Classical and Molecular Thermodynamics of Fluid Systems

This text explores the connections between different thermodynamic subjects related to fluid systems. In an innovative way, it covers the subject from first principles to the state of the art in fundamental and applied topics. Using simple nomenclature and algebra, it clarifies concepts by returning to the conceptual foundation of thermodynamics. The structural elements of classical and molecular thermodynamics of fluid systems presented cover, via examples and references, both the usefulness and the limitations of thermodynamics for the treatment of practical problems. This new edition explores recent advances in statistical associated fluid theories and contains creative end-of-chapter problems connecting the theory with real-life situations. It includes new chapters on thermodynamics of polymer solutions and molecular thermodynamics and also presents advances in the study of the activity of individual ions.

- Provides a concise structure of concepts, using simple nomenclature and algebra
- Clarifies problems usually overlooked by standard texts
- Features end-of-chapter problems to enhance the reader's understanding of the concepts
- Includes diverse topics of interest to researchers and advanced students, including elements of statistical thermodynamics, models of solutions, statistical associated fluid theory and the activity of individual ions
- Offers four appendices giving step-by-step procedures and parameters for direct use of the PRSV equation of state and the ASOG-KT group method for fugacity and activity coefficient calculations
- Features a complete set of solutions to problems throughout the book, available for download on the book's webpage under "Support Material"

This textbook is written for advanced undergraduate and graduate students studying chemical engineering and chemistry as well as for practicing engineers and researchers.

Classical and Molecular Thermodynamics of Fluid Systems

Principles and Applications

Second Edition

Juan H. Vera, Grazyna Wilczek-Vera,
Claudio Olivera-Fuentes, and Costas Panayiotou

CRC Press
Taylor & Francis Group
Boca Raton London New York

CRC Press is an imprint of the
Taylor & Francis Group, an informa business

Second edition published 2025
by CRC Press
2385 NW Executive Center Drive, Suite 320, Boca Raton FL 33431

and by CRC Press
4 Park Square, Milton Park, Abingdon, Oxon, OX14 4RN

CRC Press is an imprint of Taylor & Francis Group, LLC

First edition published by CRC Press 2017

ISBN: 978-1-032-55734-2 (hbk)
ISBN: 978-1-032-55735-9 (pbk)
ISBN: 978-1-003-43198-5 (ebk)

DOI: 10.1201/9781003431985

Typeset in Times
by codeMantra

Access the Support Material: www.routledge.com/9781032557342

Contents

SECTION II Mixtures

SECTION V Appendices

SECTION VI Solutions to End-of-Chapter Problems

Foreword

There are many good books on classical and molecular thermodynamics, some more suitable for teaching, some for research purposes, some focusing more on fundamentals, and others more on applications. The book *Classical and Molecular Thermodynamics of Fluid Systems: Principles and Applications* is a new and significantly revised version of the book of a similar title by Professor Juan H. Vera and Dr. Grazyna Wilczek-Vera. This new edition, now co-authored by four very experienced academics, is extended to include "molecular" as well as classical thermodynamics as well as applications and special topics covering a wider range of mixtures and models.

The first two sections of the book on Fundamentals and Mixtures are very similar to those of the first edition, and the emphasis is on presenting the framework of thermodynamics with very little discussion about models and theories. The latter may "come and go," while the fundamentals of thermodynamics have been in use for over 100 years. In this way, the reader is first presented with the laws of thermodynamics, and all-important concepts related to phase equilibria and models are presented only in Chapters 18 and 19, which deal with electrolytes and polymers, respectively. Chapter 19 is new and briefly presents the fundamentals of polymer thermodynamics with emphasis on the classical Flory–Huggins theory. This well-known theory is used as a tool to discuss in a coherent way a wide range of phenomena including liquid–liquid equilibria, elasticity, swelling, polymer networks, and hydrogels. The reader is, thus, getting an idea of how a rather simple approach can describe qualitatively diverse phenomena for polymer systems. The competition of entropic and enthalpic effects is discussed both in this and other chapters.

The applications section starts with a new chapter on applications of polymer solution thermodynamics, where emphasis is given on the lattice-fluid (LF) theory (and various extensions) and its applicability for, among others, phase equilibrium, blends, and glass transition calculations. Several other applications of thermodynamics are included in this section, which closes with an extended chapter on adsorption of gases on solids, where both classical and newer approaches are presented.

Among all sections of this book, the last (fourth) section on special topics is the one most heavily updated where now are presented a statistical thermodynamics approach of the generalized van der Waals theory, the generalized lattice-fluid model and associated extensions for hydrogen bonding (HB) mixtures, as well as a chapter on COnductor like Screening MOdel for Real Solvents (COSMO-RS)-type approaches and a shorter chapter on the Statistical Associating Fluid Theory (SAFT) theory. The chapter on the legacy of the van der Waals theory illustrates a variety of models which have their origin from this theory, and the connection between models is illustrated.

The LF developments are presented in detail in several of these chapters (20, 29–33). The LF approach has been combined with the group contribution concept, yielding successful models for polymer solutions, and has also been extended to HB systems with great success. It may not have been as widely used as the SAFT theory, but in many cases, the two methodologies (LF and perturbation) yield very similar results. The readers interested in exploring the whole spectrum of applications and refinements of the LF concept will find these chapters very interesting. Without doubt, the authors are fascinated by the elegance and wide use of applications of the LF concept (in its various forms), and this appears to be a favorite approach. This is also one additional aspect making this book different from others which emphasize SAFT and related theories.

The chapter on HB is very interesting, discussing in a comparative way chemical, LF, and perturbation theories, while touching upon complex phenomena like intramolecular HB, co-operativity, acid dimerization, and HB in glassy polymers. The analogies and similarities of the various HB theories are discussed as well, both in terms of functionality and results. The special role and importance of HB and how it is implemented in several modern thermodynamic models is an additional aspect which is particularly highlighted in this book. The reader will be able to appreciate why models that explicitly account for HB effects perform better than those which do not!

This book also includes a shorter but elegant SAFT chapter, which makes no attempt to reproduce the very extensive SAFT literature but focuses on the basics of original versions of the model and its molecular background rather than on applications. Several SAFT versions are briefly mentioned as well as how SAFT theory can be combined with cubic equations of state or HB terms in the context of the LF theory.

The COSMO-RS theory is presented as a pioneering piece of work, and links or similarities to the quasi-chemical theory are nicely explained. A recent development presented is the combination of COSMO-RS with an equation of state approach which can result, once broadly developed, in a promising predictive approach.

Overall, in these chapters, the interested reader can follow the progress of molecular thermodynamics during the second half of the twentieth century until now and appreciate the achievements but also the similarities and differences of the LF and perturbation theories as applied to the thermodynamics of fluids and mixtures.

A significant highlight of this book, as was the case also in the previous edition, are the two chapters on electrolyte solutions: one on fundamentals and the other on individual ion activity coefficients (IIACs). The chapter on IIACs is updated, and several recent developments are mentioned on how IIACs have been used in the development and/or validation of activity coefficient models and equations of state for electrolytes.

Finally, this book includes an updated appendix on the highly successful the Peng-Robinson-Stryjek-Vera (PRSV) equation of state, where a large parameter table of the model is included, and a discussion on the importance and influence of mixing rules.

What makes this book extra pedagogic in this new edition is also the many problems included which I believe will be of value to students, teachers, researchers, and engineers alike.

I would like to close with a personal note. I have deeply enjoyed reading this book, and I hope the readers will both enjoy and appreciate the very personal style of the authors, both on fundamentals and on applications.

Georgios M. Kontogeorgis
Professor of Applied Thermodynamics
Department of Chemical and Biochemical Engineering
Technical University of Denmark

Preface

Thinking Straight in the Age of Information Overload.

Subtitle of The Organized Mind *by Daniel J. Levitin (2014)*

Classical thermodynamics has not changed since the publication of *Classical Thermodynamics of Fluid Systems*. Thus, we repeat here what was said in its preface: "Classical thermodynamics is the human attempt to capture, with a few principles and a coherent mathematical framework, some of the natural facts that we observe in daily life. The mathematical superstructure, successful as it is, should not be confused with reality. The preface of the first edition of the classical text *Thermodynamics* by G. N. Lewis and M. Randall compared the study of this subject with a visit to an ancient cathedral. These authors said that, as with a cathedral, the perfection of thermodynamics inspires respect and awe. It is easy to forget that this perfect structure is the product of human efforts and that all the natural human imperfections, present during its construction, have been erased by the passing of time. They remarked that even the rude and loud language that could have been used by the workers at the time of construction is now silenced. One can hear only hushed voices respectfully admiring the imposing form of the final work. There are as many ways of visiting a cathedral as there are ways of studying thermodynamics. Different people search for different qualities from the same reality. Even the same person may see different details on different visits. This may explain why so many books have been written on the subject of thermodynamics. So many indeed that we feel compelled to justify why we have decided to write yet another one. The reason, as we see it, is the need to provide a unified view of seemingly independent topics of interest in the study of chemical processes. The author of a text makes choices and decides what to emphasize and for whom to write. This text is written mostly, but not exclusively, for instructors, researchers, and process engineers. It assumes knowledge of calculus and of the physical states of matter. Although not a prerequisite, an undergraduate course in chemical engineering thermodynamics or physical chemistry is desirable. This text is written for those who want to understand the connections between different thermodynamic subjects related to fluid systems. It may be used as a text for a graduate course and even by beginners seeking a strong start. In some sense, it is a look at the finished building with complete disregard for the historical steps involved in its construction. In doing so, an attempt has been made not to leave a single corner of the building unexplored, discussing the minute details that, if left unattended, may give a feeling of insecurity once at higher levels."

Up to here, we reproduced the Preface of *Classical Thermodynamics of Fluid Systems*, a basic text for the teaching of undergraduate thermodynamics courses which includes also special topics of interest for courses at a graduate level. Two deficiencies of the first edition were that it lacked end-of-chapter problems and it did not include material in polymer solutions. For this second edition, we invited Professor Claudio Olivera-Fuentes to contribute creative problems, and their solutions, to enhance the understanding of the subjects presented. In addition, in collaboration with Professor Olivera-Fuentes, who had already contributed to the first edition the tables of PRSV parameters and helped improve it in style and content, we decided to rewrite Appendix C. Considering our lack of expertise in polymer solutions, we invited Professor Costas Panayiotou to contribute two chapters on this topic, one at the basic level and the other covering applications. Thus, new Chapters 19 and 20 were included, and all other chapters up to Chapter 26 of the first edition were moved up by two in numbering. Professor Panayiotou then volunteered to write a few more chapters covering recent advances in molecular thermodynamics, and new Chapters 29 to 33 were included. Thus, Chapter 27 of the first edition became Chapter 34 of this book. Professor Panayiotou also provided end-of-chapter problems for the two chapters in polymer solutions he contributed. The incorporation of new chapters in molecular thermodynamics promoted the inclusion of related concepts in

"Adsorption of Gases in Solids" and "Elements of Statistical Thermodynamics." Thus, at this point, it became evident that the new manuscript was a different book from *Classical Thermodynamics of Fluid Systems*, and it was necessary to find a new title to emphasize its molecular thermodynamics content. In addition, solutions to all end-of-chapter problems are available as support material at the book's webpage, www.routledge.com/9781032557342. This online material contains derivations of equations and many figures, tables, and references.

As in the first edition of *Classical Thermodynamics of Fluid Systems*, in the present text of *Classical and Molecular Thermodynamics of Fluid Systems*, Section I presents, in a rational sequence, unperturbed by arbitrary approaches or temporary fashions, the structural elements of classical thermodynamics of fluid systems. Section II covers the treatment of mixtures of nonelectrolytes, aqueous solutions of electrolyte solutes and polymer solutions, with special emphasis given to the physical reality represented by equations and concepts. Section III shows both the usefulness and the limitations of thermodynamics for the treatment of practical problems related to fluid systems. Section IV discusses special topics including chapters covering advances in molecular thermodynamics and culminates with the measurement and modeling of the activity of individual ions. Section V, in addition to giving the basic equations for material balances and for work with the virial equation of state, makes available equations and parameters for the use of two simple and effective tools for the treatment of fluid mixtures of non-electrolytes: the Analytical Solution of Groups (ASOG-KT) and the PRSV Equation of State. Parameters for these two models were previously published in *Journal of Chemical Engineering of Japan* and *The Canadian Journal of Chemical Engineering*, respectively; two sources that can be difficult to access. Errors in the first page of Table D.1 of ASOG parameters of the first edition were corrected in this edition.

The purpose of the end-of-chapter problems is not to test problem-solving-ability but to expand the understanding of the subjects covered and show how, with the knowledge gained, new situations can be dealt with. Solutions to these problems are not included in the text. They can be found at the website www.routledge.com/9781032557342. The website should only be consulted after trying hard to find a solution to the problem in question. We decided not to include problems for Chapters 31–34, which deal with concepts that are still in the stage of development.

We are indebted to Professors J. Gmehling and K. Tochigi for their help in obtaining permission to reproduce the ASOG-KT parameters and to our co-authors Professors Claudio Olivera-Fuentes and Costas Panayiotou for their dedicated effort to transform our previous text *Classical Thermodynamics of Fluid Systems* into this more complete text covering *Classical and Molecular Thermodynamics of Fluid Systems*. Special thanks are due to Professor Georgios Kontogeorgis, who kindly accepted our invitation to write the Foreword for this book.

Juan H. Vera and Grazyna-Wilczek-Vera

Authors of Classical Thermodynamics of Fluid Systems

A WORD FROM THE AUTHORS OF THE SECOND EDITION

My (GW-V) initial formation in thermodynamics took place in two places: The Laboratory of Thermodynamics of Non-Electrolytes at the University of Warsaw, Poland, led by Dr. Bogusław Janaszewski, and the Institute of Technical Chemistry II at University Erlangen-Nurnberg, West Germany (the political map of Europe was different then). I am grateful to colleagues in both groups, particularly to Dr. Paweł Oracz, who contributed to my scientific growth. Special thanks are reserved for Prof. Stanisław Malanowski, supervisor of my doctoral dissertation.

I (JHV) am grateful to Alfonso Frick and Fernando Aguirre, who introduced me to the subject of this book; to my mentor and friend John M. Prausnitz, who taught me what research is all about, and to Gerald Ratcliff and Donald Patterson who received me in their research groups at McGill

University. All of them, at one time or another, transferred to me their enthusiasm for the study of thermodynamics.

I (CO-F), too, owe a debt of gratitude to Alfonso Frick and Fernando Aguirre, my professors of Thermodynamics and later colleagues at Federico Santa María Technical University in my native Valparaiso, Chile. They instilled in me the love of this subject and encouraged me to embark fifty years ago on my ongoing academic career. And of course, as always, I am grateful to my dear wife, Aminta, for her unfailing patience, support, and encouragement.

I (CP) am grateful to Juan Vera, my PhD thesis supervisor at McGill University, who taught me the beauty and strength of thermodynamics, and to Isaac Sanchez for the wonderful collaboration we had in Austin. And, as always, I am more than grateful to my wife, children, and grandchildren for being an invaluable source of joy and meaning.

We expect that by a rational understanding of the strengths and limitations of thermodynamics, the reader will be empowered to apply its principles to real-life problems. Welcome to the realm of classical and molecular thermodynamics.

Juan H. Vera
Grazyna Wilczek-Vera
Claudio Olivera-Fuentes
Costas Panayiotou

Authors

Juan H. Vera is a Professor Emeritus in the Department of Chemical Engineering, McGill University, Montreal, Canada. He earned a doctorate (Ing. Quim.) at the Universidad Tecnica Federico Santa Maria, Chile, and a Master of Science in chemical engineering at the University of California, Berkley. He co-authored the first edition of this book, a manual on copper metallurgy (in Spanish), a module in Elsevier Encyclopedia, co-edited the book *Ionic Surfactants and Aqueous Solutions: Biomolecules, Metals and Nanoparticles* (2018), and authored more than 200 refereed publications in international journals. He has an international patent on the extraction of proteins and a Canadian patent on the extraction of heavy metals.

Grazyna Wilczek-Vera is a chemist who has taught and conducted research in various areas of thermodynamics at several universities. During her more than 30-year academic career, she served as a Faculty Lecturer and the Director of Undergraduate Studies in the Department of Chemistry and as a Research Associate in the Department of Chemical Engineering at McGill University, Montreal, Canada, and as an Adjunct in the Department of Chemistry, University of Warsaw, Poland. She received the 2008 Principal's Prize for Excellence in Teaching at McGill University. She earned a PhD in chemical sciences (with distinction) and a Master of Science in chemistry (with distinction) at the University of Warsaw. She has authored 60 refereed publications in journals, given 38 conference presentations, and co-authored three books.

Claudio Olivera-Fuentes is a Professor Emeritus in the Thermodynamics and Transport Phenomena Department and Coordinator of Applied Sciences and Engineering in the Dean's Office of Research and Development at Universidad Simón Bolívar, Caracas, Venezuela. He earned a chemical engineering degree at the Universidad Técnica Federico Santa María, Chile, and an MSc in chemical engineering at the University of Manchester Institute of Science and Technology, UK. He has been an Invited Researcher at the University of Concepción, Chile, and a Visiting Scholar at the University of Pennsylvania. He has authored over 120 publications in conference proceedings and scientific journals, three book chapters, and one textbook on vector and tensor analysis applied to transport phenomena. Among other recognitions, he received the 2003 Procter & Gamble of Venezuela Award for Excellence in Teaching and the 2016 Simón Rodríguez Award for Lifetime Academic Achievement.

Costas Panayiotou is a Professor Emeritus in the Department of Chemical Engineering, Aristotle University of Thessaloniki, Greece. He earned a PhD at McGill University, Montreal, Canada. He was a Visiting Professor at the University of Texas, Austin, and at Ecole Polytechnique Federale, Lausanne, Switzerland, and he was an Otto Monsted Professor at the Technical University of Denmark (DTU). He has published more than 250 peer-reviewed articles in international journals, ten book chapters, and two textbooks, and he has served on the editorial boards of ACS and Elsevier journals.

Section I

Fundamentals

The enormous usefulness of mathematics in the natural sciences is something bordering in the mysterious and there is no rational explanation for it.*

NOTE

* Remark made by Eugene (Jano) P. Wigner (Nobel Prize in Physics, 1963) at the Richard Couvant Lecture in Mathematical Sciences, University of New York, 1959, as quoted in Wigner, E. P. 1979. *Symmetries and Reflections: Scientific Essays*. Woodbridge, CT: Ox Bow Press.

DOI: 10.1201/9781003431985-1

Section 1

Fundamentals

[illegible]

[illegible]

1 Basic Concepts and Definitions

Abstract

This chapter gives the basic concepts and definitions used in the thermodynamic study of the behavior of fluid systems. A simple system is characterized by its temperature, pressure, and composition variables. It is in equilibrium if, when isolated from its surroundings, no change is observed. A process is the change between two equilibrium states. A closed system has constant composition, and its internal energy changes due to work or heat exchanged with its surroundings. The change between two equilibrium states is independent of the path. A reversible path proceeds by infinitesimally small changes of pressure and temperature. The internal energy and the volume are state functions as they depend only on the state of the system. Heat and work are path-dependent functions. At high temperature and low pressure, the molar volume of pure compound fluids is directly proportional to the absolute temperature and inversely proportional to the pressure, so the product of the molar volume times the pressure over the absolute temperature is a constant, the universal gas constant. A system with this behavior is called "an ideal gas." The ideal gas does not condense into a liquid phase by an increase in pressure at any temperature.

Thermodynamics is an experimental science. It builds upon the results of experimental measurements, and its conclusions must be verified experimentally. On the other hand, well-established and experimentally confirmed thermodynamic models liberate us from the necessity of continuously performing experiments for each problem encountered.

This work presents a rational look into the structure of classical thermodynamics, with some insight into its molecular aspects based on simple elements of statistical thermodynamics. First, with the purpose of having a common language, it is necessary to agree on some elemental definitions and use of terms. In this work, *matter* and *energy* are considered to be primitive concepts that, similarly to space and time, do not need further definitions. In addition, we ignore the possible interconversion between matter and energy, and consider the material and energy balances to hold independently of one another.

CONCEPTS OF SYSTEM AND SURROUNDINGS

Like most disciplines, thermodynamics has its own vocabulary. A *system* is the region of the space that we decide to study. A system can have real or imaginary *boundaries* or walls. The *surroundings* are the regions of the space in contact with and affected by the system. The *universe* includes the system and its surroundings. This term should not be confused with the astronomical universe, of which we know little.

One way to introduce a topic is to consider first the most complex system that one can imagine and then, beginning from a general treatment, come down to specific cases. This work chooses instead to start building up from simple systems and add variables as the additional complexity of real problems may require.

The *simple system* chosen is fixed in space and not affected by any force field. It is homogeneous and isotropic. In addition, it is not electrically charged, it is large enough to make surface effects negligible, and it is chemically inert. As the properties of a simple system isolated from its surroundings will not change with time, such a system is said to be at a state of *equilibrium*. The *intensive state* of a simple system at equilibrium can be completely described by the values of its intensive parameters, *pressure* and *temperature*, and the values of its *composition variables*. The description of the *extensive state* of a simple system requires the specification of one

DOI: 10.1201/9781003431985-2

additional *extensive variable* as, for example, its *total mass* or its *total volume*. The ratio of the values of two extensive properties of a simple system gives a value independent of the size of the system, such as *density* or its reciprocal value, the *specific* or *molar volume*. Temperature and pressure are sometimes called *field* variables to distinguish them from specific or molar quantities. As stated above, the study of more complex systems is possible by the addition of appropriate variables.

THERMODYNAMIC PROCESS, STATE AND PATH FUNCTIONS

In a *thermodynamic process*, the system passes from one equilibrium state to another. In this process, energy and matter may be exchanged between the system and its surroundings. In the study of simple systems, we distinguish three forms of energy.

The *internal energy*, U, is the energy that a system has because it is formed by atoms and molecules in movement. *Heat* and *work* are forms of energy in transfer that can only be detected at the walls of a system during a thermodynamic process. Once the energy has been transferred to or from the system, it just shows as a change in internal energy, without distinction of the form in which it was transferred. Work, W, is defined as the energy transferred between a system and its surroundings in such a way that all net effects in the surroundings can be exactly reproduced by the effect of lifting or letting fall weights. Heat, Q, is the energy transferred in such a way that its effects on the surroundings cannot be reproduced by lifting or letting fall weights.

Energy is considered positive when it enters the system from the surroundings as heat or work.

The *internal energy of a system* is a *state function*, that is, a function that only depends on the state of the system. Thus, the internal energy change between two well-defined equilibrium states is independent of the path followed by the process.

On the other hand, heat and work are *path functions*, that is, functions whose value depends on the particular process by which energy was transferred between the system and the surroundings.

At this point, it is important to make the distinction between a system in *equilibrium* and a system in *steady state*. If heat, for example, enters and leaves the system at the same rate, the properties of the system will not change with time. This system is said to be at a steady state.

However, if such a system is *isolated from its surroundings*, its properties will evolve toward their value at equilibrium, different from the value they had at steady state.

Real boundaries or walls enclosing a simple system can be of different types. They may be *permeable* or not. Sometimes, they can be semipermeable and allow some species to go through while being nonpermeable to other species. A system enclosed by a nonpermeable wall is said to be a *closed system*, as no matter can be exchanged with the surroundings.

Real walls can be *movable* or *rigid*, thus allowing or not allowing changes in the volume of the system. Movable walls allow energy transfer between the system and the surroundings in the form of work.

Similarly, real walls can be *diathermal* or *adiabatic*. Diathermal walls allow the transfer of heat, while adiabatic walls do not. A system enclosed by nonpermeable, rigid, and adiabatic walls is said to be an *isolated system*, as it cannot exchange matter or energy with its surroundings.

CONCEPT OF A REVERSIBLE PROCESS

One of the *basic problems of classical thermodynamics* is the evaluation of the change in value of the thermodynamic properties of a system after a particular process occurs between two equilibrium states.

A way of seeing this problem more clearly is by imagining two simple systems, each isolated under its own conditions, which are enclosed together inside a nonpermeable, rigid, and adiabatic reservoir. Thus, the two subsystems are part of a composite isolated system. Under these initial conditions, no thermodynamic process is possible. The internal energy of the composite system is the sum of the internal energies of the two subsystems. As the composite system is isolated, its overall internal energy would not change if a thermodynamic process occurred between the two subsystems.

In a thought experiment, we can remove one of the internal constraints separating both subsystems. By making the wall separating both subsystems diathermal, heat will be transferred from one subsystem to the other. By making it movable, energy in the form of work will be exchanged between the subsystems. Similarly, if it is made permeable, matter will be exchanged between the subsystems. One can do these changes in any order and each of these options will cause the subsystems to evolve to different but well-defined equilibrium states.

One important point to notice here is that the change considered is between two equilibrium states of the subsystems. The rate at which the change takes place is not important. Thus, it is possible to choose an *infinitely slow process* in which all conditions are kept under control at all times and use this path for calculating the changes in the state properties of each subsystem between its initial and final equilibrium states. This imaginary infinitely slow or *quasi-static process* is referred to as a *reversible process*, and it is extremely useful for thermodynamic calculations.

COMPRESSIBLE AND INCOMPRESSIBLE FLUIDS

In the study of the thermodynamics of fluid systems, we introduce a pragmatic division between *compressible* (gases) and *incompressible* (liquids) fluids. The volume of a pure compound compressible fluid is a function of pressure and temperature, while the volume of incompressible fluids, to a good approximation, is only a function of temperature. No real fluid is totally incompressible.

Both types of fluids in a closed system can exchange energy in the form of heat with the surrounding. As we have neglected the action of force fields and surface effects, incompressible fluids in a closed system exchange energy in the form of mechanical work only in the case of stirring. Compressible fluids can also exchange energy in the form of mechanical work by movement of the walls. In open flow systems, liquids and gases can exchange energy as work by the movement of a shaft. This is called shaft work.

The simplest compressible system is the so-called ideal gas. Its equation of state, that is, the equation that gives the volumetric properties as a function of field variables temperature (T) and pressure (P), is

$$V^* = \frac{nRT}{P} \tag{1.1a}$$

or

$$v^* = \frac{V^*}{n} = \frac{RT}{P} \tag{1.1b}$$

where V^* is the total volume of the system, v^* is the molar volume of the system, n is the number of moles of gas, and R is the universal gas constant:

$$R = 8.3144598(48)\left[\frac{\text{J}}{\text{mol} \cdot K}\right] = 1545.4\left[\frac{\text{ft} \cdot \text{lbf}}{\text{lbmol} \cdot {}^{\circ}\text{R}}\right]$$

The ideal gas, in contrast with real gases, does not condense into liquid by lowering the temperature or increasing the pressure. Real gases at temperatures below a characteristic value called a critical temperature, T_c, condense into liquids by an increase in pressure above a certain value, called the

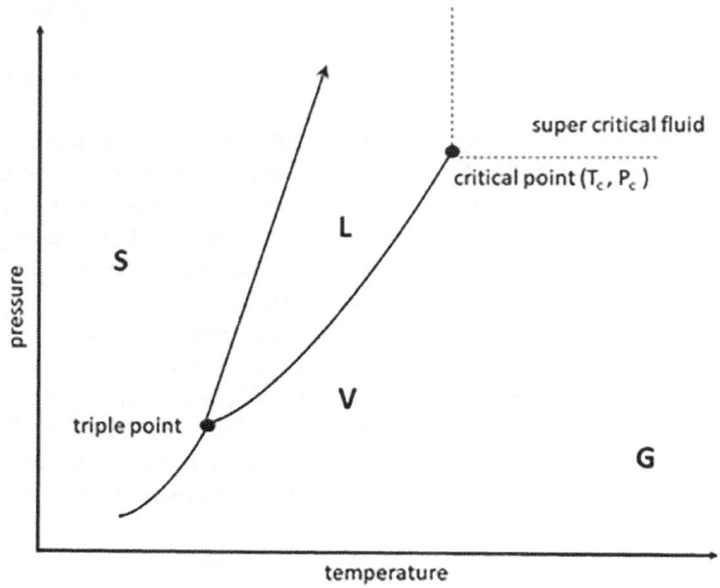

FIGURE 1.1 A typical schematic of a triple point phase diagram. S – solid, L – liquid, V – vapor, and G – gas.

saturation pressure, that is fixed for each temperature. The critical pressure, P_c, is the value of the saturation pressure of a compound at its critical temperature, above which no vapor–liquid equilibrium is possible. The ideal gas behavior is a good approximation for the behavior of real gases at low pressures and high temperatures.

The behavior of real fluids is better seen in a pressure–temperature diagram as the one shown in Figure 1.1. The saturation pressure curve, which separates the vapor (V) and liquid (L) zones in this diagram, goes from the triple point, where the vapor (V), liquid (L), and solid (S) phases coexist, to the critical point at the critical temperature and critical pressure of the fluid. At temperatures and pressures below the triple point the vapor phase is in equilibrium with the solid phase, corresponding to the sublimation phenomena. At temperatures below the triple point and pressures above the triple point pressure, there is a line separating the liquid and solid phases representing the freezing–melting phenomena. This line has a slight positive slope for most fluids, and it does not have a known upper limit. Water is one of the few exceptions for which this line has a negative slope and an increase in pressure melts the ice.

PROBLEMS

1.1 **Development of the ideal gas law**

This problem is based on the excellent papers by Jensen [1] and Laugier and Garai [2]. The ideal gas law, Equations 1.1a, b, was the end result of more than two centuries of painstaking experimental and theoretical work. Show that it is a direct consequence of combining the four earlier "laws":

Boyle and Mariotte: $PV = F_{PV}(T,n)$

Gay-Lussac: $P/T = F_{PT}(V,n)$

Charles and Gay-Lussac: $V/T = F_{VT}(P,n)$

Avogadro: $V/n = F_{VN}(T,P)$

1.2 Deviations from ideal gas behavior

Table 1.1 reproduces experimental data on volumes of gaseous propane at several temperatures and pressures. The critical properties of propane are $t_c = 206.3$ °F, $P_c = 617.4$ psia, and $v_c = 3.202$ ft³/lbmol [3]. Under what conditions does propane behave approximately as an ideal gas, for example, with less than 5% error in volume?

TABLE 1.1
Experimental Volumes of Gaseous Propane

t (°F)	100	160	220	280	340	400
T (°R)	559.67	619.67	679.67	739.67	799.67	859.67
P^{sat} (psia)	188.7	383.8				
V_L^{sat} (ft³/lbmol)	1.495	1.761				
V_V^{sat} (ft³/lbmol)	24.56	10.70				
P (psia)			Molar volume (ft³/lbmol)			
14.696	403.2	448.0	492.5	537.0	581.3	625.5
20	294.8	328.0	360.9	393.7	426.4	459.0
30	194.6	217.1	239.3	261.4	283.4	305.3
40	144.5	161.7	178.5	195.2	211.9	228.4
50	114.4	128.4	142.0	155.5	168.9	182.2
60	94.3	106.2	117.7	129.1	140.3	151.5
80	69.1	78.4	87.3	96.0	104.6	113.1
100	53.9	61.7	69.0	76.1	83.1	90.0
125	41.6	48.3	54.4	60.2	65.9	71.5
150	33.2	39.3	44.6	49.6	54.5	59.2
200		28.0	32.4	36.4	40.2	43.9
300		16.25	19.99	23.07	25.85	28.48
400			13.65	16.38	18.68	20.80
500			9.66	12.32	14.37	16.19
600			6.68	9.56(*)	11.48	13.12
800			2.272	5.98	7.85	9.29
1000			2.015	3.79	5.69	7.01
1250			1.882	2.642	4.05	5.24
1500			1.803	2.270	3.16	4.15
1750			1.749	2.099	2.706	3.45
2000			1.707	1.985	2.426	3.01
2250			1.673	1.909	2.262	2.721
2500			1.643	1.850	2.144	2.510
2750			1.618	1.804	2.055	2.367
3000			1.597	1.766	1.985	2.257
3500			1.560	1.705	1.883	2.095
4000			1.529	1.658	1.809	1.983
4500			1.503	1.618	1.750	1.901
5000			1.479	1.584	1.703	1.835
6000			1.440	1.530	1.629	1.736

Reprinted with permission from Ref. [3].
(*) Corrected from a misprint in the original.

REFERENCES

1. Jensen, W. B. 2003. The universal gas constant. *J. Chem. Educ.* 80: 731–732.
2. Laugier, A., and Garai, J. 2007. Derivation of the ideal gas law. *J. Chem. Educ.* 84: 1832–1833.
3. Reamer, H. H., Sage, B. H., and Lacey, W. N. 1949. Volumetric behavior of propane. *Ind. Eng. Chem.* 41(3): 482–484.

2 The First and Second Laws of Thermodynamics

ABSTRACT

The First and Second Laws of Thermodynamics are presented. Comparison of the internal energy change of a closed system in a real and a reversible process between two equilibrium states shows that the work lost due to irreversibility is equal to the difference between the heat exchanged in both processes. It is postulated that the differential of the reversible heat exchanged can be expressed as the product of the absolute temperature times the differential change of a state function called the entropy of the system, which is larger than the differential of the actual heat exchanged. This relation leads to the unavoidable conclusion that the entropy of an isolated system always increases when irreversible changes occur. The second law of thermodynamics implies that it is not possible to pass heat from a cold body to a hot body unless something else also occurs. Refrigerators, air conditioners, and heat pumps transfer heat from cold spaces to hot spaces but to do this it is necessary to put in work. The chapter closes by demonstrating that the second law of thermodynamics also explains why it is not possible to totally convert heat into useful work in a cyclic process.

PRINCIPLE OF CONSERVATION OF ENERGY IN A CLOSED SYSTEM

Let us consider a simple system that is closed and fixed in space. As for most practical applications, the interconversion between mass and energy occurring in atomic or nuclear processes can be safely ignored; we consider here that the principles of conservation of mass and energy can be separately stated. In this case, a change in the internal energy, U, of this simple system is due only to an exchange of energy between the system and its surroundings, be this exchange in the form of heat, Q, or work, W. According to the International Union of Pure and Applied Chemistry (IUPAC) [1] convention of signs, energy is positive when it enters the system, thus increasing its internal energy. Using the symbol Δ to indicate the difference of the final value of a function minus its initial value, we write:

$$\Delta U = Q + W \tag{2.1a}$$

The mass of a closed system is constant, so this equation can be written per unit mass using lower-case symbols for the specific values:

$$\Delta u = q + w \tag{2.1b}$$

For a process between two equilibrium states that are infinitely close one to the other, this equation takes the form:

$$du = \delta q + \delta w \tag{2.1c}$$

or

$$dU = \delta Q + \delta W \tag{2.1d}$$

DOI: 10.1201/9781003431985-3

Equations 2.1a–d are forms of the *first law of thermodynamics*. It is important to observe that the symbol d is used for the exact differential of a state function, while the symbol δ is used to indicate small quantities of heat and work exchanged during the incremental process. This difference in symbols is required to distinguish a change in the values of a state function from transfers of energy that are path dependent.

REVERSIBLE OR QUASI-STATIC PROCESSES

Real processes are difficult to describe in mathematical terms. However, we are normally interested in the change of value of a state function after a process occurs so we can imagine an alternate process that would take the system from the same initial state to the same final state as in the actual process. Thus, we consider that the change between the final and the initial equilibrium states presented by the system in the real process occurs following an infinitely slow (*reversible*) process. For a reversible process in a closed system, we write:

$$dU = \delta Q_{rev} + \delta W_{rev} \tag{2.2a}$$

or

$$du = \delta q_{rev} + \delta w_{rev} \tag{2.2b}$$

As the internal energy u is a state function, du has the same value for both the real and the reversible processes, while $\delta q_{rev} \neq \delta q$ and $\delta w_{rev} \neq \delta w$. By eliminating du from Equations 2.1c and 2.2b, we get:

$$\delta q + \delta w = \delta q_{rev} + \delta w_{rev} \tag{2.3}$$

or

$$\delta w - \delta w_{rev} = \delta q_{rev} - \delta q \tag{2.3a}$$

It is instructive, now, to visualize alternative processes of expansion or compression of a closed simple system.

Figure 2.1 depicts in a P–V diagram two states of the system characterized by the values P_1, V_1 and P_2, V_2. For a closed system, the volume is a state function of the temperature and pressure. Although one normally associates a smaller volume of a closed system with a higher pressure, the $(P_1, V_1) \rightarrow (P_2, V_2)$ process presented in the figure has been purposely drawn to show that at the higher pressure the volume of the system is larger. As the temperature axis is not shown in the diagram, the heat transfer that occurred in this process is hidden.

As shown in Figure 2.1, the same initial and final states of the system can be joined by different lines, each representing a different path or process. Each of these well-defined paths represents a different infinitely slow, that is, reversible, process.

Although there is no need to restrict the nature of the system or the shape of the vessel, it is easier to visualize it as a gas contained in a cylindrical vessel closed by a weightless movable piston that can be displaced by the difference between the external and internal forces. The cylinder and the piston have diathermal walls, and the complete set is immersed in a thermostatic air bath at a temperature we can control.

Consider that at a temperature, T, the system has a volume, V, and exerts a pressure, P, over the inside wall of the piston. The system will be in a state of equilibrium with its surroundings when the internal pressure, P, is equal to the external pressure, P_E, acting over the piston. If the piston has a weight, the pressure caused by this weight over the gas is added to the atmospheric pressure to get the value of P_E. To follow a reversible path, we may imagine that over the piston we have sand, the

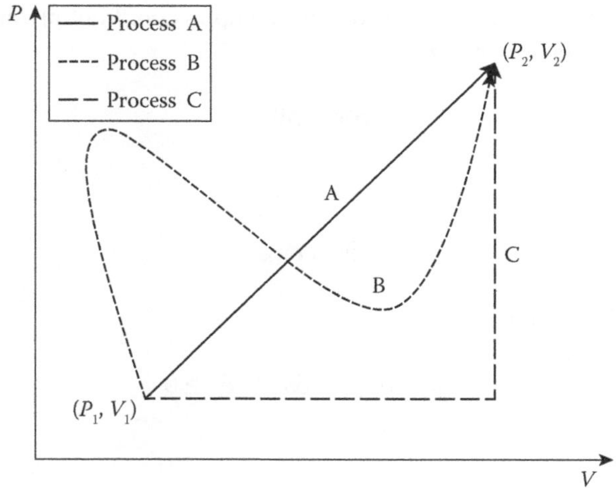

FIGURE 2.1 P–V diagram of $(P_1, V_1) \rightarrow (P_2, V_2)$ processes by three different paths: A, B, and C.

weight of which, acting over the surface area of the cylinder, plus the external atmospheric pressure, equals the internal pressure of the gas. By adding or removing sand from the top of the piston or changing slightly the temperature of the thermostatic bath, we can change the volume of the system following an arbitrarily chosen infinitely slow (reversible) path between the initial and final states that the system had in the real process.

The external force acting against the system is equal to the external pressure times the surface area of the piston, $F_E = P_E A$. An infinitesimal displacement, dL, of the piston times its surface area, A, gives the infinitesimal volume change $dV = A\ dL$ of the system. Thus, this infinitesimal volume change times the external pressure equals the work, done over the system by the external forces:

$$\delta W = -F_E dL = -P_E dV \tag{2.4}$$

A negative sign is required for dV and dL to satisfy the convention of sign for work. For compression, work is positive while the volume decreases. We can carry out a compression or an expansion process infinitely slowly by making the difference between the internal and the external pressure smaller and smaller for the small change in volume of the system. The reversible process is the imaginary limiting case when the infinitesimal expansion or compression occurs with an external pressure, P_E, equal to the *internal pressure*, P, of the system. Hence, we identify the *reversible work* with:

$$\delta W_{rev} = -PdV \tag{2.5a}$$

or

$$W_{rev} = -\int_1^2 PdV \tag{2.5b}$$

The integral in Equation 2.5b is the surface area under the corresponding P–V curve in Figure 2.1, and it differs for different reversible processes.

For the compression of a system under isothermal conditions, in a real process, the external pressure should be higher than the internal pressure. Thus, for a slight compression we write:

$$P_E \geq P$$

The equal sign applies for the imaginary reversible compression. Multiplying both sides by ($-dV$), which is a positive quantity, we obtain:

$$\delta W \geq \delta W_{rev} \tag{2.6}$$

or, per unit mass,

$$\delta w \geq \delta w_{rev} \tag{2.7}$$

Thus, from Equation 2.3a, we write:

$$\delta w - \delta w_{rev} = \delta q_{rev} - \delta q \geq 0 \tag{2.8}$$

and

$$\delta q_{rev} = \delta q + \left(\delta w - \delta w_{rev}\right) \geq \delta q \tag{2.9}$$

CONCEPT OF ENTROPY, EXTREMUM PRINCIPLE, AND CLAUSIUS INEQUALITY

The difference between the actual and the reversible work can be interpreted as the *work lost* due to the irreversibility of the process, and it is given the symbol l_w:

$$w - w_{rev} = l_w \geq 0 \tag{2.10}$$

The difference between the reversible heat and the actual heat exchanged is called the *uncompensated heat*. As shown by Equation 2.8, both of these differences are the same and are always positive.

From Equations 2.2 and 2.5, we can now write the equation of the first law for a reversible process in a closed simple system as:

$$dU = \delta Q_{rev} - PdV \tag{2.11}$$

We observe here that the reversible work is evaluated as the product of the *intensive parameter*, P, and change in the *extensive parameter*, V. A gradient in pressure is the potential for doing work, while the difference in volume is the effect shown in the system because work was done. The simple form of this equation suggests the introduction of a state function that for the reversible heat plays exactly the same role as the total volume for the reversible work. Thus, we introduce the concept of entropy through the following postulate:

There exists an extensive state function S, called the entropy of the system, such that the heat exchanged in a differential reversible process can be expressed as the product of the intensive parameter T, the absolute temperature of the system, times the change in the value of this function in the process.

A gradient in temperature is the potential for heat exchange, while the difference in entropy is the effect of the reversible heat exchange.

Thus, we write:

$$\delta Q_{rev} \equiv TdS \tag{2.12a}$$

or, per unit of mass, in specific or molar terms,

$$\delta q_{rev} \equiv Tds \tag{2.12b}$$

Consequently, for a closed system,

$$dU = TdS - PdV \tag{2.13a}$$

or

$$du = Tds - Pdv \tag{2.13b}$$

Equation 2.13a, which relates entropy as a function of state to the other thermodynamic variables, is sometimes referred to as the *differential form of the fundamental equation of thermodynamics for a closed nonreacting system.* The purpose of rewriting the equations each time in terms of specific or molar quantities is to emphasize the difference between intensive parameters as P or T and extensive state functions as U, V, or S.

There is an important distinction between the state function entropy S and the state functions internal energy, U, and volume, V. While the values of the internal energy and the volume of the system change when the system exchanges heat or work with the surroundings in a reversible process, *the value of the entropy of the system is not affected by the exchange of reversible work.* By its definition, Equation 2.12, the entropy of the system only changes due to the exchange of reversible heat that would occur in an imaginary reversible process between the same initial and final states as in the real process. In a real process, the value of the entropy of the system also changes due to the action of (real) work, but this is only because of the work lost due to the irreversibility of the process, which, according to Equation 2.9, will have an effect on the magnitude of the reversible heat exchange involved.

There are other important conclusions to be drawn from the definition of entropy. From Equations 2.9 and 2.12a,

$$dS \geq \frac{\delta Q}{T} \tag{2.14}$$

This relation, known as the *Clausius inequality*, can be converted into equality by adding to the right-hand side the term δS_{irr}, which is always positive for any irreversible process and equal to 0 for a reversible process:

$$dS = \frac{\delta Q}{T} + \delta S_{irr} \tag{2.15}$$

Thus, a reversible adiabatic process is isentropic (S=constant), due to the definition of the entropy function by Equation 2.12. If the process is irreversible and adiabatic, Equation 2.15 shows that the entropy of the system will always increase.

The entropy of a system decreases in an irreversible process if the heat lost by the system is large enough to compensate for the increase in entropy due to the irreversibility of the process. Similarly, an irreversible process can be isentropic if the heat lost in the process exactly compensates for the increase in entropy due to irreversibility. Finally, a conclusion of great importance is that if an irreversible process occurs in an isolated system, the entropy of some of the subsystems may decrease, but the entropy of other subsystems must then increase in such an amount that the entropy of the overall isolated system will always increase. As the isolated system has constant internal energy, constant volume, and constant total mass, the above statement is expressed as:

$$dS_{U,V,m} \geq 0 \tag{2.16}$$

This *extremum principle*, generally identified as the *second law of thermodynamics*, is a natural consequence of the convention of the sign assigned to heat exchanged and the definition of an entropy change by Equation 2.12, more clearly seen in the form of Equation 2.15. When two bodies exchange heat, the entropy increase of the body at the lower temperature will always be larger than the entropy decrease of the body at the higher temperature. This is so because the same amount of heat exchanged is divided by a larger value of absolute temperature in the case of the hotter body. As all real processes are irreversible, if we accept that our physical universe is an isolated system, the conclusion is that the entropy of the universe increases with each and every process that occurs. From this general statement, it can be shown that natural processes occur in one and only one direction. Thus, for example, if a hot body is put in contact with a cold body, heat will always flow from the hot body to the cold body and never in the opposite direction unless there are other changes in the surroundings.

Confirmation of Agreement with the Directional Nature of Heat Flow

Consider a system, isolated from its surroundings, which is formed by two subsystems that can only exchange energy between each other in the form of heat. Designating the total system by t and the subsystems by α and β, we have:

$$dS^t_{U,V,m} = \left(dS^\alpha + dS^\beta\right) \geq 0$$

If there is no work done, according to Equation 2.1c the heat exchanged in a real process is equal to the change in internal energy, and thus it is independent of the nature of the path followed. The same amount of energy can be exchanged between the two subsystems following a reversible path. Hence,

$$\frac{\delta Q^\alpha}{T^\alpha} + \frac{\delta Q^\beta}{T^\beta} \geq 0$$

As the total system is isolated, it follows that, $\delta Q^\alpha = -\delta Q^\beta$, or

$$\left[\frac{1}{T^\alpha} - \frac{1}{T^\beta}\right]\delta Q^\alpha \geq 0$$

If $\delta Q^\alpha \geq 0$, this means that heat enters the system α. Hence, to satisfy the above inequality, necessarily, $T^\beta \geq T^\alpha$ and vice versa. The process will continue until both temperatures are equal.

Heat can be transferred from a "cold" system to a "hot" system, but for this to occur, other changes in the surroundings, causing an overall increase in entropy of the universe, must take place. If a system decreases its entropy in an irreversible process, one or more parts of its surroundings increase their entropy in a larger amount, so the net change of entropy of the universe is always positive. This directional evolution of the universe may be referred to as the *arrow of time.*

Heat can be passed from a cold source to a hot sink by a cyclic process, as it happens in a refrigerator, in a home air conditioner, or in a heat pump. *In a cyclic process, after exchanging energy in the form of heat or work with the surroundings, the system returns to its initial state at the end of each cycle.* Thus, all net changes that mark that a process has occurred are to be found in the surroundings. We observe here that a compressor or pump transfers energy in the form of work to the system, and thus its work is assigned a positive sign, while a turbine extracts energy from the system, and thus its work, with respect to the system, is assigned a negative sign. Cyclic processes, employing a working fluid, are commonly used not only to transfer heat from a cold source to a hot sink but also to produce useful work from heat energy. Both of these cases, which provide examples of the practical use of the concept of entropy, are considered below. For the analysis of cyclic devices, we are not concerned about the temperature of the circulating fluid. All we need to know is that for the fluid to receive heat from a source, its highest temperature should be below the temperature of the source, and to give heat to a sink, its lowest temperature should be above the temperature of the sink. These temperature differences required for heat transfer are causes of further irreversibility.

REFRIGERATORS, AIR CONDITIONERS, AND HEAT PUMPS

Most refrigerators, air conditioners (including dehumidifiers), and heat pumps use the *vapor–compression refrigeration cycle* shown in Figure 2.2.

In each cycle, the working fluid passes through four units. It enters the refrigeration unit as a liquid, receives an amount of heat q_L from the cold source at a constant low temperature T_L, and it evaporates. The vapor is then compressed in an adiabatic compressor, raising its pressure and temperature. The hot compressed vapor is sent to a heat exchanger where it condenses giving an amount of heat q_H to a hot sink at a constant high temperature T_H. The liquid is then passed through an adiabatic valve or through an adiabatic turbine to produce some work and closes the cycle reentering the refrigeration unit. To visualize this, in a home refrigerator the cold source is the interior of the fridge, while the air exterior to the refrigeration unit serves as a hot sink.

According to the first law, Equation 2.2,

$$\Delta u_{cycle} = q_{net} + w_{net} = 0$$

After each cycle, there is no change in internal energy, or in entropy, for the circulating fluid and the energy received by the fluid in the form of work and as heat from the cold source is passed as heat to the hot sink:

$$|q_H| = |q_L| + |w_{net}| \tag{2.17}$$

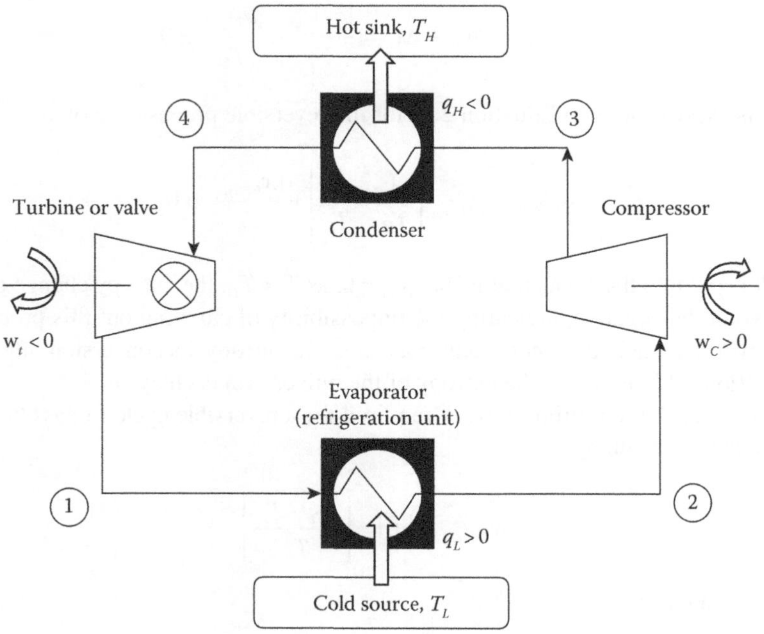

FIGURE 2.2 Refrigeration device (signs for heat and work refer to the circulating fluid). Note that T_H and T_L refer to the temperatures of the hot sink and cold source, respectively, and not to the temperatures of the circulating fluid.

Equation 2.17 shows that the absolute value of the heat released by a refrigerator to the surroundings is larger than the absolute value of heat received from the cold source. This is why air conditioners and heat pumps have their high-temperature heat exchanger outside the space that is being cooled. Otherwise, the unit would act as a heater. A refrigerator or a dehumidifier heats the room where it operates. For the surroundings, the signs for heat and work are reversed with respect to the signs of these quantities for the working fluid; consequently, by using their absolute values the above relation is valid for both the circulating fluid and the surroundings. The performances of the devices used for transferring heat from a cold source to a hot sink are measured by the magnitude of ratio of the value of the desired effect to the value of the required duty. This ratio is known as the coefficient of performance (*COP*), and with the exception of heat pumps, it takes the form:

$$(COP)=\frac{|q_L|}{|w_{net}|} \tag{2.18}$$

The comparison of performance of different devices of the same kind is better done by referring their (*COP*) to a common basis, namely, the (*COP*) of a reversible device operating between the same temperatures of the source and the sink of heat:

$$(COP)_{rev}=\frac{|q_L|_{rev}}{|w_{net}|_{rev}}$$

This value is obtained recalling that there is no entropy change due to the action of reversible work. Thus, the only places where an entropy change can be found at the end of a cycle are the cold source and the hot sink. According to the second law of thermodynamics, for the reversible process to be possible, the following condition must hold:

$$\Delta s_{U,V,m}=\Delta s_H+\Delta s_L=\frac{|q_H|_{rev}}{T_H}-\frac{|q_L|_{rev}}{T_L}\geq 0 \tag{2.19a}$$

Combining this expression with Equation 2.17 for the reversible process, we obtain:

$$\Delta s_{U,V,m}=|q_L|_{rev}\left[\frac{1}{T_H}-\frac{1}{T_L}\right]+\frac{|w_{net}|_{rev}}{T_H}\geq 0 \tag{2.19b}$$

If no net work is put into the cycle, that is, $|w_{net}|_{rev}=0$, as $T_L<T_H$, the entropy change of the universe in one cycle would be negative, indicating the impossibility of carrying out this process. Although the reversible work, as such, does not produce a change in entropy, its conversion into heat released, shown in Equation 2.17, increases the entropy of the universe in each cycle.

From Equation 2.19, the minimum work required for a reversible cycle to pass heat from a cold source to a hot sink is given by:

$$|w_{net}|_{rev}=|q_L|_{rev}\left[\frac{T_H-T_L}{T_L}\right]$$

Rearranging, we obtain:

$$(COP)_{rev}=\frac{|q_L|_{rev}}{|w_{net}|_{rev}}=\left[\frac{T_L}{T_H-T_L}\right] \tag{2.20}$$

Therefore, the second law efficiency of a refrigerating device is given by:

$$\text{Refrigerator Second Law Efficiency} = \left[\frac{|q_L|}{|w_{net}|}\right]_{actual} \frac{T_H - T_L}{T_L} \tag{2.21}$$

For a heat pump, on the other hand, the desired effect is the heat delivered to the hot sink, and the coefficient of performance is defined as:

$$(COP) = \frac{|q_H|}{|w_{net}|} \tag{2.22}$$

Hence, repeating the same algebra, we obtain:

$$(COP)_{rev} = \frac{|q_H|_{rev}}{|w_{net}|_{rev}} = \left[\frac{T_H}{T_H - T_L}\right] \tag{2.23}$$

and

$$\text{Heat Pump Second Law Efficiency} = \left[\frac{|q_H|}{|w_{net}|}\right]_{actual} \frac{T_H - T_L}{T_H} \tag{2.24}$$

HEAT ENGINES

One of the major steps made by humanity toward modern society was the conversion of heat into useful work using cyclic processes. Heat coming from wood, coal, oil, natural or liquefied gas, or even atomic energy can not only be used to replace human effort in doing work but also to generate electrical power. Electricity can then be transmitted long distances to move engines and give light and heat. The second law, in this case, says that it is impossible to convert *all* the heat into work in a cyclic process. This conclusion comes directly from the considerations given above.

Let us discuss a reversible process to see this effect. As the reversible work does not contribute to entropy changes, the single effect observed when converting all the heat into work in a cycle is that the hot source gives up an amount of heat, q_H, at a constant high temperature T_H, thus decreasing its entropy. This decrease in entropy would be the change in entropy of the universe caused by the process, thus making it impossible. On the other hand, if part of the heat received by the working fluid from the hot source were given away to a cold sink, increasing the entropy of the cold sink by a larger amount, the process would become possible.

A *cyclic heat engine*, as shown in Figure 2.3, uses a working fluid operating in a cycle that is effectively the reverse of the one described for the vapor–compression refrigeration devices.

In this case, a liquid is vaporized in a high-temperature, high-pressure evaporator and passed through a turbine to produce work. The low-pressure vapor coming from the turbine is sent to a lower-temperature condenser. The liquid from the condenser is then pumped back to the evaporator.

We consider that the turbine and the pump operate reversibly and adiabatically, and thus do not change the entropy of the operating fluid or the entropy of its surroundings. Similarly, we consider that the hot source gives an amount of heat, q_H, at a constant high temperature, T_H, and the cold sink receives an amount of heat, q_L, at a constant lower temperature, T_L.

Equation 2.17 expresses the first law for the working fluid in one cycle. As necessarily some heat must be lost to a cold sink to make the process feasible, it is of interest to know what would be the maximum possible efficiency of such a cyclic engine. As the heat from the hot source is the one that has a cost and the work obtained is the desired effect, the *thermal efficiency* of a cycle, η_{th}, is defined as the ratio of the net work obtained to the amount of heat received:

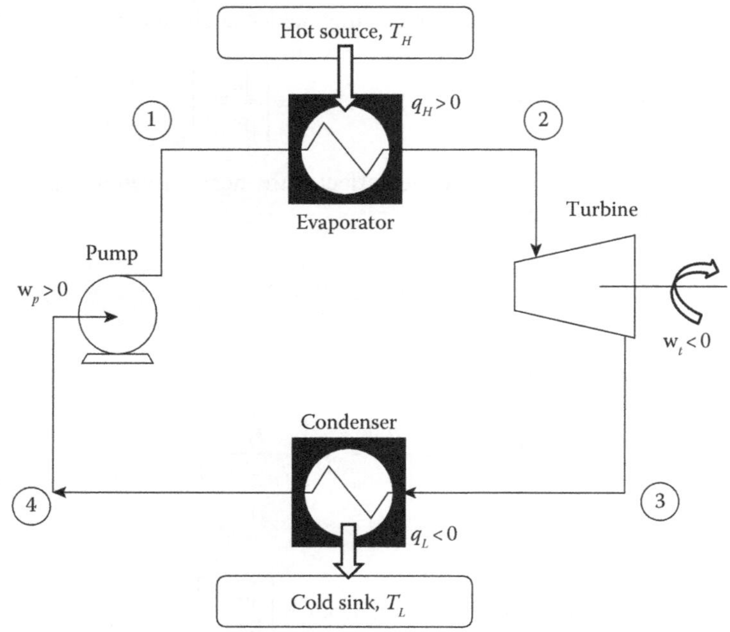

FIGURE 2.3 Heat engine (signs for heat and work refer to the circulating fluid). Note that T_H and T_L refer to the temperatures of the hot source and cold sink, respectively, and not to the temperatures of the circulating fluid.

$$\eta_{th} = \frac{|w|}{|q_H|} \tag{2.25}$$

Combining with Equation 2.17, we obtain:

$$\eta_{th} = 1 - \frac{|q_L|}{|q_H|}$$

Hence, the greater the amount of heat passed to the cold sink, the lower the efficiency. As the second law indicates that the value of q_L cannot be zero, the changes in entropy of the hot source and cold sink are used to determine the limiting value for the ratio $|q_L/q_H|$.

For this cycle, in a reversible engine it is necessary to have:

$$\Delta s_{U,V,m} = \Delta s_H + \Delta s_L = -\frac{|q_H|_{rev}}{T_H} + \frac{|q_L|_{rev}}{T_L} \geq 0$$

or

$$\frac{|q_L|_{rev}}{|q_H|_{rev}} \geq \frac{T_L}{T_H}$$

Thus, the maximum possible efficiency of a cyclic thermal engine is:

$$\eta_{th} \leq \left(1 - \frac{T_L}{T_H}\right) \tag{2.26}$$

As an illustration, we consider that the temperature of the hot source is about 700 K and heat is released to the atmosphere, which is at about 300 K. The maximum possible efficiency of such a reversible (imaginary) thermal engine will not exceed 57%. As the temperature of the atmosphere is not a controllable variable, one way to increase the efficiency of the cyclic engine is to increase the temperature of the hot source. Therefore, in practice, the circulating fluid is evaporated, and the vapor is then superheated in contact with a high-temperature hot source.

An interesting aspect to ponder is whether it would be possible to have a device that converts all the heat into work. The answer is that a noncyclic reversible device may seem to approach this total conversion but will never reach this limit.

Consider a gas contained in a cylinder–piston device that has a heater at its base. The complete system, including the movable piston, is thermally insulated from the surroundings. As heat is added to the system, the gas expands and moves the piston against the atmospheric pressure. At first, it would seem that all heat added was converted into work. However, as the volume of the gas has changed in the process, while the pressure has been kept constant, its temperature has also changed (did it increase or decrease?). So, some of the energy given by the heater is accumulated by the gas. The state of the gas at the end of the process is not the same as at the beginning. Any attempt to return the gas to its initial state would require to perform work on the system or release heat.

PROBLEMS

2.1 **Work performed in an ideal gas expansion**

A long horizontal cylinder fitted with a piston contains 0.1 mol of an ideal gas at a pressure of 5 bar. The piston is initially held in place by a stop, which is then released allowing the piston to move and the gas to expand against the external pressure of 1 bar. If the whole process takes place at 298 K constant temperature, calculate (a) the work done by the gas; (b) the work performed on the external atmosphere; and (c) the lost work, and where it goes.

2.2 **General efficiency of thermal cycles**

Consider a fluid undergoing the arbitrary reversible cycle shown in the $T-S$ diagram of Figure 2.4. Along the cycle, the entropy of the fluid moves between a minimum value S_A and a maximum value S_B. In the upper section of the cycle, going from state A to state B, the entropy increases as the fluid absorbs heat from one or more thermal sources, not at a constant temperature. The opposite happens in the lower section of the cycle, where the fluid returns from state B to state A by releasing heat to one or more thermal sinks, again at variable temperature.

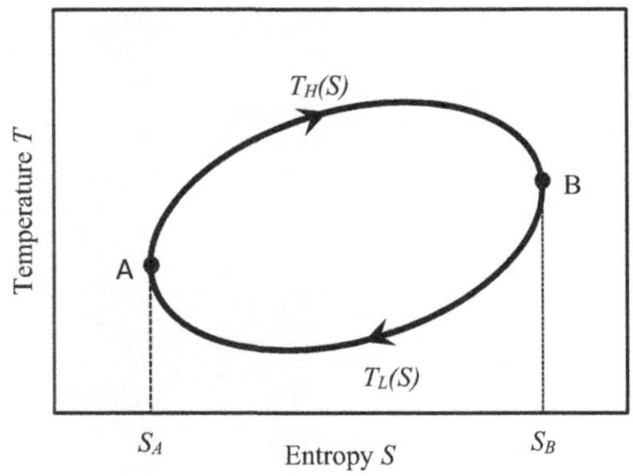

FIGURE 2.4 A general thermal cycle with non-isothermal heat input and output.

a. Prove that the thermal efficiency of the cycle depends only on the *average* temperatures at which heat is accepted and rejected.
b. If the maximum high temperature available is T_{Hi}, and the minimum low temperature available is T_{Lo}, what is the most efficient cycle possible?

REFERENCE

1. International Union of Pure and Applied Chemistry, Physical and Biophysical Chemistry Division. 2007. *Quantities, Units and Symbols in Physical Chemistry*, 3rd ed. Cambridge, UK: RSC Publishing, section 2.11.

3 Conservation of Energy in an Open Flow System
Definition of Enthalpy

ABSTRACT

Chapter 3 presents one of the most useful practical applications of the energy balances and introduces the state function enthalpy as the sum of the internal energy of a unit of mass and the product of the acting pressure times its molar volume. The energy balance of a unit of mass moving inside a duct considers the possible changes in kinetic, potential, and internal energy of the unit of mass, plus the work involved to move it in and out of the duct, done by the pressure at both ends, the work received from any intermediate engine and the energy exchanged with the surroundings as heat. For generality, the energy balance is considered first for a system in unsteady state and then limited to the case of steady state. Two common applications are discussed in detail: the mechanical energy balance commonly used in engineering studies of piping systems, and the use of the enthalpy function for the study of closed systems.

In a completely rigorous line of thought, it can be argued that principles of classical thermodynamics are only applicable to systems at equilibrium. In fact, there is a separate branch of thermodynamics dedicated to nonequilibrium cases. However, being aware of the approximation involved, there is one simple application of the energy balance that is worth discussing here.

CLOSED SYSTEM MOVING IN SPACE

First, consider the case of a closed system of mass, m, moving in space. In this case, we distinguish three components of its total energy, E: its internal energy, U; its kinetic energy, E_K, with respect to an external point of reference; and its potential energy, E_P, with respect to a fixed plane of reference. Thus,

$$E = U + E_K + E_P \tag{3.1}$$

or, per unit mass,

$$e = u + e_k + e_p \tag{3.2}$$

OPEN FLOW SYSTEM AT UNSTEADY STATE

Now, let us consider a control volume in which fluid enters by several streams, i, and fluid leaves by several streams, j. Per unit time, indicated by a dot over the symbol, in each stream i enters a mass $\delta\dot{m}_i$ of fluid and in each stream j exits a mass $\delta\dot{m}_j$. In addition, per unit time, the system exchanges an amount of heat, $\delta\dot{Q}$, and an amount of work, $\delta\dot{W}$, with the surroundings. Thus, the change of energy in the control volume, per unit time, is:

$$d\dot{E} = \sum_{in}(u + e_K + e_P)_i\,\delta\dot{m}_i - \sum_{out}(u + e_K + e_P)_j\,\delta\dot{m}_j + \delta\dot{Q} + \delta\dot{W} \tag{3.3}$$

DOI: 10.1201/9781003431985-4

For the work, we can distinguish two components: the work associated with a moving shaft, or shaft power $\delta \dot{W}_s$, and the work associated with the fluid entering in each stream against a pressure P_i and with the fluid going out in each stream against a pressure P_j. Per unit mass, the work associated with the movement of the fluid is equal to the product of its specific volume and the prevailing pressure at the point of interest. This can be visualized considering that the force is the pressure times the surface area over which it acts, and when multiplying by the displacement necessary to move a unit of mass, the product of the surface area times the displacement gives the specific volume of the fluid. Hence,

$$\delta \dot{W} = \delta \dot{W}_s + \left(\sum_{in} (Pv)_i \, \delta \dot{m}_i - \sum_{out} (Pv)_j \, \delta \dot{m}_j \right) \tag{3.4}$$

Combining both equations, we write:

$$d\dot{E} = \sum_{in} (h + e_K + e_P)_i \, \delta \dot{m}_i - \sum_{out} (h + e_K + e_P)_j \, \delta \dot{m}_j + \delta \dot{Q} + \delta \dot{W}_s \tag{3.5}$$

DEFINITION OF ENTHALPY

In Equation 3.5, we have used the symbol h for the sum of the internal energy u and the product Pv:

$$h = u + P\text{v} \tag{3.6a}$$

or

$$H = U + PV \tag{3.6b}$$

All terms on the right-hand side of Equation 3.6 *are a function of the state of the system. Thus, the function* h *is also a state function, and it is called enthalpy. Obviously, the definition of enthalpy is independent of the movement of a fluid, so it is equally valid for a closed system.*

OPEN FLOW SYSTEM AT STEADY STATE

One important application of Equation 3.5 is the case of a steady state in which the mass and energy of the system, as well as all flows of mass and energy, are constant with time. In this case, we remove the symbol δ that was introduced to account for possible variations of the flows with time. Thus, setting $d\dot{E}=0$ in Equation 3.5, for a system at steady state, we write:

$$\sum_{out} (h + e_K + e_P)_j \, \dot{m}_j - \sum_{in} (h + e_K + e_P)_i \, \dot{m}_i = \dot{Q} + \dot{W}_s \tag{3.7}$$

In this case, the total mass entering the system is equal to the total mass leaving the system:

$$\sum_{in} \dot{m}_i = \sum_{out} \dot{m}_j \tag{3.8}$$

A common situation encountered in engineering is to have a system with only one stream entering and one stream leaving, as the one shown in Figure 3.1. For this case, at steady-state operation, we write:

$$\dot{m}_i = \dot{m}_j = \dot{m} \tag{3.9}$$

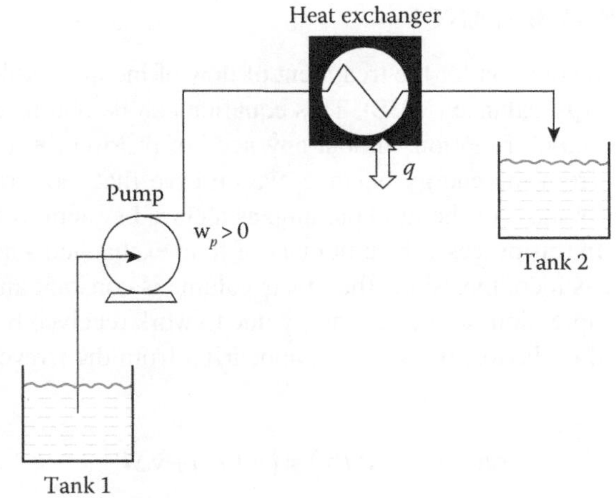

FIGURE 3.1 Open flow system.

Hence, the energy balance equation takes the form:

$$(\Delta h + \Delta e_K + \Delta e_P)\dot{m} = \dot{Q} + \dot{W}_s \tag{3.10}$$

where Δ indicates the difference between the value of the property at the exiting stream minus the value of the property at the intake stream. Dividing both sides by $\dot{m}$,

$$\Delta h + \Delta e_K + \Delta e_P = q + w_s \tag{3.11}$$

All quantities in this equation are expressed per unit mass of fluid. The kinetic energy term, written in terms of the linear velocity $\dot{v}$ of the fluid, that is, the volumetric flow rate over the surface area of the duct in which it flows, takes the form[1]:

$$\Delta e_K = \frac{(\dot{v}_{out})^2 - (\dot{v}_{in})^2}{2} \tag{3.12}$$

Similarly, the potential energy term, written in terms of the elevation of the point over an arbitrarily set reference plane, Z, takes the form:

$$\Delta e_P = g(Z_{out} - Z_{in}) \tag{3.13}$$

where *g* is the acceleration of gravity. These two terms are usually negligible in comparison with the enthalpy change, and they are normally omitted. Equation 3.7 without these terms is sometimes referred to as the *enthalpy balance*. For the case of negligible contribution of the kinetic and potential energy terms, Equation 3.11 reduces to:

$$\Delta h = q + w_S \tag{3.14}$$

Equation 3.14 has a direct similitude with Equation 2.1, the first law for closed systems. It is particularly useful for the analysis of compressors and turbines dealing with gases or steam, and also in the study of distillation columns, evaporators, and concentration tank operations.

MECHANICAL ENERGY BALANCE

Finally, a commonly used equation for the treatment of flow of incompressible fluids in pipes is the so-called mechanical energy balance (MEB). This equation can be obtained from fluid mechanics by integration of the equation of motion without any need of passing through the energy balance. To obtain this equation from the energy balance, we observe that the term Δu is the change in internal energy of a unit of mass of the fluid passing, as a closed system, from the state at the inlet to the state at the outlet. In this process, the unit mass of fluid exchanged a net amount of heat q. As the fluid considered here is incompressible, the molar volume is constant and there is no reversible work of expansion or compression, so the net energy due to work received by the system, according to Equation 2.10, is equal to the lost work, $w = l_w$, appearing from the irreversibility effects caused by friction:

$$\Delta h = \Delta u + \Delta(P\text{v}) = (q + l_w) + \text{v}\Delta P \tag{3.15}$$

Replacing Equations 3.12, 3.13, and 3.15 into Equation 3.11, we write the MEB as:

$$\text{v}(P_{out} - P_{in}) + \frac{\dot{\upsilon}_{out}^2 - \dot{\upsilon}_{in}^2}{2} + g(Z_{out} - Z_{in}) - w_s + l_w = 0 \tag{3.16a}$$

or

$$\text{v}\Delta P + \frac{\Delta\dot{\upsilon}^2}{2} + g\Delta Z - w_s + l_w = 0 \tag{3.16b}$$

As stated above, the linear velocity of the fluid is equal to the volumetric flow rate of the fluid divided by the surface area of the duct. Thus, with the exception of l_w, all terms in this equation are measurable and the value of l_w can be back-calculated. This has been done for many typical flow configurations of practical interest, and engineering correlations are available to calculate the value of l_w for different designs.

While the energy balance for an open flow system, Equation 3.11, is valid for both compressible and incompressible fluids, the MEB as written, Equation 3.16, is only valid for incompressible fluids. The fluid mechanics of compressible fluids is complex, and it is better treated through equations pertaining to transport phenomena. The fact that the molar volume of compressible fluids is a function of not only temperature but also pressure creates a completely new reality when the fluid flows.

At ordinary (room) temperatures and pressures, real gases, with the exception of hydrogen, helium, and neon, cool down upon expansion. The three latter gases experience the same effect, but only at much lower temperatures. The steady flow characteristics of a compressible fluid in a horizontal duct of a constant cross-surface area depend on whether the velocity at the entrance is subsonic or supersonic. When the gas enters at subsonic velocity, the fluid expands as the pressure drops and the temperature decreases. The entropy and velocity increase until reaching the velocity of sound at which the entropy is at its maximum. If this maximum length is exceeded, chocking occurs and the entrance conditions cannot be maintained. Chocking is a discontinuity; there is an abrupt increase in pressure, temperature, and density that reduces the downstream velocity to the subsonic range.

Supersonic flow at the entrance of the duct can be accomplished with a special nozzle. In this case, there is a compression of the gas as it flows. Thus, there is an increase in pressure, density, temperature, and entropy down the duct. The velocity decreases. Stable flow conditions can only be maintained if the length of the duct is less than or equal to a critical length at which a sonic discharge velocity is produced while the entropy is at its maximum value. If the duct is longer, a shock wave will be produced with an abrupt increase in pressure, temperature, and density that reduces the downstream velocity to subsonic values. A further increase of the duct length will cause the shock wave to move back toward the entrance, traveling at a velocity larger than sonic. When it reaches the entrance, only subsonic velocities will exist in the duct. Beyond this, the shock wave will move into the nozzle feeding the duct.

The thermodynamics of compressible fluid flow requires elements of equations of state, which is covered in Chapter 5. A more detailed thermodynamic treatment of the adiabatic flow of gases is presented in Chapter 23.

USE OF THE ENTHALPY FUNCTION IN CLOSED SYSTEMS

Although the enthalpy function defined by Equation 3.6 is particularly useful for the treatment of open flow systems, it is also applicable to the study of closed systems undergoing changes at constant pressure. If the pressure of the closed system is constant, according to Equation 2.4 the expansion or compression work exchanged with the surroundings is directly given by the product of the pressure times the change in volume of the system. Thus, in this case, we can write Equation 2.2 as:

$$du_P = \delta q + \delta w_P = \delta q - P d\text{v}$$

$$d(u + P\text{v})_P = \delta q$$

$$dh_P = \delta q$$

Thus, for a constant pressure process in a closed system, we write:

$$\Delta h_P = q \tag{3.17}$$

As the heat exchanged in a process is a measurable quantity, Equation 3.17 provides values for enthalpy changes in mixing processes.

PROBLEMS

3.1 **Energy balance in a pump**

A centrifugal pump, operated by an electric motor, pumps water at 25°C at a rate of 8 m^3/h. At steady state, the motor consumes 0.94 kW of electric power. The pressure difference measured between the suction and the discharge of the pump is 310.5 kPa. Consider that the suction and the discharge pipes are of the same diameter and the two points where the pressures are measured are at the same height. The density of liquid water can be taken as 1000 kg/m^3.

a. Calculate the efficiency of the motor-pump unit. The efficiency is defined as the ratio between the "net power" received by the fluid and the electric power consumed by the motor.
b. Calculate the rate of entropy increase of the universe.
c. How much the answers to the above questions would change if the diameters of the suction and discharge pipes are 4.4 and 4.0 cm, respectively, and the difference in height between the two points where the pressures are measured is 20 cm, the discharge being higher than the suction? The pressure difference measured between the suction and the discharge of the pump is 310.5 kPa.

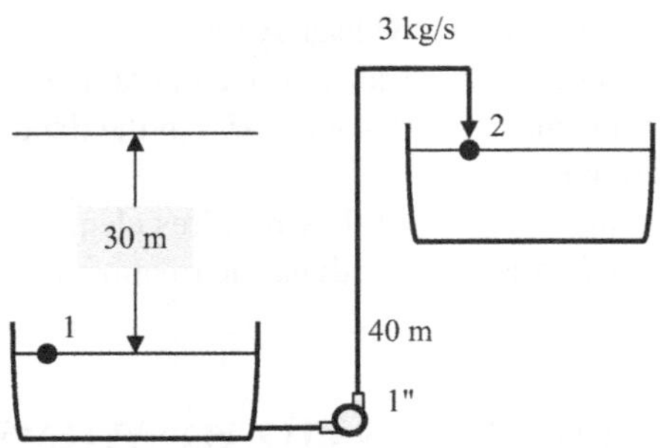

FIGURE 3.2 Hydraulic circuit for Problem 3.2.

Hint: The net work received by the fluid is equal to the work done by the pump over the fluid minus the work lost by the fluid in the pump.

3.2 **Energy balance in a hydraulic circuit**

3 kg/s of water at 10°C are pumped from storage tank 1 to raised tank 2 as shown in Figure 3.2. The water level in tank 2 is 30 m above the level in tank 1. The pipe is 40 m long, and its diameter is 1″. Since the pipe is not insulated, heat is transferred from the environment, which is at 25°C to the flowing water, at a rate of 210 W/m of pipe length. The nominal power of the pump is 3 HP and its efficiency is 90%. The density of liquid water can be taken as 1000 kg/m^3. Calculate

a. The lost work.
b. The enthalpy change of water between tanks 1 and 2.
c. The rate of entropy increase of the "universe."

3.3 **Energy balance in a pipe**

Water at 25°C flows in a horizontal ¾″ diameter, schedule 40, steel pipe. Two pressure measuring devices are installed in the pipe at 4.91 m distance one from the other. At a flow rate of 3.28 m^3/h, the pressure drop between the two points is 22.56 kPa, and at a flow rate of 6.16 m^3/h, the pressure drop is 76.22 kPa. A ¾″ nominal diameter, schedule 40, steel pipe has a 2.03 cm internal diameter.

Calculate the rate of increase of the entropy of the universe per meter of pipe in both cases, if there is no heat exchange between the pipe and the surroundings.

NOTE

1 From equations for momentum transport, it has been shown [1] that when there is a well-developed velocity profile in a cross section area, the kinetic energy term has the form $\Delta e_k = \dfrac{1}{2}\Delta\dfrac{\dot{\upsilon}^3_{ave}}{\dot{\upsilon}_{ave}}$. For most cases, the kinetic energy term represents a small contribution; thus, Equation 3.12 is commonly used.

REFERENCE

1. Bird, R. B. 1957. The equations of change and the macroscopic mass, momentum, and energy balances. *Chem. Eng. Sci.* 6: 123–181.

4 The Algebra of State Functions
The Helmholtz and Gibbs Functions

Abstract

Two important additional state functions are added to the internal energy U and enthalpy H already discussed. These are the Helmholtz A and the Gibbs G functions, defined as the difference of U and H with the product of the absolute temperature times the entropy, respectively. For an open system, the total differential of these state functions is expressed as the sum of the partial derivative with respect to their independent variables, times the differential of the independent variable. The all-important chemical potential of compound i in a mixture is defined as the partial derivative of the state function with respect to the number of moles of compound i, at constant values of the other independent variables. The fact that for state functions the order of differentiation with respect to two variables is immaterial is used to obtain Maxwell's relations which are the stepping stone to relate thermodynamic functions with measurable properties. The presentation of other useful mathematical relations for partial derivatives completes the chapter.

Until now, we have encountered four state functions of a system characterized by the measurable variables: absolute pressure, P; absolute temperature, T; and the composition of the system expressed by the number of moles of each compound present, n_i. These state functions are the total volume, V; the internal energy, U; the entropy, S; and the enthalpy, H. Two more state functions are used in thermodynamics. These are the *Helmholtz function*:

$$A = U - TS \tag{4.1}$$

and the *Gibbs function*:

$$G = H - TS \tag{4.2}$$

THERMODYNAMIC FUNCTIONS IN OPEN SYSTEMS: THE CHEMICAL POTENTIAL

The systems studied in the previous chapters were closed, so changes in composition were not considered. Thus, the dependence of state functions on the number of moles of each compound was left implicit. State functions have exact differentials, meaning that the value obtained by integration of their differentials between two states is independent of the path followed. We observe that Equation 2.13 suggests that the natural variables for the function U, internal energy, in a differential expansion should be S and V. If we write the total differential of the internal energy, considering its dependence on composition, we have:

$$dU = \left(\frac{\partial U}{\partial S}\right)_{V,n} dS + \left(\frac{\partial U}{\partial V}\right)_{S,n} dV + \sum_i^c \left(\frac{\partial U}{\partial n_i}\right)_{S,V,n_{j\neq i}} dn_i \tag{4.3}$$

DOI: 10.1201/9781003431985-5

The number of moles of each compound may change because the system is open to the surroundings and mass enters or leaves the system, or because there are chemical reactions transforming moles of some compounds into moles of some other compounds. For a closed nonreacting system, the last term of the right-hand side cancels out; thus, comparing with Equation 2.13, we conclude that:

$$T = \left(\frac{\partial U}{\partial S}\right)_{V,n} \tag{4.4}$$

and

$$-P = \left(\frac{\partial U}{\partial V}\right)_{S,n} \tag{4.5}$$

In addition, for simplicity, we introduce the symbol μ_i for the partial derivative of the total internal energy with respect to n_i at constant $n_{j\neq i}$, S, and V, and call it the *chemical potential*:

$$\mu_i \equiv \left(\frac{\partial U}{\partial n_i}\right)_{S,V,n_{j\neq i}} \tag{4.6}$$

Hence, we rewrite Equation 4.3 as:

$$dU = TdS - PdV + \sum_i^c \mu_i dn_i \tag{4.7}$$

Equation 4.7 is sometimes referred to as the *differential form of the fundamental equation of thermodynamics*. For a system of constant composition, Equation 4.7 reduces to Equation 2.13. The composition of the system can change because it is open or because chemical reactions occur.

For the enthalpy of a system with changes in composition during a process, from Equation 3.6,

$$dH = dU + PdV + VdP \tag{4.8}$$

Thus, using Equation 4.7, we can write:

$$dH = TdS + VdP + \sum_i^c \mu_i dn_i \tag{4.9}$$

Again, from the expression of the exact differential of H, in terms of S, P, and n, we write:

$$dH = \left(\frac{\partial H}{\partial S}\right)_{P,n} dS + \left(\frac{\partial H}{\partial P}\right)_{S,n} dP + \sum_i^c \left(\frac{\partial H}{\partial n_i}\right)_{S,P,n_{j\neq i}} dn_i \tag{4.10}$$

So, we obtain:

$$T = \left(\frac{\partial H}{\partial S}\right)_{P,n} \tag{4.11}$$

$$V = \left(\frac{\partial H}{\partial P}\right)_{S,n} \tag{4.12}$$

$$\mu_i = \left(\frac{\partial H}{\partial n_i}\right)_{S,P,n_{j\neq i}} \tag{4.13}$$

A similar exercise with Equation 4.1 for A and Equation 4.2 for G gives:

$$-S = \left(\frac{\partial A}{\partial T}\right)_{V,n} \tag{4.14}$$

$$-P = \left(\frac{\partial A}{\partial V}\right)_{T,n} \tag{4.15}$$

$$\mu_i = \left(\frac{\partial A}{\partial n_i}\right)_{T,V,n_{j\neq i}} \tag{4.16}$$

$$-S = \left(\frac{\partial G}{\partial T}\right)_{P,n} \tag{4.17}$$

$$V = \left(\frac{\partial G}{\partial P}\right)_{T,n} \tag{4.18}$$

$$\mu_i = \left(\frac{\partial G}{\partial n_i}\right)_{P,T,n_{j\neq i}} \tag{4.19}$$

Thus, to gain perspective, we group the expressions involving the same property as:

$$T = \left(\frac{\partial U}{\partial S}\right)_{V,n} = \left(\frac{\partial H}{\partial S}\right)_{P,n} \quad ;\text{ (4.4) and (4.11)}$$

$$-P = \left(\frac{\partial U}{\partial V}\right)_{S,n} = \left(\frac{\partial A}{\partial V}\right)_{T,n} \quad ;\text{ (4.5) and (4.15)}$$

$$V = \left(\frac{\partial H}{\partial P}\right)_{S,n} = \left(\frac{\partial G}{\partial P}\right)_{T,n} \quad ;\text{ (4.12) and (4.18)}$$

$$-S = \left(\frac{\partial A}{\partial T}\right)_{V,n} = \left(\frac{\partial G}{\partial T}\right)_{P,n} \quad ;\text{ (4.14) and (4.17)}$$

In the above differentiations, the numbers of moles of all the compounds are held constant, so the same relations can be written in terms of the molar properties of the system.

For the chemical potential, Equations 4.6, 4.13, 4.16, and 4.19 give:

$$\mu_i \equiv \left(\frac{\partial U}{\partial n_i}\right)_{S,V,n_{j\neq i}} = \left(\frac{\partial H}{\partial n_i}\right)_{S,P,n_{j\neq i}} = \left(\frac{\partial A}{\partial n_i}\right)_{T,V,n_{j\neq i}} = \left(\frac{\partial G}{\partial n_i}\right)_{P,T,n_{j\neq i}}$$

We observe that in all cases, the natural or "canonical" variables for U are S and V, for H are S and P, for A are T and V, and for G are P and T. The main use of the above relations is for the derivation of the so-called *Maxwell's relations*, which are the stepping stones for the evaluation of the thermodynamic functions from measurable variables.

MAXWELL'S RELATIONS

For a state function of several variables, the order of partial differentiation with respect to two of the variables is immaterial. This is, in fact, a property of all functions having an exact differential. Hence, in general, for a function X of variables x_1 and x_2, among other variables that are held constant, we write:

$$\frac{\partial}{\partial x_1}\left[\left(\frac{\partial X}{\partial x_2}\right)_{x1}\right]_{x2} = \frac{\partial}{\partial x_2}\left[\left(\frac{\partial X}{\partial x_1}\right)_{x2}\right]_{x1}$$

Thus, taking the variables V and S for the function U, we have:

$$\frac{\partial}{\partial V}\left[\left(\frac{\partial U}{\partial S}\right)_{V,n}\right]_{S,n} = \frac{\partial}{\partial S}\left[\left(\frac{\partial U}{\partial V}\right)_{S,n}\right]_{V,n}$$

or from Equations 4.4 and 4.5,

$$\left[\frac{\partial T}{\partial V}\right]_{S,n} = -\left[\frac{\partial P}{\partial S}\right]_{V,n} \tag{4.20}$$

Similarly, from the state function H, with variables P and S, from Equations 4.11 and 4.12, we write:

$$\left[\frac{\partial T}{\partial P}\right]_{S,n} = \left[\frac{\partial V}{\partial S}\right]_{P,n} \tag{4.21}$$

From A, with variables T and V, from Equations 4.14 and 4.15, we get:

$$\left[\frac{\partial P}{\partial T}\right]_{V,n} = \left[\frac{\partial S}{\partial V}\right]_{T,n} \tag{4.22}$$

and from G, with variables P and T, from Equations 4.17 and 4.18,

$$-\left[\frac{\partial S}{\partial P}\right]_{T,n} = \left[\frac{\partial V}{\partial T}\right]_{P,n} \tag{4.23}$$

In addition to these "classic" Maxwell's relations, which are particularly useful to relate the thermodynamic functions with the measurable variables P, T, and V, there are another two that are important for μ_i, the chemical potential. From Equations 4.17 and 4.19,

$$\frac{\partial}{\partial T}\left[\left(\frac{\partial G}{\partial n_i}\right)_{T,P,n_{j\neq i}}\right]_{P,n} = \frac{\partial}{\partial n_i}\left[\left(\frac{\partial G}{\partial T}\right)_{P,n}\right]_{T,P,n_{j\neq i}}$$

$$\left[\frac{\partial \mu_i}{\partial T}\right]_{P,n} = -\left[\frac{\partial S}{\partial n_i}\right]_{T,P,n_{j\neq i}} \tag{4.24}$$

and from Equations 4.18 and 4.19,

$$\left[\frac{\partial \mu_i}{\partial P}\right]_{T,n} = \left[\frac{\partial V}{\partial n_i}\right]_{T,P,n_{j\neq i}} \tag{4.25}$$

We observe here that the evaluation of the chemical potential function, μ_i, from the Gibbs function G by Equation 4.19, and its temperature and pressure derivatives at constant composition, from Equations 4.24 and 4.25, respectively, involves the differentiation of an extensive property with respect to the number of moles of compound i, keeping constant the temperature, pressure, and number of moles of all other compounds different from i. This particular form of partial derivative is discussed in detail in Chapter 6.

OTHER USEFUL MATHEMATICAL RELATIONS

Additional equations can be obtained using the following three mathematical relations:

$$\left(\frac{\partial X}{\partial Y}\right)_Z = \frac{1}{\left(\frac{\partial Y}{\partial X}\right)_Z} \quad \text{Inversion} \tag{4.26}$$

$$\left(\frac{\partial X}{\partial Y}\right)_Z = \left(\frac{\partial X}{\partial W}\right)_Z \left(\frac{\partial W}{\partial Y}\right)_Z \quad \text{Chain rule} \tag{4.27}$$

$$\left(\frac{\partial X}{\partial Y}\right)_Z = -\left(\frac{\partial X}{\partial Z}\right)_Y \left(\frac{\partial Z}{\partial Y}\right)_X \quad \text{Change of variable} \tag{4.28}$$

Finally, a particularly useful transformation for any property of a closed system considered a function of V and T is obtained from the exact differential:

$$dX = \left(\frac{\partial X}{\partial V}\right)_{T,n} dV + \left(\frac{\partial X}{\partial T}\right)_{V,n} dT$$

For a change of temperature at constant pressure, we obtain:

$$\left(\frac{\partial X}{\partial T}\right)_{P,n} = \left(\frac{\partial X}{\partial V}\right)_{T,n} \left(\frac{\partial V}{\partial T}\right)_{P,n} + \left(\frac{\partial X}{\partial T}\right)_{V,n} \tag{4.29}$$

In order to fix ideas with respect to the difference between state and path functions, we consider here a simple example. It is strongly suggested that you try to solve the example before reading the solution.

Example

Consider the following two differentials:

$$df = \left(9x^2y^2 - 2y + 7\right)dx + \left(6x^3y - 2x\right)dy$$

and

$$\delta w = \left(9x^2y^2 - 2y + 7\right)dx + \left(6x^3y + 2x\right)dy$$

1. Test whether these differentials are exact or not.
2. For both cases, evaluate the line integral from the point ($x=1$, $y=3$) to the point ($x=2$, $y=6$) along the following two paths:

$$\text{Path 1: } y = 3x; \text{ Path 2: } y = x^2 + 2$$

3. Classify the functions f and w in thermodynamic terms as either a path or a state function. Decide whether the value of the line integral of the first of these functions should be designated by f or Δf; equally, decide whether the line integral of the second function should be designated by w or Δw.

Solution

1. In order to decide whether the differentials of the functions are exact, we verify whether the functions preceding the differentials of the variables x and y meet Maxwell's conditions. For clarity, we consider the following general form:

$$dF = \left(\frac{\partial F}{\partial x}\right)_y dx + \left(\frac{\partial F}{\partial y}\right)_x dy$$

An exact differential should meet Maxwell's relation of the form:

$$\frac{\partial}{\partial y}\left[\left(\frac{\partial F}{\partial x}\right)_y\right]_x = \frac{\partial}{\partial x}\left[\left(\frac{\partial F}{\partial y}\right)_x\right]_y$$

For the function f,

$$\frac{\partial}{\partial y}\left[9x^2y^2 - 2y + 7\right]_x = 18x^2y - 2 \text{ and } \frac{\partial}{\partial x}\left[6x^3y - 2x\right]_y = 18x^2y - 2$$

Thus, df is an exact differential. For the function w, however, the partial differential with respect to y of the coefficient of dx is the same as above, while the partial differential with respect to x of the coefficient of dy is:

$$\frac{\partial}{\partial x}\left[6x^3y + 2x\right]_y = 18x^2y + 2$$

indicating that the expression for δw is not an exact differential.

2. First, we verify that both paths of integration pass through the two extreme points of the integration. After this verification, we proceed with the integration of both functions following path 1. For this, we replace $y=3x$ and $dy=3dx$ in the expressions for df and δw. After reducing terms, we obtain:

$$df = \left(135x^4 - 12x + 7\right)dx \text{ and } \delta w = \left(135x^4 + 7\right)dx$$

Thus, integrating between $x=1$ and $x=2$,

$$\int_1^2 df = 826 \text{ and } \int_1^2 \delta w = 844$$

For the integration following path 2, we replace $y=x^2+2$ and $dy=2x\,dx$ in the expressions for df and δw. After reducing terms, we obtain:

$$df = 3\left(7x^6 + 20x^4 + 10x^2 + 1\right)dx \text{ and } \delta w = \left(21x^6 + 60x^4 + 38x^2 + 3\right)dx$$

Thus, integrating between $x=1$ and $x=2$,

$$\int_1^2 df = 826 \text{ and } \int_1^2 \delta w = 844\frac{2}{3}$$

As expected from the test of the validity of Maxwell's equation, the value resulting from integration of the function f is independent of the path, while the value of the integration of function w is path dependent.

3. In thermodynamic terms, f is a state function like U, H, S, A, or G, and its integral should be designated by Δf, while w is a path function, and its integral should be designated by w.

PROBLEMS

4.1 **An open system containing a single component**

For a closed system, the fundamental Equation 2.13a in terms of total energy, entropy, and volume can also be written as Equation 2.13b in terms of the respective intensive (per mole or unit mass) variables,

$$\left[dU = TdS - PdV \quad \Leftrightarrow \quad du = Tds - Pdv\right]_n$$

Consider, however, an open system composed of a single pure species. For $c = 1$, Equation (4.7) becomes:

$$dU = TdS - PdV + \mu dn$$

What is required for Equation 2.13b to be still valid in this case?

4.2 **Jacobians and the Maxwell relations**

The Jacobian of two functions $\mathcal{F}(x,y)$ and $\mathcal{G}(x,y)$ is the determinant symbolized and defined as:

$$J(\mathcal{F},\mathcal{G}) \equiv \frac{\partial(\mathcal{F},\mathcal{G})}{\partial(x,y)} \equiv \begin{vmatrix} \left(\frac{\partial \mathcal{F}}{\partial x}\right)_y & \left(\frac{\partial \mathcal{F}}{\partial y}\right)_x \\ \left(\frac{\partial \mathcal{G}}{\partial x}\right)_y & \left(\frac{\partial \mathcal{G}}{\partial y}\right)_x \end{vmatrix} = \left(\frac{\partial \mathcal{F}}{\partial x}\right)_y \left(\frac{\partial \mathcal{G}}{\partial y}\right)_x - \left(\frac{\partial \mathcal{F}}{\partial y}\right)_x \left(\frac{\partial \mathcal{G}}{\partial x}\right)_y$$

As a result of this definition, the mathematical properties of Jacobians combine those of derivatives and determinants. It can be shown in particular that any partial derivative can be interpreted as a Jacobian with one dependent variable used also as independent,

$$\frac{\partial(x,w)}{\partial(x,y)} = \frac{\partial(w,x)}{\partial(y,x)} = \left(\frac{\partial w}{\partial y}\right)_x ; \quad \frac{\partial(w,x)}{\partial(x,y)} = \frac{\partial(x,w)}{\partial(y,x)} = -\left(\frac{\partial w}{\partial y}\right)_x$$

and also that Jacobians obey the equivalent of the chain rule of differentiation,

$$\frac{\partial(\mathcal{F},\mathcal{G})}{\partial(w,z)} = \frac{\partial(\mathcal{F},\mathcal{G})}{\partial(x,y)} \cdot \frac{\partial(x,y)}{\partial(w,z)}$$

Using these properties, prove that the four Maxwell relations for constant n, Equations 4.20–4.23, can be written as the single formula $J(T,S) = J(P,V)$.

5 Calculation of Changes in the Value of Thermodynamic Properties

Abstract

Equations for the calculation of differences in enthalpy and entropy in solids and liquids are presented. They use the measurable coefficients of isothermal compressibility and of thermal expansion and the specific heats at constant pressure and constant volume. The Clausius–Clapeyron equation, used to calculate the difference in enthalpy and entropy for phase changes, is derived. Equations of state requiring the knowledge of the critical constants are used to calculate changes in the gas phase. The reduced temperature is defined as the ratio of the absolute temperature to the critical temperature and the reduced pressure is defined as the ratio of the absolute pressure to the critical pressure. Equations for changes in the properties of the ideal gas are given and the method for preparation of tables and plots of the thermodynamic properties of real gases is discussed.

MEASURABLE PROPERTIES

In this chapter, we consider the calculation of changes in internal energy, enthalpy, and entropy for a closed system undergoing a process between an initial equilibrium state at a pressure P_1 and temperature T_1 and a final equilibrium state at P_2 and T_2. The internal energy, enthalpy, and entropy are nonmeasurable properties. A preliminary consideration of Equations 4.7 and 4.9 shows that in order to make any progress, we must first eliminate the entropy from these equations.

The pressure, temperature, composition, and volume of a closed system are *measurable properties*. Two additional measurable properties are the *coefficient of thermal expansion*,

$$\alpha = \frac{1}{V}\left(\frac{\partial V}{\partial T}\right)_{P,n} = \frac{1}{\mathrm{v}}\left(\frac{\partial \mathrm{v}}{\partial T}\right)_{P,n} \tag{5.1}$$

and the *coefficient of isothermal compressibility*,

$$\kappa_T = -\frac{1}{V}\left(\frac{\partial V}{\partial P}\right)_{T,n} = -\frac{1}{\mathrm{v}}\left(\frac{\partial \mathrm{v}}{\partial P}\right)_{T,n} \tag{5.2}$$

As the volume of a closed system decreases with an increase in pressure at constant temperature, in order to have positive values for the coefficient of isothermal compressibility, a negative sign is used on the right-hand side of Equation 5.2. The heat exchanged between a closed system and its surroundings, during a process carried out under controlled conditions, is also measurable. Considering that the heat required to change the temperature of a closed system by 1° depends on its initial temperature, it is necessary to refer to an infinitesimal change in temperature and write, per unit mass,

$$c \equiv \lim_{\Delta T \to 0}\left[\frac{\delta q}{\Delta T}\right] \tag{5.3}$$

The quantity c, so defined, is called the *specific heat* and is a function not only of temperature but also of the path followed by the process. Two paths are commonly used. For a constant pressure

 DOI: 10.1201/9781003431985-6

(isobaric) process, the external pressure, P_E, and the internal pressure, P, of the system are the same and constant. Thus, from Equations 2.11 and 3.6,

$$\delta q_P = du + P_E dv = du + d(Pv) = dh_P$$

Once the path of the process is defined as isobaric, the heat exchanged with the surroundings becomes equal to the change in the state function enthalpy and the relation for a reversible process between the same two states is applicable. From Equation 4.9 at constant pressure and number of moles,

$$\delta q_P = \delta q_{rev,P} = dh_P = T\ ds_P \tag{5.4}$$

An important conclusion from this relationship is that for a closed system undergoing an isobaric process, the *specific heat at constant pressure*, at a particular temperature T, can be written as:

$$c_P = \left(\frac{\partial h}{\partial T}\right)_{P,n} = T\left(\frac{\partial s}{\partial T}\right)_{P,n} \tag{5.5}$$

The subscript n is kept as a reminder that this equation is valid for a closed system. Thus, as the heat exchanged to change by a small increment the temperature of a closed homogeneous system is measurable, the value of the specific heat can be experimentally obtained.

Similarly, by defining a constant volume path for the process, the heat exchanged between the system and its surroundings from Equations 2.2 and 2.3 takes the form:

$$\delta q_v = \delta q_{rev,v} = du_v = Tds_v \tag{5.6}$$

Thus, the *specific heat at constant volume* can then be written as:

$$c_v = \left(\frac{\partial u}{\partial T}\right)_{v,n} = T\left(\frac{\partial s}{\partial T}\right)_{v,n} \tag{5.7}$$

Usually, the specific heat at constant volume is not measured independently, but if needed, it is obtained from the exact relation:

$$c_v = c_P - \frac{Tv\alpha^2}{\kappa_T} \tag{5.8}$$

The derivation of this equation is given at the end of this chapter, but it is strongly recommended that the reader attempts to obtain it before seeing how it is done.

CALCULATION OF PROPERTY CHANGES FOR SOLID AND LIQUID PHASES

Once experimental values for the specific heat at constant pressure have been measured and tabulated, for most practical applications solids and liquids can be safely considered to be incompressible and the small effects caused by volume changes with pressure are ignored. Enthalpy changes can be calculated by integration of rearranged Equation 5.5, where superscript c indicates a "condensed phase":

$$\Delta h^c = \int_{T_1}^{T_2} c_P dT \tag{5.9}$$

Using the arithmetic average of the c_P values at T_1 and T_2, we write:

$$\Delta h^c \approx \overline{c_P}\Delta T \tag{5.10}$$

For the change in internal energy, if ever needed, although the correction for pressure is usually negligible, we may use:

$$\Delta u^c = \Delta h^c - \mathrm{v}\Delta P \tag{5.11}$$

For the calculation of entropy changes, from Equation 5.5,

$$\Delta s^c = \int_{T_1}^{T_2} \frac{c_P}{T} dT \tag{5.12}$$

or, considering an average value for specific heat at constant pressure,

$$\Delta s^c \approx \overline{c_P} \ln \frac{T_2}{T_1} \tag{5.13}$$

The superscript c has been used in these equations as a reminder that these expressions are only applicable to condensed, incompressible phases.

CALCULATION OF PROPERTY CHANGES FOR PHASE CHANGE OF A PURE COMPOUND

If the process is isobaric and isothermal, from Equation 5.4 we write:

$$q_{P,T} = \Delta h_{P,T} + T\Delta s_{P,T}$$

A particular process that occurs at constant pressure and temperature is the *change of phase for a pure compound*, from either solid to liquid, liquid to gas, or solid to gas. As the *phases in equilibrium are said to be "saturated,"* instead of indicating by subscripts that the pressure and temperature are kept constant, a superscript s is used to indicate the change in the property between saturated phases. Thus, we write:

$$\Delta s^s = \frac{\Delta h^s}{T_s} \tag{5.14}$$

As shown by Equation 5.4, the enthalpy change in a phase change is equal to the heat required, which is also a measurable quantity. For this reason, the enthalpy change of vaporization, for example, is simply called the *heat of vaporization*. Similarly, we use the terms *heat of condensation*, *heat of sublimation*, *heat of melting*, and so forth. Once the enthalpy of phase change is measured at a particular temperature, the entropy change can be calculated with Equation 5.14 and the internal energy change, if ever required, can be obtained from Equation 3.6. However, once the enthalpy change is measured at one temperature, it is desirable to have some way of calculating it at different temperatures using properties that are more easily measured. The treatment that follows considers the phase change from saturated liquid (sL) to saturated vapor (sV), but the equations obtained are applicable to any two phases in equilibrium. Rearranging Equation 5.14, we write:

$$h^{sV} - T_s s^{sV} = h^{sL} - T_s s^{sL} \tag{5.14a}$$

or, from Equation 4.2,

$$g^{sV} = g^{sL} \tag{5.14b}$$

For an incremental temperature change from T_s to $(T_s + dT_s)$,

$$g^{sV} + dg^{sV} = g^{sL} + dg^{sL}$$

By comparison between the last two equations, we get:

$$dg^{sV} = dg^{sL}$$

From the general expression for the exact differential of the Gibbs energy, for a closed system we have:

$$dg = \left(\frac{\partial g}{\partial T}\right)_{P,n} dT + \left(\frac{\partial g}{\partial P}\right)_{T,n} dP$$

Using Equations 4.17 and 4.18, we obtain:

$$dg = -sdT + \text{v}dP$$

As we know from the physical phenomenon of phase change, once the temperature of equilibrium changes, the saturation pressure of the system changes accordingly. Thus, considering the saturation values for pressure as a function of temperature as independent variable, and eliminating the subscript s for the independent variable T, for the change in the saturation equilibrium condition we write:

$$-s^{sV}dT + \text{v}^{sV}dP^s = -s^{sL}dT + \text{v}^{sL}dP^s$$

Rearranging we get:

$$\frac{dP^s}{dT} = \frac{\Delta s^s}{\Delta \text{v}^s} \tag{5.15a}$$

Combining this expression with Equation 5.14,[1]

$$\frac{dP^s}{dT} = \frac{\Delta h^s}{T\Delta \text{v}^s} \tag{5.15b}$$

For work with volumetric properties, it is usual to introduce the definition of the *compressibility factor*, z:

$$z \equiv \frac{P\text{v}}{RT} \tag{5.16}$$

In this definition, which is valid for any physical state of matter, R is the so-called gas constant and v is the molar volume. Thus, introducing the compressibility factor we write:

$$\frac{dP^s}{dT} = \frac{\Delta h^s}{RT^2\Delta z^s}P^s$$

or

$$\frac{d\ln P^s}{dT} = \frac{\Delta h^s}{RT^2\Delta z^s} \tag{5.15c}$$

This exact relation is known as the *Clausius–Clapeyron equation*, and it is valid for any phase change. For the particular case of vapor–liquid equilibrium, from experimental data it has been found that the ratio between the heat of vaporization and the change due to vaporization of the compressibility factor is a very weak function of temperature. Hence, we can write:

$$\frac{\Delta h^s}{R\Delta z^s} = B + DT + ET^2 + \ldots$$

where B, D, E, …, are empirical constants that can be evaluated from experimental data for each pure compound. Thus,

$$d\ln P^s = \left[\frac{B}{T^2} + \frac{D}{T} + E + \ldots\right]dT$$

The indefinite integral of this equation gives:

$$\ln P^s = A - \frac{B}{T} + D\ln T + ET + \ldots \tag{5.17a}$$

where A is the integration constant. This semi-empirical correlation for the vapor pressure of pure compounds is known as the *Riedel equation.* A simple empirical equation for the vapor pressure applicable over a limited temperature range, closely related to the Riedel equation, is the *Antoine equation.* It has the form:

$$\log_{10} P^s = A_A - \frac{B_A}{t + C_A} \tag{5.17b}$$

where the temperature t is in degrees Celsius and the constants A_A, B_A, and C_A are also empirically evaluated.

Rewriting Equation 5.15c, we get:

$$\Delta h^s = RT^2\Delta z^s\frac{d\ln P^s}{dT} \tag{5.18}$$

The fact that this exact equation is verified using experimental data can be considered among the best evidence of the usefulness of the definition of entropy. The derivation of this equation was possible thanks to the introduction of the concept of entropy by a postulate. After some algebra, the entropy was eliminated, and the final result, Equation 5.18, confirmed by the experiment supports the theory. In addition, this exact equation coupled with a semi-empirical correlation like the Riedel equation provides a valuable tool to evaluate enthalpy and entropy changes of vaporization. For approximate calculations for the case of vapor–liquid equilibrium, Equation 5.15b is often simplified. Two parameters used to characterize the volumetric behavior of fluids are the reduced temperature, T_r, and the reduced pressure, Pr, defined by:

$$T_r = \frac{T}{T_c} \tag{5.19}$$

and

$$P_r = \frac{P}{P_c} \tag{5.20}$$

where T_c and P_c are the critical temperature and critical pressure of the fluid, respectively.

For $T_r \geq 0.8$ the empirical relation of Haggenmacher gives a reasonable approximation for the difference in the compressibility factors of the saturated phases. This equation has the form:

$$\Delta z^s \approx \left[1 - \frac{P_r}{T_r^3}\right]^{0.5}$$

For $T_r \leq 0.8$, the molar volume of the liquid is orders of magnitude smaller than the molar volume of the vapor. Thus, the value of the liquid compressibility factor is negligibly small compared with its value for the vapor. Hence, considering the vapor phase as an ideal gas,

$$\Delta z^s \approx 1$$

USE OF EQUATIONS OF STATE FOR A PURE COMPOUND OR FOR A MIXTURE OF CONSTANT COMPOSITION

Due to their compressible nature, gases are the real problem for the calculation of the value of property changes. In this case, we need to use equations of state (EOSs). Any thermodynamic function in terms of its canonical variables can be used to define the state of the system. For example, an expression for H in terms of P, S, and the set of number of moles n, as suggested by Equation 4.10, is an EOS. For practical purposes, however, unless otherwise specified, *the term* EOS *refers to the relation of the volume of the system as a function of temperature, pressure, and composition.* These are all measurable properties. Moreover, for our purposes, we only consider either a pure compound or a mixture of fixed composition.

Depending on the variable that it expressed explicitly, there are two forms of EOS commonly used. These are the *volume explicit form*:

$$\mathrm{v} = \mathrm{v}(P, T)$$

and the *pressure explicit form*:

$$P = P(\mathrm{v}, T)$$

The symbol used in the right-hand side of the above expressions indicates that the property shown outside the brackets is a function of the variables in brackets.

CASE OF INDEPENDENT VARIABLES *V* AND *T*, *P* = *P*(*V*, *T*)

For a closed system, the exact differential of the internal energy, u, in terms of temperature and volume as variables takes the form:

$$du = \left(\frac{\partial u}{\partial T}\right)_{\mathrm{v},n} dT + \left(\frac{\partial u}{\partial \mathrm{v}}\right)_{T,n} d\mathrm{v}$$

The first partial derivative of the right-hand side is equal to the measurable specific heat at constant volume, Equation 5.7. For the second partial derivative, from the fundamental equation (4.7),

$$du = Tds - Pd\text{v}$$

From here we write:

$$\left(\frac{\partial u}{\partial \text{v}}\right)_{T,n} = T\left(\frac{\partial s}{\partial \text{v}}\right)_{T,n} - P$$

In addition, from the Maxwell relation, Equation 4.22,

$$\left(\frac{\partial s}{\partial \text{v}}\right)_{T,n} = \left(\frac{\partial P}{\partial T}\right)_{\text{v},n}$$

Thus, we finally can write for du,

$$du = c_\text{v} dT + \left[T\left(\frac{\partial P}{\partial T}\right)_{\text{v},n} - P\right] d\text{v} \tag{5.21}$$

For the calculation of enthalpy changes, in this case we use:

$$dh = du + d(P\text{v}) \tag{5.22}$$

To calculate changes in entropy, we follow a similar path and write:

$$ds = \left(\frac{\partial s}{\partial T}\right)_{\text{v},n} dT + \left(\frac{\partial s}{\partial \text{v}}\right)_{T,n} d\text{v}$$

For the first partial derivative we use Equation 5.7, and for the second one we use the Maxwell relation, Equation 4.22, as used above; thus,

$$ds = \frac{c_\text{v}}{T} dT + \left(\frac{\partial P}{\partial T}\right)_{\text{v},n} d\text{v} \tag{5.23}$$

Integration of these equations between two equilibrium states will give the value of the property change as a function of T_1, T_2, v_1, and v_2. However, the variables normally measured are T_1, P_1, T_2, and P_2. Thus, before doing any calculation of a property change, it is necessary to use the EOS and evaluate v_1 using T_1, P_1, and also evaluate v_2 using T_2, P_2. The direct path between the two equilibrium states presents an additional problem. The specific heat of the gas is a function of temperature and pressure, and the latter dependence is usually unknown. However, as the value of a property change is independent of the path followed between the two equilibrium states considered, the calculations are done using a three-step path. First, the gas is taken at constant temperature from T_1, P_1 to T_1, P^*, where P^* is a pressure low enough that the gas behaves as an ideal gas. In the second step, the temperature is changed from T_1 to T_2, at constant pressure P^*. Finally, in the third step, the gas is taken from T_2, P^* to T_2, P_2. This path has the advantage that the specific heat capacities of different gases as an ideal gas are known as a function of temperature.

Case of Independent Variables P and T, $V = V(P, T)$

For a closed system,

$$dh = \left(\frac{\partial h}{\partial T}\right)_{P,n} dT + \left(\frac{\partial h}{\partial P}\right)_{T,n} dP$$

From Equation 4.9, in molar terms for a closed system,

$$dh = Tds + \mathrm{v}dP$$

Thus, for the second differential of the right-hand side we write:

$$\left(\frac{\partial h}{\partial P}\right)_{T,n} = T\left(\frac{\partial s}{\partial P}\right)_{T,n} \mathrm{v}$$

and using the Maxwell relation, Equation 4.23, we eliminate the term containing the entropy and write:

$$\left(\frac{\partial h}{\partial P}\right)_{T,n} = \mathrm{v} - T\left(\frac{\partial \mathrm{v}}{\partial T}\right)_{P,n}$$

In terms of measurable properties, the first differential of the right-hand side is given by Equation 5.5. Hence, finally,

$$dh = c_P dT + \left[\mathrm{v} - T\left(\frac{\partial \mathrm{v}}{\partial T}\right)_{P,n}\right] dP \tag{5.24}$$

As before, for du we use directly

$$du = dh - d(P\mathrm{v}) \tag{5.25}$$

For entropy change with independent variables P and T,

$$ds = \left(\frac{\partial s}{\partial T}\right)_{P,n} dT + \left(\frac{\partial s}{\partial P}\right)_{T,n} dP$$

The first differential of the right-hand side is given by Equation 5.5, and the second one is given by the same Maxwell relation, Equation 4.23, used above to obtain Equation 5.24; thus,

$$ds = \frac{c_P}{T} dT - \left(\frac{\partial \mathrm{v}}{\partial T}\right)_{P,n} dP \tag{5.26}$$

Again, in this case, when calculating the change in the value of a thermodynamic property between a state at P_1, T_1 and a state at P_2, T_2, it is necessary to do a three-step calculation. First, the gas is taken at constant temperature from T_1, P_1 to T_1, P^*, where P^* is a pressure low enough that the gas behaves as an ideal gas. Then the temperature is changed from T_1 to T_2, at constant pressure P^*, and finally, the gas is taken from T_2, P^* to T_2, P_2.

CHANGES IN THE VALUES OF THE THERMODYNAMIC PROPERTIES OF THE IDEAL GAS

The ideal gas behavior gives a good approximation to the behavior of nonpolar gases, such as nitrogen, oxygen, or air, at room or higher temperature and at around atmospheric pressure. The compressibility factor of an ideal gas is unity, $z^* = 1$, and Equation 5.16 takes the form:

$$P^* = \frac{RT}{\mathrm{v}^*} \tag{5.27}$$

The asterisk is used here to indicate ideal gas behavior. We observe that there is a subtle difference between Equation 1.1 and Equation 5.27. In Equation 1.1, the molar volume of the ideal gas is evaluated at a pressure, P, and a temperature, T. As the pressure in this case is an independent variable, there is no need to mark it with an asterisk. In Equation 5.27, the volume of the ideal gas and the temperature are the independent variables, so the value of the pressure obtained is only valid for the case of an ideal gas, and it is marked with an asterisk. For the change in internal energy of the ideal gas from Equation 5.21,

$$du^* = c_V^* dT + \left[T\left(\frac{\partial P^*}{\partial T}\right)_{V,n} - P^* \right] dv^*$$

But, from Equation 5.27,

$$\left[T\left(\frac{\partial P^*}{\partial T}\right)_{V,n} - P^* \right] = 0$$

so

$$du^* = c_V^* dT \tag{5.28a}$$

Thus, the internal energy of the ideal gas is only a function of temperature. From Equations 3.6a and 5.27, the enthalpy of the ideal gas can be written as:

$$h^* = u^* + P^* v^* = u^* + RT$$

or

$$dh^* = du^* + RdT = \left(c_V^* + R\right) dT$$

Thus, the enthalpy of the ideal gas is only a function of temperature. In addition, from Equation 5.5,

$$dh^* = c_P^* \, dT \tag{5.28b}$$

Hence,

$$c_P^* = c_V^* + R \tag{5.29}$$

This result can also be obtained by applying Equation 5.8 for the ideal gas. For the entropy change of the ideal gas, from Equation 5.26 we write:

$$ds^* = \frac{c_P^*}{T} dT - \frac{R}{P} dP \tag{5.30}$$

Thus, the entropy of the ideal gas is a function of both temperature and pressure.

ADIABATIC REVERSIBLE COMPRESSION OR EXPANSION OF THE IDEAL GAS

As an approximation, we consider here constant values of the specific heats of the ideal gas, $\overline{c}_P^*$ and $\overline{c}_V^*$. An adiabatic reversible thermodynamic process is isentropic. Thus, from Equation 5.30 we write:

$$\overline{c}_P^* d \ln T = R d \ln P$$

Integrating between a state at P_1, T_1 and a state at P_2, T_2, and rearranging, we obtain:

$$\frac{T_2}{T_1} = \left(\frac{P_2}{P_1}\right)^{\frac{R}{\bar{c}_P^*}} = \left(\frac{P_2}{P_1}\right)^{\frac{\gamma-1}{\gamma}} \tag{5.31}$$

In Equation 5.31, we have defined:

$$\gamma \equiv \frac{\bar{c}_P^*}{\bar{c}_V^*} \tag{5.32}$$

Combining Equation 5.31 with the ideal gas equation, Equation 1.1,

$$\frac{T_2}{T_1} = \left(\frac{v_1}{v_2}\right)^{\gamma-1} \tag{5.33}$$

Finally, combining Equations 5.31 and 5.33,

$$P_1 v_1^{\gamma} = P_2 v_2^{\gamma} \tag{5.34}$$

DEVIATIONS FROM THE IDEAL GAS BEHAVIOR

For nonpolar gases, such as oxygen, nitrogen, carbon dioxide, or air, at normal ambient temperature or higher, and at pressures below 5 atm, the following truncated forms of the virial equation give a satisfactory approximation of their volumetric behavior:

$$z = 1 + \frac{B}{v} + \frac{C}{v^2} + \ldots \tag{5.35}$$

and

$$z = 1 + \frac{B}{RT}P + \frac{C - B^2}{(RT)^2}P^2 + \ldots \tag{5.36}$$

In Equations 5.35 and 5.36, B and C are the second and third virial coefficients, which are usually given as a function of the reduced temperature, defined by Equation 5.19, by expressions of the form:

$$B = \frac{RT_c}{P_c}\left(B^{(0)} + \omega B^{(1)}\right) \tag{5.37}$$

where ω is the so-called Pitzer's acentric factor, defined by

$$\omega = -\log_{10}\left(P^s \text{ at } T_r = 0.7\right) + \log_{10} P_c - 1.000 \tag{5.38}$$

Further discussion of the virial EOS is presented in Chapter 27 and Appendix B. Larger deviations from the ideal behavior are treated with cubic EOSs. One such equation is discussed in Chapter 25 and Appendix C.

PREPARATION OF TABLES AND PLOTS OF THERMODYNAMIC PROPERTIES

Tables and plots of thermodynamic properties are constructed on the basis of extensive collections of values of the measurable properties. At constant temperature, molar volumes are measured as a function of pressure. Values of the specific heats, vapor pressure, and heats of vaporization are independently measured and tested for consistency. Once the data are available, it is necessary to assign a zero value to the nonmeasurable properties, like enthalpy and entropy. Normally, the property for the saturated liquid at a well-defined temperature is assigned the zero value. When combining information from different sources, it is imperative to control that they are all using the same reference point. Otherwise, a displacement of the relative scale is required. Once the zero is fixed, the enthalpy of the saturated vapor at that equilibrium temperature is evaluated using the exact Equation 5.15b, and the entropy using Equation 5.14. From these points, the (relative) values of the properties at any state in the vapor or liquid phases can be obtained using the equations for a single phase discussed above. For precise work, once volumes of the fluid have been measured as a function of pressure at constant temperature, the data are correlated using extended empirical virial EOSs of the form:

$$P = \frac{RT}{\mathrm{v}}\left[1 + \sum_{k=1}^{n} \frac{C_k}{\mathrm{v}^k}\right] \tag{5.39}$$

For a pure compound, the adjustable parameters C_k are only a function of temperature.

Derivation of Equation 5.8

We start from the definition (Equation 5.7):

$$c_\mathrm{V} = T\left(\frac{\partial s}{\partial T}\right)_{\mathrm{v},n}$$

From the transformation (Equation 4.29),

$$c_\mathrm{V} = T\left[\left(\frac{\partial s}{\partial T}\right)_{P,n} - \left(\frac{\partial s}{\partial \mathrm{v}}\right)_{T,n}\left(\frac{\partial \mathrm{v}}{\partial T}\right)_{P,n}\right]$$

But from Equation 5.5,

$$c_P = T\left(\frac{\partial s}{\partial T}\right)_{P,n}$$

Thus,

$$c_\mathrm{V} = c_P - T\left[\left(\frac{\partial s}{\partial \mathrm{v}}\right)_{T,n}\left(\frac{\partial \mathrm{v}}{\partial T}\right)_{P,n}\right]$$

From Equations 4.22 and 4.28, we get:

$$\left(\frac{\partial s}{\partial \mathrm{v}}\right)_{T,n} = \left(\frac{\partial P}{\partial T}\right)_{\mathrm{v},n} = -\frac{\left(\frac{\partial \mathrm{v}}{\partial T}\right)_{P,n}}{\left(\frac{\partial \mathrm{v}}{\partial P}\right)_{T,n}}$$

so

$$c_v = c_P + \frac{T\left[\left(\dfrac{\partial v}{\partial T}\right)_{P,n}\right]^2}{\left(\dfrac{\partial v}{\partial P}\right)_{T,n}}$$

Using the definitions of α (Equation 5.1) and κ_T (Equation 5.2),

$$c_v = c_P - \frac{Tv\alpha^2}{\kappa_T}$$

For the ideal gas,

$$\frac{Tv^*\alpha^{*2}}{\kappa_T^*} = R$$

Thus,

$$c_P^* - c_v^* = R$$

PROBLEMS

5.1 **Hydraulic circuit, revisited**

In Problem 3.2, we found the lost work in a simple hydraulic circuit, and we used it to compute the rate of entropy increase of the "universe" (actually just the water plus its immediate surroundings), based on an estimated average temperature of the flowing water. From Equations 5.10 and 5.13, show what the true mean temperature should have been, and recalculate with it the rate of entropy generation.

5.2 **Phase equilibria of argon**

In the vicinity of the solid-liquid-vapor triple point of argon, its vaporization and sublimation pressures [1] may be approximated, respectively, by:

$$\ln P^{vap} = 13.847183 - \frac{805.76406}{T}$$

$$\ln P^{sub} = 15.629480 - \frac{955.13089}{T}$$

with P in kPa and T in K.

a. Determine the temperature and pressure of the triple point.
b. Estimate the heats of vaporization, sublimation, and melting at the triple point.
c. Knowing *additionally* that the melting pressure is given in the same units by:

$$P^{mel} = 334602 + 3993.41T$$

and that the heat of vaporization at the triple point is actually $\Delta h^{vap} = 6539.9$ kJ/kmol, repeat the calculation of the enthalpies of sublimation and melting. Explain any differences from the answers in (b).

5.3 **Differential forms of the Helmholtz and Gibbs energy functions**

From Equations 4.14–4.17, we know that for a pure component or a mixture of constant composition (omitting for simplicity the subscript n from all derivatives):

$$\left(\frac{\partial a}{\partial T}\right)_v = -s; \quad \left(\frac{\partial a}{\partial v}\right)_T = -P$$

$$\left(\frac{\partial g}{\partial T}\right)_P = -s; \quad \left(\frac{\partial g}{\partial P}\right)_T = v$$

Derive the differential forms for $a(T,P)$ and $g(T,v)$.

5.4 **Coal to diamonds (and back)**

Graphite (G) and diamond (D) are two distinct solid phases of carbon, differing in crystalline structure.

a. Show that graphite is the stable phase at 310 K and 100 kPa, and estimate the pressure that must be exerted to change it into diamond at that temperature. Relevant thermodynamic data for both forms of carbon at 298.15 K and 100 kPa are [2,3]

	Density (kg/m³)	Enthalpy (kJ/kmol)	Gibbs Energy (kJ/kmol)
Graphite	2200	0(*)	0(*)
Diamond	3513	1895	2900

(*) Conventionally assigned reference state.

b. How can diamonds therefore exist at room temperature and pressure? Why don't they turn into graphite? Are diamonds truly forever?

5.5 **The vortex tube**

In the remarkable device known as the *Hilsch, Ranque–Hilsch* or *vortex tube*, compressed gas (normally air) at temperature T_E enters tangentially to the cylinder and acquires a rotational motion which causes it to separate into a cold fraction x at temperature T_L and a hot fraction $1-x$ at temperature T_H that exit to the atmosphere through opposite ends of the tube (Figure 5.1). Assuming that the tube has no moving parts, operates adiabatically and reversibly, develop the equations for the "cold cut" x and the temperature ratios T_L/T_E, T_H/T_E as functions of the ratio of inlet to outlet pressure. Consider air as an ideal gas with constant specific heats.

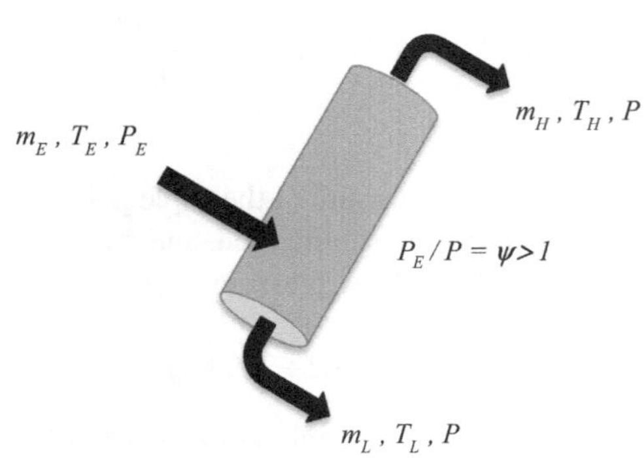

FIGURE 5.1 Schematic representation of a vortex tube.

5.6 **Coefficient of isentropic compressibility and velocity of sound in a fluid**

Like the coefficient of isothermal compressibility defined in Equation 5.2, the coefficient of *isentropic* compressibility is defined as:

$$\kappa_s = -\frac{1}{V}\left(\frac{\partial V}{\partial P}\right)_{S,n} = -\frac{1}{v}\left(\frac{\partial v}{\partial P}\right)_s$$

Show that $\kappa_s = \kappa_T \dfrac{c_v}{c_P}$

NOTE

1 For liquid–solid equilibrium at a temperature T, as for most compounds $v^{sL} > v^{sS}$, an increase in pressure increases the melting temperature, so at the temperature T, liquid freezes to solid. For a few compounds, including water, $v^{sL} < v^{sS}$. Thus, for these compounds an increase in pressure lowers the melting temperature, and at the temperature T, solid melts. This effect makes ice skating easier, as a liquid phase is formed under the pressure of a skate.

REFERENCES

1. Tegeler, Ch., Span, R., and Wagner, W. 1999. A new equation of state for argon covering the fluid region for temperatures from the melting line to 700 K at pressures up to 1000 MPa. *J. Phys. Chem. Ref. Data* 28(3): 779–850.
2. Haynes, W. M. (Ed.). 2017. *CRC Handbook of Chemistry and Physics*, 97th ed. Boca Raton, FL: CRC Press.
3. Wagman, D. D., Evans, W. H., Parker, V. B., Schumm, R. H., Halow, I., Balley, S. M., Churney, K. L., and Nuttall, R. L. 1982. The NBS tables of chemical thermodynamic properties. *J. Phys. Chem. Ref. Data* 11(Supplement 2): 1–392.

Section II

Mixtures

$$\phi_i = \gamma_i \phi_i^{0}$$*

NOTE

* Equation from Wilczek-Vera, G., and Vera, J. H. 1989. A consistent method to combine PRSV EOS with excess Gibbs functions. *Fluid Phase Equilib.* 51: 197–208.

DOI: 10.1201/9781003431985-7

6 Partial Molar Properties and Property Changes by Mixing

ABSTRACT

The concept of partial molar properties of compounds in mixtures is introduced. The all-important Gibbs–Duhem equation and Maxwell's equations in terms of partial molar properties are derived. An ideal mixture is defined as the one formed isothermally without change in volume and without heat effects, which implies that the partial molar volumes and enthalpies are equal to the values of the molar properties of the pure compounds. Partial molar properties are expressed as a function of mole fractions and plots for binary systems are presented.

PARTIAL MOLAR PROPERTIES AND THE GIBBS–DUHEM EQUATION

For a single-phase system, the thermodynamic state functions *V*, *U*, *S*, *H*, *A*, and *G* are all *extensive* and can be expressed in terms of the state variables *P*, *T* and composition. Volume is a measurable property and provides a more visual sense of algebra. However, the treatment is general and expressions similar to Equation 4.3, but with independent variables *P* and *T*, can be written for all other extensive functions *U*, *S*, *H*, *A*, or *G*. Thus, for the total differential of any of the state functions, taking the volume as an example, we write:

$$dV = \left(\frac{\partial V}{\partial T}\right)_{P,n} dT + \left(\frac{\partial V}{\partial P}\right)_{T,n} dP + \sum_{k}^{c}\left(\frac{\partial V}{\partial n_k}\right)_{T,P,n_{j\neq k}} dn_k \tag{6.1}$$

For the particular case of the total volume, the partial derivatives with respect to temperature and pressure are *measurable properties* and are given a symbol and a name. In both of these derivatives, the numbers of moles of all compounds are kept constant, so they refer to a closed system. Therefore, we can consider these two terms per unit mass. For an open system, the third term on the right-hand side represents the effect on volume due to a change in the numbers of moles of the components of the system. For simplicity, we use the symbol $\overline{\mathrm{v}}_i$ for the partial derivative of the total property *V* with respect to the number of moles of component *i*, at constant temperature, constant pressure, and constant number of moles of all other compounds, different from *i*:

$$\overline{\mathrm{v}}_i \equiv \left(\frac{\partial V}{\partial n_i}\right)_{T,P,n_{j\neq i}} \tag{6.2}$$

Similar relations can be written for the partial molar properties of *U*, *S*, *H*, and *G*. The derivatives have dimensions. The derivative of an extensive variable with respect to the change in another extensive variable, as in Equation 6.2, gives a value that is independent of the size of the homogeneous system. If the number of moles of each component of the mixture is multiplied by a factor λ, at constant temperature and pressure, the volume V_λ of the system formed by λ times the moles of each species will be λ times the volume *V* of the original system:

$$\lambda V(T,P,n_1,n_2,\ldots,n_c) = V_\lambda(T,P,\lambda n_1,\lambda n_2,\ldots,\lambda n_c) \tag{6.3}$$

DOI: 10.1201/9781003431985-8

For clarity, in Equation 6.3 we have shown in parentheses the variables for each function. From here on, however, this detail will be omitted. To demonstrate the intensive character of $\overline{v}_i$, we differentiate Equation 6.3 with respect to the number of moles of compound i, n_i at constant temperature, pressure, and moles of all other compounds different from i:

$$\lambda\left(\frac{\partial V}{\partial n_i}\right)_{T,P,n_{j\neq i}} = \left(\frac{\partial V_\lambda}{\partial \lambda n_i}\right)_{T,P,\lambda n_{j\neq i}}\left(\frac{\partial \lambda n_i}{\partial n_i}\right) = \left(\frac{\partial V_\lambda}{\partial \lambda n_i}\right)_{T,P,n_{j\neq i}}\lambda$$

or simplifying by dividing by λ:

$$\left(\frac{\partial V}{\partial n_i}\right)_{T,P,n_{j\neq i}} = \left(\frac{\partial V_\lambda}{\partial \lambda n_i}\right)_{T,P,\lambda n_{j\neq i}} \tag{6.4}$$

We conclude that the value of $\overline{v}_i$ as defined by Equation 6.2 remains unchanged by a change in the size of the system at constant composition, temperature, and pressure. If now we differentiate Equation 6.3 with respect to a change of the arbitrary size factor λ,

$$\left(\frac{\partial \lambda V}{\partial \lambda}\right)_{T,P,n} = \sum_k^c \left(\frac{\partial V_\lambda}{\partial \lambda n_k}\right)_{T,P,\lambda n_{j\neq k}}\left(\frac{\partial \lambda n_k}{\partial \lambda}\right) = \sum_k^c \left(\frac{\partial V}{\partial n_k}\right)_{T,P,\lambda n_{j\neq k}} n_k$$

Thus, we write:

$$V = \sum_k^c n_k \overline{v}_k \tag{6.5}$$

This equation can also be written in molar terms as:

$$\mathrm{v} = \sum_k^c x_k \overline{v}_k \tag{6.6}$$

In Equation 6.6, we have used the definition of the mole fraction of the component i:

$$x_i = \frac{n_i}{\sum_k n_k} \tag{6.7}$$

Equation 6.6 clearly shows that $\overline{v}_i$, as defined by Equation 6.2, is the contribution of compound i to the molar property of the mixture, in this case the molar volume. Thus, $\overline{v}_i$ is known as the *partial molar volume* of compound i in the mixture, and it is a function of temperature, pressure, and composition. Equations 6.2, 6.5, and 6.6 are general forms, not limited to the volume of the system, and with the appropriate change of symbols, they directly apply to U, S, H, A, or G. From Equation 6.5, we then write the general relation:

$$dV = \sum_k^c n_k d\overline{v}_k + \sum_k^c \overline{v}_k dn_k$$

On the other hand, from Equations 6.1 and 6.2, we write:

$$dV = \left(\frac{\partial V}{\partial T}\right)_{P,n} dT + \left(\frac{\partial V}{\partial P}\right)_{T,n} dP + \sum_k^c \overline{v}_k dn_k$$

Combining the two equations above, we write:

$$\sum_{k}^{c} n_k d\overline{v}_k = \left(\frac{\partial V}{\partial T}\right)_{P,n} dT + \left(\frac{\partial V}{\partial P}\right)_{T,n} dP \tag{6.8}$$

or, in molar terms,

$$\sum_{k}^{c} x_k d\overline{v}_k = \left(\frac{\partial v}{\partial T}\right)_{P,n} dT + \left(\frac{\partial v}{\partial P}\right)_{T,n} dP \tag{6.9}$$

In Equations 6.8 and 6.9, we have not introduced Equations 5.1 and 5.2, which, being particular for volume, do not have a counterpart for the nonmeasurable thermodynamic properties *U*, *S*, *H*, and *G*. The all-important Equation 6.9 is known as the *Gibbs–Duhem equation. It shows that the partial molar properties of the compounds in a mixture are not all independent.*

Observing that the two derivatives of the right-hand side are taken at constant composition, we combine Equations 6.6 and 6.9 and write:

$$\sum_{k}^{c} x_k d\overline{v}_k = \sum_{k}^{c} x_k \left(\frac{\partial \overline{v}_k}{\partial T}\right)_{P,n} dT + \sum_{i}^{c} x_k \left(\frac{\partial \overline{v}_k}{\partial P}\right)_{T,n} dP$$

Rearranging,

$$\sum_{k}^{c} x_k \left[d\overline{v}_k - \left(\frac{\partial \overline{v}_k}{\partial T}\right)_{P,n} dT - \left(\frac{\partial \overline{v}_k}{\partial P}\right)_{T,n} dP \right] = 0$$

As this equation is valid for any fixed composition, *independent of the particular set of mole fractions considered, for a mixture of fixed composition* a valid solution is that the term in brackets is identically equal to zero; thus, we write:

$$d\overline{v}_i = \left(\frac{\partial \overline{v}_i}{\partial T}\right)_{P,n} dT + \left(\frac{\partial \overline{v}_i}{\partial P}\right)_{T,n} dP \tag{6.10}$$

Equation 6.10 is the general Gibbs–Duhem equation in terms of partial molar properties *for a mixture of fixed composition.* As an example, it is written here for the partial molar volume, but it also applies to $\overline{u}$, $\overline{h}$, $\overline{s}$, and $\overline{g}$.

MAXWELL RELATION APPLIED TO PARTIAL MOLAR PROPERTIES

As discussed in Chapter 4, for a state function of several variables, the order of partial differentiation with respect to two variables is immaterial. Thus, this property of all functions having an exact differential can be combined with the definition of partial molar property, Equation 6.2, to obtain:

$$\frac{\partial}{\partial n_i}\left[\left(\frac{\partial V}{\partial n_j}\right)_{T,P,n_{k\neq j}}\right]_{T,P,n_{k\neq i}} = \frac{\partial}{\partial n_j}\left[\left(\frac{\partial V}{\partial n_i}\right)_{T,P,n_{k\neq i}}\right]_{T,P,n_{k\neq j}}$$

or

$$\left(\frac{\partial \overline{v}_j}{\partial n_i}\right)_{T,P,n_{k\neq i}} = \left(\frac{\partial \overline{v}_i}{\partial n_j}\right)_{T,P,n_{k\neq j}} \tag{6.11}$$

Equation 6.11, not often found in literature, shows again that the partial molar properties of different compounds in a mixture are not independent of each other. One important point to keep in mind is that when the expressions for the partial molar properties of all compounds in a mixture are obtained from a single expression for the total property, these expressions always satisfy the Gibbs–Duhem equation and also Maxwell's relations. This is an important conclusion not properly emphasized in the literature.

GENERALITY OF THE RELATIONS BETWEEN THERMODYNAMIC PROPERTIES

The derivation of Equation 6.10 from Equation 6.9 shows that any equation in terms of molar properties can be directly written in terms of partial molar properties if the algebra does not include a change in composition and the relation between the functions is linear in composition. As a simple example, consider the definition of enthalpy by Equation 3.6a:

$$h = u + P\text{v} \tag{3.6a}$$

$$\sum_k^c x_k \bar{h}_k = \sum_k^c x_k \bar{u}_k + P \sum_k^c x_k \bar{\text{v}}_k \quad \text{or} \quad \sum_k^c x_k \left(\bar{h}_k - \bar{u}_k - P\bar{\text{v}}_k \right) = 0$$

For a mixture of fixed composition, with the same argument used to derive Equation 6.10, we obtain:

$$\bar{h}_i = \bar{u}_i + P\bar{\text{v}}_i \tag{6.12}$$

A similar algebra can be followed to derive other relations involving partial molar properties from the equation relating the molar properties of a simple system.

LIMITING VALUES OF THE PARTIAL MOLAR PROPERTIES

Partial molar properties at particular values of P and T are functions of the composition of a homogeneous mixture. For liquid mixtures, but not for gaseous mixtures, they can be considered to be independent of pressure. Considering the general form of Equation 6.6, it is evident that as the mixture becomes more and more concentrated in one of the compounds, the value of the partial molar property of that particular component will get closer and closer to the value of the property for that pure compound. Thus, taking the volume as an example, we write:

$$\lim_{x_i \to 1} \bar{\text{v}}_i = \text{v}_i^0 \tag{6.13}$$

In Equation 6.13, we have introduced the superscript 0 to denote the value of a property for a pure compound. Equations similar to Equation 6.13 can be written for all other partial molar properties. However, on the other extreme of composition, that is, when a compound i gets more and more diluted in a mixture, the value of its partial molar property will get further and further away from the value of the property for the pure compound. The limiting value at infinite dilution is certainly characteristic of the composition of the solvent mixture. The partial molar volume of methanol at infinite dilution in water is not the same as the partial volume of methanol at infinite dilution in ethanol and is different from its value in a mixture of ethanol and water. Thus, in this case, it is necessary to specify the composition of the solvent, and we write:

$$\lim_{x_i \to 0} \bar{\text{v}}_i = \bar{\text{v}}_i^{\infty,r} \tag{6.14}$$

In Equation 6.14, the superscript ∞ indicates infinite dilution, and the superscript r is a reminder of the need to specify the composition of the solvent mixture. In practice, when the composition of the solvent is understood from the physical situation, the latter subscript is omitted.

PARTIAL MOLAR PROPERTIES AS A FUNCTION OF MOLE FRACTIONS

The use of the volume as an example to derive expressions for partial molar properties is particularly useful in obtaining a physical sense of the meaning of a partial molar property at infinite dilution. Consider, for example, the case that the molar volume v of a binary mixture is measured as a function of composition, expressed by the mole fraction of component 1, x_1. In a plot of v versus x_1, at $x_1=0$, $\mathrm{v} = \mathrm{v}_2^0$, and at $x_1=1$, $\mathrm{v} = \mathrm{v}_1^0$. However, it is not immediate where to find $\overline{\mathrm{v}}_1$ or $\overline{\mathrm{v}}_2$. For generality, consider a multicomponent mixture containing c compounds for which the molar volume at a pressure P and a temperature T is known as a function of the mole fraction of $(c-1)$ components, which are independent variables. As the mole fractions add to unity, one of these values, say x_i, is not an independent variable. Hence, in terms of independent variables we write:

$$\mathrm{v} = \mathrm{v}(x_1, x_2, \ldots, x_{i-1}, x_{i+1}, \ldots, x_c)$$

In the variables for the function above, we have omitted P and T, which are kept constant in the derivation of the partial molar property, and also x_i, which is selected to be the dependent mole fraction. From Equation 6.2,

$$\overline{\mathrm{v}}_i \equiv \left(\frac{\partial V}{\partial n_i}\right)_{T,P,n_{j\neq i}} = \left(\frac{\partial n\mathrm{v}}{\partial n_i}\right)_{T,P,n_{j\neq i}} = \mathrm{v}\left(\frac{\partial n}{\partial n_i}\right)_{n_{j\neq i}} + n\left(\frac{\partial \mathrm{v}}{\partial n_i}\right)_{T,P,n_{j\neq i}}$$

However, $\left(\frac{\partial n}{\partial n_i}\right)_{n_{j\neq i}} = 1$, and applying the chain rule,

$$\left(\frac{\partial \mathrm{v}}{\partial n_i}\right)_{T,P,n_{j\neq i}} = \sum_{k\neq i}\left(\frac{\partial \mathrm{v}}{\partial x_k}\right)_{\substack{T,P,x_r\\ r\neq i,r\neq k}}\left(\frac{\partial x_k}{\partial n_i}\right)_{k\neq i}.$$

Moreover, in the latter expression,

$$\left(\frac{\partial x_k}{\partial n_i}\right)_{k\neq i} = -\frac{x_k}{n}$$

Thus, finally,

$$\overline{\mathrm{v}}_i = \mathrm{v} - \sum_{k\neq i} x_k\left(\frac{\partial \mathrm{v}}{\partial x_k}\right)_{\substack{T,P,x_r\\ r\neq i,r\neq k}} \tag{6.15}$$

Equation 6.15 is useful for the derivation of partial molar properties from an expression of the molar property as a function of mole fractions.

EQUATIONS AND PLOTS FOR BINARY MIXTURES

For a binary mixture, from Equation 6.15 we can write for component 2:

$$\overline{\mathrm{v}}_2 = \mathrm{v} - x_1\left(\frac{\partial \mathrm{v}}{\partial x_1}\right)_{T,P} \tag{6.16a}$$

where no mole fraction is kept constant for the differentiation. If x_1 changes, x_2 must change accordingly. For component 1, we write:

$$\bar{v}_1 = v - x_2\left(\frac{\partial v}{\partial x_2}\right)_{T,P}$$

or, considering $x_1 = 1 - x_2$ and $\left(\frac{\partial v}{\partial x_2}\right)_{T,P} = \left(\frac{\partial v}{\partial x_1}\right)_{T,P}\left(\frac{\partial x_1}{\partial x_2}\right) = -\left(\frac{\partial v}{\partial x_1}\right)_{T,P}$,

$$\bar{v}_1 = v + (1 - x_1)\left(\frac{\partial v}{\partial x_1}\right)_{T,P} \tag{6.16b}$$

A simple geometrical exercise helps to give a physical sense of the partial molar properties and their values at infinite dilution. From Equations 6.16a and 6.16b,

$$\left(\frac{\partial v}{\partial x_1}\right)_{T,P} = \frac{(\bar{v}_1 - v)}{(1 - x_1)} = \frac{(v - \bar{v}_2)}{x_1} \tag{6.17}$$

Thus, drawing a tangent line to the curve representing v versus x_1, the partial molar volumes are located at the points where the tangent line cuts the corresponding vertical axis at $x_1 = 0$ and $x_1 = 1$. A tangent line at one of the extremes of the curve representing v versus x_1 will give the corresponding pure compound molar volume at that extreme of the line and the value of the partial molar volume of the compound at infinite dilution at the other extreme of the line (Figure 6.1).

Once the partial molar volumes of the components of a binary mixture are known, they can be plotted as a function of composition. A typical example is shown in Figure 6.2.

The curve for $\bar{v}_1$ versus x_1 runs from $\bar{v}_1^{\infty}$ at $x_1 = 0$ to v_1^0 at $x_1 = 1$, and we observe that it reaches zero slope at the latter end. This necessary condition is a result of the restrictions imposed by the Gibbs–Duhem equation. From Equation 6.9, for a binary mixture at constant pressure and temperature,

$$[x_1 d\bar{v}_1 + x_2 d\bar{v}_2 = 0]_{T,P}$$

Thus, for a change in composition, taking x_1 as independent variable,

$$\left[x_1\frac{d\bar{v}_1}{dx_1} + x_2\frac{d\bar{v}_2}{dx_1} = 0\right]_{T,P}$$

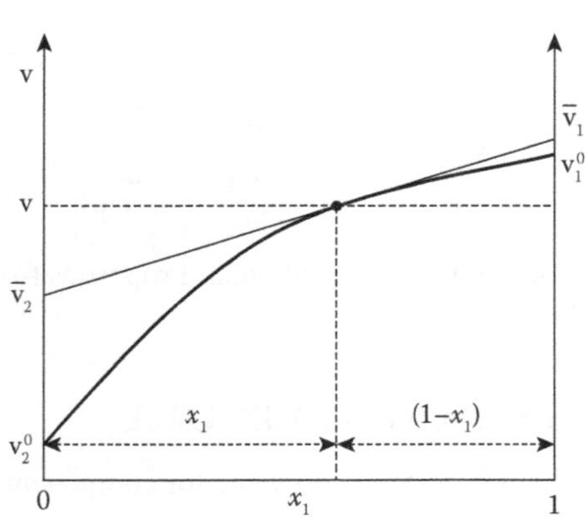

FIGURE 6.1 Determination of partial molar properties.

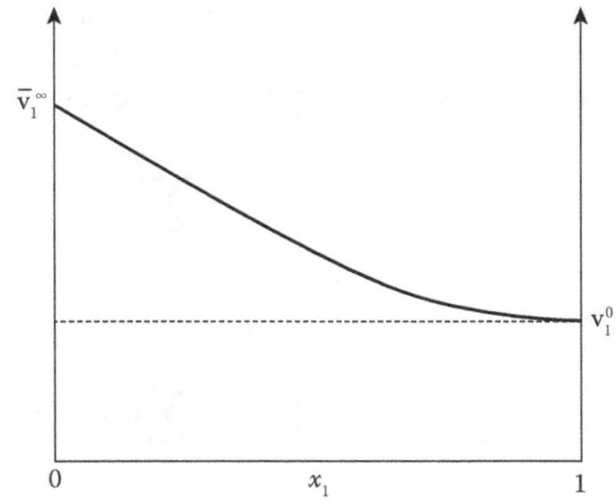

FIGURE 6.2 Partial molar volume as a function of concentration.

For the limit $x_1 \to 1, x_2 \to 0$,

$$\lim_{x_1 \to 1} \left[\frac{d\bar{v}_1}{dx_1} \right]_{T,P} = 0$$

This limiting behavior is valid for volume or enthalpy but not for entropy or entropy-related functions, like the Gibbs or Helmholtz functions or the chemical potential. In Chapter 7, attention is focused on what is perhaps the most important of all partial molar properties: the chemical potential.

PROPERTY CHANGES BY MIXING

In general, when pure compounds at pressure P and temperature T are mixed to form a mixture at P and T, the molar property of the mixture is different from the weighted addition of the molar properties of the compounds. When mixed, the compounds contribute to the molar properties of the mixture by their partial molar property values. Thus, for example, the volume change of mixing is:

$$\Delta v_m = \sum_k^c x_k \left(\bar{v}_k - v_k^0 \right) = v - \sum_k^c x_k v_k^0 \tag{6.18}$$

As discussed in the presentation of Equation 5.4, for a process happening in a closed system, the heat exchanged with the surroundings is equal to the enthalpy change. Thus, the enthalpy change of mixing of compounds under these conditions is properly called *heat of mixing*:

$$\Delta h_m = \sum_k^c x_k \left(\bar{h}_k - h_k^0 \right) = h - \sum_k^c x_k h_k^0 \tag{6.19}$$

Similar expressions can be written for the change of other thermodynamic properties upon mixing.

IDEAL MIXTURES

For the treatment of fluid mixtures, in gaseous or liquid phases, it is convenient to distinguish between ideal and nonideal behavior. A mixture that forms without heat effect and without volume change at constant pressure and temperature is called an *ideal mixture*. This measurable physical evidence indicates that the molecules mix without interfering with each other; that is, a mole of component k occupies the same volume in the mixture as it would as a pure substance. We may say that the molecules do not "see" each other as being different species:

$$\Delta v_m^{id} = \sum_k^c x_k\left(\overline{v}_k^{id} - v_k^0\right) = v_m^{id} - \sum_k^c x_k v_k^0 = 0 \tag{6.20}$$

and

$$\Delta h_m^{id} = \sum_k^c x_k\left(\overline{h}_k^{id} - h_k^0\right) = h_m^{id} - \sum_k^c x_k h_k^0 = 0 \tag{6.21}$$

With the same argument used to derive Equation 6.10, in the composition range that Equations 6.20 and 6.21 are valid, we can write:

$$\overline{v}_k^{id} = v_k^0 \tag{6.22}$$

and

$$\overline{h}_k^{id} = h_k^0 \tag{6.23}$$

Mixtures formed from compounds that are close to each other in a chemical homologous series approximate closely this behavior. For example, mixtures of benzene and toluene, methanol and ethanol, or acetic acid and propionic acid form with negligible volume change or heat effect. Chapter 15 discusses a generalization of the concept of ideal mixture. As shown there, the entropy change and the Gibbs energy change, when forming an ideal mixture, are different from zero.

PROBLEMS

6.1 **Partial molar volumes and volume change of mixing**

The compressibility factor z of gaseous mixtures of trifluoromethane (CHF_3, component 1) and tetrafluoromethane (CF_4, component 2) at 0°C (273.15 K) and 300 psia (2068.4 kPa) changes with composition as reported in [1]:

y_1	0.0	0.193	0.410	0.607	0.8407	1.0
z	0.8945	0.8860	0.8643	0.8361	0.7862	0.7348

Compute the volume change of mixing and estimate the partial molar volumes as functions of composition.

6.2 **Partial volumes per unit mass**

When the *density* ρ of a certain binary liquid mixture at 20°C and 1 atm is plotted against the *mass fraction* w_1 of component 1, and a tangent to the resulting curve is drawn at the point where $\rho = 0.9486$ g/cm³, this straight line intersects the vertical axis $w_1 = 0$ at $\rho = 1.0080$ g/cm³, and the vertical axis $w_1 = 1$ at $\rho = 0.8280$ g/cm³. What are the *partial specific volumes* of the two components at the composition of the point of tangency?

6.3 **Partial molar properties as a function of *unconstrained* molar fractions**

It is *physically* impossible to change *only one* mole fraction, for example, x_i in a multicomponent mixture, because by definition all mole fractions sum to unity, therefore *at least one other* mole fraction, for example, x_k must also change, as in the *constrained* derivatives in Equation 6.15.

a. How can the change of only two of the mole fractions be accomplished in practice, and how does the corresponding change in the mixture volume v relate to the partial molar volumes?
b. Usually, however, experimentally determined properties of mixtures at some temperature and pressure are correlated by analytical functions of mole fractions, for example, $v(x_1,...,x_c)$, and it is *mathematically* possible (and quite licit!) to obtain from these the partial derivatives with respect to *one* mole fraction keeping *all other* mole fractions constant. How can the partial molar properties be derived from these *unconstrained* partial derivatives?

6.4 **Partial properties of a binary regular mixture**

Use the results of Example 6.3 to derive expressions for the partial molar volumes in binary mixtures that exhibit "regular" behavior according to the symmetric Margules formula:

$$v = x_1 v_1^0 + x_2 v_2^0 + A x_1 x_2$$

where v_1^0, v_2^0, A may all be functions of temperature and pressure.

6.5 **Molar concentration and mass density of an ideal solution**

How do the molar concentration (moles per unit volume) and mass density of an ideal solution depend on its mole and mass fractions, respectively?

6.6 **Partial molar volumes and density of a ternary mixture**

In the ternary system (1) 1,4-dioxane+(2) methyl acetate+(3) n-butanol at 298.15 K and 100 kPa, the partial molar volumes of components 1 and 2 are given by:

$$\overline{v}_1 = v_1^0 + a_{12}x_2 + a_{13}x_3 - f(x_1, x_2, x_3)$$

$$\overline{v}_2 = v_2^0 + a_{12}x_1 + a_{23}x_3 - f(x_1, x_2, x_3)$$

where f is the function:

$$f = a_{12}x_1x_2 + a_{13}x_1x_3 + a_{23}x_2x_3$$

The pure-component molar volumes are $v_1^0 = 85.669$, $v_2^0 = 79.896$, $v_3^0 = 91.940$ cm³/mol, and the model parameters are $a_{12} = 0.0044746$, $a_{13} = 1.0643$, $a_{23} = 1.5956$. Find the molar density of a mixture having mole fractions $x_1 = 0.20$, $x_2 = x_3 = 0.40$.

REFERENCE

1. Lange, H. B., Jr., and Stein, F. P. 1970. Volumetric behavior of a polar-nonpolar gas mixture: Trifluoromethane-tetrafluoromethane system. *J. Chem. Eng. Data* 15(1): 56–61.

7 The Chemical Potential and the Gibbs–Helmholtz Equation

Abstract

This short chapter presents the identification of the chemical potential as the partial molar Gibbs energy and its variation with temperature and pressure. It shows that the partial derivative of the dimensionless Gibbs energy is equal to the negative value of the dimensionless enthalpy over the absolute temperature, a relation also valid for the partial molar properties and known as the Gibbs–Helmholtz equation. The Gibbs–Duhem and the Maxwell equations are written in terms of the chemical potentials.

CHEMICAL POTENTIAL AS A PARTIAL MOLAR PROPERTY

Due to the importance of the chemical potential in thermodynamics, we recapitulate here some of the equations related to it. Comparison of Equations 4.19 and 6.2 written for the Gibbs function G shows that what was defined as the chemical potential by Equation 4.19 is shown by Equation 6.2 to be the partial molar Gibbs energy:

$$\mu_i \equiv \bar{g}_i = \left(\frac{\partial G}{\partial n_i}\right)_{P,T.n_{j\neq i}} \tag{7.1}$$

Thus, using the general form of Equation 6.6, we write:

$$g = \sum_k^c x_k \bar{g}_k = \sum_k^c x_k \mu_k \tag{7.2a}$$

or, alternatively,

$$G = \sum_k^c n_k \bar{g}_k = \sum_k^c n_k \mu_k \tag{7.2b}$$

Notice that *the chemical potential is not the partial molar property for the internal energy, the enthalpy, or the Helmholtz energy.* Equations 4.6, 4.13, and 4.16 show that for these functions, the differentiation with respect to the number of moles of compound i is not taken at constant pressure and temperature, as required by the definition of a partial molar property.

CHEMICAL POTENTIAL DEPENDENCE ON TEMPERATURE AND PRESSURE

The change of the chemical potential with temperature and pressure is given by Equations 4.24 and 4.25, which can be rewritten in terms of partial molar properties as:

$$\left[\frac{\partial \mu_i}{\partial T}\right]_{P,n} = -\left[\frac{\partial S}{\partial n_i}\right]_{T,P,n_{j\neq i}} = -\bar{s}_i \tag{7.3}$$

DOI: 10.1201/9781003431985-9

and

$$\left[\frac{\partial \mu_i}{\partial P}\right]_{T,n} = \left[\frac{\partial V}{\partial n_i}\right]_{T,P,n_{j\neq i}} = \overline{\mathrm{v}}_i \tag{7.4}$$

GIBBS–HELMHOLTZ EQUATION

With the same procedure used for the derivation of Equation 6.12, from Equation 4.2 in molar terms, we write:

$$\mu_i \equiv \overline{g}_i = \overline{h}_i - T\,\overline{s}_i \tag{7.5}$$

Equation 7.5 is a key relation to derive the all-important *Gibbs–Helmholtz equation*. For this purpose, we consider the following mathematical identity:

$$\left[\frac{\partial\left(\mu_i/RT\right)}{\partial T}\right]_{P,n} \equiv -\frac{\mu_i}{RT^2} + \frac{1}{RT}\left(\frac{\partial \mu_i}{\partial T}\right)_{P,n}$$

Thus, using Equation 7.3 and combining with Equation 7.5, we write:

$$\left[\frac{\partial\left(\mu_i/RT\right)}{\partial T}\right]_{P,n} = -\frac{1}{RT^2}\left[\mu_i + T\overline{s}_i\right] \equiv -\frac{\overline{h}_i}{RT^2} \tag{7.6a}$$

Equation 7.6a is the Gibbs–Helmholtz equation in terms of partial molar properties. As the partial derivative is taken at constant composition, this expression can be directly written in terms of molar or total properties. In the latter case, it takes the form:

$$\left(\frac{\partial\left(G/RT\right)}{\partial T}\right)_{P,n} \equiv -\frac{H}{RT^2} \tag{7.6b}$$

GIBBS–DUHEM EQUATION IN TERMS OF CHEMICAL POTENTIALS

Although all equations valid for partial molar properties can be directly written in terms of the chemical potential, due to its importance we rewrite Equations 6.9 and 6.10 in this form here. From the general form of Equation 6.9, *mutatis mutandis*, we write the all-important *Gibbs–Duhem equation in terms of chemical potentials*:

$$\sum_k^c x_k d\mu_k = \left(\frac{\partial g}{\partial T}\right)_{P,n} dT + \left(\frac{\partial g}{\partial P}\right)_{T,n} dP$$

or, using Equations 4.17 and 4.18, in molar terms,

$$\sum_k^c x_k d\mu_k = -s\ dT + \mathrm{v}\ dP \tag{7.7a}$$

This equation is also conveniently written as:

$$\sum_{k}^{c} n_k d\mu_k = -S\ dT + V\ dP \tag{7.7b}$$

Equation 7.7b corresponds to Equation 6.8, the Gibbs–Duhem equation, written in terms of the chemical potentials.

Furthermore, using the same procedure employed for the derivation of Equation 6.10, from Equation 7.7a, *for a mixture of fixed composition*, independent of the particular set of mole fractions considered, we obtain:

$$d\mu_i = -\overline{s}_i\ dT + \overline{v}_i\ dP \tag{7.8}$$

This equation is one of the stepping stones for the calculation of chemical potentials. At constant pressure and temperature, it may seem to indicate that the chemical potential of a compound i in a mixture is constant and independent of composition. On the other hand, we know that like any partial property, chemical potential is a function of the composition of the mixture. The fact is that Equation 7.8 was derived *for the case of fixed composition*, no matter what composition it would be. Thus, the composition dependence is hidden in the functionality of the partial molar properties $\overline{v}_i$ and $\overline{s}_i$. Therefore, the constant value of the chemical potential at constant pressure and temperature depends on the composition of the system. This point is of utmost importance.

From the recognition that the chemical potential is the partial molar Gibbs energy, one can write Equation 6.15 for the chemical potential as:

$$\mu_i = g - \sum_{k \neq i} x_k \left(\frac{\partial g}{\partial x_k} \right)_{\substack{T,P,x_r \\ r \neq i, r \neq k}} \tag{7.9}$$

Equations can be transformed in so many ways that sometimes it is difficult to recognize them. One good example is the case of the *Euler's or integral form of the fundamental equation*, Equation 7.10. Combining Equation 4.2 with Equations 7.2b and 3.6b, we write:

$$\sum_{k}^{c} n_k \mu_k = (U + PV) - TS$$

or, rearranging,

$$U = TS - PV + \sum_{k}^{c} n_k \mu_k \tag{7.10}$$

It is interesting to observe that from the three terms contributing to the total internal energy in Equation 7.10, the enthalpy and the Helmholtz functions retain only two of the three pieces of information contained in the internal energy:

$$H = U + PV = TS + \sum_{k}^{c} n_k \mu_k$$

$$A = U - TS = -PV + \sum_{k}^{c} n_k \mu_k$$

The Gibbs energy, on the other hand, retains only the compositional contribution, and thus it is the appropriate function to use when studying molecular interactions:

$$G = H - TS = U + PV - TS = \sum_{k}^{c} n_k \mu_k$$

It is also interesting to consider the differential form of Equation 7.10:

$$dU = TdS + SdT - PdV - VdP + \sum_{k}^{c} n_k d\mu_k + \sum_{k}^{c} \mu_k dn_k$$

or

$$dU = \left[TdS - PdV + \sum_{k}^{c} \mu_k dn_k \right] + \left[SdT - VdP + \sum_{k}^{c} n_k d\mu_k \right]$$

Comparison of this differential with Equation 4.7 shows that three terms from the second bracket of the right-hand side cancel out. These are the terms considered in the Gibbs–Duhem equation (7.7b).

MAXWELL RELATION IN TERMS OF THE CHEMICAL POTENTIAL

As the chemical potential is the partial molar Gibbs energy, from Equation 6.11 we write:

$$\left(\frac{\partial \mu_j}{\partial n_i} \right)_{T,P,n_{k \neq i}} = \left(\frac{\partial \mu_i}{\partial n_j} \right)_{T,P,n_{k \neq j}} \tag{7.11}$$

This relation is of particular use for the study of electrolyte solutions, and it is seldom emphasized in literature.

PROBLEMS

7.1 **Enthalpy and Gibbs energy of an ideal gas as functions of temperature**

The enthalpy and Gibbs energy of nitrogen, as a hypothetical ideal gas, are conventionally assigned the value zero at 298.15 K and 1 bar (the *standard state*). The ideal gas heat capacity at constant pressure between 50 and 1000 K is given by the polynomial [1]:

$$\frac{c_P^*}{R} = a_0 + a_1 T + a_2 T^2 + a_3 T^3 + a_4 T^4$$

with $a_0 = 3.539$, $a_1 = -0.261 \times 10^{-3}$, $a_2 = 0.007 \times 10^{-5}$, $a_3 = 0.157 \times 10^{-8}$, $a_4 = -0.099 \times 10^{-11}$, and R the universal gas constant. Compute the enthalpy and Gibbs energy of nitrogen as an ideal gas at 500 K and 5 bar.

7.2 Reciprocal changes of chemical potentials

Show that $\left(\frac{\partial \mu_i}{\partial \mu_j}\right)_{T,P,\mu_{k \neq i,j}} = -\frac{x_j}{x_i}$

7.3 Pressure as a fundamental function

Suppose that an equation is available for pressure P as a function of temperature T and chemical potentials $\mu_1,\ldots,\mu_c$ of a homogeneous multicomponent mixture. Is this a *fundamental equation*, that is, can all other thermodynamic properties of the mixture be derived from a knowledge of $P(T,\mu_1,\ldots,\mu_c)$?

REFERENCE

1. Poling, B. E., Prausnitz, J. M, and O'Connell, J. P. 2001. *The Properties of Gases and Liquids*, 5th ed. New York: McGraw-Hill.

8 The Principles of Physical and Chemical Equilibrium

Abstract

The extremum principle of a maximum of entropy at equilibrium is extended to enthalpy, the Helmholtz function, and more importantly, to the Gibbs function, G. In the case of phase equilibrium without chemical reactions in a closed system, the minimum in the value of the Gibbs function implies the equality of the chemical potential of a compound in the different phases in which it is present. In the case of chemical equilibrium in a single-phase closed system, it implies that the sum of the products of the stoichiometric coefficients times the corresponding chemical potentials is null. A change in the number of moles of a compound is equal to the stoichiometric coefficient of compound i times the change of a parameter known as reaction coordinate, which characterizes the advance of the reaction.

In this chapter, we discuss the conditions for equilibrium in multicomponent closed systems that can be formed by one or more phases (open subsystems) in direct contact. The number of moles of each compound present in each subsystem can change for one of the following reasons:

1. The subsystem is open and exchanges matter with its surroundings.
2. The subsystem is closed and undergoes chemical reactions.
3. The subsystem is open and undergoes chemical reactions.

The first case is a typical problem of phase equilibrium in which two nonreacting phases of different compositions are put in direct contact in an overall closed system. The second case is typical of a study of chemical equilibrium. The third case is when two reacting phases are put in direct contact and allowed to reach phase and chemical equilibrium. These three cases will be discussed separately.

Before going into detail, it is necessary to agree on a change in the nomenclature. For the study of partial molar properties in mixtures of constant composition, it was convenient to use the subscript n to indicate that the numbers of moles did not change. For a closed system undergoing chemical reactions, the numbers of moles of the compounds will change, although the total mass of the system will be constant. Thus, to indicate a closed system from here on, we use the subscript m.

EXTREMUM PRINCIPLE FOR FUNCTIONS OTHER THAN ENTROPY

In Chapter 2, Equation 2.16 expressed in mathematical terms the extremum principle in terms of entropy. It is desirable to have this general principle expressed in terms of U, A, H, and G. For this, combining Equations 2.1d and 2.15 for an actual process in a closed system,

$$dU = TdS - T\delta S_{irr} + \delta W \tag{8.1}$$

For a process at constant entropy and at constant volume, the first and third terms of the right-hand side are null. As the second term is always negative for a real process, the internal energy decreases and it is a minimum at equilibrium. Thus, the mathematical expression of the extremum principle in terms of internal energy for a process at constant entropy and volume in a closed system takes the form:

$$dU_{S,V,m} \leq 0 \tag{8.2}$$

DOI: 10.1201/9781003431985-10

By a similar argument, using Equations 4.1 and 3.6b, respectively, it is shown that in a closed system, A is at minimum at equilibrium for a process at constant temperature and volume, while H is at minimum for a process at constant entropy and volume:

$$dA_{T,V,m} \leq 0 \tag{8.3}$$

and

$$dH_{S,V,m} \leq 0 \tag{8.4}$$

Due to its importance for G, we repeat the argument here. From Equations 3.6b and 4.2 for a closed system, at constant temperature and pressure, we write:

$$dG_{T,P,m} = [dU + PdV - TdS]_{T,P,m}$$

Combining this equation with Equation 8.1, and observing that at constant pressure the product PdV is equal to the actual irreversible work $\delta W = -P_E dV$, we write:

$$dG_{T,P,m} = -TdS_{T,P,m} \leq 0 \tag{8.5}$$

Thus, for a process at constant temperature and pressure in a closed system, the Gibbs energy always decreases and it has its minimum value at equilibrium for the particular set of values of pressure and temperature. The key point to notice here is that once the system is at equilibrium, it is totally immaterial by which process it attained the equilibrium state at T, P, and m. In mathematical terms, for the condition of G being a minimum with respect to the variation of a generalized independent variable X *at equilibrium* we write:

$$\left(\frac{\partial G}{\partial X}\right)_{T,P,m} = 0 \tag{8.6}$$

We are now in a position to present the analysis of the three possible processes by which a subsystem, forming a part of a closed system, can change its number of moles.

PHASE EQUILIBRIUM WITHOUT CHEMICAL REACTIONS

We consider here two nonreacting open subsystems α and β, in direct contact, forming part of an overall closed system of mass, m, which are left to reach equilibrium at a temperature, T, and a pressure, P. As the internal energy, the entropy, and the volume of the two subsystems in contact are additive, and the pressure and temperature of both subsystems at equilibrium are the same, the Gibbs energy of the overall closed system is the sum of the Gibbs energies of the subsystems. At equilibrium,

$$G = G^{\alpha} + G^{\beta}$$

Thus, for a differential change in the masses of the subsystems at constant temperature and pressure we have:

$$dG_{T,P,m} = dG^{\alpha}_{T,P} + dG^{\beta}_{T,P}$$

For each subsystem, from Equation 7.2b,

$$dG_{T,P} = \sum_k^c n_k d\mu_k + \sum_k^c \mu_k dn_k$$

According to the Gibbs–Duhem equation, Equation 7.7b, at constant pressure and temperature, the first term of the right-hand side is null. Thus, with all generality for each subsystem, we write:

$$dG_{T,P} = \sum_k^c \mu_k dn_k \tag{8.7}$$

For the overall closed system at constant pressure and temperature, adding the differential Gibbs energy changes for each subsystem and noting that necessarily $dn_i^\beta = -dn_i^\alpha$, we have:

$$dG_{T,P,m} = \sum_k^c \left(\mu_k^\alpha - \mu_k^\beta\right) dn_k^\alpha$$

From this expression and Equation 8.6, for a differential change in the number of moles of any of the compounds i at equilibrium,

$$\left(\frac{\partial G}{\partial n_i^\alpha}\right)_{T,P,m} = \mu_i^\alpha - \mu_i^\beta = 0$$

or

$$\mu_i^\alpha = \mu_i^\alpha \text{ for a compound } i \text{ in all the phases in which it is present} \tag{8.8a}$$

If there are more than two phases present in the closed system, a similar analysis applied to each pair of phases in direct constant, for the case of π phases, leads to:

$$\mu_i^\alpha\left(T,P,x^\alpha\right) = \mu_i^\beta\left(T,P,x^\beta\right) = \ldots = \mu_i^\pi\left(T,P,x^\pi\right) \tag{8.8b}$$

As indicated, the chemical potentials of compound i in all phases in which it is present are equal and are at the same temperature and pressure, but in each phase, they are at the particular composition of the phase.

We observe that the equality of the chemical potential of a pure compound in two phases in equilibrium was deduced previously from direct application of the first and second laws. According to Equation 7.2a, for a pure compound the chemical potential is equal to the molar Gibbs energy, and as shown by Equation 5.14b for any two phases of a pure compound in equilibrium, the molar Gibbs energies are equal.

CHEMICAL EQUILIBRIUM IN A SINGLE-PHASE CLOSED SYSTEM

In a single-phase closed system undergoing chemical reactions at constant temperature and pressure, we write Equation 8.7 as:

$$dG_{T,P,m} = \sum_k^c \mu_k dn_k \tag{8.7a}$$

where in this case, the subscript m has been added to the left-hand side to emphasize that this is valid for a closed system.

In a chemical reaction, the number of moles of *reactants*, R, consumed and the number of moles of *products*, P, produced are related by *stoichiometric coefficients*, λ:

$$\sum_{j}^{reactants} \lambda_j R_j = \sum_{k}^{products} \lambda_k P_k$$

Thus, the change in the number of moles of any two reactants is proportional to the ratio of their stoichiometric coefficients. The same happens for any two products. On the other hand, the ratio of the change in the number of moles of a reactant and a product is in the ratio of their stoichiometric coefficients affected with a minus sign, because the number of moles of reactant decreases as the number of moles of product increases. Thus, it is simpler to define the *stoichiometric number*, v_i, as:

$$\nu_i = \lambda_i \text{ for products}$$

and

$$\nu_i = -\lambda_i \text{ for reactants}$$

and rewrite the stoichiometric equation in the form:

$$\sum_{k}^{products} \nu_k P_k + \sum_{j}^{reactants} \nu_j R_j = 0$$

Hence, for any two compounds, reactants or products,

$$\frac{dn_1}{dn_2} = \frac{\nu_1}{\nu_2}$$

or, equivalently, for any two compounds participating in the reaction, reactants or products,

$$\frac{dn_1}{v_1} = \frac{dn_2}{v_2} = \ldots = d\xi$$

The parameter ξ is known as the *reaction coordinate*. Therefore, we write:

$$dn_i = v_i d\xi \tag{8.9}$$

Equation 8.7a can be written as:

$$dG_{T,P,m} = \left(\sum_{k}^{c} \nu_k \mu_k \right) d\xi$$

or, at equilibrium, from Equation 8.6,

$$\left(\frac{\partial G}{\partial \xi} \right)_{T,P,m} = \sum_{k}^{c} \nu_k \mu_k = 0 \tag{8.10}$$

This condition for chemical equilibrium applies to all reactions that can occur in a closed system (Figure 8.1).

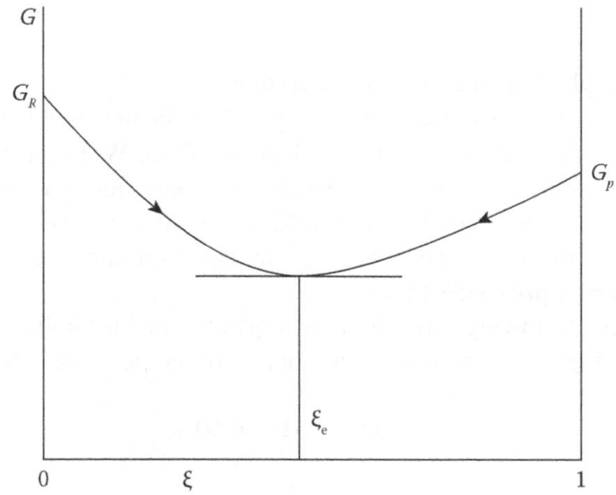

FIGURE 8.1 Gibbs function of a closed system in which a chemical reaction takes place.

Although the chemical potential for compound i at equilibrium has a single value for all possible reactions, the stoichiometric number of compound i in different reactions is not necessarily the same. Thus, the *thermodynamic condition for chemical equilibrium* is more precisely written in the form:

$$\sum_{k}^{c} \nu_{k,r}\mu_k = 0 \tag{8.11}$$

In this expression, the subscript r indicates that there is one such relation for each *independent reaction* that can occur in the system.

The *number of independent chemical reactions* possible (NIR) is obtained by writing the equation of formation of all compounds considered to be present, starting from their constituent elements. By algebraic combination of these imaginary reactions of formation, all elements, which are considered not to be present as elements, are eliminated. The set of equations resulting from this process gives the number of independent reactions possible. Although the set of final equations depends on the algebra, the number of such equations should be the same for all sets. In general, this number is larger or equal to the number of compounds and elements present as elements minus the number of constituent elements. It is highly recommended to eliminate the elements not present as elements in an organized way by selecting one equation containing one such element and using it to eliminate that particular element from all other equations.

PHASE AND CHEMICAL EQUILIBRIUM IN A CLOSED SYSTEM

In this case, Equations 8.8 and 8.11 will apply simultaneously. As the reactions will be at equilibrium in each of the phases present, with the same stoichiometric numbers and the same values for the chemical potentials of the compounds, the chemical equilibrium is conveniently solved in the phase where calculations are simpler. This is usually the gas phase, when present.

As shown in this chapter, the conditions for phase equilibrium in simple systems or chemical equilibrium in reacting systems are the direct consequence of the minimum in the Gibbs energy at equilibrium. This condition, in turn, is a direct consequence of the increase in entropy of the system due to the irreversibility of real processes. The generality of the thermodynamic treatment presented up to this point is further discussed in Chapter 10.

PROBLEMS

8.1 **Extremum principles for unconstrained systems**

In Equations 8.2–8.5, the extremum principles have been stated for systems under constant conditions of the appropriate independent variables. What form do these equations adopt if such restrictions are relaxed, for example, how does internal energy U behave if the system boundaries are diathermal, deformable, and permeable? And what is the equivalent extremum principle for the fundamental pressure equation introduced in Problem 7.3?

8.2 **Phase equilibria of a pure substance**

Suppose that the following three equations give the molar Gibbs energy, g kJ/mol of a certain material M in different states as a function of temperature T K at 1 atm:

$$g^{(\alpha)} = 607.9 - 0.09650\ T$$

$$g^{(\beta)} = 751.0 - 0.2148\ T$$

$$g^{(\gamma)} = 624.7 - 0.1160\ T$$

which correspond to the solid, liquid, and vapor phases, although not necessarily in that order. Determine which equation pertains to each state, and find the melting, sublimation, and boiling temperatures, if they exist. What can you conclude about the triple (solid–liquid–vapor) point of this material?

8.3 **Stability of chemical equilibrium**

While the condition given in Equation 8.10 and illustrated in Figure 8.1 is *necessary* for chemical equilibrium, it is not *sufficient*, because a zero derivative can also arise at a maximum rather than a minimum Gibbs energy, indicating an *unstable* state. What additional condition is required for Equation 8.10 to actually describe a minimum? Show that such condition is fulfilled by any reaction that takes place in an ideal mixture in which the chemical potentials of the components are given by $\mu_i^{id}(T,P,x) = g_i^0(T,P) + RT \ln x_i$, where $g_i^0(T,P)$ is the molar Gibbs energy of the pure component at the same temperature and pressure of the mixture.

9 The Phase Rule and Duhem's Theorem

ABSTRACT

The phase rule states that the number of degrees of freedom available in a closed system without chemical reactions is equal to the number of compounds present, minus the number of phases, plus two. This number is reduced by the number of independent reactions if reactions are present. Duhem's theorem fixes in two the number of degrees of freedom for the case of a flash of a liquid mixture without chemical reactions. Additional constraints may reduce further the number of degrees of freedom.

When solving a problem of phase or chemical equilibrium, it is necessary to know if there is enough information or it would be necessary to specify some variables before evaluating the remaining variables. As in any multivariable system of equations, if the number of variables is equal to the number of independent equations, the system can be solved. If there are more equations than variables, the system is overdetermined and, in the general case, no solution exists. On the other hand, if there are fewer relations than variables, the system is underdefined. The difference between the number of variables and the number of relations between them gives the available degrees of freedom (*DF*), that is, the number of variables that need to be specified in order to solve for all others.

For simplicity, we discuss separately the case of a closed nonreacting system with two or more phases in equilibrium and the case of a closed reacting system in which phases are in physical and chemical equilibrium.

PHASE RULE FOR NONREACTING SYSTEMS

The chemical potential of a compound in a homogeneous mixture is a function of temperature, pressure, and the composition of the mixture. Thus, in mathematical terms, each equality of the chemical potential of compound i included in Equation 8.8a or 8.8b represents one equation relating the variables of both phases in contact. These variables are the mole fractions x of the compounds in each phase, the temperature, T, and the pressure, P.

Thus, if π phases are present, Equation 8.8 indicates that there are $(\pi - 1)$ independent equations per compound relating to the variables involved. For c compounds, the total number of equations available is $c\,(\pi - 1)$.

With respect to the variables involved, we recall that the mole fractions in each phase add to unity, so there are only $(c-1)$ independent mole fractions for a phase with c components. Assuming that all c compounds are present in all the phases, as the pressure and the temperature are the same for all phases, the total number of independent variables is $[(c-1)\,\pi + 2]$.

We recall here that in the case we are considering, there are no chemical reactions; thus,

$$DF = \left[(c-1)\,\pi + 2\right] - c(\pi - 1)$$

or

$$DF = c - \pi + 2 \tag{9.1}$$

DOI: 10.1201/9781003431985-11

If one compound is not present in one phase, that is, if the value of one composition variable is known to be zero, there is one variable less and also one equation less, as the equality of chemical potentials of that compound does not enter into the count. Thus, the number of degrees of freedom remains unchanged.

PHASE RULE FOR REACTING SYSTEMS

In this case, we consider a single phase. If there is a number of independent reactions (*NIR*) that are possible in the system, the number of variables will not change, but the number of relations between the variables will increase by exactly the number *NIR* of independent equilibrium conditions (Equation 8.10). Hence, in this case,

$$DF = \left[(c-1)\ \pi + 2\right] - \left[c\ (\pi - 1) + NIR\right]$$

$$DF = (c - NIR) - \pi + 2 \tag{9.2}$$

If one reaction goes to completion at equilibrium and some reactants that were initially in the system are totally consumed, both the reaction and the compounds totally consumed can be ignored.

COMMON CASES OF APPLICATION OF THE PHASE RULE

It is useful to keep in mind the number of variables that it is necessary to specify to define completely the intensive state of a system in equilibrium. For a pure compound or for a mixture of known composition in one phase, the values of two intensive variables are required. One single intensive variable completely defines the state of a pure compound in two phases. For a binary system in two phases, there are two degrees of freedom. Thus, if the composition of one of the phases is known, either the pressure or the temperature can be fixed arbitrarily, but not both. In chemical engineering calculations for vapor–liquid equilibrium in binary systems, if the liquid composition is known, the calculation of the vapor phase composition is called a "bubble" calculation. Similarly, the calculation of the liquid phase composition from information on the vapor phase composition is called a "dew" calculation. If the pressure is fixed and the temperature is unknown, the problem is further indicated as a T calculation, and vice versa for the pressure calculation. Thus, a *bubble P* calculation is the case of knowing the liquid composition and the temperature and having to obtain the vapor phase composition and pressure.

DUHEM'S THEOREM

One interesting case in chemical engineering is the "flashing" of a liquid mixture of known composition in a low-pressure drum to obtain a liquid phase and a vapor phase in equilibrium. This practical process is run in a continuous way, and the mass flow rate and composition of the feed are known. The *Duhem theorem* gives the number of intensive variables necessary to fix, so the mass flow rates and compositions of the vapor and liquid product streams are uniquely determined. The surprising result is that even requiring an extra extensive variable to determine the extensive state of each phase, the number of intensive variables that is necessary to fix is only two, independent of the number of components of the mixture. Thus, if the composition and mass flow rate of the feed are known, fixing the temperature and pressure in the drum completely determines the system. Duhem's theorem can be demonstrated by a simple count of the number of variables and relations involved. In addition to the intensive variables considered by the phase rule, it is necessary to know one additional extensive variable per phase present to define the extensive state of each of them.

Thus, the total number of variables in this case for π phases is:

$$(\text{Number of variables}) = \left[(c-1)\ \pi + 2\right] + \pi = c\pi + 2$$

On the other hand, in the case considered, there are also additional relations between the variables. In the absence of chemical reactions, at steady state, each compound that enters in the feed must leave the system in the outflow streams. The equations relating the masses of each compound in the three streams are known as material balances. There are as many material balances as compounds in the system. Hence,

$$(\text{Number of relations}) = (\pi - 1)c + c = c\pi$$

Thus, $DF=2$.

PHASE RULE WITH ADDITIONAL CONSTRAINTS

Additional constraints not considered in the cases discussed above will discount degrees of freedom. If a pure compound is present as a phase in a system of c compounds, this will discount $(c-1)$ variables. In some complex cases, it is better to obtain the degrees of freedom by direct counting of variables and restrictions. For example, a rigid and semi-permeable membrane separating two phases in equilibrium will cause the pressure to not be the same in both phases, and some compositions may have a zero value in some of the phases. Even if a compound is present in more than one phase but cannot permeate through the membrane, the equality of potentials would not apply for that compound and one relationship is lost.

PROBLEMS

9.1 **Triple point of a pure substance**

How many degrees of freedom does the triple point of a pure substance have? How many triple points does a pure substance have?

9.2 **Duhem's theorem for reacting systems**

According to Equations 8.9 and 8.10, the equilibrium molar amounts and compositions of a reacting system at a given temperature and pressure depend solely on the reaction coordinate ξ. Show that if the initial amounts of all components are known, only two more variables need to be specified to completely determine the equilibrium state, whatever the number of components. Can this result be extended to systems with multiple reactions?

10 Generality of the Thermodynamic Treatment for More Complex Systems

ABSTRACT

Chapter 10 is a preview of how to handle thermodynamic problems requiring the consideration of more variables than those specified in Chapter 2 for the treatment of simple systems. Two end-of-chapter problems of phase equilibrium across curved surfaces are included in this unusual exploratory chapter. Detailed material for systems including charged particles or ions is treated in Chapter 18 and for systems including surface effects is considered in Chapter 22. The important point stressed in this chapter is that the basic concepts do not change with the introduction of more variables necessary for the treatment of nonsimple systems.

Simple systems discussed in the first seven chapters were formed by nonreacting compounds, did not include electrically charged species, and were considered to be large enough to make surface effects negligible. Chapter 8 introduced a first generalization of the thermodynamic treatment and included the discussion of reacting systems. Thus, we accept now that the internal energy of the system can change by the exchange of heat with the surroundings, by expansion or compression work due to the action of external forces, and also by the change in the number of moles of the components due to the fact that either the system is open, or it undergoes chemical reactions, or both.

In this section, we discuss the possibility of having work other than expansion or compression work in a thermodynamic reversible process between two equilibrium states of the system. Equation 4.7 had the form:

$$dU = TdS - PdV + \sum_{k}^{c} \mu_k \, dn_k \tag{4.7}$$

If there is a reversible work $\delta \hat{W}_{rev}$, different from expansion work, it needs to be incorporated. Thus, we write the generalized form of Equation 4.7 as:

$$d\hat{U} = TdS - PdV + \delta \hat{W}_{rev} + \sum_{k}^{c} \mu_k \, dn_k \tag{10.1}$$

In Equation 10.1, a special symbol is used for the internal energy as a reminder that the system under discussion is not a simple system. Otherwise, for all practical consequences, the algebra done for simple systems will remain unchanged. For example, the relation of the internal energy with the enthalpy of such a system follows Equation 3.6, written as:

$$\hat{H} = \hat{U} + PV$$

Similarly, the Gibbs energy of the nonsimple system follows the definition given in Equation 4.2, namely,

DOI: 10.1201/9781003431985-12

$$\hat{G} = \hat{H} - TS = \hat{U} + PV - TS$$

The differential form of this expression,

$$d\hat{G} = d\hat{U} + PdV + VdP - TdS - SdT$$

combined with Equation 10.1, gives:

$$d\hat{G} = -SdT + VdP + \delta\hat{W}_{rev} + \sum_{k}^{c} \mu_k \, dn_k \tag{10.2}$$

In most practical applications, the pressure and temperature are kept constant and we write:

$$d\hat{G}_{T,P} = \delta\hat{W}_{rev} + \sum_{k}^{c} \mu_k \, dn_k \tag{10.3}$$

We consider two examples of application of Equation 10.3 here.

EQUILIBRIUM INVOLVING IONS

Ions are *charged particles* produced by dissolving salts in water or other appropriate solvent. The word *ion* means "traveler" in Greek. Dissolving sodium chloride in water, for example, and applying an electric potential on two electrodes immersed in the solution, negatively charged chloride species and positively charged sodium species will travel toward the appropriate electrodes. Sodium ions will travel toward the negatively charged electrode, or *cathode*, and chloride ions toward the positively charged electrode, or *anode*. Thus, these positively and negatively charged ions are called *cations* and *anions*, respectively. As an electric current will pass from one electrode to the other when a salt is in solution, the salts generating ions are called *electrolytes*.

Electrochemical systems are complex and the thermodynamic description of electrochemical processes is far from simple. The problem starts with the difficulty of defining unambiguously the electrostatic potential of an electrolyte solution. As with the internal energy or the enthalpy, in thermodynamics, we normally do not know the "value" of a property, but we are satisfied with being able to determine the difference in its value for a particular process or condition.

Thus, to proceed, we accept that a phase α of an electrolyte solution has an electrostatic potential Φ^{α}, with respect to an arbitrary zero. This potential is determined by the concentrations of electrolytes present, the temperature, and to a minor extent for liquid solutions, the pressure of the system. The extra work required in this case is the electrical work necessary to move a charge $\delta\Omega_i$ from a state at zero potential to a state at potential Φ of the solution. The addition of an ion to the solution adds energy to its internal energy. As with the expansion or compression work, the thermodynamic work is the work done by the external force; hence,

$$\delta\hat{W}_{rev,i} = \Phi\delta\Omega_i$$

The units used for the potential are volts. As the charge is measured in coulombs, that is, joules per volt, the work is expressed in joules. Each ion i has a characteristic charge Z_i. Sodium ion has charge +1, chloride ion has charge −1, and so on. The Faraday constant F is equal to 96484 J/(V mol), which is the product of the charge of an electron, that is, 1.6022×10^{-19} J/V, times Avogadro's number, that is, $6.0220 \times 10^{23}\,\text{mol}^{-1}$. Hence, the charge of dn_i moles of ion i is:

$$\delta\Omega_i = Z_i F dn_i$$

and the electric work per dn_i moles of ion i is:

$$\delta\hat{W}_{rev,i} = \Phi Z_i F dn_i$$

Introducing this result in Equation 10.3, we write:

$$d\hat{G}_{T,P} = \sum_k^c (\mu_k + \Phi Z_k F)\, dn_k$$

The *electrochemical potential* of ion i is then defined by:

$$\hat{\mu}_i = \mu_i + \Phi Z_i F \tag{10.4}$$

Thus, the above equation reduces to the familiar form of Equation 8.7, namely,

$$d\hat{G}_{T,P} = \sum_k^c \hat{\mu}_k \; dn_k \tag{10.5}$$

and in total parallel with Equation 7.1, we write:

$$\hat{\mu}_i = \left(\frac{\partial \hat{G}}{\partial n_i}\right)_{P,T,\Phi,n_{j\neq i}} \tag{10.6}$$

From here on, *mutatis mutandis*, the algebra is not any different than the one for noncharged species. For *phase equilibrium of ions in solution*, the equilibrium condition takes the form:

$$\hat{\mu}_i^{\alpha} = \hat{\mu}_i^{\beta} = \ldots = \hat{\mu}_i^{\pi} \tag{10.7}$$

Writing this equation, in terms of chemical potentials, for an ion in two phases, we get:

$$\mu_i^{\alpha} + \Phi^{\alpha} Z_i F = \mu_i^{\beta} + \Phi^{\beta}\; Z_i F$$

or

$$\mu_i^{\alpha} - \mu_i^{\beta} = \left[\Phi^{\beta} - \Phi^{\alpha}\right] Z_i F$$

This expression is better written as:

$$\frac{1}{Z_i}\left[\mu_i^{\alpha} - \mu_i^{\beta}\right] = \Delta\Phi\; F \tag{10.8}$$

Equation 10.8, in which $\Delta\Phi$ is known as the *Donnan potential*, shows that the right-hand side is independent of the ion considered, and consequently, different ions in solutions at equilibrium meet the condition:

$$\frac{1}{Z_i}\left[\mu_i^{\alpha} - \mu_i^{\beta}\right] = \frac{1}{Z_j}\left[\mu_j^{\alpha} - \mu_j^{\beta}\right] \tag{10.9}$$

This relation is known as the *Donnan equilibrium* condition for ions in solutions in contact through membranes.

Similarly, the case of "chemical equilibrium" for a single closed phase is found when a salt is dissolved in an appropriate solvent such as water, and it dissociates into its ions. A chemical equilibrium is established between the undissociated salt in solution and its ions. Strong electrolytes, like NaCl or KOH, dissociate completely, while weak electrolytes like alanine or acetic acid, will dissociate only partially. Another case of ionic equilibrium is metal in equilibrium with its ions in an acid solution. For all these cases, the condition of chemical equilibrium, obtained by the same algebra as the one used to obtain Equation 8.10 for the case of nonelectrolytes, gives:

$$\sum_k \nu_k \hat{\mu}_k = 0 \tag{10.10}$$

Using Equation 10.4, we write:

$$\sum_k \nu_k \mu_k + F\Phi \sum_k \nu_k Z_k = 0$$

In this equation, the second summation is null, as it is the difference of the charges of all the cations and the charges of all the anions. The sign of the charges is implicit in Z_i:

$$\sum_k \nu_k Z_k = 0 \tag{10.11}$$

Equation 10.11 is known as the *electroneutrality condition for all solutions of electrolytes.* Thus, finally for chemical equilibrium of electrolytes in solution, we write:

$$\sum_k \nu_k \mu_k = 0 \tag{10.12}$$

Equation 10.12 is identical to Equation 8.11 for nonelectrolytes. The important point to observe here is that in Equation 10.9 for phase equilibrium and also in Equation 10.12 for chemical equilibrium in electrolyte solutions, the loosely defined phase electrostatic potential Φ cancels out. In some sense, this potential has played the same role as the entropy in Chapter 5; it was used as a stepping stone to derive equations of practical use. The use of these relations will be further explored in Chapter 18 dealing with electrolyte solutions.

ADSORPTION OF CHEMICAL SPECIES ON A SURFACE

As another example of the generality of the thermodynamic treatment, we discuss here the basic principle of equilibrium in the case of *adsorption of a solute from a solution over a solid surface.* The solution can be in the gas or liquid phase. As in the case of the previous example, we will not go into the details of the application here.

In the case of adsorption equilibrium, there are three phases present: a fluid phase that can be a solution or a pure compound, a solid phase, and an adsorbed phase. While the first two phases are volumetric in nature, the adsorbed phase is considered to be a two-dimensional phase covering the surface of the solid. At equilibrium, for any chemical species present in the fluid phase and in the adsorbed phase, the chemical potential of the compound will be the same in both phases:

$$\mu_i^{ads} = \mu_i$$

The symbol without superscript in this expression refers to the chemical potential in the fluid phase. The work of adsorption is the work to extend the adsorbing surface area A of the adsorbent to receive more adsorbate. The surface tension σ is defined as the force per unit of frontal surface length exerted by the surface; thus, the adsorption work to increase the adsorbed area by a differential dA is:

$$\delta \hat{W}_{rev} = \sigma dA$$

Introducing this expression into Equation 10.3, for the adsorbed phase we obtain:

$$d\hat{G}^{ads}_{T,P} = \sigma dA + \sum_{k}^{c} \mu_k \, dn_k^{ads} \tag{10.13}$$

Here, using the equality of chemical potentials for compound i, the superscript *ads* is omitted. We now use the same method employed to derive Equation 6.5, namely, the fact that if the number of moles of *all* compounds adsorbed is multiplied by a factor λ, the area covered and the value of the Gibbs function will also be changed by the same factor. Thus,

$$d\left[\lambda \hat{G}^{ads}_{T,P}\right] = \sigma d\left[\lambda A\right] + \sum_{k}^{c} \mu_k \, d\left[\lambda n_k^{ads}\right]$$

Differentiation of this expression with respect to λ gives:

$$\hat{G}^{ads}_{T,P} = \sigma A + \sum_{k}^{c} \mu_k \, n_k^{ads}$$

Thus,

$$d\hat{G}^{ads}_{T,P} = \sigma dA + A d\sigma + \sum_{k}^{c} \mu_k \, dn_k^{ads} + \sum_{k}^{c} n_k^{ads} \, \mu_k$$

Comparison of this expression with Equation 10.13 shows that, necessarily,

$$A d\sigma + \sum_{k}^{c} n_k^{ads} \, d\mu_k = 0 \tag{10.14}$$

Equation 10.14 is the Gibbs–Duhem equation for the adsorbed phase.

In the work of adsorption, it is common to define the *Gibbs adsorption*, Γ_i, as the number of moles of compound i adsorbed per unit surface area of adsorbent.

$$\Gamma_i = \frac{n_i^{ads}}{A} \tag{10.15}$$

Combining Equations 10.14 and 10.15, we obtain:

$$d\sigma = -\sum_{k}^{c} \Gamma_k \, d\mu_k \tag{10.16}$$

Variations of Equation 10.16 are commonly used in adsorption studies. For example, for the case of adsorption on a solid surface of a solute 2 from a solution in a solvent 1, in which the solvent is *not* adsorbed by the solid surface, that is, $\Gamma_1 = 0$, from Equation 10.16 we get:

$$\Gamma_2 = -\frac{d\sigma}{d\mu_2} \tag{10.17}$$

This expression is normally used as the starting point in physical chemistry texts.

As a final example of the large number of applications to be found by extension of the basic thermodynamic treatment, we present here one expression useful for the study of small systems where the surface effects become important.

Consider a droplet of a solution in a chamber containing vapors of the solvent and air. The number of moles of solute in the droplet is constant, while the number of moles of solvent can be changed by evaporation or condensation of solvent in the droplet. In this case, as there is no change in the number of moles of solute, Equation 10.13 takes the form:

$$d\hat{G}^{ads}_{T,P} = \sigma_s dA + \mu_s dn_s$$

The subscript s has been used to indicate that the surface tension in this case is that of the solvent in the droplet. Assuming a spherical droplet of radius r, a change in the surface area of sphere is related to a change in its volume by:

$$dA = 2\frac{dV}{r}$$

In addition, for a dilute solution the partial molar volume of the solvent can be approximated by its pure compound molar volume v_s and we can write:

$$dV = \mathrm{v}_s dn_s$$

Thus,

$$d\hat{G}^{ads}_{T,P} = \left[2\sigma_s \frac{\mathrm{v}_s}{r} + \mu_s\right] dn_s$$

and the chemical potential of the solvent in the droplet, following Equation 10.6, is:

$$\hat{\mu}_s = \left(\frac{\partial \hat{G}}{\partial n_s}\right)_{P,T,\sigma_s} = \left[2\sigma_s \frac{\mathrm{v}_s}{r} + \mu_s\right] \tag{10.18}$$

Clearly, the change in the chemical potential of the solvent is important in small systems, and as the radius of the droplet increases, the difference of chemical potentials becomes negligible.

The purpose of this chapter has been to show some selected examples in which the same standard ideas of the first nine chapters can be applied to discuss more complex systems. We have followed the developments to the point where the equations obtained are normally used as a starting point for applications.

PROBLEMS

10.1 **Equilibrium between a liquid drop and a vapor**

A small spherical drop of a pure liquid is surrounded by a large amount of vapor of that same fluid. Because of the large area-to-volume ratio of the drop, the energy associated with surface tension makes a significant contribution to the total energy of the liquid. Derive the criteria for equilibrium between the drop and the vapor. Show that the pressure inside the drop is greater than that of the vapor outside and find its dependence on the drop radius r.

10.2 **Dew points of a pure vapor**

Methanol vapor is compressed isothermally at 30°C until it begins to condense into liquid drops that grow in size as the pressure increases. Using the results of the previous example, estimate the pressure at which the vapor is in equilibrium with drops of radius (a) 10 nm, (b) 1 μm, and (c) 0.1 mm. At 30°C, the molar volume of liquid methanol is 0.0410 m^3/kmol, its saturation pressure is 21.914 kPa [1], and its surface tension is 21.69 mN/m [2]. For simplicity, assume that the liquid phase is incompressible, and the vapor phase behaves as an ideal gas.

REFERENCES

1. Lemmon, E. W., Bell, I. H., Huber, M. L., and McLinden, M. O. 2023. Thermophysical properties of fluid systems. In *NIST Chemistry WebBook, NIST Standard Reference Database Number 69*, eds. P. J. Linstrom and W. G. Mallard. Gaithersburg, MD: National Institute of Standards and Technology. Available at https://doi.org/10.18434/T4D303 (accessed July 2, 2024).
2. Mulero, A., Cachadiña, I., and Parra, M. I., 2012. Recommended correlations for the surface tension of common fluids. *J. Phys. Chem. Ref. Data* 41(4): 043105.

11 Ideal Gas and Ideal Gas Mixtures

Abstract

This chapter presents the expression of the chemical potential for a compound in an ideal gas mixture in terms of pressure, temperature, and composition. It is observed that when mixing pure compounds ideal gases at the same temperature and pressure, the mixture forms without volume change and heat effect. However, there are entropy, Helmholtz and Gibbs energy change

CHEMICAL POTENTIAL OF A COMPOUND IN AN IDEAL GAS MIXTURE

As discussed in Chapter 8, the study of phase and chemical equilibrium requires knowledge of the chemical potentials of the components of a mixture at the temperature, T, of the system. A convenient starting point to calculate chemical potentials is Equation 7.8, which indicates that at constant pressure and temperature the chemical potentials of the compounds in the mixture have fixed values. This result is in agreement with the conclusion obtained in Chapter 9, from the application of the phase rule to a single-phase multicomponent mixture of fixed composition. This system has two degrees of freedom, and by fixing the temperature and pressure, the system is completely specified. We also recall that, in principle, in Equation 7.8 the composition dependence of the chemical potential is hidden in the functionality of the partial molar properties $\overline{v}_i$ and $\overline{s}_i$. Thus, at constant temperature, Equation 7.8 reads:

$$d\mu_{i,T} = \overline{v}_i dP \tag{11.1}$$

This equation would be directly applicable if the partial molar volume of compound i in the mixture as a function of pressure could be obtained from Equation 6.2 or 6.15. This would be the case if an equation of state (EOS) giving the volume of the mixture as a function of composition and pressure at the temperature of interest were available. In practice, the use of available volume-explicit EOSs is restricted to gases of nonpolar pure compounds at low pressure and relatively high temperatures. Thus, a better way to proceed is to start the analysis considering the case of a compound in a multicomponent ideal gas mixture for which the volumetric behavior can be described without approximations. From Equation 1.1 or 5.27, for n moles of mixture, we write:

$$V^* = \frac{\left[\sum_k^c n_k\right] RT}{P}$$

The partial molar volume of compound i in an ideal gas mixture, from Equation 6.2, is:

$$\overline{v}_i^* \equiv \left(\frac{\partial V^*}{\partial n_i}\right)_{T,P,n_{j\neq i}} = \frac{RT}{P} \tag{11.2}$$

DOI: 10.1201/9781003431985-13

Comparison of Equation 11.2 with Equation 5.27 shows that the partial molar volume of a component of an ideal gas mixture at temperature, T, and pressure, P, is equal to the molar volume of the pure compound as an ideal gas under the same conditions of pressure and temperature:

$$\overline{v}_i^* = v_i^{0,*} = \frac{RT}{P}$$

This is true for an ideal gas mixture only. We observe that n_i moles of pure compound i as ideal gas, at the temperature and pressure of the mixture, occupy a total volume that is different from the total volume of the gas mixture:

$$V_i^{0,*}(P,T) = \frac{n_i RT}{P} \neq V^*$$

In an ideal gas, molecules are mere points in space and do not interact in any way with one another. Each compound i in the ideal gas mixture occupies the whole volume V^*. If this is the case, as all other components of the ideal mixture also exert pressure on the walls, the pressure exerted on the wall of the container by a compound i cannot be the total pressure P. Both conditions cannot hold simultaneously. The n_i moles of compound i in the ideal gas mixture exert a *partial pressure* p_i^* on the walls of the container when the total volume of the mixture is V^*. Thus,

$$p_i^* = \frac{n_i RT}{V^*} = \frac{n_i}{\left[\sum_k^c n_k\right]} P = y_i P$$

In this equation, we have used the symbol y_i for the mole fraction of compound i in a gas phase mixture in exact parallel with Equation 6.7. Addition over all components of the mixture gives the total pressure P exerted by the ideal gas mixture. For compound i in a real gas mixture, that is, a gas mixture whose volumetric behavior does not follow Equation 5.27, we define its partial pressure by:

$$\overline{p}_i \equiv y_i P \tag{11.3}$$

Two points need clarification here. First, the partial pressure of a compound in a *real* gas mixture is not the pressure that compound i would exert on the wall if it occupies the same volume at the temperature T. This is only true for a component of an ideal gas mixture. Second, the partial pressure is not a partial molar property. The partial pressure of a component in a gas mixture is simply defined by Equation 11.3. Returning to the ideal gas case, combining Equations 11.1 and 11.2, we write:

$$d\mu_{i,T}^* = \frac{RT}{P} dP = RTd\ln P$$

Integrating between two pressures P_1 and P_2 for a compound as ideal gas at temperature T, we write:

$$\mu_i^*(T,P_2) = \mu_i^*(T,P_1) + RT\ln\frac{P_2}{P_1}$$

It is conventional to consider state 1 as the pure compound ideal gas at the temperature T of the system and at *one-unit pressure*, whatever the units of pressure being used. Hence, for the pure compound as ideal gas at pressure P and temperature T of the mixture, we write:

$$\mu_i^{0,*}(T,P) = \mu_i^{0,*}(T,1) + RT\ln P$$

and for the compound i, in an ideal gas mixture at total pressure P and temperature T,

$$\mu_i^*(T,P,y) = \mu_i^{0,*}(T,1) + RT\ln(y_i P) \tag{11.4}$$

PROPERTY CHANGES IN THE MIXING OF IDEAL GASES

For the process of mixing pure compounds as ideal gases at a pressure P and a temperature T, to form an ideal gas mixture at the same pressure and temperature in a closed system, by analogy to Equations 6.18 and 6.19, for the molar Gibbs energy change, we write:

$$\Delta g_m^* = \sum_{k=1}^{c} y_k\left(\overline{g}_k^* - g_k^{0,*}\right)$$

Because according to Equation 7.1 the chemical potential is the partial molar Gibbs energy, from Equation 11.4,

$$\Delta g_m^* = \sum_{k=1}^{c} y_k\left(\mu_k^* - \mu_k^{0,*}\right) = RT\sum_{k=1}^{c} y_k\left[\ln(y_k P) - \ln P\right]$$

or

$$\Delta g_m^* = RT\sum_{k=1}^{c} y_k \ln y_k \tag{11.5}$$

The Gibbs–Helmholtz equation, Equation 7.6b, written in terms of molar property changes, has the form:

$$\left(\frac{\partial\left(\Delta g_m/RT\right)}{\partial T}\right)_{P,n} \equiv -\frac{\Delta h_m}{RT^2}$$

Thus, as the right-hand side of Equation 11.5 becomes independent of temperature after division by RT:

$$\Delta h_m^* = \sum_{k=1}^{c} y_k\left(\overline{h}_k^* - h_k^{0,*}\right) = 0 \tag{11.6}$$

In addition, as Equations 11.2 and 5.27 show that the partial molar volume of each component in the mixture is equal to its pure compound molar volume:

$$\Delta v_m^* = \sum_{k=1}^{c} y_k\left(\overline{v}_k^* - v_k^{0,*}\right) = 0 \tag{11.7}$$

Two more conclusions can be reached, for the constant pressure mixing process,

$$\Delta u_m^* = \sum_{k=1}^{c} y_k\left(\overline{u}_k^* - u_k^{0,*}\right) = \Delta h_m^* - P\Delta v_m^* = 0 \tag{11.8}$$

and, finally, as the process is isothermal,

$$\Delta g_m^* = \Delta h_m^* - T\Delta s_m^*$$

Thus,

$$\Delta s_m^* = -R\sum_{k=1}^{c} y_k \ln y_k \tag{11.9}$$

Note that when forming an ideal gas mixture at pressure P and temperature T in a closed system, starting with the pure compounds as ideal gases at the same pressure and temperature, the process occurs without any volume change or heat effect. However, the entropy does change. As discussed in Chapter 6, the fact that the process is isochoric and isenthalpic indicates that the ideal gas mixture is ideal. On the other hand, ideal mixtures are not necessarily in the ideal gas state. They can be a real gas mixture, a liquid, or even a solid mixture. An additional important conclusion is that for the ideal gas case, the chemical potential composition dependence shows in the argument of the logarithmic term of Equation 11.4. This observation is the basis for the treatment of real mixtures discussed in Chapter 12.

PROBLEMS

11.1 **Specific heat of air**

Standard air (i.e., dry, and free from carbon dioxide and other trace constituents) is defined as a mixture of 78.12% nitrogen (molar mass $M_{N2} = 28.01348$), 20.96% oxygen ($M_{O2} = 31.9988$), and 0.92% argon ($M_{Ar} = 39.948$) by moles [1,2]. Table 11.1 gives values of the specific heats of these three gases at zero pressure (ideal gas state) obtained from the NIST WebBook [3].

Calculate the specific heat of air as an ideal gas at these temperatures, and compare the results to Bridgeman's equation $c_V^* = 0.1714 + 2.3\times10^{-5}t - 1.0\times10^{-8}t^2$, where c_V^* is in cal/(g°C) and t is in °C [4].

11.2 **Mixing of two streams of moist air**

The discipline of *psychrometry* is concerned with the thermodynamic and transport properties of *moist air*, mixtures of air and water vapor, and their use in domestic and industrial processes such as air conditioning and evaporative water cooling. At the

TABLE 11.1
Ideal Gas-Specific Heat of the Components of Standard Air

	c_P^* (J·mol^{-1} K^{-1})		
T (K)	N_2	O_2	Ar
275	29.117	29.277	20.786
285	29.120	29.316	20.786
295	29.124	29.361	20.786
305	29.129	29.410	20.786
315	29.135	29.465	20.786
325	29.142	29.525	20.786
335	29.151	29.589	20.786
345	29.161	29.658	20.786
355	29.173	29.731	20.786
365	29.186	29.809	20.786
375	29.202	29.890	20.786

pressures and temperatures typical of such applications, the system {air + water vapor} is conventionally assumed to behave as an ideal gas mixture, with air considered as a pure pseudo-component, and the vapor contents, volume, and enthalpy of the mixture are given per unit mass of dry air.

a. Derive the equations for the specific volume and enthalpy of moist air on a dry air basis.
b. Calculate the temperature, humidity, and volumetric flow rate of the air stream obtained in an air conditioning system that operates at 101 kPa by adiabatic mixing of 0.708 m^3/s outside air carrying 0.0168 kg H_2O/kg dry air at 35°C, and 1.18 m^3/s recycled air with 0.0109 kg H_2O/kg dry air at 27°C.

11.3 **Entropy generation in the mixing of two ideal gas streams**

Write an expression for the partial molar and partial mass entropy of a component in an ideal gas mixture, and use it to calculate the entropy generation in the mixing process of Problem 11.2(b).

11.4 **Displacement of chemical equilibrium**

According to Equation 8.10 and Figure 8.1, the general criteria for chemical reaction equilibrium are:

$$\left(\frac{\partial G}{\partial \xi}\right)_{T,P,m} = \sum_i^c \nu_i \mu_i \begin{cases} < 0 & \text{reaction proceeds "to the right" } (\xi \text{ increases}) \\ = 0 & \text{reaction stops (equilibrium)} \\ > 0 & \text{reaction proceeds "to the left" } (\xi \text{ decreases}) \end{cases}$$

Consider a chemical reaction taking place *in an ideal gas mixture* at constant temperature and pressure. After equilibrium has been reached, the system is perturbed by adding or removing an amount δn_k of one of the components. How does the system respond to this perturbation? Apply your conclusions to the particular case of the ammonia synthesis, $N_2 + 3H_2 \leftrightarrow 2NH_3$.

REFERENCES

1. National Oceanic and Atmospheric Administration, National Aeronautics and Space Administration, and United States Air Force 1976. *U.S. Standard Atmosphere, 1976* (NOAA – S/T 76-1562). Washington, DC: US Government Printing Office.
2. Jacobsen, R. T., Penoncello, S. G., Beyerlein, S. W., Clarke, W. P., and Lemmon, E. W. 1992. A thermodynamic property formulation for air. *Fluid Phase Equilib.* 79: 113–124.
3. Lemmon, E. W., Bell, I. H., Huber, M. L., and McLinden, M. O. 2023. Thermophysical properties of fluid systems. In *NIST Chemistry WebBook, NIST Standard Reference Database Number 69*, eds. P. J. Linstrom and W. G. Mallard. Gaithersburg MD: National Institute of Standards and Technology. Available at https://doi.org/10.18434/T4D303 (accessed June 29, 2022).
4. Bridgeman, O. C. 1929. The Joule-Thomson effect and heat capacity at constant pressure for air. *Phys. Rev.* 34: 527–533.

12 Equilibrium in Terms of Fugacity and Activity

Abstract

Based on the expressions for the chemical potentials of components of an ideal gas mixture, Lewis's definitions of fugacity, fugacity coefficients, and activity are introduced. Equations for phase equilibrium are written in terms of fugacities and for chemical equilibrium in terms of activities. The equation for the temperature dependence of the equilibrium constant of a reaction is directly obtained from application of the Gibbs–Helmholtz equation.

DEFINITION OF FUGACITY AND FUGACITY COEFFICIENT

For the treatment of mixtures in gaseous or liquid phases for which a pressure explicit equation of state would be available with the adjustable parameters dependent on composition, one could think of using Equation 11.1 to calculate chemical potentials. In fact, the partial molar volume, as defined by Equation 6.2, can be then evaluated using the following transformation:

$$\bar{v}_i \equiv \left(\frac{\partial V}{\partial n_i}\right)_{T,P,n_{j\neq i}} = -\left(\frac{\partial P}{\partial n_i}\right)_{T,V,n_{j\neq i}} \left(\frac{\partial V}{\partial P}\right)_{T,n} = \left(\frac{\partial P}{\partial n_i}\right)_{T,V,n_{j\neq i}} V\kappa_T$$

In the first equality to the right, after the identity, we have applied Equation 4.28, and in the second equality, we have applied the definition of the coefficient of isothermal compressibility, Equation 5.2. Although this treatment is feasible, it is cumbersome and discouraging. One alternative, which will be developed in Chapter 13, is to evaluate the chemical potential from the Helmholtz function using Equation 4.16. We include these considerations here to emphasize the ingenious step taken by G. N. Lewis by introducing the definition of fugacity. This step is comparable to the introduction of the concepts of entropy or of chemical potential. G. N. Lewis proposed to start from what was known for ideal gas mixtures for which the isothermal differential form of Equation 11.4 is:

$$d\mu_{i,T}^{*}(P,y) = RT\ d\ln[y_i P] \tag{12.1}$$

In this expression, appearing naturally from the treatment of the ideal gas mixture, the composition dependence of the chemical potential is in the argument of the logarithmic term. In a major leap forward, Lewis defined a function of composition, temperature, and pressure f_i, called *fugacity*, which for any component i in a mixture, real or ideal, gives the isothermal change of the chemical potential of compound i by the relation:

$$d\mu_{i,T}(P,x) \equiv RT\ d\ln f_i(x,T,P) \tag{12.2}$$

In this identity, which is valid for any phase gas or liquid, for generality, the mole fraction has been designated by x. From Equation 11.4 for compound i in an ideal gas mixture and in the pure compound state, we have, respectively:

$$f_i^{*} = x_i P \tag{12.3a}$$

DOI: 10.1201/9781003431985-14

and

$$f_i^{0,*} = P \tag{12.3b}$$

As the pressure decreases tending to zero, any system *at equilibrium* approaches the behavior of the ideal gas system, and the limiting value of the fugacity tends to the value of the "partial pressure" of the component in the ideal gas mixture; hence,

$$\lim_{P\to 0} f_i = x_i P \tag{12.4}$$

This limiting behavior suggests the convenience of defining the *fugacity coefficient*, ϕ_i, of compound i by:

$$\phi_i(T,P,x) \equiv \frac{f_i(T,P,x)}{x_i P} \tag{12.5}$$

and we rewrite Equation 12.4 as:

$$\lim_{P\to 0} \phi_i = 1 \tag{12.6}$$

The relations for a pure compound equivalent to Equations 12.5 and 12.6 are:

$$\phi_i^0(T,P) \equiv \frac{f_i^0(T,P)}{P} \tag{12.7}$$

and

$$\lim_{P\to 0} \phi_i^0 = 1 \tag{12.8}$$

Integrating Equation 12.2 between the state of compound i as pure compound ideal gas at one unit pressure and its state in a mixture at pressure P, after rearrangement, gives:

$$\mu_i(T,P,x) \equiv \mu_i^{0,*}(T,1) + RT\ln f_i(T,P,x) \tag{12.9}$$

PHASE EQUILIBRIUM IN TERMS OF FUGACITIES

Thus, for phase equilibrium of compound i in phases α and β, the equality of chemical potentials, $\mu_i^\alpha = \mu_i^\beta$, takes the form:

$$\mu_i^{0,*}(T,1) + RT\ln f_i^\alpha\left(T,P,x^\alpha\right) = \mu_i^{0,*}(T,1) + RT\ln f_i^\beta\left(T,P,x^\beta\right)$$

or, simply,

$$f_i^\alpha\left(T,P,x^\alpha\right) = f_i^\beta\left(T,P,x^\beta\right) \tag{12.10}$$

Thus, *for phase equilibrium calculations, with all generality, the equality of fugacities replaces the equality of chemical potentials.* At this point, one may question the advantage of having defined the fugacity and fugacity coefficient of a compound in a mixture, in comparison with the direct use of chemical potentials. One advantage is encountered in the treatment of dilute solutions. As shown by Equation 12.1 for an ideal gas mixture, as the concentration of compound i goes to zero, its chemical potential goes to minus infinity, while according to Equation 12.3, its fugacity goes to zero. On the other extreme, as the compound gets more concentrated in the ideal gas mixture, its fugacity goes to the value of the pressure of the system. This physical interpretation of fugacity helps in making estimates for iterative calculations. Other advantages will become apparent. Isothermal integration of Equation 12.2 for compound i between some arbitrarily chosen state (a) and its state in a mixture at pressure P gives:

$$\mu_i(T,P,x)=\mu_i^{(a)}\left(T,P^{(a)},x^{(a)}\right)+RT\ln\frac{f_i(T,P,x)}{f_i^{(a)}\left(T,P^{(a)},x^{(a)}\right)} \tag{12.11}$$

One possibility is to consider the arbitrary state as the pure compound at the temperature of the system and at a well-defined real or imaginary state characterized by its pressure $P^{(\theta)}$. This particular reference state is called the *standard state* of the compound. In this case, we write:

$$\mu_i(T,P,x)=\mu_i^{(\theta)}\left(T,P^{(\theta)}\right)+RT\ln\frac{f_i(T,P,x)}{f_i^{(\theta)}\left(T,P^{(\theta)}\right)} \tag{12.12}$$

DEFINITION OF ACTIVITY

The ratio of fugacities of the right-hand side of Equation 12.12 is called the *activity of compound* i *with respect to the standard state* θ. This function is normally used in the study of chemical equilibrium:

$$a_{i,\theta}(T,P,x)\equiv\frac{f_i(T,P,x)}{f_i^{(\theta)}\left(T,P^{(\theta)}\right)} \tag{12.13}$$

Depending on the problem studied, it may be more convenient to use other units for measuring concentration and other standard states. A value of the activity of a compound without a clear specification of the standard state used for its evaluation is meaningless.

In simplified nomenclature, we rewrite Equation 12.12 as:

$$\mu_i=\mu_i^{(\theta)}+RT\ln a_{i,\theta} \tag{12.12a}$$

CHEMICAL EQUILIBRIUM IN TERMS OF ACTIVITY

For the case of chemical equilibrium, then, we write the condition of equilibrium, Equation 8.11, as:

$$\sum_k^c \nu_k\mu_k^{(\theta)}+RT\sum_k^c \nu_k\ln a_{k,\theta}=0$$

or

$$\sum_k^c \nu_k\ln a_{k,\theta}=-\frac{\sum_k^c \nu_k\mu_k^{(\theta)}}{RT} \tag{12.14a}$$

For all practical purposes, the above expression is the final result needed. Observe that due to the implicit sign of the stoichiometric numbers, the sum of the right-hand side is the sum of the standard values of the molar Gibbs energies of the products minus the sum of the standard values of the molar

Gibbs energies of the reactants; that is, this sum represents the *standard Gibbs energy change of reaction*. From Equation 7.2b, we write:

$$\Delta G_T^{\theta} \equiv \sum_k^c \nu_k \mu_k^{(\theta)} \tag{12.14b}$$

where the subscript T indicates that the value of the Gibbs energy change of reaction is at the temperature of the system and the superscript θ indicates that this value is based on well-defined standard states for the chemical potentials. In addition, as the coefficients multiplying the logarithms can be written as exponents of the argument of the logarithms and the sum of logarithms is equal to the logarithm of the product of the terms, we get:

$$\sum_k^c \nu_k \ln a_{k,\theta} = \sum_k^c \ln a_{k,\theta}^{\nu_k} = \ln\left[\prod_k a_{k,\theta}^{\nu_k}\right]$$

For simplicity, the symbol K_T is used for the product in the above expression, and noting that the stoichiometric numbers are positive for products and negative for reactants, we revert to the use of the stoichiometric coefficients that are always positive and write:

$$K_T \equiv \prod_k a_{k,\theta}^{\nu k} = \frac{\prod_k^{products} a_{Pk,\theta}^{\lambda_{Pk}}}{\prod_j^{reactants} a_{Rj,\theta}^{\lambda_{Rj}}} \tag{12.15}$$

This expression is called the *thermodynamic equilibrium constant*. Thus, Equation 12.14a takes the simple form:

$$\ln K_T = -\frac{\Delta G_T^{\theta}}{RT} \tag{12.16}$$

ENTHALPY CHANGE OF REACTION

The consistent use of Equation 12.16 requires that the same standard state for a compound is used in the calculation of the activities for Equation 12.15 and for the calculation of the *standard Gibbs energy change of reaction* in Equation 12.14b. Because in practice these two terms are calculated independently, as a reminder, we have kept θ as a subscript for the activities and as a superscript for the Gibbs energy change.

Finally, the temperature dependence of the equilibrium constant is obtained by combining Equation 12.16 with the Gibbs–Helmholtz equation, Equation 7.6b:

$$\left(\frac{\partial \ln K_T}{\partial T}\right)_{P,n} \equiv \frac{\Delta H_T^{\theta}}{RT^2} \tag{12.17}$$

where

$$\Delta H_T^{\theta} \equiv \sum_k^c \nu_k h_k^{(\theta)} \tag{12.18}$$

is the *standard enthalpy change of reaction* at the temperature of the system.

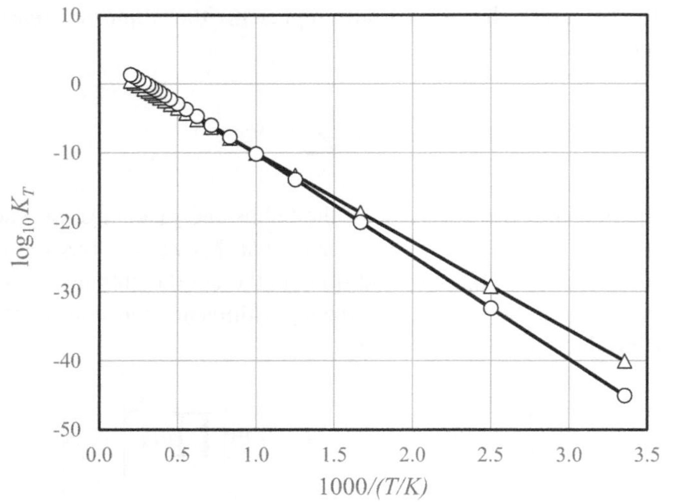

FIGURE 12.1 Equilibrium constants for the dissociation of H_2O (triangles) and CO_2 (circles).

PROBLEMS

12.1 **Physical interpretation of chemical potentials at infinite dilution**

It is mentioned in the text that in an ideal gas mixture, as the concentration of compound i goes to zero, its chemical potential goes to minus infinity. Moreover, from Equations 12.5 and 12.9,

$$\mu_i(T,P,x) = \mu_i^{0,*}(T,1) + RT\ln\left[\phi_i(T,P,x)x_iP\right]$$

it can be seen that this limiting behavior is characteristic of all real mixtures, whether gaseous or not. How can this universal limit be interpreted in physical terms?

12.2 **Further standard property changes of reaction**

Develop expressions for the standard entropy and standard heat capacity changes of reaction in terms of the equilibrium constant.

12.3 **Equilibrium constants as a function of temperature**

Figure 12.1 is a plot of data from [1] for the equilibrium constant against temperature for two dissociation reactions in the gas phase:

$$H_2O \leftrightarrow H_2 + \frac{1}{2}O_2$$

$$CO_2 \leftrightarrow CO + \frac{1}{2}O_2$$

What can be inferred from the shape of these lines?

REFERENCE

1. Reynolds, W. C., and Colonna, P. 2018. *Thermodynamics: Fundamentals and Engineering Applications*. Oxford: Cambridge University Press.

13 Calculation of Fugacities from Equations of State

ABSTRACT

The fugacity of a compound in a mixture is conveniently expressed as the product of its mole fraction, times the pressure and times a term known as fugacity coefficient. The advantage of introducing the concept of fugacity coefficient is that this term is directly obtainable from equations of state which either give the molar volume as a function of temperature, pressure, and composition or give the pressure as a function of temperature, molar volume, and composition. Final expressions for the fugacity coefficient are presented for both cases.

The determination of the value of the fugacity of a compound as a function of temperature, pressure, and mixture composition is crucial for the study of physical equilibrium, as indicated by Equation 12.10. Similarly, the values of activity required by Equation 12.14a for the study of chemical equilibrium can be obtained directly from Equation 12.13 once the values of fugacities are known. The key equation for the calculation of the values of fugacities is Equation 12.5, written as:

$$f_i = y_i \phi_i P \tag{13.1}$$

For a pure compound, Equation 12.7 takes the form:

$$f_i^0 = \phi_i^0 P \tag{13.2}$$

For generality, in these equations, we have not indicated the independent variables for each function. As the above equations are perfectly general and apply to any kind of mixture, for the mole fraction of compound i we use either the symbol y_i or x_i, as required. Since for a pure compound the fugacity of saturated liquid is equal to the fugacity of the saturated vapor, for this particular case, the method for calculating fugacity coefficients discussed here is directly applicable to a saturated liquid phase. For liquid compounds at pressures different than the saturation pressure, a correction must be applied. In this chapter, we consider the evaluation of fugacities from equations of state (EOSs) that use temperature, pressure, and composition as independent variables, $\mathrm{v}=\mathrm{v}\ (T, P, x)$, and from EOSs using temperature, molar volume, and composition as independent variables, $P=P\ (T, \mathrm{v}, x)$.

USE OF VOLUME-EXPLICIT EOSs: $v=v\ (T, P, x)$

Combining Equations 11.1 and 12.2, we obtain:

$$d \ln f_i = \frac{\overline{\mathrm{v}}_i}{RT} dP \tag{13.3}$$

For compound i, in an ideal gas mixture at the same temperature T, pressure P, and composition x of the real mixture,

$$d \ln f_i^* = \frac{\overline{\mathrm{v}}_i^*}{RT} dP$$

DOI: 10.1201/9781003431985-15

or, using Equations 11.2 and 12.3a,

$$d\ln[x_i\ P] = \frac{1}{P}dP \tag{13.4}$$

Subtracting Equation 13.4 from Equation 13.3 and using Equation 12.5, we obtain:

$$d\ln\phi_i = \frac{1}{RT}\left[\overline{v}_i - \frac{RT}{P}\right]dP$$

Integrating this expression from zero pressure to P and remembering that as $P \to 0$, $\phi_i \to 1$,

$$\ln\phi_i(T,P,x) = \frac{1}{RT}\int_0^P\left[\overline{v}_i - \frac{RT}{P}\right]dP \tag{13.5}$$

For compound i in an *ideal gas* mixture, from Equation 11.2,

$$\phi_i^* = 1 \tag{13.6}$$

In addition, writing Equation 13.5 for a pure compound i at temperature T and pressure P,

$$\ln\phi_i^0(T,P) = \frac{1}{RT}\int_0^P\left[v_i^0 - \frac{RT}{P}\right]dP \tag{13.7}$$

Although Equation 13.7 is of limited practical use due to the lack of EOSs with pressure as an independent variable, it is of considerable theoretical interest. The fact that we do not have expressions in terms of these independent variables does not imply that this relation is not valid in general terms. One interesting variation of this equation is obtained by introducing the definition of the compressibility factor, Equation 5.16,

$$\ln\phi_i^0(T,P) = \int_0^P\frac{\left[z_i^0 - 1\right]}{P}dP \tag{13.7a}$$

In addition to its applicability to a pure compound, this equation is equally applicable to a mixture of constant composition. As discussed in Chapter 9, a single-phase mixture of constant composition behaves as a pure compound and has two degrees of freedom. Thus, for a mixture of constant composition, we can write:

$$\ln\phi_m(T,P) = \int_0^P\frac{\left[z_m - 1\right]}{P}dP \tag{13.7b}$$

Equation 13.7b is a *de facto* definition of the fugacity coefficient of a gas mixture of constant composition. Taking the analogy one step further, and interpreting the mixture of constant composition in a single phase as a new pure compound, we can define the fugacity of a gas mixture by:

$$\phi_m \equiv \frac{f_m}{P} \tag{13.8}$$

Returning to the case of a compound in a mixture and subtracting Equation 13.7 from Equation 13.5, we write:

$$\ln \frac{\phi_i(T,P,x)}{\phi_i^0(T,P)} = \frac{1}{RT}\int_0^P \left[\bar{v}_i - v_i^0\right] dP \tag{13.9}$$

An immediate conclusion obtained from this equation is that, using Equation 6.22, for compound i in an ideal mixture, be it gas, liquid, or solid, the right-hand side is null and *the fugacity coefficient of the compound in the ideal mixture at any composition is equal to the fugacity coefficient of the pure compound at the same pressure and temperature:*

$$\phi_i^{id}(T,P,x) = \phi_i^0(T,P) \tag{13.10}$$

Thus, for a compound in an ideal mixture, from Equations 13.1 and 13.10,

$$f_i^{id}(T,P,x) = x_i\phi_i^0(T,P)P \tag{13.11}$$

It is important to note that as per Equation 13.9, Equations 13.10 and 13.11 are valid only if the ideal mixture behavior holds in all the range from zero pressure to the pressure P of the mixture. Equation 13.11 is especially useful for first estimates for the fugacity in iterative calculations.

USE OF PRESSURE-EXPLICIT EOSs: $P = P(T, v, x)$

EOSs having as independent variables the volume, the temperature, and the composition of the fluid are the most common ones in practice. It is unfortunate that the derivation of equations for this case is the more elaborate one. Although algebra can be done by computer applications, it is important to understand the principles involved in the derivation of equations, and thus we will deal with algebra following a conceptual approach.

First, we observe that in this case, the integration required needs to be done over volume. For the real system of fixed composition x, at volume V and temperature T, the pressure P is given by Equation 5.16, written as:

$$P = z\frac{\left(\sum n_k\right)RT}{V} \tag{5.16a}$$

For a system as ideal gas at the same total volume, temperature, and composition, according to Equation 5.27 the pressure will have a value P^* given by:

$$P^* = \frac{\left(\sum n_k\right)RT}{V} \tag{5.27a}$$

Thus,

$$P^* = \frac{P}{z}$$

Independent of the set of variables considered, Equation 12.9 is written as:

$$\mu_i(T,P,x) = \mu_i(T,v,x) = \mu_i^{0,*}(T,1) + RT\ln f_i$$

while for the ideal gas system, it has the form:

$$\mu_i^*(T,P^*,x)=\mu_i^*(T,v,x)=\mu_i^{0,*}(T,1)+RT\ln(x_iP^*)=\mu_i^{0,*}(T,1)+RT\ln(x_iP)-RT\ln z$$

Subtracting the ideal gas equation from the equation for the real system, using Equation 12.5 and rearranging,

$$\ln\phi_i=\frac{\mu_i-\mu_i^*}{RT}-\ln z \qquad (13.12)$$

In this equation, we have not indicated the variables for each function, as the two possibilities are clearly shown in the two preceding equations. The next step is to obtain the chemical potential using Equation 4.16. For this, from Equation 4.15,

$$dA_T=-PdV$$

and, at the same value of the total volume,

$$dA_T^*=-P^*dV=-\frac{\left(\sum_k n_k\right)RT}{V}dV$$

$$d\left[A_T-A_T^*\right]=-\left[P-\frac{\left(\sum_k n_k\right)RT}{V}\right]dV$$

Thus, integrating between zero pressure, or infinite volume, and the pressure and total volume of the system, and observing that at the lower integration limit both integrands are equal to zero, we write:

$$A_T-A_T^*=-\int_\infty^V\left[P-\frac{\left(\sum_k n_k\right)RT}{V}\right]dV=+\int_V^\infty\left[P-\frac{\left(\sum_k n_k\right)RT}{V}\right]dV$$

This equation combined with Equation 4.16 gives:

$$\mu_i-\mu_i^*=\int_V^\infty\left[\left(\frac{\partial P}{\partial n_i}\right)_{T,V,n_{j\neq i}}-\frac{RT}{V}\right]dV$$

Hence, Equation 13.12 can be written as:

$$\ln\phi_i=\frac{1}{RT}\int_V^\infty\left[\left(\frac{\partial P}{\partial n_i}\right)_{T,V,n_{j\neq i}}-\frac{RT}{V}\right]dV-\ln z \qquad (13.13)$$

We emphasize that in this equation, the compressibility factor z refers to the mixture and that the partial derivative of the analytical equation for pressure with respect to the number of moles of compound i is unrelated to any partial molar quantity. As EOSs are normally written in terms of the molar volume, it is convenient to rewrite this equation as:

$$\ln \phi_i = \frac{1}{RT}\int_{v}^{\infty}\left[n\left(\frac{\partial P}{\partial n_i}\right)_{T,V,n_{j\neq i}} - \frac{RT}{v}\right]dv - \ln z \tag{13.13a}$$

However, *the differentiation of the analytical expression for the pressure with respect to the number of moles of compound* i *must be done at constant total volume* V *and* not *constant molar volume* v = V/n.

For the derivation of an expression for a pure compound fugacity coefficient from an EOS using as independent variables pressure, temperature, and composition, it is simpler to start with the differential form of Equation 13.7a:

$$d\ln\phi = [z-1]\frac{dP}{P}$$

From Equation 5.15, we obtain:

$$dP = \frac{RT}{v}dz - z\frac{RT}{v^2}dv$$

Thus, dividing by *P*, using again Equation 5.16 and rearranging,

$$\frac{dP}{P} = \frac{dz}{z} - \frac{dv}{v}$$

Hence, we can now write:

$$d\ln\phi = [z-1]\left[\frac{dz}{z} - \frac{dv}{v}\right] = dz - d\ln z - [z-1]\frac{dv}{v}$$

Integrating this expression from zero pressure, at which both the fugacity coefficient and the compressibility factor are unity, and the molar volume goes to infinity, to the pressure and volume of the system,

$$\ln\phi = [z-1] - \ln z - \int_{\infty}^{v}[z-1]\frac{dv}{v} \tag{13.14}$$

This expression is useful to calculate the fugacity coefficient of a pure compound or of a mixture of constant composition.

PROBLEMS

13.1 **Fugacity coefficients of a pure fluid from the virial equation of state**

The virial equation of state is a series expansion in powers of molar density or volume ("Leiden series"):

$$z = 1 + B\rho + C\rho^2 + \ldots = 1 + \frac{B}{v} + \frac{C}{v^2} + \ldots$$

or alternatively in powers of pressure ("Berlin series")

$$z = 1 + \frac{B}{RT}P + \frac{C - B^2}{(RT)^2}P^2 + \ldots$$

In Appendix B, the expression for the fugacity coefficient from this second virial form is given as:

$$\ln\phi = \frac{B}{RT}P + \frac{C - B^2}{2(RT)^2}P^2 + \ldots$$

a. Derive the analogous formula for the fugacity coefficient given by the Leiden series.
b. As explained in Chapter 12, fugacity coefficients approach unity for any system at equilibrium in the ideal gas limit as pressure (or density) goes to zero. Are there, however, any other conditions under which a real fluid behaves as an ideal gas in the restricted sense of having fugacity coefficient equal to unity? Develop your analysis for both virial expansions, using the empirical correlations for the virial coefficients given in Appendix B, Equations B.2–B.7. Consider only the special case of a "simple" fluid with acentric factor $\omega = 0$.

13.2 **Fugacity coefficients in a multicomponent mixture from the virial (Leiden) equation of state**

Derive an expression for the fugacity coefficient of a component in a gaseous mixture described by the truncated Leiden virial expansion of the preceding Problem.

13.3 **Critical fugacity coefficients from cubic equations of state**

The simplest cubic equations of state (CEOS), frequently used in engineering practice, can be written in general as:

$$P = \frac{RT}{\mathrm{v} - b} - \frac{a}{(\mathrm{v} + mb)(\mathrm{v} + nb)}$$

or, in dimensionless form, as:

$$1 = \frac{1}{z - B_P} - \frac{A_P}{(z + mB_P)(z + nB_P)}$$

with

$$A_P \equiv \frac{aP}{(RT)^2}, \quad B_P \equiv \frac{bP}{RT}$$

The fluid-specific parameters a and b may be functions of temperature (a always, b almost never in modern CEOS), but do not depend on pressure or volume.

a. Derive an expression for the fugacity coefficient of a fluid described by this type of CEOS.
b. Two celebrated versions of the above are the Redlich-Kwong (RK) CEOS [1] defined by $m = 0,\ n = 1$, and the Peng-Robinson (PR) CEOS [2] defined by $m = 1 - \sqrt{2}$, $n = 1 + \sqrt{2}$, an important variant of which, in terms of the temperature dependence of a, is the PRSV equation described in Appendix C. Compute the fugacity coefficients predicted by these two models at the critical point, where the dimensionless parameters A and B and the compressibility factor z take the values [3]:

$$A_{Pc} = (1 + B_{Pc} - z_c)^3, \quad B_{Pc} = \frac{1}{\left[(1+m)^{1/3} + (1+n)^{1/3}\right]^3}, \quad z_c = \frac{1 + (1 - m - n)B_{Pc}}{3}$$

REFERENCES

1. Soave, G. 1972. Equilibrium constants from a modified Redlich-Kwong equation of state. *Chem. Eng. Sci.* 27: 1197–1203.
2. Peng, D.-Y., and Robinson, D. B. 1976. A new two-constant equation of state. *Ind. Eng. Chem. Fundam.* 15: 59–64.
3. Hernández, J., Olivares, J. F., and Olivera-Fuentes, C. 2023. The 'Lewis line': A proposed new ideal curve of fluids. *Fluid Phase Equilib.* 568: 113755.

14 Fugacity of a Mixture and of Its Components

ABSTRACT

This chapter explores the idea of treating a mixture of constant composition as it were a pure compound. It is shown that the logarithm of the fugacity coefficient of a compound is just a partial molar property of the logarithm of the fugacity coefficient of the mixture. This allows to write the Gibbs–Duhem equation in terms of fugacity coefficients and in terms of fugacities, indicating that in both cases, these terms are not all independent.

FUGACITY AND FUGACITY COEFFICIENT OF A MIXTURE

The definitions of the fugacity coefficient and fugacity of a mixture by Equations 13.7b and 13.8, respectively, bring up an interesting extension to the idea of treating a mixture of constant composition as a pseudo–pure compound. For a pure compound, the chemical potential is equal to the molar Gibbs energy of the compound. Thus, we can rewrite Equation 12.9 as:

$$g = RT\ln f + g^{*}(T,1)$$

In this equation, we have omitted all distinctions between pure compound and mixture of constant composition. Translating this equation to the latter case, we would write:

$$g_m = RT\ln f_m + g_m^{*}(T,1) \tag{14.1}$$

On the other hand, for a compound i in the mixture, from Equation 12.9,

$$\mu_i = RT\ln f_i + \mu_i^{*}(T,1)$$

and according to Equation 7.2a,

$$g_m = \sum_k^c x_k\bar{g}_k = \sum_k^c x_k\mu_k$$

or

$$g_m = RT\sum_k^c x_k \ln f_k + RT\sum_k^c x_k\mu_k^{*}(T,1)$$

 DOI: 10.1201/9781003431985-16

Hence, from Equations 14.1 and 7.2a,

$$g_m = RT\ln f_m + g_m^*(T,1) = RT\sum_k^c x_k \ln f_k + RT\sum_k^c x_k\mu_k^*(T,1)$$

or

$$\ln f_m = \sum_k^c x_k \ln f_k - \frac{1}{RT}\left[g_m^*(T,1) - \sum_k^c x_k\mu_k^*(T,1)\right]$$

But, from Equation 11.5 we know that:

$$\left[g_m^*(T,1) - \sum_k^c x_k\mu_k^*(T,1)\right] = RT\left[\sum_k^c x_k \ln x_k\right]$$

So, we finally write:

$$\ln f_m = \sum_k^c x_k \ln \frac{f_k}{x_k} \tag{14.2}$$

and, as an additional step, subtracting ln P from both sides, we obtain:

$$\ln \phi_m = \sum_k^c x_k \ln \phi_k \tag{14.3}$$

FUGACITY COEFFICIENTS AS PARTIAL PROPERTIES

Comparison of Equations 14.2 and 14.3 with Equation 6.6 indicates that (f_i/x_i) behaves as a partial molar property for ln f_m, and ln ϕ_k behaves as a partial molar property for ln ϕ_m. Thus, if ln ϕ_m is obtained from an equation of state for the mixture using Equation 13.7b, then, following Equation 6.2, the fugacity coefficient for component i of the mixture can be obtained as:

$$\ln \phi_i \equiv \left(\frac{\partial n \ln \phi_m}{\partial n_i}\right)_{T,P,n_{j\neq i}} \tag{14.4}$$

or, following Equation 6.15,

$$\ln \phi_i = \ln \phi_m - \sum_{k\neq i} x_k \left(\frac{\partial \ln \phi_m}{\partial x_k}\right)_{\substack{T,P,x_r \\ r\neq i,\ r\neq k}} \tag{14.5}$$

GIBBS–DUHEM EQUATION IN TERMS OF FUGACITY COEFFICIENTS

The recognition that the logarithm of the fugacity of a compound in a mixture behaves as a partial molar property of the logarithm of the fugacity of the mixture has other useful implications. First, the fugacity coefficients of the components in a mixture are not all independent, but they are related by the *Gibbs–Duhem equation*, Equation 6.9, written as:

$$\sum_k^c x_k d\ln \phi_k = \left(\frac{\partial \ln \phi_m}{\partial T}\right)_{P,n} dT + \left(\frac{\partial \ln \phi_m}{\partial P}\right)_{T,n} dP \tag{14.6}$$

and also, from Equation 6.11,

$$\left(\frac{\partial \ln \phi_j}{\partial n_i}\right)_{T,P,n_{k\neq i}} = \left(\frac{\partial \ln \phi_i}{\partial n_j}\right)_{T,P,n_{k\neq j}} \tag{14.7}$$

The derivatives of the right-hand side of Equation 14.6, giving the variation of the logarithm of the fugacity coefficient with temperature and pressure, can be obtained from Equation 14.1 written as:

$$\ln \phi_m = \frac{g_m - g_m^*(T,1)}{RT} - \ln P^\#$$

For the derivative with respect to pressure, using Equation 4.18 we obtain:

$$\left(\frac{\partial \ln \phi_m}{\partial P}\right)_{T,n} = \frac{\mathrm{v}_m}{RT} - \frac{1}{P} = \frac{z_m - 1}{P} \tag{14.8}$$

and for the derivative with respect to temperature, using Equation 7.6b,

$$\left(\frac{\partial \ln \phi_m}{\partial T}\right)_{P,n} = -\frac{\left[h_m - h_m^*(T,1)\right]}{RT^2} \tag{14.9}$$

In this latter equation, which is seldom used, as most cases of study are under isothermal conditions, $h_m^*(T,1)$ can be written simply as $h_m^*(T)$ since the enthalpy of the ideal gas is independent of pressure. Equations 14.8 and 14.9 are also valid for the variation of the fugacity coefficient of a pure compound with pressure and temperature. For a compound i in a mixture, by the same argument used to derive Equation 6.10, from Equation 14.6 we obtain:

$$d \ln \phi_k = \left(\frac{\partial \ln \phi_k}{\partial T}\right)_{P,n} dT + \left(\frac{\partial \ln \phi_k}{\partial P}\right)_{T,n} dP \tag{14.10}$$

and applying the same argument to Equations 14.8 and 14.9,

$$\left(\frac{\partial \ln \phi_i}{\partial P}\right)_{T,n} = \frac{\overline{\mathrm{v}}_i}{RT} - \frac{1}{P} \tag{14.11}$$

and for the derivative with respect to temperature, using Equation 7.6,

$$\left(\frac{\partial \ln \phi_i}{\partial T}\right)_{P,n} = -\frac{\left[\overline{h}_i - h_i^*(T,1)\right]}{RT^2} \tag{14.12}$$

GIBBS–DUHEM EQUATION IN TERMS OF FUGACITIES

Again, in this case, Equation 14.12 finds limited use. Sometimes it is desirable to have the *Gibbs–Duhem equation in terms of fugacities* instead of fugacity coefficients. This form is better obtained from Equation 14.6. Replacing Equation 12.5 on the left-hand side and Equations 14.8 and 14.9 on the right-hand side of Equation 14.6, we obtain:

$$\sum_k^c x_k d \ln f_k - \sum_k^c x_k d \ln x_k - \left(\sum_k^c x_k\right) d \ln P = -\frac{\left[h_m - h_m^*(T,1)\right]}{RT^2} dT + \left(\frac{\mathrm{v}_m}{RT} - \frac{1}{P}\right) dP$$

In this expression, it is interesting to notice that:

$$\sum_{k}^{c} x_k d\ln x_k = \sum_{k}^{c} x_k \frac{dx_k}{x_k} = \sum_{k}^{c} dx_k = d\left(\sum_{k}^{c} x_k\right) = 0$$

and, in addition,

$$\left(\sum_{k}^{c} x_k\right) d\ln P = \frac{1}{P} dP$$

Thus,

$$\sum_{k}^{c} x_k d\ln f_k = -\frac{\sum_{k}^{c} x_k\left[\bar{h}_k - h_k^{0,*}(T,1)\right]}{RT^2} dT + \frac{\sum_{k}^{c} x_k \bar{v}_k}{RT} dP \tag{14.13}$$

Equation 14.13 is the general form of the Gibbs–Duhem equation in terms of fugacities. Grouping all terms of Equation 14.13 under a single summation on the left-hand side,

$$\sum_{k}^{c} x_k \left[d\ln f_k + \frac{\left[\bar{h}_k - h_k^{0,*}(T,1)\right]}{RT^2} dT - \frac{\bar{v}_k}{RT} dP \right] = 0$$

For this equation to be valid in all of the composition range, necessarily,

$$d\ln f_i = -\frac{\left[\bar{h}_i - h_i^{0,*}(T,1)\right]}{RT^2} dT + \frac{\bar{v}_i}{RT} dP \tag{14.14}$$

From Equation 14.14, then,

$$\left(\frac{\partial \ln f_i}{\partial P}\right)_{T,n} = \frac{\bar{v}_i}{RT} \tag{14.15}$$

and

$$\left(\frac{\partial \ln f_i}{\partial T}\right)_{P,n} = -\frac{\left[\bar{h}_i - h_i^{0,*}(T,1)\right]}{RT^2} \tag{14.16}$$

Equations 14.15 and 14.16 can also be obtained directly from Equations 14.11 and 14.12, respectively. For a pure compound, Equations 14.15 and 14.16 take the form:

$$\left(\frac{\partial \ln f_i^0}{\partial P}\right)_{T} = \frac{v_i^0}{RT} \tag{14.17}$$

and

$$\left(\frac{\partial \ln f_i^0}{\partial T}\right)_{P} = -\frac{\left[h_i^0 - h_i^{0,*}(T,1)\right]}{RT^2} \tag{14.18}$$

It is important to emphasize that the expressions for fugacity and fugacity coefficients for all compounds forming a mixture must be derived from the same equation of state. This ensures that these expressions are thermodynamically consistent and satisfy both the Gibbs–Duhem equation and Maxwell's relations.

To close, we consider a particular case of Equation 14.13 that is of practical importance. Most studies of liquid mixtures are performed under isothermal conditions, and the first term of the right-hand side of this equation cancels out. In addition, for liquid mixtures the second term of the right-hand side is negligible because the volume of the liquid phase is orders of magnitude smaller than the product RT.

Hence, *for isothermal studies of liquid mixtures, the Gibbs–Duhem equation in terms of fugacities* is usually written as:

$$\sum_k^c x_k d \ln f_k = 0 \tag{14.19}$$

PROBLEMS

14.1 **Fugacity coefficients in a multicomponent mixture from the virial (Berlin) equation of state**

As stated in Appendix B, for the range of pressures where the second virial term suffices, the fugacity coefficient of component i is given by Equation B.11,

$$\ln \phi_i = \left(2\sum_k y_k B_{ik} - B_m \right) \frac{P}{RT}$$

whereas the fugacity coefficient of the mixture as a whole should be given by the correspondingly truncated Equation B.8,

$$\ln \phi_m = \frac{B_m P}{RT}$$

Use Equation 14.5 to derive the first equation from the second one, that is, to obtain $\ln \phi_i$ as the partial property for $n \ln \phi_m$.

14.2 **More on fugacity coefficients as partial properties**

For any extensive property E, the partial properties defined as:

$$\bar{e}_i \equiv \left(\frac{\partial E}{\partial n_i} \right)_{T,P,n_{j\neq i}}$$

fulfill *two conditions:* (I) They add up to the mixture property,

$$\sum_k x_k \bar{e}_k = \frac{E}{n}$$

and (II) They obey the Gibbs–Duhem equation,

$$\sum_k x_k \left(d\bar{e}_k \right)_{T,P} = 0$$

Thus, (I) is a *necessary but not sufficient* condition for a set of component functions to constitute partial properties; one must also *prove* that they satisfy the Gibbs–Duhem relation.

a. Suppose that an expression of the form:

$$\frac{E}{n} = \sum_k x_k \varepsilon_k (T,P,x)$$

has been found for a certain extensive property E. What can be inferred about the relation between the functions ε_k and the partial properties $\bar{e}_k$?

b. Prove that the component fugacity coefficients defined by Equation 13.5 are indeed partial properties.

14.3 **Fugacity coefficients in a binary gas mixture**

Experimental data reported in [1] for the vapor–liquid equilibrium of ethyl acetate ("EtAc," component 1)+benzene ("Bz," component 2) at 50°C are shown in Table 14.1. Estimate the fugacity coefficients of both components in the vapor phase at each composition. Since the equilibrium pressures are low, use Equations B.1–B.3, B.9, and B.11 with the mixing rules given in Appendix B. Can the vapor mixture be assumed to behave as an ideal gas at these conditions?

TABLE 14.1
Vapor–Liquid Equilibrium of EtAc+Bz at 50°C [1]

x_1	y_1	P (kPa)
0.1497	0.1641	36.996
0.3346	0.3519	37.509
0.4580	0.4696	37.690
0.5697	0.5783	37.825
0.6289	0.6342	37.940
0.6708	0.6746	37.972
0.7170	0.7182	38.000
0.7599	0.7616	38.053
0.7629	0.7646	38.053
0.8162	0.8162	38.077
0.8545	0.8549	38.102
0.8637	0.8637	38.108
0.8976	0.8976	38.124

NOTE

This expression can be written as $RT \ln\phi_m = g_m(T,P,x) - \left[g_m^*(T,1,x) + RT \ln P\right] = g_m(T,P,x) - g_m^*(T,P,x) = g_m^{res}(T,P,x)$, where the residual value of the molar Gibbs energy $g_m^{res}(T,P,x)$ is the contribution of the intermolecular forces to the property of the fluid.

REFERENCE

1. Linék, J., Procházka, K., and Wichterle, I. 1972. Liquid-vapour equilibrium. LIV. The systems ethyl acetate-benzene, ethyl acetate-toluene and ethyl acetate-ethylbenzene. *Coll. Czech. Chem. Commun.* 37: 3010–3014.

15 Fugacities, Activities, and Activity Coefficients in Liquid Mixtures of Nonelectrolytes

Abstract

In the first half of the last century, equations of state were only able to represent gas mixtures. So, liquid mixtures were dealt with "Raoult's Law." This chapter shows that the correction of this "Law" by the introduction of activity coefficients is directly related to treatment with fugacity coefficients. Solvents approach ideal behavior at high solute dilution and solutions are formed without volume (and heat) effects. Thus, for the definition of activity, Lewis's convention uses pure compound liquid as the standard state, while Henry's convention, useful for solutes in dilute solutions, uses the state of the solute at infinite dilution. The use of molality instead of mole fraction as a measure of concentration of solutes is discussed.

LEWIS'S CONVENTION FOR ACTIVITY COEFFICIENTS

The methods to calculate fugacities described in Chapter 13 assumed the availability of equations of state valid from zero pressure to the pressure of the system at the temperature and composition of the system. While for systems in the gas phase, this is a reasonable assumption, for condensed mixtures, this is not always the case. The volumetric behavior of an aqueous solution of sugar, for example, is not easily described from the ideal gas to the liquid phase using a single equation of state. However, the fact that we do not have an analytical expression to handle the algebra does not mean that the thermodynamic relations between the thermodynamic properties are not valid. Although in this chapter we develop relations of general use, we concentrate mostly on cases when the pure compounds forming a mixture exist as liquids at the temperature T and pressure P of the system. Hence, using Equation 13.9,

$$\ln\frac{\phi_i(T,P,x)}{\phi_i^0(T,P)}=\frac{1}{RT}\int_0^P\left[\overline{v}_i-v_i^0\right]dP \quad (13.9)$$

For convenience, we define:

$$\ln\gamma_{i,0}(T,P,x)\equiv\frac{1}{RT}\int_0^P\left[\overline{v}_i-v_i^0\right]dP \quad (15.1)$$

and using Equations 12.5 and 12.7, we write:

$$\gamma_{i,0}(T,P,x)=\frac{\phi_i(T,P,x)}{\phi_i^0(T,P)}=\frac{f_i}{x_if_i^0} \quad (15.2a)$$

or

$$f_i\,(T,P,x)=x_i\gamma_{i,0}(T,P,x)f_i^0(T,P) \quad (15.2b)$$

DOI: 10.1201/9781003431985-17

Or dividing both sides by x_iP and using Equations 13.1 and 13.2,

$$\phi_i = \gamma_{i,0}\phi_i^0 \tag{15.2c}$$

Equation 15.2c is an important link between the thermodynamic treatments based on equation of state and those based on excess functions, and it has been used to open Section II, "Mixtures," of this book.

In addition, *the activity of compound* i *with respect to the standard state of pure compound at the same pressure and temperature of the system*, indicated by the subscript 0, according to Equation 12.13, is:

$$a_{i,0}(T,P,x) \equiv \frac{f_i\,(T,P,x)}{f_i^0\,(T,P)} = x_i\gamma_{i,0} \tag{15.3}$$

The fact that the mole fraction of compound i multiplied by the factor $\gamma_{i,0}$, defined by Equation 15.1, gives its activity suggests calling $\gamma_{i,0}$ the activity coefficient of compound i. This convention, in which both the activity coefficient and the activity are unity for the pure compound, is known as the *Lewis convention*. Other useful conventions are discussed in Chapter 16. In preparation for the more general treatment of ideal mixtures presented in Chapter 16, in Chapter 12 we introduced the general symbol θ as a way to identify the particular standard state chosen for the analytical treatment of equations. Comparison of Equation 15.3 with Equation 12.13 shows that Lewis's convention refers to the particular case when the standard state used for the definition of the activity of compound i is chosen as the pure compound at the temperature and pressure of the system.

Although we have indicated temperature, pressure, and composition as the variables for the activity coefficient, its definition by Equation 15.1 indicates that the main variables are composition and temperature only. The integration of the right-hand side of Equation 15.1 can be divided into the integration from pressure zero to the saturation pressure of the compound and the integration from the saturation pressure to the pressure of the system. The first of these integrals is independent of the pressure of the system, while for the second integral, the volumes refer to liquid volumes that, for all practical purposes, are independent of pressure.

As discussed in the derivation of Equation 13.11, for compound i in an ideal mixture from zero pressure to the pressure P of the system, the right-hand side of Equation 15.1 is null and the activity coefficient of the compound in the ideal mixture at any composition is equal to unity,

$$\gamma_{i,0}^{id}(T,P,x) = 1 \tag{15.4}$$

Thus, in Lewis's convention, for a compound in an ideal mixture, Equation 15.3 gives:

$$f_i^{id,0}(T,P,x) = x_i f_i^0(T,P) \tag{15.5}$$

For the activity coefficient of a compound in an actual solution, in Lewis's convention, we write more formally,

$$\lim_{x_i \to 1} \gamma_{i,0} = 1 \tag{15.6}$$

Deviations from Lewis's ideal behavior are said to be positive if the activity coefficient is larger than unity and negative if the activity coefficient is smaller than unity. The approximation of ideal mixture is good for the solvent in dilute solutions and for mixtures of close members of the same homologous series at any concentration. In these cases, we will have negligible heat and volume effects by mixing the compounds. As the difference in the hydrocarbon chain length increases between members of the same homologous series, weak negative deviations appear. Negative deviations are also found in mixtures of components from different homologous series when either hydrogen bonding or electron donor–acceptor complexes can be formed between the compounds. However, *most mixtures of compounds from different homologous series present positive deviations from ideality. Positive deviations increase with the difference in polarity between the compounds.*

The fact that dilute mixtures are formed with negligible volume change or heat effects suggests a generalization of the concept of ideal mixture, as discussed below.

HENRY'S CONVENTION FOR THE ACTIVITY COEFFICIENT OF A SOLUTE

Lewis's convention for the activity coefficient is a convenient one when the compounds in the mixture all exist as pure liquids at the temperature of the system. When one of the compounds is solid at the temperature of the system, Lewis's convention can be used but needs some additional considerations. Let us consider a case in which *the solvent, component 1, is almost pure and the solute, component 2, is at high dilution.* For the solvent, in Lewis's convention,

$$f_1^{dil} \cong x_1 f_1^0 \tag{15.7}$$

From the isothermal form of Equation 14.13, as given by Equation 14.19, neglecting pressure effects for liquid mixtures,

$$x_1 d\ln f_1^{dil} + x_2 d\ln f_2^{dil} = 0$$

Thus, as f_1^0 is independent of composition,

$$d\ln f_2^{dil} = -\frac{x_1}{x_2} d\ln x_1 = -\frac{dx_1}{x_2} = \frac{dx_2}{x_2} = d\ln x_2$$

or

$$d\ln\left(\frac{f_2^{dil}}{x_2}\right) = 0$$

This result indicates that in the dilute region, the ratio of the fugacity of the solute to its mole fraction is a constant independent of composition. Designating this constant value by $H_{2,1}$, known as *Henry's constant for solute 2 dissolved in solvent 1*, we write:

$$f_2^{dil} = x_2 H_{2,1} \tag{15.8}$$

This expression, in fact, defines an ideal behavior for the solute, and we can define an *activity coefficient in Henry's convention*, $\gamma_{i,H}$, which is unity at infinite dilution of the solute. Thus, for a concentrated solution of the solute i in solvent r, we write:

$$f_i = x_i \gamma_{i,H} H_{i,r} \tag{15.9}$$

with

$$\lim_{x_i \to 0} \gamma_{i,H} = 1 \tag{15.10}$$

We observe that for the definition of the activity coefficient in Henry's convention, it is particularly important to identify the nature of the solvent r in which the solute is at infinite dilution. The value of Henry's constant depends on the solvent in which the solute is at infinite dilution. For a particular

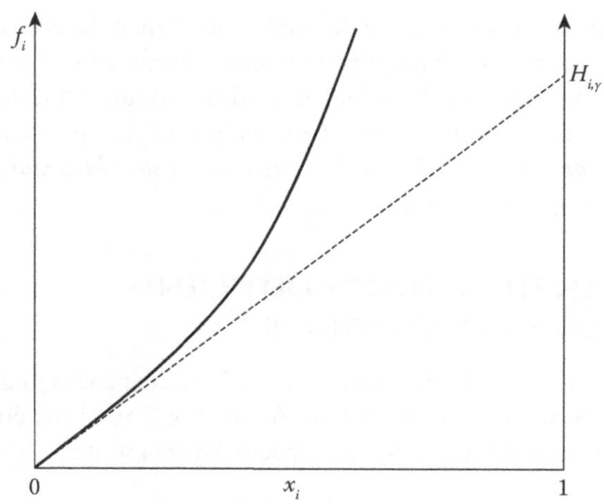

FIGURE 15.1 Determination of Henry's constant.

solvent r, the value of Henry's constant is then the limiting value of the ratio of the fugacity of the compound to its mole fraction as the mole fraction tends to zero. Because the ratio tends to a limit of zero over zero at infinite dilution, applying the L'Hopital theorem, we obtain:

$$H_{i,r} = \lim_{x_i \to 0} \frac{f_i}{x_i} = \lim_{x_i \to 0} \frac{df_i}{dx_i} \tag{15.11}$$

Thus, Henry's constant is the value of the slope of the fugacity as a function of mole fraction for the solute at high dilution in solvent r. In a graphical representation of the fugacity of the solute as a function of its mole fraction, the value of Henry's constant would be found where the tangent to the fugacity curve at its origin, $x_i = 0$, cuts the vertical axis at $x_i = 1$ (Figure 15.1).

GENERALIZED DEFINITION OF ACTIVITY

As both conventions have much in common, we can write a single generalized expression for both of them as:

$$f_i = x_i \gamma_{i,\theta} f_i^{\theta} \tag{15.12}$$

Thus, the activity of compound i with respect to the generalized standard state θ of the compound is given by Equation 12.13, written as:

$$a_{i,\theta} \equiv \frac{f_i}{f_i^{\theta}} = x_i \gamma_{i,\theta} \tag{15.13}$$

For a component i in an ideal mixture, in these two conventions,

$$f_i^{id,\theta} = x_i f_i^{\theta} \tag{15.14}$$

and the activity of the compound in the ideal mixture is equal to the mole fraction and becomes unity for the pure compound in its standard state.

The main difference between both conventions is that in the Lewis convention, the reference state used to normalize the activity coefficient to unity is the same as the standard state that gives unit activity for the compound in a real or ideal mixture, while in Henry's convention, this is not the case. In Henry's convention, the reference state used to normalize the activity coefficient to unity is the state of the solute at infinite dilution in the particular solvent, while the standard state that gives unit activity is the imaginary state of the solute in an ideal mixture (in Henry's sense), extrapolated to a mole fraction of the solute equal to unity. *Thus, in general we distinguish between a reference state used to normalize the activity coefficient to unity and a standard state in which the activity of the compound in an ideal mixture would be unity.*

RELATIONS BETWEEN THE ACTIVITY COEFFICIENTS IN LEWIS'S AND HENRY'S CONVENTIONS

For the case of a compound that can be a solute or a solvent, depending on the situation, it may be convenient to use Lewis's or Henry's convention. As the fugacity of the compound is independent of the convention used to define the activity coefficient, we can write:

$$f_i = x_i \gamma_{i,0} f_i^0 = x_i \gamma_{i,H} H_{i,r}$$

or

$$\ln \gamma_{i,0} = \ln \gamma_{i,H} + \ln \frac{H_{i,r}}{f_i^0} \tag{15.15}$$

Taking the limit of this expression when the mole fraction of component i tends to zero in the particular solvent (state of infinite dilution), using Equations 15.10 and 15.14 and rearranging,

$$\ln H_{i,r} = \ln f_i^0 + \lim_{x_i \to 0} \left(\ln \gamma_{i,0} \right) \tag{15.16}$$

This equation is important because it allows us to evaluate Henry's constant from a knowledge of the fugacity of the pure compound, Equation 13.2, and the value of the activity coefficient in Lewis's convention at infinite dilution in the particular solvent. Replacing Equation 15.16 back into Equation 15.15,

$$\ln \gamma_{i,H} = \ln \gamma_{i,0} - \lim_{x_i \to 0} \left(\ln \gamma_{i,0} \right) \tag{15.17}$$

Similarly, taking the limit of Equation 15.15 when x_i tends to unity, using Equation 15.6 and rearranging,

$$\ln \gamma_{i,0} = \ln \gamma_{i,H} - \lim_{x_i \to 1} \left(\ln \gamma_{i,H} \right) \tag{15.18}$$

This equation is not particularly useful, and it is included here for completeness only. The important point is that having analytical expressions for the activity coefficient of the compounds as a function of composition in one convention, it is possible to obtain the analytical expressions in the other convention.

GENERALIZED EQUATIONS FOR AN IDEAL MIXTURE

The Gibbs energy change obtained by forming an ideal mixture starting from all its components in their respective standard states can be written as:

$$\Delta g_m^{id,\theta} = \sum_1^c x_k \left(\bar{g}_k^{id,\theta} - g_k^\theta \right) = RT \sum_1^c x_k \ln \frac{f_k^{id,\theta}}{f_k^\theta} = RT \sum_1^c x_k \ln x_k \tag{15.19}$$

For the enthalpy change of mixing in forming the ideal mixture, then, from Equation 7.6,

$$\Delta h_m^{id,\theta} = \sum_1^c x_k\left(\bar{h}_k^{id,\theta} - h_k^{\theta}\right) = -RT^2\left(\frac{\partial\left(\Delta g_m^{id,\theta}/RT\right)}{\partial T}\right)_{P,n} = 0$$

Thus,

$$\bar{h}_k^{id,\theta} = h_k^{\theta} \tag{15.20}$$

Similarly, from Equation 4.18, as Equation 15.17 indicates that $\Delta g_m^{id,\theta}$ is independent of pressure, we write:

$$\Delta v_m^{id,\theta} = \sum_1^c x_k\left(\bar{v}_k^{id,\theta} - v_k^{\theta}\right) = \left(\frac{\partial \Delta g_m^{id,\theta}}{\partial P}\right)_{T,n} = 0$$

Thus,

$$\bar{v}_k^{id,\theta} = v_k^{\theta} \tag{15.21}$$

In addition, we can also write:

$$\Delta u_m^{id,\theta} = \sum_1^c x_k\left(\bar{u}_k^{id,\theta} - u_k^{\theta}\right) = \Delta h_m^{id,\theta} - P\Delta v_m^{id,\theta} = 0$$

from where

$$\bar{u}_k^{id,\theta} = u_k^{\theta}$$

Recapitulating, a generalized ideal mixture, real or hypothetical, is formed from its pure compounds at their respective standard state without any heat effect or volume change:

$$\Delta h_m^{id,\theta} = 0 \tag{15.22}$$

and

$$\Delta v_m^{id,\theta} = 0 \tag{15.23}$$

while from Equation 4.2, the entropy change of forming the ideal mixture from the individual compounds in their standard states is:

$$\Delta s^{id,\Theta} = \frac{\Delta h^{id,\Theta} - \Delta g^{id,\Theta}}{T}$$

$$\Delta s_m^{id,\theta} = -R\sum_1^c x_k \ln x_k \tag{15.24}$$

PROPERTIES OF IDEAL MIXTURES IN LEWIS'S CONVENTION

In Lewis's convention, ideal mixtures will form starting from the pure compounds at the same pressure and temperature of the mixture without changes in volume or heat effects. Thus, all the above equations are valid. As studies of liquid mixtures in the Lewis convention are the most common, we repeat these equations here:

$$\Delta v_m^{id,0} = \sum_k x_k \left(\bar{v}_k^{id,0} - v_k^0 \right) = 0$$

and this relation implies that:

$$\bar{v}_i^{id,0} = v_i^0$$

Similarly, from

$$\Delta h_m^{id,0} = \sum_k x_k \left(\bar{h}_k^{id,0} - h_k^0 \right) = 0$$

we obtain $\bar{h}_i^{id,0} = h_i^0$

For the Gibbs energy change,

$$\Delta g_m^{id,0} = RT \sum_k x_k \ln \frac{f_k^{id,0}}{f_k^0} = RT \sum_k x_k \ln \frac{x_k f_k^0}{f_k^0} = RT \sum_k x_k \ln x_k$$

$$\Delta s^{id,0} = \frac{\Delta h^{id,0} - \Delta g^{id,0}}{T} = -R \sum_k x_k \ln x_k$$

As these equations are just particular cases of Equations 15.22–15.24, there is no need to assign them a particular equation number here.

USE OF MOLALITY AS A MEASURE OF CONCENTRATION

Molality, that is, the moles of solute per 1 kg of solvent, is a convenient way to express the concentration of dilute solutions. In this case, the fugacity of solute i at a molality m_i is written in Henry's convention as:

$$f_i = \tilde{m} \gamma_{i,H}^{(m)} H_{i,r}^{(m)} \tag{15.25}$$

In this expression, $\tilde{m}$ is the dimensionless molality of the solute, that is, the value of its molality divided by 1 [mol of i/1 kg of solvent]. This later value is the concentration of the standard state for the ideal solution in Henry's sense. In addition, we have introduced the superscript (m) in order to distinguish the activity coefficient and the standard fugacity used here from those used in Henry's convention with mole fraction as a measure of concentration. Again here, taking the limit of zero molality and applying L'Hôpital's theorem, we obtain:

$$H_{i,r}^{(m)} = \lim_{\tilde{m} \to 0} \frac{f_i}{\tilde{m}} = \lim_{\tilde{m} \to 0} \frac{df_i}{d\tilde{m}} \tag{15.26}$$

Conversion between the two sets of activity coefficients in Henry's convention is direct. As the fugacity is independent of the system chosen to express concentration, we have:

$$f_i = x_i \gamma_{i,H} H_{i,r} = \tilde{m}_i \gamma_{i,H}^{(m)} H_{i,r}^{(m)}$$

Thus,

$$\ln \gamma_{i,H} = \ln\left(\frac{\tilde{m}_i}{x_i}\right) + \ln \gamma_{i,H}^{(m)} + \ln\left(\frac{H_{i,r}^{(m)}}{H_{i,r}}\right) \tag{15.27}$$

Simple algebra shows that:

$$x_i = \frac{\tilde{m}_i}{\tilde{m}_i + \sum_{k \neq 1} \tilde{m}_k + 1/M_s}$$

or

$$\frac{\tilde{m}_i}{x_i} = \frac{1}{M_s} + \tilde{m}_i + \sum_{k \neq i} \tilde{m}_k$$

where M_s is the molar mass of the solvent in kg/mol. Thus, as the normalization of the activity coefficient in Henry's convention is taken at the limit when the number of moles of i, n_i, tends to zero in pure solvent, the dimensionless molalities of all other possible solutes are identically zero and the expression above tends to the constant value of the first term on the right-hand side, while $\ln \gamma_{i,H}$ and $\ln \gamma_{i,H}$ tend to zero as n_i goes to zero. Hence,

$$\frac{H_{i,r}^{(m)}}{H_{i,r}} = M_s \tag{15.28}$$

Equation 15.28 gives a simple relation between both values of Henry's constants. Replacing these terms back in Equation 15.27 and rearranging, we finally write:

$$\ln \gamma_{i,H} = \ln \gamma_{i,H}^{(m)} + \ln\left(1 + M_s \sum_k m_k\right) \tag{15.29}$$

where this time the sum is over all solutes, including solute i.

PROBLEMS

15.1 **Ideality of a virial gas mixture**

Derive the requirements for a gaseous mixture that obeys the truncated virial equation of state, as described in Appendix B and used in Problem 14.3, to be an ideal solution in the Lewis convention.

15.2 **Lewis and Henry activity coefficients of a binary liquid mixture**

Consider again the experimental vapor-liquid equilibrium data of the ethyl acetate (1) + benzene (2) system at 50°C in Problem 14.3. Calculate the activity coefficients of both components in the liquid phase

a. In the Lewis convention
b. In the Henry convention

The vapor pressures of the pure components are given by the Antoine equations [1].

$$\log_{10} P_1^s/\text{mmHg} = 7.10233 - \frac{1245.239}{t/°\text{C} + 217.911}$$

$$\log_{10} P_2^s/\text{mmHg} = 6.90565 - \frac{1211.033}{t/°\text{C} + 220.790}$$

15.3 **Activity coefficients of aqueous sucrose solutions**

Robinson et al. [2] measured the vapor pressure lowering ΔP of sucrose solutions, relative to the value of 23.756 mmHg for pure water, as a function of molality at 25°C. Their reported values are shown in Table 15.1.

a. Compute the Lewis activity coefficients of water at these compositions and temperature. Assume that the vapor phase behaves as an ideal gas, and neglect pressure effects on the coefficients.
b. From the above results, derive the Henry activity coefficients of sucrose, both in terms of mole fractions and of molalities. The molar masses (molecular weights) for water (1) and sucrose (2) are $M_1 = 18.015$, $M_2 = 342.30$ g/mol.

TABLE 15.1
The Vapor Pressure Lowering ΔP of Sucrose Solutions, Relative to the Value of 23.756 mmHg for Pure Water, as a Function of Molality at 25°C [2]

m_2 (mol/kg)	ΔP (mm/kg)	m_2 (mol/kg)	ΔP (mm/kg)
0.1	0.043	2.5	1.294
0.2	0.087	3.0	1.602
0.3	0.131	3.5	1.921
0.5	0.222	4.0	2.250
0.7	0.315	4.5	2.581
1.0	0.462	5.0	2.916
1.5	0.721	5.5	3.256
2.0	0.999	6.0	3.602

REFERENCES

[1] Linék, J., Procházka, K., and Wichterle, I. 1972. Liquid-vapour equilibrium. LIV. The systems ethyl acetate–benzene, ethyl acetate–toluene and ethyl acetate-ethylbenzene. *Coll. Czech. Chem. Commun.* 37: 3010–3014.

[2] Robinson, R. A., Smith, P. K., Smith, E. R. B. 1942. The osmotic coefficients of some organic compounds in relation to their chemical constitution. *Trans. Faraday Soc.* 38: 63–70.

16 Activity Coefficients and Excess Properties

Abstract

The second half of the last century was an active period in experimental measurement and modeling of the activity coefficients as a function of composition and temperature in liquid mixtures. As these coefficients are not all independent, tests of consistency of the measured values were developed based on the satisfaction of the Gibbs–Duhem equation. For modeling purposes, it was found useful to define the excess Gibbs energy as the difference between the Gibbs energy of a mixture and that of an ideal mixture at the same composition and temperature. Similar excess properties were defined for enthalpy, volume, etc. The chapter explores the relations of excess properties to the activity and fugacity coefficients of the compounds forming the mixture.

To evaluate the activity coefficients of components of a liquid mixture of nonelectrolytes from experimental data, it is necessary to know the value of the fugacity of the pure compounds in the liquid state at the pressure and temperature of the system.

EVALUATION OF THE FUGACITY OF A PURE COMPOUND LIQUID AT *P* AND *T* OF THE SYSTEM

For a pure compound that exists as a saturated liquid at the temperature T of the system, from Equation 14.17, we write:

$$\ln\left(\frac{f_i^{0,L}}{f_i^{0,s}}\right) = \frac{1}{RT}\int_{P_i^s}^{P} v_i^{0,L}\,dP \approx \frac{v_i^{0,L}\left(P-P_i^s\right)}{RT}$$

or

$$f_i^{0,L} = f_i^{0,s}\exp\left[\frac{v_i^{0,L}\left(P-P_i^s\right)}{RT}\right]$$

But from Equation 13.2,

$$f_i^{0,s} = \phi_i^{0,s}P_i^s$$

Thus,

$$f_i^{0,L} = \phi_i^{0,s}\exp\left[\frac{v_i^{0,L}\left(P-P_i^s\right)}{RT}\right]P_i^s \tag{16.1}$$

The fugacity coefficient of a pure compound, saturated at the temperature, T, of the system, can be calculated from an equation of state for the saturated vapor phase using either Equation 13.7 or Equation 13.13. The molar liquid volume can be obtained from data for liquid densities, and the saturation pressure can be obtained from an equation of the type of Equation 5.17. The exponential

DOI: 10.1201/9781003431985-18

term in Equation 16.1 is usually referred to as the *Poynting correction*. Considering a typical molar liquid volume of 100 mL and a temperature of 300 K, the Poynting correction for 10 atm of difference between the pressure, P, of the system and the saturation pressure is about 1.05. For 100 atm of pressure difference, the Poynting correction is about 1.5. In the rare cases of negative deviations from ideality, in which the pressure of the system is lower than the saturation pressure, P_i^s, of the compound, the Poynting correction is a value slightly smaller than unity.

On the other hand, experimental evidence shows that the compressibility factor of pure compounds is less than unity at reduced temperatures below unity. Thus, according to Equation 13.7a, the fugacity coefficients at saturation for nonpolar compounds are less than unity. Approximate estimates based on generalized correlations suggest that at a reduced temperature of 0.7, the fugacity coefficient is slightly less than 1; it is about 0.8 at a reduced of temperature 0.8, and decreases to about 0.75 at a reduced temperature of 0.9, reaching 0.6 near the critical point. At moderate conditions, and also at high pressure and reduced temperatures close to unity, both the fugacity coefficient and the Poynting correction tend to compensate. At reduced temperatures below 0.8 and moderate pressure, their product is less than unity, and at high pressure and low temperatures, their product is higher than unity. As a first approximation, then, the fugacity of the pure liquid is closer to the saturation pressure of the pure compound than to the total pressure of the system.

$$f_i^{0,L} \approx P_i^s$$

Similar equations can be written for a compound that exists as a saturated solid at the temperature, T, of the system.

EVALUATION OF ACTIVITY COEFFICIENTS FROM PHASE EQUILIBRIUM STUDIES

The key equations for the study of phase equilibria are obtained from the equality of fugacities at equilibrium, Equation 12.10. In the case of vapor–liquid equilibrium, for the fugacity of the compound in the vapor phase we use Equation 13.1, and for the fugacity of the compound in the liquid phase, we use Equations 15.2b and 16.1,

$$y_i\phi_i^V P = x_i\gamma_{i,0}\phi_i^{0,s}\exp\left[\frac{v_i^{0,L}\left(P-P_i^s\right)}{RT}\right]P_i^s \tag{16.2}$$

In Equation 16.2, we have used the conventional symbols y_i and x_i for the mole fraction in the vapor and liquid phases, respectively. In this equation, all quantities other than the activity coefficient can be determined independently. The fugacity coefficients are obtained from equations of state, while the compositions, the pressure, and the temperature are measured directly. The density and saturation pressure of the pure compound are obtained from the literature. For practical work, the following shorthand notation is useful:

$$y_iP = x_i\gamma_{i,0}\left(P_i^s\right)' \tag{16.2a}$$

In Equation 16.2a, we have used the following definition for a "corrected" value of the vapor pressure of the pure compound:

$$\left(P_i^s\right)' \equiv \left(\frac{\phi_i^{0,s}}{\phi_i^V}\right)\exp\left[\frac{v_i^{0,L}\left(P-P_i^s\right)}{RT}\right]P_i^s \tag{16.3}$$

As discussed in the paragraph following Equation 16.1, at reduced temperatures below 0.8 and pressures of a few atmospheres, at which ϕ_i^V is nearly equal to 1, the whole factor correcting the value of the vapor pressure is close to unity. Thus, as the normal boiling point of organic liquids is normally below a reduced temperature of 0.7, the correction factor is usually ignored and Equation 16.2 is simply used as:

$$y_i P = x_i \gamma_{i,0} P_i^s \tag{16.4}$$

This expression is known as the modified Raoult's equation. For historical reasons, Equation 16.4 with $\gamma_{i,0} = 1$ is called Raoult's law.

Although there are analytical methods based on group contributions to estimate the values of the activity coefficients in a liquid mixture, their evaluation comes from experimental measurements carried out on binary systems. Complete sets of data have been collected and are available in the literature. For the case of two compounds that *exist as liquids at the temperature of the system*, each liquid pure compound is degassed until its vapor pressure coincides with the known value at the temperature of interest. Then, one compound is added to a thermostated cell and the other compound is added in steps, registering the total pressure once the temperature is stabilized after each addition. Both the vapor and the liquid phases are then analyzed, and the activity coefficient of each compound is determined from Equation 16.2 or, at moderate conditions, Equation 16.4. We observe that the activity coefficient of each compound in the mixture can be evaluated independently. The problem is, however, that the activity coefficients are not all independent. For thermodynamic consistency, they should satisfy the Gibbs–Duhem equation, as discussed below.

For calculation of phase equilibrium, Equation 16.2 can be used without further assumptions if the compound i exists as liquid at the temperature of the system. Otherwise, as discussed in Chapter 13, the use of an equation of state is preferred.

For the case of liquid–liquid equilibrium, in which a compound i is present in two liquid phases, α and β, at mole fractions x_i^α and x_i^β, the activity coefficients $\gamma_{i,0}^\alpha$ and $\gamma_{i,0}^\beta$ are obtained from measurements in an isothermal cell containing the two degassed liquid phases in equilibrium with a vapor phase. Equation 16.2 applies to the vapor–liquid equilibrium of each liquid phase with the vapor phase. In addition, as the pure compound fugacities are the same for both liquid phases, from Equation 12.10 we write:

$$x_i^\alpha \gamma_{i,0}^\alpha = x_i^\beta \gamma_{i,0}^\beta \tag{16.5}$$

Equation 16.5, which is also satisfied by the trivial solution $x_i^\alpha = x_i^\beta$, has been of common use for liquid–liquid equilibrium calculations. An alternative method for these calculations is presented in Chapter 20.

GIBBS–DUHEM EQUATION IN TERMS OF ACTIVITY COEFFICIENTS IN LEWIS'S CONVENTION

From Equation 15.2, we write:

$$\ln f_i = \ln x_i + \ln \gamma_{i,0} + \ln f_i^0$$

Combining this expression with the Gibbs–Duhem equation in terms of fugacities, Equation 14.13, we obtain:

$$\sum_k^c x_k d\ln x_k + \sum_k^c x_k d\ln \gamma_{k,0} + \sum_k^c x_k d\ln f_k^0 = -\frac{\sum_k^c x_k\left[\bar{h}_k - h_k^{0,*}(T)\right]}{RT^2} dT + \frac{\sum_k^c x_k \bar{v}_k}{RT} dP$$

As shown in the derivation of Equation 14.13, the first term on the left-hand side of this expression is null. The third term on the left-hand side, according to Equation 14.13, has the form:

$$\sum_k^c x_k d \ln f_k^0 = -\frac{\sum_k^c x_k\left[h_k^0 - h_k^{0,*}(T)\right]}{RT^2} dT + \frac{\sum_k^c x_k \mathrm{v}_k^0}{RT} dP$$

Thus,

$$\sum_k^c x_k d \ln \gamma_{k,0} = -\frac{\sum_k^c x_k\left[\bar{h}_k - h_k^0\right]}{RT^2} dT + \frac{\sum_k^c x_k\left[\bar{\mathrm{v}}_k - \mathrm{v}_k^0\right]}{RT} dP \quad (16.6)$$

Equation 16.6 is the Gibbs–Duhem equation in terms of activity coefficients in Lewis's convention. As in previous cases, we can now group all the terms under a single summation. For this summation to be null at all concentrations, the factor multiplying the mole fractions must be null:

$$d \ln \gamma_{i,0} = -\frac{\left[\bar{h}_i - h_i^0\right]}{RT^2} dT + \frac{\left[\bar{\mathrm{v}}_i - \mathrm{v}_i^0\right]}{RT} dP \quad (16.7)$$

The change of the activity coefficient with temperature is not of much interest, as the activity coefficients are usually measured at the temperature of interest. For the pressure variation, we obtain:

$$\left(\frac{\partial \ln \gamma_{i,0}}{\partial P}\right)_T = \frac{\left[\bar{\mathrm{v}}_i - \mathrm{v}_i^0\right]}{RT} \quad (16.8)$$

Although, due to the incompressibility of the liquid phase, for most practical situations the activity coefficients are quite insensitive to changes in pressure, for completeness, we note here that the pressure chosen for its definition is rather arbitrary. In the treatment above, we have chosen the pressure P of the system.

In some cases, the standard state is chosen as the saturated liquid pure compound. In this case, the fugacity of the pure compound, f_i^{0s}, depends only on temperature. From Equation 15.2, it is clear that as the fugacity of the compound is independent of the choice of standard state, a change of the latter will require a change of the activity coefficient to a value $\gamma_{i,0s}$ satisfying the relation:

$$\gamma_{i,0s} f_i^{0s}\left(T, P_i^s\right) = \gamma_{i,0} f_i^0(T, P)$$

Thus,

$$d \ln \gamma_{i,0s} = d \ln \gamma_{i,0} + d \ln f_i^0(T,P) - d \ln f_i^{0s}\left(T, P_i^s\right)$$

From this equality, using Equations 16.7 and 14.17 and observing that the last term on the right-hand side is independent of pressure, we obtain:

$$\left(\frac{\partial \ln \gamma_{i,0s}}{\partial P}\right)_T = \frac{\bar{\mathrm{v}}_i}{RT} \quad (16.9)$$

The fact that for liquid mixtures the typical magnitude of the terms on the right-hand side of both Equations 16.8 and 16.9 is of the order of 10^{-3} indicates that the activity coefficients are quite insensitive to pressure changes.

The Gibbs–Duhem equation, Equation 16.6 in Lewis's convention, is used for testing the thermodynamic consistency of values of activity coefficients obtained in isothermal experiments. For these conditions in binary systems, Equation 16.6 takes the form:

$$x_1 d\ln\gamma_1 + x_2 d\ln\gamma_2 = \frac{\Delta v_m}{RT}dP \tag{16.10}$$

Thus, for a change in composition,

$$x_1\frac{d\ln\gamma_{1,0}}{dx_1} + x_2\frac{d\ln\gamma_{2,0}}{dx_1} = \frac{\Delta v_m}{RT}\frac{dP}{dx_1}$$

where we have used Equation 6.18 for the volume change of mixing. This stringent test of slopes has not found much application in practice and alternative procedures are of common use.

In the paragraph under Equation 14.18, we remarked that when all fugacity coefficients are obtained from a single equation of state, they satisfy the Gibbs–Duhem equation. Following the same pattern, it is convenient to combine all the independently measured activity coefficients in a single function, as discussed in the next section. When a single analytical expression for the excess Gibbs energy is used to reduce the experimental values of the activity coefficients of all components of the mixture, the smoothed values produced necessarily satisfy the Gibbs–Duhem equation.

EXCESS GIBBS ENERGY AND OTHER EXCESS FUNCTIONS IN ANY CONVENTION

For generality, we use here the symbol θ to indicate the standard state chosen to define the ideal mixture behavior. For the particular case of the Lewis convention, the symbol 0 replaces θ. The excess Gibbs energy function, G_θ^E, is defined as the difference between the Gibbs energy of the actual mixture and the Gibbs energy of the ideal mixture, with respect to well-defined standard states of the compounds, both mixtures taken at the same composition, pressure, and temperature:

$$G_\theta^E \equiv G(T,P,x) - G^{id,\theta}(T,P,x) \tag{16.11a}$$

or, in molar terms,

$$g_\theta^E \equiv g(T,P,x) - g^{id,\theta}(T,P,x) \tag{16.11b}$$

Following the same procedure used in the derivation of Equation 6.10, we write for the partial molar excess properties,

$$\bar{g}_{i,\theta}^E \equiv \bar{g}_i - \bar{g}_i^{id,\theta} \tag{16.11c}$$

Analogous relations hold for the excess volume, excess enthalpy, excess entropy, and so forth.

From Equation 4.2, we write:

$$g_\theta^E \equiv h_\theta^E - Ts_\theta^E \tag{16.12a}$$

In addition, following again the same method used to derive Equation 6.10, relations between partial molar excess properties are obtained. For example,

$$\bar{g}_{i,\theta}^E \equiv \bar{h}_{i,\theta}^E - T\bar{s}_{i,\theta}^E \tag{16.12b}$$

and, from Equation 7.6, the Gibbs–Helmholtz form,

$$h_\theta^E = -RT^2\left(\frac{\partial\left(g_\theta^E/RT\right)}{\partial T}\right)_{P,n} \tag{16.13a}$$

$$\bar{h}_{i,\theta}^E = -RT^2\left(\frac{\partial\left(\bar{g}_{i,\theta}^E/RT\right)}{\partial T}\right)_{P,n} \tag{16.13b}$$

As discussed in Chapter 26, these expressions are useful in the development of models for the representation of mixtures.

ACTIVITY COEFFICIENTS AND THEIR RELATION TO EXCESS FUNCTIONS IN ANY CONVENTION

The activity coefficients are directly related to the excess Gibbs energy. From Equation 16.9c, we write:

$$\bar{g}_{i,\theta}^E \equiv \bar{g}_i - \bar{g}^{id,\theta} = \mu_i - \mu_i^{id,\theta} = RT\ln\frac{f_i}{f_i^{id,\theta}}$$

With the purpose of having general equations, not limited to Lewis's convention, we rewrite Equations 15.2 and 15.9 as:

$$f_i = x_i\gamma_{i,\theta}f_i^\theta \tag{16.14}$$

Also, from Equation 15.14,

$$f_i^{id,\theta} = x_i f_i^\theta \tag{16.15}$$

Thus,

$$\bar{g}_{i,\theta}^E = RT\ln\gamma_{i,\theta} \tag{16.16}$$

From this expression and Equation 16.13b,

$$\bar{h}_{i,\theta}^E = -RT^2\left(\frac{\partial\ln\gamma_{i,\theta}}{\partial T}\right)_{P,n} \tag{16.17}$$

and from Equations 6.12b, 16.16, and 16.17,

$$\bar{s}_{i,\theta}^E = -R\ln\gamma_{i,\theta} - RT\left(\frac{\partial\ln\gamma_{i,\theta}}{\partial T}\right)_{P,n} \tag{16.18}$$

Of these expressions, the most important is Equation 16.16. It shows that *the logarithm of the activity coefficient is the partial molar property of the excess Gibbs energy over the product* RT. Thus, according to Equations 6.5 and 6.2, respectively, we write:

$$\frac{g_\theta^E}{RT} = \sum_k x_k\ln\gamma_{k,\theta} \tag{16.19}$$

and

$$\ln \gamma_{i,\theta} = \left(\frac{\partial \left(n g_\theta^E / RT \right)}{\partial n_i} \right)_{T,P,n_{j \neq i}} \tag{16.20}$$

or, alternatively,

$$\ln \gamma_{i,\theta} = \frac{g_\theta^E}{RT} - \sum_{k \neq i} x_k \left(\frac{\partial \left(g_\theta^E / RT \right)}{\partial x_k} \right)_{T,P,xj;j \neq i,j \neq k} \tag{16.21}$$

Equations 16.19 and 16.20, or Equation 16.21 if preferred, are the key equations for smoothing experimentally measured data of activity coefficients and producing sets of values of these activity coefficients that necessarily satisfy the Gibbs–Duhem equation. Individually measured activity coefficients are smoothed using a parametric form for the left-hand side of Equation 16.19. This function can be suggested by a model of solutions or be a simple polynomial form. Once adjustable parameters for the excess Gibbs energy function have been fitted to minimize the deviation of the fit from the experimental data, values of activity coefficients are generated using Equation 16.20 or 16.21. These smoothed values necessarily satisfy the Gibbs–Duhem equation.

THERMODYNAMIC CONSISTENCY OF ACTIVITY COEFFICIENTS IN LEWIS'S CONVENTION

As discussed above, results obtained from the measurement of the individual activity coefficients in binary mixtures can be made to be thermodynamically consistent by obtaining the values of the excess Gibbs function using Equation 16.19, fitting these resulting values with an analytical function containing a few adjustable parameters, and using Equation 16.20 to generate "smoothed" consistent values of the activity coefficients. More rigorous studies prefer to test directly how consistent are the values of the individual activity coefficients obtained from the experiments at each composition. Two tests based on the combination of the Gibbs–Duhem equation with the definition of excess Gibbs energy have been developed for this purpose. Equation 16.19 in the Lewis convention for a binary system takes the form:

$$\frac{g_0^E}{RT} = x_1 \ln \gamma_{1,0} + x_2 \ln \gamma_{2,0}$$

Then,

$$d\left(\frac{g_0^E}{RT} \right) = x_1 d \ln \gamma_{1,0} + x_2 d \ln \gamma_{2,0} + \left(\ln \gamma_{1,0} \right) dx_1 + \left(\ln \gamma_{2,0} \right) dx_2$$

The first two terms on the right-hand side of this expression are the terms appearing in the Gibbs–Duhem equation, Equation 16.10; thus, for isothermal data, considering that $dx_2 = -dx_1$, we write:

$$d\left(\frac{g_0^E}{RT} \right) = \frac{\Delta v_m}{RT} dP + \left[\ln \left(\frac{\gamma_{1,0}}{\gamma_{2,0}} \right) \right] dx_1$$

or

$$\ln \left(\frac{\gamma_{1,0}}{\gamma_{2,0}} \right) = \frac{d\left(g_0^E / RT \right)}{dx_1} - \frac{\Delta v_m}{RT} \frac{dP}{dx_1} \tag{16.22}$$

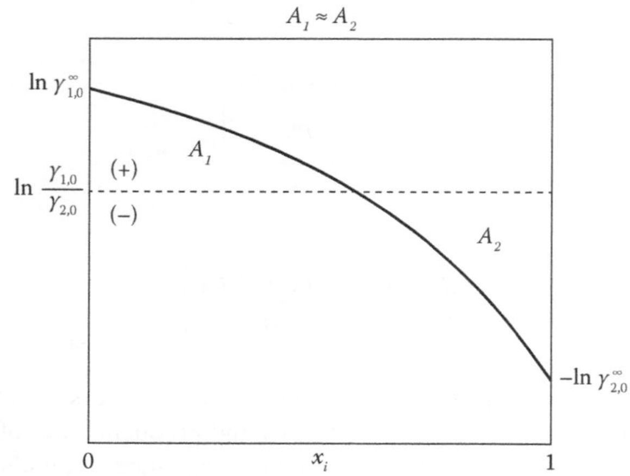

FIGURE 16.1 Consistency test for activity coefficients.

This differential test is less stringent than the one based on Equation 16.10, but still, only high-precision data can satisfy it. Data are normally tested by the integral form of this equation. Integrating between $x_1=0$ and $x_1=1$, neglecting the small contribution of the second term on the right-hand side, and noting that in both limits the value of g_0^E is zero,

$$\int_0^1 \ln\left(\frac{\gamma_{1,0}}{\gamma_{2,0}}\right) dx_1 = 0 \qquad (16.23)$$

Activity coefficients that do not satisfy this integral test, sometimes referred to as the area test, are normally rejected. Equation 16.23 indicates that the surface area under the curve $\ln(\gamma_{1,0}/\gamma_{2,0})$ in a plot versus x_1 should add to zero. Thus, the surface area under this curve will have a section above the horizontal axis and an equivalent section below the horizontal axis (Figure 16.1).

With respect to Equation 16.23, one interesting feature to notice is that in the plot of $\ln(\gamma_{1,0}/\gamma_{2,0})$ versus x_1, at the limit $x_1 \to 0$, $\gamma_{2,0} \to 1$ and $\gamma_{1,0}$ tends to a large value at infinite dilution that we denote by $\gamma_{1,0}^{\infty}$. Similarly, at the limit $x_1 \to 1$, $\gamma_{1,0} \to 1$ and $\gamma_{2,0}$ tends to a large value at infinite dilution denoted by $\gamma_{2,0}^{\infty}$. Thus, the curve of $\ln(\gamma_{1,0}/\gamma_{2,0})$ versus x_1 has a value $\ln \gamma_{1,0}^{\infty}$ at $x_1=0$, crosses the x-axis at some intermediate composition, and reaches the value $\left[-\ln \gamma_{2,0}^{\infty}\right]$ at $x_1=1$. This feature of the function $\ln(\gamma_{1,0}/\gamma_{2,0})$ will be used in the study of azeotropy. In Chapter 17, we first discuss when a mixture will split in two or more phases and when it will be stable in a single phase.

EXCESS GIBBS ENERGY IN LEWIS'S CONVENTION IN TERMS OF FUGACITY COEFFICIENTS

Due to its interest in the development of mixing rules for equations of state, we digress briefly here and observe that by combining Equation 16.19 for the Lewis convention with the activity coefficients given by Equation 15.2a, we obtain:

$$\frac{g_0^E}{RT} = \sum_k x_k \ln \gamma_{k,0} = \sum_k x_k \ln \frac{\phi_k}{\phi_k^0} = \sum_k x_k \ln \phi_k - \sum_k x_k \ln \phi_k^0$$

and using Equation 14.3, we write:

$$\frac{g_0^E}{RT} = \ln \phi_m - \sum_k x_k \ln \phi_k^0 \tag{16.24}$$

Equation 16.24 is an equivalent form of Equation 15.2c.

PROBLEMS

16.1 **Consistency test of activity coefficients of a binary system**
Use the area test of Equation 16.23 and Figure 16.1 to evaluate the thermodynamic consistency of the Lewis activity coefficients of the liquid mixture of ethyl acetate (1)+benzene (2) at 50°C obtained in Problem 15.2 from vapor–liquid equilibrium data.

16.2 **More on the area test of activity coefficients of a binary system**
Assume that experimentally determined isothermal Lewis activity coefficients of a binary liquid mixture have been correlated by a polynomial of adequate degree,

$$\ln \frac{\gamma_{1,0}}{\gamma_{2,0}} = \mathcal{P}(x_1) = \sum_{k=o} c_k x_1^k$$

Derive the corresponding expressions for the excess Gibbs energy and the individual activity coefficients. What happens if the data are not fully consistent? Use the results of Problem 16.1 to illustrate the analysis.

16.3 **Correlation of activity coefficients of a binary system**
Re-evaluate the experimental data of Problem 16.1. This time, use the procedure described in the text below Equation 16.21, that is, smooth the excess Gibbs energy values by fitting with an appropriate model and use this to generate consistent values of the activity coefficients.

17 Mixture Behavior, Stability, and Azeotropy

ABSTRACT

This chapter presents what to expect when mixing two compounds in liquid state. Based on a classification of pure compounds proposed by Berg in 1969, hints are given to anticipate if the mixture would have ideal behavior or would present positive or negative deviations from ideality in Lewis's sense. Mixtures exhibiting positive deviations are endothermic and those presenting negative deviations are exothermic. The cases of separation of the mixture in two liquid phases and the case of the mixture presenting azeotropic behavior receive special attention. Some typical phase equilibrium diagrams are presented to give a better physical sense of mixture behavior.

BASIC RELATIONS

As compounds mix to some degree in the liquid phase, the Gibbs energy of the system decreases toward a minimum at equilibrium. The Gibbs energy change obtained by forming a mixture starting from all the pure compounds can be written as:

$$\Delta g_m \equiv g - g^0 = g - \sum_k x_k \mu_k^0 \tag{17.1}$$

The terms of Equation 17.1 can be written more explicitly as:

$$\Delta g_m = \sum_1^c x_k \left(\bar{g}_k - g_k^0\right) = \sum_1^c x_k \left(\mu_k - \mu_k^0\right) = RT \sum_1^c x_k \ln \frac{f_k}{f_k^0} \tag{17.2}$$

For a binary system,

$$\frac{\Delta g_m}{RT} = x_1 \ln \frac{f_1}{f_1^0} + x_2 \ln \frac{f_2}{f_2^0} \tag{17.3}$$

Thus, expressing the fugacities in terms of the pure compound standard state, that is, defining the ideal behavior in Lewis's sense, from Equation 15.2 we write:

$$\frac{\Delta g_m}{RT} = \left(x_1 \ln x_1 + x_2 \ln x_2\right) + \left(x_1 \ln \gamma_{1,0} + x_2 \ln \gamma_{2,0}\right) \tag{17.4a}$$

The first bracket of the right-hand side gives the contribution of the ideal mixture, Equation 15.19, and as both mole fractions are less than unity, its value is always negative. This suffices to prove that an ideal mixture is always a stable mixture. The second bracket gives the contribution of nonideality of the mixture, and it can be identified with the excess Gibbs energy in Lewis's sense, Equation 16.19. Thus,

$$\frac{\Delta g_m}{RT} = \frac{\Delta g_m^{id,0}}{Rt} + \frac{g_0^E}{RT} \tag{17.4b}$$

DOI: 10.1201/9781003431985-19

POSITIVE AND NEGATIVE DEVIATIONS FROM IDEAL BEHAVIOR IN LEWIS'S SENSE

For negative deviations from ideality in Lewis's sense, $\gamma_{i,0} < 1$, the second bracket of the right-hand side of Equation 17.4a is also negative and makes the mixture still more stable than an ideal solution. For positive deviations from ideality in Lewis's sense, $\gamma_{i,0} > 1$, the second bracket of the right-hand side of Equation 17.4a is positive. If this contribution is large enough to overcome the contribution of the ideal mixture term, it produces instability, that is, liquid–liquid phase separation. A "good solvent" for a given compound should give negative deviations, ideal mixing, or at most weak positive deviations.

Briefly stated, ideal mixtures are formed without volume change or heat effects. In the case of positive deviations from ideality in Lewis's sense, there is a decrease in temperature and an increase in volume in adiabatic mixing. The opposite happens in the case of negative deviations in Lewis's sense. In molecular terms, molecules that "dislike" each other take more space among themselves and produce a cold environment. Molecules that "like" each other come closer together and create a warm milieu.

Although molecular effects are complex and resist generalizations, it is possible to establish some rules of thumb and be prepared to find exceptions. For the case of organic compounds, there are two main characteristics of their molecules that help in anticipating the behavior of their mixtures:

1. Length and shape of the hydrocarbon skeleton.
2. Functional groups present.

The effect of the functional group is so dominant that we classify the organic molecules in homologous families. In this respect, water can be considered the shortest alcohol, that is, an alcohol without a hydrocarbon chain. When mixing compounds, the group–group interactions govern the physical and chemical behavior of the mixture. Some group–group interactions are so strong as to cause a chemical reaction and the formation of new compounds. Good examples are the cases of groups COOH and OH, COO and H_2O, and CO and NH_2. We restrict our discussion here to non-reactive cases.

Large differences in the lengths of the hydrocarbon chains tend to give negative deviations. These are important in hydrocarbon–hydrocarbon systems, but they tend to be overshadowed by the larger effect of functional groups when these are present. The effect of the shape of the hydrocarbon skeleton is minor, and in most practical situations, it can be ignored. However, the effect of unsaturation in hydrocarbon structures may be important since unsaturated aliphatic compounds and also aromatic compounds can participate in electron donor–acceptor complexes and thus favor mixing.

BERG CLASSIFICATION OF LIQUIDS AND MIXTURE BEHAVIOR

Berg [1] proposed a classification of liquids in five groups that we extend here to six by differentiating between saturated and unsaturated hydrocarbons. The classification is based mostly on the capacity of the molecules to form hydrogen bonds and electron donor–acceptor complexes, but it also roughly orders the compounds on a scale of decreasing polarity. The modified Berg's classification is as follows:

- *Group 1*: Molecules forming tridimensional networks of H bonds. These are highly associated liquids, such as water, polyalcohols, polyphenols, hydroxyacids, amino alcohols, and amides.
- *Group 2*: Molecules having both active hydrogen atoms and electron donor atoms, such as O, N, and F. These are associated liquids, such as nitro compounds and nitriles with

hydrogen atoms in the α position (i.e., in the adjacent C atom) with respect to the N atom, nitromethane and acetonitrile, phenols, acids (including HCN, HF, HCl, and HNO_3), alcohols, and primary and secondary amines.

- *Group 3*: Molecules having electron donor atoms (O, N, and F) but not having an active hydrogen atom. Nitriles and nitro compounds without α-hydrogen, esters, ketones, aldehydes, esters, tertiary amines, and so forth, are in this group.
- *Group 4*: Molecules having an active hydrogen atom but not an electron donor atom, such as O, N, or F. These are polychlorinated compounds with Cl and H in the same or adjacent carbon atoms.
- *Group 5*: Unsaturated hydrocarbons, aliphatic or aromatic.
- *Group 6*: Saturated hydrocarbons.

From group 3 down, mixtures of compounds from the same group are close to ideal with slight positive deviations. Within the same group, positive deviations increase as the difference between the molecules increases. For example, a mixture of two ketones will be more ideal than a mixture of one ketone and one nitrile. Long hydrocarbon chains in both molecules of the same group tend to reduce the nonideality. If the molecules have similar vapor pressures, their mixtures can exhibit maximum-pressure azeotropes.

When mixing compounds of different groups from group 3 down, there are two cases deserving special mention. Members of group 3 mixed with members of group 4 will always give negative deviations from ideality. If their vapor pressures are similar, they can exhibit minimum-pressure azeotropes. Second, polar solvents from group 3 (and also from group 2), when mixed with members of group 5, give electron donor–acceptor complexes and thus favor solubility.

Groups 1 and 2 contain the most difficult "personalities" among the chemical species. When members of group 1 or 2 are mixed with members of group 4, 5, or 6 they always exhibit positive deviations that can be strong enough to cause liquid phase separation. When members of groups 1 and 2 are mixed with members of group 3, there can be either weak positive or weak negative deviations. And this is also the situation when mixing members of group 1 among themselves, members of group 2 among themselves, and members of group 1 with members of group 2. Negative deviations that lead to azeotropic behavior are found in mixtures of water with strong acids or other associating liquids of group 2 and also in mixtures of organic acids or phenols with amines or alcohols.

While negative deviations from ideality in Lewis's sense produce stable mixtures, strong positive deviations may cause instability, leading to the separation of the mixture into two liquid phases. Typical examples of a binary system formed by two partially miscible compounds are the cases of butanol and water and the more extreme case of petroleum and water. However, even in the case of strong positive deviations, two liquid compounds are miscible to a minor extent. Starting from one of the liquid compounds pure (solvent), it is always possible to dissolve in it a small concentration of the second liquid compound (solute). However, if the concentration of the solute is increased above a certain limit, the liquid mixture may split into two separate liquid phases. Each liquid phase will be rich in one of the components and lean in the other. In ternary or higher systems, the additional components will distribute between the two liquid phases formed.

PARTIAL MISCIBILITY

In the mixing process of two completely miscible liquid compounds, or in the miscibility regions of two partially miscible liquid compounds, that is, in the range where the mixture is stable, the Gibbs energy decreases toward equilibrium; thus, for the formation of a stable binary mixture the Gibbs energy change is necessarily negative,

$$\Delta g_m < 0$$

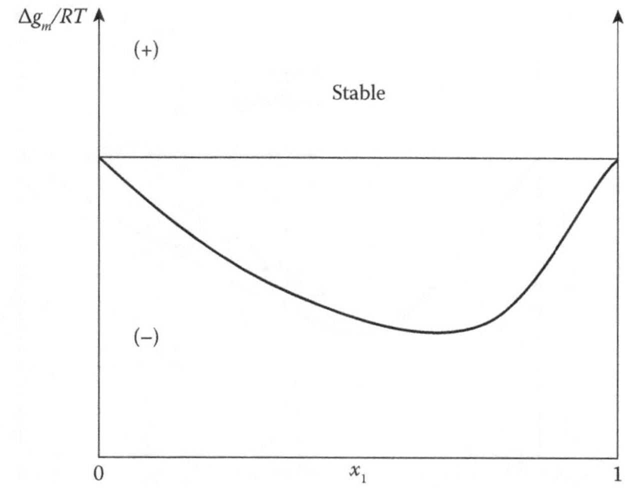

FIGURE 17.1 Stable system.

Figure 17.1 presents a plot ($\Delta g_m/RT$) versus x showing that for a system in which the two compounds are totally miscible in all the composition range, the slope of the curve $\Delta g_m/RT$ versus x increases continuously with an increase in x. Thus,

$$\frac{d}{dx}\left[\frac{d\Delta g_m / RT}{dx}\right]_T \geq 0$$

So, the criterion for the stability of a binary mixture can be written as:

$$\left[\frac{\partial^2(\Delta g_m / RT)}{\partial x^2}\right]_T \geq 0 \tag{17.5}$$

where x refers to x_1 or to x_2. When Equation 17.5 is satisfied over all of the concentration range, the compounds are miscible in all proportions. Binary systems having strong positive deviations from ideality in Lewis's sense may exhibit immiscibility if Equation 17.5 fails to be satisfied. If this happens, the mixture will be unstable and split into two phases. This behavior is depicted in Figure 17.2, where between points 1 and 2, the slope of the curve of ($\Delta g_m/RT$) versus x decreases with composition. In this case, the Gibbs energy of the system is less when there are two phases present instead of one.

For a binary liquid mixture, Equation 17.1 takes the form:

$$\Delta g_m \equiv g - x_1\mu_1^0 - x_2\mu_2^0 \tag{17.1a}$$

Taking the second derivative with respect to either x_1 or x_2, for the condition of stability we obtain:

$$\left[\frac{\partial^2 \Delta g_m}{\partial x^2}\right]_T = \left[\frac{\partial^2 g}{\partial x^2}\right]_T \geq 0 \tag{17.6}$$

This result shows a very important conclusion; that is, both the plots of Δg_m versus x and g versus x convey the same information regarding a possible instability of the system. If at any concentration range, the second derivative of the function becomes negative, that is, the curve presents an inflection point, the mixture becomes unstable and separates into two phases.

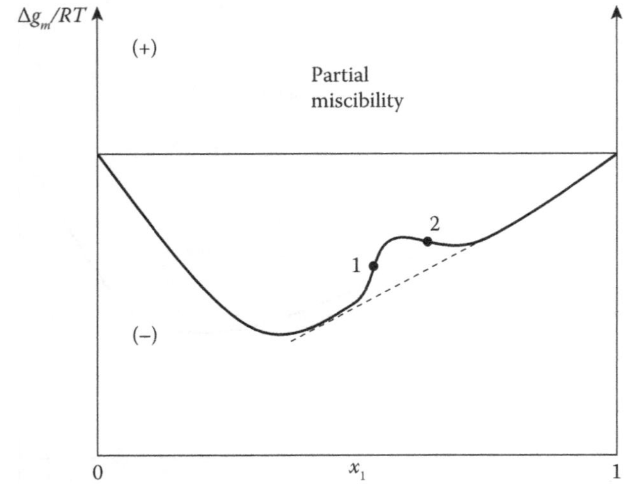

FIGURE 17.2 System showing instability gap *k*.

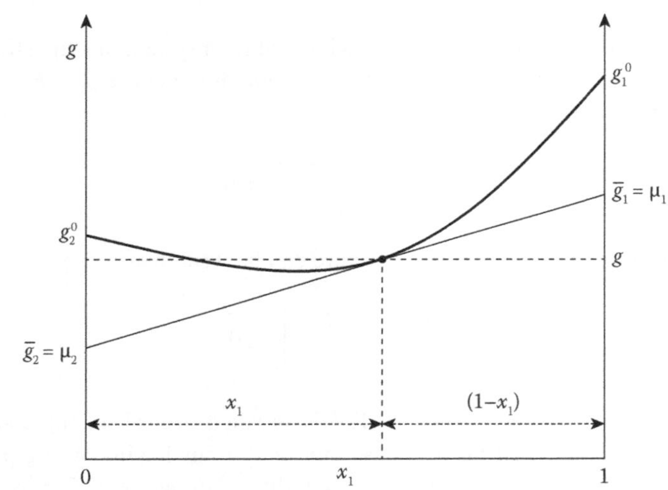

FIGURE 17.3 Stable binary system.

In Chapter 6, we discussed the relations between molar properties and partial molar properties, taking the molar volume as an example, but observing that the conclusions were also valid for all other molar and partial molar quantities. Hence, by analogy to what was shown in Figure 6.1, Figure 17.3 shows that a tangent line to the curve of g versus x intercepts the vertical axis located at $x_1=0$ at a value equal to the chemical potential of compound 2, and the vertical axis located at $x_1=1$ at a value equal to the chemical potential of compound 1 at the particular composition where it touches the curve. If the curve of g versus x does not have any inflection point, the system presents complete miscibility in all the composition range. In this case, the slope of the curve with respect to x_1 will continuously increase with respect to an increase in x_1 or with respect to an increase of x_2. However, if Equation 17.6 is not satisfied in some composition range, as shown in Figure 17.4, the tangent will touch the curve in two points representing two liquid phases of different compositions having identical values for the chemical potential of compound 1 and also identical values for the chemical potential of compound 2. Thus, if the condition for stability is not satisfied, the mixture will split into two liquid phases of different compositions.

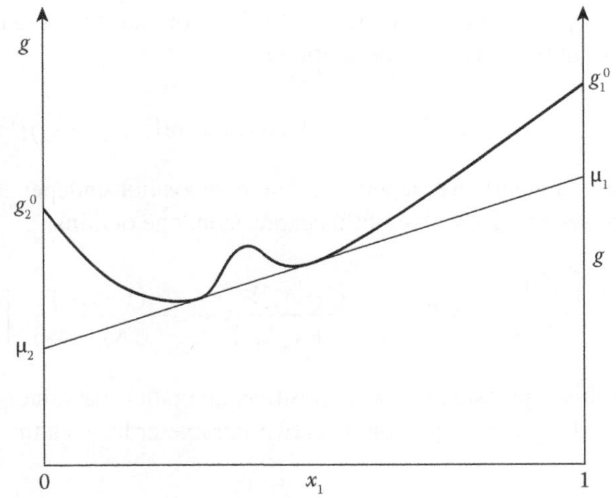

FIGURE 17.4 System showing instability gap.

The points of contact of the tangent line with the curve for the molar Gibbs energy, describing the compositions of the liquid phases at equilibrium, are called binodal points, and the inflection points are called spinodal points. For modeling purposes, it is of interest to relate stability to the excess Gibbs energy in the Lewis convention. From Equations 17.4b and 17.5, we conclude that to show *instability* in some range of composition, the following condition must be met:

$$\frac{\partial^2 \Delta g_m}{\partial x^2} = \frac{\partial^2 \Delta g_m^{id,0}}{\partial x^2} + \frac{\partial^2 g^E}{\partial x^2} < 0$$

or

$$\frac{\partial^2 g^E}{\partial x^2} < -\frac{\partial^2 \Delta g_m^{id,0}}{\partial x^2}$$

Taking the second derivative of the contribution of the ideal mixture, Equation 15.19, with respect to either x_1 or x_2, we obtain:

$$\frac{\partial^2 \Delta g_m^{id,0}}{\partial x^2} = \frac{RT}{x_1 x_2}$$

As the right-hand side of this expression is always positive, we confirm that the ideal mixture is stable in all of the composition range. The most interesting conclusion, however, is obtained when we combine this later result with that obtained for the second derivative of the excess Gibbs energy and write that for *instability*,

$$\frac{\partial^2 g^E}{\partial x^2} < -\frac{RT}{x_1 x_2} \tag{17.7}$$

Equation 17.7 shows that in order to represent liquid–liquid separation in a binary system, that is, in order to represent *instability*, an excess Gibbs energy function must have a negative value of its second derivative *in at least part of the composition range*. This is equivalent to say that in some part of the composition range, excess Gibbs energy function must present two inflection points in which its derivative changes sign. Notably, the two-parameter expression of Wilson [2], which is one of the most successful for the representation of vapor–liquid equilibrium, fails to meet this condition, and thus it is unable to represent liquid–liquid equilibrium. In fact, for a binary system, the Wilson equation for the excess Gibbs energy can be written as:

$$g^E{}_{Wilson} = -RT\left[x_1 \ln(x_1 + x_2\Lambda_{12}) + x_2 \ln(x_1\Lambda_{21} + x_2)\right]$$

where Λ_{12} and Λ_{21} are temperature-dependent but composition-independent adjustable binary parameters. Taking the second derivative of this expression, one obtains:

$$\frac{\partial^2\left[g^E\right]_{Wilson}}{\partial x^2} = RT\left[x_1\frac{(1-\Lambda_{12})^2}{(x_1 + x_2\Lambda_{12})^2} + x_2\frac{(1-\Lambda_{21})^2}{(x_1\Lambda_{21} + x_2)^2}\right]$$

The right-hand side of this expression is always positive, no matter the values of the binary parameters Λ_{12} and Λ_{21}. Hence, the Wilson equation in its two-parameter form cannot satisfy Equation 17.7.

CONDITION FOR STABILITY OF LIQUID MIXTURES

Particularly useful forms of the stability condition are those written in terms of the fugacities of the compounds. For clarity, before giving the final forms, we present three alternative formulations of the Gibbs–Duhem equation for a binary liquid mixture under isothermal conditions. From Equation 14.19, for this case, we write:

$$x_1 d \ln f_1 + x_2 d \ln f_2 = 0$$

and for an infinitesimal change in composition of the mixture,

$$x_1\frac{d \ln f_1}{dx_1} + x_2\frac{d \ln f_2}{dx_1} = 0 \tag{14.19a}$$

Replacing x_2 by $(1-x_1)$ and rearranging, we write:

$$x_1\left[\frac{d \ln f_1}{dx_1} - \frac{d \ln f_2}{dx_1}\right] = \frac{d \ln f_2}{dx_2} \tag{14.19b}$$

while replacing x_1 by $(1-x_2)$ in Equation 14.19a, we obtain:

$$x_2\left[\frac{d \ln f_1}{dx_1} - \frac{d \ln f_2}{dx_1}\right] = \frac{d \ln f_1}{dx_1} \tag{14.19c}$$

Taking the derivative of Equation 17.3 with respect to x_1,

$$\frac{d(\Delta g_m / RT)}{dx_1} = (\ln f_1 - \ln f_2) + \left(x_1\frac{d \ln f_1}{dx_1} + x_2\frac{d \ln f_2}{dx_1}\right) - \left(\ln f_1^0 - \ln f_2^0\right)$$

We recognize here that the second bracket of the right-hand side is null, according to Equation 14.19a above, and the third bracket is a constant, independent of composition. Thus, for the second derivative, the only important term is the first bracket. Hence,

$$\frac{d^2(\Delta g_m / RT)}{dx_1^2} = \frac{d \ln f_1}{dx_1} - \frac{d \ln f_2}{dx_1}$$

Comparison of this result with inequality (Equation 17.5) shows that the *condition for stability* can be written as:

$$\left(\frac{d\ln f_1}{dx_1} - \frac{d\ln f_2}{dx_1}\right) > 0 \tag{17.8a}$$

or, equivalently, as $dx_1 = -dx_2$,

$$\left(\frac{d\ln f_1}{dx_1} + \frac{d\ln f_2}{dx_2}\right) > 0 \tag{17.8b}$$

Combining the inequality in Equation 17.8a with Equations 14.19b and 14.19c of the Gibbs–Duhem equation written above, for stability it is necessary that:

$$\frac{d\ln f_1}{dx_1} > 0 \tag{17.9}$$

and

$$\frac{d\ln f_2}{dx_2} > 0 \tag{17.10}$$

According to the inequalities in Equations 17.9 and 17.10, in a stable liquid mixture the fugacity of a compound always increases with an increase in its concentration.

AZEOTROPY

For the purposes of this discussion, we start with Equation 16.4, which is valid at moderate conditions of temperature and pressure. Adding the equations for both compounds in a binary mixture, we obtain:

$$P = x_1\gamma_{1,0}P_i^s + x_2\gamma_{2,0}P_2^s \tag{17.11}$$

For an ideal mixture, then,

$$P^{id} = x_1P_i^s + x_2P_2^s = P_2^s + \left(P_1^s - P_2^s\right)x_1$$

Thus, at constant temperature, in a plot of P versus x_1, the equilibrium pressure of an ideal mixture will be a straight line. On the other hand, if the activity coefficients are larger than unity, that is, for positive deviations from ideality, the equilibrium pressure will be a curve above the straight line. If the activity coefficients are smaller than unity, that is, negative deviations from ideality, the equilibrium pressure will be a curve below the straight line. In an ideal solution, the different kinds of molecules do not distinguish each other, and they mix without heat effects or volume change. In the case of positive deviations, the two compounds attract each other with forces that are weaker than those existing in an ideal solution. Similarly, in the case of negative deviations, the two compounds attract each other with forces that are stronger than those existing in an ideal solution. There are extreme cases in which at constant temperature the equilibrium pressure of the mixture passes through a maximum for positive deviations or a minimum for negative deviations. The composition at which this extremum occurs is called the *azeotropic point* of the mixture. An azeotropic mixture behaves as a pure compound and at constant pressure boils at constant temperature. From Equation 14.13, for the vapor phase of a binary mixture in an isothermal system,

$$y_1 d \ln f_1^V + y_2 d \ln f_2^V = \frac{v_m^V}{RT} dP$$

or for an incremental change of composition in the liquid phase,

$$y_1 \frac{d \ln f_1^V}{dx_1} + y_2 \frac{d \ln f_2^V}{dx_1} = \frac{v_m^V}{RT} \frac{dP}{dx_1}$$

Similarly, for the liquid phase,

$$x_1 \frac{d \ln f_1^L}{dx_1} + x_2 \frac{d \ln f_2^L}{dx_1} = \frac{v_m^L}{RT} \frac{dP}{dx_1}$$

Although for the liquid phase the right-hand side can be considered null and neglected, we keep the term here, as this simplification is not applicable for the gas phase, and it makes no difference for the treatment that follows. Because at equilibrium the fugacities of a compound in the vapor and liquid phases for each compound are equal, we omit the superscripts *V* and *L*. Thus, taking the difference between these two equations, we write:

$$(y_1 - x_1) \frac{d \ln f_1}{dx_1} + (y_2 - x_2) \frac{d \ln f_2}{dx_1} = \left(\frac{\Delta v}{RT} \right) \frac{dP}{dx_1}$$

As the mole fractions in both the liquid phase and the vapor phase add to unity, we express the mole fractions of compound 2 in terms of compound 1 and write:

$$(y_1 - x_1) \left[\frac{d \ln f_1}{dx_1} - \frac{d \ln f_2}{dx_1} \right] = \left(\frac{\Delta v}{RT} \right) \frac{dP}{dx_1}$$

At a maximum or a minimum of the equilibrium pressure, the term on the right-hand side is equal to zero, and according to Equation 17.8a, the square bracket on the left-hand side is always positive for a stable liquid mixture; thus, necessarily,

$$y_1^{az} = x_1^{az}$$

As the composition of the vapor phase becomes equal to the composition of the liquid phase, at the azeotropic point the mixture behaves as a pure compound and boils at a constant temperature. In fact, the word *azeotrope* comes from the Greek term (*zeo*="to boil," *trope*="something changed," a="no," which means "no change on boiling"), precisely indicating this behavior. A liquid mixture that has an azeotropic point but its initial composition is different from the azeotropic composition will evolve toward the azeotropic composition as it boils, and once this composition is reached, it will continue boiling at constant temperature as if it were a pure compound. Boiling at constant pressure, a maximum-pressure azeotropic mixture has a boiling point that is lower than the boiling points of the pure compounds, that is, it presents a minimum in the representation of the boiling temperature versus composition. The reverse is obviously true for a mixture presenting a minimum-pressure azeotropic behavior.

ACTIVITY COEFFICIENTS AND AZEOTROPY

At the azeotropic point, as the composition of the vapor phase is the same as the composition of the liquid phase, we write Equation 16.4 as:

$$P^{az} = \gamma_{i,0}^{az} P_i^s$$

or

$$\gamma_{i,0}^{az} = \frac{P^{az}}{P_i^s}$$

Thus, for a binary system that presents an azeotropic point,

$$\ln \frac{\gamma_{1,0}^{az}}{\gamma_{2,0}^{az}} = \ln \frac{P_2^s}{P_1^s}$$

As discussed after the presentation of Equation 16.23, the term $\ln(\gamma_{1,0}/\gamma_{2,0})$ has the value $\gamma_{1,0}^{\infty}$ at $x_1 = 0$ and $\left[-\ln \gamma_{2,0}^{\infty}\right]$ at $x_1 = 1$. Thus, to have an azeotropic point, the logarithm of the ratio of the vapor pressures of the compounds should be within these limits. If it happens that

$$-\ln \gamma_{2,0}^{\infty} \leq \ln \frac{P_2^s}{P_1^s} \leq \ln \gamma_{1,0}^{\infty} \tag{17.12}$$

the binary system will present an azeotropic point. For systems presenting negative deviations from ideality, the inequality signs get reversed.

From a thermodynamic point of view, azeotropic behavior is due to a combination of nonideality of the liquid mixture and nearly equal pure compound vapor pressure values (or boiling point temperatures) of the components of the mixture.

HETEROAZEOTROPES

When two compounds are partially miscible or almost immiscible in the liquid phase, vapor–liquid equilibrium is established between the two liquid phases, say phase α rich in compound 1 and phase β rich in compound 2, and a single vapor phase. The heterogeneous liquid phase will then be in equilibrium with a vapor phase of fixed composition and will boil at a constant temperature that is lower than the boiling points of either of the compounds. This fact is in agreement with the behavior of homoazeotropes originated by positive deviations from ideality in Lewis's sense. At equilibrium, the fugacity of each of the compounds in the vapor phase will be equal to its fugacity in the liquid phase in which it is diluted, and to its fugacity in the liquid phase in which it is concentrated. For convenience, we choose the latter. In phase α, the mole fraction and the activity coefficient of compound 1 are both close to unity, and the same is true for compound 2 in phase β. Thus, using Equation 16.2a, we write:

$$y_1^{az} P^{az} = x_1^{\alpha} \gamma_1^{\alpha} \left(P_1^s\right)' = \left(1 - x_2^{\alpha}\right) \gamma_1^{\alpha} \left(P_1^s\right)'$$

and

$$y_2^{az} P^{az} = x_2^{\beta} \gamma_2^{\beta} \left(P_2^s\right)' = \left(1 - x_1^{\beta}\right) \gamma_2^{\beta} \left(P_2^s\right)'$$

On the right-hand side of both equations above, we have opted for expressing the composition of each phase in terms of the solubility of compound 1 in phase β, x_1^{β}, and the solubility of compound 2 in phase α, x_2^{α}. Adding over both compounds,

$$P^{az} = \left(1 - x_2^{\alpha}\right) \gamma_1^{\alpha} \left(P_1^s\right)' + \left(1 - x_1^{\beta}\right) \gamma_2^{\beta} \left(P_2^s\right)' \tag{17.13}$$

Hence,

$$y_1^{az} = \frac{\left(1-x_2^\alpha\right)\gamma_1^\alpha\left(P_1^s\right)'}{\left(1-x_2^\alpha\right)\gamma_1^\alpha\left(P_1^s\right)' + \left(1-x_1^\beta\right)\gamma_2^\beta\left(P_2^s\right)'}$$

or

$$y_1^{az} = \frac{\left(P_1^s\right)'}{\left(P_1^s\right)' + \left[\left(1-x_1^\beta\right)\gamma_2^\beta / \left(1-x_2^\alpha\right)\gamma_1^\alpha\right]\left(P_2^s\right)'} \tag{17.14}$$

with a similar expression for y_2^{az}.

The assumption of total immiscibility, that is, $x_2^\alpha = 0$ and $x_1^\beta = 0$, provides first approximations that usually are quite close to the final results:

$$P^{az} \approx P_1^s + P_2^s$$

and

$$y_i^{az} \approx \frac{P_i^s}{P_1^s + P_2^s}$$

Considering these two approximate expressions, it is easy to see that as the vapor pressures are a function of temperature, the composition of the azeotropic vapor phase is a function of the boiling temperature of the heteroazeotrope. In addition, it is also seen that the boiling temperature of the heteroazeotrope can be obtained using expressions for the temperature dependence of the vapor pressures of the pure compounds and iterating values of temperature until the azeotropic equilibrium pressure equals the prevailing pressure of the system. The normal boiling point of the azeotropic mixture is obtained when the prevailing pressure is set equal to 1 atm. Obviously, with the appropriate correction factors, these conclusions are also applicable to Equations 17.13 and 17.14.

PHASE EQUILIBRIUM DIAGRAMS

Although with the advent of computers graphical methods have lost their importance as tools to solve problems, plots retain their educational value. Good figures help us to think, retain concepts, and even program a computer. For binary vapor–liquid equilibrium, typical diagrams depict in a single figure the curves of the equilibrium pressure P versus x_1 and versus y_1, at a fixed temperature T, or the curves of the equilibrium temperature T versus x_1 and versus y_1 at a fixed pressure P. A typical vapor–liquid equilibrium diagram for the isothermal case is presented in Figure 17.5, and the isobaric case is shown in Figure 17.6.

For a mixture of overall composition z_1, points b and d are the bubble and dew points, respectively. In both figures, the solid line is the bubble line, that is, the line joining the bubble points, and the segmented line is the dew line, that is, the line joining the dew points. For pressures above the bubble line in Figure 17.5, the mixture is in liquid phase, and for pressures below the dew line, the mixture is in vapor phase.

Decompressing at constant temperature a liquid mixture of overall composition z_1, the bubble point pressure is the value of the pressure at which the first bubble of vapor appears. A further decrease in pressure will cause an increase in the vapor phase with the corresponding decrease of the liquid phase. At the dew point, the last drop of liquid phase disappears. Starting with the vapor phase and increasing the pressure at constant temperature, at the dew point the first drop of liquid appears, and at the bubble point, the last bubble of vapor disappears.

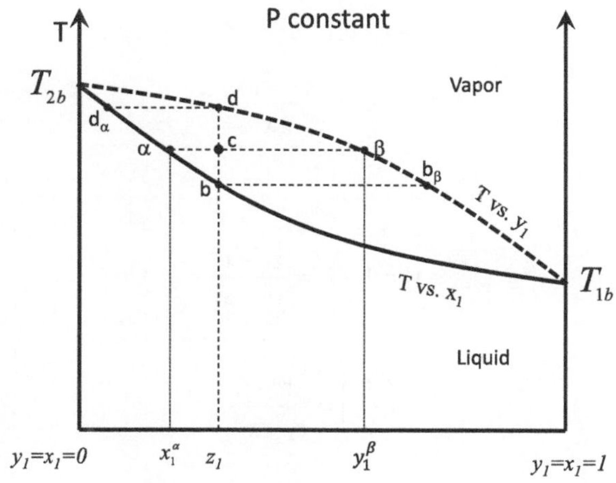

FIGURE 17.5 Isothermal vapor–liquid equilibrium diagram.

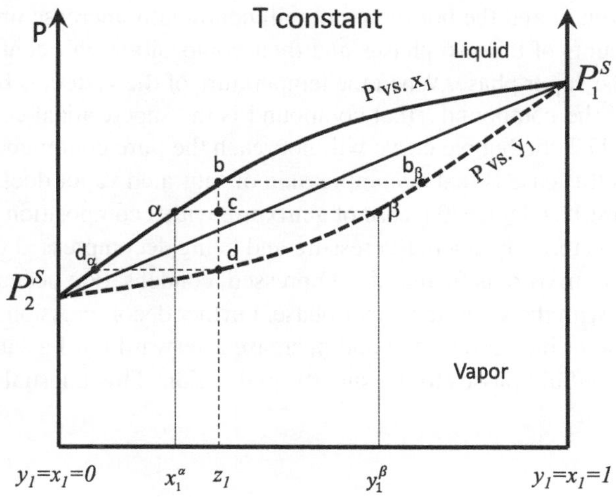

FIGURE 17.6 Isobaric vapor–liquid equilibrium diagram.

Similarly, for temperatures below the bubble line in Figure 17.6, the mixture is in liquid phase, and for temperatures above the dew line, the mixture is in vapor phase. Changes in temperature at constant pressure will result in the appearance or disappearance of one of the phases at the dew and bubble points. The horizontal line joining the equilibrium compositions of coexisting liquid and vapor phases at fixed temperature and pressure is called a tie line. Thus, in Figures 17.5 and 17.6, the line $b - b_\beta$ is the bubble point tie line and the line $d_\alpha - d$ is the dew point tie line for a mixture of overall composition z_1. At an intermediate point between the bubble point and the dew point of a mixture, such as point c in the figures, a saturated liquid phase coexists at equilibrium with a saturated vapor phase. Taking as a basis of calculation 1 mole of mixture of composition z_1, at point c the system will be formed by L moles of liquid phase of composition x_1^α and V moles of vapor phase of composition y_1^β with $(L+V)=1$. Since in 1 mole of mixture of composition z_1 there are z_1 moles of compound 1, by material balance of compound 1,

$$z_1 = Vy_1^\beta + Lx_1^\alpha = (1-L)\,y_1^\beta + Lx_1^\alpha$$

or

$$L = \frac{y_1^\beta - z_1}{y_1^\beta - x_1^\alpha}$$

Hence,

$$V = 1 - L = \frac{z_1 - x_1^\alpha}{y_1^\beta - x_1^\alpha}$$

$$\left(\frac{L}{V}\right) = \frac{y_1^\beta - z_1}{z_1 - x_1^\alpha} \tag{17.15}$$

As can be seen in Figures 17.5 and 17.6, the relative amounts of phases L and V are inversely proportional to the segments of the tie line joining the point at the mixture composition z_1 with the corresponding equilibrium compositions x_1^α and y_1^β. This is the so-called *lever rule*. If the temperature of the mixture is increased at a constant pressure P, as shown in Figure 17.6, the liquid will start boiling at point b and the boiling point will continually increase until point d is reached. Both the relative amounts of the two phases and their composition will continuously change until all the mixture is in the vapor phase. When the temperature of the system is higher than the critical temperature of one of the compounds, that compound is in "supercritical condition." In this case, as depicted in Figure 17.7, the bubble curve will not reach the pure compound 1 axis. One interesting effect observed in this case is that decompression of saturated vapor does not always produce a superheated gas. Figure 17.7 depicts the case of a mixture with a composition slightly to the right of critical point c, which is initially at a high pressure and being decompressed until a saturated vapor phase is reached. If the mixture is further decompressed, condensation occurs, and a saturated liquid phase will coexist with the saturated vapor phase. Further decompression will cause the relative amount of liquid phase to increase at first and decrease afterward until a saturated vapor phase is reached again. Then the fluid passes to the superheated region. This unusual behavior is known as retrograde condensation.

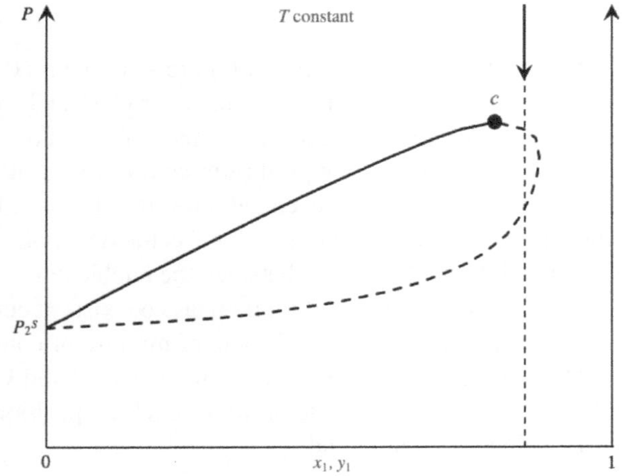

FIGURE 17.7 Isothermal vapor–liquid equilibrium at $T > T_c$ of component 1.

In Figures 17.5 and 17.6, we observe that the compositions of the liquid and vapor phases in equilibrium are never equal to each other. Not all binary mixtures behave this way. As discussed before, binary mixtures having strong positive deviations from ideality in Lewis's sense may present a maximum in the curve of P versus x_1 at a fixed temperature T. At this maximum pressure, $y_1 = x_1$, the "azeotropic point," the mixture boils at constant pressure and temperature, as if it were a pure compound. Figures 17.8 and 17.9 illustrate maximum boiling pressure (minimum boiling temperature) behavior. Similarly, binary mixtures having strong negative deviations from ideality in Lewis's sense, that is, mixtures of compounds forming either a hydrogen bond or an electron donor–acceptor complex, may present a minimum in the equilibrium pressure at constant temperature, that is, a maximum boiling temperature at constant pressure.

Liquid mixtures exhibiting strong positive deviations from ideality in Lewis's sense eventually may show (liquid–liquid) phase separation. As liquid–liquid equilibrium concentrations are a weak

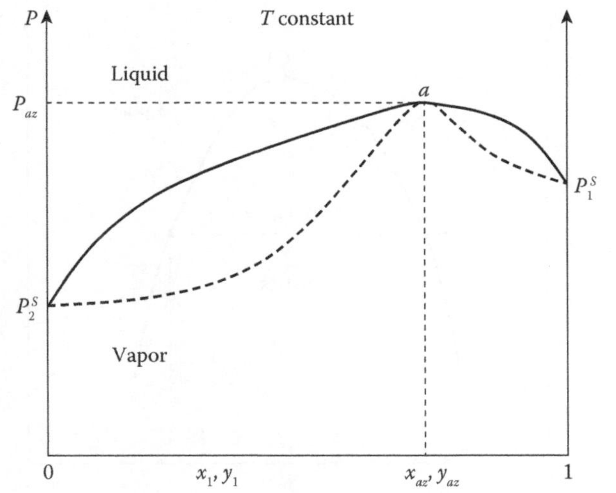

FIGURE 17.8 Isothermal azeotropic system.

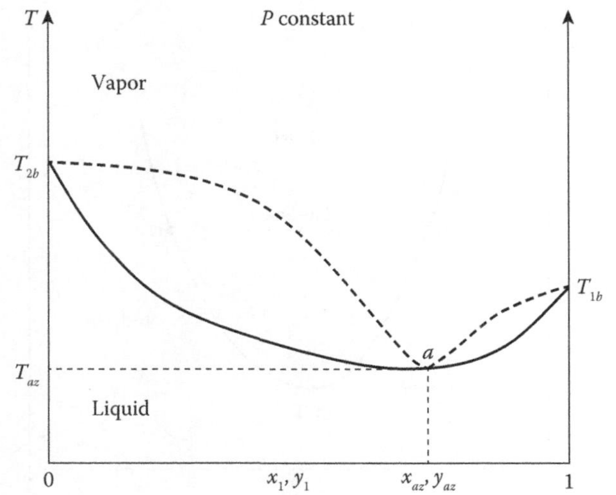

FIGURE 17.9 Isobaric azeotropic system.

function of pressure, temperature versus mole fraction, $T-x$, diagrams are preferred to $P-x$ diagrams. In this case, the equilibrium pressure is seldom of concern.

Figure 17.10 shows the binodal curve, that is, the mutual solubility curve joining the binodal points described in Figure 17.4, as a function of temperature for the most common case of a system presenting an upper consolute or upper critical solution temperature (UCST). For this kind of system, the compounds are totally miscible at high temperatures, and at low temperatures two liquid phases coexist at equilibrium. The less common case of a system showing a lower critical solution temperature (LCST) is shown in Figure 17.11. Some rare systems, like the nicotine–water system, present a closed solubility loop exhibiting a lower and a UCST. At the critical point, the second and third derivatives of the molar Gibbs energy with respect to the mole fraction are equal to zero. Similarly to vapor–liquid equilibrium, for liquid–liquid equilibrium of a mixture of an overall composition z_1, the amounts L_α and L_β of the phases present are in the proportion given by the lever rule.

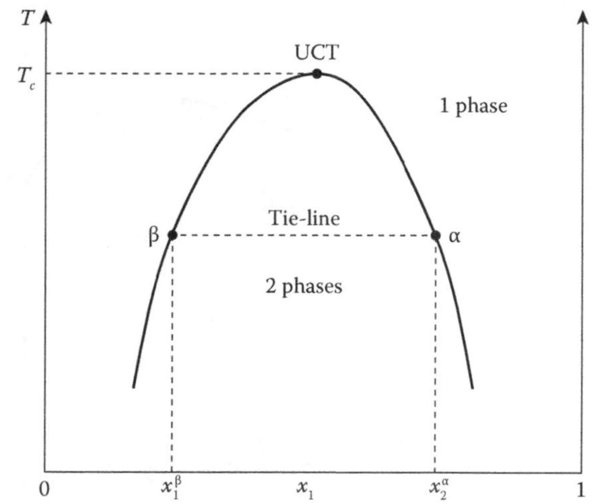

FIGURE 17.10 Two-phase isobaric system with UCST.

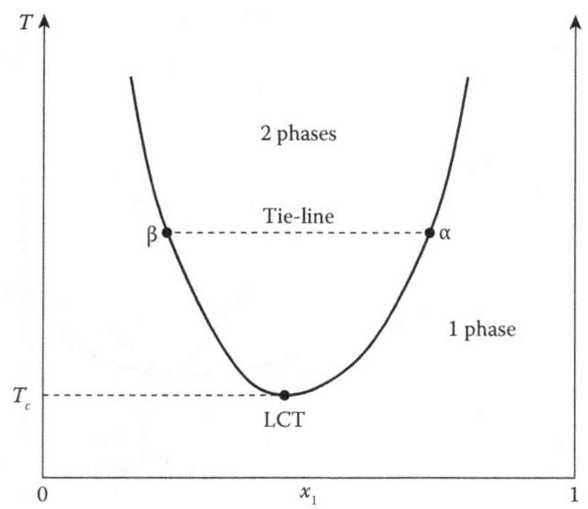

FIGURE 17.11 Two-phase isobaric system with LCST.

The solubilities x_1^β and x_2^α are fixed once the temperature and the pressure are fixed. If a vapor phase coexists with the two liquid phases, fixing the temperature or the pressure, according to the phase rule, all other variables are fixed.

Figure 17.12 depicts the case of two liquid phases in equilibrium with a vapor phase at a temperature T and equilibrium pressure P_{eq}, which is of interest in steam distillation. At equilibrium, the compositions of the three phases are fixed at x_1^β, x_2^α, and y_1^{az}. If this three-phase system exchanges heat with its surroundings, neither the temperature, the pressure, nor the compositions of the phases will change while there are three phases present. Only the relative amount of the phases will change until one of the phases disappears. This state of the system is called a heteroazeotropic point. If at constant pressure the temperature increases, one liquid phase will disappear, and if the temperature decreases, the vapor phase will disappear. The vapor phase in equilibrium with two liquid phases is always an azeotropic phase, in the sense that its composition will not change until one of the phases disappears. For all practical purposes, a diagram at constant pressure of two solid solutions S_α and S_β in equilibrium with a liquid phase L looks identical to Figure 17.12. For this case, the point of coexistence of the three phases is called a eutectic point instead of an azeotropic point.

Figure 17.13 shows the case of solid–liquid equilibrium when an intermediate compound (1–2) is formed from pure compounds 1 and 2.

The case presented in Figure 17.13 considers that there are no solid solutions of either compound 1 or compound 2 with the intermediate compound (1–2). In cases of having more than one intermediate compound, the diagram becomes more complex.

In all the cases discussed above, we have considered positive deviations from ideality in Lewis's sense. These are by far the most common cases found in practice. Figures 17.14 and 17.15 present typical P versus x_1 diagrams for negative deviations with no minimum-pressure azeotrope and with a minimum-pressure azeotrope, respectively. Binary systems formed by two volatile compounds showing negative deviations from ideality in Lewis's sense never split into two liquid phases. If the solute is nonvolatile, at high concentration it may saturate the solution and separate into a solid phase.

For ternary systems, triangular diagrams are used. These diagrams are particularly useful in cases of liquid–liquid equilibria of two highly immiscible solvents containing a common solute.

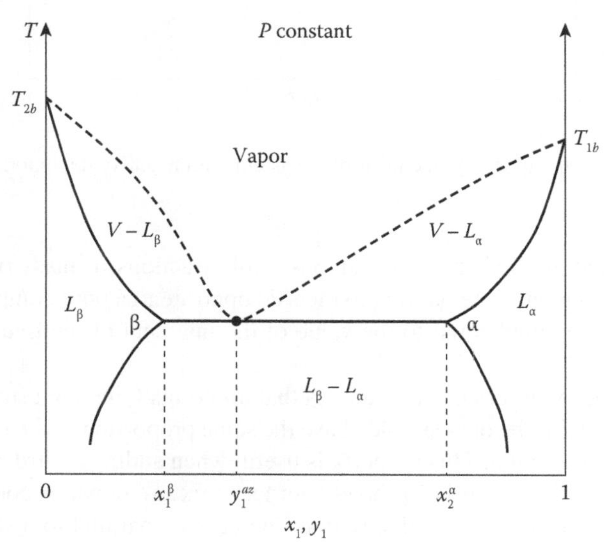

FIGURE 17.12 Isobaric heteroazeotrope.

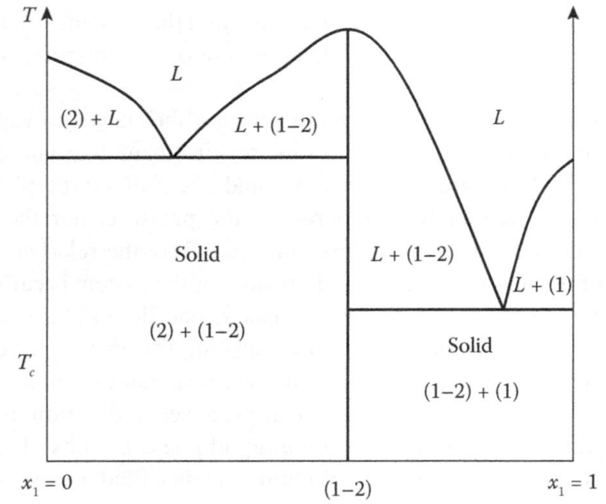

FIGURE 17.13 Isobaric solid–liquid equilibrium with the formation of an intermediate.

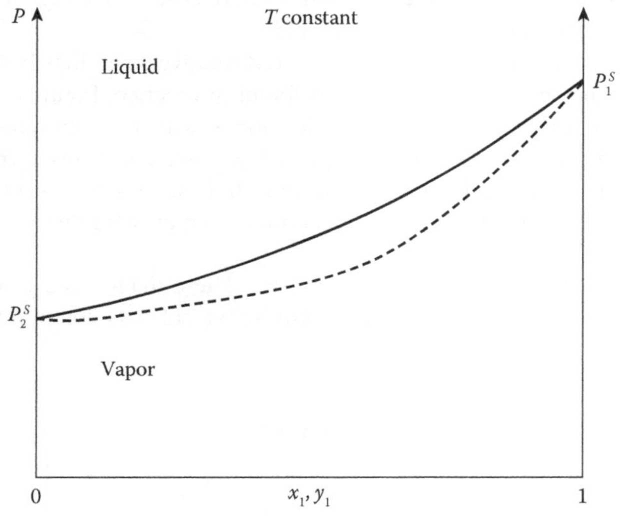

FIGURE 17.14 Isothermal vapor–liquid equilibrium system with negative deviations from ideality.

Figure 17.16 shows how to read the compositions (mole fractions or mass fractions) in a ternary diagram. All points lying on a line parallel to the side opposite to a pure compound vertex have the composition in that compound equal to the value of the intercept of the line with the axis of that compound.

An important property of ternary diagrams is that all points lying on a straight line that joins a pure compound vertex with the opposite side have the same proportion of the other two compounds, as shown in the side intercepted. This property is useful when adding a third compound to a binary mixture. The dotted line in Figure 17.16 shows a ternary mixture in which compounds B and C are in proportion 1:2. At any point of the diagram, a line passing parallel to a side opposite to a particular vertex will intercept the side corresponding to the compound of that vertex at the compound composition in the mixture.

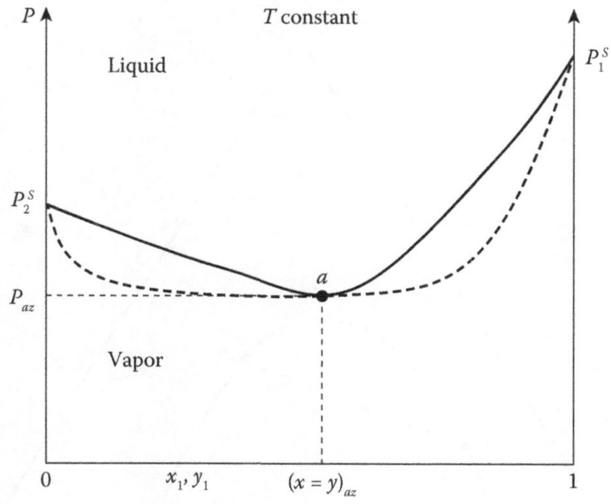

FIGURE 17.15 Isothermal azeotropic vapor–liquid equilibrium system with negative deviations from ideality.

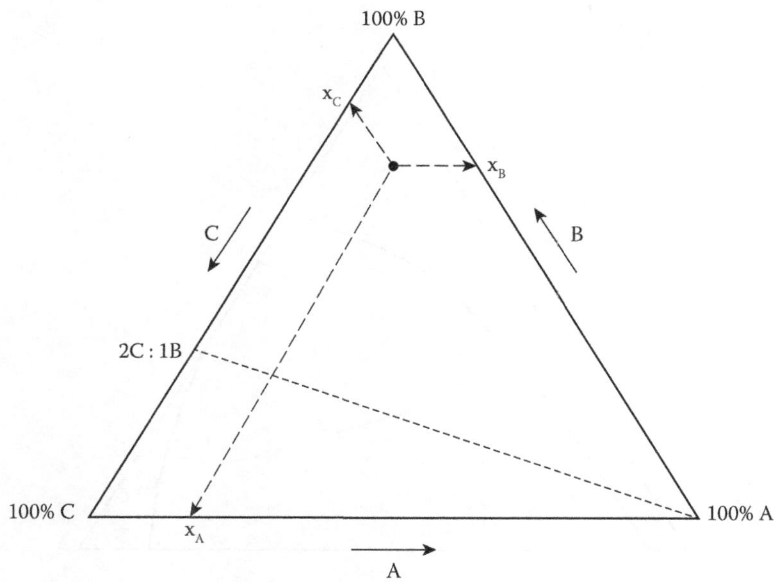

FIGURE 17.16 Reading a triangular diagram.

Figure 17.17 shows the case of solvents A and C, which have an immiscibility gap, and a solute B that is miscible with both solvents in all proportions. The binodal curve and one tie line are shown. Figure 17.18 shows the rare case of two partially miscible pairs.

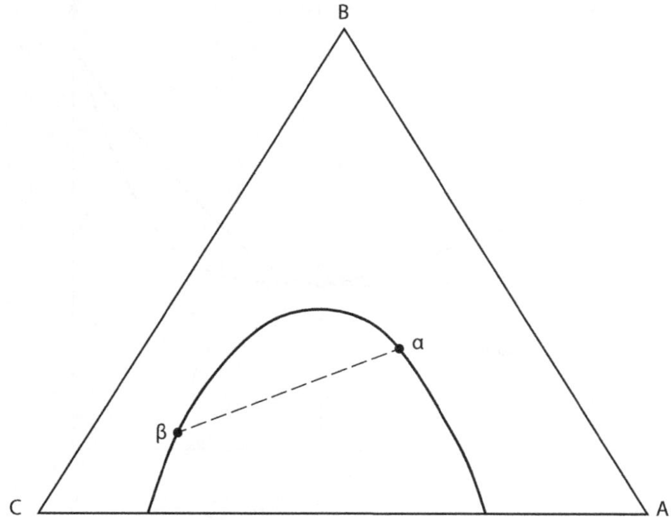

FIGURE 17.17 One immiscibility gap.

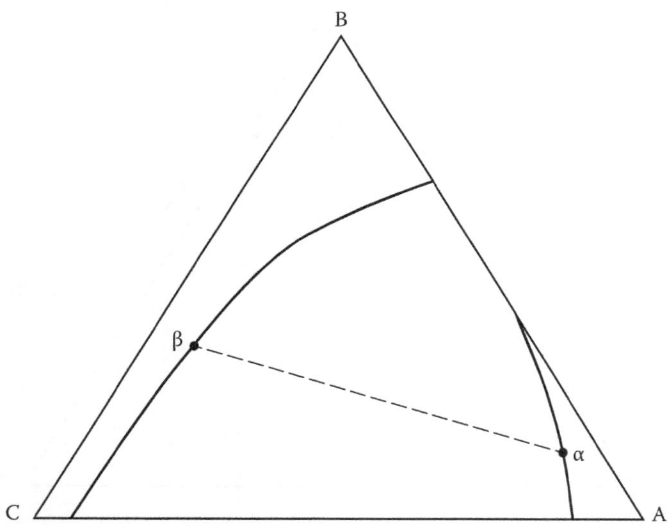

FIGURE 17.18 Two miscibility gaps.

REFERENCES

1. Berg, L. 1969. Selecting the agent for distillation processes. *Chem. Eng. Prog.* 65(9): 52–57.
2. Wilson, G. M. 1964. Vapor-liquid equilibrium. XI. A new expression for the excess free energy of mixing. *J. Am. Chem. Soc.* 86(2): 127–130.

PROBLEMS

17.1 **Stability and phase separation of a binary liquid mixture**

The simplest class of nonideal binary liquid mixtures is the regular solution (two-suffix Margules) model with excess Gibbs energy given by:

$$\frac{g_0^E}{RT} = A_{12}x_1x_2$$

a. What values of A_{12} make the mixture unstable, that is, cause it to split into two liquid phases?
b. Plot the Gibbs energy of mixing versus composition for A_{12} equal to 100%, 150%, and 200% of the instability threshold value. On the graph, locate the compositions of the two phases in equilibrium, and verify your findings with numerical calculations.

17.2 **A ternary azeotropic system**

The ternary system n-hexane (1)+benzene (2)+isopropanol (3) forms a minimum temperature azeotrope of composition $x_1^{az} = 0.6175$, $x_2^{az} = 0.0600$, $x_3^{az} = 0.3225$ at 62.30°C and 760 mmHg [1]. From this data, calculate the constants in the model:

$$\frac{g_0^E}{RT} = A_{12}x_1x_2 + A_{13}x_1x_3 + A_{23}x_2x_3$$

Is the binary system n-hexane+isopropanol azeotropic at this same temperature? If so, calculate the azeotropic pressure and composition. The vapor pressures P^s (atm) for T (K) are given by [1]:

$$\ln P^s = C_1 + \frac{C_2}{C_3 + T} + C_4T + C_5 \ln T$$

	C_1	C_2	C_3	C_4	C_5
(1) n-hexane	113.283	−7151.50	0.0	0.01995	−17.00
(2) benzene	133.313	−8026.29	0.0	0.02393	−20.29
(3) isopropanol	8.719	−1873.00	−140.55	0.0	0.0

17.3 **Liquefaction of air**

In the bestselling science-fiction novel *The Martian* [2], an astronaut is marooned on Mars after his crewmates, believing him dead, leave him behind in their emergency escape from the planet. In order to survive in the harsh Martian environment, the protagonist and narrator must appeal to all his technical expertise and creativity. In one passage of the book, he describes the operation of an atmospheric regulator that controls the composition of the air by freeze separation of the gases: if there is too much oxygen, the air is collected in a tank and cooled to 90 K. This condenses the oxygen to liquid, but leaves the nitrogen still gaseous, because its condensation temperature is 77 K.

Discuss the thermodynamic basis for the above separation process. Draw the phase diagram for vapor–liquid equilibrium of the system N_2 (1)+O_2 (2), assuming that these fluids form ideal mixtures, and use it to describe the changes as gaseous air is cooled to its dew point and below. For simplicity, neglect the possible presence of other gases in the air. The vapor pressures of both components are given in reduced variables $P_r^s = P^s / P_c$, $T_r = T / T_c$ by [3,4]:

$$\ln P_r^s = \frac{a(1-T_r)+b(1-T_r)^{3/2}+c(1-T_r)^m+d(1-T_r)^n}{T_r}$$

	T_c (K)	P_c (bar)	a	b	c	D	m	n
N_2	126.20	33.944	−6.11102	1.2189	−0.69366	−1.89893	2.5	5
O_2	154.58	50.43	−6.28275	1.73619	−1.81349	−0.0253645	3	6

17.4 **Vapor–liquid–liquid equilibrium of two immiscible liquids**

Construct the isothermal $P-x-y$ phase diagram of n-heptane (1)+water (2) at 300 K, assuming that these two liquids are completely immiscible. The vapor pressures are given by [1]:

$$\ln P_r^s = \frac{A(1-T_r)+B(1-T_r)^{1.5}+C(1-T_r)^3+D(1-T_r)^6}{T_r}$$

	T_c (K)	P_c (bar)	A	B	C	D
(1) n-heptane	540.3	27.4	−7.67468	1.37068	−3.53620	−3.20243
(2) water	647.3	221.2	−7.76451	1.45838	−2.77580	−1.23303

17.5 **Ternary liquid–liquid equilibrium**

In Problem 17.2, the parameters $A_{12} = 0.4745$, $A_{13} = 1.949$, $A_{23} = 2.294$ of an excess Gibbs energy model were obtained for the ternary system n-hexane (1)+benzene (2)+isopropanol (3) from its azeotropic composition and pressure at 62.30°C. It was pointed out that the value of A_{23} meant that the benzene+isopropanol binary should exhibit partial miscibility, most likely an unrealistic prediction. Assume nevertheless that a fictitious ternary system exists that is well represented by this model with the above parameters. Construct the phase diagram of the system at 62.30°C and a pressure high enough that no vaporization takes place, that is, the liquid mixture is above its dewpoint pressure at all compositions.

17.6 **Equilibrium between two phases of identical composition**

It is shown in this chapter that when the composition of a binary system in vapor–liquid equilibrium at certain temperature and pressure is the same in both phases, the vapor pressure is a maximum or a minimum at that temperature, and the boiling temperature is a maximum or a minimum at that pressure. Prove that the same is true for the equilibrium between any two phases of identical composition, whatever the number of components and their physical state.

REFERENCES FOR PROBLEMS

1. Govindaswamy, S., Andiappan, A. N., and Lakshmanan, S. M. 1976. Isobaric vapour-liquid equilibrium data for the ternary and sub-binary systems containing n-hexane(1)-benzene(2)-isopropanol(3), *J. Chem. Eng. Jpn* 9(5): 345–349.
2. Weir, A. 2014. *The Martian*, New York: Crown Publishing.
3. Reid, R. C., Prausnitz, J. M., and Poling, B. E. 1987. *The Properties of Gases and Liquids*, 4th ed. New York: McGraw-Hill.
4. Poling B. E., Prausnitz J. M., and O'Connell, J. P. 2001. *The Properties of Gases and Liquids*, 5th ed. New York: McGraw-Hill.

18 The Thermodynamics of Aqueous Electrolyte Solutions

Abstract

It is shown that the activity of water in electrolyte solutions behaves differently and thus it needs to be treated differently than in nonelectrolytes. To see why, get two glasses of water, one with sugar and one with salt, a desk lamp, an extension cord, a knife, pliers, and rubber gloves. With the knife separate the two wires of the extension cord, cut one of them, and peel the ends. Put the gloves on, plug the lamp turned on to the extension cord and the cord to electricity. Immerse the cut ends of the cable first into the water with sugar and then into the water with salt.

As discussed in Chapter 10, when a salt is dissolved in water or in other appropriate solvents, the molecules dissociate into ions. In aqueous solutions, strong electrolytes, that is, those formed from a strong acid neutralized with a strong base, will dissociate almost completely into ions, while weak electrolytes will dissociate only partially. In media of lower dielectric constant than water, such as furfural, acetonitrile, alcohols, chloroacetic acid, dioxane, acetone, and acetic acid, or in their mixtures with water, conductivity measurements show that all electrolytes are increasingly weak; that is, they are partially associated, as the solvent moves down in the scale of dielectric constants. Thus, the classification of strong electrolytes as strong acids, bases, and their salts (chlorides, fluorides, sulfates of sodium, potassium, magnesium, copper, zinc, etc.) is only valid in aqueous solution. On the other hand, weak electrolytes such as acetic acid or chloroacetic acid in concentrated aqueous solutions can associate to such a high degree as to change the properties of water as solvent. The dielectric constant of air is so low that there are no ions present in the vapor phase over a solution of a volatile electrolyte, all molecules are fully associated. In mercury or sodium lamps, ions exist in the vapor phase under a voltage difference and in the absence of air.

BASIC RELATIONS

With these considerations in mind, without specifying the extent of the actual dissociation in aqueous solution, for 1 mole of an electrolyte E that in total dissociation would give ν_+ cations C^{Z+} and ν_- anions A^{Z-}, we write:

$$C_{\nu+}A_{\nu-} = \nu_+ C^{Z+} + \nu_- A^{Z-}$$

For a single electrolyte, the electroneutrality condition reduces to:

$$\nu_+ Z_+ + \nu_- Z_- = 0 \quad (18.1)$$

In this equation, the sign of the charge is implicit in Z_i. For clarity, it is better to have the signs of the charges explicit and write:

$$\nu_+ Z_+ = \nu_- |Z_-| \quad (18.1a)$$

DOI: 10.1201/9781003431985-20

Variations of this relation that are often used in the literature without further explanation may be confusing at first, so we write some of them in detail:

$$\frac{1}{Z_+} = \frac{\nu_+}{\nu_-|Z_-|} = \frac{\nu_+}{\nu_+ Z_+} \tag{18.1b}$$

$$\frac{1}{|Z_-|} = \frac{\nu_-}{\nu_+ Z_+} \tag{18.1c}$$

One important form is obtained by multiplying Equation 18.1a first by $|Z_-|$ to obtain:

$$\nu_+|Z_+Z_-| = \nu_-(Z_-)^2$$

and then multiplying Equation 18.1a by Z_+ to obtain:

$$\nu_+(Z_+)^2 = \nu_-|Z_+Z_-|$$

Taking the difference of these two expressions and rearranging, we get:

$$|Z_+Z_-| = \frac{\nu_+(Z_+)^2 + \nu_-(Z_-)^2}{\nu} \tag{18.1d}$$

with

$$\nu \equiv \nu_+ + \nu_- \tag{18.2}$$

According to Equation 12.15, the equilibrium constant for the ionic dissociation in terms of activities takes the form:

$$K_T = \frac{a_+^{\nu_+} a_-^{\nu_-}}{a_E} \tag{18.3}$$

For the dissociation of the electrolyte E, according to Equation 12.16, the value of the equilibrium constant is obtained from the standard Gibbs energy change:

$$K_T = \exp\left[\frac{\mu_E^\theta - \nu_+\mu_+^\theta - \nu_-\mu_-^\theta}{RT}\right] \tag{18.4}$$

Some treatments of electrolyte solutions have proposed to use mole fractions as a measure of composition. For all practical purposes, the use of molality is simpler and gives a better range of values. As an example, the solubility of common salt (NaCl) in water at 298 K is 360 g kg^{-1} of water or 6.16 moles per 55.51 moles of water. Thus, at saturation, that is, the maximum concentration of salt possible at this temperature, the mole fraction of each ion is 0.100, while the molality is 6.16. Having decided to use molality as the measure of concentration, the next step is to choose the standard states for the activity coefficients to be used in the evaluation of the activities by:

$$a_i = \tilde{m}_i \gamma_i = \nu_i \tilde{m} \gamma_i \tag{18.5}$$

Thus, for the cation,

$$a_+ = \tilde{m}_+ \gamma_+ = \nu_+ \tilde{m} \gamma_+ \tag{18.5a}$$

and for the anion,

$$a_- = \tilde{m}_-\gamma_- = \nu_-\tilde{m}\gamma_- \tag{18.5b}$$

In this expression, as the activities are dimensionless, $\tilde{m}_i$ is the dimensionless molality of the ion i and $\tilde{m}$ is the dimensionless molality of the electrolyte solute; that is, the value of the molality divided by 1 [mole of i/1 kg of solvent].

Similarly, to the case of the use of molality for nonelectrolytes discussed in Chapter 15, the reference state for the activity coefficient of the ions is their state at infinite dilution, and their standard state is the ideal solution in Henry's sense at 1[mole of i/1 kg of solvent]. At the reference state, the activity coefficient of an ion is normalized to unity:

$$\lim_{m_i \to 0} \gamma_i = 1 \tag{18.6}$$

MEAN IONIC ACTIVITY COEFFICIENT

In a single salt solution, at the standard state, the activity of an ion is equal to unity (dimensionless). This is so because in its standard state, the ion is in an ideal solution at unit molality. The normalization of the activity coefficients of the ions to unity at their state in an infinitely dilute solution is of great importance. At this state, the presence of any other ion is immaterial, be it a co-ion or a counterion. Thus, the same condition is valid independently of the nature of the electrolyte generating the ion. With this normalization, although its value is not known, the standard state potential of an ion in solution is fixed and well defined, and it is independent of where the ion came from. The standard state for the electrolyte is chosen so that the constant K_T in Equation 18.4 is equal to unity:

$$\mu_E^\theta = \nu_+\mu_+^\theta + \nu_-\mu_-^\theta$$

Again here, the value of μ_E^θ for the electrolyte is not known, but we know that for each electrolyte it has a fixed and well-defined value depending only on the temperature and the pair of ions forming the electrolyte. Hence, from Equation 18.3 we write:

$$a_E = a_+^{\nu_+} a_-^{\nu_-} = \left(\tilde{m}_+\gamma_+\right)^{\nu_+} \left(\tilde{m}_-\gamma_-\right)^{\nu_-} = \tilde{m}^{\nu}\gamma_\pm^{\nu}\left(\nu_+^{\nu_+}\nu_-^{\nu_-}\right) \tag{18.7}$$

where the mean ionic activity coefficient of the electrolyte $\gamma_\pm$ is defined as:

$$\gamma_\pm^{\nu} \equiv \gamma_+^{\nu_+}\gamma_-^{\nu_-} \tag{18.8}$$

with $\nu \equiv \nu_+ + \nu_-$, as defined by Equation 18.2.

At infinite dilution, by normalization of the activity coefficients of the ions, we have:

$$\lim_{m \to 0} \gamma_\pm = 1 \tag{18.9}$$

OSMOTIC COEFFICIENT

For a single electrolyte aqueous solution at constant temperature, neglecting pressure effects, the Gibbs–Duhem equation, which relates the changes in a_E with the changes in the activity of water a_W, takes the form:

$$n_E d\ln a_E + n_W d\ln a_W = 0$$

where

$n_E = \tilde{m}$ is the number of moles of salt, and the number of moles of water is:

$$n_W = \frac{1000\,(\mathrm{g})}{M_W^{cgs}\,(\mathrm{g/mol})}$$

in cgs system of units.

In the SI units n_W takes the form:

$$n_W = \frac{1\,(\mathrm{kg})}{M_W^{SI}\,(\mathrm{kg/mol})}$$

Thus,

$$\widetilde{m}\,d\ln a_E + \frac{1000}{M_W^{cgs}}d\ln a_W = 0 \quad \text{or} \quad \widetilde{m}\,d\ln a_E + \frac{1}{M_W^{SI}}d\ln a_W = 0 \tag{18.10}$$

The activity of water is sometimes given in terms of the osmotic coefficient of the solution, which is defined as:

$$\varphi \equiv -\frac{1000}{M_W^{cgs}\sum_j(\nu\tilde{m})_j}\ln a_W \quad \text{or} \quad \varphi \equiv -\frac{1}{M_W^{SI}\sum_j(\nu\tilde{m})_j}\ln a_W \tag{18.11}$$

For an aqueous solution of a nonvolatile electrolyte, the activity of water is obtained directly by measuring the vapor pressure, P, of the solution at the temperature of interest. Then, from Equation 16.4:

$$a_W = x_W\gamma_W = \frac{P}{P_W^S} \tag{18.12}$$

Here, P_W^S is the vapor pressure of pure water at the temperature of the system. For work at high pressure, the correction factors appearing in Equation 16.2a should be included in Equation 18.12. It is of interest to obtain the relations between the osmotic coefficient of a single electrolyte solution, φ, and the mean ionic activity coefficient of the electrolyte. For a single electrolyte in solution, Equation 18.11 takes the form:

$$\ln a_W \equiv -\frac{M_W^{cgs}\nu\tilde{m}}{1000}\varphi \quad \text{or} \quad \ln a_W \equiv -M_W^{SI}\nu\tilde{m}\varphi \tag{18.11a}$$

Then,

$$d\ln a_W \equiv -\frac{M_W^{cgs}\nu\tilde{m}}{1000}d\varphi - \frac{M_W^{cgs}\nu}{1000}\varphi\,d\tilde{m} \quad \text{or} \quad d\ln a_W \equiv -M_W^{SI}\nu\tilde{m}\,d\varphi - M_W^{SI}\nu\varphi\,d\tilde{m}$$

Combining this expression with the Gibbs–Duhem equation, Equation 18.10, we get:

$$d\ln a_E = \nu d\varphi + \nu\,\varphi\frac{d\tilde{m}}{\tilde{m}}$$

From Equation 18.7,

$$d\ln a_E = \nu\frac{d\tilde{m}}{\tilde{m}} + \nu d\ln \nu_{\pm}$$

Thus, equating these two expressions and rearranging, the relation between the osmotic coefficient of a single electrolyte solution and the mean ionic activity coefficient takes the form:

$$d \ln \gamma_{\pm} = d\varphi + \frac{(\varphi - 1)}{\tilde{m}} d\tilde{m} \tag{18.13}$$

This differential relation can be used to obtain the mean ionic activity coefficient in terms of the osmotic coefficient and vice versa. Integrating between the limit at infinite dilution, where the mean ionic activity coefficient and the osmotic coefficient tend to unity, and a molality m,

$$\ln \gamma_{\pm} = (\varphi - 1) + \int_{0}^{\tilde{m}} \frac{(\varphi - 1)}{\tilde{m}} d\tilde{m} \tag{18.14}$$

Because in Equation 18.14 the molality appears as a ratio, for simplicity, the tilde differentiating it from its dimensional value is sometimes dropped. Rearranging Equation 18.13, we write:

$$\tilde{m} d \ln \gamma_{\pm} = \tilde{m} d\varphi + \varphi d\tilde{m} - d\tilde{m} = d(\tilde{m}\varphi) - d\tilde{m}$$

or

$$d(\tilde{m}\varphi) = \tilde{m} d \ln \gamma_{\pm} + d\tilde{m}$$

Integrating between the same limits as before, and rearranging,

$$\varphi = 1 + \frac{1}{\tilde{m}} \int_{0}^{\tilde{m}} \tilde{m} d \ln \gamma_{\pm} \tag{18.15}$$

Again, as the molality appears as a ratio in this expression, sometimes the difference between the molality and its dimensionless form is ignored.

NEED FOR CONSIDERING THE EXISTENCE OF INDIVIDUAL IONS

One natural question arising at this point is why, for the thermodynamic treatment of strong electrolytes, one should go through all the trouble of considering independent ions in solution instead of defining an activity coefficient for the electrolyte by the relation:

$$\hat{a}_E = \tilde{m}\gamma_E$$

The idea here is not to ignore the real existence of ions, which is evident from conductivity measurements, but to explore whether it is absolutely necessary to consider their existence for the thermodynamic treatment when it would be much simpler to consider the electrolyte as a single entity. The answer to this question is found by observing the behavior of the solution of an electrolyte in the very dilute region.

If we ignore the presence of ions and consider only the presence of molecules, from the above relation, in the very dilute region, when $\gamma_E \to 1$ and $\hat{a}_E^{dil} \approx \tilde{m}$, we get the following relation:

$$\frac{d \ln \hat{a}_E^{dil}}{d\tilde{m}} = \frac{1}{\tilde{m}}$$

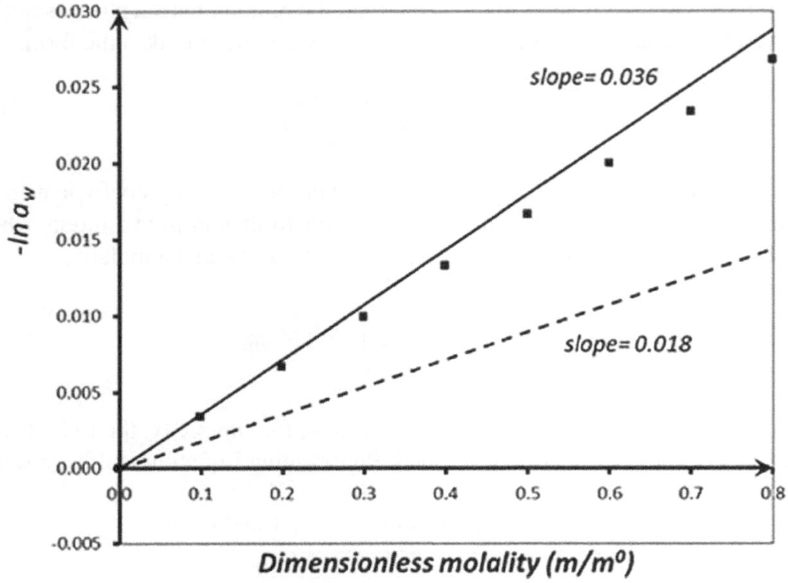

FIGURE 18.1 Comparison of the experimental dependence of $-\ln a_W$ as a function of molality for aqueous NaCl solution at 25°C with calculated lines: ——, calculated line with assumed presence of ions; ------, calculated line with the assumption of no ions present; ■, experimental data [1]. (Reprinted and partially modified from Wilczek-Vera, G. and Vera J. H. Peculiarities of the Thermodynamics of Electrolyte Solutions: A Critical Discussion. *Can. J. Chem. Eng.* 81:70–79, 2003. With permission from Wiley & Sons.)

In addition, using the Gibbs–Duhem equation, Equation 18.10, we can calculate the slope of $\ln \hat{a}_W^{dil}$ versus the molality of salt $\tilde{m}$:

$$\frac{d\ln \hat{a}_W^{dil}}{d\tilde{m}} = -\frac{M_W^{cgs}}{1000} \qquad or \qquad \frac{d\ln \hat{a}_W^{dil}}{d\tilde{m}} = -M_W^{SI}$$

On the other hand, if the presence of the ions is considered for thermodynamic treatment, in the dilute region, that is, when $\gamma_\pm \to 1$, Equation 18.7 gives:

$$a_E^{dil} = \tilde{m}^\nu \left(\nu_+^{\nu_+} \nu_-^{\nu_-}\right)$$

and from Equation 18.11 we obtain

$$\frac{d\ln \hat{a}_W^{dil}}{d\tilde{m}} = -\frac{\nu M_W^{cgs}}{1000} \qquad \text{or} \qquad \frac{d\ln \hat{a}_W^{dil}}{d\tilde{m}} = -\nu M_W^{SI}$$

Figure 18.1 shows a plot of ($-\ln a_W$) versus $\tilde{m}$ in the dilute region for NaCl ($\nu=2$) in an aqueous solution at 298.2 K. The slope of the line is 0.036, as predicted by the consideration of the ions, and not 0.018, as would be the case ignoring the ions in the thermodynamic approach.

BEHAVIOR OF THE ACTIVITY OF INDIVIDUAL IONS

At high dilution in a one-salt solution, the ions of one kind are not affected by any other ion present in the solution. One reasonable question is whether, as the solution gets concentrated, the activity coefficient of an ion in the presence of a particular counterion will be the same or different from the activity coefficient of the same ion in the presence of a different counterion, both solutions at

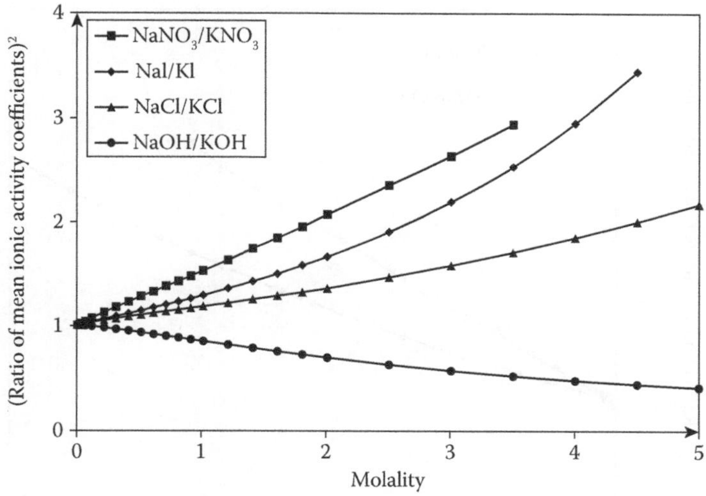

FIGURE 18.2 Purported ratios of $\gamma_{Na^+}/\gamma_{K^+}$ calculated for different salts. (Experimental data [2]; Plot: reprinted and partially modified with permission from Wilczek-Vera, G. and Vera J.H.: Peculiarities of the thermodynamics of electrolyte solutions: a critical discussion. *Can. J. Chem. Eng.* 2003. 81. 70–79. Copyright Wiley-VCH Verlag GmbH & Co. KGaA.)

the same concentration and temperature. This question can be stated more clearly by considering Equation 18.8 for solutions at the same concentration of two salts with the same anion, but with different cation. Consider, for example, a solution of a sodium salt with an anion Y_ and a potassium salt with the same anion, both solutions at the same concentration and temperature. If the activity coefficients of the anion were independent of the cation present, we would have:

$$\left[\frac{\left(\gamma_{\pm}^{2}\right)_{NaY}}{\left(\gamma_{\pm}^{2}\right)_{KY}}\right] \equiv \left[\frac{\gamma_{Na^+}\gamma_{Y^-}}{\gamma_{K^+}\gamma_{Y^-}}\right] = \left[\frac{\gamma_{Na^+}}{\gamma_{K^+}}\right]?$$

If this were true, the ratio of the mean ionic activity coefficients of two electrolytes with a common ion would have the same value independent of the nature of the common ion.

Figure 18.2 shows the values of the left-hand side of this equation for aqueous solutions of electrolytes of the same cation with different anions at 298.2 K. Figure 18.3 shows the values of the left-hand side of this equation for aqueous solutions of electrolytes of the same anion with different cations at 298.2 K. These results show that at the same concentration, the activity coefficient of an ion is strongly dependent on the nature of the counterion present. At high dilution, the activity coefficient of an ion tends to be independent of the counterion, but as the solution becomes more concentrated, the ion behaves differently depending on its counterion.

DEBYE–HÜCKEL THEORY FOR DILUTE AQUEOUS ELECTROLYTE SOLUTIONS

Up to this point, we have made a deliberate effort to avoid introducing models or theories that are not part of the mathematical structure of thermodynamics. Models and theories change with time; the mathematics supporting thermodynamic treatment of fluid systems does not. However, as a tool to discuss equations useful in the thermodynamic treatment of aqueous electrolyte solutions, we introduce here the expressions obtained by Debye and Hückel [3] for the treatment of dilute electrolyte solutions. This theory has stood the test of time. Debye and Hückel developed a model by considering the mathematics of inserting charged particles (ions) into a solvent. In their model, based on what is called the MacMillan–Mayer framework, the solvent is not a component of the mixture,

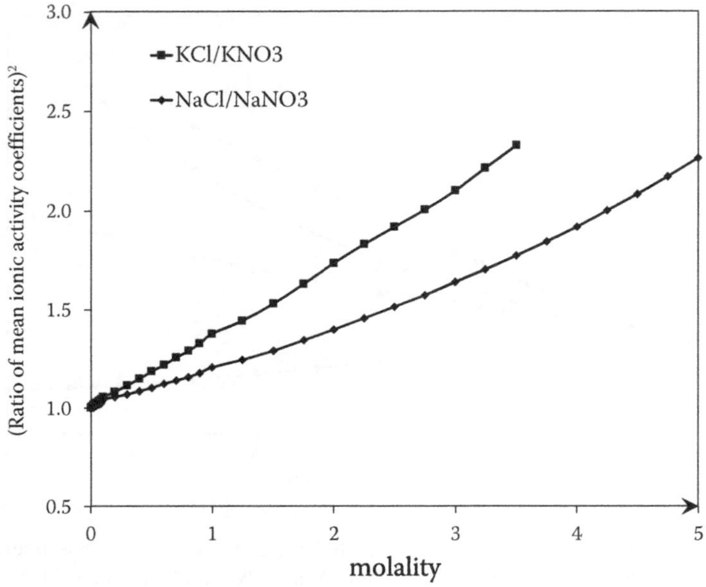

FIGURE 18.3 Purported ratios of $\gamma_{Cl^-}/\gamma_{NO_3^-}$ calculated for different salts. (Experimental data [2]; Plot: reprinted and partially modified with permission from Wilczek-Vera, G. and Vera J.H.: Peculiarities of the thermodynamics of electrolyte solutions: a critical discussion. *Can. J. Chem. Eng.* 2003. 81. 70–79. Copyright Wiley-VCH Verlag GmbH & Co. KGaA.)

but it is just a background medium supporting the ions. This restriction presents a problem for the thermodynamic treatment of electrolyte solutions in mixed solvents, but for all practical purposes, it can be ignored in the work of aqueous solutions or other single-solvent systems. An excellent analysis of the derivations of the Debye–Hückel equations can be found in [4].

In practical work, it has become usual to employ an extended form of the Debye–Hückel expression known as the Hückel equation [5] which for the activity coefficient of a single ion in a single solvent takes the form:

$$\ln\gamma_i = -\frac{A_D Z_i^2 I^{1/2}}{1+B_{Di}I^{1/2}} + b_i I \tag{18.16}$$

where A_D is a theoretical parameter that depends on the solvent and the temperature of the system. Although in principle B_{Di} is a parameter related to the size of the ion, in practice it is usually set equal to unity. The parameter b_i is an adjustable parameter. The ionic strength I, written in terms of (dimensionless) molality, is defined as:

$$I = \frac{1}{2}\sum_{j=1}^{all\ ions}\tilde{m}_j Z_j^2 \tag{18.17}$$

For a single electrolyte in solution, the ionic strength is given by:

$$I = \frac{\tilde{m}}{2}\sum_{j=1}^{all\ ions}\nu_j Z_j^2 \tag{18.17a}$$

For a 1:1 electrolyte, such as HCl or NaCl, the ionic strength is numerically equal to the (dimensionless) molality. The Debye–Hückel type of expression for the mean ionic activity coefficient of the electrolyte obtained from Equation 18.8, assuming the same value of B_D for the cation and the anion, can be written as:

$$\ln\gamma_\pm = -\frac{A_D|Z_+Z_-|I^{1/2}}{1+B_DI^{1/2}} + bI \tag{18.18}$$

with

$$b = \frac{(\nu_+b_+ + \nu_-b_-)}{\nu}$$

The absolute value of the product of the charges of the ions in Equation 18.18 arises from the relation of electrical neutrality for the electrolyte, Equation 18.1d. From Equation 18.10, the activity coefficient for water in a single electrolyte aqueous solution obtained from the Debye–Hückel theory (DHT) is:

$$\ln\gamma_W = \frac{2A_DM_W^{cgs}}{(10B_D)^3}\left[\left(1+B_DI^{1/2}\right) - \frac{1}{\left(1+B_DI^{1/2}\right)} - 2\ln\left(1+B_DI^{1/2}\right)\right] + \ln\left[1+\frac{\nu M_W^{cgs}\tilde{m}}{1000}\right] - \frac{\nu M_W^{cgs}\tilde{m}}{1000}\left[1+\frac{bI}{2}\right]$$

or

$$\ln\gamma_W = \frac{2A_DM_W^{SI}}{(B_D)^3}\left[\left(1+B_DI^{1/2}\right) - \frac{1}{\left(1+B_DI^{1/2}\right)} - 2\ln\left(1+B_DI^{1/2}\right)\right] + \ln\left[1+\nu M_W^{SI}\tilde{m}\right] - \nu M_W^{SI}\tilde{m}\left[1+\frac{bI}{2}\right] \tag{18.19}$$

LIMITING VALUES OF THE MEAN IONIC ACTIVITY COEFFICIENT AND THE OSMOTIC COEFFICIENT AT HIGH DILUTION

One immediate application of the Debye–Hückel equation is to obtain explicitly the limiting value of the osmotic coefficients as the ionic strength goes to zero. As it can be seen from the definition of the osmotic coefficient by Equation 18.11, and more clearly from the case of a single electrolyte in solution, as the molality of the solute goes to zero, the activity of water goes to unity and its logarithm goes to zero, so no clear limiting behavior for the osmotic coefficient is obtained. At high dilution, the limiting behavior obtained from the Debye–Hückel equation takes the form:

$$\ln\gamma_\pm^\infty = -A_D|Z_+Z_-|I^{1/2} \tag{18.20}$$

Thus, according to Equation 18.17a, for a single electrolyte in solution,

$$\ln\gamma_\pm^\infty = -A_D|Z_+Z_-|\left[\frac{1}{2}\sum_{j=1}^{all\ ions}\nu_jZ_j^2\right]^{\frac{1}{2}}\tilde{m}^{1/2} = -A_\gamma\tilde{m}^{1/2} \tag{18.20a}$$

We have indicated by a subscript γ in the Debye–Hückel constant A that for any electrolyte other than type 1:1, the value of this constant is modified in this step. This fact is normally ignored in texts. Inserting Equation 18.20a into Equation 18.15, we observe that in the dilute region, the osmotic coefficient is given by:

$$\varphi^{\infty} = -\frac{A_\gamma}{\tilde{m}} \int_0^{\tilde{m}} \tilde{m} d\left(\tilde{m}^{1/2}\right) + 1 = -\frac{A_\gamma}{3} \tilde{m}^{1/2} + 1$$

Hence,

$$\lim_{\tilde{m} \to 0} \varphi = 1 \tag{18.21}$$

As discussed in Chapter 20, this limiting behavior of the osmotic coefficient is well known for the case of the osmotic pressure in the study of nonelectrolyte solutions, but it is not immediately obvious for aqueous electrolyte solutions. In addition, for electrolyte aqueous solutions, this limiting behavior seems to create a problem with the value of the integrand at the lower limit of the integral in Equation 18.14:

$$\ln \gamma_{\pm} = (\varphi - 1) + \int_0^{\tilde{m}} \frac{(\varphi - 1)}{\tilde{m}} d\tilde{m} \tag{18.14}$$

At the limit of infinite dilution, the integrand seems to be undefined. However, the behavior of the osmotic coefficient in the dilute region just obtained shows that:

$$\lim_{\tilde{m} \to 0} \left[\frac{\varphi - 1}{\tilde{m}^{1/2}}\right] = -\frac{A_\gamma}{3} \tag{18.22}$$

This result suggests that a change in the denominator of the integrand of Equation 18.14 from $\tilde{m}$ to $\tilde{m}^{1/2}$ solves the problem. This is achieved by noting that:

$$\frac{d\left(\tilde{m}^{1/2}\right)}{d\tilde{m}} = \frac{1}{2} \tilde{m}^{-1/2} = \frac{1}{2} \frac{\tilde{m}^{1/2}}{\tilde{m}}$$

or

$$2 \frac{d\left(\tilde{m}^{1/2}\right)}{\tilde{m}^{1/2}} = \frac{d\tilde{m}}{\tilde{m}}$$

Hence, Equation 18.14 is written as:

$$\ln \gamma_{\pm} = (\varphi - 1) + 2 \int_0^{\tilde{m}} \frac{(\varphi - 1)}{\tilde{m}^{1/2}} d\tilde{m}^{1/2} \tag{18.14a}$$

In this form, at the lower limit of integration the integrand has a well-defined value given by Equation 18.22. For this reason, experimentally measured values of the osmotic coefficient or of the mean ionic activity coefficient are fitted with advantage as a function of $\tilde{m}^{1/2}$ instead of $\tilde{m}$.

INDIRECT MEASUREMENT OF THE MEAN IONIC ACTIVITY COEFFICIENT

Equation 18.14a is the basic relation for calculating the mean ionic coefficient from the osmotic coefficient data. The vapor pressure of water (or solvent) is measured under isothermal conditions for different molalities of a single electrolyte in solution. The activity of water is then obtained from Equation 18.12, and the osmotic coefficient is calculated from Equation 18.11a. The values of the osmotic coefficient are then correlated as a function of molality, usually in terms of $\tilde{m}^{1/2}$, and the smooth function is used with Equation 18.14a to obtain values of the mean ionic activity coefficient $\gamma_\pm$ as a function of molality.

DIRECT ELECTROCHEMICAL MEASUREMENT OF THE MEAN IONIC ACTIVITY COEFFICIENT

The mean ionic activity coefficient of an electrolyte in aqueous solution can be obtained by measuring the potential difference between the responses of an ion-selective electrode (ISE) sensitive to the cation and an ISE sensitive to the anion. These ISEs, of which the pH meter is a precursor, are commercially available for most ions of practical interest. The voltage response $E_{i,k}$ of an ISE sensitive to ion i at molality $\tilde{m}_{i,k}$ in an aqueous solution is given by the Nernst equation:

$$E_{i,k} = E_{i,0} + \frac{RT}{Z_i F} \ln\left(a_{i,k}\right) \tag{18.23}$$

In this equation, other than the usual symbols R, T, Z_i, and $a_{i,k}$ for the gas constant, the absolute temperature, the charge of the ion, and the activity of ion i at molality m_k, $E_{i,0}$ [V] is the standard electrode potential and F is the Faraday constant, equal to $9.64853321233100184 \times 10^4$ [C/mol]. As the ISEs are transient instruments, the standard electrode potential has a constant value only over relatively short lapses of time. If an ISE is immersed in the same solution on different days, the standard electrode potential will have different values. Hence, either measurements are carried out in continuous runs in short time periods, or if carried out on different days, the electrodes should be recalibrated daily as explained below. For clarity, we write the Nernst equation for the cation and for the anion as:

$$E_{+,k} = E_{+,0} + \frac{RT}{Z_+ F} \ln\left(a_{+,k}\right)$$

and

$$E_{-,k} = E_{-,0} - \frac{RT}{|Z_-| F} \ln\left(a_{-,k}\right)$$

If both electrodes are immersed in the same solution at the same time and their responses are measured directly one against the other in a voltmeter, the voltage difference is:

$$\Delta E_k = \Delta E_0 + \frac{RT}{F} \ln\left[a_{+,k}^{\frac{1}{Z_+}} a_{-,k}^{\frac{1}{|Z_-|}}\right] \tag{18.24a}$$

Now, using Equations 18.1b and 18.1c, we write:

$$\left[a_{+,k}^{\frac{1}{Z_+}} a_{-,k}^{\frac{1}{|Z_-|}}\right] = \left[a_{+,k}^{\nu_+} a_{-,k}^{\nu_-}\right]^{\frac{1}{\nu_+ Z_+}}$$

But, according to Equation 18.7,

$$a_+^{\nu_+} a_-^{\nu_-} = \tilde{m}^{\nu} \gamma_\pm^{\nu} \left(\nu_+^{\nu_+} \nu_-^{\nu_-}\right)$$

so,

$$\Delta E_k = \Delta E_0 + \frac{RT}{\left(\nu_+ Z_+\right) F} \ln\left[\tilde{m}^{\nu} \gamma_\pm^{\nu} \left(\nu_+^{\nu_+} \nu_-^{\nu_-}\right)\right] \tag{18.24b}$$

The only residual problem is the evaluation of ΔE_0. This is usually done by measuring several points in the dilute region where the mean ionic activity coefficient can be obtained from Equation 18.18 approximated as:

$$\ln \gamma_{\pm}^{dil} = -\frac{A_D |Z_+ Z_-| I^{1/2}}{1 + I^{1/2}}$$

The rewarding result is that the mean ionic activity coefficient values for different electrolytes obtained by the direct electrochemical measurements agree with the values obtained by measurement of the osmotic coefficient [6].

ELECTROCHEMICAL MEASUREMENT OF THE MEAN IONIC ACTIVITY COEFFICIENT WITH INDEPENDENT MEASUREMENTS FOR THE CATION AND THE ANION

Although, as discussed above, the mean ionic activity coefficient can be measured connecting both ISEs, for the cation and for the anion, directly to a voltmeter, it is also instructive to consider the case of measuring each voltage response separately against a single-junction reference electrode, as shown in Figure 18.4. The reference electrode is immersed in an internal standard solution that leaks into the sample solution, creating a liquid junction potential at the surface of contact between the two solutions. This contact between the sample solution and the reference solution is necessary to close the electric circuit. In this case, the voltage response of the electrodes for an ion i takes the form:

$$E_{i,k}^{R} = E_{i,0}^{R} + \frac{RT}{Z_i F} \ln\left(a_{i,k}\right) + E_{J,k} \tag{18.25}$$

Thus, we write for the cation and for the anion:

$$E_{+,k}^{R} = E_{+,0}^{R} + \frac{RT}{Z_+ F} \ln\left(a_{+,k}\right) + E_{J,k}$$

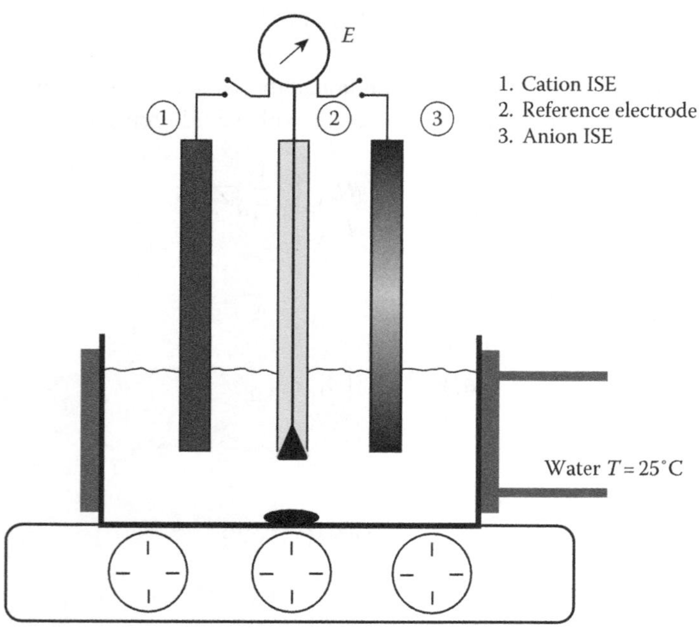

FIGURE 18.4 Schematics of the experimental setup for measurements of ionic activities.

$$E^R_{-,k} = E^R_{-,0} - \frac{RT}{|Z_-|F}\ln\left(a_{-,k}\right) + E_{J,k}$$

We have used a superscript R to indicate that in this case, the voltage response is measured against a reference electrode and its value will be different from the value obtained in the previous experimental setup. The additional term, $E_{J,k}$, is due to the junction potential generated by the contact of the sample solution with the reference solution that happens at the tip of the reference electrode. This junction potential will have the same value for the responses of both ISEs connected to the same reference electrode in simultaneous measurements. Taking the difference of both independent readings, we obtain:

$$\Delta E^R_k = \Delta E^R_0 + \frac{RT}{F}\ln\left[a_{+,k}^{\frac{1}{Z_+}} a_{-,k}^{\frac{1}{|Z_-|}}\right] \tag{18.24c}$$

Equation 18.24c is formally identical to Equation 18.24a and by calibration of the value of ΔE^R_0 will yield values for the mean ionic activity coefficients of the electrolyte. Again, the rewarding result is that the mean ionic activity coefficient values for different electrolytes obtained by this approach agree with the values obtained by the direct electrochemical procedure and with the values obtained from measurements of the osmotic coefficient [6].

ACTIVITY OF A SECOND SOLUTE IN AN AQUEOUS ELECTROLYTE SOLUTION

The direct measurement of the mean ionic activity coefficient of an electrolyte using ISEs opens the possibility of determining the activity of a water-soluble solute dissolved in an electrolyte aqueous solution. According to Equation 7.11, for a solute A in an aqueous solution with an electrolyte E,

$$\left[\frac{\partial \mu_A}{\partial \tilde{m}_E}\right]_{T,\tilde{m}_A} = \left[\frac{\partial \mu_E}{\partial \tilde{m}_A}\right]_{T,\tilde{m}_E}$$

Considering

$$\mu_A = \mu_A^\theta + RT\ln\left(\tilde{m}_A \gamma_A\right)$$

and

$$\mu_E = \mu_E^\theta + RT\ln\left(\tilde{m}_E^{\nu} \gamma_\pm^{\nu} \nu_+^{\nu_+} \nu_-^{\nu_-}\right)$$

we obtain

$$\left[\frac{\partial \ln \gamma_A}{\partial \tilde{m}_E}\right]_{T,\tilde{m}_A} = \nu\left[\frac{\partial \ln \gamma_\pm}{\partial \tilde{m}_A}\right]_{T,\tilde{m}_E}$$

Thus, integrating between 0 and $\tilde{m}_E$, at constant m_A,

$$\ln\left[\frac{\gamma_A^{(2)}}{\gamma_A^{(1)}}\right] = \nu\int_0^{\tilde{m}_E}\left[\frac{\partial \ln \gamma_\pm}{\partial \tilde{m}_A}\right]_{T,\tilde{m}_E} d\tilde{m}_E \tag{18.26}$$

In this expression, the superscripts (1) and (2) indicate the value of the activity coefficient of the solute A in the absence of electrolyte and in the presence of electrolyte, respectively, both at the same molality of solute A. The former value should be available from independent measurements, usually from the isopiestic method in the case of nonelectrolytes. To obtain the activity coefficient of the solute A in the presence of the electrolyte, measurements are made of the mean ionic activity coefficient of the electrolyte at fixed values of the molality of the electrolyte changing the concentration of the solute A. These results are then correlated by a function of the form:

$$\ln\left[\frac{\gamma_{\pm}^{(2)}}{\gamma_{\pm}^{(1)}}\right] = \sum_{k=1} C_k m_A^k$$

Usually, no more than three or four terms are required on the right-hand side of this equation. Again here, the superscripts (1) and (2) indicate the value of the activity coefficient of the electrolyte in the absence of solute A and in the presence of solute A, respectively, both at the same molality of the electrolyte. The former value should be available from independent measurements. The values of the coefficients C_k are correlated as a function of the molality of the electrolyte:

$$C_k = \sum_{j=1} C_{kj} m_E^j$$

The rest is simple algebra necessary to obtain the value of the right-hand side of Equation 18.26 and get values of $\gamma_A^{(2)}$.

AQUEOUS SOLUTIONS OF WEAK ELECTROLYTES

For weak electrolytes, only a fraction α of the electrolyte added to the solution will ionize so that the activity of the electrolyte that can be obtained from the activity of water, by means of the Gibbs–Duhem equation, is an effective value. In this case, with a degree of dissociation α, the molality of the ion i in solution would be $\nu_i\tilde{m}\alpha$ and its activity would be $\gamma_i\nu_i\tilde{m}\alpha$, while for the undissociated electrolyte in solution, the activity would be $\gamma_E\tilde{m}(1-\alpha)$. The dissociation constant from Equation 18.3 is:

$$K_T^{dis} = \frac{a_+^{\nu_+} a_-^{\nu_-}}{a_E}$$

Replacing and rearranging,

$$\frac{\alpha}{(1-\alpha)^{\frac{1}{\nu}}} = \frac{1}{\gamma_{\pm}}\left[\frac{\gamma_E K_T^{dis}}{\tilde{m}^{\nu-1}\nu_+^{\nu_+}\nu_-^{\nu_-}}\right]^{\frac{1}{\nu}}$$

A weak electrolyte has a weak dissociation, so the left-hand side is close to α. For a 1:1 weak electrolyte in a dilute solution, $\gamma_E \approx 1$,

$$\alpha \approx \frac{1}{\gamma_{\pm}}\left[\frac{K_T^{dis}}{\tilde{m}}\right]^{\frac{1}{2}} \tag{18.27}$$

This approximation is interesting as a way of understanding the effect of additional ions on the dissociation of the weak electrolyte. If an electrolyte that does not contain the cation or the anion of the weak electrolyte is added to the solution, the ionic strength will increase, and thus the value of the mean ionic activity coefficient will decrease, resulting in an increase of the degree of dissociation

α. If the added electrolyte contains either the cation or the anion of the weak electrolyte, the activity of the common ion will increase, and as the dissociation constant does not change, the activity of the undissociated electrolyte must increase; that is, the degree of dissociation decreases. This is the so-called common ion effect.

CHARGED ORGANIC MOLECULES

Amino acids are organic compounds having an amino group, $-NH_2$, and a carboxylic group, $-COOH$, attached to a hydrocarbon chain. Without entering into the chemistry of amino acids, with the sole interest in their behavior in aqueous electrolyte solutions, we observe that most of the common amino acids at pH above 9.4 have the carboxyl group negatively charged, $-COO^-$, and at pH below 2.2, the amino group is positively charged, $-NH_3^+$. At a pH between 2.2 and 9.4, the amino acid has both a negative carboxyl group and a positive amino group. This state with zero net charge is known as the zwitterionic form, from the German word *zwitter*, meaning "hybrid." At the midpoint between the two limiting values of pH, the amino acid has a weak positive charge in the amino group and a weak negative charge in the hydroxyl group that closely compensate. This state is known as the isoelectric point, and the pH is designated as the pI of the amino acid. The interesting point is that at the isoelectric point, the amino acid presents a minimum in its solubility in water. Amino acids associate in chains known as peptides, and peptides associate in larger molecules known as proteins, of which enzymes are a special kind presenting catalytic properties. In aqueous solutions, the presence of electrolytes, such as $(NH_4)_2SO_4$, for example, affects the solubility of proteins. In a dilute electrolyte solution, the protein is more soluble than in pure water, but if the concentration of the electrolyte is increased beyond a certain point, the protein is less soluble than in pure water. These behaviors are known as the salting-in and salting-out of the protein by changing the concentration of the electrolyte.

A different kind of charged organic molecules are the so-called ionic liquids. These are organic salts that are in liquid phase at temperatures below the boiling point of water. This temperature limit is rather arbitrary, but it comes from the possibility of using these ionic liquids as solvents for green chemistry synthesis. Although ionic liquids are rather viscous, they have low volatility and high thermal stability properties that make them recyclable.

EXPERIMENT TO DETERMINE THE CHARGE OF A PROTEIN

The molecules of proteins behave as polyions in aqueous solutions and are too large to pass through some membranes, while other ions can. When one such polyelectrolyte is in an aqueous solution on one side of a membrane, and if some electrolyte formed by small ions, for example, NaCl, and water can freely pass through the membrane, we are in the presence of the Donnan equilibrium described in Chapter 10. Due to the condition of electroneutrality on both sides of the membrane and the inability of a polyion to migrate, ions are at different concentrations on each side of the membrane. This causes a difference in electric potential to arise between the two phases. As an illustration of the chemical equilibrium between ionic species separated by a permeable membrane, we discuss here an experiment designed for the determination of the charge of the sodium salt of an anionic protein. There are many kinds of membranes. Thus, the first step is to decide what kind of membrane is appropriate for the purpose. Some membranes are permeable only to the solvent; some are more permeable to cations than to anions or vice versa. It is known that biological membranes behave differently when they are part of a living tissue than when they are not. For this reason, the consideration of equilibrium in living biological systems may be a very poor assumption. Generally, it has been observed that single-charged ions, such as Na^+ or Cl^-, permeate through membranes more easily than polycharged ions, such as Ba^{2+} or SO_4^{2-}; hence, for this experiment NaCl is chosen as the electrolyte. The membrane is chosen to be permeable to both the sodium and the chloride

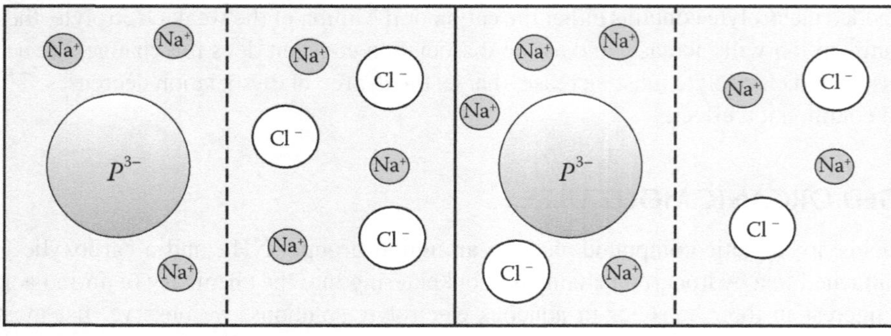

FIGURE 18.5 The system protein+electrolyte separated by an electrolyte-permeable membrane: (a) immediately after loading and (b) after the equilibrium is established.

ions. Figure 18.5a presents the experimental setup in which the same weight of deionized water is charged on both sides of the membrane, a molality m_P of the protein is charged to the left-hand side of the cell, and a molality m of sodium chloride is charged to the right side. After equilibration, a molality $m_{-,R}$ of Cl^- is measured by titration of the solution in the right-hand side of the membrane (Figure 18.5b).

The dissociation equilibrium of the sodium salt of the anionic protein P with a negative charge $|Z_P|$ is given by:

$$P\text{Na}_{|Z_P|} = P^{-|Z_P|} + |Z_P|\,\text{Na}^+$$

The charge balance of the solution on the left-hand side of the membrane where there is the polyion plus chloride and sodium ions is:

$$|Z_P| m_P + m_{-,L} = m_{+,L} \tag{18.28}$$

or

$$|Z_P| = \frac{m_{+,L} - m_{-,L}}{m_P} \tag{18.28a}$$

The molality of chloride ions on the right-hand side of the membrane after $m_{-,L}$ moles of chloride ion migrated to the left-hand side of the membrane is:

$$m_{-,R} = m - m_{-,L}$$

Hence, the charge balance on the right-hand side of the membrane gives:

$$m_{+,R} = m_{-,R} = m - m_{-,L} \tag{18.29}$$

To obtain the molality of sodium in the left-hand side of the membrane, we apply the expression for the Donnan equilibrium, Equation 10.9, with the chemical potentials of the ions given by Equation 12.12a, and write:

$$\left[\frac{a_{+,L}}{a_{+,R}}\right]^{\frac{\nu_+}{\nu_-}} = \left[\frac{a_{-,R}}{a_{-,L}}\right] \tag{18.30}$$

For NaCl, $\nu_+ = \nu_- = 1$, so we write:

$$\left[\frac{m_{+,L}\gamma_{+,L}}{m_{+,R}\gamma_{+,R}}\right] = \left[\frac{m_{-,R}\gamma_{-,R}}{m_{-,L}\gamma_{-,L}}\right]$$

Although in complete thermodynamic rigor, the dimensionless molalities should be used for activities, in order to simplify the nomenclature, we are ignoring this requirement here. Thus, using the definition of the mean ionic activity coefficient,

$$m_{+,L} = \left[\frac{m_{+,R}m_{-,R}}{m_{-,L}}\right]\left[\frac{\gamma_{\pm,R}}{\gamma_{\pm,L}}\right]^2$$

Replacing in this expression the molalities of sodium and chloride on the right-hand side of the membrane given by Equation 18.29, we write:

$$m_{+,L} = \frac{\left(m - m_{-,L}\right)^2}{m_{-,L}}\left[\frac{\gamma_{\pm,R}}{\gamma_{\pm,L}}\right]^2 \quad (18.31)$$

Thus, $|Z_P|$ is directly obtained from Equation 18.28a. The mean ionic activity coefficient of sodium chloride on the right-hand side of the membrane can be obtained using the Debye–Hückel expression at low concentrations of NaCl or more complex correlations for concentrated solutions. On the other hand, the mean ionic activity coefficient of sodium chloride on the protein side of the membrane is not easily evaluated because the charged protein increases the ionic strength of the solution on that side of the membrane. If we assume that the mean activity coefficients of NaCl have the same value on both sides of the membrane, from Equation 18.31,

$$m_{+,L}m_{-,L} = \left(m - m_{-,L}\right)^2$$

Replacing back $m_{+,L}$ from Equation 18.28,

$$\left(|Z_P|\, m_P + m_{-,L}\right)m_{-,L} = \left(m - m_{-,L}\right)^2$$

and expanding and rearranging,

$$m_{-,L} = \frac{m}{\left(2 + \dfrac{|Z_P|\, m_P}{m}\right)}$$

From this approximate expression, it can be seen that less than half of the NaCl added to the right-hand side permeated to the left-hand side. Thus, the contribution to the ionic strength of the charged protein on the left-hand side is compensated to some extent by the smaller concentration of NaCl, and the ratio of the activity coefficients in Equation 18.31 can be set to unity as a good approximation.

As a final comment on Donnan equilibrium, it is interesting to note that although the solutions at both sides of the membrane are electrically neutral, the solution at the left-hand side has more charges than the right-hand side solution. This difference in charge, which produces the difference in potential $\Delta\Phi$ in electric potential appearing in Equation 10.8, is due to the presence of charged organic molecules. In the absence of these molecules, the concentration of cations and concentration of anions would be the same on both sides of the permeable membrane to both ions.

Determination of the Charge of Biomolecules by Complexation with Ionic Surfactants

Although not requiring thermodynamic principles, an alternative to the experiment based on Donnan equilibrium to determine the charge of organic molecules, is their complexation with an ionic surfactant [7–9]. Controlled addition of a monovalent ionic surfactant to an aqueous solution containing a purified biomolecule, forms a complex which floats over the solution. A commonly used monovalent surfactant is sodium bis(2-hexylethyl) sulfo succinate, known as AOT surfactant. As shown in Figure 18.6 for the case of complexation of lysozyme with AOT, the moles of surfactant added do not distribute randomly in the number of biomolecules available but concentrate in a few ones and when one of these few molecules has its valences saturated it floats as a complex over the solution. Thus, when no more biomolecules are complexed, it indicates that the charge of each of the complexed molecules has been saturated and this directly gives the charge number of the biomolecule.

ELECTROLYTE SOLUTIONS IN SOLVENTS OTHER THAN WATER

Recently, there has been increased interest in the study of electrolyte solutions in solvents other than water. For these developments it is important to keep in mind that the DHT and its extensions are based on the McMillan–Mayer framework in which the solvent is just a background for the long-range interactions between ions. Thus, the first thermodynamic treatment of solutions of electrolytes in solvents different from water and in mixed solvents focused on converting the equations from the McMillan–Mayer to the Lewis–Randall framework. This conversion was discussed by Cabezas and O'Connell [10] and by Haynes and Newman [11]. The extension of this treatment to mixed solvents is presented in [12]. An alternative view has been presented by Breil and Mollerup [13], who maintain that "the McMillan-Mayer framework is superfluous when the thermodynamics of the electrolyte solutions is described by the Helmholtz energy functions."

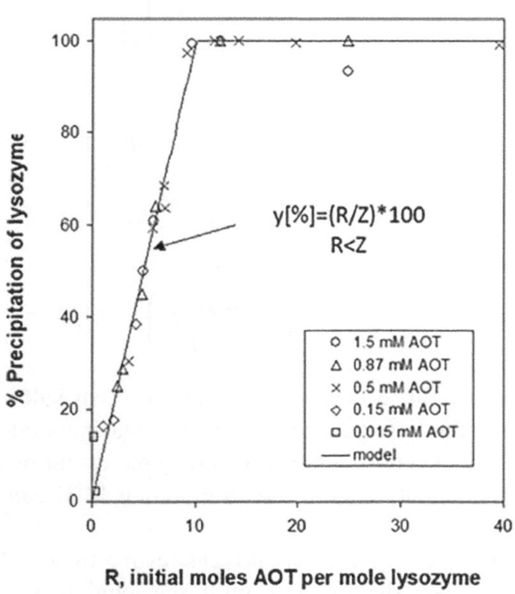

FIGURE 18.6 Precipitation of lysozyme as a function of the amount of AOT added. Z is the charge of the macromolecule; in this case $Z=10$. For $R>Z$, $y=100\%$.

In fact, using the proper value of the Debye–Hückel constant from the DHT is directly applicable to electrolyte solutions in solvents other than water. From physics, this constant is given as:

$$A_\gamma = \left[\frac{N_A^2}{8\pi} \left(\frac{e^2}{\varepsilon_0 RT} \right)^{3/2} \right] \left(\frac{2d_s}{D_s^3} \right)^{1/2} \tag{18.32}$$

The term in square brackets on the right-hand side of Equation 18.32 is independent of the nature of the solvent. In this term, N_A, R, and T have their usual meaning as Avogadro's number, the gas constant, and the absolute temperature, while e and ε_0 are the charges of the electron and the vacuum permittivity. The density d_s and the dielectric constant D_s of the solvent change for solvents other than water. Recent computer simulation studies of aqueous electrolyte solutions [14], have proposed the use of a concentration-dependent dielectric constant, which should be evaluated independently. This topic is discussed further in [15,16]. For deeper understanding of the topics covered, see Chapter 34.

CONCLUSIONS

In this chapter, we have presented a general view and the main concepts necessary for the understanding of the behavior of aqueous electrolyte solutions. This is just the starting point for more detailed studies. One case of interest, for example, is the measurement of the activity of individual ions, which is discussed in Chapter 34.

REFERENCES

1. Robinson, R. A., and Stokes, R. H. 1959. *Electrolyte Solutions*, 2nd ed. London: Butterworths Scientific, Appendix 8.10, Table 1.
2. Hamer, W. J., and Wu, Y.-C. 1972. Osmotic coefficients and mean activity coefficients of uni-univalent electrolytes in water at 25° C. *J. Phys. Chem. Ref. Data*. 1: 1047–1099.
3. Debye, P., and Hückel, E. 1923. Zur Theorie der Elektrolyte. I. Gefrierpunktserniedrigung und verwandte Erscheinungen. *Phys. Z.* 24: 185–206.
4. Silva, G. M., Liang, X., and Kontogeorgis, G. M. 2022. On the derivations of the Debye-Hückel equations. *Mol. Phys.* 120(10): e2064353 (22 pages). https://www.tandfonline.com/doi/full/10.1080/00268976.2022.2064353.
5. Silva, G. M., Liang, X., and Kontogeorgis, G. M. 2022. The true Hückel equation for electrolyte solutions and its relation with the Born term. *J. Mol. Liq.* 368(Part A): 120554. https://doi.org/10.1016/j.molliq.2022.120554.
6. Wilczek-Vera, G., and Vera, J. H. 2003. Peculiarities of the thermodynamics of electrolyte solutions: A critical discussion. *Can. J. Chem. Eng.* 81: 70–79.
7. Wilczek-Vera, G., Dumortier Shin, Y., Rodil, E., and Vera, J. H. 2018. Corrigendum to *Biochem. Eng. J.*, 17 (2004) 91–97 by Y.O. Shin, E. Rodil, and J.H. Vera. *Biochem. Eng. J.* 134: 57–59.
8. Wilczek-Vera, G., Dumortier Shin, Y., Rodil, E., and Vera, J. H, 2018. Corrigendum to Biotechnology Progress 2003, 19, 928-935 by Y.O. Shin, M.E. Weber, and J.H. Vera. *Biotech. Progr.* 134: 57–59.
9. Ionic surfactants and aqueous solutions. In *Biomolecules, Metals and Nanoparticles*, 2018, eds. G. Wilczek-Vera, and J. H. Vera. Berlin, Boston, MA: De Gruyter, pp. 353–355.
10. Cabezas, H., and O'Connell, J. P. 1993. Some uses and misuses of dilute solution thermodynamic models. *Ind. Eng. Chem. Res.* 32: 2892–2904.
11. Haynes, C. A., and Newman, J. 1998. On converting from the McMillan-Mayer framework. I. Single-solvent system. *Fluid Phase Equilib.* 45: 255–268.
12. Curtis, R. A., Blanch, H. W., Prausnitz, J. M., and Newman, J. 2001. McMillan-Mayer solution thermodynamics for a protein in a mixed solvent. *Fluid Phase Equilib.* 192: 131–153.
13. Breil, M. P., and Mollerup, J. M. 2006. The McMillan-Mayer framework and the theory of electrolyte solutions. *Fluid Phase Equilib.* 242, 129–135.

14. Valisko, M., and Boda, D. 2023. Resurrection of Hückel's idea: Decoupling ion-ion and ion-water terms in activity coefficients via state-dependent dielectric constant. *Fluid Phase Equilib.* 572(15): 113826–113847.
15. Silva, G. M., Liang, X., and Kontogeorgis, G. M. 2023. How to account for the concentration dependency of relative permittivity in the Debye-Hückel and Born equations. *Fluid Phase Equilib.* 566: 113671. https://doi.org/10.1016/j.fluid.2022.113671.
16. Novak, N., Kontogeorgis, G. M., Castier, M., and Economou, J. G. 2024. Theoretical considerations on single and mixed solvent electrolyte solutions. *Fluid Phase Equilib.* 576: 113924. https://doi.org/10.1016/j.fluid.2023.113924.

PROBLEMS

18.1 Conversion of units

A. For the case of a single solute of molecular weight $M_i^{cgs}\left[\frac{\text{g}}{\text{mol}}\right]$ or $M_i^{SI}\left[\frac{\text{kg}}{\text{mol}}\right]$ in solution, the concentration can be expressed in variety of units. The units most commonly used in thermodynamics of electrolytes are molality $m_i\left[\frac{\text{moles of solute } i}{\text{kg of solvent}}\right]$ and molarity $c_i\left[\frac{\text{moles of solute } i}{\text{L of solution}}\right]$. Derive the interconversion of molarity to molality and vice versa in cgs units. Repeat this exercise in SI units.

B. Using Excel and the data from Table 18.1, calculate the molality of aqueous solutions of H_2SO_4 for the different mass % of acid in water

$$M_w(\text{H}_2\text{SO}_4)\left[\frac{\text{g}}{\text{mol}}\right] = 98.079$$

18.2 Mean ionic activity coefficients

Using data from results from the previous problem (Table 18.1) and Table 18.2, calculate the mean ionic activity coefficients for H_2SO_4 in aqueous solution in the molarity scale at 25°C. Remember that the $d_{i\,solution}^{25°\text{C}} = 0.9973 d_{i\,solution}^{20°\text{C}}$

TABLE 18.1
Concentration Data for Aqueous Solution of H_2SO_4 at 20°C [1a]

Mass (%)	$d_{i\,solution}^{cgs}$ (g/cm³)	c_i (mol/L)
0.5	1.0016	0.051
1	1.0049	0.102
10	1.0661	1.087
20	1.1398	2.324
30	1.2191	3.729
40	1.3028	5.313
50	1.3952	7.113
60	1.4987	9.168
70	1.6105	11.494
80	1.7272	14.088
90	1.8144	16.649
98	1.8361	18.346
100	1.8305	18.663

TABLE 18.2
Mean Ionic Activity Coefficients for H_2SO_4 at 25°C [1]

m_i	$\gamma_\pm (H_2SO_4)$ [1b]	m_i	$\gamma_\pm (H_2SO_4)$ [1c]
0	1	0	1
0.1	0.2655	0.001	0.804
0.2	0.209	0.002	0.74
0.3	0.1826	0.005	0.634
0.4		0.01	0.542
0.5	0.1557	0.02	0.445
0.6		0.05	0.325
0.7	0.1417	0.1	0.251
0.8		0.2	0.195
0.9		0.5	0.146
1	0.1316	1	0.125
		2	0.119
		5	0.197
		10	0.527
		15	1.077
		20	1.701

18.3 **Debye–Hückel constant**

The original derivation of the limiting Debye–Hückel law [2–4] is based on the linearized Poisson–Boltzmann equation and was derived in the P–V–T system:

$$\log(\gamma_\pm) = -Az_+z_-\sqrt{I} \quad \text{where} \quad A = \frac{\sqrt{2}e_0F^2}{\ln(10)\,8\pi(\varepsilon_r\varepsilon_0RT)^{3/2}}, \quad I = \frac{1}{2}\sum_i z_i^2 c_i$$

A. Using the values presented in Table 18.3, calculate the value of *A* and its units at 25°C according to the equation above.

By what factor has this value of A to be multiplied if the concentration of the electrolyte solution is expressed in:

B. mol/L

C. mmol/L

D. mol/kg

E. How will these values change if natural instead of decimal logarithm is used?

TABLE 18.3
Basic Model Parameters at 25°C [1d]

Parameter	Notation	Value	Unit
Elementary charge	e_0	1.602×10^{-19}	C
Faraday constant	F	96485.33	C/mol
Gas constant	R	8.3145	J/K/mol
Permittivity of vacuum	ε_0	8.85×10^{-12}	F/m
Relative permittivity of water	ε_r	78.408*	
Temperature	T	298.15	K

**Note:* Literature values of relative permittivity of water vary from 78.3 to 78.46. Here the value reported in [1e] is used.

18.4 **Osmotic coefficient**

Using data from Table 18.2, calculate the osmotic coefficient for H_2SO_4 at 25°C in the concentration range of 0–20 m.

18.5 **Donnan equilibrium**

Derive Equation 18.30. Note: If you get it upside down, it means that you forgot something. Check before consulting the solution.

REFERENCES FOR PROBLEMS

1. John R. Rumble, ed. 2023. *CRC Handbook of Chemistry and Physics*, 104th edn (Internet Version 2023). Boca Raton, FL: CRC Press/Taylor & Francis.
   a. "Concentrative Properties of Aqueous Solutions: Density, Refractive Index, Freezing Point Depression, and Viscosity at 20°C".
   b. "Activity Coefficients of Acids, Bases, and Salts at 25°C as a Function of Concentration (in Molality)".
   c. "Mean Activity Coefficients of Electrolytes at 25°C as a Function of Concentration".
   d. "CODATA RECOMMENDED VALUES OF THE FUNDAMENTAL PHYSICAL CONSTANTS: 2018".
   e. "TABLE 1. Thermophysical Properties of H_2O from the Triple Point to 100°C at P=1 bar (0.100MPa)".
2. Debye, P., and Hückel, E. 1923. Zur Theorie der Elektrolyte. I. Gefrierpunktserniedrigung und verwandte Erscheinungen. *Physikalische Zeitschrift* 24(9): 185–206.
3. Manov, G. G., Bates, R. G., Hamer, W. J., and Acree, S. F. 1947. Values of the constants in the Debye-Hückel equation for activity coefficients. *J. Am. Chem. Soc.* 65(9): 1765–1767. https://doi.org/10.1021/ja01249a028.
4. Silva, G. M., Liang, X., and Kontogeorgis, G. M. 2022. On the derivations of the Debye–Hückel equations. *Mol. Phys.* https://doi.org/10.1080/00268976.2022.2064353.

19 The Thermodynamics of Polymer Solutions

Abstract

This chapter emphasizes the difference between a polymer solution and a solution of a nonsegmented solute. A polymer chain in solution may take different spaghetti-like conformations. In good solvents the chain is more extended, in bad solvents, it is more compact. In the limit, it separates from the solvent. Liquid–liquid equilibria and fractionation of polydisperse polymers are discussed. The elasticity and swelling of cross-linked polymer network structures and gels are also discussed.

INTRODUCTION

Polymers are molecules consisting of many (πολυ) parts (μερη) or *repeating units* typically coming from the *monomer* (*μόνο μέρος, one part*) used in the *polymerization or* synthesis of polymers from their monomers. When the number of repeating units is small, say ten or twenty, the molecule is called an *oligomer* (*ολίγο, few*). When this number is high, the molecule is a *macromolecule* (*μακρο, large*) or, simply, *polymer.* Polymers with very high numbers of repeating units and molecular weights of the order of 1 million daltons are called *high polymers.*

Although the same principles of thermodynamics apply, due to the long chains of the solute and their random crumbled shapes, polymer solutions present many peculiarities. The macromolecular character of polymers deprives polymers of vapor pressure and heat of vaporization – properties typically used for the thermodynamic characterization of low molecular weight substances. In addition, most polymers at ambient conditions are found in a nonequilibrium glassy or a semi-crystalline state. This state is of primordial importance not only for the usefulness of the polymer as a material but also for its processing with solvents. It has a significant impact in numerous technological areas, such as polymer membrane processes, sol-gel processes, polymer fractionation, polymer foaming, production of insulation materials and high-impact microcellular materials, aerogels, nanofibers, macroporous scaffolds for tissue engineering, etc.

When a typical polymer chain with a backbone of carbon atoms enters a solution, it may take a large variety of shapes or conformations, different from the extended zig-zag chain or the fully extended *contour length* (total length of C–C bonds). In solution, the chain is rather crumpled and takes a much more compact conformation of a *random coil,* as shown in Figure 19.1. The bulk of the repeating units of this random coil may be visualized as being enclosed in a sphere of radius, R_g, called *radius of gyration*, which depends very much on the nature of the solvent.

Good solvents increase R_g as the polymer tends to maximize its contact with the solvent. *Bad solvents* decrease R_g as the polymer avoids the contacts with the solvent and prefers the contacts with its own repeating units. The critical conditions, where monomer–monomer attraction just counterbalances monomer–monomer repulsion, are referred to as Θ-conditions. When the solvent is too bad, the polymer solution phase separates in a very dilute phase and in a very concentrated one.

The average geometric characteristics of the chain conformations are obtained by applying *random-flight* or *random-walk* statistics in a Gaussian chain of n segments. The average distance between the two chain ends $\sqrt{\langle r_n^2 \rangle}$ is an important and representative measure of the conformation compactness. For very long chains $\sqrt{\langle r_n^2 \rangle / 6} = \sqrt{\langle R_g^2 \rangle}$. Figure 1 shows how the end-to-end distance,

DOI: 10.1201/9781003431985-21

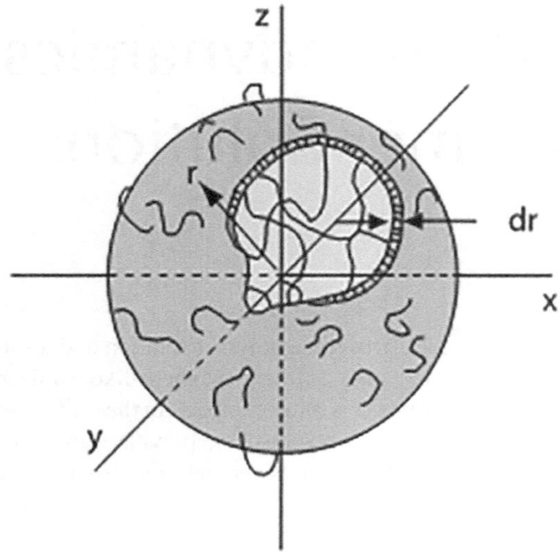

FIGURE 19.1 Schematic representation of the chain conformation in solution. If one end of the randomly coiled chain is located at the center of the axes, in an isotropic system the other end of the chain may be found anywhere on a spherical shell of radius r and thickness dr [1] (fig. 3.11).

r_n, of a chain with n C–C bonds, of length l each, may be visualized. The average end-to-end distance of this polymer chain is given by Equation 19.1:

$$\langle r_n^2 \rangle = C_n n l^2 \tag{19.1}$$

where C_n is the *steric factor* or the Flory characteristic ratio of the chain conformation. For the freely jointed chain, $C_n = 1$ but for actual chains $C_n \geq 2$.

A factor more useful in practice is the *steric hindrance coefficient*, α, which is the ratio of the end-to-end distances of the conformation at the *unperturbed*, or Θ – condition, over the end-to-end distance of the freely rotating chain:

$$\alpha = \sqrt{\bar{r}_\Theta^2 / \bar{r}_f^2} \tag{19.2}$$

Thus, in a very dilute polymer solution, the isolated polymer chains may be viewed as spherical random coils moving freely in the surrounding solvent, as shown in Figure 19.2. In good solvents, the size of the globules of the polymer chains is significantly larger than the size of the above ideal unperturbed conformations. As the concentration increases, it reaches a semi-dilute region around a critical concentration value, c_c, where the polymer spheres or globules touch each other and the free movement through the solvent starts becoming hindered. Above this concentration region, the globules fuse with their neighbors and the polymer chains penetrate each other and ultimately form a homogenous concentrated solution or nearly pure polymer melt. The chains exhibit also the above ideal unperturbed conformations in the case of nearly pure polymer melt and, thus, their root-mean-square end-to-end distance scales with their degree of polymerization or their number of segments, n, as $\sqrt{\langle r_n^2 \rangle} \propto n^{1/2}$. Although we mainly focus on concentrated or semi-concentrated polymer solutions, some interesting aspects of the very dilute solutions are also discussed.

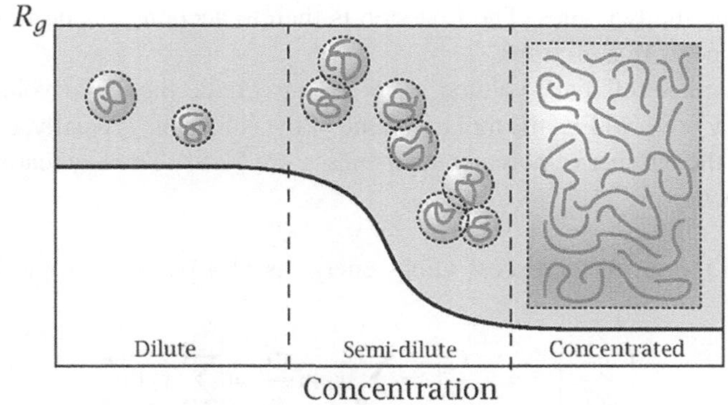

FIGURE 19.2 Schematic representation of the three main regions of polymer solutions according to their polymer concentration. Each spherical particle or globule encloses nearly all segments of one single polymer chain. In good solvents, the chains are expanded and the radius of gyration, R_g, or the size of the spheres in the very dilute solution (left) may be significantly larger compared to the corresponding average size in the region of the concentrated solutions (right) resembling Θ – conditions. In the semi-dilute or cross-over regime (center), R_g fluctuates between large and small. Adapted from Ref. [2] (fig. 33.21).

CONCENTRATED POLYMER SOLUTIONS

Although polymer solutions could present complex behavior, in relatively concentrated solutions in nonpolar solvents the thermodynamic treatment is rather simple.

Considering Equation 16.12a,

$$g_\Theta^E = h_\Theta^E - Ts_\Theta^E \qquad (16.12a),$$

if there are no strong interactions between the functional groups forming the molecules and the predominant effect is due to the difference in the chain lengths between the polymer and the solvent, the mixing will be almost athermal. There will not be a significant temperature change when mixing the polymer and the solvent at the same temperature. Thus, for *athermal mixing*, as an approximation, we write:

$$g_\Theta^E \approx -Ts_\Theta^E$$

$$h_\Theta^E = \sum x_i \bar{h}_i^E \approx 0$$

So, $\bar{h}_i^E \approx 0$, and from Equation 16.17,

$$\bar{h}_i^E = -RT^2\left(\frac{\partial \ln \gamma_{i,\Theta}}{\partial T}\right) \approx 0$$

Hence, $\gamma_{i,\Theta}$ is independent of temperature. Clearly the nonideality of the mixing process in this case is due to an entropic effect.

The Regular Solution Theory and the Flory–Huggins Model

The **R**egular **S**olution **T**heory (RST) [3] and the **F**lory–**H**uggins (FH) model [4,5], for which theoretical justifications are presented in Chapter 27, are very well suited to be the starting point of polymer solution thermodynamics. The first step is then to account for the molecular size differences of the components in a mixture.

In RST, a standard or reference volume, v_0, is used to divide the molar volume of component j into segments, $r_j = v_j^L / v_0$. In FH, the molar volume of the solvent is, typically, used as the reference volume, giving for the number of polymer segments, $r_p = v_p^L / v_s^L$. The key features of RST and FH model are the following:

A. For athermal mixtures, the excess Gibbs energy is of a purely entropic character and is given by:

$$\left(\frac{g_0^E}{RT}\right)_{FH} = -\frac{Ts_0^E}{R} = \sum_j x_j \ln\frac{\Phi_j}{x_j} = \sum_j x_j \ln\frac{r_j}{r} \tag{19.3}$$

where $r = \sum_k x_k r_k$ and the volume or segment fraction, Φ_j, is given by:

$$\Phi_j = \frac{x_j r_j}{\sum_k x_k r_k} = \frac{x_j r_j}{r} \tag{19.4}$$

B. The activity coefficient obtained from Equation 19.3 is given by:

$$(\ln\gamma_i)_{FH} = \left(1 - \frac{\Phi_i}{x_i} + \ln\frac{\Phi_i}{x_i}\right) = \left(1 - \frac{r_i}{r} + \ln\frac{r_i}{r}\right) \tag{19.5}$$

which, in a binary polymer–solvent mixture takes the following form, when the solvent is treated as mono-segmental ($r_s = 1$):

$$(\ln\gamma_s)_{FH} = \left(1 - \frac{1}{r_p}\right)\Phi_p + \ln\left(\frac{\Phi_s}{x_s}\right) \tag{19.5a}$$

The solvent activity, a_s, takes then the form:

$$(\ln a_s)_{FH} = (\ln\gamma_s x_s)_{FH} = \left(1 - \frac{1}{r_p}\right)\Phi_p + \ln(1 - \Phi_p) \tag{19.6}$$

C. When the intermolecular interactions are not negligible, the above equations should be completed by the enthalpic or the "interaction" term. For regular solutions, RST gives the following complete form for a binary mixture in terms of the solubility parameters, δ_i:

$$\left(\frac{g_0^E}{RT}\right)_{RST} = \left(x_1 \ln\frac{\Phi_1}{x_1} + x_2 \ln\frac{\Phi_2}{x_2}\right) + \left(x_1 v_1^L + x_2 v_2^L\right)\Phi_1\Phi_2\frac{(\delta_1 - \delta_2)^2}{RT} \tag{19.7}$$

Notably, Equation 19.7 cannot account for negative excess enthalpies. The corresponding FH equation is:

$$\left(\frac{g_0^E}{RT}\right)_{FH} = \left(x_s \ln\frac{\Phi_s}{x_s} + x_p \ln\frac{\Phi_p}{x_p}\right) + \left(x_s + x_p r_p\right)\Phi_s\Phi_p\chi_F \tag{19.8}$$

The FH interaction parameter, χ_F, is a dimensionless quantity defined as:

$$\chi_F = \frac{z}{2}\frac{\varepsilon_{ss} + \varepsilon_{pp} - 2\varepsilon_{sp}}{RT} = \frac{z\Delta\varepsilon_{sp}}{2RT} = \frac{v_s^L B}{RT} \tag{19.9}$$

where ε_{ij} is the interaction energy of the contact *i-j* and $z\Delta\varepsilon_{sp}/2$ the interchange energy, that is, the average energy change upon replacement of a solvent molecule by a polymer segment. B is the same energy per solvent molar volume. Thus, χ_F may take, both, positive and negative values, which makes Equation 19.8 more versatile than Equation 19.7. In fact, for data correlation purposes, χ_F is often treated as an adjustable binary parameter.

D. The complete forms of the activity coefficients are obtained from the corresponding equations for the excess Gibbs energy. From Equation 19.7 we obtain:

$$\left(\ln\gamma_1\right)_{RST} = \left(1 - \frac{\Phi_1}{x_1} + \ln\frac{\Phi_1}{x_1}\right) + \frac{v_1^L\Phi_2^2}{RT}\left(\delta_1 - \delta_2\right)^2 \tag{19.10}$$

and, from Equation 19.8, we obtain ($r_s = r_l = 1$):

$$\left(\ln\gamma_1\right)_{FH} = \left(1 - \frac{\Phi_1}{x_1} + \ln\frac{\Phi_1}{x_1}\right) + \Phi_2^2\chi_F = \left(1 - \frac{1}{r} + \ln\frac{1}{r}\right) + \Phi_2^2\chi_F \tag{19.11}$$

Similarly, the complete form of Equation 19.6 is:

$$\left(\ln a_s\right)_{FH} = \frac{\mu_s - \mu_s^0}{RT} = \left(1 - \frac{1}{r_p}\right)\Phi_p + \ln\left(1 - \Phi_p\right) + \Phi_p^2\chi_F \tag{19.12}$$

Equation 19.12 is often used for the determination of χ_F from experimental vapor pressure data over a polymer solution. Polymers are not volatile and, thus, the activity of the solvent may be equated to the ratio p_s/p_s^0 of the vapor pressure over the polymer solution, p_s, and the corresponding vapor pressure, p_s^0 of pure solvent, or $a_s = p_s/p_s^0$. The above set of equations is sufficient for some first calculations of thermodynamic quantities in polymer solutions.

LIQUID–LIQUID EQUILIBRIA AND THE FH MODEL

Polymer solutions exhibit both UCST (Upper Critical Solution Temperature) and LCST (Lower Critical Solution Temperature) behavior. The simple FH model can handle only UCST behavior. The LCST behavior requires somewhat more advanced models and will be examined in a later section.

As stated in Chapter 17, the stability criterion of a binary system is Equation 17.6,

$$\left[\frac{\partial^2 \Delta g_m}{\partial x_2^2}\right]_T = \left[\frac{\partial^2 g}{\partial x_2^2}\right]_T \geq 0 \tag{17.6}$$

If these second-order derivatives become negative (the g or Δg_m vs. x curve shows an inflection point) the system phase separates. When there is phase splitting or liquid–liquid equilibrium (LLE), there are two inflection points, as shown in the upper part of Figure 19.3. At the critical point these two inflection points coincide, and the following equation holds:

$$\left[\frac{\partial^2(\Delta g_m)}{\partial x_2^2}\right]_T = \left[\frac{\partial^3(\Delta g_m)}{\partial x_2^3}\right]_T = 0 \text{ (critical point)} \tag{19.13}$$

Recalling that $\Delta\mu_1 = \left(\partial\Delta g_m / \partial n_1\right)_{T,P}$, a more convenient equation for the critical point is:

$$\left[\frac{\partial\mu_1}{\partial\Phi_2}\right]_T = \left[\frac{\partial^2\mu_1}{\partial\Phi_2^2}\right]_T = 0 \tag{19.14}$$

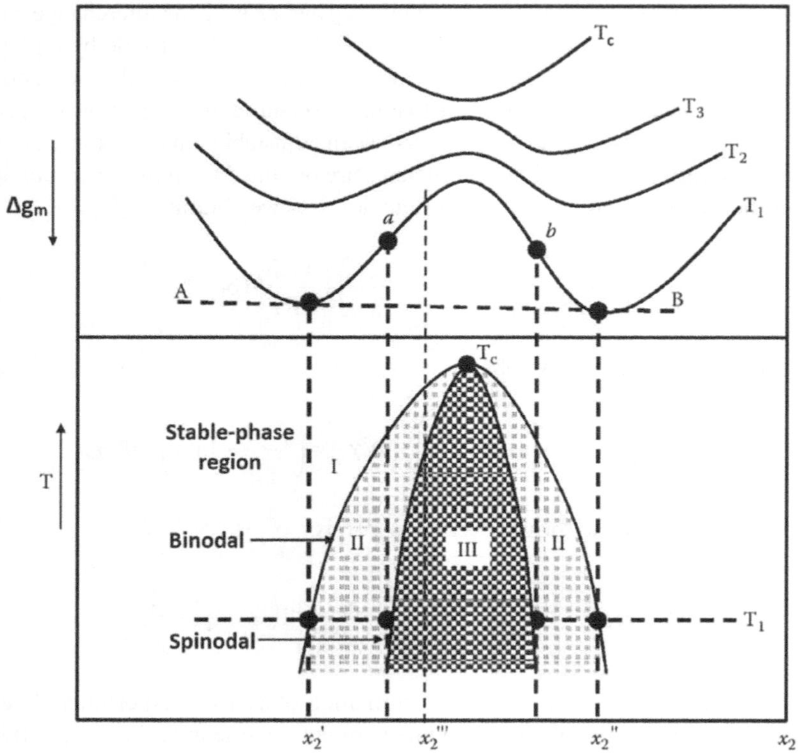

FIGURE 19.3 Upper part: Schematic presentation of the Δg_m curves vs. mole fraction, x_2. Curves of temperatures T_1, T_2, and T_3 correspond to partially miscible systems. Curve T_c corresponds to the critical temperature (UCST). Above temperature T_c the system is fully miscible. By drawing a common tangent, like line AB, the two contact points are projected, in the lower part, on the $T-x$ plane and form the binodal curve (solid line). The inflection points of curves T_1, T_2, and T_3, like points a and b, are projected to the instability/metastability or spinodal curve (dotted line). Thus, there are three regions in the $T-x$ plane. The region I outside the binodal is the stable region of full miscibility. Region II between binodal and spinodal is the metastable region. Region III inside the spinodal curve is the instability region. Adapted from Ref. [6] (fig. 8.2) with permission.

Applying Equation 19.14 in Equation 19.8, from the first derivative we obtain:

$$\frac{1}{1-\Phi_{2,cr}} - \left(1 - \frac{1}{r_p}\right) - 2\Phi_{2,cr}\chi_{cr} = 0 \tag{19.15}$$

where the subscript cr indicates critical point. And from the second derivative:

$$\frac{1}{\left(1-\Phi_{2,cr}\right)^2} - 2\chi_{cr} = 0 \tag{19.16}$$

From the last two equations, we obtain for the critical volume fraction:

$$\Phi_{2,cr} = \frac{1}{1+\sqrt{r_p}} \approx \frac{1}{\sqrt{r_p}} \tag{19.17}$$

and for the critical value of the FH χ parameter:

$$\chi_{cr} = \frac{1}{2}\left(1+\frac{1}{\sqrt{r_p}}\right)^2 = \frac{1}{2}+\frac{1}{\sqrt{r_p}}+\frac{1}{2r_p} \tag{19.18}$$

Thus, for high polymers ($r_p \to \infty$), we have χ_{cr}=0.5. Bad solvents for a polymer have χ_F values near 0.5 or higher. Good solvents have χ_F values less than 0.5. Negative χ_F values imply a very good solvent for the polymer. The typical temperature dependence of χ_F is given by the following equation:

$$\chi_F = a + b/T \tag{19.19}$$

Thus, as the temperature increases, the solvent becomes better for the polymer. This general behavior is the basis for the fractionation of polydisperse polymers, namely, mixtures of polymers of the same repeating unit but of different numbers of repeating units per polymer chain.

Fractionation of Polydisperse Polymers

Typical polymerizations produce polydisperse polymers and the need often arises for their fractionation into polymer samples of narrower molecular weight variation. This fractionation may be rationally designed by controlling the goodness of the solvent for the polymer. As discussed in the previous section, there are two ways to cause phase separation in a homogenous polymer solution: by changing the temperature or by adding a nonsolvent (or, bad solvent). Both ways increase the χ_F parameter and, thus, the fractionation may be quantified in terms of the χ_F parameter variation.

The number of polymer segments, r_p, is analogous and often close to the number of the repeating units of the polymer or its degree of polymerization (DP_n) and analogous to the molecular weight of the chain molecule. Equation 19.18 then, is a good guide for the solvent selection that best suits the fractionation of the polymer with the given r_p. Figure 19.4 shows graphically how the critical χ_F varies by varying DP_n of the very same polymer. When χ_F is lower than 0.5, all polymer fractions are fully dissolved by the solvent and the solution is homogenous. By increasing χ_F to 0.51, the system phase separates and, practically, only the polymer chains with DP_n lower than 10,000 remain dissolved in the solvent. The polymer-rich phase with the long chains is removed and their cleaned fraction is obtained. Further addition of bad solvent in the supernatant solution or increase of χ_F to ca. 0.522 leaves dissolved only the chains with DP_n less than 2000 while a further increase to ca. 0.531, leaves dissolved only the chains with DP_n less than 1000. In each such step, the removed phase is cleaned, typically, by solvent evaporation, and the corresponding fraction is obtained.

One may follow the reverse of the above process starting from a bad solvent and gradually decrease χ_F by the addition of a good solvent or by the increase of temperature. In this case one recovers first the low DP_n chains from the supernatant solution and gradually the chains of increasing DP_n.

DILUTE POLYMER SOLUTIONS

The above FH model is valid for (semi-) concentrated solutions but fails for dilute solutions. Yet, with some simple changes, it becomes a useful tool for the description of key aspects of the thermodynamic behavior of dilute polymer solutions.

In very dilute solutions, Φ_p is nearly zero. Starting from Equation 19.12, we may expand the term $\ln(1-\Phi_p)$ in a Taylor series as follows:

$$\ln(1-\Phi_p) = -\Phi_p - \Phi_p^2/2 - \Phi_p^3/3\ldots \tag{19.20}$$

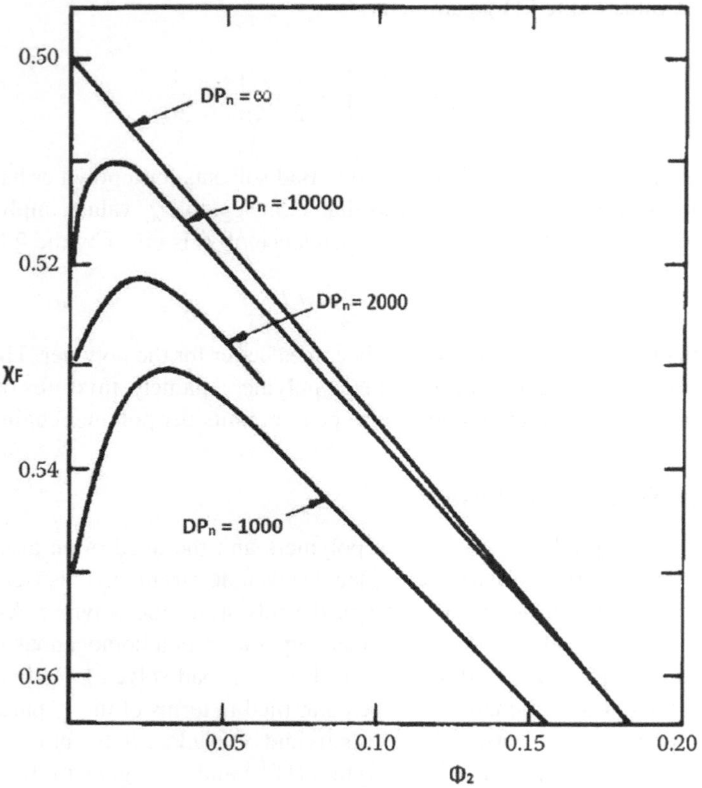

FIGURE 19.4 Plot of χ_F vs. polymer volume fraction, Φ_2 for various degrees of polymerization, DP_n. Adapted from Ref. [6] (fig. 8.3) with permission.

Retaining the first two terms only and replacing in Equation 19.12 we obtain:

$$\mu_s - \mu_s^0 = -RT\left[\frac{\Phi_p}{r_p} + \left(\frac{1}{2} - \chi_F\right)\Phi_p^2\right] \tag{19.21}$$

This equation is often written in a slightly different form by making use of the following two identities: $r_p = \text{v}_p^L / \text{v}_s^L$ and $\Phi_p = c_p \text{v}_{sp,p} = c_p\, \text{v}_p^L / M_p$, where c_p is the concentration of the solution (kg/L), $V_P = \text{v}_p^L$ and $V_s = \text{v}_s^L$ are the molar volumes of polymer and solvent, respectively, and $\text{v}_{sp,p}$ is the specific volume of the polymer. From their ratio, we obtain:

$$\frac{\Phi_p}{r_p} = \frac{c_p \text{v}_s^L}{M_p} \tag{19.22}$$

Replacing in Equation 19.21, we obtain the very useful equation:

$$\mu_s - \mu_s^0 = -RT\left[\frac{c_p \text{v}_s^L}{M_p} + \text{v}_{sp,p}^2\left(\frac{1}{2} - \chi_F\right)c_p^2\right] \tag{19.23}$$

Equation 19.23 becomes particularly useful when combined with the equation for the osmotic pressure, Equation 22.27:

$$\pi = \frac{RT}{\text{v}_s^L}(-\ln a_s) = -\frac{RT}{\text{v}_s^L}\left(\frac{\mu_s - \mu_s^0}{RT}\right) = -\frac{\mu_s - \mu_s^0}{\text{v}_s^L} \tag{19.24}$$

Replacing in Equation 19.23 we obtain:

$$\pi v_s^L = RT\left[\frac{c_p v_s^L}{M_p} + v_{sp,p}^2\left(\frac{1}{2} - \chi_F\right)c_p^2\right]$$

or

$$\frac{\pi}{c_p} = \frac{RT}{M_p} + RT\frac{v_{sp,p}^2}{v_s^L}\left(\frac{1}{2} - \chi_F\right)c_p \tag{19.25}$$

or

$$\frac{\pi}{RTc_p} = \frac{1}{M_p} + \frac{v_{sp,p}^2}{v_s^L}\left(\frac{1}{2} - \chi_F\right)c_p \tag{19.26}$$

Equation 19.26 is used for the determination of the (number-average) molecular weight, M_p, of the polymer from osmotic pressure measurements. A plot of π/RTc_p vs. c_p gives a straight line. The constant term is equal to $1/M_p$ and the slope is equal to the osmotic second virial coefficient, B:

$$B = \frac{v_{sp,p}^2}{v_s^L}\left(\frac{1}{2} - \chi_F\right) = \frac{\left(v_p^L\right)^2}{v_s^L M_p^2}\left(\frac{1}{2} - \chi_F\right) \tag{19.27}$$

As observed, B is a measure of polymer–solvent interactions. Positive B (or $\chi_F<0.5$) indicates a good solvent for the polymer, while negative B values indicate bad solvents for the same polymer. When $B=0$, the solution is considered ideal, and the osmotic pressure obeys the classical van't Hoff equation:

$$\frac{\pi}{c_p} = \frac{RT}{M_p} \quad \left(B = 0, \chi_F = 0.5\right) \tag{19.28}$$

The experimental osmotic pressure data in the above plot may not fall on a straight line with increasing polymer concentration. In these cases, as with gases, the osmotic virial equation of polymer solutions may be written in alternative more extensive ways containing higher-order virial coefficients:

$$\frac{\pi}{RTc_p} = \frac{1}{M_p} + Bc_p + B_3c_p^2 + \dots \tag{19.29}$$

or

$$\frac{\pi}{RTc_p} = \left(\frac{1}{M_p}\right)\left(1 + \Gamma_2 c_p + \Gamma_3 c_p^2 + \dots\right) \tag{19.30}$$

For good solvents ($\chi_F<0.5$), $\Gamma_3=(\Gamma_2/2)^2$ and Equation 19.30 takes the following form:

$$\frac{\pi}{c_p} = \left(\frac{RT}{M_p}\right)\left(1 + \frac{1}{2}\Gamma_2 c_p\right)^2 \tag{19.31}$$

Obviously, with Equation 19.31, the above plot is not linear, that is, Equation 19.31 may describe departures from linearity, as does Equation 19.30. One may still handle the experimental data and make them fall on a straight line by taking the square roots of both sides of Equation 19.31:

$$\sqrt{\frac{\pi}{c_p}} = \sqrt{\frac{RT}{M_p}}\left(1 + \frac{1}{2}\Gamma_2 c_p\right) \tag{19.32}$$

As observed, a plot of $\sqrt{\pi / c_p}$ vs c_p will lead to a straight line and, as previously, the constant term will give the molecular weight of the polymer and the slope will give the second virial coefficient $\Gamma_2 = BM_p$. Positive slopes will indicate good solvents for the polymer. Zero slope implies zero second virial coefficient. The solution is again called ideal, and the solvent is considered to be a Θ – solvent, in which the polymer conformations assume the unperturbed dimensions, as discussed above.

EXCLUDED VOLUME AND FLORY–KRIGBAUM MODEL

The above FH model deals with flexible polymer chains. But how flexible or stiff a chain is, depends on the length scale of observation. If the length scale is very small, all chains are rather stiff. If the length scale is large, the chains become flexible. Thus, for very long chains there is always a length scale for which the polymer chain behaves as a freely jointed chain obeying the random-flight Equation 19.1 with $C_n = 1$, which is now written as:

$$\langle r_n^2 \rangle = n_e l_e^2 \tag{19.33}$$

The effective number of segments, n_e, is different from n and the effective or *Kuhn length*, l_e, of each of these segments is also different from the bond length, l, of Equation 19.1. But, since the chain is now a freely jointed chain, its *contour length*, l_c, is the length of the fully stretched chain or $l_c = n_e l_e$. Replacing in Equation 19.33, we see that the Kuhn length is just the ratio of $\langle r_n^2 \rangle$ and the contour length. In flexible chains $l_e \approx l$ while in stiff chains $l_e \gg l$. Just to give an idea of the stiffness of real polymers, the Kuhn length of polyethylene is 1.24 nm (while the C–C bond length is 0.154 nm). For the more flexible poly(ethylene oxide) $-(CH_2O)_n-$, the Kuhn length is 0.34 nm, while for the stiff double helix of DNA, in which $l = 0.35$ nm, the Kuhn length varies from 30 to 100 nm.

Apart from the chain stiffness, the conformation of a real polymer chain is also dictated by the fact that two chain segments cannot occupy the same site in the volume of the system. Thus, in reality, the chain conformation is obtained not by a random flight or *random walk* but rather by a *self-avoiding walk (saw)*, which tends to expand the chain conformation. This brings into the discussion the important concept of *excluded volume*.

The excluded volume in polymers refers, precisely, to the idea that one part of a long chain molecule cannot occupy space that is already occupied by another part of the same molecule. It is a volume correction relative to the ideal behavior of the chain at Θ conditions, as mentioned above. This is analogous to the definition of the second virial coefficient of gases by Equation 27.15 relative to the ideal gas behavior (although the Θ-temperature is more alike the Boyle point of gases). The excluded volume encompasses both repulsive and attractive contributions. Thus, if $U(r)$ is the intermolecular potential for the polymer segment interactions, the excluded volume, v_{ex}, for the interacting segments is obtained from the following defining equation (cf. Equation 27.15 for comparison):

$$\mathrm{v}_{ex} = -2\pi \int \left\{ \exp\left[-U(r)/kT\right] - 1 \right\} r^2 dr \tag{19.34}$$

At high temperatures, the attractive contribution of the potential is negligible and, thus, the excluded volume is dominated by the repulsive contribution and is proportional to the second virial coefficient of the polymer in the solution. In contrast, at very low temperatures the attractive contribution predominates and the polymer collapses. The polymer chain behaves ideally at the intermediate temperature, Θ, where the chain conformation takes the unperturbed dimensions dictated by the chain flexibility or chain stiffness.

As mentioned already, the direct consequence of the excluded volume on the chain conformation is its expansion relative to the ideal behavior. Focusing on a single coil at high temperatures well above Θ, its free energy is the interplay of two opposing factors, the energetic one arising from the interaction potential dominated by the segment repulsions that tend to expand the coil and the entropic or elasticity one, due to its most probable random coil compact conformation, that opposes this expansion. The final coil dimensions will be determined by minimization of this free energy with respect to its radius. Thus, accounting for the excluded volume, the end-to-end distance of a chain of n segments, scales (that is, leaving minor details. varies) with n as:

$$\sqrt{\langle r_n^2 \rangle}_{saw} \propto \left(Bnl^2\right)^{3/5} = Cn^{3/5} \tag{19.35}$$

This equation should be contrasted with the equations of random-walk statistics (cf. Equation 19.1), in which the end-to-end distance scales with n as $n^{1/2}$ rather than $n^{3/5}$. Both expressions are approximations.

The scaling of the end-to-end distance with the polymer size is generalized as $\sqrt{\langle r_n^2 \rangle} \propto n^{\nu}$, where the exponent ν is a characteristic parameter of the polymer chain and function of the physical space available to the polymer chain, often expressed as $\nu = 3/(d+2)$, where d is the dimensionality of the system. On a flat surface $d=2$, in a three-dimensional volume $d=3$. Thus, for a chain of $n=10{,}000$, in a very dilute solution in a good solvent $n^{3/5}=251$, while in somewhat more concentrated solutions and in the ideal case of Θ conditions (cf. Equation 19.1) $n^{1/2}=100$. Thus, in a good solvent, the chain expands significantly when $\nu=3/5=0.6$ (swollen chain) compared to the case of $\nu=0.5$ (ideal chain). For two-dimensional systems, $\nu=0.75$, which describes very well the conformations of adsorbed chains on surfaces, such as lipid surfaces.

The above originates from the seminal work of Flory and Krigbaum [7] on dilute polymer solutions. In contrast to the FH model, Flory and Krigbaum have derived the entropy contribution of the polymer chain to the chemical potential via an enumeration of probable chain arrangements by excluding sites already occupied by segments of the chain. Their result is Equation 19.36:

$$\mu_s - \mu_s^0 = -RT\mathrm{v}_s^L\left[\frac{c_p}{M_p} + \frac{1}{2}\frac{N_A \mathrm{v}_{ex}}{M_p^2}c_p^2\right] \tag{19.36}$$

Substituting in Equation 19.24 for the osmotic pressure, we obtain:

$$\pi = RT\left(\frac{c_p}{M_p} + \frac{1}{2}\frac{N_A \mathrm{v}_{ex}}{M_p^2}c_p^2\right) \quad \text{or} \quad \frac{\pi}{RTc_p} = \frac{1}{M_p} + \frac{1}{2}\frac{V_{ex}}{M_p^2}c_p \tag{19.37}$$

where V_{ex} is the molar excluded volume. By comparing Equation 19.37 with Equation 19.27, we may now define the second virial coefficient as follows:

$$B = \frac{1}{2}\frac{V_{ex}}{M_p^2} \tag{19.38}$$

This equation shows that the excluded volume is directly proportional to the second virial coefficient and with a rather equivalent content.

Flory and Krigbaum have considered the last term in Equation 19.21 to represent the excess chemical potential of the solvent for very long polymer chains, or

$$\left(\mu_s - \mu_s^0\right)^E = \Delta h_s^E - T\Delta s_s^E = -RT\left(\frac{1}{2} - \chi\right)\Phi_p^2 \tag{19.39}$$

and expressed the enthalpic and the entropic contributions to the excess chemical potential as:

$$\Delta h_s^E = RT\kappa\Phi_p^2 \quad \text{and} \quad \Delta s_s^E = R\psi\Phi_p^2 \tag{19.40}$$

Replacing in Equation 19.39, we obtain:

$$\psi - \kappa = \left(\frac{1}{2} - \chi_F\right) \tag{19.41}$$

At Θ conditions, $\chi_F = ½$. Thus, at $T = \Theta$ we also have $\psi = \kappa$ or $\Delta h_s^E = T\Delta s_s^E$. The temperature Θ may, then, be defined as follows:

$$\Theta = T\frac{\kappa}{\psi} \tag{19.42}$$

With this definition, Equation 19.39 may be written as:

$$\left(\mu_s - \mu_s^0\right)^E = -RT\psi\left(1 - \frac{\Theta}{T}\right)\Phi_p^2 \tag{19.43}$$

As observed with Equation 19.43, at $T = \Theta$ the excess chemical potential becomes zero, or the activity coefficient of the solvent is equal to one and the polymer conformation has the unperturbed dimensions. This is what is called *ideal behavior* in this context. At temperatures higher than Θ, the polymer swells and shrinks at temperatures lower than Θ.

Experimental LLE data, like those of Figure 19.4, may be used to determine Θ. Combining Equation 19.18 with Equations 19.41 and 19.42, we obtain:

$$\psi\left(\frac{\Theta}{T_c} - 1\right) = \frac{1}{\sqrt{r_p}} + \frac{1}{2r_p} \tag{19.44}$$

Equation 19.44 may be written, alternatively, as ($r_p = DP_n$):

$$\frac{1}{T_c} = \frac{1}{\Theta}\left\{1 + \frac{1}{\psi}\left(\frac{1}{\sqrt{DP_n}} + \frac{1}{2DP_n}\right)\right\} \tag{19.45}$$

By plotting $1/T_c$ vs $\left(\frac{1}{\sqrt{DP_n}} + \frac{1}{2DP_n}\right)$ we obtain a straight line. From the intercept we obtain Θ and from the slope we obtain ψ. Again, for very long chain molecules, the temperature Θ is equal to the UCST. Table 19.1 shows values of Θ for some common polymers.

Flory has defined the *linear expansion coefficient*, relative to the unperturbed chain conformation, as follows:

$$\alpha = \sqrt{\langle R_g^2 \rangle} \Big/ \sqrt{\langle R_g^2 \rangle}_\Theta \tag{19.46}$$

and derived still another useful equation for Θ:

$$\alpha^5 - \alpha^3 = 2C\psi\left(1 - \frac{\Theta}{T}\right)M_p^{1/2} \tag{19.47}$$

C is a system parameter. Combining Equations 19.43 and 19.47, one may obtain equations interconnecting the excess chemical potential with the expansion coefficient of the coil.

The above FH model and its variations, including the Flory–Krigbaum one, although very useful for the interpretation of numerous thermodynamic aspects of polymer solutions, also has severe limitations, one of the most notable being its inability to account for the LCST behavior of polymer solutions and polymer mixtures. This inability arises, mainly, from the fact that the FH model cannot account for the volumetric behavior of polymer solutions, that is, the effect of external conditions of temperature, pressure, and composition on the density of the polymer mixture. It also cannot account for strong specific interactions, like hydrogen bonding, which may also lead to LCST behavior. Thus, in the rest of this chapter and in the next chapter we focus on approaches that may account for the first effect, namely, on the equation-of-state (EOS) approach. The effect of hydrogen bonding will be examined in Chapter 32.

THE LCST BEHAVIOR AND THE EQUATION-OF-STATE APPROACH TO POLYMER SOLUTIONS

So far, we have dealt with, mainly, nonpolar polymer – solvent mixtures, in which the solubility was decreasing with temperature decrease and was leading to the UCST behavior. As discussed in Chapter 17 and shown schematically in Figure 19.3, the negative value of the mixing Gibbs energy, $\Delta g^E = \Delta h^E - T\Delta s^E$, is a necessary but not sufficient condition for the stability of the mixture, and phase separation may occur. In these systems, both Δs^E and Δh^E are typically positive. As the temperature is lowered, the negative term, $[-T\,\Delta s^E]$, diminishes and Δg^E may become positive leading to phase separation.

The values of Θ and ψ in Table 19.1 show that the entropic contribution, ψ, at the Θ temperature in the first six systems is positive, while for the seventh it is negative. But at Θ temperature $\psi = \kappa$. As the temperature increases, the first six systems become more stable due to the negative $[-T\,\Delta s^E]$ term. In contrast, in the seventh system, this term is positive and increases with temperature. When it overcomes the negative Δh^E term, the system phase separates exhibiting LCST behavior. This particular system is stable at lower temperatures due to the *hydrogen-bonding* interactions between the poly(acrylic acid) and the base solvent dioxane. As the temperature increases the strength of this hydrogen-bonding interaction diminishes and the system becomes unstable and phase separates. Figure 19.5 shows schematically the simultaneous UCST and the LCST behavior. LCST behavior is also very common in polymer systems, even in nonpolar polymer solutions, but for a different reason.

In the case of ordinary low molar mass compounds, the phase stability classification is well-established in the following scheme [8]:

1. UCST:	$\Delta h^E>0;\ \Delta s^E>0.$	(a) $\Delta V_m>0;$	(b) $\Delta V_m<0.$
2. LCST:	$\Delta h^E<0;\ \Delta s^E<0.$	–	(b) $\Delta V_m<0.$

Systems exhibiting LCST behavior with volume expansion upon mixing ($\Delta V_m > 0$) are not known. Polymer solution behavior conforms to the above scheme and as shown below, it is common for polymer solutions to exhibit both UCST and LCST behaviors (cf. Figure 19.5). This implies, according

TABLE 19.1
Typical Values of Θ and ψ for Common Polymers

Polymer	Solvent	ψ	Θ (K)
Polyethylene (PE)	Nitrobenzene	1.09	503
Polystyrene (PS)	Cyclohexane	0.79	307
Polystyrene (PS)	Methylcyclohexane	0.61	342.3
Polyisobutylene (PIB)	Di-iso-butyl ketone	0.65	333.1
Poly(methyl methacrylate) (PMMA)	4-Heptanone	0.61	305
Poly(dimethyl siloxane) (PDMS)	Chlorobenzene	0.19	255
Poly(acrylic acid) (PAA)	Dioxane	–0.31	302.2

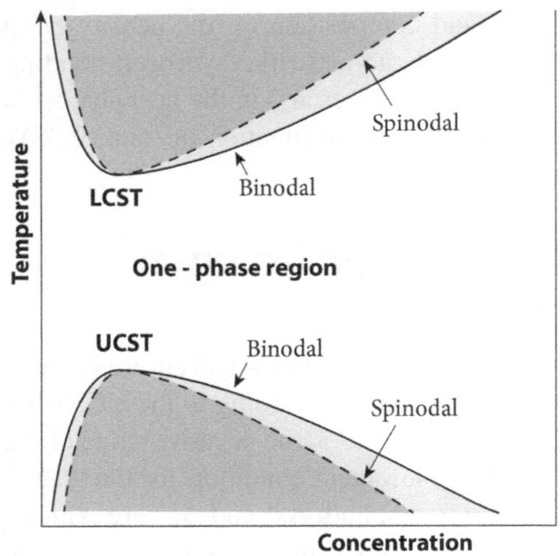

FIGURE 19.5 Schematic presentation of UCST and LCST behavior of polymer solutions. Above LCST and below UCST the system phase separates.

to the above scheme, that mixing enthalpies and entropies must change sign with temperature. Such a change may not be needed in volumes of mixing.

In polymer solutions, the LCST behavior is determined experimentally by heating the solution in a sealed glass tube up to temperatures often approaching the critical temperature of the solvent. As the temperature increases the liquid solvent expands much faster than the polymer but the expansion of the solvent is restricted due to the presence of the polymer leading to a drastic reduction of its partial molar volume relative to its molar volume, or to a negative excess volume. This reduction implies an entropy reduction which, when significant, will lead to phase separation and is referred to as LCST behavior of *volumetric* or *free-volume* origin. This asymmetry of thermal expansivity and compressibility of solvent and polymer cannot be accounted for by the plain FH model [4,5], and this brings in the discussion of *the equation-of-state* (*EOS*) approach.

There are numerous EOS approaches of varying degrees of approximation to the thermodynamics of polymer solutions, today. Some of the most widely used EOS models, till today, were reviewed already by Lambert et al. [9] and were divided into four categories:

- *cell* models,
- *lattice-fluid* models,
- *cell-hole* models, and
- *tangent-sphere* or *perturbation-theory* models.

The most representative of the cell models is the **F**lory–**P**atterson–**P**rigogine (FPP) model which was a significant advancement in polymer solution thermodynamics and paved the way for numerous refinements. The central role of free volume on the LCST behavior of polymer systems was initially and systematically explored by this FPP model. The FPP approach is a model of the liquid state and will be discussed in Chapter 29 along with the popular and versatile **L**attice-**F**luid (LF) model, which is a model of both the liquid and the gaseous state. The cell-hole models combine elements of cell and LF models and often show improved correlation of volumetric data of polymers, but they are models of the liquid state only.

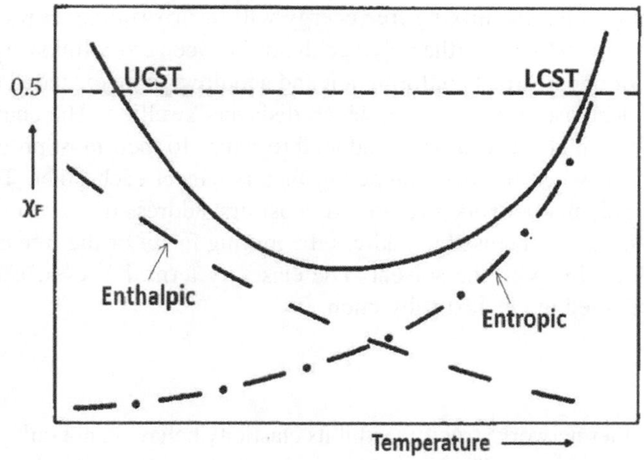

FIGURE 19.6 The variation of χ_F (solid line) with temperature according to the EOS approach. The enthalpic term (- - -) predominates at low T leading to UCST behavior. At high T the entropic (free-volume) term (-. -.) predominates leading to LCST behavior. Adapted with permission from Ref. [10] (fig. 1b).

The tangent-sphere models attract very much interest today and seem quite promising. The family of the PHCT (**P**erturbed **H**ard **C**hain **T**heory) models will be also discussed in Chapter 29 and the family of SAFT (**S**tatistical **A**ssociating **F**luid **T**heory) models will be discussed in Chapter 33.

The χ_F parameter variation with temperature, which is shown schematically in Figure 19.6, is more or less common to the above EOS approaches. As shown in the figure, at lower temperatures, the enthalpic term predominates while at higher temperatures it is the entropic or free-volume term that predominates and leads to the LCST behavior.

Besides the above models, two lattice models that attract much interest today are the LCT (**L**attice **C**luster **T**heory) of Freed and coworkers, and the BGY (**B**orn–**G**reen–**Y**von) model of Lipson and coworkers which uses the BGY integral-equation approach as an extension of the LF approach. Both models will be touched upon in Chapter 30.

To the above, one should add the **M**olecular **D**ynamics (MD) and **M**onte-**C**arlo (MC) simulation approaches. There is much progress and very promising applications of MD and MC simulations of pure polymers and polymer mixtures, although, the statement of Binder et al. [11] that "the study of phase equilibria in polymer solutions by simulation methods is still in its infancy" remains, to some extent, a rather valid one. However, today's intense efforts for an efficient combination and unification of programming frameworks of artificial intelligence with MD simulations produce much hope for significant advances in the field.

So far, the polymer molecules have been considered as independent linear chains. When the polymer chains are interconnected extensively, the thermodynamic behavior changes drastically and this is discussed in the next section.

SWELLING OF POLYMER NETWORK STRUCTURES

Crosslinking is the process of formation of covalent bonds which hold together portions of different polymer chains. When the polymer chains are extensively cross-linked, a three-dimensional random network structure is produced which is not soluble in any solvent. In contact with a solvent or its vapor, this network structure may expand or swell to a varying degree depending on the goodness or affinity of the solvent for the polymer. There are, in fact, two competing factors which determine this degree of swelling at equilibrium, the *mixing* and the *elasticity* ones. Depending on the polymer–solvent affinity, the mixing free energy will favor swelling. However, as the network structure swells, the *network chains* (the polymer chains between cross-links) are forced to abandon their most probable or unperturbed conformation and acquire a more extended (less probable) one, causing an entropy decrease in the system, which disfavors swelling. The chains in the extended conformations acquire an elastic character and tend to return to their most probable conformation. Equilibrium is reached when the two competing factors cancel each other. Thus, to discuss the thermodynamics of polymer network swelling we must first address these two antagonistic factors.

The above FH model may be used to address the mixing factor or the free energy change upon mixing the network chains with the solvent. The elasticity term, however, is a new one and the essentials of it are recalled in the next subsection.

Rubber Elasticity

The rubber-like polymer network structure exhibits elasticity behavior, not only in the swollen state, but in its pure (dry) state as well. Thus, a dry rubber band or specimen of volume, V, of the network structure will be considered subject to an elongation force, f, in the x direction until equilibrium is reached at the external temperature T and pressure P conditions. The force exerted elongates the specimen by dL producing an additional work $dW = fdL$. Assuming the deformation process is reversible, the first law of thermodynamics for the specimen gives:

$$dU = dq + dW = TdS - PdV + fdL \tag{19.48}$$

Differentiating the Helmholtz free energy, $A = U - TS$, and substituting for dU from Equation 19.48, gives:

$$dA = -SdT - PdV + fdL \tag{19.49}$$

from which we obtain:

$$f = \left(\frac{\partial A}{\partial L}\right)_{T,V} = \left(\frac{\partial U}{\partial L}\right)_{T,V} - T\left(\frac{\partial S}{\partial L}\right)_{T,V} \tag{19.50}$$

This equation may be written in the following more useful alternative form (cf. Problem 19.4):

$$f = \left(\frac{\partial U}{\partial L}\right)_{T,V} + T\left(\frac{\partial f}{\partial T}\right)_{P,dL} \tag{19.51}$$

These are the two central equations of the thermodynamics of rubber elasticity. In direct analogy with the ideal gas, the internal energy of an ideal rubber does not change upon deformation. According, then, to Equation 19.50, the force opposing its elongation is of a pure entropic character. On the other hand, Equation 19.51 explains a most peculiar facet of rubber elasticity and tells that the (positive) force required to keep the elongation constant, increases as the temperature increases!

As mentioned already, the elasticity contribution to the free energy change of network deformation is of purely entropic character and arises from the departure of the conformation of network chains from their unperturbed conformation. A simple way to obtain an expression for the elasticity contribution to the free energy change, ΔA_{el}, is to consider conformations of Gaussian network chains of nearly equal size subject to an *affine* deformation in the rubber-like network specimen (displacement of cross-links linear in the macroscopic specimen strain) (cf. Problem 19.5). If L_{x0} is the initial length of the rectangular parallelepiped specimen in the x direction and L_x the length after deformation, the corresponding extension ratio is, $\alpha_x = L_x/L_{x0}$, and similarly for the other directions. With this notation, the obtained expression for ΔA_{el} is:

$$\Delta A_{el} = -T\Delta S_{el} = \frac{n_c kT}{2}\left(\alpha_x^2 + \alpha_y^2 + \alpha_z^2 - 3\right) \tag{19.52}$$

where n_c is the number of network chains.

Equation 19.52 is basic to the molecular theories of rubber-like elasticity and is used to obtain the elastic equations of state for any type of deformation. In the typical case of elongation, the deformation occurs at essentially constant volume ($\alpha_x \alpha_y \alpha_z = 1$). Thus, if the network is stretched in the x direction by the amount $\alpha_x = \alpha > 1$, it would have its perpendicular dimensions compressed by the amounts $\alpha_y = \alpha_z = \alpha^{-1/2} < 1$. Accordingly, Equation 19.52 becomes:

$$\Delta A(T) = \left(\frac{n_c kT}{2}\right)\left[\alpha^2 + \frac{2}{\alpha} - 3\right] \text{(elongation)} \tag{19.53}$$

Combining with Equation 19.50 (where $L = \alpha L_0$), the elastic force is given by:

$$f = \left(\frac{\partial \Delta A}{\partial L}\right)_{T,V} = \left(\frac{n_c kT}{L_0}\right)\left[\alpha - \frac{1}{\alpha^2}\right] \tag{19.54}$$

and the engineering stress, σ, by:

$$\sigma = \frac{f}{A_0} = \frac{n_c kT}{V_0}\left(\alpha - \frac{1}{\alpha^2}\right) = E_{el}\left(\alpha - \frac{1}{\alpha^2}\right) \tag{19.55}$$

A_0 is the undeformed cross-sectional area and $V_0 = L_0 A_0$ is the volume of the specimen. E_{el} is referred to as the *modulus of elasticity* of the rubber-like network. Again, the peculiarity of this modulus of elasticity is the increase with increasing temperature, implying that the material becomes stronger or less elastic with increasing temperature!

Flory [12,13] has used statistical mechanics for the calculation of the entropy of deformation, based on the consideration of the entropy of formation of the network from a corresponding set of independent chains and obtained the following alternative equation for the free energy change:

$$\Delta A_{el} = \frac{n_c kT}{2}\left[\alpha_x^2 + \alpha_y^2 + \alpha_z^2 - 3 - \ln(\alpha_x \alpha_y \alpha_z)\right] \tag{19.56}$$

For an ideal rubber the volume does not change upon deformation and the logarithmic term in Equation 19.56 is just zero. However, on swelling the network, the volume changes and the above logarithmic term is no longer zero, although often rather small. Thus Equation 19.56 is preferred (over Equation 19.52) for the study of swelling including the general case of anisotropic swelling. For the more typical case of isotropic swelling ($\alpha_x = \alpha_y = \alpha_z = \alpha$) the equation is further simplified to:

$$\Delta A_{el} = \frac{3n_e kT}{2}\left(\alpha^2 - 1 - \ln \alpha\right) \tag{19.57}$$

In fact, Flory is replacing n_c by the *effective number*, n_e, of network chains, that is, chain strands between two cross-linkages and members of internal closed loops contributing to the network elasticity. Chain ends and other defects do not contribute to this elasticity.

If n_p is the number of primary uncrosslinked polymer chains and ν_0 is the total number of cross-links in the network, the effective number of cross-links is just $\nu_e = \nu_0 - n_p$. The number n_e of effective network chains will be twice this quantity, that is,

$$n_e = 2\nu_e = 2\nu_0\left(1 - n_p / \nu_0\right) \tag{19.58}$$

If no defects were present, we should have an "idealized" network with $n_e = 2\nu_0$. If ρ is the density of the dry network and M_n is the (number-average) molar mass of the primary polymer chains, the number n_p is given by:

$$\frac{n_p}{V_0} = \frac{\rho N_A}{M_n} \tag{19.59}$$

where N_A is Avogadro's number and V_0 the volume of the dry polymer network. Similarly, the following equation could be written:

$$\frac{2\nu_0}{V_0} = \frac{\rho N_A}{M_c} \tag{19.60}$$

where M_c should be considered as the molar mass of the idealized network chain, in which loose ends and defects are absent. Combining the last three equations, we obtain:

$$n_e = 2\nu_0\left(1 - \frac{2M_c}{M_n}\right) = \frac{\rho V_0}{M_c}\left(1 - \frac{2M_c}{M_n}\right)N_A \tag{19.61}$$

This equation, along with Equation 19.57, will be used in the next subsection on swelling.

Swelling and Collapse of Polymer Network Structures: Hydrogels

As seen above, when the polymer network swells, it is also deformed but, this time, the volume changes until equilibrium is reached. At equilibrium, the chemical potential of the solvent in the swollen structure, μ_s, and in its pure state, μ_s^0, are equal. The chemical potential in the swollen structure consists of two terms [12,14], $\mu_{s,mix}$ and $\mu_{s,el}$, the mixing and the elasticity ones, respectively. Thus, at equilibrium, the following equation applies:

$$\mu_{s,mix} + \mu_{s,el} = \mu_s^0 \quad \text{or}$$

$$\frac{\mu_{s,mix} - \mu_s^0}{RT} = -\frac{\mu_{s,el}}{RT} \tag{19.62}$$

The left-hand side of this equation may be well approximated by Equation 19.12 of the FH model for polymer–solvent mixtures. For the right-hand side, the elasticity-free energy change, Equation 19.57, must be differentiated with respect to the number of solvent molecules, n_s. Prior to this, it should be recalled that the extent of swelling or the swelling ratio of the polymer network may be expressed in terms of the extension ratio, α, as follows:

$$\alpha^3 = \frac{V}{V_0} = \frac{V_0 + n_s V_s}{V_0} \tag{19.63}$$

V_s is the molecular volume of the solvent and V_0 is the initial volume of the unswollen polymer network. Flory and Rehner [14] have considered that the primary polymer chains were cross-linked in their bulk pure (melt) state and, thus, V_0 was set equal to the volume of the plain polymer network or, $\alpha^3 = 1/\Phi_p$. Combining Equations 19.57 and 19.63, the following equation for the elastic part of the chemical potential of the solvent is obtained:

$$\mu_{s,el} = \left(\frac{\partial \Delta A_{el}}{\partial n_s}\right)_{T,P} = \left(\frac{\partial \Delta A_{el}}{\partial \alpha}\right)_{T,P}\left(\frac{\partial \alpha}{\partial n_s}\right)_{T,P} = \left\{\frac{3n_e kT}{2}\left[2\alpha - \frac{1}{\alpha}\right]\right\}\left\{\frac{1}{3\alpha^2}\frac{V_s}{V_0}\right\} \tag{19.64}$$

Substituting from Equation 19.61, Equation 19.64 may be conveniently re-expressed in terms of the polymer volume fraction, Φ_p, as follows:

$$\frac{\mu_{s,el}}{RT} = n_e \frac{V_s}{V_0}\left\{\left[\Phi_p^{1/3} - \frac{\Phi_p}{2}\right]\right\} = \frac{\rho V_s}{M_c}\left(1 - \frac{2M_c}{M_n}\right)\left[\Phi_p^{1/3} - \frac{\Phi_p}{2}\right] \tag{19.65}$$

Thus, replacing Equations 19.12 and 19.65 in Equation 19.62, the following equation for the swelling at equilibrium is obtained:

$$\Phi_p + \ln\left(1 - \Phi_p\right) + \Phi_p^2 \chi_F = \rho V_s \left(\frac{2}{M_n} - \frac{1}{M_c}\right)\left[\Phi_p^{1/3} - \frac{\Phi_p}{2}\right] \tag{19.66}$$

This is a very important equation and may be used in a variety of ways, depending on what is known and what is sought to be estimated. If the FH's parameter, χ_F, and the molar mass of the network chains, M_c, are known, the equation may be used to predict the degree of the network swelling by the solvent. This degree of swelling may often be large, of the order of several hundreds, especially, in the case of superabsorbent materials. If the degree of swelling is measured, Equation 19.66 may be used to estimate M_c or the degree of crosslinking.

For the derivation of Equation 19.66 it was considered that the cross-links were introduced in the dry polymer network or dry *gel*. The case of introducing cross-links in polymer chains, already in solution, may be considered as well [15]. In this case, the swollen gel state should be distinguished from the *relaxed* state where the cross-links were introduced, and Equation 19.66 should be modified accordingly. If w_p is the weight of the polymer, the volume of the dry polymer is $V_p = w_p/\rho$. The volume fraction of the polymer in the swollen gel of volume V_{sw} is $\Phi_{p,sw} = V_p/V_{sw}$. Similarly, the volume fraction of the polymer in the relaxed state of volume V_r is $\Phi_{p,r} = V_p/V_r$. In this case, then, Equation 19.66 is changed to Equation 19.67 [15]:

$$\Phi_{p,sw} + \ln\left(1 - \Phi_{p,sw}\right) + \chi_F \Phi_{p,sw}^2 = \rho V_s \left(\frac{2}{M_n} - \frac{1}{M_c}\right)\left[\left(\frac{\Phi_{p,sw}}{\Phi_{p,r}}\right)^{1/3} - \frac{1}{2}\frac{\Phi_{p,sw}}{\Phi_{p,r}}\right]\Phi_{p,r} \tag{19.67}$$

Equations 19.66 and 19.67 are key equations for the study and technological development of gels.

When the solvent is water, the gels are referred to as *hydrogels*. Hydrogels are attractive materials for biomedical applications, such as controlled drug delivery and tissue engineering. These applications exploit not only the swelling capacity of hydrogels but also their controlled *collapse*, that is, their shrinkage and abrupt reduction of their swelling ratio under the influence of external stimuli, such as temperature, pH, co-solvents, etc.

Poly(N-isopropyl acrylamide) (PNIPAM) and a number of its derivatives are examples of hydrophilic polymers soluble in water at ambient conditions. Their aqueous solutions exhibit an LCST at around 35°C, which is close to the physiological temperature of human bodies. Above LCST, the aqueous solution phase separates. If PNIPAM is cross-linked, it swells significantly with water at ambient conditions. By heating this hydrogel, it collapses when the temperature reaches the LCST region of the aqueous (uncrosslinked) polymer solution. If the hydrogel were loaded with a drug or other active compound, it would release it upon collapse, and this is the basis for various biomedical applications.

The origin of this LCST behavior of hydrogels is different from that of the LCST discussed in the previous sections. Hydrophilic polymers, like PNIPAM, exhibit strong hydrogen-bonding interactions with water and form homogenous aqueous solutions. At higher temperatures, however, these interactions weaken, the polymer solubility diminishes and the solution phase separates.

The plain FH model with a constant χ_F parameter is not adequate to describe hydrogel collapse unless χ_F is expressed as an empirical function of composition and temperature. Swollen rubbers or gels in the absence of such strong intermolecular interactions may still collapse upon temperature increase as the LCST of the volumetric origin of the (uncrosslinked) polymer solution is approached. The collapse temperature of hydrogels also changes as the external pressure changes. The appropriate molecular thermodynamic approach for the phase behavior of polymer solutions and gels requires, then, not only the EOS features discussed in connection with Figure 19.6, but also efficient accounting for the strong hydrogen-bonding interactions, as discussed in Chapter 32, and, eventually, other strong electrostatic interactions.

PROBLEMS

19.1 **Average end-to-end distance of a polymer chain**

An important and representative measure of the conformation compactness of a polymer chain is the average distance between the two chain ends, usually written as $\sqrt{\langle r_n^2 \rangle}$, obtained by applying random-walk statistics in a Gaussian chain of n segments. This average end-to-end distance is given by Equation 1:

$$\bar{r}_n^2 = \int_0^\infty r^2 w(n,r)dr \tag{1}$$

where the *radial probability density*, *w*(*r*, *n*), of the freely jointed Gaussian chain is given by Equation 2:

$$w(n,r) = 4\pi r^2 \left(\frac{3}{2\pi n l^2}\right)^{3/2} \exp\left(-\frac{3r^2}{2nl^2}\right) \tag{2}$$

Show that by replacing Equation 2 in Equation 1 and integrating, the following remarkably simple equation is obtained for the freely jointed chain:

$$\sqrt{\langle r_n^2 \rangle} = l\sqrt{n} \tag{3}$$

19.2 **Estimation of parameter χ_F**

Polyisobutylene (PIB) of molar mass of 1,000,000 Da, and density of 0.915 g/cm³, is in thermodynamic equilibrium with propane vapors at 35°C, where the propane vapor pressure is 9050 mmHg and the density is 0.490 g/cm³. The concentration of absorbed propane in the mass of PIB at various partial pressures of propane is given in the following Table 19.2 [16].

Use the data of the table and obtain χ_F. Make also a graphical estimation of χ_F.

19.3 **Osmometry**

A typical osmometer consists of two compartments for the polymer solution and the pure solvent, which are separated by a semi-permeable membrane. The osmotic pressure is measured by the rise, *h*, of the solution in a glass tube attached on the compartment with

TABLE 19.2
Partial pressures of propane over PIB solution at 35°C [16]

P_1 (mmHg)	496	941	1446	1452
c, g C_3H_8/g PIB	0.0061	0.0116	0.0183	0.0185

the polymer solution. The following data were obtained for a polymer solution in a solvent of density $\rho = 0.850$ g/cm³ at 25°C (Table 19.3):

Estimate the molecular weight of the polymer and the second virial coefficient.

19.4 **The elastic force of a rubber**

Derive Equation 19.51 by using the Gibbs free energy of the deformed rubber specimen.

19.5 **Calculation of the entropy of rubber deformation**

The end-to-end distances of the stretched and unstretched chains of the figure may be calculated by the Gaussian distribution function:

$$W_n(x,y,z) = \left(\frac{3}{2\pi\langle r^2\rangle_0}\right)^{3/2} \exp\left(-\frac{3r^2}{2\langle r^2\rangle_0}\right) \quad (1)$$

where, $r^2 = x^2 + y^2 + z^2$, and $\langle r^2\rangle_0$ represents the dimensions of the free chains as unperturbed by excluded volume effects. Apply Equation 1 and calculate the entropy of deformation of a rubber-like polymer network, assuming that the n_c network chains are of a nearly equal length and that the strain-induced displacements of the cross-links are linear (*affine*) in the macroscopic strain of the rubber.

TABLE 19.3
Osmometric rise data

c (g/dL)	*h* (cm)
0.352	0.770
0.682	1.881
1.039	3.221
1.462	5.681
1.966	9.623

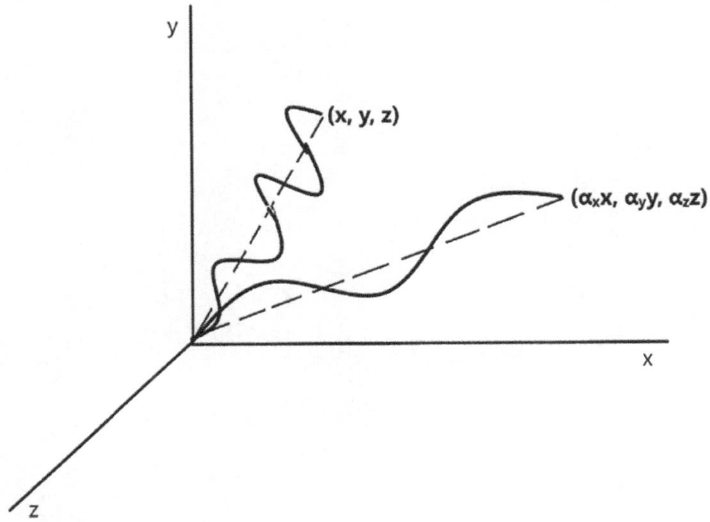

FIGURE 19.7 Schematic presentation of the network chain deformation.

REFERENCES

1. Hiemenz, P. C. 1977. *Principles of Colloid and Surface Chemistry.* New York: Marcel Dekker. Out of print.
2. Dill, K. A., and Bromberg, S. 2010. *Molecular Driving Forces. Statistical Thermodynamics in Biology, Chemistry, Physics, and Nanoscience*, 2nd ed. London: Garland Science (Taylor and Francis Group).
3. Hildebrand, J., and Scott, R. L. 1962. *Regular Solutions.* Englewood Cliffs, NJ: Prentice Hall.
4. Flory, P. J. 1942. Thermodynamics of high polymer solutions. *J. Chem. Phys.* 10: 51–61.
5. Huggins, M. L. 1942. Thermodynamic properties of solutions of long-chain compounds. *Ann. N.Y. Acad. Sci.* 43: 1–32.
6. Cowie, J. M. G., and Arrighi, V. 2007. *Polymers: Chemistry & Physics of Modern Materials*, 3rd ed. Boca Raton, FL: CRC Press.
7. Krigbaum, W. R., and Flory, P. J. 1953. Statistical mechanics of dilute polymer solutions. IV. Variation of the osmotic second coefficient with molecular weight. *J. Am. Chem. Soc.* 75: 1775–1784.
8. Prigogine, I., and Defay, R. 1954. *Chemical Thermodynamics.* London: Longmans.
9. Lambert, S. M., Song, Y., and Prausnitz, J. M. 2000. Equations of state for polymer systems. In *Equations of State for Fluids and Fluid Mixtures*, eds. J. V. Sengers, R. F. Kayser, C. J. Peters, and H. J. White Jr.. London, IUPAC / Elsevier B.V. pp. 523–588.
10. Patterson, D. 1969. Free-volume and polymer solubility: A qualitative view. *Macromolecules* 2: 672–679.
11. Binder, K., Mognetti, B., Paul, W., Virnau, P., and Yelash, L. 2011. Computer simulations and coarse-grained molecular models predicting the equation of state of polymer solutions. In *Polymer Thermodynamics. Polymer Containing Liquid Mixtures*, eds. B. A. Wolf, B. A., and Enders, S. Berlin: Springer-Verlag, pp. 523 – 588.
12. Flory, P. J. 1953. *Principles of Polymer Chemistry.* Ithaca, NY: Cornell University Press.
13. Treloar, L. R. G. 1975. *The Physics of Rubber Elasticity*, 3rd edn. Oxford: Clarendon Press.
14. Flory, P. J., and Rehner, J. 1943. Statistical mechanics of cross-linked polymer networks II. Swelling. *J. Chem. Phys.* 11: 512–550.
15. Peppas, N. A., and Barr-Howell, B. D. 1986. Characterization of the cross-linked structure of hydrogels. In *Hydrogels in Medicine and Pharmacy. Vol. I, Fundamentals*, ed. N. A. Peppas. Boca Raton, FL: CRC Press, pp. 27–56.
16. Prager, S., Bagley, E., and Long, F. A. 1953. Equilibrium sorption data for polyisobutylene-hydrocarbon systems. *J. Am. Chem. Soc.* 75: 2742–2745.

Section III

Applications

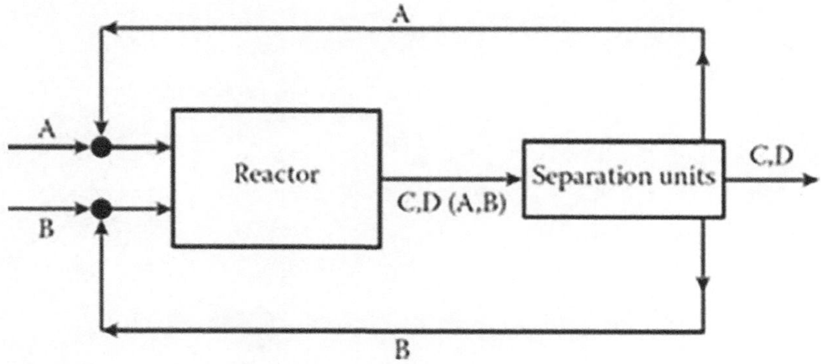

Going beyond the equilibrium conversion.

DOI: 10.1201/9781003431985-22

20 Applications of Polymer Solution Thermodynamics

Abstract

This chapter discusses, first, the phase transitions of polymers, especially the glass transition. The essentials of the Lattice-Fluid equation-of-state model are presented, subsequently. The model is applied to phase equilibrium calculations in a variety of polymer–gas, polymer–solvent, polymer–polymer, polymer–supercritical fluid systems and in nonequilibrium glassy polymer systems.

INTRODUCTION

This chapter concentrates on applications of the Lattice-Fluid (LF) model [1–4], which is a natural extension of the Flory–Huggins model, discussed in Chapter 19. The LF model, enhanced by the incorporation of the free-volume feature, applies to both the liquid and the gas state and thus, it can be used for phase equilibria calculations, including polymer–supercritical fluid equilibria. The enhanced model can incorporate strong specific interactions, such as hydrogen bonding, including three-dimensional hydrogen-bonding networks, cooperative interactions, and intramolecular interactions. More details of these applications are given in Chapters 29–32. The LF model, extended to the nonequilibrium glassy state, significantly broadens its range of applications for the design of novel polymer-based materials. Its essentials are presented below. In view of the interest in glassy polymers, the basics of the phase transitions of polymers, primarily, solid polymers, is discussed first. These transitions are essential for understanding the thermodynamics of polymer mixtures.

PHASE TRANSITIONS OF POLYMERS

Nearly perfect crystalline polymers, or polymer monocrystals, may be obtained under carefully controlled conditions. The most common occurrence is that of semi-crystalline polymers, typically ranging from 5% to over 90% degree of crystallinity. In semi-crystalline polymers, crystalline microphases of varying sizes are either dispersed in an amorphous matrix or are interconnected/separated by amorphous phases. Due to this dispersion, semi-crystalline polymers do not melt at a single temperature but have a range of melting temperatures. When the molten polymer is cooled down, the crystallization does not occur at the melting temperature T_m, but at a lower temperature, T_{mc}. However, a most crucial peculiarity of polymers is their crystallization by *chain folding* into *lamellae* or thin sheets, which typically may lead to monocrystals in solution or to *spherulites* in pure polymer melt. Melting and crystallization transitions are attributed to corresponding enthalpy and entropy changes, often expressed per repeated unit or per unit mass or unit volume. The presence of impurities leads to a polymer melting-point depression. If the impurity is a sorbed solvent B, the melting-point depression of polymer A is often given by:

$$\frac{1}{T_m}-\frac{1}{T_m^0}=-\frac{R}{\Delta H_u}\frac{V_u}{V_B}\left(\Phi_B-\chi_F\Phi_B^2\right) \tag{20.1}$$

where T_m and T_m^0 are the melting points of the mixture and of the pure polymer, respectively. ΔH_u and V_u are the melting enthalpy and volume, respectively, per repeating unit of the polymer, V_B and Φ_B are the molar volume and the volume fraction of the solvent. This equation provides the information necessary to obtain the Flory–Huggins χ_F parameter of the polymer/solvent pair.

DOI: 10.1201/9781003431985-23

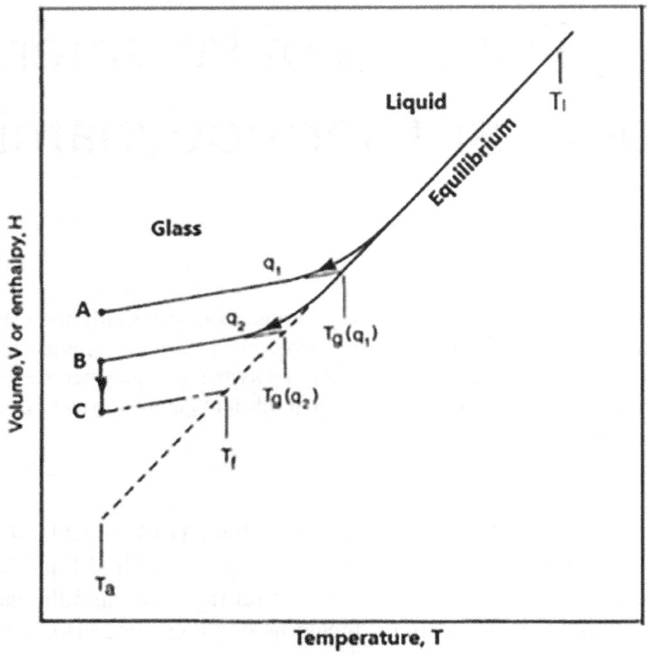

FIGURE 20.1 Variation of polymer enthalpy or volume at glass transition for two rates of cooling, q_1 and $q_2 < q_1$, leading to the corresponding glass transition temperatures ($T_{g1} > T_{g2}$). Reprinted with permission from Ref. [5, fig. 3.1].

Amorphous polymers do not melt or crystallize. They exhibit one or more important phase transitions of significantly different character, the most notable being the *glass transition* in which upon heating an amorphous polymer up to the *glass transition temperature*, T_g, it undergoes a phase transition from the glassy to the rubbery state. In terms of mechanical strength, this transition implies a fall of three orders of magnitude in the elasticity modulus from the relatively strong glassy polymer to the elastic rubber. Thus, T_g is important for practical applications of the polymer. However, the glass transition does not imply any discontinuity in the volume or the enthalpy of the polymer. This behavior is shown schematically in Figure 20.1.

As observed in Figure 20.1, upon cooling the initial liquid/melt at T_c, there is a change in the slopes dV/dT or dH/dT at the glass transition temperature. The geometrical determination of the T_g is also shown in the figure. As shown in the figure, T_g varies with the cooling rate. The higher the cooling rate, the higher the T_g. This is an indication that the glass transition is not an equilibrium phase transition, but it has a kinetic character. If the glassy polymer at point *B* is annealed at the annealing temperature, T_a, a relaxation process takes place leading to the glassy state *C*, which has a lower specific volume (glass densification) as well as lower enthalpy content. As shown in Figure 20.1, a fictitious temperature, T_f, may be attributed to state *C*.

Gee [6] conducted a systematic study on the character of the glass transition in oligomers and polymers. If the transition were an equilibrium phenomenon, the following equalities would apply (first and fourth Ehrenfest equations):

$$\frac{dT_g}{dP} = \frac{\Delta\kappa_T}{\Delta\alpha} = \frac{TV\Delta\alpha}{\Delta C_p} \tag{20.2}$$

where $\Delta\alpha$, $\Delta\kappa_T$, and ΔC_p are the change from the liquid to the glassy state of the thermal expansion coefficient, the isothermal compressibility, and the heat capacity, respectively. Careful measurements on polystyrene samples gave the following results [6]:

$$\frac{dT_g}{dP} = 3{,}100\left[\frac{\text{K}}{\text{N/m}^2}\right], \quad \frac{\Delta\kappa_T}{\Delta\alpha} = 7{,}100\left[\frac{\text{K}}{\text{N/m}^2}\right], \quad \frac{TV\Delta\alpha}{\Delta C_p} = 3{,}600\left[\frac{\text{K}}{\text{N/m}^2}\right]$$

The large discrepancy of the second ratio from the other two is a further indication that the polymer glass transition is not an equilibrium phase transition. The glass transition does not exhibit a discontinuous change in quantities which are first-order derivatives of Gibbs energy, G, such as the volume, the entropy, or the enthalpy. It exhibits discontinuous changes in quantities which are second-order derivatives of G, such as the heat capacity or the thermal expansion coefficient. As the extent of these changes is not what would be expected for an equilibrium transition, it is often referred to as a *quasi-second-order phase transition.*

The phase transition upon cooling a semi-crystalline polymer is shown schematically in Figure 20.2. Starting from a temperature well above the melting point of the polymer, a discontinuous change in the volume or the enthalpy takes place at the crystallization temperature, T_{mc}. Below T_{mc} the polymer is in a semi-crystalline/glass state. Annealing it for a certain time at T_{mc}, a relaxed denser structure is obtained, as shown in the figure. In the same figure it is also shown the phase transition in the absence of crystallization. Depending on the duration of the annealing process, glasses of various degrees of relaxation structures are obtained. If further annealing of the relaxed glass occurs at temperature T_a, the glass undergoes crystallization with a further reduction in enthalpy or volume.

Of much interest is the effect of pressure on the glass transition. Experimental volumetric (PVT) data [8] over a broad range of temperatures and pressures up to 800 bar are shown in Figure 20.3. The starting point was again the equilibrated liquid/melt at temperatures well above the glass transition. The glass was formed by cooling isobarically from the melt at various pressures, using a cooling rate of 5°C/h. As observed, the pressure has a significant effect not only on the density of the glass but also on the glass transition temperature. The higher the pressure, the higher the value of T_g obtained with the same cooling rate.

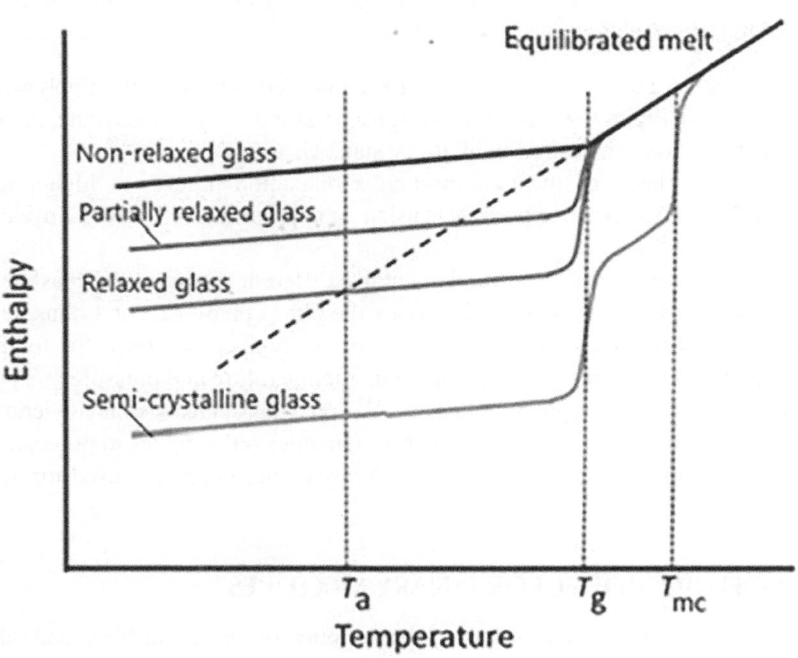

FIGURE 20.2 Schematic representation of phase transitions in semi-crystalline polymer and annealing at the glass transition temperature, T_g. Adapted with permission from Schawe, J.K.E., and Wrana, C. 2020. Competition between structural relaxation and crystallization in the glass transition range of random copolymers. *Polymers* 12(8): 1778, fig.5. Copyright 2023. MDPI. Reprinted with permission from Ref. [7, fig. 5].

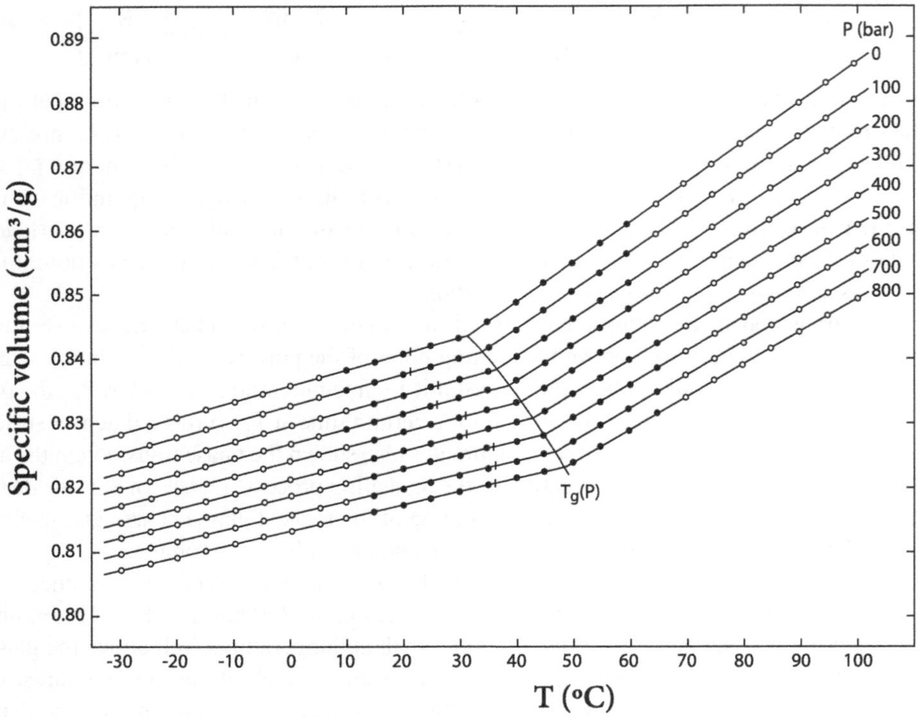

FIGURE 20.3 The specific volume of poly(vinyl acetate) (PVAc) over a range of temperatures and pressures [8]. The glass was formed by cooling isobarically from the melt at various pressures, using a cooling rate of 5°C/h. The effect of pressure on the glass transition temperature is shown by the line $T_g(P)$. Reprinted with permission from Ref. [8, fig. 4].

This effect of pressure on T_g may be used for the controlled densification of polymer glasses. As shown schematically in Figure 20.4, one may design a pressure–temperature route, or cycle, in order to obtain a densified glass at a desired final glassy state, G.

Thus, besides the glass densification through a relaxation process at higher temperatures, shown in Figure 20.2, an alternative process is using an appropriate P–T route or cycle as shown in Figure 20.4.

It is important to keep in mind the fundamentally different nature of the glass and the liquid state. Polymer melts or liquids at equilibrium obey the Gibbs phase rule and, thus, their densities and all other basic thermodynamic quantities are uniquely specified once the temperature and pressure are specified. In contrast, the specification of temperature and pressure of a polymer glass does not specify a unique corresponding density. Various glass densities may be compatible with a given T, P pair. This implies that standard thermodynamics only applies to polymers above their glass transition temperature. Nonequilibrium thermodynamics should be used for describing the glassy state.

THE LATTICE-FLUID MODEL FOR BINARY MIXTURES

The LF model has been extensively used for the description of phase equilibria and related properties in a variety of polymer systems. The next section reviews some representative applications. As most of the systems studied are binary mixtures, this section gives the working LF equations for these mixtures. The generalized formalism for multicomponent mixtures is presented in Chapter 29. The equation of state (EOS) (cf. Equation 29.37) remains identical in form for all cases:

$$\tilde{P}+\tilde{\rho}^2+\tilde{T}\left[\ln\left(1-\tilde{\rho}\right)+\left(1-1/r\right)\tilde{\rho}\right]=0$$

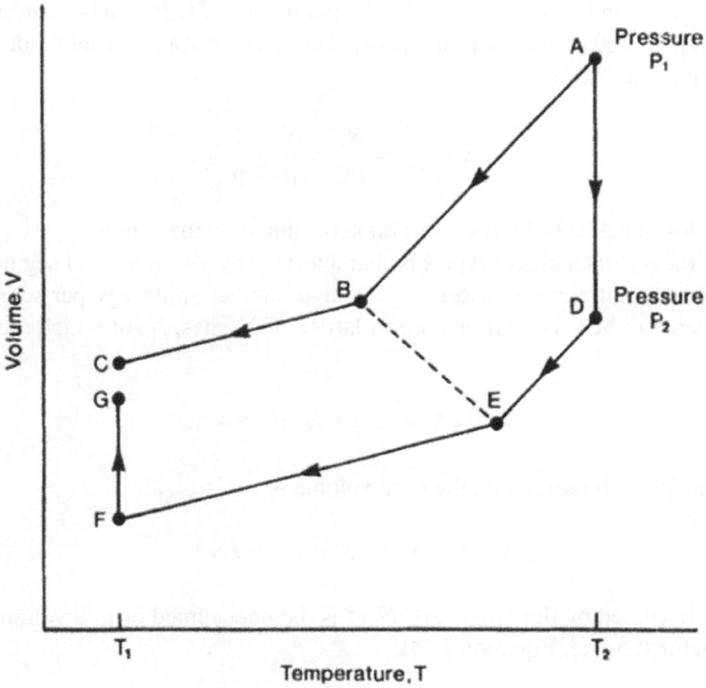

FIGURE 20.4 Schematic illustration of the pressure–temperature cycle used to obtain densified glasses. The line *BE* corresponds to the line $T_g(P)$ of Figure 20.3. The line *FG* is the last step of the isothermal pressure release until the final glassy state, *G*. Adapted with permission from Ref. [9, fig. 1a].

or

$$EOS = \frac{\tilde{P}\tilde{v}}{\tilde{T}} + \frac{\ln(1-\tilde{\rho})}{\tilde{\rho}} + \frac{\tilde{\rho}}{\tilde{T}} + 1 - \frac{1}{r} = 0 \quad \text{(at equilibrium)} \tag{20.3}$$

$\tilde{P}$ and $\tilde{T}$ are a reduced pressure and temperature, respectively, defined as follows (cf. Equation 29.36):

$$\tilde{P} = \frac{P}{P^*} = \frac{Pv^*}{\varepsilon^*}, \qquad \tilde{T} = \frac{T}{T^*} = \frac{RT}{\varepsilon^*}, \tag{20.4}$$

with ε^* given by:

$$\varepsilon^* = \phi_1\varepsilon_1^* + \phi_2\varepsilon_2^* - \phi_1\theta_2 RTX_{12} \tag{20.5}$$

and X_{ij} given by:

$$X_{ij} = \frac{\varepsilon_i^* + (s_i / s_j)\varepsilon_j^* - 2\sqrt{(s_i / s_j)}\varepsilon_{ij}^*}{RT} \tag{20.6}$$

with

$$\varepsilon_{ij}^* = \zeta_{ij}\sqrt{\varepsilon_i^*\varepsilon_j^*} \tag{20.7}$$

The parameters ε*, v*, and r or, alternatively, the parameters T^*, P^*, and ρ*, are the three scaling constants or EOS parameters of each liquid [1–4]. The reduced volume $\tilde{v}$ and reduced density $\tilde{\rho}$ are defined as (cf. Equation 29.28):

$$\tilde{v} = \frac{V}{V^*} = \frac{N_r}{rN} = \frac{\rho^*}{\rho} = \frac{1}{\tilde{\rho}} \tag{20.8}$$

where V^* is the close-packed or hard-core volume occupied by the r-mers.

In LF, each of the N_k molecules of type k is characterized by a number r_k of segments of hard-core volume, v_k^*, of s_k external contacts and of ε_k^* average interaction energy per segment. These segments are considered to be arranged on a quasi-lattice of N_r sites, N_0 of which are empty, and in a binary mixture:

$$N_r = r_1N_1 + r_2N_2 + N_0 = rN + N_0 \tag{20.9}$$

If v* is the volume of each lattice site, the total volume is:

$$V = (rN + N_0)v^* = V^* + N_0v^* \tag{20.10}$$

V^* is the volume occupied by the r-mers and N_0v^* is the unoccupied or free volume. Segment fractions in LF are defined by (cf. Equation 19.4):

$$\phi_1 = \frac{x_1r_1}{r} = 1 - \phi_2 \tag{20.11}$$

and the surface-area fractions by

$$\theta_1 = \frac{x_1r_1s_1}{x_1r_1s_1 + x_2r_2s_2} = \frac{x_1r_1s_1}{rs} = 1 - \theta_2 \tag{20.12}$$

The segment close-packed volume, v*, is given by:

$$v^* = \phi_1v_1^* + \phi_2v_2^* \tag{20.13}$$

When v_1^* differs significantly from v_2^* the following alternative mixing and combining rules are often used [3]:

$$v^* = \phi_1^2v_1^* + \phi_2^2v_2^* + 2\phi_1\phi_2v_{12}^* \tag{20.14}$$

with

$$v_{12}^* = \xi_{12}\left(\frac{\left(v_1^*\right)^{1/3} + \left(v_2^*\right)^{1/3}}{2}\right)^3 \tag{20.15}$$

ξ_{12} is a binary correction factor close to unity. The chemical potential of component 1 in the mixture is given by:

$$\frac{\mu_1}{RT} = \ln\phi_1 + \left(1 - \frac{1}{r}\right)\phi_2 + r_1\tilde{\rho}X_{12}\theta_2^2 + r_1(\tilde{v} - 1)\ln(1 - \tilde{\rho}) + \ln\frac{\tilde{\rho}}{\omega_1} - \frac{r_1\tilde{\rho}}{\tilde{T}_1} + r_1\frac{\tilde{P}\tilde{v}}{\tilde{T}}\frac{v_1^*}{v^*}D_{12} - (DR)(EOS) \tag{20.16}$$

When Equation 20.13 is used for v*, $D_{12}=1$. When Equation 20.14 is used, D_{12} is given by:

$$D_{12} = 2\frac{\phi_1 v_1^* + \phi_2 v_{12}^*}{v^*} - 1 \tag{20.17}$$

At equilibrium the last term in Equation 20.16 is zero since the EOS term is equal to zero (cf. Equation 20.3) while $DR = rN\tilde{v}\,\partial\tilde{\rho}/\partial N_1$. We will come back to this term in a later section on the thermodynamics of the glassy polymer state.

The corresponding equation for the chemical potential of pure compound 1 is:

$$\frac{\mu_1^0}{RT} = r_1(\tilde{v}_1 - 1)\ln(1-\tilde{\rho}_1) + \ln\frac{\tilde{\rho}_1}{\omega_1} - \frac{r_1\tilde{\rho}_1}{\tilde{T}_1} + r_1\frac{\tilde{P}_1\tilde{v}_1}{\tilde{T}_1} \tag{20.18}$$

The volume fraction, Φ_i, required to obtain an equation for the Flory–Huggins parameter, χ_F, is defined as:

$$\Phi_1 = \frac{N_1 r_1 v_1^*}{rNv^*} = \frac{\phi_1 v_1^*}{v^*} = 1 - \Phi_2 \tag{20.19}$$

Combining Equations 20.16 and 20.18 with the equation for χ_F (cf. Equation 19.12):

$$\frac{\mu_1 - \mu_1^0}{RT} = \ln\Phi_1 + \left(1 - \frac{r_1 v_1^*}{r_2 v_2^*}\right)\Phi_2 + \chi_F\Phi_2^2 \tag{20.20}$$

the following equation is obtained for χ_F at low pressures:

$$\chi_F = \left\{ \begin{bmatrix} \ln\dfrac{\phi_1}{\Phi_1} + (\phi_2 - \Phi_2) + r_1\tilde{\rho}X_{12}\theta_2^2 + r_1(\tilde{v}-1)\ln(1-\tilde{\rho}) - \\ -r_1(\tilde{v}_1 - 1)\ln(1-\tilde{\rho}_1) + \ln\dfrac{\tilde{\rho}}{\tilde{\rho}_1} - \dfrac{r_1(\tilde{\rho}-\tilde{\rho}_1)}{\tilde{T}_1} \end{bmatrix} \right\} \Bigg/ \Phi_2^2 \tag{20.21}$$

Following Flory's practice [10], an entropic correction term, $-T\Delta s_{12}$, is often added to the interchange energy, $\Delta\varepsilon_{12^*}$, which ultimately affects only the chemical potential (and the χ_F parameter). Thus, the term containing X_{12} in Equation 20.21 should be replaced by the term:

$$r_1\theta_2^2\left(\Delta\varepsilon_{12}^* - T\Delta s_{12}\right)/RT = r_1\theta_2^2(X_{12} - \Delta s_{12}/R) = r_1\theta_2^2(X_{12} + Q_{12})$$

where X_{12} and Q_{12} are dimensionless, but the product RQ_{12} has units of entropy. Q_{12} is normally set equal to zero.

Heats and volumes of mixing are obtained from Equations 20.22 and 20.23 (cf. Equations 29.55 and 29.56):

$$H^E = U^E + PV^E = rN\left[\sum_k \tilde{\rho}_k\phi_k\varepsilon_k^* - \tilde{\rho}\varepsilon^*\right] + PV^E \tag{20.22}$$

where

$$V^E = rN\left[\tilde{v}v^* - \sum_k \tilde{v}_k\phi_k v_k^*\right] \tag{20.23}$$

In the limit of infinite dilution and at low pressures, the heat of mixing takes the following form:

$$\Delta H^{\infty} = \frac{RT}{\rho_2^* v_2^*}\left\{\frac{\tilde{\rho}_2 - \tilde{\rho}_1}{\tilde{T}_2} + \tilde{\rho}_1 \frac{s_2}{s_1} X_{12} + \kappa_{T1} P_1^* \tilde{\rho}_1^2 \psi_1\right\} \tag{20.24}$$

where

$$\psi_1 = \tilde{\rho}_1\left(\frac{1}{\tilde{T}_1} - \frac{1}{\tilde{T}_2} + \frac{s_2}{s_1} X_{12}\right) + \frac{1}{r_2} - \frac{1}{r_1} \tag{20.25}$$

ρ_2^* is the close-packed density of component 2: $\rho_2^* = M_2 / V_2^*$.

The phase stability condition, Equation 17.5, applied in Chapter 19 to the Flory–Huggins model, may also be applied to the LF model for both the UCST and LCST. By setting $g = G/rN$, the phase stability condition (spinodal inequality) may be written as:

$$\partial^2 g / \partial \phi_1^2 > 0 \text{ or}$$

$$\tilde{\rho}\left[2\frac{\theta_1 \theta_2^2}{\phi_1 \phi_2^2} X_{12} + \kappa_T P^* \tilde{T} \psi^2 - \frac{P\tilde{v}^2}{RT} W\right] < \frac{1}{r_1 \phi_1} + \frac{1}{r_2 \phi_2} \tag{20.26}$$

where

$$\psi = \tilde{\rho}\left[\frac{1}{\tilde{T}_1} - \frac{1}{\tilde{T}_2}\left(\theta_2^2 - \theta_1^2 \frac{s_2}{s_1}\right) X_{12}\right] + \frac{P\tilde{v}}{RT}\left(v_1^* - v_2^* + \frac{\phi_1 - \phi_2}{2} W\right) + \frac{1}{r_2} - \frac{1}{r_1} \tag{20.27}$$

$W = \dfrac{d^2 v^*}{d\phi_1^2}$, is equal to zero, when Equation 20.13 is used, and $W = 2\left(v_1^* + v_2^* - 2v_{12}^*\right)$, when Equation 20.14 is used instead.

When the entropic correction is added to the chemical potential, the term with X_{12} in the spinodal inequality Equation 20.26 should be replaced by the term $2\dfrac{\theta_1 \theta_2^2}{\phi_1 \phi_2^2}\left(X_{12} + Q_{12}\right)$ and, similarly, in Equation 20.27, X_{12} should be replaced by $\left(X_{12} + Q_{12}\right)$.

Gas solubilities in polymer matrices are often treated with the classical Henry's law constant, $H = y_1 P/c_1$, or the weight fraction Henry's law constant, defined by:

$$H_w = \lim_{y_1, w_1 \to 0} \frac{y_1 P}{w_1} \tag{20.28}$$

y_1 is the mole fraction of the solute (gas) in the vapor phase and c_1 and w_1 are its concentration (mass/unit-volume) and weight fraction, respectively, in the polymeric liquid phase. The LF equation for H_w is:

$$H_w = \frac{RT\rho_2}{M_1} \exp\left\{r_1\left[1 + \tilde{\rho}_2 X_{12} + (\tilde{v}_2 - 1)\ln(1 - \tilde{\rho}_2) - \frac{\tilde{\rho}_2}{\tilde{T}_1}\right]\right\} \tag{20.29}$$

Since polymers are not volatile, it is not possible to measure their vapor pressure or heat of vaporization that would give a direct measure of the strength of their intermolecular forces. This information is typically obtained indirectly using "probe" molecules. In this regard, an important experimental technique is inverse gas chromatography (IGC). In fact, a large portion of the experimental gas solubilities in polymer matrices is obtained via IGC. The measured quantity is the specific retention volume, V_g, which is related to Henry's constant through the equation:

$$V_g = \frac{1}{H}\left(\frac{RT}{\rho_2 M_1}\right) = \frac{1}{\rho_2}\exp\left\{-r_1\left[1+\tilde{\rho}_2 X_{12}+(\tilde{v}_2-1)\ln(1-\tilde{\rho}_2)-\frac{\tilde{\rho}_2}{\tilde{T}_1}\right]\right\} = V_g^0\frac{T^0}{T} \qquad (20.30)$$

where V_g^0 is the standard or corrected retention volume to the standard reference temperature, $T^0 = 273.15$ K.

APPLICATIONS OF THE LF MODEL

In all applications of the above equations found in the literature, characteristic LF scaling constants of solvents and polymers are needed. Compilations of these constants are available in the literature [1,2,11]. The scaling constants of solvents are usually obtained by simultaneous correlation of EOS properties, such as orthobaric densities, vapor pressures, heats of vaporization, etc. For polymers, their volumetric or *PVT* properties are normally used for the determination of the scaling constants.

Alternative constants may be obtained from the above defining equations, such as, $v^* = (RT^*/P^*)$, $\rho^* = 1/v^*_{sp}$, etc.

Several polymer–solvent and polymer–polymer systems have been studied with the LF model [3,4,11–13]. In Ref. [3], the LF model was used to calculate heats of mixing at infinite dilution, volumes of mixing, and χ_F parameter or activity coefficients over the full range of composition. The calculations were done in two different ways. In the first way, one adjustable binary parameter (ζ_{12}) was used while s_1/s_2 and ξ_{12} were set equal to unity and Q_{12} was set equal to zero. Following Flory's practice [10], the ζ_{12} parameter was determined from the heats of mixing at infinite dilution. Once the parameters were evaluated as detailed above, the rest of the calculations are rather predictions. In the second way, the literature values for s_1/s_2 were used while ξ_{12} and Q_{12} were also treated as adjustable parameters. This second way is, in essence, a test of the flexibility of the model to describe simultaneously all properties with available experimental information. The results give a good measure of the model's flexibility and versatility.

Binodal and spinodal curves in liquid–liquid equilibria (UCST and LCST behaviors) were also calculated with the model. The shape of binodal curves in polymer mixtures depends strongly on the molecular weight distribution, that is on the polydispersity of the polymers. The role of polydispersity on the phase equilibrium description with the LF model is not discussed here. Further information can be found in references [14] and [15]. Important characteristics of liquid–liquid equilibria in polymer solutions can be examined by application of the spinodal inequality condition (Equation 20.26). Figure 20.5 presents a comparison between the experimental [16] LCST behavior and the LF calculations for mixtures of poly(vinyl methyl ether) (PVME) and nearly monodisperse polystyrene (PS) samples of molecular weights varying from 10,000 to 200,000. The simple version of the LF model ($\xi_{12} = s_1/s_2 = 1.0$, $Q_{12} = 0.0$) was used for the calculation of binodal and spinodal curves. Binodal and spinodal curves are very sensitive to the value of the ζ_{12} binary parameter. All calculations were done by setting $\zeta_{12} = 1.00466$.

Gas solubilities in polymer matrices are well described by the LF model, even in its plain version ($s_1/s_2 = 1$, $Q_{12} = 0$). In Figure 20.6, the predicted IGC retention volumes, V_g, of 27 hydrocarbons in poly(1-butene) at 373 K are compared with the experimental values [11,17,18]. The average error is around 13%. Higher accuracy can be obtained by adjusting the binary parameter, ζ_{12}. It should be underlined that Henry's law can be used in different ways for reporting gas solubilities. Quite often the alternative form $P_1 = K_P V_1^0$ is used, where V_1^0 is the solubility of the solute (cm^3 of solute gas per g of polymer at 273.15 K and 1 atm). The relation of K_P to H_w is $K_P = H_w M_1/22{,}414$. A variety of gas-polymer systems have been studied with the LF model [11–13].

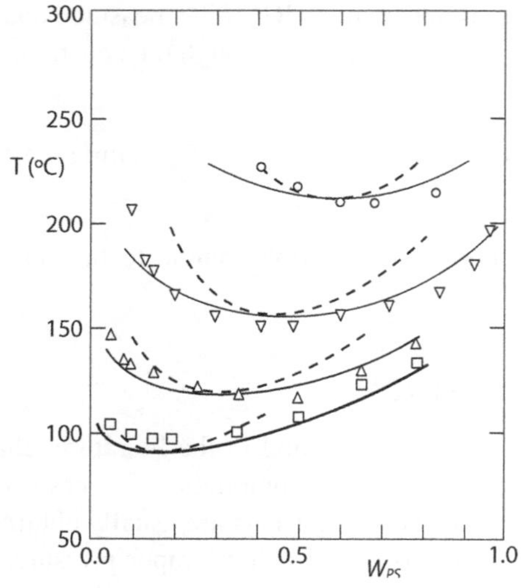

FIGURE 20.5 Comparison of experimental [16] and calculated phase equilibria for mixtures of PVME with PS of varying molecular weights. The molecular weight for PVME is 51,500 and for PS is: 10,000 (circles), 20,400 (rev. triangles), 51,000 (triangles), and 200,000 (squares). W_{ps} is the weight fraction of PS. Solid lines are calculated binodals by the LF model, while dashed lines are the corresponding spinodals. Reprinted with permission from Ref [3, fig. 11].

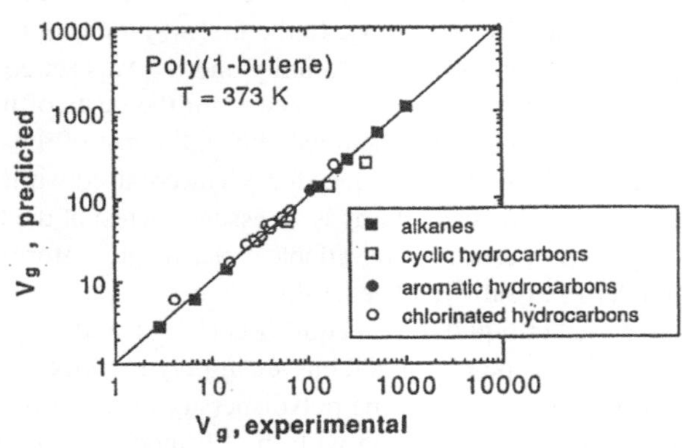

FIGURE 20.6 Comparison of experimental [18] and predicted specific retention volume for 27 hydrocarbon probes in poly(1-butene) at 373 K [17]. Reprinted with permission from Ref. [17, fig. 2].

Many polymer systems exhibiting hydrogen-bonding interactions have been studied with extended variants of LF model (LFHB and NRHB). These more elaborate models are discussed in Chapter 32.

THE LF MODEL AND THE GLASS TRANSITION OF POLYMERS

In order to study glassy polymers, Equation 20.16 or 20.18 need to be considered with focus on the term containing the chain-flexibility/symmetry parameter, ω_i.

In the frame of the Flory–Huggins and the LF model, this parameter is expressed by:

$$\omega_i = \delta_i / 2e^{r_i - 1} \tag{20.31}$$

The flexibility term, δ_i, is a characteristic temperature-dependent property of each polymer and is given by [19,20]:

$$\delta_i = z\left(\frac{z-2}{f_i}\right)^{(r_i-2)f_i}\left(\frac{1}{1-f_i}\right)^{(r_i-2)(1-f_i)} \tag{20.32}$$

The term dividing δ_i in Equation 20.31 is a correction factor arising from the excluded volume of the chain in the system. The "flex factor," f_i, is used to indicate the $f_i(r_i-2)$ bonds of the chain, which are in high-energy or flexed states, and the $(1-f_i)(r_i-2)$ bonds, which are in the low-energy states. If $\Delta\varepsilon_{gi}$ is the energy change upon bond flexing (as an example, from a "trans" to a "gauche" conformation) of component i, the total energy of a binary mixture is:

$$U = -rN\tilde{\rho}\varepsilon^* + N_1(r_1-2)f_1\Delta\varepsilon_{g1} + N_2(r_2-2)f_2\Delta\varepsilon_{g2} \tag{20.33}$$

and the flex factor is given by:

$$f_i = \frac{(z-2)\exp(-\Delta\varepsilon_{g1}/RT)}{1+(z-2)\exp(-\Delta\varepsilon_{g1}/RT)} \tag{20.34}$$

In the plain LF formalism ($s_1/s_2=1$; $v^* = \phi_1 v_1^* + \phi_2 v_2^*$), the system entropy is given by:

$$S = -rNR\left\{\begin{aligned} &(\tilde{v}-1)\ln(1-\tilde{\rho}) + \frac{\ln\tilde{\rho}}{r} + \frac{\phi_1}{r_1}\ln\frac{\phi_1}{r_1} + \frac{\phi_2}{r_2}\ln\frac{\phi_2}{r_2} + \\ &1 + \frac{\ln(2/Z)-1}{r} + \frac{\phi_1}{r_1}(r_1-2)\left[\ln(1-f_1) - f_1\frac{\Delta\varepsilon_{g1}}{RT}\right] + \\ &\frac{\phi_2}{r_2}(r_2-2)\left[\ln(1-f_2) - f_2\frac{\Delta\varepsilon_{g2}}{RT}\right] \end{aligned}\right\} \tag{20.35}$$

The only unknown parameter in Equation 20.35 is $\Delta\varepsilon_{gi}$. In order to determine it, it is necessary to recall that the glassy state is, essentially, a frozen liquid or amorphous state and then, adopt *Gibbs – Di Marzio criterion*, which states that the system entropy approaches zero as the system temperature approaches the glass transition temperature, T_g, upon cooling. In practice, ordinary solvents are considered fully flexible molecules and their $\Delta\varepsilon_{gi}$ is set equal to zero. The corresponding parameter for the polymer is obtained by zeroing S at $T=T_g$. Otherwise, the phase equilibrium calculations are done as described in the previous section.

Of much interest is the description of the phase equilibrium of polymer–solvent systems down to the glass transition region and over a broad range of external pressure and temperature. When the solvent is a supercritical fluid, such as carbon dioxide, these phase equilibrium studies are of much technological interest for the production of nano-foamed, micro-foamed and macro-foamed or cellular polymeric materials.

In order to understand the peculiarity of the phase diagrams of glassy polymer solutions, it is necessary to first recall the effect of pressure and solvent concentration on the glass transition temperature (T_g). Figure 20.7 shows the effect of pressure on the glass transition temperature of pure poly(methyl methacrylate) (PMMA). An increase in pressure reduces the free volume and restricts the free motion of the chain, leading to a quasi-frozen state, thus, increasing the value of T_g. Figure 20.7 is a representative diagram showing that the increase in pressure increases the T_g of the polymer. A similar effect was seen in Figure 20.3.

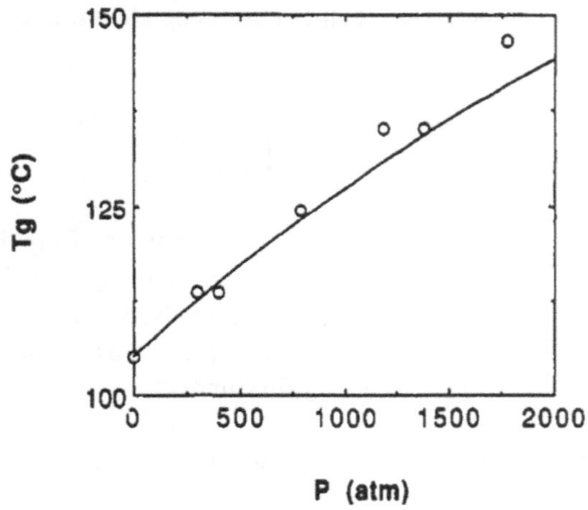

FIGURE 20.7 The effect of pressure on the glass transition temperature of PMMA. Symbols are experimental data [21]. Line is the correlation with the LF model. Reprinted with permission from Ref. [20, fig. 9].

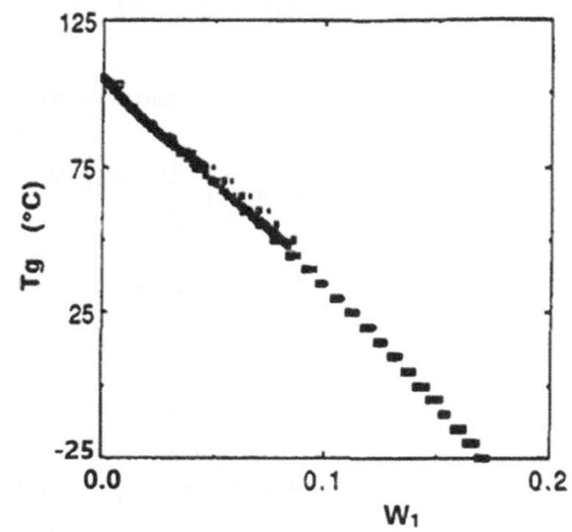

FIGURE 20.8 The glass transition depression with solvent concentration for various ζ_{12} and ε_1^* values. Reprinted with permission from Ref. [20, fig. 3].

Figure 20.8 shows the effect of solvent concentration on the glass transition temperature of the polymer. As seen in this figure, there is a nearly linear T_g depression with solvent composition. This diagram is, again, representative for the T_g depression of polymers with solvents of various interaction energies, ε_1^* and/or ζ_{12} binary interaction parameters.

As observed in Figures 20.7 and 20.8, the effect of pressure on T_g is the opposite of the effect of solvent composition on T_g. It is, then, of interest to see what the combined effect of pressure and solvent composition would be on the glass transition temperature. As shown in Figure 20.9, there are four distinct types of T_g versus P diagrams depending on the solvent solubility in the polymer matrix. The scaling constants for solvent and polymer in Figure 20.9 are those of CO_2 and PMMA ($T_g = 105°C$) and the solubility varies by varying the ζ_{12} parameter. The four types of diagrams in Figure 20.9 are general ones and are not an exclusive behavior of the CO_2-PMMA system.

Of particular interest are the three types of diagrams with pressure maxima presented in Figure 20.9. Considering the diagram of type IV as an example, at an external pressure of 20 atm and at 120°C, the system is in the liquid state. By cooling down to a temperature of 75°C, the system

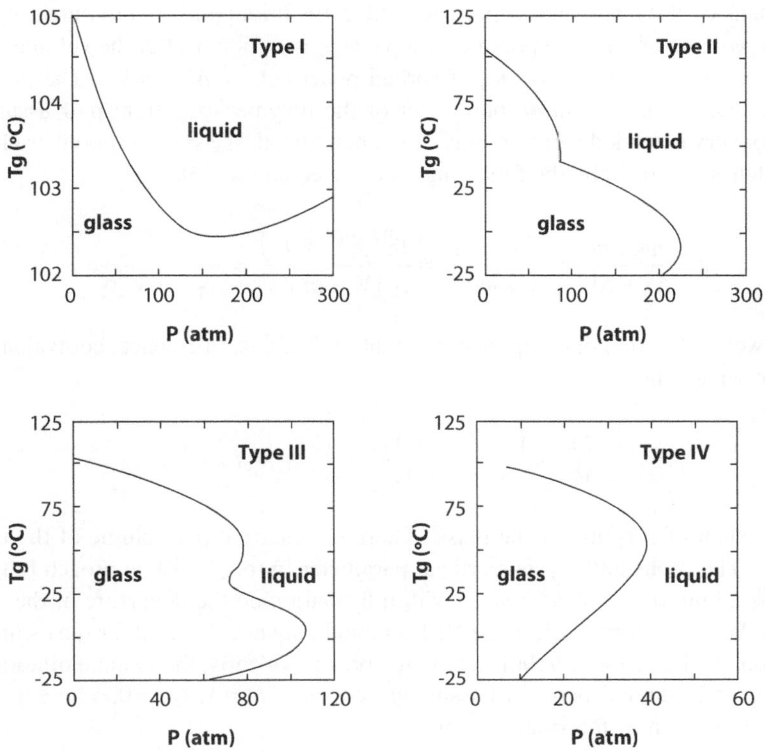

FIGURE 20.9 The four types of T_g behavior as a function of pressure predicted by the LF model. Reprinted with permission from Ref [20, fig. 2].

enters the glassy state and remains in it down to ca 0°C. Cooling further down, the system reenters the liquid state. This peculiar phenomenon is called *retrograde vitrification*. These diagrams are very much used in the production of foams in the liquid state and stabilization in the glassy state, for example by rapid depressurization and removal of the supercritical solvent. However, sorption in the glassy state is not a thermodynamic-equilibrium phenomenon and its description requires some further modification of the LF model presented in the next section.

THE NONEQUILIBRIUM LATTICE-FLUID (NELF) MODEL

As discussed above, when an amorphous polymer is found at a temperature well below its glass transition temperature, T_g, the polymer is considered to be in a (semi) frozen state. Since such a state is not a thermodynamic one, two state variables are no longer sufficient for a full description of the polymer state. One of the most interesting applications of the LF model is its extension to the study of gas solubility in glassy polymer matrices via a nonequilibrium thermodynamic approach. This extension was initiated by Doghieri and Sarti [22], who introduced an *order* parameter or *internal state* parameter to account for the departure of the state from equilibrium. Their approach requires information at every set of external conditions. In what follows, an alternative semi-predictive approach [23] is presented, which introduces a characteristic "dilation" constant in the LF model for the thermodynamic calculations in nonequilibrium glassy polymer systems.

Since the glassy state is not an equilibrium one, its density cannot be obtained from the EOS. It must be given as external information or as an *order parameter* of the nonequilibrium state. Let ρ_2^0

be the (initial) density of the pure glassy polymer and let the initial mass of our pure polymer sample be m_2. Thus, the volume of our pure polymer sample is $V_0=m_2/\rho_2^0$ cm^3. Let the volume change upon sorption be proportional to the mass, m_1, of sorbed penetrant, or $\Delta V=m_1V_{12}$. The proportionality factor, V_{12}, is a characteristic *dilation* parameter of the polymer–penetrant pair assumed to be a constant over not very extended sorption intervals. Then, the glassy system (polymer plus dissolved penetrant gas) density is given by the following defining equation [23]:

$$\rho=\frac{m_2+m_1}{V_0+\Delta V}=\frac{\rho_2^0V_0+m_1}{V_0+m_1V_{12}}=\frac{\rho_2^0\left(\rho_2^0V_0+m_1\right)}{\rho_2^0\left(V_0+m_1V_{12}\right)}=\frac{\rho_2^0}{w_2+w_1V_{12}\rho_2^0} \tag{20.36}$$

where w_i is the weight fraction of component i. Equation 20.36 is, in essence, equivalent to defining the volume dilation as follows:

$$\frac{\Delta V}{V_0}=\frac{V-V_0}{V_0}=\frac{m_1V_{12}}{V_0}=\frac{m_1}{V_0}V_{12}=s_1V_{12} \tag{20.37}$$

where s_1 is the extent of sorption or the mass of sorbed penetrant per volume of the initial (pure) polymer sample. The "constant" V_{12} is a crucial parameter in this NELF approach [23] and it may be considered as a kind of *order parameter* with information on the departure of the system from nonequilibrium. With this information, the NELF model may now be used for the estimation of the extent of sorption, m_1/V_0, or the weight fraction, w_1, or, equivalently, the volume dilation.

For the LF model adopted here, in its simple version ($s_1/s_2=1$, $Q_{12}=0$, $v^*=\phi_1v_1^*+\phi_2v_2^*$), the Gibbs free energy is given by (cf. Equation 29.51):

$$\frac{G}{RT}=rN\left\{-\frac{\tilde{\rho}}{\tilde{T}}+\frac{\tilde{P}\tilde{v}}{\tilde{T}}+(\tilde{v}-1)\ln(1-\tilde{\rho})+\frac{1}{r}\ln\tilde{\rho}+\sum_{k=1}^{t}\frac{\phi_k}{r_k}\ln\frac{\phi_k}{\omega_k}\right\} \tag{20.38}$$

However, care must be exercised on the way the derived thermodynamic quantities are obtained from Equation 20.38 since the system state is, now, a nonequilibrium one. The chemical potential of the penetrant is given by (cf. Equation 20.16):

$$\begin{aligned}\frac{\mu_1}{RT}=\frac{1}{RT}\left(\frac{\partial G}{\partial N_1}\right)_{T,P,m_2}&=\ln\phi_1+\left(1-\frac{1}{r}\right)\phi_2+r_1\tilde{\rho}X_{12}\phi_2^2+r_1(\tilde{v}-1)\ln(1-\tilde{\rho})\\&+\ln\frac{\tilde{\rho}}{\omega_1}-\frac{r_1\tilde{\rho}}{\tilde{T}_1}+r_1\frac{\tilde{P}\tilde{v}}{\tilde{T}}\frac{v_1^*}{v^*}-(DR)(EOS)\end{aligned} \tag{20.39}$$

but, now, the EOS term (cf. Equation 20.3) is no longer zero and the last term in Equation 20.39 reads:

$$(DR)(EOS)=rN\tilde{v}\frac{\partial\tilde{\rho}}{\partial N_1}EOS=\left\{r_1\frac{w_1}{\phi_1}\frac{w_2\rho^*}{\rho_2^*}\left[\frac{\tilde{\rho}}{\tilde{\rho}_2^0}\left(1-V_{12}\rho_2^0\right)+\frac{\rho_2^*}{\rho_1^*}-1\right]\right\}\left\{\frac{\tilde{P}\tilde{v}}{\tilde{T}}+\left[\frac{\ln(1-\tilde{\rho})}{\tilde{\rho}}+\frac{\tilde{\rho}}{\tilde{T}}+1\right]-\frac{1}{r}\right\} \tag{20.40}$$

where use was made of Equation 20.41 for ρ^* arising from the conservation of hard-core volumes in the mixture (cf. Problem 20.2):

$$\frac{1}{\rho^*}=\frac{w_1}{\rho_1^*}+\frac{w_2}{\rho_2^*} \tag{20.41}$$

The extent of sorption or penetrant solubility is obtained by equating the (equilibrium) chemical potential of the pure fluid penetrant given by Equation 20.18 with its chemical potential in the glassy system given by Equation 20.39.

A comprehensive set of experimental data to test the NELF model was provided by Fleming and Koros [24] with solubilities of carbon dioxide in glassy bisphenol A polycarbonate (PC) at 35°C. In addition to solubilities, volume dilation data were also provided for both unconditioned (as received) and conditioned (in CO_2 atmosphere at 900 psi and at 35°C) polymer. In order to simplify the subsequent calculations, the segment volume, v_1^*, of the penetrant was set equal to $v_2^*=11.75\,cm^3/mol$. The remaining scaling constants were obtained from correlation of EOS data [25] for CO_2 leading to $T_1^*=405$ K and $\rho_1^*=1210\,kg/m^3$. The density of the unconditioned sample was set equal to 1.200 g/cm³ [24]. The conditioned sample of initial volume equal to 1 cm³ had shown a volume equal to 1.008 cm³ after removal of the sorbed CO_2. Thus, the density of the conditioned sample was set equal to 1.200/1.008 = 1.190 g/cm³. These measurable quantities are the key parameters for the NELF calculations below. V_{12} is the intermediate order parameter interconnecting volume dilation, ρ, and penetrant solubilities. No adjustable parameters are used in the calculations. From the equality of chemical potential of penetrant, the value of ϕ_1 or w_1 is obtained but for the concentration it is required to know, or to calculate, the volume (dilation).

Figure 20.10 presents a comparison between the experimental solubilities of CO_2 in the "as received" glassy PC sample and in the conditioned PC sample at 35°C. The calculations were made considering the volume dilation, $\Delta V/V^0$, known [24] (which is equivalent to saying that V_{12} is known – cf. Equation 20.37). Similarly, the calculated volume dilations are obtained by considering the gas solubilities known.

Of much interest is the simultaneous calculation of sorption on an unconditioned sample and desorption from a conditioned PC sample. This is shown in Figure 20.11, where a comparison is made between experimental [24] and calculated gas solubilities (considering volume dilations known) and volume dilations (considering gas solubilities known). In view of the nonequilibrium character of the system, the comparison in Figure 20.11 is quite satisfactory.

With these results, it is to be expected that the NELF model could be a nearly predictive tool if a universal relation were found between V_{12} and the extent of sorption m_1/V_0. At a given pressure, one would obtain m_1/V_0 from the above equality of chemical potentials and, from the universal relation, V_{12}, which would give directly the volume dilation, s_1V_{12}, or the glassy system density. Similarly, one would obtain the extent of sorption or penetrant solubility from the volume dilation.

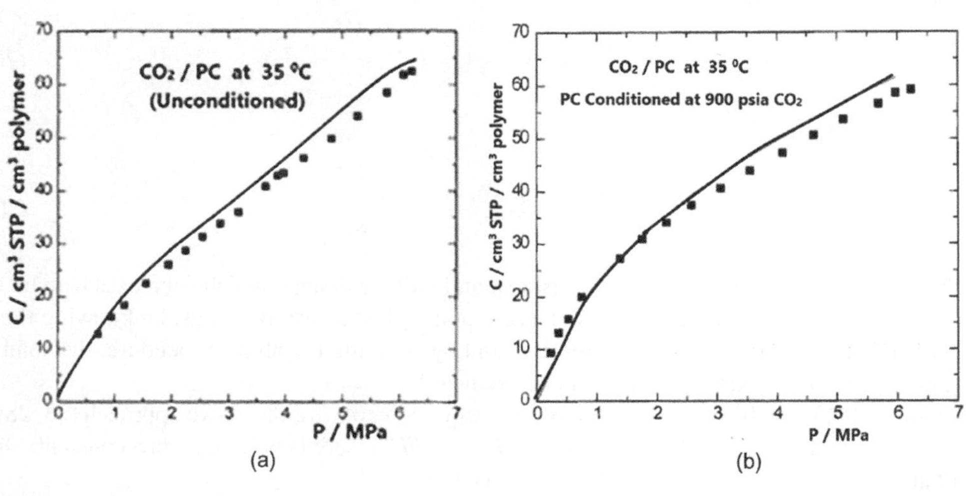

FIGURE 20.10 Experimental [24] (symbols) and calculated gas solubilities of CO_2 in unconditioned (a) and conditioned (b) PC at 35°C by considering the excess volume known. Reproduced, with permission, from Ref. [23, fig. 5 and 6].

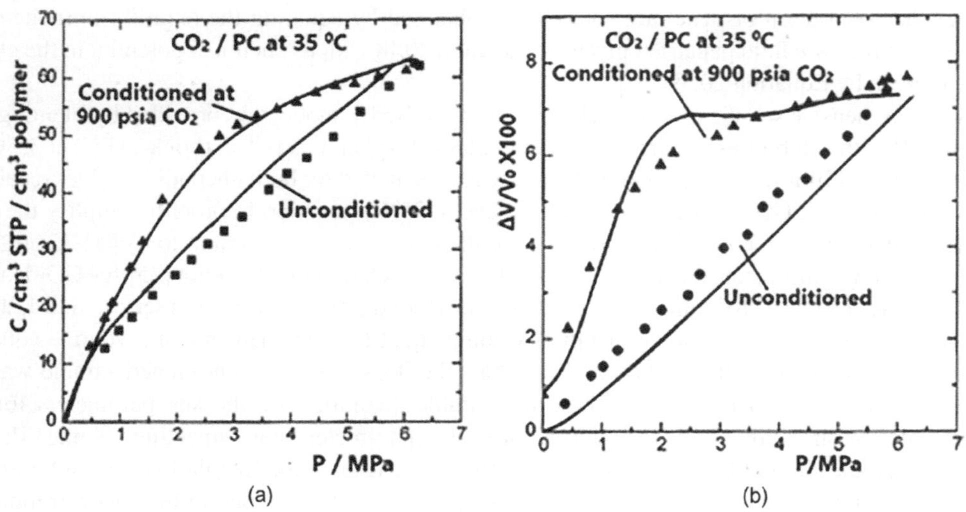

FIGURE 20.11 Experimental [24] (symbols) and calculated (lines) gas solubilities on unconditioned and conditioned PC sample, considering volume dilations known (a) and volume dilations on the same sample by considering gas solubilities known (b). Reproduced with permission from Ref. [23, figs. 7 and 9].

Instead of Equation 20.36, Doghieri and Sarti introduced an (apparent) polymer density, ρ_2, as order parameter and from mass balance they have set $\rho = \rho_2/w_2$. In this case, DR in Equation 20.40 is given by (cf. Problem 20.2):

$$DR = \frac{r_1 w_1 \rho^*}{\phi_1 \rho_1^*} = \frac{r_1 v_1^*}{\phi_2 v_2^*}\frac{\tilde{\rho}_2}{\tilde{\rho}} = \frac{r_1 v_1^*}{v^*} \quad \text{(Doghieri – Sarti)} \tag{20.42}$$

However, in this case, the density ρ_2 must be given at every point. Later, Baschetti et al. [26] introduced a "swelling" constant, k_{sw}, analogous to the above dilation constant V_{12} and turned also their NELF approach into a semi-predictive model for the solubility in glassy polymers at any penetrant gas partial pressure, P_1 [27]. Their empirical relation, based on a review of available experimental volume dilation data, has the form:

$$\rho_2(P_1) = \rho_2^0(1 - k_{sw}P_1) \tag{20.43}$$

or, equivalently [26]:

$$\frac{\Delta V}{V_0} = \frac{\rho_2^0 - \rho_2}{\rho_2} = \frac{k_{sw}P_1}{1 - k_{sw}P_1} \tag{20.44}$$

The "swelling" constant, k_{sw}, is a characteristic quantity of the system, which decreases with increasing temperature and is expected to approach zero at the glass transition. Thus, by knowing the volume dilation at any penetrant partial pressure, and by applying the above procedure, the solubility of the penetrant in the glassy polymer can be predicted.

Equation 20.43 may be obtained directly from Equation 20.37 of the above approach [23, 28] and Henry's law, as shown in Problem 20.3. In fact, $k_{sw} = V_{12}/H$, where H is Henry's law constant. On the other hand, combining Equations 20.37 and 20.39 gives:

$$\frac{\Delta V}{V_0} = \frac{m_1}{V_0}V_{12} = \frac{k_{sw}P_1}{1 - k_{sw}P_1} \quad \text{or} \quad s_1 = \frac{m_1}{V_0} = \frac{P_1/H}{1 - k_{sw}P_1} \tag{20.45}$$

Thus, the knowledge of k_{sw} enables the prediction of volume dilation, m_1V_{12}, at any pressure and, thus, ρ. Then, from the equality of chemical potentials, the penetrant solubility, or m_1, is obtained. The second form on the right-hand side of Equation 20.45 also provides a basis or link with the widely used dual-mode sorption model [29] at low pressures. However, in practice, neither k_{sw} nor V_{12} are constants over a broad range of penetrant pressures or extends of sorption and care must be exercised in their use. Often, also, the binary interaction parameter ζ_{12} (cf. Equation 20.6) needs to be adjusted for satisfactory calculations. For its simplicity, the combination of Equations 20.39, 20.42, and 20.44 seems to be the preferred implementation of this NELF model.

The Doghieri and Sarti (DS) approach does not use the above LF formalism (referred to as NELF model, for distinction). DS approach uses an entirely equivalent off-lattice LF formalism, discussed in Chapter 29, in which, among other changes, the role of interchange energy, $\Delta\varepsilon_{12}$* is replaced by the interchange of the reducing pressures, ΔP_{12}* [2,11]. Their approach is alternatively known in the literature as the NETGP (Non-Equilibrium Thermodynamics of Glassy Polymers). Both, the NETGP and the above NELF approach may be adapted to alternative EOS frameworks, such as the LFHB, NRHB, and SAFT models discussed in Chapters 32 and 33.

GLASS TRANSITION AND HYDROGEN BONDING

The previous sections presented the role of the entropy equation in accounting for the glass transition and the thermodynamics of the glassy state. The role of this equation is further enhanced by the hydrogen-bonding (HB) contribution in associated polymer systems. Hydrogen bonding is discussed in Chapter 32. For the purposes of the present chapter, it is considered that the equations for all basic thermodynamic quantities consist of two contributions, the HB contribution due to hydrogen-bonding interactions and the *physical* contribution due to all other types of intermolecular interactions.

As with the other thermodynamic quantities, the entropy equation consists of two parts, the physical, S_P, and the HB contribution, S_H. Equation 20.35 is used for S_P. The hydrogen-bonding contribution, derived in Chapter 32, will be used below.

In the frame of the Gibbs and DiMarzio approach to the glass transition [21], the full expression for the flexibility term, δ, (essentially, the number of internal configurations available to the semi-flexible chain of the polymer) is given by [21]:

$$\ln\delta = \ln(z/2) - (r-2)\ln(1-f) + f(r-2)\frac{\Delta\varepsilon_g}{RT} \tag{20.46}$$

where the fraction f of flexed bonds is given by Equation 20.34. It should be recalled that the "flex" energy, $\Delta\varepsilon_g$, is a characteristic parameter of each polymer and is calculated usually by zeroing the entropy, S, at the glass transition temperature of the polymer at ambient pressure [19,30]. With known flex energies, the T_g's are estimated.

An interesting application is the effect of HB on the glass transition behavior of polymer mixtures [30]. For a mixture of self-associated polymer 1 (d_1 donor sites, a_1 acceptor sites) with cross-associated polymer 2 (a_2 acceptor sites), S_H is given by (cf. Equation 32.83)

$$\frac{S_H}{RrN} = -\nu_H + \frac{\phi_1 d_1}{r_1}\ln\frac{\phi_1 d_1}{\phi_1 d_1 - r_1\nu_H} + \frac{\phi_1 a_1}{r_1}\ln\frac{\phi_1 a_1}{\phi_1 a_1 - r_1\nu_{11}} + \frac{\phi_2 a_2}{r_2}\ln\frac{\phi_2 a_2}{\phi_2 a_2 - r_2\nu_{12}} \tag{20.47}$$

where, $r\nu_H$ is the total number of hydrogen bonds per mol, $r\nu_{11}$ and $r\nu_{12}$ are the corresponding numbers of hydrogen bonds of type donor 1–acceptor 1 and donor 1–acceptor 2, respectively.

The experimental T_g's of the mixture of copolymer styrene-co-vinyl-phenol (SVPh60) and poly(iso-butyl methacrylate) (PIBMA) and the calculated T_g's with and without the HB contribution

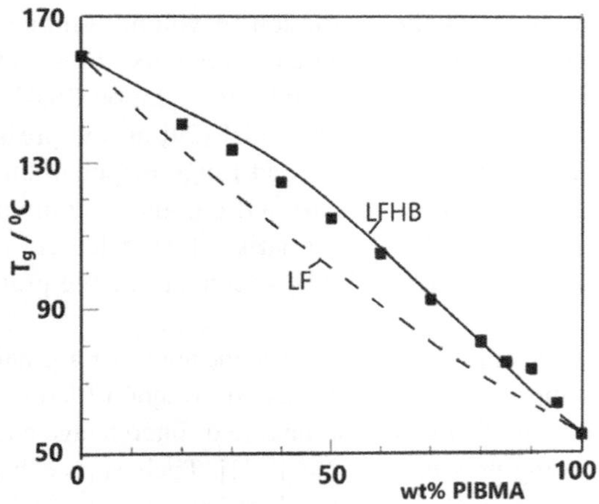

FIGURE 20.12 Glass transition temperatures of the mixture SVPh60/PIBMA. Symbols are experimental data [30]. Solid line was calculated with the LFHB model. Dashed line is the non-hydrogen-bonding LF contribution to T_g. Reproduced with permission from Ref. [30, fig. 1].

are compared in Figure 20.12. As seen in this figure, the effect of HB is rather significant in this mixture. In fact, it is the HB contribution which explains the unusual positive deviation of mixture T_g's from linearity with composition.

There are not many EOS models of polymer solution thermodynamics that can be used to perform as many types of calculations as does the LF model and certainly there is probably none as simple as the LF model with this capacity. However, the LF model is an approximate mean-field interaction model bearing all shortcomings of the mean-field approximation and is a model with limited predictive capacity. One of the severe shortcomings of the plain LF model is its limited capacity to describe the behavior of systems interacting with strong specific forces such as hydrogen bonding. As indicated above, this has been corrected and the improved LFHB model is discussed in Chapter 32 along with its extension to the NRHB (Non-Randomness with Hydrogen Bonding) model.

REFERENCES

1. Sanchez, I. C., and Lacombe, R. H. 1976. An elementary molecular theory of classical fluids. Pure fluids. *J. Phys. Chem.* 80: 2352–2362.
2. Sanchez, I. C., and Lacombe, R. H. 1978. Statistical thermodynamics of polymer solutions. *Macromolecules* 11: 1145–1156.
3. Panayiotou, C. 1987. Lattice-fluid theory of polymer solutions. *Macromolecules* 20: 861–871.
4. Panayiotou, C., and Sanchez, I. C. 1991. Hydrogen bonding in fluids: An equation of-state approach. *J. Phys. Chem.* 95: 10090–10097.
5. Hutchinson, J. M. 1997. *Relaxation processes and physical aging.* In *The Physics of Glassy* Polymers, eds. R. N. Haward, and R. J. Young. Dordrecht: Springer, pp. 85–153.
6. Gee, G. 1970. The glassy state in polymers. *Contemp. Phys.* 11(4): 313–334.
7. Schawe, J. K. E., and Wrana, C. 2020. Competition between structural relaxation and crystallization in the glass transition range of random copolymers. *Polymers* 12(8): 1778.
8. McKinney, J. E., and Goldstein, M. 1974. PVT relationships for liquid and glassy poly(vinyl acetate). *J. Res. Nat. Bur. Std.* 78A: 331–353.
9. McKinney, J. E., and Simha, R. 1977. Thermodynamics of the densification process for polymer glasses. *J. Res. Nat. Bur. Std.* 81A: 283–297.
10. Eichinger, B. E., and Flory, P. J. 1968. Thermodynamics of polymer solutions Part 1. Natural rubber and benzene. *Trans. Faraday Soc.* 64: 2035–2052.

11. Sanchez, I. C., and Panayiotou, C. 1994. Equations of state thermodynamics of polymer and related solutions. In *Models for Thermodynamic and Phase Equilibria Calculations*, ed. S. Sandler. New York: Marcel Dekker Inc., p. 191.
12. Panayiotou, C. 1986. Thermodynamics of gas solubility in polymeric liquids. *Makromol. Chem.* 181: 2867–2882.
13. Panayiotou, C. 1987. Thermodynamics of random copolymer mixtures. *Makromol. Chem.* 188: 2733–2743.
14. Taimoori, M., and Panayiotou, C. 2003. The non-random distribution of free volume in fluids: Polydisperse polymer systems. *Fluid Phase Equilib.* 205: 249–265.
15. Panayiotou, C. 2003. The QCHB model of fluids and their mixtures. *J. Chem. Thermodyn.* 35: 349–381.
16. Nishi, T., and Kwei, T. K. 1975. Cloud point curves for poly(vinyl methyl ether) and monodisperse polystyrene mixtures. *Polymer* 16(4): 285–290.
17. Sanchez, I. C., and Rodgers, P. A. 1990. Solubility of gases in polymers. *Pure Appl. Chem.* 62(11): 2107–2114.
18. Du, Q., Hattam, P., and Munk, P. 1990, Inverse gas chromatography 7. Polymer-solvent interactions of hydrocarbon polymers. *J. Chem. Eng. Data* 35(3): 367–371.
19. Gibbs, J. H., and Di Marzio, E. A. 1958. Nature of the glass transition and the glassy state. *J. Chem. Phys.* 28(3): 373–383.
20. Condo, P. D., Sanchez, I. C., Panayiotou, C. G., and Johnston, K. P. 1992. Glass transition behavior including retrograde vitrification of polymers with compressed fluid diluents. *Macromolecules* 25: 6119–6127.
21. Olabisi, O., and Simha, R. 1975. Pressure-volume-temperature studies of amorphous and crystallizable polymers. I. Experimental. *Macromolecules* 8: 206–210.
22. Doghieri, F., and Sarti, G. C. 1996. Nonequilibrium lattice fluids: A predictive model for the solubility in glassy polymers. *Macromolecules* 29: 7885–7896.
23. Boudouris, D., and Panayiotou, C. 1998. On the solubility of gas molecules in glassy polymers and the nonequilibrium approach. *Macromolecules* 31: 7915–7920.
24. Fleming, G. K., and Koros, W. J. 1990. Carbon dioxide conditioning effects on sorption and volume dilation behavior for bisphenol A-polycarbonate. *Macromolecules* 23: 1353–1360.
25. Vargaftik, N. B. 1975. *Handbook of Physical Properties of Liquids and Gases*, 2nd ed. Washington, DC: Hemisphere Publishing.
26. Baschetti, G. M., Doghieri, F., and Sarti, G. C. 2001. Solubility in glassy polymers: Correlations through the non-equilibrium lattice fluid model. *Ind. Eng. Chem. Res.* 40(14): 3027–3037.
27. Galizia, M., Stevens, K. A., Smith, Z. P., Paul, D. R., and Freeman, B. D. 2016. Non-equilibrium lattice fluid modeling of gas solubility in HAB-6FDA polyimide and its thermally rearranged analogs. *Macromolecules* 49: 8768–8779.
28. Panayiotou, C. 2024, Thermodynamics of the glassy polymer state: Equilibrium and non-equilibrium aspects. *Polymers* 16: 298. https://doi.org/10.3390/polym16020298
29. Paul, D. R. 2014. Dual mode sorption model. In *Encyclopedia of Membranes*, eds. E. Drioli, and L. Giorno. Berlin: Springer, pp. 1–2.
30. Prinos, J., and Panayiotou, C. 1995. Glass-transition temperatures in hydrogen-bonded polymer mixtures. *Polymer* 36(6): 1223–1227.

PROBLEMS

20.1 Gas sorption in polymers and Henry's law

Prove Equation 20.30, starting from Equation 20.29 and the definition of Henry's law constant.

20.2 The order parameter in glassy polymers

Doghieri and Sarti [1] have proposed to set the density ρ_2 of the glassy polymer as the order parameter in gas sorption studies. Show that this option leads to Equation 20.42.

20.3 The change of glassy polymer density with gas sorption

Baschetti et al. [2] have proposed the empirical Equation 20.43 for the prediction of the change of polymer density ρ_2 with the partial pressure of gas penetrant. Show how this equation could be derived from Henry's law and Boudouris [3] Equation 20.37.

REFERENCES FOR PROBLEMS

1. Doghieri, F., and Sarti, G. C. 1996. Nonequilibrium lattice fluids: A predictive model for the solubility in glassy polymers. *Macromolecules* 29: 7885–7896.
2. Baschetti, G.M., Doghieri, F., and Sarti, G. C. 2001. Solubility in glassy polymers: Correlations through the non-equilibrium lattice fluid model. *Ind. Eng. Chem. Res.* 40(14): 3027–3037.
3. Boudouris, D., and Panayiotou, C. 1998. On the solubility of gas molecules in glassy polymers and the nonequilibrium approach. *Macromolecules* 31: 7915–7920.

21 The Thermodynamics of Chemical Reactions

Abstract

The basic definitions of terms and variables used in the study of chemical reactions are introduced and the false idea that the compositions at chemical equilibrium give the maximum conversion attainable is dispelled. Down-to-earth examples are presented to illustrate a simple method for determining the number of independent reactions. The Solvay process to produce soda ash from brine and lime is presented as an example of how an indirect method can be used to carry out a seemingly impossible reaction. Ways of calculating the equilibrium compositions for reactions in the gas and liquid phases are discussed.

In Section I, we reviewed the fundamentals, while in Section II, Chapters 5–17, we studied the behavior of mixtures. In this chapter, the second specifically considering an application of the thermodynamics of mixtures, attention is centered on a reacting system. As usual in thermodynamics, we do not consider the kinetics of the process but, as discussed in Chapters 8 and 12, we only pay attention to the final state of the system at equilibrium, after reactions have occurred.

STATEMENT OF THE PROBLEM AND BASIC DEFINITIONS

To clarify the applicability of chemical reaction equilibrium, let us state (as shown in Figure 21.1) that some reactions get so displaced toward the products that, for all practical purposes, there is no need to perform equilibrium calculations. This is the case, for example, of the combustion of many organic compounds to yield carbon dioxide and water. Considering that the main purpose of these reactions is usually the generation of heat, their study is presented in Chapter 21 dealing with heat effects.

The study of chemical reaction equilibrium is important when the final state consists of a mixture of both reactants and products in significant amounts. In this case, as at equilibrium, there is no

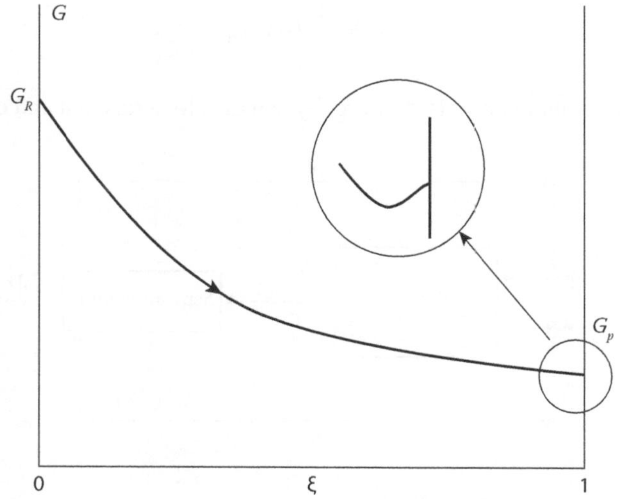

FIGURE 21.1 Gibbs function of a closed system in which a chemical reaction is almost complete.

DOI: 10.1201/9781003431985-24

net reaction, and close to equilibrium the rate of reaction tends to slow down; the design of reactors is based on kinetic rather than thermodynamic considerations. Moreover, although for this case the calculation of the composition of a reacting mixture at equilibrium gives the maximum possible conversion that can be expected at the conditions of operation, in principle it is possible to carry out a reaction in such a way that all the reactants fed to the system are completely converted into products.

Consider the case shown in Figure 21.2 in which reactants *A* and *B* enter a continuous reactor to give an output stream of products *C* and *D* plus unreacted *A* and *B*. The output stream is then sent to a separation system from which products *C* and *D* are obtained while the unreacted *A* and *B* compounds are recycled to the reactor. The net effect is that *A* and *B* enter the overall system and products *C* and *D* leave the overall system.

Thus, one may ask why to worry about chemical equilibrium when even with a very low conversion it is possible to convert all the reactants to products. The answer is that the size of the reactor and the number and complexity of the separation units depend on the amount of chemicals that need to be recycled. The energy consumed also depends on how efficient the reactor is. The higher the conversion that can be obtained in the reactor in a single pass, the better.

This maximum conversion is limited by the composition attainable at equilibrium. Thus, the study of chemical equilibrium is important as a way to establish the maximum possible *extent of reaction*, ξ, in a discontinuous batch reactor for a given feed composition at specified conditions of temperature and pressure. No matter how fast or how slow the kinetics of the reaction is, the value of the extent of reaction at equilibrium will determine the maximum conversion of each reactant in the reactor, that is, the maximum possible yield of products at the operating conditions. The *conversion* χ_i *of a reactant i*, for which the initial and final numbers of moles are n_{i0} and n_i, respectively, is defined as:

$$\chi_i = \frac{n_{i0} - n_i}{n_{i0}} \tag{21.1}$$

Integrating Equation 8.9 between ($n_{i,0}$, ξ=0) and (n_i, ξ) and rearranging, we write:

$$n_i = n_{i0} + \nu_i \xi \tag{21.2}$$

and combining this result with Equation 21.1, we have:

$$\chi_i = \frac{1}{n_{i0}}(-\nu_i)\xi \tag{21.3}$$

where the *stoichiometric number*, ν_i, is intrinsically negative for a reactant and positive for a product.

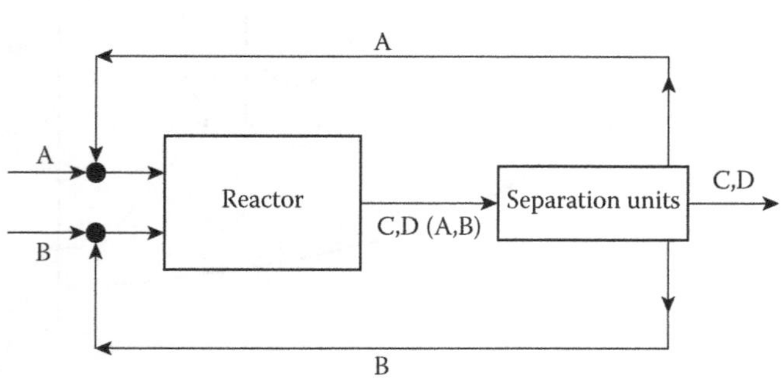

FIGURE 21.2 Diagram of a continuous reactor.

Hence, the conversion of reactant i in a closed system at equilibrium is determined by the equilibrium value of the extent of reaction, ξ.

If the actual conversion obtained in such a closed system is much less than the equilibrium conversion, then kinetic factors are limiting the advancement of the reaction and the use of a catalyst should be considered. On the other hand, if the actual conversion is close to the equilibrium conversion, then thermodynamic factors are limiting and the attention should be directed to the separation and recycling of products.

In the case of a single reaction, for a given value of the extent of reaction and a knowledge of the initial number of moles of the chemical species, n_{i0}, the mole fractions of the compounds are obtained from Equation 21.2 as:

$$y_i = \frac{n_{i0} + \nu_i \xi}{n_0 + \xi \Delta \nu} \tag{21.4}$$

with

$$n_0 = \sum_k n_{k0} \tag{21.5}$$

and, as discussed in Chapter 8, the change in the number of moles of the stoichiometric reaction,

$$\Delta \nu = \sum_k \nu_k = \sum_{products} \lambda_P - \sum_{reactants} \lambda_R \tag{21.6}$$

where λ_k is the coefficient of compound k in the stoichiometric equation for the reaction. If there is more than one independent reaction, Equations 21.2–21.4 are directly generalized to:

$$n_i = n_{i0} + \sum_r \nu_{i,r} \xi_r \tag{21.2a}$$

$$\chi_i = \frac{1}{n_{i0}} \left(-\sum_r \nu_{i,r} \xi_r \right) \tag{21.3a}$$

and again

$$y_i = \frac{n_{i0} + \sum_r \nu_{i,r} \xi_r}{n_0 + \sum_r \xi_r \Delta \nu_r} \tag{21.4a}$$

In Equations 21.4 and 21.4a, we have used the symbol y_i for the mole fraction, but obviously, similar expressions apply to the mole fraction x_i of a compound in a liquid mixture.

While the reaction proceeds, one of the reactants may be totally consumed. This reactant will then be the *limiting reactant* fixing the extent of reaction at a value ξ_L. This can be calculated by setting the number of moles of this reactant to zero in Equation 21.2. Thus,

$$\xi_L = -\frac{n_{L,0}}{\nu_L}$$

In some practical cases, one of the reactants is much more expensive than the others, and it is convenient to use it as the limiting reactant.

For those cases in which the attainment of chemical equilibrium is important, it is desirable to determine the value of the extent of reaction ξ at equilibrium conditions when the reactants are charged in their stoichiometric amounts to a batch reactor, that is, to a system that it is closed after the reactants are loaded.

DETERMINATION OF THE NUMBER OF INDEPENDENT REACTIONS

Equations 21.4 and 21.4a show that the first step to get the equilibrium composition is to know how many *independent reactions* are possible, and then determine the value of the extent of reaction at equilibrium for each of the independent reactions. Although there are different ways of determining the number of possible independent reactions (one of them purely mathematical based on the rank of the matrix of stoichiometric numbers), we present here a method centered on the reactions of formation. This approach is more intuitive and connects well with the conceptual steps used for evaluation of the equilibrium constant.

As described in Chapter 8, in this method, the number of independent reactions is obtained by writing the stoichiometric equations of formation, from its constituent elements, of each chemical species considered to be present in the system. Then all the elements considered not to be present as elements, that is, in their elemental form, are eliminated in a systematic way. The final number of independent equations resulting from this process must be the same no matter the path chosen to eliminate the elements not present as elements. This procedure is illustrated in the examples that follow.

Each independent reaction has an equilibrium value of the extent of reaction, ξ_r. The solution of the equilibrium problem for any of the sets of independent reactions gives a unique set of values ξ_r, that is, a unique set of equilibrium concentrations.

The important point is that it is not necessary to know which are the independent reactions that happen in reality. If the equilibrium compositions are obtained for the set of reactions generated by the above procedure, these will be the equilibrium compositions of the compounds in the real system. When there is any question about the possible presence of a particular element in its elemental form in the equilibrium mixture, it is safe to assume that the particular element is present as element. If it were not present, the final result would show that at equilibrium, its "presence" is negligible and it could have been ignored from the start. The only price to pay for a wrong guess is the additional work of solving some extra equations. The same is true when it is not known what products will be obtained from multiple reactions. It is wise to include as many products as one could assume to be present, do the additional work, and from the results see which products could have been left out to begin with. As in most cases of applications, ideas are better clarified through examples.

Example 21.1: Catalytic Oxidation of Ammonia

In the catalytic oxidation of ammonia with air to produce nitric acid, the following chemical species are present in the reactor(s): NH_3, O_2, N_2, NO, NO_2, H_2O, and HNO_3. Determine the number of independent reactions for this system considering that (1) all chemical species participate in the reaction and (2) under the reaction conditions, N_2 can be considered inert.

Solution

Reactions of formation (marked as f) of compounds considered to be present:

$$f_1.\ N_2+3H_2=2NH_3$$

$$f_2.\ N_2+O_2=2NO$$

$$f_3.\ N_2+2O_2=2NO_2$$

$$f_4.\ 2H_2+O_2=2H_2O$$

$$f_5.\ H_2+N_2+3O_2=2HNO_3$$

Elements forming the compounds: N, H, O
Elements present as elements: N_2, O_2
Elements not present as elements: H_2
Thus, with reaction f_1, H_2 is eliminated from reactions f_4 and f_5, and reaction f_1 is then eliminated. The final reactions are
1. $N_2+O_2=2NO$
2. $N_2+2O_2=2NO_2$
3. $2N_2+6H_2O=4NH_3+3O_2$
4. $2N_2+9O_2+2NH_3=6HNO_3$

As all elements not present in their elemental form have been eliminated, there are four independent reactions ($r=4$). As an exercise, one can use equation f_4 to eliminate H_2 from f_1 and f_5 and see that although a different set of equations is obtained, again $r=4$.
In practice, it is known that N_2 does not participate in the reactions. N_2 is one of the elements forming the compounds, but if it had not been added as an inert, it would not be produced in the reactor and it would not be present as an element in the final products. The situation of nitrogen in this case is similar to the situation of hydrogen. Clearly, if N_2 is replaced by argon in the mixture with oxygen, as a way to "dilute" the chemicals as N_2 does, nothing will change in the reactions. In this case, argon will not appear in the reactions of formation of the chemical compounds, and thus it would not be necessary to eliminate it from equations in which it is not present. Therefore, as N_2, being one of the elements forming the compounds, would not appear as an element in the final products, it is necessary to eliminate it from Equations 1–4 above. Eliminating N_2 using Equation 1, the final reactions in this case are
1. $O_2+2NO=2NO_2$
2. $6H_2O+4NO=4NH_3+5O_2$
3. $7O_2+4NO+2NH_3=6\ HNO_3$

The result then is $r=3$.
According to the technical literature, the reactions that actually take place are
$4NH_3+5O_2=6H_2O+4NO$ (reaction 2 above, reversed)
$O_2+2NO=2NO_2$ (reaction 1 above)
and

$$H_2O+3NO_2=NO+2HNO_3$$

The third reaction is obtained by a linear combination of reaction 3, with the two other.

Example 21.2: Production of 1,3-Butadiene

In the production of 1,3-butadiene by dehydrogenation of n-butane, it is expected to find the following chemical species: n-butane, hydrogen, 1,3-butadiene ($CH_{2=CH}$–CH=CH2 or 1,3-C_4H_6), and the side products: 1-butene ($CH_{2=CH}$–CH_2–CH_3 or 1-C_4H_8), 2-butene (CH_3–CH=CH–CH_3 or 2-C_4H_8), 1,2-butadiene ($CH_{2=C=CH}$–CH_3 or 1,2-C_4H_6), 1-butyne (CH≡C–CH_2–CH_3 or 1-C_4H_6), and 2-butyne (CH_3–C≡C–CH_3 or 2-C_4H_6). Determine the number of independent reactions to be considered when (1) only the above compounds are present at equilibrium and (2) N_2 is added as an inert to minimize side reactions.

Solution

Reactions of formation of compounds considered to be present:

$$f_1.\ 4C+5H_2=C_4H_{10}$$

$$f_2.\ 4C+3H_2=1,3\text{-}C_4H_6$$

$$f_3.\ 4C+4H_2=1\text{-}C_4H_8$$

$$f_4.\ 4C+4H_2=2\text{-}C_4H_8$$

$$f_5.\ 4C+3H_2=1,2\text{-}C_4H_6$$

$$f_6.\ 4C+3H_2=1\text{-}C_4H_6$$

$$f_7.\ 4C+3H_2=2\text{-}C_4H_6$$

Elements forming the compounds: C, H
Element present as element: H_2
Element not present as element: C
Use equation f_1 above to eliminate C from all the rest:
1. $C_4H_{10}=1,3\text{-}C_4H_6+2H_2$
2. $C_4H_{10}=1\text{-}C_4H_8+H_2$
3. $C_4H_{10}=2\text{-}C_4H_6+H_2$
4. $C_4H_{10}=1,2\text{-}C_4H_6+2H_2$
5. $C_4H_{10}=1\text{-}C_4H_6+2H_2$
6. $C_4H_{10}=2\text{-}C_4H_6+2H_2$

As no other elimination is required, $r=6$.
Since N_2 is an inert and it does not participate in the reactions, $r=6$.

Example 21.3: Isomerization of 1,3-Butadiene

In the isomerization of 1,3-butadiene to obtain 1,2-butadiene, it is expected that we will also find 1-butyne and 2-butyne in the products. Determine the number of independent reactions to be considered.

Solution

Reactions of formation:

$$f_1.\ 4C+3H_2=1,3\text{-}C_4H_6$$

$$f_2.\ 4C+3H_2=1,2\text{-}C_4H_6$$

$$f_3.\ 4C+3H_2=1\text{-}C_4H_6$$

$$f_4.\ 4C+3H_2=2\text{-}C_4H_6$$

Elements forming the compounds: C, H
No elements present as elements
Thus, with reaction f_1 eliminate C and H_2 in all other equations. This eliminates one reaction; then, $r=3$.

THERMODYNAMIC EQUILIBRIUM CONSTANT IN TERMS OF THE EQUILIBRIUM COMPOSITIONS

To simplify matters, we consider here the case of a single reaction. As discussed in Chapter 12, once the stoichiometry of the reaction is decided, the thermodynamic equilibrium constant, K_T, is a function of temperature only, and it is related to the activities of the compounds by:

$$K_T \equiv \frac{\prod_k a_{Pk,\theta}^{\lambda_{Pk}}}{\prod_j a_{Rj,\theta}^{\lambda_{Rj}}} \tag{12.15}$$

The dependence of the value of K_T on the stoichiometry of the equation, as written, is clearly shown by the presence of the stoichiometric coefficients in Equation 12.15. Thus, giving a numerical value of the *Thermodynamic equilibrium constant*, without specifying the stoichiometry of the equation, is meaningless. On the other hand, once the stoichiometric equation has been specified and the same stoichiometric coefficients are consistently used in all calculations, the compositions at equilibrium are independent of how the stoichiometric equation was written in the first place. This fact is better demonstrated with one simple example.

Consider a typical reaction,

$$\lambda_A A + \lambda_B B \leftrightarrow \lambda_C C + \lambda_D D$$

for which the thermodynamic equilibrium constant takes the form:

$$K_T \equiv \frac{a_C^{\lambda_C} a_D^{\lambda_D}}{a_A^{\lambda_A} a_B^{\lambda_B}}$$

According to Equation 12.16, the value of the thermodynamic equilibrium constant in this case will be obtained from the Gibbs energies of the compounds in their standard states as:

$$\ln K_T = \frac{\left[\left(\lambda_A G_A^0 + \lambda_B G_B^0\right) - \left(\lambda_C G_C^0 + \lambda_D G_D^0\right)\right]}{RT}$$

An independent thinker may decide to work with the chemical equation multiplied by a factor *F*, which can even be a negative number, thus reversing the sense of the reaction. For the reaction written as:

$$(F\lambda_A)A + (F\lambda_B)B \leftrightarrow (F\lambda_C)C + (F\lambda_D)D$$

the equilibrium constant takes the form:

$$K_T^* \equiv \frac{a_C^{F\lambda_C} a_D^{F\lambda_D}}{a_A^{F\lambda_A} a_B^{F\lambda_B}} = \left(\frac{a_C^{\lambda_C} a_D^{\lambda_D}}{a_A^{\lambda_A} a_B^{\lambda_B}}\right)^F$$

where the asterisk, used as a superscript on the left-hand side, indicates that the numerical value of this thermodynamic equilibrium constant will be different from the previous one. In fact, the value of this new thermodynamic equilibrium constant will be obtained as:

$$\ln K_T^* = \frac{\left[\left(F\lambda_A G_A^0 + F\lambda_B G_B^0\right) - \left(F\lambda_C G_C^0 + F\lambda_D G_D^0\right)\right]}{RT}$$

$$= \frac{\left[\left(\lambda_A G_A^0 + \lambda_B G_B^0\right) - \left(\lambda_C G_C^0 + \lambda_D G_D^0\right)\right]F}{RT}$$

or

$$\frac{1}{F}\ln K_T^* = \ln K_T$$

So, this new numerical value will give:

$$K_T^* = [K_T]^F = \left(\frac{a_C^{\lambda_C} a_D^{\lambda_D}}{a_A^{\lambda_A} a_B^{\lambda_B}} \right)^F$$

As shown by this result, a change in the way of writing the stoichiometric equation does not change the relation between the activities, that is, the relation between the equilibrium compositions. Hence, once the stoichiometric equation has been decided, it is necessary to keep the same stoichiometric coefficients for all calculations in order to make sense of the numerical value of the thermodynamic equilibrium constant.

Another general point that deserves mentioning is that the sign of the standard Gibbs energy change only indicates that if it is negative, the equilibrium constant is larger than unity, and if it is positive, the equilibrium constant is smaller than unity for the equation as written. The larger the value of the equilibrium constant, the more the equilibrium composition is displaced toward the products. The sign of the standard Gibbs energy change should not be interpreted as meaning that the reaction as written is spontaneous or not. As this value is obtained considering the reactants and the products in their standard state, and not the conditions of the reaction, any conclusion regarding the spontaneity of the reaction is unreliable. In addition, even when the conditions could be favorable for a spontaneous reaction, the reaction may not happen due to the need for activation energy to initiate it. An example of this is the reaction of oxygen with the stoichiometric amount of hydrogen to form water. The standard Gibbs energy change to form liquid water is –237.14 kJ/mol of water, and to form water vapor is –228.61 kJ/mol of water. This reaction with an extremely favorable equilibrium constant will not proceed unless initiated by a spark or another means. Once started, however, it will proceed with explosive speed.

A reaction with an unfavorable equilibrium constant can be displaced toward the products by adding an excess of one of the reactants that is either easy to separate and recover or inexpensive. Moreover, even a reaction with an unfavorable value of the equilibrium constant can be forced to proceed toward completion by continuously withdrawing one of the products of the reaction.

SOLVAY PROCESS

A good example of a successful procedure to carry out a difficult reaction is the Solvay process to produce soda ash, Na_2CO_3, starting from brine, $NaCl_{(aq)}$, and lime, $CaCO_3$.

The desired reaction is:

$$2\text{NaCl} + \text{CaCO}_3 \rightarrow \text{Na}_2\text{CO}_3 + \text{CaCl}_2, \quad \Delta\text{G}^0 = +96.9 \ (\text{kJ / molNa}_2\text{CO}_3)$$

which has a positive standard Gibbs energy change. Common salt, NaCl, and lime are inexpensive raw materials, while soda ash is a valued commodity with multiple industrial applications and calcium chloride, $CaCl_2$, is used as a road salt.

In 1791, Leblanc proposed a method to carry out the reaction with the help of sulfuric acid. The Leblanc process was replaced by the more economical Solvay process, proposed in 1861. Variations of the Solvay scheme have been proposed, but they do not change the basic feature of the process presented in Figure 21.3, together with the reactions in each unit.

This process promotes the reaction using ammonia, NH_3. The key steps happen in the ammonization unit, in which the brine is mixed with ammonia, becoming alkaline, and in the Solvay unit, where the alkaline brine is contacted with carbon dioxide gas, CO_2, to form sodium bicarbonate, $NaHCO_3$. In an alkaline environment, the sodium bicarbonate is less soluble than NaCl, and it is withdrawn from the unit as a precipitate. The precipitate is finally calcined to give sodium carbonate.

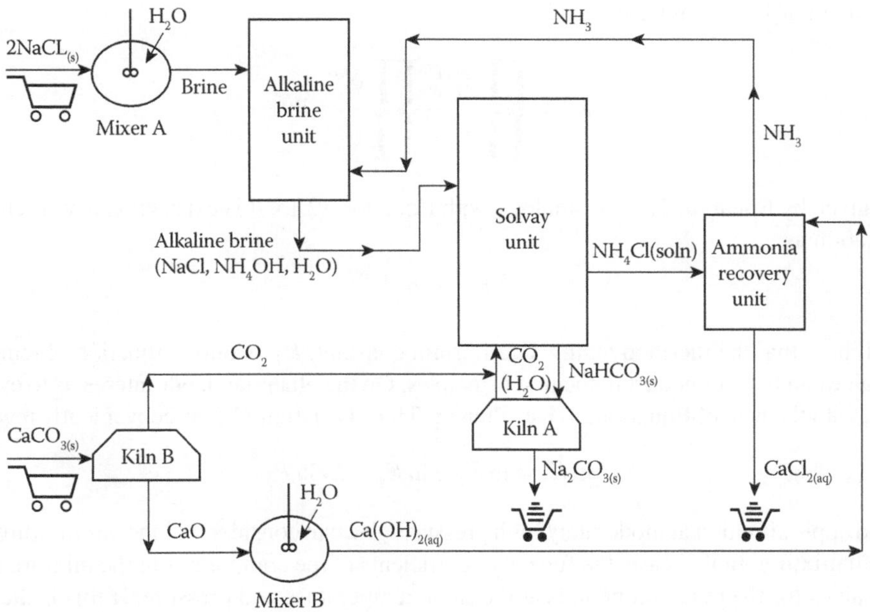

FIGURE 21.3 Solvay process.

The standard process needs a small makeup of ammonia to compensate for losses. If ammonia is easily available, it can be fed directly to the process and the unit of ammonia recovery can be eliminated. In this case, the by-products would be ammonium chloride and calcium hydroxide, both having useful applications. Moreover, if the carbon dioxide is obtained from flue gas of combustion, the whole feed of calcium carbonate can be eliminated and the only by-product will be ammonium chloride. In any case, this process shows that a reaction with a very unfavorable equilibrium constant can be carried almost to completion.

Once the value of thermodynamics in the study of chemical reactions is properly understood, the next step is to relate the thermodynamic equilibrium constant to the composition of the reacting mixture at equilibrium. For this, we recall that the activity of a compound is given as:

$$a_{i,\theta} \equiv \frac{f_i}{f_i^{(\theta)}} \tag{12.13}$$

REACTIONS IN THE GAS PHASE

For a reaction in the gas phase, the fugacity of compound i is given by:

$$f_i = y_i \phi_i P \tag{13.1}$$

and the standard state is the pure compound i as ideal gas at one unit of pressure, normally taken as 1 bar. Thus, $f_i^{(\theta)} = 1$ (unit of pressure used to measure the pressure P), and from Equation 12.13, remembering that by dividing by the unit of pressure the symbol P represents a dimensionless quantity, we write:

$$a_{i,\theta} = y_i \phi_i P$$

Hence, Equation 12.15 is rewritten as:

$$K_T = \frac{\prod_k y_{Pk,\theta}^{\lambda_{Pk}} \prod_k \phi_{Pk,\theta}^{\lambda_{Pk}}}{\prod_j y_{Rj,\theta}^{\lambda_{Rj}} \prod_j \phi_{Rj,\theta}^{\lambda_{Rj}}} P^{\Delta\nu} \tag{21.7}$$

with $\Delta\nu$ given by Equation 21.6. By analogy with Equation 12.15, this expression is written in shorthand notation as:

$$K_T = K_y K_\phi P^{\Delta\nu} \tag{21.7a}$$

We recall here that the thermodynamic equilibrium constant, K_T, is only a function of temperature, and its value can be obtained from the first principles. On the other hand, our interest is to evaluate K_y and then, with the help of Equation 21.4, evaluate ξ. Thus, Equation 21.7a is conveniently rewritten as:

$$\ln K_y = \ln K_T - \ln K_\phi - \Delta\nu \ln P \tag{21.7b}$$

As a good approximation at moderately high pressures, we can consider that the gas mixture behaves as an ideal mixture. In this case, the fugacity coefficients of the compounds in the mixture are equal to their values for the pure compounds at the same temperature and pressure. If this is the case, K_ϕ and all of the right-hand side of Equation 21.7b are independent of composition. Furthermore, at low pressure we can consider that the gas mixture behaves as an ideal gas, in which case the fugacity coefficients are all equal to unity and Equation 21.7b reduces to:

$$\ln K_y^{id} = \ln K_T - \Delta\nu \ln P \tag{21.8}$$

where the superscript *id* has been used as a reminder that this form considers an ideal gas phase. This simplified form separates the effect of temperature and pressure, showing more clearly that if the number of moles increases by the reaction, an increase in pressure will decrease the value of K_y; that is, it will favor the displacement of the reaction toward the reactants, and vice versa. This is a form of the Le Chatelier principle.

At very high pressure, when the assumption of ideal gas mixture is not valid, an iterative procedure may be required to correct for the composition dependence of K_ϕ.

REACTIONS IN THE LIQUID PHASE

Solutions of electrolytes in water require special consideration, and its discussion was covered in Chapter 18. For nonelectrolyte liquids and solids, the standard state is usually taken as the pure compound in its natural state at 1 bar and 298.15 K. For solutes, sometimes molality is used as a measure of composition and the standard state is the solute in a hypothetical ideal solution, in Henry's sense, at unit molality and 298.15 K. In the case of using mole fraction as a measure of composition, the most common case for mixtures of nonelectrolytes, the activity of compound i is directly given by Equation 15.13,

$$a_{i,\theta} = x_i \gamma_{i,\theta} \tag{15.13}$$

and Equation 12.15 takes the simple form,

$$K_T = \frac{\prod_k x_{Pk}^{\lambda_{Pk}} \prod_k \gamma_{Pk,\theta}^{\lambda_{Pk}}}{\prod_j x_{Rj}^{\lambda_{Rj}} \prod_j \gamma_{Rj,\theta}^{\lambda_{Rj}}} = K_x K_\gamma \tag{21.9}$$

where we have used the same shorthand notation used in Equation 21.7a for reactions in the gas phase. Thus, for reactions in liquid phase, we write:

$$\ln K_x = \ln K_T - \ln K_\gamma \tag{21.9a}$$

Even when in appearance this form is simpler than Equation 21.7b, in practice its use may get complicated by a variation in the value of K_γ, which may even be a function of the initial composition of the liquid mixture.

This observation is better illustrated with one exaggerated example. Consider a mixture in which acetic acid and ethyl alcohol are the reactants and water and ethyl acetate are the products:

$$(\text{Acetic acid}) + (\text{Ethyl alcohol}) \Leftrightarrow (\text{Ethyl acetate}) + (\text{Water})$$

While the thermodynamic equilibrium constant is only a function of temperature, the value of K_γ will be different if the initial mixture is formed with a drop of ethyl acetate in a large amount of water or with a drop of acid in a large amount of alcohol. In the first case, the activity coefficient of water (one of the products) will be close to unity, while the activity coefficients of ethyl acetate, acid, and alcohol will be close to their values at infinite dilution in water. In the second case, the activity coefficient of alcohol (one of the reactants) will be close to unity, while the activity coefficients of water, acetic acid, and ethyl acetate will be close to their values at infinite dilution in ethyl alcohol. Although this case is exaggerated, it shows that the value of K_γ depends on the composition of the feed. In addition, the value of K_γ cannot be accurately calculated due to the scarcity of data on activity coefficients in reacting mixtures. In the simple example considered here, it would be required to know the values of the activity coefficients of all four participants in the reaction as a function of composition of the quaternary mixture. One practical way of proceeding is to measure the compositions at equilibrium under typical conditions of operation and knowing the value of the thermodynamic equilibrium constant K_T, back-calculate K_γ. In this case, as the activity coefficients are normally a weak function of temperature, for small variations in the initial composition of the mixture, the value of K_γ can be assumed to be constant. An example of the application of this idea is presented later in the chapter.

REACTIONS WITH COMPOUNDS IN THE SOLID PHASE

In some cases, a compound (elemental or nonelemental) is present in a solid state at equilibrium with other compounds, either solid, liquid, or gas. In these cases of heterogeneous reactions, Equation 12.15 is used with the activities expressed in different forms for different compounds.

For the compounds in the vapor phase, the activities are obtained from Equation 13.1, as:

$$a_{i,\theta} = y_i \phi_i P$$

For the compounds in the liquid phase, the activities are obtained from Equation 15.13 as:

$$a_{i,\theta} = x_i \gamma_{i,\theta}$$

For those compounds that are pure solids, Equation 12.13 is used as:

$$a_{i,\theta} = \frac{f_i}{f_i^{(\theta)}\left(1\ \text{bar}, T\right)} = \exp\frac{v_i^s (P-1)}{RT}$$

where P must be in bars. The molar volumes of solids are very small when compared with the ratio $RT/(P-1)$; thus, for pure solids, as a good approximation,

$$a_{i,\theta} = 1$$

For this reason, in many situations, the activities of pure solid compounds do not show in the right-hand side of Equation 12.15. However, if the pure solid is not an element in its natural (elemental) state at 1 bar and 298.15 K, the value of the Gibbs energy of formation of the solid must be included in the calculation of K_T, as discussed below. On the other hand, if the pure solid is an element in its natural elemental form, by convention, the Gibbs energy of formation is taken as zero.

CALCULATION OF THE THERMODYNAMIC EQUILIBRIUM CONSTANT

In any case, for reactions in a gas phase, in a liquid phase, or in the presence of a solid phase, it is necessary to get the value of the thermodynamic equilibrium constant, K_T, at the temperature, T, of the system. This is done in two steps. In order to use standard thermodynamic tables, first its value is obtained at the temperature $T_0=298.15$ K, and then the value at the temperature, T, of the system is calculated.

The first step is to write the standard reaction to be considered. In the standard reaction, the stoichiometric amounts of reactants are converted into the stoichiometric amounts of products. We then assume that each of the compounds participating in the stoichiometric reaction is formed from a reaction of its elements in their natural elemental state at 298.15 K and 1 bar. These are imaginary reactions that are used only for tabulation purposes. The Gibbs energies and the enthalpies of the elements in their standard states are arbitrarily set equal to zero. *Lange's Handbook of Chemistry* [1] gives values of the standard Gibbs energies of formation, $\Delta G^0_{f,i,T_0}$, and the standard enthalpies of formation, $\Delta H^0_{f,i,T_0}$, of different pure compounds, i. These values correspond, respectively, to the standard chemical potential and the standard enthalpy of the compound in the nomenclature used in Chapter 12.

From Equation 12.14, the standard Gibbs energy change for the stoichiometric reaction at 298.15 K is then:

$$\Delta G^\theta_{T_0} \equiv \sum_k^c \nu_k \Delta G^0_{f,k,T_0} \equiv \sum_k^c \nu_k \mu^\theta_k \quad (21.10)$$

and from Equation 12.18, the standard enthalpy change for the stoichiometric reaction at 298.15 K is:

$$\Delta H^\theta_{T_0} \equiv \sum_k^c \nu_k \Delta H^0_{f,k,T_0} \equiv \sum_k^c \nu_k h^\theta_k \quad (21.11)$$

From Equation 12.16, the value of the thermodynamic equilibrium constant at $T_0=298.15$ K can be obtained directly as:

$$\ln K_{T_0} = -\frac{\Delta G^\theta_{T_0}}{RT_0} \quad (21.12)$$

The value of the thermodynamic equilibrium constant at a temperature T different from T_0 is then obtained by integration of Equation 12.17 as:

$$\ln K_T = \ln K_{T_0} + \int_{T_0}^{T} \frac{\Delta H^\theta_T}{RT^2} dT \quad (21.13)$$

Again, here Equation 21.11 gives the standard enthalpy change of reaction at T_0=298.15 K and the value of the standard enthalpy change at a temperature T is required. From Equation 5.5, we write:

$$\Delta C_P^\theta = \left(\frac{\partial \Delta H_T^\theta}{\partial T}\right)_{P,n}$$

with

$$\Delta C_P^\theta \equiv \sum_k^c \nu_k c_{P,k} \tag{21.14}$$

Thus, by integration we obtain:

$$\Delta H_T^\theta = \Delta H_{T_0}^\theta + \int_{T_0}^{T} \left(\Delta C_P^\theta\right) dT \tag{21.15}$$

Equation 21.11, using the heats of formation of the compounds at 298.15 K, gives the value of the first term of the right-hand side of Equation 21.15, while the integral is evaluated using information on the specific heats, $c_{P,k}$, of the compounds. Different correlations for specific heats use different series as a function of temperature. For generality, we use here the following series in which particular correlations may set to zero some of the coefficients:

$$c_{P,k} = c_{0,k} + c_{1,k}T + c_{2,k}T^2 + c_{3,k}T^3 + c_{4,k}T^{-2} \tag{21.16}$$

Thus, from Equation 21.14,

$$\Delta C_P^\theta = \Delta C_0 + \Delta C_1 T + \Delta C_2 T^2 + \Delta C_3 T^3 + \Delta C_4 T^{-2}$$

with

$$\Delta C_0 \equiv \sum_k^c \nu_k c_{0,k}$$

and similar expressions for ΔC_1, ΔC_2, ΔC_3, and ΔC_4. Although the algebra is trivial, for the purposes of discussion we give here the final forms obtained after integration. From Equation 21.15,

$$\Delta H_T^\theta = \Delta H_{T_0}^\theta + (\Delta C_0)(T - T_0) + \frac{(\Delta C_1)}{2}\left(T^2 - T_0^2\right) + \frac{(\Delta C_2)}{3}\left(T^3 - T_0^3\right) + \frac{(\Delta C_3)}{4}\left(T^4 - T_0^4\right) - (\Delta C_4)\left(\frac{1}{T} - \frac{1}{T_0}\right) \tag{21.17}$$

From Equation 21.13, then,

$$\ln K_T = \ln K_{T_0} - \frac{\Delta H_0}{R}\left(\frac{1}{T} - \frac{1}{T_0}\right) + \frac{\Delta C_0}{R}\ln\frac{T}{T_0} + \frac{\Delta C_1}{2R}\left(T - T_0\right) + \frac{\Delta C_2}{6R}\left(T^2 - T_0^2\right) + \frac{\Delta C_3}{12R}\left(T^3 - T_0^3\right) + \frac{\Delta C_4}{2R}\left(\frac{1}{T^2} - \frac{1}{T_0^2}\right) \tag{21.18}$$

with

$$\Delta H_0 = \Delta H^{\theta}_{T_0} - \Delta C_0 T_0 - \frac{\Delta C_1}{2} T_0^2 - \frac{\Delta C_2}{3} T_0^3 - \frac{\Delta C_3}{4} T_0^4 + \frac{\Delta C_4}{T_0} \tag{21.19}$$

where $\Delta H^{\theta}_{T_0}$ is given by Equation 21.11.

Two simplifications of the above equations are common. If the specific heats of all compounds are considered to be constant, that is, independent of temperature, and as a good approximation they are evaluated at their mean value in the temperature range from T_0 to T, then Equation 21.18 takes the form:

$$\ln K_T = \ln K_{T_0} - \frac{\Delta H_0}{R}\left(\frac{1}{T} - \frac{1}{T_0}\right) + \frac{\Delta \bar{C}}{R} \ln \frac{T}{T_0}$$

with

$$\Delta H_0 = \Delta H^{\theta}_{T_0} - \Delta \bar{C} T_0$$

Thus, combining both results,

$$\ln K_T = \ln K_{T_0} - \frac{\Delta H^{\theta}_{T_0}}{R}\left(\frac{1}{T} - \frac{1}{T_0}\right) + \frac{\Delta \bar{C}}{R}\left(\frac{T_0}{T} - 1 + \ln \frac{T}{T_0}\right)$$

The last term of the right-hand side of this equation normally is negligible in comparison with the other two terms, so the approximated form is simply written as:

$$\ln K_T = \ln K_{T_0} - \frac{\Delta H^{\theta}_{T_0}}{R}\left(\frac{1}{T} - \frac{1}{T_0}\right)$$

For an exothermic reaction, for example, $\Delta H^{\theta}_{T_0} \leq 0$, a plot of ($\ln K_T$) versus ($1/T$) will show a straight line with a positive slope, where the high temperature range is to the left of the diagram. At higher temperatures, the value of the equilibrium constant decreases for an exothermic reaction, indicating that the reaction displaces toward the reactants. The reverse is valid for an endothermic reaction. This result, like the approximation shown for Equation 21.7b, is an example of the Le Chatelier principle. The reaction displaces its equilibrium in a sense that tends to counteract the effect of a change in temperature or pressure.

EXAMPLES OF EVALUATION OF THE THERMODYNAMIC EQUILIBRIUM CONSTANT

The numerical value of the thermodynamic equilibrium constant depends on the way that the stoichiometric equation is written. Thus, giving a value of the equilibrium constant without specifying the stoichiometric equation is meaningless. On the other hand, once the stoichiometry has been specified, the same stoichiometric coefficients must be used for all calculations. If this is done, the equilibrium compositions obtained are independent of how the stoichiometric equation was written.

In order to evaluate the equilibrium constant at T_0=298.15 K, we need the value of $\Delta G^{\theta}_{T_0}$ and use Equation 21.11. Complete tables of standard heats and Gibbs energy changes for the reactions of formation of pure compounds at 298.15 K and 1 bar are available in the literature [1].

As for the determination of the number of independent reactions in a system, the reactions of formation of the compounds present in the equilibrium mixture are the basic pieces of information needed to study the chemical equilibria between different species. To evaluate K_T at a temperature different from 298.15 K, as indicated by Equations 21.18 and 21.19, we need information on the specific heats of the compounds and on the standard heats of reaction $\Delta H^{\theta}_{T_0}$, which again, according to Equation 21.11, is obtained from tables for the standard heat of formation.

The standard heats and standard Gibbs energy of formation of the elements in their natural (elemental) states at 1 bar and 298.15 K are conventionally assigned a zero value. This convention is particularly important since it provides a basis to express the enthalpy and Gibbs energy of all compounds on the same basis.

Example of a Heterogeneous Reaction

Calculate the standard enthalpy change and the standard Gibbs energy change at 1 bar and 298.15 K, for the following reaction:

$$Mg(OH)_{2(s)} + 2HCl_{(g)} = MgCl_{2(s)} + 2H_2O_{(g)}$$

Note that it is important to specify per mole of which compound the value of $\Delta H^{\theta}_{T_0}$ or $\Delta G^{\theta}_{T_0}$ is given. As different compounds have different stoichiometric coefficients, once the stoichiometric equation is specified, it is necessary to keep it unchanged to make sense of numerical values of $\Delta H^{\theta}_{T_0}$ and $\Delta G^{\theta}_{T_0}$.

We write first the reactions of formation of 1 mole of all nonelemental species starting from the constituent elements in their standard states. From *Lange's Handbook of Chemistry* [1], we get the values of $\Delta H^0_{f,T_0}$ and $\Delta G^0_{f,T_0}$ (see Table 21.1).

The overall reaction R is obtained by adding reaction f_3 plus twice reaction f_4 and subtracting reaction f_1 plus twice reaction f_2. After performing this operation, all elements cancel out and we recover the overall reaction: $R=f_3+2f_4-(f_1+2f_2)$. The same operation should be carried out with the values of $\left(\Delta H^B_{f,T_0}\right)$ in order to get the overall enthalpy change of the reaction:

$$\Delta H^{\theta}_{T_0} = [-641.3 + 2(-241.826)] - [(-924.7) - 2(-92.31)] = -16.13 \text{kJ/molMgCl}_2$$

Thus, the reaction is exothermic. Similarly, for the standard Gibbs energy change of reaction,

$$\Delta G^{\theta}_{T_0} = [(-591.8) + 2(-228.61)] - [(-833.7) - 2(-95.3)] = -24.72 \text{kJ/molMgCl}_2$$

TABLE 21.1
Molar Enthalpies and Molar Gibbs Energies for the Reactions of Formation

No.	Reactions of Formation	$\Delta H^0_{f,T_0}$ [1] kJ/mol	$\Delta G^0_{f,T_0}$ [1] kJ/mol
f_1	$O_{2(g)}+H_{2(g)}+Mg_{(s)}=Mg(OH)_{2(s)}$	–924.7	–833.7
f_2	$\frac{1}{2}H_{2(g)}+\frac{1}{2}Cl_{2(g)}=HCl_{(g)}$	–92.31	–95.30
f_3	$Cl_{2(g)}+Mg_{(s)}=MgCl_{2(s)}$	–641.3	–591.8
f_4	$H_{2(g)}+\frac{1}{2}O_{2(g)}=H_2O_{(g)}$	–241.826	–228.61

Source: Data taken from Lange, N. A., and Speight, J. G. 2005. *Lange's Handbook of Chemistry*. New York: McGraw-Hill.

In principle, the standard reaction has a favorable equilibrium. Now we can calculate the value of the thermodynamic equilibrium constant at T_0=298.15 K for the reaction.

From Equation 21.12,

$$\ln K_{T_0} = -\frac{(-24.72\times 10^3)}{(8.314)(298.15)}\left[\frac{\text{J}/\text{mol}}{\left(\text{J}/\text{mol·K}\right)K}\right] = +9.97_2$$

$$K_{T_0} = 2.14_3 \times 10^4$$

The very large value of K_T indicates that at equilibrium the reaction is almost totally displaced toward the products (see Figure 21.1).

Example of a Reaction in the Gas Phase

The Haber process produces ammonia in a batch reactor operating at 100 bar and 450°C with a 50% excess over the stoichiometric proportion of N_2 with respect to H_2.

As a first step toward understanding the necessity of operating at the above conditions, we calculate the value of the thermodynamic equilibrium constant at 25°C and 450°C for the stoichiometric reaction written as:

$$N_2 + 3H_2 = 2NH_3$$

For this calculation, we distinguish useful from useless information. The thermodynamic equilibrium constant is only a function of temperature. The pressure and the proportion of the chemical species are totally immaterial. From *Lange's Handbook of Chemistry* [1], for the formation of 1 mole of ammonia,

$$\Delta H^0_{f,T_0} = -45.94\left(\frac{\text{kJ}}{\text{mol NH}_3}\right)$$

$$\Delta G^0_{f,T_0} = -16.4\left(\frac{\text{kJ}}{\text{mol NH}_3}\right)$$

Thus, for the equation as written, considering a zero value for the elements in their elemental form,

$$\Delta H^\theta_{T_0} = -91.88\left(\frac{\text{kJ}}{\text{mol N}_2}\right)$$

$$\Delta G^\theta_{T_0} = -32.8\left(\frac{\text{kJ}}{\text{mol N}_2}\right)$$

From Equation 21.12,

$$\ln K_{T_0} = -\frac{-32.8\times 10^3}{8.314\times 298.15} = 13.23$$

$$K_{T_0} = 5.58\times 10^5$$

This value is so high that, from the thermodynamics point of view, it would seem that the reaction should be complete at room temperature. Unfortunately, a very stable triple bond in the nitrogen molecule requires the use of a metal catalyst in the Haber process. The reaction will not proceed without the catalyst, and this catalyst only works at temperatures in the 400°C–500°C range.

This explains why we should operate at 450°C. Let us then calculate K_T at 450°C=723.15 K:

$$\ln K_T = 13.23 - \frac{\left(-91.88\times10^3\right)}{8.314}\left(\frac{1}{723.15}-\frac{1}{298.15}\right) = -8.55$$

$$K_T = 1.93\times10^{-4}$$

Thus, at this temperature, the value of the thermodynamic equilibrium constant decreased by nine orders of magnitude. As the catalyst requires this high temperature, some other way is needed to improve the conversion of the reaction. At a high pressure, according to the principle of Le Chatelier, the reaction tends to displace from 4 moles of reactants to 2 moles of product. The value of the thermodynamic equilibrium constant does not change with pressure, but the all-important value of K_y will change. Thus, now we need to calculate the value of K_y at 100 bar.

Using the approximation of Equation 21.7b, that sets $K_\phi=1$, it is immaterial that the reactor is operating with a 50% excess of N_2 over the stoichiometric proportion. For the reaction, as written, $\Delta\nu=-2$; thus, at 100 bar, as a first approximation,

$$\ln K_y^* = -8.53-(-2)\ln(100) = 0.677$$

and

$$K_y^* = 1.97$$

The increase of pressure increases the value of K_y by four orders of magnitude, but it also makes the whole process very expensive. Note that the value of P used in the above equation must be in bars. If P had been given in any other units, it would have been mandatory to convert it to bars. (Why?)

To evaluate K_ϕ, we recall the convention introduced in Equation 21.7:

$$K_\phi = \frac{\phi_{NH_3}^2}{\phi_{N_2}\phi_{H_2}^3}$$

As a first approximation, we can assume an ideal mixture of real gases, and consider $\phi_k = \phi_k^0$. In this case, the initial composition of the mixture has no effect on the value of K_y. At very high pressure, the use of equations of state with mixing rules may be required to calculate the values of the fugacity coefficients. In such a case, an iterative method can be needed and the initial composition of the mixture will slightly affect the value of K_ϕ.

Example of a Reaction in Liquid Phase

For the following reaction at equilibrium at 298.15 K,

$$(\text{Acetic acid [A]}) + (\text{Ethanol [E]}) = (\text{Ethyl acetate [EA]}) + (\text{Water [W]})$$

the value of K_T can be obtained using data from Tables 1.56 and 2.53 of *Lange's Handbook of Chemistry* [1], and the values of $\Delta G^0_{f,T_0}$ for A, E, EA, and W are –390.2, –174.8, –332.7, and –237.14 kJ/mol, respectively. Thus,

$$\Delta G^\theta_{T_0} = [(-237.14)+(-332.7)]-[(-390.2)+(-174.8)] = -4.8 \text{ kJ/mol}$$

In this case, it is not necessary to specify per mole of which compound $\Delta G^\theta_{T_0}$ is given:

$$\ln K_{T_0} = \frac{-(-4.8)\times 10^3}{8.314\times 298.15} = 1.952$$

$$K_{T_0} = 7.05$$

In practical work, it is sometimes useful to define a "constant" K_C in terms of concentrations by:

$$K_C \equiv \frac{C_C^{\lambda_C} C_D^{\lambda_D}}{C_A^{\lambda_A} C_B^{\lambda_B}}$$

This so-called constant, which is not really a constant, can be directly related to the thermodynamic equilibrium constant by considering that the mole fraction of compound i can be written as:

$$X_i = \frac{n_i}{\sum n_k} = \frac{n_i}{v}\frac{v}{\sum n_k} = c_i(\rho_m)^{-1}$$

where ρ_m is the molar density of the mixture at equilibrium. Thus,

$$\ln K_x = \ln K_c - \Delta \nu \ln \rho_m$$

and from Equation 21.9a, ln K_c is directly related to K_T. Notably for reactions without a change in the number of moles, that is, reactions for which $\Delta\nu=0$, as in the case of the reaction of acetic acid with ethanol, $K_c=K_x$.

EXPRESSION OF K_X AS A FUNCTION OF THE CONVERSION OF ONE OF THE REACTANTS

It is of interest to relate K_C to the conversion of one of the reactants. For the general reaction,

$$\lambda_A A + \lambda_B B \leftrightarrow \lambda_C C + \lambda_D D$$

defining the conversion in terms of reactant B, from Equation 21.1,

$$n_{B0} - n_B = \chi_B n_{B0}$$

or

$$n_B = n_{B0}(1-\chi_B) = n_{B0} - \chi_B n_{B0}$$

For reactant A,

$$n_{A0} - n_A = \frac{\lambda_A}{\lambda_B}(n_{B0} - n_B)$$

or

$$n_A = n_{A0} - \frac{\lambda_A}{\lambda_B}\chi_B n_{B0}$$

For products C and D,

$$n_C = n_{C0} + \frac{\lambda_C}{\lambda_B}\chi_B n_{B0}$$

$$n_D = n_{D0} + \frac{\lambda_D}{\lambda_B}\chi_B n_{B0}$$

The total number of moles at equilibrium then is:

$$n_T = n_0 + \chi_B n_{B0}\left(\frac{\lambda_C}{\lambda_B} + \frac{\lambda_D}{\lambda_B} - \frac{\lambda_A}{\lambda_B} - 1\right) = n_0 + \frac{\chi_B n_{B0}}{\lambda_B}\Delta\nu \quad \text{with} \quad n_0 = \sum_k n_{0k}$$

Therefore,

$$K_x = \frac{\left[n_{C0} + \frac{\lambda_C}{\lambda_B}\chi_B n_{B0}\right]^{\lambda_C}\left[n_{D0} + \frac{\lambda_D}{\lambda_B}\chi_B n_{B0}\right]^{\lambda_D}}{\left[n_{A0} - \frac{\lambda_A}{\lambda_B}\chi_B n_{B0}\right]^{\lambda_A}\left[n_{B0} - \chi_B n_{B0}\right]^{\lambda_B}}\left[n_0 + \frac{n_{B0}}{\lambda_B}\chi_B\Delta\nu\right]^{-\Delta\nu}$$

For the reaction of acetic acid with ethanol, assuming that initially there are no products present and remembering that $\Delta\nu = 0$,

$$K_x = \frac{\chi_B^2}{[r - \chi_B][1 - \chi_B]}$$

or

$$(K_x - 1)\chi_B^2 - K_x(r+1)\chi_B + K_x r = 0$$

where

$$r = n_{A,0} / n_{B,0}$$

A laboratory experiment at 298.15 K using $r = 1$ gave a conversion of 2/3; thus,

$$K_x = K_c = \frac{(2/3)^2}{[1 - 2/3][1 - 2/3]} = 4$$

TABLE 21.2
Comparison of Calculated Conversion Values χ_B with the Experimental Ones

$r = n_{A,0}/n_{B,0}$	$\chi_{B,\ observed}$	$\chi_{B,\ calculated}$
0.05	0.050	0.049
0.08	0.078	0.078
0.18	0.171	0.171
0.28	0.226	0.258
0.33	0.293	0.298
0.50	0.414	0.423
1.00	0.665	0.667
2.24	0.876	0.864
8.00	0.966	0.967

As a result, from Equation 21.9 we obtain $K_\gamma = K_T/K_x = 1.76$. More importantly, assuming K_x is constant, that is, K_γ is constant, it is possible to estimate the conversion at other values of r. Table 21.2 compares calculated values with experimental values of the conversion measured in the lab for different values of r.

This exercise based on real data demonstrates that with the help of the thermodynamic theory, one can get reasonable estimates in practical situations.

CHEMICAL EQUILIBRIUM WITH MORE THAN ONE INDEPENDENT REACTION

As indicated by Equations 21.2a–21.4a, when there is more than one independent reaction, for each independent reaction, there is an extent of reaction ξ_r. The number of equations for equilibrium constants and the number of unknown extents of reaction are equal to the number of independent reactions, and any method to solve a set of nonlinear equations can be used to solve the problem. The technique required to solve the set of nonlinear equations can be quite sophisticated, and the discussion of such techniques is beyond the scope of this text. Here, we will only mention that as an alternative to solving the set of equations of the equilibrium constants, it is possible to directly minimize the Gibbs energy of the system constrained by the material balances of the elements. While the number of moles of each compound and the total number of moles of the system change when a chemical reaction occurs, the amount of each element present in the system is conserved. A mathematical technique to solve a constrained minimum makes use of the Lagrange multipliers.

REFERENCE

1. Lange, N. A., and Speight, J. G. 2005. *Lange's Handbook of Chemistry*. New York: McGraw-Hill.

PROBLEMS

21.1 **Number of independent reactions**

When calcium carbonate ($CaCO_3$) is mixed with carbon (C) and heated up, it decomposes to form calcium oxide (CaO), carbon monoxide (CO), and carbon dioxide (CO_2). Write a set of independent reactions for this process if these are the only species present at equilibrium.

21.2 **Heat of reaction and equilibrium constants as functions of temperature**

A frequently available correlation for the temperature dependence of standard heat capacities is the polynomial form

$$c_{P,k} = c_{0,k} + c_{1,k}T + c_{2,k}T^2 + c_{3,k}T^3 + c_{4,k}T^4 = \sum_{j=0}^{4} c_{j,k}T^j$$

Derive the equations that replace Equations 21.17 and 21.18 for ΔH_T^θ and $\ln K_T$ as functions of temperature in this case.

21.3 **Equilibrium composition in the Haber reaction**

Calculate the equilibrium composition of the ammonia synthesis (Haber) reaction at 450°C and 100 bar, from an initial mixture of N_2 and H_2 with 50% excess N_2 over the stoichiometric proportion. Assume that the gas phase is an ideal mixture of nonideal gases that obey the virial equation (Berlin series) truncated at the second coefficient, given by Equations B.1–B.3 of Appendix B. Do not assume constant heat of reaction. Pure component data are given in Table 21.3. The correlation for ideal gas heat capacities is the one given in Problem 21.2.

21.4 **Esterification reaction in liquid phase**

Acetic acid, CH_3COOH (A, component 1) and ethanol, CH_3CH_2OH (E, component 2) in stoichiometric (1:1) molar proportions are reacted in liquid phase at 350 K to produce ethyl acetate, $CH_3COOCH_2CH_3$ (EA, component 3) and water, H_2O (W, component 4). Calculate the equilibrium composition, and the minimum pressure required to avoid the formation of a vapor phase. In Table 21.4 are given the enthalpy and Gibbs energy of formation of the liquids at 298.15 K and 1 bar, vapor pressures at 350 K, and coefficients for heat capacity in J/(mol·K) as a function of temperature in K, according to the correlation in Problem 21.2.

TABLE 21.3
Pure Component Properties for $N_{2(g)}$, $H_{2(g)}$, $NH_{3(g)}$

	$N_{2(g)}$	$H_{2(g)}$	$NH_{3(g)}$
$\Delta H^0_{f,298}$ (J/mol)	0	0	−45,940
$\Delta G^0_{f,298}$ (J/mol)	0	0	−16,410
T_c (K)	126.2	33.25	405.4
P_c (kPa)	3398	1297	11,353
ω	0.037	−0.216	0.257
$c_{0,k}$/R	2.883	3.539	4.238
$10^3 c_{1,k}$/R	3.681	−0.261	−4.215
$10^5 c_{2,k}$/R	−0.772	0.007	2.041
$10^8 c_{3,k}$/R	0.692	0.157	−0.2126
$10^{11} c_{4,k}$/R	−0.213	−0.099	0.761

Source: Data taken from Appendix C in Poling, B. E., Prausnitz, J. M., and O'Connell, J. P. 2001. *The Properties of Gases and Liquids*, 5th ed. New York: McGraw-Hill.

TABLE 21.4
Pure Component Properties of CH_3COOH ($A_{(l)}$), Ethanol, CH_3CH_2OH ($E_{(l)}$), $CH_3COOCH_2CH_3$ ($EA_{(l)}$), H_2O ($W_{(l)}$)

	Acetic Acid $A_{(l)}$	Ethanol $E_{(l)}$	Ethyl Acetate $EA_{(l)}$	Water $W_{(l)}$
$\Delta H^0_{f,298}$ (J/mol)	–484,500	–276,980	–478,800	–312,400
$\Delta G^0_{f,298}$ (J/mol)	–389,000	–173,680	–332,400	–237,129
P^s (kPa)	24.177	95.637	100.15	41.645
$c_{0,k}$	139.64	102.64	226.23	276.37
$c_{1,k}$	–0.3208	–0.13693	–0.6248	–2.0901
$c_{2,k}$	8.958E-04	–3.0341E-05	1.472E-03	8.125E-03
$c_{3,k}$	0	2.0386E-06	0	–1.4116E-05
$c_{4,k}$	0	0	0	9.3701E-09

Source: Data taken from the online DIPPR® 801 Database, https://dippr.aiche.org/SampleDb.

21.5 **Simultaneous gas-phase reactions**

Methane gas (CH_4) is burned with the stoichiometric amount of oxygen. The reaction products at 2500 K and 1 bar are CO_2 and H_2O, which in turn undergo the dissociation reactions (cf. Problem 12.3).

Reaction	$\ln K_T$
$CO_{2(g)} \leftrightarrow CO_{(g)} + \frac{1}{2}O_{2(g)}$	$10.250 - \frac{33908}{T/K}$
$H_2O_{(g)} \leftrightarrow H_{2(g)} + \frac{1}{2}O_{2(g)}$	$6.7185 - \frac{29609}{T/K}$

Find the equilibrium composition of the mixture, assuming complete combustion of the methane.

21.6 **Reactions with solid phases**

Find the equilibrium pressure and composition for the process described in Problem 20.1 (calcination of limestone), carried out at 1273 K from an initial equimolar mixture of C(s) and $CaCO_3$ (s). The equilibrium constants are

$$C_{(s)} + CaCO_{3(s)} \leftrightarrow CaO_{(s)} + 2\ CO_{(g)} \qquad K_{T1} = 492.7 \tag{1}$$

$$2\ CO_{(g)} \leftrightarrow C_{(s)} + CO_{2(g)} \qquad K_{T2} = 0.007936 \tag{2}$$

21.7 **Oxidation of silver**

Polishing the silverware is frequently depicted as one of the duties of the servants, especially the butler, in stately manors. This chore is necessary because silver oxidizes in contact with air, losing its shine as it becomes tarnished with a surface layer of silver oxide according to the reaction:

$$4\ Ag_{(s)} + O_{2(g)} \leftrightarrow 2\ Ag_2O_{(s)}$$

TABLE 21.5
Pure Component Properties for $Ag_{(s)}$, $Ag_2O_{(s)}$, $O_{2(g)}$

	$\Delta H^0_{f,298}$	s^0_{298}	c_0	$10^3 c_1$	$10^{-6} c_2$	$10^6 c_4$
$Ag_{(s)}$	0	42.7	24.22	2.74	0	2.84
$Ag_2O_{(s)}$	−31,000	121.3	59.33	40.79	−0.46	0
$O_{2(g)}$	0	205.1	29.15	6.48	−0.18	−1.02

Source: Data taken from Binnewies, M., and Milke, E. 2002. *Thermochemical Data of Elements and Compounds*. Weinheim: Wiley-VCH.

a. What are the maximum allowable pressure and composition of an oxygen-containing gas mixture to avoid the formation of silver oxide at 298.15 K? Does your answer depend on the number of other components, all of them inert gases? Analyze the process in terms of the Phase Rule.
b. Usual recipes for removing the tarnish include rubbing the silverware with toothpaste, baking soda (optionally mixed with sea salt or lemon juice), or of course a variety of commercial silver polish products. By removing the oxide, all of these methods have the drawback of throwing away part of the silver. Can the reaction be reversed, and the oxide reduced back to silver simply by heating the pieces?

Table 21.5 gives pure component data on standard heats of formation, entropies, and coefficients of heat capacity for Equation 21.16 with $c_{3,k} = 0$. Units are J, mol, K.

22 The Thermodynamics of Equilibrium-Based Separation Processes

Abstract

As applied thermodynamics received a major impulse from the development of petroleum refining followed by the development of the industry of plastics, the chapter starts introducing the basic concepts of distribution coefficient and selectivity, necessary for evaluating the goodness of separation agents. The thermodynamically based processes of azeotropic and extractive distillation employed for difficult separations are discussed in detail. Step-by-step schemes for bubble, dew point, and flash vapor–liquid calculations are presented. The chapter closes with notes on calculations of liquid–liquid, solid–liquid, and equilibrium through membranes.

Separation processes usually represent a major component in the cost of chemical products, as most industries need to purify chemicals to be fed to reactors and separate the reaction products obtained. Some industries, such as petroleum refining, are mostly a sequence of separation steps.

Separation processes can be classified into two broad categories: equilibrium-stage processes and rate-controlled processes. Only the first of these categories is in the realm of thermodynamics.

Although equilibrium is seldom attained, thermodynamics is of help in the design of equilibrium-stage processes. In these processes, a phase containing the compound to be separated is contacted with a second phase which has a much lower concentration of that compound. The two phases are left to equilibrate; the second phase is removed, and the desired compound is recovered from it. Sometimes, the first phase is contacted again with another amount of (fresh) second phase. This sequence of contacts, in different equilibrium stages, is the basis of the design of separation processes discussed in this chapter. The key problem is to evaluate the relative proportions of the compounds when the phases contacted are in equilibrium.

DISTRIBUTION COEFFICIENT AND SELECTIVITY

Figure 22.1 depicts a typical separation unit in which two phases, designated generically by α and β, are in direct contact at a pressure, P, and a temperature, T, common to both phases.

Phase α can be a liquid phase and phase β a vapor phase, both α and β can be liquid phases, or one of the phases can be solid and the other either liquid or gas. Our interest is to determine the relative amounts of the compounds in each phase in the system at equilibrium. Mole fractions are normally used as a measure of composition. In the case of vapor–liquid equilibrium, the symbols x_i and y_i are used for the mole fractions of compound i in the liquid phase and in the vapor phase, respectively. For liquid–liquid equilibrium, the two liquid phases are designated as phases L_1 and L_2. The distribution coefficient K_i of a compound i between the two phases α and β is defined as:

$$K_i^{\beta\alpha} \equiv \frac{x_i^{\beta}}{x_i^{\alpha}} \tag{22.1}$$

DOI: 10.1201/9781003431985-25

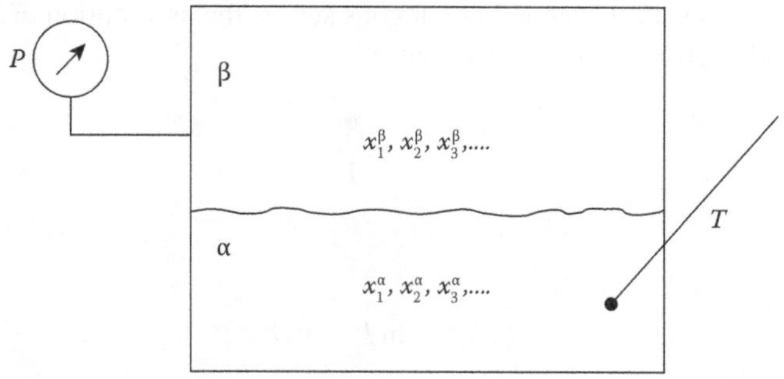

FIGURE 22.1 Two-phase system unit.

In vapor–liquid equilibrium, the distribution coefficient is called the *K*-factor, the superscripts α and β are eliminated, and we write:

$$K_i \equiv \frac{y_i}{x_i} \tag{22.1a}$$

A value of K_i larger than unity, $K_i > 1$, indicates that compound *i* tends to concentrate in phase β. In vapor–liquid equilibrium nomenclature, for "light" compounds, that is, compounds that tend to concentrate in the vapor phase, the *K*-factor is larger than unity, and for "heavy" compounds, that is, compounds that tend to concentrate in the liquid phase, the *K*-factor is less than unity. In the separation of a multicomponent mixture, the comparison of the values of the distribution coefficients for two compounds *i* and *j* shows which of the components is selectively taken out from phase α by phase β. The selectivity α_{ij} is defined as:

$$\alpha_{ij} = \frac{K_i^{\beta\alpha}}{K_j^{\beta\alpha}} \tag{22.2}$$

In vapor–liquid equilibrium, the selectivity is called relative volatility, and it is written as:

$$\alpha_{ij} = \frac{K_i}{K_j} = \frac{(y_i / x_i)}{(y_j / x_j)} \tag{22.2a}$$

The thermodynamic treatment of a phase equilibrium problem always starts by equating the fugacities, or the chemical potentials, of each compound in all phases in which the compound is present. The values of the distribution coefficient and the selectivity are then obtained solving the resulting system of equations. For a vapor–liquid equilibrium of compounds that are present in significant amounts in both phases, using Equation 16.2a, we write:

$$K_i \equiv \frac{y_i}{x_i} = \gamma_{i,0} \frac{(P_i^s)'}{P} \tag{22.3}$$

At moderate pressures, the corrections to the vapor pressure can be safely ignored, and we write:

$$K_i \approx \gamma_{i,0} \frac{P_i^s}{P} \tag{22.3a}$$

For mixtures of compounds of the same homologous series, the assumption of ideal solution in the Lewis sense is a good approximation, so we write:

$$K_i^{id} = \frac{P_i^s}{P}$$

or

$$\ln K_i^{id} = \ln P_i^s - \ln P$$

For the case of the vapor–liquid equilibrium of a light compound (j) and a heavy compound (i), say methane dissolved in hexane, the light compound is predominantly in the vapor phase, and we may use Henry's convention for the solute. Thus, using Equation 13.1 for the vapor phase and Equation 15.9 for the solute j in the liquid solvent i, we write:

$$K_j = \gamma_{j,H} \frac{H_{j,i}}{\phi_j P}$$

Considering the case of an ideal solution in Henry's sense, at moderate pressure, we write:

$$\ln K_j^{id} = \ln H_{j,i} - \ln P$$

Hence, for both the solvent i and the solute j, a plot of the logarithm of the distribution coefficient versus the logarithm of the pressure will give a straight line with a slope of –1 in the diluted region. A typical plot of this kind is shown in Figure 22.2.

At P=1 (one unit of pressure), the value of ln $H_{j,i}$, in which $H_{j,i}$ has the same units as P, can be found in the scale of ln K_j. At low pressure, the logarithms of the distribution coefficients for the light and heavy compounds are a linear function of the logarithm of the pressure. At higher pressures, the curves bend due to nonidealities, and the two curves join at the mixture critical point.

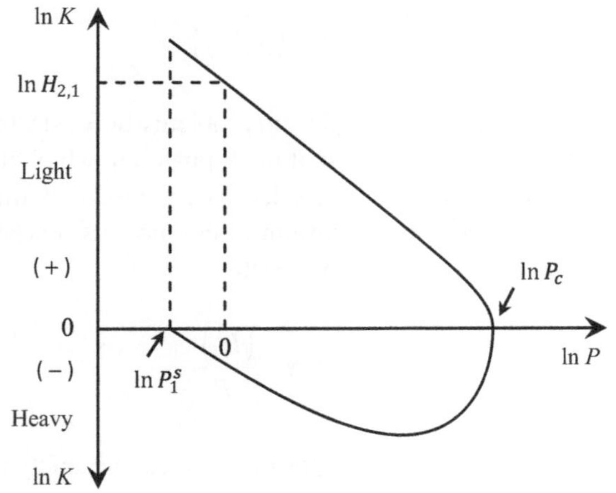

FIGURE 22.2 Logarithm of the distribution coefficient as a function of pressure.

For the case of two liquid mixtures α and β in equilibrium, using Equation 16.5, we write:

$$K_i \equiv \frac{x_i^{\beta}}{x_i^{\alpha}} = \frac{\gamma_{i,0}^{\alpha}}{\gamma_{i,0}^{\beta}} \tag{22.4}$$

and

$$\alpha_{ij} = \frac{\left(\gamma_{i,0}^{\alpha} / \gamma_{i,0}^{\beta}\right)}{\left(\gamma_{j,0}^{\alpha} / \gamma_{j,0}^{\beta}\right)}$$

If both mixtures were ideal in Lewis's sense, the composition of both phases would be the same and there would not be any phase separation. As discussed in Chapter 17, mixtures separate into two liquid phases only in the case of strong positive deviations from ideality in Lewis's sense. If this is the case, as the mole fractions of compound i are inversely proportional to the values of its activity coefficients in both phases, the phase in which the activity coefficient is smaller will have a larger mole fraction of compound i. If two sparsely miscible liquid pure compounds are put in contact, one of the compounds will be almost pure in one liquid phase and at high dilution in the other. Say that the compounds are water (1) and benzene (2), the aqueous phase is phase α and the organic phase is phase β. In Lewis's convention, the activity coefficient of compound 1 in phase α will be close to unity and in phase β it will be a value much larger than unity. In order to work with activity coefficients of the order of unity, the ideal behavior in Henry's sense can be used for the compound at high dilution. In this convention, at low concentration its activity coefficient will be close to unity. For component 2, the reverse will hold. Thus, considering ideal behavior in Lewis's sense for the solvent and in Henry's sense for the solute, we write:

$$x_1^{\alpha}\left(P_1^S\right)' = x_1^{\beta} H_{1,\beta}$$

$$K_1^{id} \equiv \frac{x_1^{\beta}}{x_1^{\alpha}} = \frac{\left(P_1^S\right)'}{H_{1,\beta}}$$

Similarly for compound 2,

$$K_2^{id} = \frac{x_2^{\alpha}}{x_2^{\beta}} = \frac{\left(P_2^S\right)'}{H_{2,\alpha}}$$

The value of x_1^{β} is the solubility of water in the organic phase, and the value of x_2^{α} is the solubility of benzene in the aqueous phase. If, in addition, the two liquid phases α and β are at equilibrium with a vapor phase,

$$y_1\phi_1^0 P = x_1^{\alpha}\left(P_1^s\right)' = x_1^{\beta} H_{1,\beta}$$

$$y_2\phi_2^0 P = x_2^{\alpha} H_{2,\alpha} = x_2^{\beta}\left(P_2^s\right)'$$

In this case, one can define two different K-factors for each compound depending on the phase chosen. At moderate conditions, we write:

$$y_1 P = x_1^{\beta} H_{1,\beta}$$

$$y_2 P = x_2^{\beta} \left(P_2^s\right)'$$

Adding and rearranging, we obtain:

$$x_1^{\beta} = \frac{P - \left(P_2^s\right)'}{H_{1,\beta} - \left(P_2^s\right)'}$$

A similar relation can be written for x_2^{α}. These equations can be used to estimate Henry's constant if the solubilities are small. Normally, the solubility is measured at atmospheric pressure and at the temperature required for a practical application. Thus, it is necessary to correct the value of Henry's constant for its use at pressures of interest. For simplicity, consider that compound 2 is at high dilution in solvent 1. From Equation 15.8, in Henry's convention,

$$f_2^{dil} = x_2 H_{2,1}$$

or

$$\ln f_2^{dil} = \ln x_2 + \ln H_{2,1}$$

Recalling Equation 14.15, we write for the case of compound 2 at high dilution,

$$\left(\frac{\partial \ln H_{2,1}}{\partial P}\right)_{T,n} + \left(\frac{\partial \ln f_2^{dil}}{\partial P}\right)_{T,n} = \frac{\overline{v}_2^{\infty}}{RT}$$

and integrating between 1 atm and the pressure P, we obtain:

$$\ln H_{2,1}(P) = \ln H_{2,1}(1 \text{ atm}) + \frac{\overline{v}_2^{\infty}(P-1)}{RT} \tag{22.5}$$

This result gives a good approximation for most practical applications.

One last case of interest is the one of a compound at high dilution in two immiscible solvents in contact. This is, for example, the case of a small amount of liquid butanol added to a two-liquid-phase water–benzene system. Calling the solute, butanol, component 1, and the aqueous and organic phases α and β, respectively, we write:

$$x_1^{\alpha} H_{1,\alpha} = x_1^{\beta} H_{1,\beta}$$

$$K^{id} = \frac{x_1^{\beta}}{x_1^{\alpha}} = \frac{H_{1,\alpha}}{H_{1,\beta}}$$

Due to the simplifications made, in this case K^{id} is a function of temperature only. When nonidealities of the phases become important, the K-factor becomes a function of the composition of the phases. For completeness, we mention here that in practical liquid–liquid equilibrium work, sometimes an empirical distribution coefficient is used in terms of concentration units (mol/dm³),

$$K_{i,practical} = \frac{c_i^{\beta}}{c_i^{\alpha}}$$

This empirical term includes the ratio of the molar densities of the phases in the value of K.

For organic mixtures, Berg rules, discussed in Chapter 17, are of help in the design of separation processes. Mixtures of organic liquids with water can sometimes be separated by the addition of salts. This is the case, for example, of a mixture of methanol and water when K_2CO_3 is added as shown schematically in Figure 22.3.

The effect is usually referred to as "salting out" the mixture. Similar unusual effects are obtained by the addition of water-soluble polymers to water. For a polymer with a hydrocarbon skeleton to be water soluble, it must have active groups able to combine with water. The hydrocarbon "tail" and the water will keep away from each other as much as they can. In a practical situation, the hydrophobic hydrocarbon tail will go to the surface and tail-out, while the active hydrophilic group will face the water body. This configuration will obviously affect the surface properties of water (surface tension), and such a compound is properly called a surfactant. At larger concentrations of the surfactant, the surface will get too crowded and eventually, at a so-called critical micelle concentration (CMC), the surfactant molecules will form "drops" inside the body of water. These drops, called micelles, will have all the hydrocarbon tails facing inside the drop, with the active groups at the surface of the micelle facing the water. The nonpolar, hydrocarbon microphase thus formed can dissolve organic soluble materials. At a high micellar concentration of an aqueous phase in contact with an organic phase, the structure can reverse to give a continuous organic phase with water drops surrounded by the active groups. These so-called reverse micelles can be used to "trap"

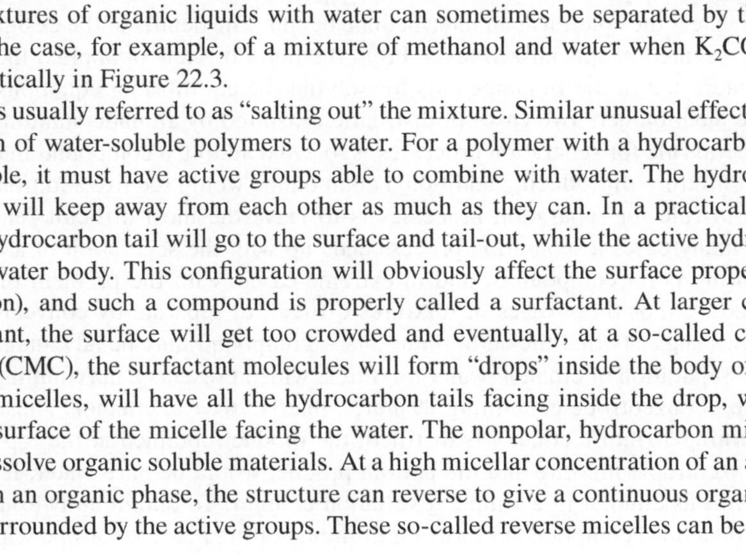

FIGURE 22.3 Ternary system of methanol–water–K_2CO_3.

water-soluble solutes in organic phases for their eventual separation. The formation of micelles and reverse micelles is a clear example of local composition in mixtures and has major potential for the development of separation techniques in bioprocesses.

AZEOTROPIC AND SOLVENT DISTILLATION

The study of vapor–liquid equilibrium is of special importance because it is estimated that about 90% of the industrial separations are done by distillation. In this case, the separating agent is heat. A distillation column consists of a reboiler and either a packed or a plate column on top of the reboiler where there is partial condensation returning the heavy compound by gravity down to the reboiler and collecting the light compound at the top of the column. As the separation is never complete, the top product is condensed and a part of it is refluxed to the column. The fresh feed normally enters the column at an intermediate height. The details of the design of distillation columns are presented in specialized texts. From the point of view of applied thermodynamics, we are only interested in the fundamentals for solving the equilibrium equations. Before going into detail, we mention here two cases of distillation inspired by an understanding of thermodynamics. A general rule for separation processes is to avoid adding a compound that is not present in the original mixture. Introducing additional compounds would require additional separations. Another general rule for separation processes is to leave the most difficult separation for the end. Thus, in many cases a separation process ends up with the separation of a binary mixture with close boiling point compounds, and in extreme cases, with the problem of separating an azeotropic mixture. For these cases of mixtures difficult to separate by conventional methods, thermodynamics suggests that the addition of a new compound may be advantageous. A classic example is the separation of ethanol–water mixtures, which presents a maximum pressure, minimum boiling point azeotrope containing 89 mol%, that is 96 wt%, ethanol. Thus, if an aqueous mixture containing 10 mol% ethanol is distilled, the product obtained at the top of the column would be the azeotropic mixture and the bottom product would be pure water. It is not possible to obtain anhydrous ethanol in a simple distillation column. To obtain anhydrous ethanol, it is necessary to add a third compound to form a minimum boiling point azeotrope with water having a boiling point that is lower than the boiling point of ethanol. Considering the modified Berg classification of liquids and their mixtures, benzene, toluene, and cyclohexane are good candidates to form a heteroazeotrope with water that, upon condensation of the vapor, will separate into two liquid phases. Cyclohexane is normally preferred. Thus, in a first column, the ethanol–water azeotrope is separated from the excess water and sent as a feed to a second column, to which cyclohexane is fed close to the top of the column in the amount required to give the vapor phase composition of the azeotropic mixture water–cyclohexane. In this second column, anhydrous ethanol is obtained at the bottom and the azeotropic mixture of water–cyclohexane is obtained as the top product, condensed, and separated into an aqueous phase and a cyclohexane phase. The cyclohexane, together with fresh cyclohexane to make up for any losses, is recycled close to the top of the second column. The aqueous phase containing a small contamination of cyclohexane is sent as a feed to a third column, where the cyclohexane is obtained at the top and the water is obtained at the bottom.

A different application of thermodynamics involving the addition of a compound not previously present in the mixture is the case of solvent distillation, also called extractive distillation. In this case, a compound that is a good solvent for the heavier component in a binary mixture difficult to separate is added at the top of the distillation column to "wash" that component down. Normally, the solvent is a heavier compound of the same homologous series of the heavy component to be "washed down." As indicated by Berg's classification, compounds of the same homologous series form solutions presenting close to ideal behavior, and thus they are good solvents. There are cases, however, in which the two compounds forming the mixture of difficult separation are of a similar homologous series. In this case, a compound of a different homologous series that is a particularly good solvent for one of the compounds in the initial mixture can be used. A typical example of this case of solvent distillation is the use of furfural, a polar compound, to separate 1-butene, an unsaturated hydrocarbon, from isobutane, a saturated hydrocarbon. Furfural and 1-butene form an electron-donor–acceptor complex. In solvent distillation, the separating agent is fed to the top of the column, while in azeotropic distillation, it is fed by the middle of the column. In the latter case, columns have a larger diameter in their lower section. These physical aspects of the equipment clearly reflect the different thermodynamic requirements for proper use of the separating agent.

VAPOR–LIQUID EQUILIBRIUM

In this section, we discuss in some detail a few examples of vapor–liquid equilibrium calculations that are of practical interest. For the vapor phase, invariably the fugacity of each compound i is expressed using Equation 13.1. For the fugacity of a compound in the liquid phase, we may use Equation 12.5 and write Equation 12.10 as:

$$x_i \phi^L_{i(T,P,x)} P = y_i \phi_{i(T,P,y)} P \tag{22.6}$$

or

$$x_i \phi^L_{i(T,P,x)} = y_i \phi_{i(T,P,y)}$$

This is the so-called ϕ approach, in which the fugacities of the compounds are expressed in terms of the fugacity coefficient in each phase. For the vapor phase, we use simply the symbol ϕ_i, while for the liquid phase, we use ϕ_i^L. In general, all fugacity coefficients of the vapor phase must be calculated from a single equation of state (EOS) and all fugacity coefficients of the liquid phase must be calculated from a single EOS. The equations of state do not need to be the same for both phases, although usually the same (cubic) EOS is used. For the vapor phase at low pressures, for example, we may find it convenient to set all values $\phi_i = 1$ (ideal gas behavior), while for the liquid phase, even the assumption of an ideal mixture, $\phi_i^L = \phi_i^{0,L}$, could be a poor one.

For the compounds in the liquid phase, we may express the fugacities in terms of activity coefficients and write Equation 16.2 as:

$$x_i \gamma_{i,0(T,P,x)} \left(P_i^s\right)^{\#} = y_i \phi_{i(T,y)} P \tag{22.7}$$

For generality, in Equation 22.7 we preferred to keep the vapor phase correction to ideality separate from the liquid phase correction. From Equations 16.2 and 16.3 then,

$$\left(P_i^s\right)^{\#} = \phi_i^0 \exp\left[\frac{v_i^{0L}\left(P - P_i^s\right)}{RT}\right] P_i^s \tag{22.8}$$

In this treatment, called the $\gamma - \phi$ approach, all activity coefficients for the compounds in the liquid phase should originate from a single expression for the excess Gibbs energy of the mixture, and all fugacity coefficients for the compounds in the vapor phase from a single EOS. Analytical solution of groups (ASOG) [1,2] is a particularly useful method to obtain activity coefficients for vapor–liquid equilibrium calculations with $\gamma-\phi$. This method is discussed in Chapter 28 and in Appendix D.

Alternatively, for the compounds in the liquid phase, we can use Equation 15.9 if the compound is diluted and write:

$$x_i \gamma_{i,H(T,P,x)} H_{i,r} = y_i \phi_{i(T,P,y)} P \tag{22.9}$$

BUBBLE AND DEW POINT CALCULATIONS

For a "dew pressure" problem, all the values of the vapor phase mole fractions y_i ($i=1, 2, \ldots, c$) and the temperature T are known. The values of the liquid phase mole fractions x_i ($i=1, 2, \ldots, c$) and the pressure P are to be calculated. Similarly, for a "bubble temperature" problem, all compositions of the liquid phase and pressure are known, and the values of the vapor phase mole fractions and temperature are to be calculated. The unknown variable, pressure or temperature, is the one defining the problem to be solved, while the use of "dew" or "bubble" indicates the phase in which the compositions are known. Because in both kinds of problems the composition of one of the phases and either the pressure or the temperature is known, we can present a generalized treatment. Let us call α the phase for which all compositions X_i^α are known. This can be either the liquid (bubble point calculation) or the vapor (dew point calculation) phase. Let us designate by ξ_A the known intensive variable, T or P, and by ξ_B the unknown variable (P or T). Thus, the known variables are $X_i^\alpha\left(i=1, 2, \ldots, c\right)$ and ξ_A, and the unknown variables are $X_i^\beta\left(i=1, 2, \ldots, c\right)$ and ξ_B. In addition, as Equations 22.6, 22.7, and 22.9 are symmetrical, we generalize them under the form:

$$X_i^\alpha \Phi^\alpha_{i\left(\xi_A, \xi_B, X^\alpha\right)} = X_i^\beta \Phi^\beta_{i\left(\xi_A, \xi_B, X^\beta\right)}. \tag{22.10}$$

The terms included in Equation 22.10, for each type of bubble or dew point calculation, are presented in Table 22.1.

The basic idea behind the calculation scheme is shown in Flowchart 22.1.

In step 1, the known variables are read. In step 2, the missing value (T or P) is estimated, and then in step 3, preliminary values of all Φ_i^α are obtained. In the first pass by step 4, the unknown phase β is considered to behave like an ideal mixture so that $\Phi_i^\beta = \Phi_i^{id}$ at T and P. This is a very poor approximation for dew point calculations and probably a not-so-poor approximation for bubble point calculations at low pressure. In any case, for the first pass this approximation is totally justified. In subsequent passes, the values of Φ_i^β are calculated with the new value of ξ_B obtained in step 13 and the last values X_i^β obtained in step 11. The unknown compositions X_i^β are calculated in step 7 from the general form of Equation 22.10.

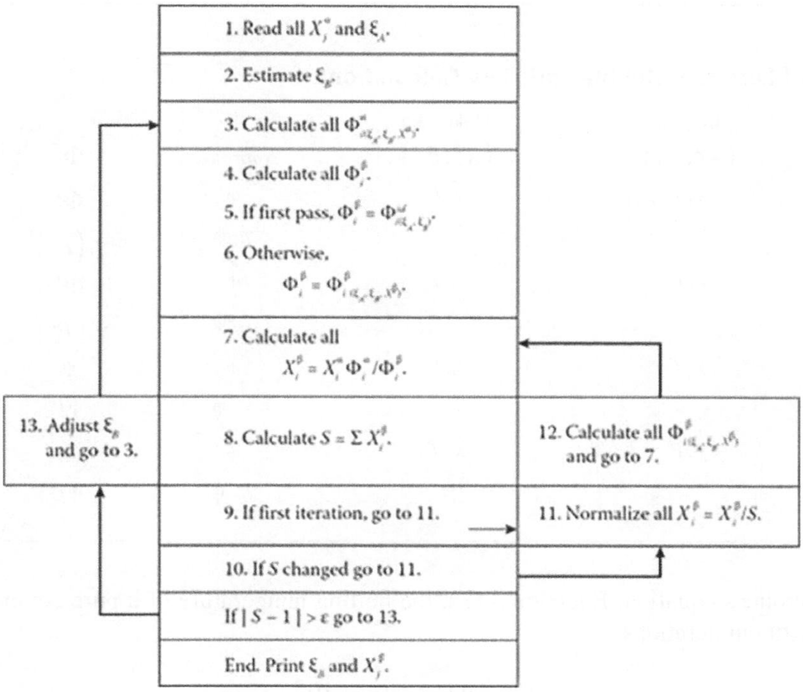

FLOWCHART 22.1 General scheme for vapor–liquid calculation for a tolerance $\left|\sum x_i^\beta - 1\right| \leq \varepsilon$.

Since, based on previous assumptions, the values $X_1^\beta, X_2^\beta, X_3^\beta$, and so forth, are calculated independently, they will not add to unity as they should. For this reason, their sum is calculated in step 8, and if it is the first iteration, the values of Φ_i^β need to be recalculated. In step 11, the X_i^β values are normalized, and in step 12, the value of Φ_i^β is calculated. The normalization step is very important because the functions to compute Φ_i^β, either by an EOS or a g^E function, are designed to work with normalized values of composition. Steps 3–10 are then repeated until no further change in S is observed. Then the value of S is compared with unity, and if it differs by less than a previously specified error ε, the result is accepted. If the error is larger than ε, the variable ξ_B is adjusted in step 13, and the process is repeated. The missing information in Flowchart 22.1 is how to estimate ξ_B (P or T), and how to adjust its value in consecutive iterations. For the initial estimates, we assume that both phases behave ideally. From Equation 16.4, assuming ideal gas phases,

$$P = \sum x_i \gamma_i P^s_{i(T)} \tag{22.11}$$

and assuming ideal gas and liquid phases,

$$P = \frac{1}{\sum \left(\frac{y_i}{P^s_{i(T)}} \right)} \tag{22.12}$$

Equation 22.11 is useful for bubble point calculations, and Equation 22.12 for dew point calculations. When the temperature is known, the pressure is directly estimated. When the pressure is known, the estimation of temperature requires trial and error. However, since only an estimate is required, it is enough to consider a weighted average ($T = \Sigma x_i T_b$ or $T = \Sigma y_i T_b$) of the boiling temperatures of the components at the given pressure.

TABLE 22.1
Summary of Different Bubble and Dew Calculations

Problem	Known (X_i^{α}, ξ_A)	Unknown (X_i^{β}, ξ_B)	Approach	Φ_i^{α}	Φ_i^{β}
Bubble P	x_i, T	y_i, P	ϕ	Φ_i^L	ϕ_i
			$\gamma-\phi$	$\gamma_i\left(P_i^s\right)^{\#}$	$\phi_i P$
Bubble T	x_i, P	y_i, T	ϕ	Φ_i^L	ϕ_i
			$\gamma-\phi$	$\gamma_i\left(P_i^s\right)^{\#}$	$\phi_i P$
Dew P	y_i, T	x_i, P	ϕ	ϕ_i	ϕ_i^L
			$\gamma-\phi$	$\phi_i P$	$\gamma_i\left(P_i^s\right)^{\#}$
Dew T	y_i, P	x_i, T	ϕ	ϕ_i	ϕ_i^L
			$\gamma-\phi$	$\phi_i P$	$\gamma_i\left(P_i^s\right)^{\#}$

From Antoine's equation, Equation 5.17b, the boiling temperature of a pure compound can be calculated without iterations:

$$t_b = \left[B/\left(A - \log_{10} P\right)\right] - C \tag{5.17c}$$

Thus, a rough estimate of ξ_B (T or P) can be made in step 2.

For the adjustment of ξ_B in step 13, the following rules of thumb are used for the first adjustment:

1. If $\Sigma y_i > 1$, the estimated P was too low, or the estimated T was too high. Thus, in a bubble point calculation, $P = 1.05P$ or $T = 0.95T$ is used for the second iteration of the outside loop if $S > 1$. Conversely, if $S < 1$, $P = 0.95P$ or $T = 1.05T$ is used for the second iteration. Before continuing, it is necessary to make $S_{\text{OLD}} = S$ and $\xi_{\text{OLD}} = \xi$.
2. If $\Sigma x_i > 1$, the estimated P was too high, or the estimated T was too low. Thus, for the second iteration in a dew point calculation, $P = 0.95P$ or $T = 1.05T$ is used if $S > 1$. Conversely, $P = 1.05P$ or $T = 0.95T$ is used for the second iteration if $S < 1$. Before continuing, make $S_{\text{OLD}} = S$ and $\xi_{\text{OLD}} = \xi$.

Starting from the second iteration, the adjustment can be made using Newton's method in incremental form. This method uses the information of the two previous iterations. For ξ_{OLD}, the value of the sum is S_{OLD}, and for ξ, it is S. For ξ_{NEW}, it is desired to have the sum equal 1. Thus,

$$\frac{\xi_{\text{NEW}} - \xi}{(1-S)} = \frac{\xi - \xi_{\text{OLD}}}{S - S_{\text{OLD}}}$$

or

$$\xi_{\text{NEW}} = \xi + \frac{\xi - \xi_{\text{OLD}}}{S - S_{\text{OLD}}}(1-S)$$

Although there are many alternative and accelerating methods, the above general guidelines are enough to solve most vapor–liquid equilibrium problems of practical interest. There is one particular case, however, that leads itself to a much simpler method of solution. This is the common case of a bubble point calculation at low-pressure and subcritical conditions for all compounds. In this case, using the $\gamma-\phi$ approach, Equation 22.7 reduces to:

$$x_i \gamma_{i(T,P,x)} P_i^s = y_i \phi_{i(T,P)}^o P \tag{22.13}$$

Since the x values and either T or P are known, the calculation of each y_i can be made once P or T is specified. In the general terms of Flowchart 22.1, Φ_B is not a function of X_β. In this case, from step 6 the calculation goes to step 9, eliminating steps 7, 8, 11, and 12. Exactly the same can be said in the ϕ approach, if it is used for this case.

FLASH CALCULATIONS

For flash calculations, the approach is different. In this case, the feed flow rate, F, the temperature, T, the pressure, P, and the overall composition of the mixture, z_i, are known. The mixture is separated into a vapor with flow rate V and composition y_i, and a liquid stream with flow rate L and composition x_i.

There are many approaches to solve this problem. C. J. King (1971) [3] recommends the approach of Rachford and Rice (1952) [4]. This approach is simple and particularly suited for Newton's method of numerical solution. Before carrying out the solution of a flash problem, it is advisable to verify that the mixture of composition z_i^{a1} is indeed in the two-phase region at the conditions of P and T. This is done by performing one bubble pressure and one dew pressure calculation at the desired temperature. To satisfy the flash condition, the value of pressure P should be between the bubble pressure and the dew pressure of the mixture. Once this is verified, the calculation can proceed as follows. The material balance (see Appendix A) gives:

$$F = L + V$$

and for each compound i,

$$z_i F = x_i L + y_i V$$

In addition, from the equilibrium condition,

$$K_i = \frac{y_i}{x_i}$$

Combining the first two equations to eliminate L and the resulting equation with the third to eliminate y_i, after rearrangement we obtain:

$$x_i = \frac{z_i}{1+(V/F)(K_i-1)} \tag{22.14}$$

and then

$$y_i = K_i x_i = \frac{K_i z_i}{1+(V/F)(K_i-1)} \tag{22.15}$$

The function to be zeroed according to Rachford and Rice is:

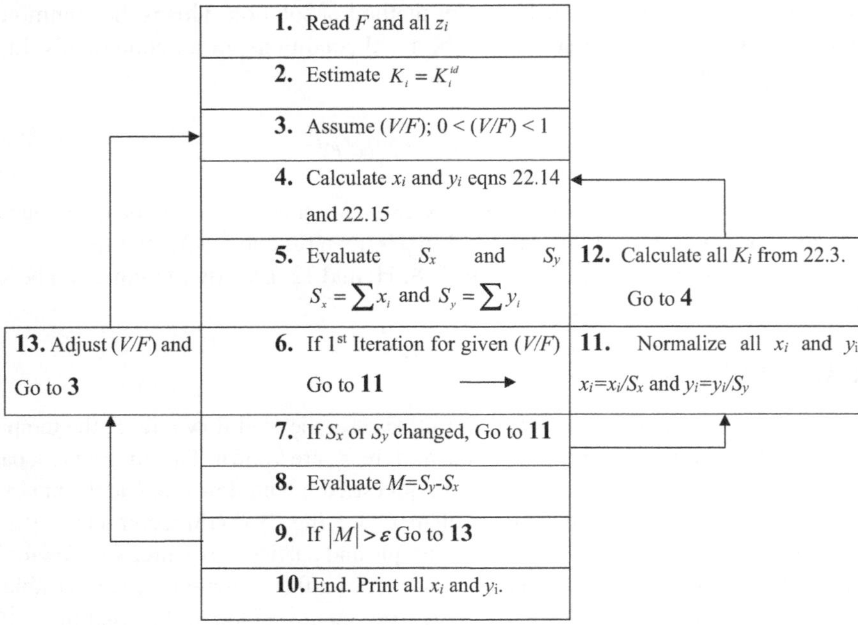

FLOWCHART 22.2 Iteration scheme for flash calculations.

$$M(V/F) = \sum y_i - \sum x_i$$

Combining with Equations 22.14 and 22.15,

$$M(V/F) = \sum \frac{z_i(K_i - 1)}{1 + (V/F)(K_i - 1)} \tag{22.16}$$

The variable of iteration is (V/F). Its value must always be between 0 and 1, and the flash temperature should be between the bubble and the dew temperatures of the mixture. The problem is solved when $M(V/F)=0$, within some predetermined accuracy ε. A scheme of solution is presented in Flowchart 22.2. It is assumed here that a decision has been previously made to use the $\gamma-\phi$ approach with a given EOS for the vapor phase and a given excess Gibbs energy function for the liquid phase. The adjustment of (V/F) can be made by the incremental form of Newton's method as:

$$\frac{(V/F)_{\text{NEW}} - (V/F)}{O - M} = \frac{(V/F) - (V/F)_{\text{OLD}}}{M - M_{\text{OLD}}}$$

or

$$(V/F)_{\text{NEW}} = (V/F) - M\frac{(V/F) - (V/F)_{\text{OLD}}}{M - M_{\text{OLD}}}$$

The new value of (V/F) must be between 0 and 1. In this particular case, the Newton method can be applied in its differential form, if desired, due to the fact that from Equation 22.16 it is possible to evaluate $dM/d(V/F)$ under the assumption that all values of K_i are constant.

LIQUID–LIQUID EQUILIBRIUM

Most treatments of liquid–liquid equilibrium employed in practice make use of activity coefficients to express the fugacities of the compounds in the liquid phase. Although the model for the excess

Gibbs energy need not be the same in both liquid phases, almost invariably a single model is used for both phases. Depending on the case, either the Lewis or the Henry convention can be used for the normalization of the activity coefficients. Only in special cases is Henry's convention preferred. When Lewis's convention is used for both phases, Equation 22.4 shows that the distribution coefficient for a compound is equal to the ratio of its activity coefficients in each of the phases. Thus, the model used to calculate the activity coefficients is quite important. The most commonly used models are the nonrandom two-liquid equation [5] equation and the universal quasi-chemical [6] equation. The Wilson [7] equation, which is so flexible for vapor–liquid equilibrium calculations, cannot be used for liquid–liquid calculations. The reason for this is that, as discussed in Chapter 17, this equation is unable to satisfy the condition for liquid phase instability.

At the present stage of development, the calculation of liquid–liquid equilibrium for a ternary system, starting with parameters for the excess Gibbs energy functions of the three binaries, is at best approximate. The usual case is having a known overall composition of a ternary mixture and wanting to determine the compositions of the two liquid phases in equilibrium, if any two such phases exist. Formally, this case is similar to a flash problem, so a calculation scheme like the one shown in Flowchart 22.2 could be used, except that K_i is given by Equation 22.4 rather than by Equation 22.3. Thus, the estimate of K_i for step 2 needs to be done with a crude guess of the distribution of the third compound into the two partially miscible liquid phases. A sounder thermodynamic approach, seldom used in practice, is the following. Consider 1 mole of liquid mixture containing n_A and n_B moles of two partially miscible compounds A and B and n_C moles of solute C that distributes into the two liquid phases at the temperature of interest. After stirring the mixture and letting it equilibrate, the phase rich in compound A will have α moles and the phase rich in compound B will have $(1-\alpha)$ moles. The overall mole fractions z_i of the three compounds in the feed are known; hence, by material balance,

$$z_i = n_i^\alpha + n_i^\beta = \alpha x_i^\alpha + (1-\alpha)x_i^\beta$$

with

$$x_i^\alpha = \frac{n_i^\alpha}{\alpha}$$

and

$$\alpha = \sum n_k^\alpha$$

Thus, if n_A^α, n_B^α, and n_C^α could be evaluated, the number of moles in both phases and the mole fractions in both phases could be obtained. In addition, we know that n_i^α must be positive and it cannot be larger than z_i, so we can transfer the problem to the evaluation of three unbound variables ζ_i defined by:

$$n_i^\alpha = \frac{z_i}{1+|\zeta_i|}$$

The first estimate of the variables ζ_i is obtained from knowledge of the solubility of B in A, plus a reasonable estimate of the distribution of C between the phases. Their final values are obtained by minimization of the Gibbs energy of the system at equilibrium:

$$G = \sum n_i^\alpha \mu_i^\alpha + \sum n_i^\beta \mu_i^\beta \quad \text{with } \mu_i^\alpha = \mu_i^\theta + RT\ln\left(x_i^\alpha \gamma_i^\alpha\right) \text{ and}$$

$$\mu_i^\beta = \mu_i^\theta + RT\ln\left(x_i^\beta\gamma_i^\beta\right)$$

Rearranging, we write:

$$\tilde{G} = \frac{G - \sum z_i\mu_i^\theta}{RT} = \sum n_i^\alpha \ln\left(x_i^\alpha\gamma_i^\alpha\right) + \sum n_i^\beta \ln\left(x_i^\beta\gamma_i^\beta\right)$$

Thus, when $\tilde{G}$ is at its minimum, G also reaches its minimum and all compositions can be determined. Although the method has a sound thermodynamic base, its successful application depends strongly on the validity of the excess Gibbs energy model used to evaluate the activity coefficients of the three compounds in the two phases. In practice, most of the liquid–liquid equilibrium problems found in liquid extraction processes are solved with empirical diagrams drawn from laboratory data, but the analytical method is sometimes used as a first approximation.

SOLID–LIQUID EQUILIBRIUM

There are two simple cases of interest for a binary system. Either the solvent is solid in equilibrium with a solution, as in the case of an iceberg in an aqueous solution of salt like the ocean, or the solute is solid, as in a saturated aqueous solution of salt. It is also possible for the solid phase to be a solid solution or a hydrate (solvate), but we restrict the discussion here to a pure compound solid phase.

Let us designate by x_i the mole fraction in the liquid phase of the compound (solvent or solute) that is pure in a solid phase. At equilibrium, we can write:

$$f_i^{0,s} = x_i\gamma_i f_i^{0,L} \tag{22.17}$$

where $f_i^{0,L}$ is the fugacity of compound i in a hypothetical pure subcooled liquid at the temperature T of the system. Hence,

$$\ln x_i = \ln\frac{f_i^{0,S}}{f_i^{0,L}} - \ln\gamma_i \tag{22.18}$$

The first term on the right-hand side corresponds to the dimensionless molar Gibbs energy change of the pure compound i when passing from the state of subcooled liquid to solid at temperature T:

$$\frac{\Delta g_i}{RT} = \frac{\Delta h_i - T\Delta s_i}{RT} = \ln\frac{f_i^{0,S}}{f_i^{0,L}} \tag{22.19}$$

TABLE 22.2
Steps for the Calculations of Δh_k and Δs_k

Step k	Δh_k	Δs_k
1. Subcooled liquid at T and P is heated to the melting point T_m at P	$C^L{}_P(T_m - T)$	$C^L{}_p \ln\frac{T_m}{T}$
2. Liquid at T_m and P is solidified; Δh_m is the "heat of fusion"	$-\Delta hm$	$-\frac{\Delta h_m}{T_m}$
3. Solid is cooled from T_m to T	$C^Sp(T-T_m)$	$C^S{}_p \ln\frac{T_m}{T}$

The enthalpy and entropy changes are accounted for in the three steps shown in Table 22.2. Hence, the total changes are:

$$\Delta h = \Delta h_1 + \Delta h_2 + \Delta h_3 = \left(C_P^L - C_P^S\right)\left(T_m - T\right) - \Delta h_m$$

$$\Delta s = \Delta s_1 + \Delta s_2 + \Delta s_3 = \left(C_P^L - C_P^S\right)\ln\frac{T_m}{T} - \frac{\Delta h_m}{T_m}$$

Calling $\Delta C_P = C_P^L - C_P^S$, we write:

$$\ln\frac{f_i^{0,S}}{f_i^{0,L}} = -\frac{\Delta h_m}{R}\left(\frac{1}{T} - \frac{1}{T_m}\right) + \frac{\Delta C_P}{R}\left(\frac{T_m}{T} - 1 - \ln\frac{T_m}{T}\right) \tag{22.20}$$

Combining Equations 22.18 and 22.20, we write:

$$\ln x_i = -\frac{\Delta h_m}{RT_m}\left(\frac{T_m}{T} - 1\right) + \frac{\Delta C_P}{R}\left(\frac{T_m}{T} - 1 - \ln\frac{T_m}{T}\right) - \ln\gamma_i \tag{22.21}$$

As a first approximation, the effect of the ΔC_p term can be neglected and we write:

$$\ln x_i = -\frac{\Delta h_m}{RT_m}\left(\frac{T_m}{T} - 1\right) - \ln\gamma_i \tag{22.22}$$

As an additional approximation, we can assume ideal solution in the liquid phase, $\gamma_i = 1$, and write:

$$\ln x_i^{id} = -\frac{\Delta h_m}{RT_m}\left(\frac{T_m}{T} - 1\right) \tag{22.23}$$

When the solvent is present as a solid, T_m is the melting point, or freezing point, of the pure solvent. In this case, Δh_m refers to the heat of fusion of the pure solvent, and Equation 22.23 is used to calculate the "freezing point depression" due to the presence of a solute.

Alternatively, when the solute is present as a pure solid, both T_m and Δh_m refer to the solute and Equation 22.23 is used to calculate the "ideal solubility." More realistic values of the solubility can be obtained by using some g^E model to calculate γ_i. Sometimes, Equation 22.23 is written in a different form using:

$$\frac{\Delta h_m}{RT_m} = \frac{\Delta S_m}{R}$$

$$\ln x_i^{id} = \frac{\Delta s_m}{R}\left(1 - \frac{T_m}{T}\right) \tag{22.24}$$

SOLUBILITY OF A VOLATILE SOLID IN A COMPRESSED GAS

When a volatile solid compound is contact with a compound in a gas phase it is expected that the solubility of the solid in the gas will increase with an increase in temperature and decrease with an increase in pressure. In fact, the solubility of a solid (2) in a gas (1) increases with temperature but as the pressure increases the solubility will first decrease, and usually pass through a minimum and then increase with a further increase in pressure. Data previously reported in the literature for CO-H_2, N_2-H_2 at temperatures from 40 to 60 K and for CO_2-air from 120 to 190 K were studied by Hinkley and Reid [8] and reproduced well with the fugacity of the pure compound (2) in the solid phase given by Equation 16.1 and in the compressed vapor phase by Equation 13.2. The fugacity coefficients were obtained with the virial EOS truncated after the second virial coefficient.

For the minimum mole fraction of the volatile solid in the gas, Hinkley and Reid [8], considering that $y_1 >> y_2$, obtained:

$$y_{2,\,min} = -\frac{2\; e\; B_{12}\; P_2^s}{R\; T}$$

And for the pressure at the minimum solubility Hinkley and Reid wrote:

$$P_{at\ minimum\ y_2} = -\frac{R\; T}{B_1 + 2\; B_{12}}$$

Observing that Hinckley and Reid had simultaneously used the pressure explicit virial equation to evaluate the fugacity coefficient and the volume explicit virial equation for the expression of the pressure at the minimum solubility, Ashraf [9], using consistently the more accurate pressure explicit virial equation, rederived the equation for the pressure at the minimum solubility and obtained,

$$P_{at\ minimum\ y_2} = -\frac{R\; T}{B_1 + 2\; B_{12}}\left[1-\left(\frac{B_1}{2B_{12}}\right)^2\right]$$

Which can also be written in a simpler form as:

$$P_{at\ minimum\ y_2} = \frac{R\; T}{2\; B_{12}}\left(\frac{B_1}{2B_{12}}-1\right)$$

Although the corrected expression for the pressure at the minimum solubility does not change significantly the calculated results obtained by Hinkley and Reid, it shows the effect of using the more accurate virial form.

One interesting case of application of the idea of increasing the pressure to decrease the presence of a volatile solid in a gas phase happened in the development of electrical light bulbs. It was not easy to find a material that, passing an electric current, it could be heated to a high temperature to produce clear light. Eventually, it was found that tungsten, with a melting point of 3695 K, the highest melting point of all metals, inside an evacuated glass bulb, met all the required conditions. However, the problem was that at the high temperature of operation, tungsten evaporated and deposited on the surface of the protective glass bulb, blackening it. Irving Langmuir (Nobel Prize in Chemistry, 1932) reasoned that by putting an inert gas under pressure inside the glass bulb would solve the problem, and so it did. Thermodynamics is an experimental science so, in the absence of virial coefficients for tungsten, most probably the optimum pressure of the gas inside the glass bulb was found by trial and error.

PHASE EQUILIBRIUM THROUGH MEMBRANES

With the increasing interest in bioseparations, old concepts are recovering importance in chemical engineering thermodynamics. In this section, we briefly consider the basic principles of osmosis. The osmotic phenomena occur when a solution of a nonvolatile solute in a solvent is in contact with pure solvent through a membrane only permeable to the solvent. This can be the case of an aqueous solution of sugar, a nonionic solute, in contact with water through a semi-permeable membrane-like

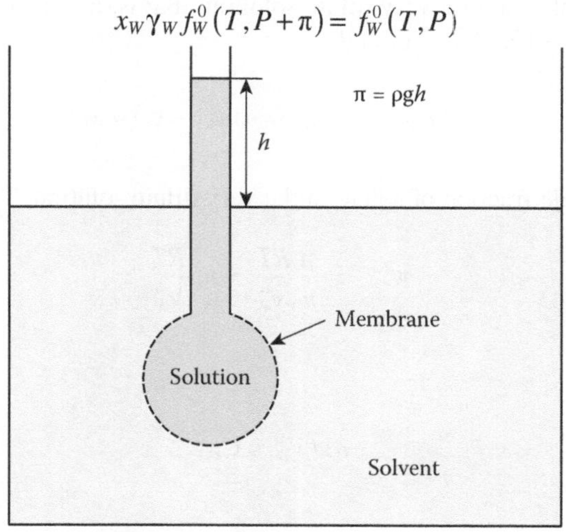

FIGURE 22.4 Osmotic equilibrium.

collodion, or of an aqueous solution of sodium chloride, an ionic solute, in contact with water through a semi-permeable membrane such as a cell wall. The physical effect caused by such a contact is an overpressure on the side of the solution with respect to the side of the pure solvent. This overpressure is called osmotic pressure, π. The osmotic pressure can be measured by means of an experiment such as the one shown in Figure 22.4.

The osmotic pressure, at a given temperature, is a function of the concentration of the solution. The thermodynamic treatment of the osmotic phenomena is simple. At equilibrium, the fugacity of the solvent in the solution at T, $(\pi+P)$ and x must be equal to the fugacity of the pure solvent (water) at T and P.

Hence,

$$f_W(T,P+\pi,x)=f_W^0(T,P)$$

$$\ln(x_W\gamma_W)+\ln\frac{f_W^0(T,P+\pi)}{f_W^0(T,P)}=0 \tag{22.25}$$

The second term of Equation 22.25 refers to the pure solvent and is directly obtained from Equation 14.17 as:

$$\ln\frac{f_W^0(T,P+\pi)}{f_W^0(T,P)}=\frac{v_W^0}{RT}\pi \tag{22.26}$$

where v_W^0 is the molar volume of pure liquid water assumed to be incompressible. Thus,

$$\pi=\frac{RT}{v_W^0}(-\ln a_W) \tag{22.27}$$

where we have used the definition of activity coefficient:

$$a_W\equiv x_W\gamma_W$$

If we consider the special case of a very dilute solution, that is, almost pure water on the solution side of the membrane, $\gamma_W \approx 1$ and $a_W \approx x_W$ and

$$-\ln a_W \approx -\ln x_W \approx -\ln(1-x_s) \approx x_s$$

where $x_s \approx n_s/n_W$ is the mole fraction of solute in the very dilute solution. Thus,

$$\pi^{dilute} = \frac{n_s RT}{n_W v_W^0} = n_s \frac{RT}{V_W^0} \tag{22.28}$$

or with

$$n_s / V_W^0 = \tilde{C}_s$$

$$\pi^{dilute} = \tilde{C}_s RT$$

Equation 22.28 is the well-known van't Hoff equation for dilute solutions. For concentrated solutions, ($-\ln a_W$) will differ from (n_s/n_W) and Equation 22.27 is used to define the osmotic coefficient φ:

$$(-\ln a_W) \equiv \frac{n_s}{n_W} \varphi \tag{22.29}$$

Thus,

$$\pi = \frac{n_s RT}{V_W^0} \varphi \tag{22.30}$$

RECAPITULATION OF CONCEPTS

The solution to a phase equilibrium problem always starts from the equality of the fugacity for each compound in the different phases, in which it is present at equilibrium, at a given temperature T and pressure P. How do we express the fugacity in each case is a matter of convenience. Equations of state are almost invariably used for the vapor phase. Simple equations of state, such as the virial pressure expansion truncated after the second virial coefficient, are particularly useful for low-pressure, moderate-temperature vapor–liquid equilibrium calculations with the $\gamma-\phi$ approach. In this case, for the liquid phase, the fugacity of each compound is expressed in terms of activity coefficients derived from a single expression for the excess Gibbs energy of the mixture. If the experimental data for activity coefficients are missing, they can be predicted using a group contribution method, like ASOG [1,2].

For high-pressure calculations, and even for low-pressure calculations, the use of cubic equations of state has become popular in the last decades. They are discussed in Chapter 25, and the details for application of the Peng–Robinson–Stryjek–Vera (PRSV) EOS [10,11] are given in Appendix C. Other more complex equations of state have also been used.

For liquid–liquid equilibria and for solid–liquid equilibria, the fugacities are almost invariably expressed in terms of activity coefficients.

The techniques to solve phase equilibrium problems change somewhat from problem to problem, but they are all based on the solution of the material balances and the equality of the fugacities at equilibrium.

NOTE

1 Note that the symbol z_i is used here for the composition of the feed. It should not be confused with the compressibility factor.

REFERENCES

1. Kojima, K., and Tochigi, K. 1979. *Prediction of Vapor-Liquid Equilibria by the ASOG Method*. Tokyo: Kodansha-Elsevier.
2. Tochigi, K., and Gmehling, J. 2011. Determination of ASOG parameters – Extension and revision. *J. Chem. Eng. J. Jpn*. 44: 304–306.
3. King, C. J. 1971. *Separation Processes*. New York: McGraw-Hill.
4. Rachford Jr., H. H., and Rice, J. D., 1952. Technical Note 136: Procedure for use of electronic digital computers in calculating flash vaporization hydrocarbon equilibrium. *Pet. Trans. AIME*. 195: 327–328.
5. Renon, H., and Prausnitz, J. M. 1969. Derivation of the three-parameter Wilson equation for the excess Gibbs energy of liquid mixtures. *AIChE J*. 15(5): 785–787.
6. Abrams, D. S., and Prausnitz, J. M. 1975. Statistical thermodynamics of liquid mixtures. New expression for the excess Gibbs energy of partly or completely miscible systems. *AIChE J*. 21(1): 116–128.
7. Wilson, G. M. 1964. Vapor-liquid equilibrium. XI. A new expression for the excess free energy of mixing. *J. Am. Chem. Soc*. 86(2): 127–130.
8. Hinckley, R. B., and Reid, R. C. 1964. Rapid estimation of the minimum solubility of solids in gases. *AIChE J*. 10(3): 416–417.
9. Ashraf, F. A., and Vera, J. H. 1980. On the derivation of the equations for the estimation of the minimum solubility of 'solids' in compressed gases, *Chem. Eng. Commun*. 4: 563–567.
10. Stryjek, R., and Vera, J. H. 1986. PRSV: An improved Peng-Robinson equation of state for pure compounds and mixtures. *Can. J. Chem. Eng*. 64(2): 323–333.
11. Stryjek, R., and Vera, J. H. 1986. PRSV2: A cubic equation of state for accurate vapor–liquid equilibria calculations. *Can. J. Chem. Eng*. 64: 820–826.

PROBLEMS

22.1 **Simultaneous phase and chemical equilibria**

Substance A decomposes according to the reaction $A \leftrightarrow B + C$. At 400 K and 3 bar, an equilibrium is reached at which a liquid phase of molar composition $x_A = 0.30$, $x_B = 0.40$, $x_C = 0.30$ coexists with a vapor phase of composition $y_A = 0.10$, $y_B = 0.30$, $y_C = 0.60$. Assuming that this system obeys Raoult's law, calculate

a. The reaction equilibrium constant at 400 K, taking as standard states (a1) the ideal gases at 1 bar; (a2) the pure liquids at 1 bar.
b. The pressure range over which the system at chemical equilibrium exists in two phases (liquid and vapor) at 400 K, if the reaction starts from pure A.
c. The phase compositions as functions of pressure at 400 K.

22.2 **Vapor–liquid–liquid equilibrium**

Dimethyl ether (1, DME) and water (2, H_2O) are partially miscible in liquid phase. At 373.26 K and 2848 kPa, a triple equilibrium exists between two liquid ("alpha" and "beta") phases and a vapor phase, with compositions [1]

$$x_1^\alpha = 0.104\ ,\ x_1^\beta = 0.755\ ,\ y_1 = 0.936$$

Consider the simple empirical two-parameter models for the Gibbs energy of binary systems,

$$[\text{Redlich}-\text{Kister}]\ \frac{g_0^E}{RT} = x_1 x_2 \left[B + C(x_1 - x_2)\right]$$

$$[\text{van Laar}] \quad \frac{g_0^E}{RT} = \frac{(A_{12}x_1)(A_{21}x_2)}{A_{12}x_1 + A_{21}x_2}$$

a. Compute the parameters of both models from the $\alpha-\beta$ equilibrium compositions on the triple line.
b. Use the parameters thus obtained to predict the complete $P-x-y$ diagram of $\alpha-$ vapor and $\beta-$ vapor equilibrium at 373.26 K.
c. Compute the parameters of both models from the $\alpha-$ vapor equilibrium compositions on the triple line and use them to predict only the $P-x-y$ diagram of $\alpha-$ vapor equilibrium at 373.26 K. Compare with the results of (a) and (b).
d. Same as (c), but for the $\beta-$ vapor equilibrium.
e. Is it possible to describe the three-phase equilibrium with a single set of parameters?
f. The mutual solubility of DME and H_2O increases at higher pressures. At 373.26 K, the liquid mixture achieves full miscibility (a critical solution point) at 50,800 kPa, $x_1^\alpha = x_1^\beta = 0.35$ [1]. Can the $P-x^\alpha-x^\beta$ diagram be predicted using the parameters found in (a)?

Because of the temperature and pressures involved and the strong non-ideality of the system, neither the original nor the modified Raoult's law is valid. Obtain the vapor fugacities from the truncated virial expansion with Tsonopoulos's correlations (Appendix B), including the Poynting correction factors for the liquids. Relevant pure-component properties are given in Table 22.3.

22.3 ***K*-factors and vapor–liquid equilibrium from the PRSV equation of state**

Use the PRSV EOS (Appendix C) to predict the vapor–liquid equilibria of ethane (1)+n-heptane (2) at 450 K. Draw the $P-x-y$ and $\ln K-\ln P$ (Figure 22.2) diagrams. Pure-component data are given in Table 22.4. Take the binary interaction parameter for this mixture as $\delta_{12} = 0.020$ [3].

22.4 **Solid solubility in a gas phase**

Solids may have low but nonzero vapor (sublimation) pressures. A volatile solid in contact with a gas therefore evaporates to some extent, until it reaches solid–gas equilibrium as required by Equation 12.10.

TABLE 22.3
Pure-Component Properties

	T_c (K)	P_c (kPa)	v_c (m³/kmol)	ω	v_L (m³/kmol)	P^s (kPa)
(1) DME	400.1	5370	0.1700	0.2002	0.09326	3294
(2) H_2O	647.1	22064	0.05595	0.3449	0.01871	101.7

Source: Data taken from the DIPPR® 801 Database [2] or calculated at 373.26 K from equations therein.

TABLE 22.4
Pure-Component Properties from Appendix C

	T_c (K)	P_c (kPa)	ω	κ_1
(1) Ethane	305.43	4978.76	0.09781	0.01181
(2) n-Heptane	540.10	2735.75	0.35022	0.02325

a. Show that, at a given temperature, the solubility (mole fraction) of a solid (component 1) in a non-condensable gas (component 2) reaches either a minimum or a maximum at the pressure at which the partial molar volume of the vaporized solid in the gas phase equals its molar volume in the solid state.
b. Predict the solubility of solid carbon dioxide ("dry ice") in gaseous nitrogen at 150 K and pressures from 1 to 1400 bar. Use the PRSV EOS (Appendix C) for the vapor/gas phases. Set the binary interaction parameter $\delta_{12} = 0$ in the mixing rule of Equation C.68. At 150 K, the sublimation pressure of CO_2 is 842.6 Pa, and the molar volume of the virtually incompressible solid is 27.08 cm³/mol. Verify that your results agree with (a).

REFERENCES FOR PROBLEMS

1. Pozo, M. E., and Streett, W. B. 1984. Fluid phase equilibria for the system dimethyl ether/water from 50 to 220°C and pressures to 50.9 MPa. *J. Chem. Eng. Data* 29: 324–329.
2. Design Institute for Physical Properties. DIPPR® 801 Sample Database. https://dippr.aiche.org/SampleDb (accessed April 3, 2023).
3. Ahmed, T. 2010. *Working Guide to Vapor-Liquid Phase Equilibria Calculations*. Burlington, MA: Gulf Professional Publishing.

23 Heat Effects

Abstract

Heat effects are both roots of the development of thermodynamics and a signal of nonideality in the behavior of mixtures. Thus, the chapter starts with the use of steam as a heating agent, and it continues with the study of heats of mixing. Special attention is given to the use of enthalpy–concentration data for technical calculations. Solved examples of calculation of heat effects in chemical reactions are presented.

HEAT EFFECTS IN PURE COMPOUNDS

Engineers have prepared tables of thermodynamic properties for most compounds of practical interest. In these tables, the values of the molar or specific volume, v, enthalpy, h, and entropy, s, are given as a function of temperature and pressure. To use these tables, one must first determine the physical state of the compound at the known values of T and P. The key piece of information is the value of the vapor pressure, P_s (also called saturation pressure), at the temperature of interest. This value is normally reported in the first column of the tables, together with the values of v^s, h^s, and s^s for both the liquid and the vapor. If the pressure of interest is higher than the saturation pressure, the compound is in the state of compressed liquid. In this case, as for all practical purposes, liquids are incompressible, the difference in pressure is ignored, and the values of the thermodynamic properties of the liquid compound are considered to be equal to those it has as saturated liquid. Only for work at high pressure may corrections be needed. In the case that the pressure of interest is lower than the saturation pressure, the compound is in the gas phase and its properties are those reported for that pressure. Usually, interpolation between two values of pressure from the tables is required. Sometimes, it is also necessary to interpolate between two values of temperature before proceeding with the interpolation of pressure. In technical language, a compound in the gas phase is referred to as a superheated vapor, meaning that its temperature is higher than the saturation temperature corresponding to its pressure.

Figure 23.1 depicts a schematic of an isotherm in a pressure–enthalpy (P–h) diagram.

In the special case that the pressure corresponds exactly with the vapor pressure at the temperature of interest, the compound can be in one of the following states: as saturated liquid, as a system of saturated liquid in contact with saturated vapor, as wet saturated vapor, or as dry saturated vapor. The wet saturated vapor is a saturated vapor phase containing small droplets of saturated liquid. All these cases, generically known as a "saturated mixture," will have different values of the molar or

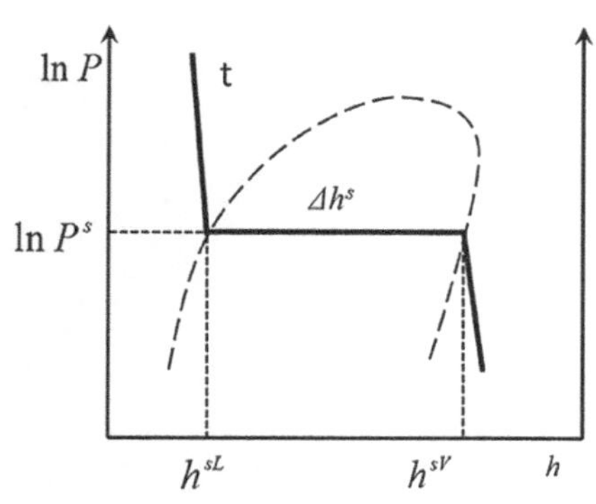

FIGURE 23.1 A schematic pressure–enthalpy diagram for a liquid–vapor system.

DOI: 10.1201/9781003431985-26

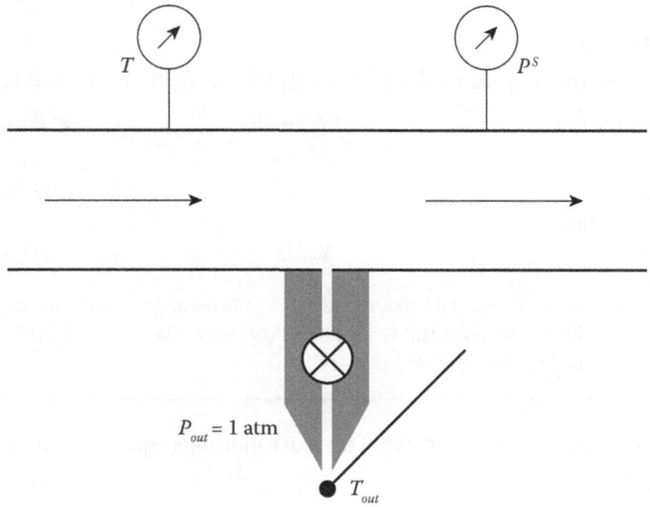

FIGURE 23.2 Adiabatic throttling calorimeter.

specific volume, enthalpy, or entropy. In the case of using saturated steam as a heating medium, it is necessary to determine in which of the above cases is the system. For heating purposes, saturated steam is used instead of saturated (boiling) water or superheated steam. This is because the heat of condensation of steam is orders of magnitude larger than the heat that can be obtained from cooling a single-phase fluid.

In order to determine the content of dry saturated steam in a saturated mixture, engineers have designed a device called an adiabatic throttling calorimeter (Figure 23.2).

Knowing the temperature of the saturated steam to be used for heating, the saturation pressure and the enthalpies of the saturated liquid and the saturated vapor phases, h^{sL} and h^{sV}, are readily obtained from tables. If the saturated mixture contains a mass fraction χ of dry saturated steam, the specific enthalpy of the saturated mixture is:

$$h_m^s = (1-\chi)h^{sL} + \chi^{h^{sV}}$$

$$= h^{sL} + \chi\left(h^{sV} - h^{sL}\right)$$

$$= h^{sL} + \chi\Delta h^s$$

or

$$\chi = \frac{h_m^s - h^{sL}}{\Delta h^s} \tag{23.1}$$

The mass fraction χ of dry saturated steam is called the *quality of the steam.* The value of h_m^s is obtained by purging a small stream of saturated mixture from the steam pipe into the atmosphere. As shown in Figure 23.2, this is done through an insulated pipe having a high-pressure valve. The temperature of the steam at the exit of the valve is measured after a short time of purging. When steady state is reached and the temperature t_{out} stabilizes, h_{out} can be determined knowing P_{atm} and t_{out}. From the first law for an open flow system (Chapter 3), neglecting kinetic and potential energy

TABLE 23.1
Vapor and Liquid Enthalpies of the Saturated Steam Mixture

Pressure (MPa)	h^{sV} (kJ/mol)	h^{sL} (kJ/mol)
1.171	50.134	14.293
1.455	50.263	15.098
Interpolating		
1.4	50.238	14.942

Source: Green, D. W., and Perry, R. H., *Perry's Chemical Engineers' Handbook*, 8th ed., McGraw-Hill Professional Publishing, Blacklick, OH, 2007, Table 2-305, pp. 2-413 to 2-414.

changes, as there is no heat exchanged with the surroundings and no shaft work involved, from Equation 3.14 we get:

$$\Delta h_{pipe}^{in} = 0$$

or

$$h_{out} = h_m^s$$

Thus, the quality of the steam, χ, is evaluated from Equation 23.1, and it is usually expressed in percent.

Example 23.1: Determination of Steam Quality

The manometer installed in a steam pipe indicates a pressure of 14 bar at the point where the saturated steam mixture passes the entrance of an adiabatic throttling calorimeter. After opening the throttling valve of the calorimeter, the temperature of the superheated steam leaving the calorimeter is 400 K. The outside pressure is 1 atm (1 atm = 1.013 bar). Determine the quality of the steam in the pipe.

Table 2–305 of Perry's handbook, pages 2–413 to 2–414 [1], provides information presented in Table 23.1. As the tabulated values of pressure are in megapascals, we need to convert the pressure data to proper units:

$$14\ [\text{bar}] = 1.4\ [\text{MPa}] \text{ and } 1\ [\text{atm}] = 0.1\ [\text{MPa}]$$

For the value of h_{out} at 400 K and 0.1 MPa, we read 49.189 [kJ/mol].

Thus,

$$\chi = \frac{h_m^s - h^{sL}}{\Delta h^s} = \frac{49.189 - 14.942}{50.238 - 14.942} = 0.97$$

The quality of steam is 97%.

HEAT EFFECTS IN MIXTURES

Except in the rare case of ideal solutions, each time that different chemical species are mixed, there are heat effects. If the mixing process is adiabatic, the temperature of the final mixture will be different from the mixed mean temperature, which we could calculate from knowledge of the masses, the temperatures, and the specific heats of the pure compounds' initial states. If the mixing process is carried out under isothermal conditions in a thermostatic bath, heat will be exchanged with the surroundings.

As it is always done in thermodynamics, we classify first the main cases encountered in practice and then systematize the experimental information in order to be able to treat new situations. There are two broad categories of mixing processes: (a) mixing processes in which there is no chemical reaction between the species mixed and (b) mixing processes in which chemical reactions occur. In this second category, two heat effects are superimposed: the heat effect due to the chemical reactions and the heat effect due to the purely physical interactions of mixing. The heat effects due to reactions are much larger than the heat effects due to purely physical interactions, so when a reaction occurs, the latter can be ignored. We refer to the first case as the study of heat of mixing and the second case as the study of heat of reaction. For the cases of mixing of the compounds, either in an open flow system without shaft work, Equation 3.14, or in a closed system at constant pressure, Equation 3.17, we write:

$$q = \Delta h \tag{23.2}$$

In the rare cases when a mixing process occurs in a closed system at constant volume, there is no work done, and from Equation 2.3, the heat effect is equal to the change in internal energy, not the change in enthalpy. However, for liquid mixtures the difference between internal energy and enthalpy is totally negligible, and frequently we retain the form of Equation 23.2. We recall that the difference in enthalpy is always taken as the final enthalpy minus the initial enthalpy, and that q is positive when heat enters the system. Conventionally then, heat evolved is indicated with a negative sign.

Heat of Mixing, Δh_m

The definition of the heat of mixing requires that the final mixture and all the pure compounds before mixing be at the same temperature and pressure. We recall that the effect of pressure is negligible in the case of liquids. For the molar or specific heat of mixing at a constant temperature T, we write:

$$q = \Delta h_m = h - \sum_i x_i h_i^0 \tag{23.3}$$

where h is the enthalpy of the mixture at temperature T and pressure P, and h_i^0 are the enthalpies of the pure compounds at the same conditions of temperature and pressure of the mixture.

In the literature, sometimes the heat of mixing is referred to as the excess enthalpy of the mixture. In fact, as discussed in Chapter 6, an ideal mixture is formed without heat effects, so for an ideal mixture in Lewis's sense, the second term of the right-hand side of Equation 23.3 corresponds to the enthalpy of the ideal mixture, and then by analogy with Equation 16.11b we write:

$$q = \Delta h_m = h - h^{id} = h^E \tag{23.4}$$

When forming a real mixture, starting with all pure compounds at the same temperature, the temperature of the mixture will have a different value. Considering that the definition of the heats of mixing, Equation 23.3, requires that the final mixture be at the same temperature as the initial temperature of the pure compounds before mixing, we can think of the process as occurring in two steps. First, we mix adiabatically the pure compounds, all at an initial common temperature T. In this step, there is no heat exchanged with the surroundings and the mixture will have an enthalpy h_{ad}. From Equation 23.3, with $q=0$,

$$h_{ad} = \sum_i x_i h_i^0$$

Thus, in an adiabatic mixing the mixture has the same enthalpy as the sum of the input enthalpies of the pure compounds, but unless it is an ideal mixture, its temperature will be different from the temperature of the pure compounds before mixing. In a second step then, the mixture is heated or cooled to bring its temperature back to T. In this step, the system exchanges with the surroundings an amount of heat q, per mole or per unit mass of the mixture:

$$q = h - h_{ad}$$

Thus, for this second step the amount of heat exchanged between the system and the surroundings is exactly that given by Equation 23.3. When the enthalpy of the mixture is larger than the sum of the enthalpies of the pure compounds at the same conditions of temperature and pressure, the temperature of the system after the adiabatic step is lower than the initial temperature of the pure compounds, that is, $\Delta h_m > 0$, then $q > 0$. Thus, heat must be supplied to the system to keep the temperature constant. In this case, we say that the process is endothermic. The opposite case is referred to as an exothermic mixing process.

For the mixing of compounds in the gas phase, the heat of mixing is usually negligibly small and can be ignored in most practical situations. In fact, for any ideal mixture in Lewis's sense, gas, liquid, or solid, the heat of mixing is exactly zero. Thus, although perfectly general, most of the treatment that follows is intended for *liquid* mixtures.

To fix ideas, let us consider the enthalpy of an ideal binary liquid mixture, in Lewis's sense,

$$h^{id,0} = x_1 h_1^0 + x_2 h_2^0 = h_2^0 + \left(h_1^0 - h_2^0\right) x_1$$

In this case, the enthalpy of the mixture is a linear function of composition, and on a plot of h versus x_1, it will be represented as a straight line joining the points ($x_1 = 0$, $h = h_2$) and ($x_1 = 1$, $h = h_1$). In this case, the mixture is formed without heat effects. On the other hand, Equation 23.3 tells us that for an endothermic case, $\Delta h_m > 0$, $q > 0$, and the curve of h versus x goes above the ideal straight line, as shown in Figure 23.3, and heat must be supplied to the system to keep the temperature constant. This is the most common case for organic mixtures, and it corresponds with positive deviations from

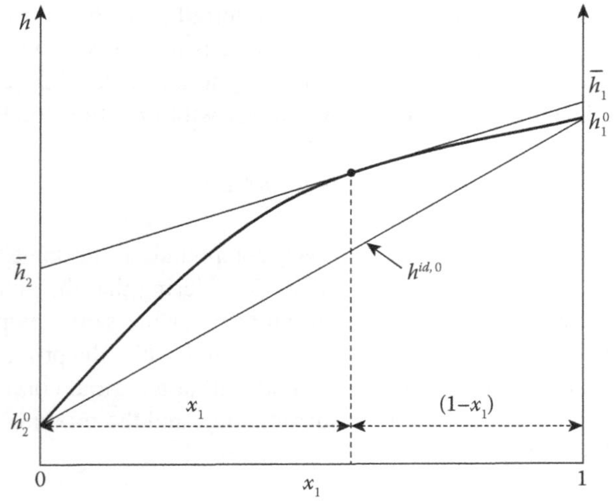

FIGURE 23.3 Enthalpy of binary liquid mixture for the endothermic case.

ideality in Lewis's sense. The more dissimilar the compounds are, the larger the positive deviations and the heat required to keep the temperature constant. Only a few binary mixtures presenting association between the two dissimilar compounds show exothermic behavior corresponding to negative deviations from ideality in Lewis's sense. Alcohol–amine systems or the system acetone–chloroform, for example, are cases in which the curve of h versus x falls below the ideal-mixture straight line, as well as the aqueous solutions of strong acids (e.g., H_2SO_4) and hydroxides (e.g., NaOH).

For completeness, in Figure 23.3 we have also shown the locations of the partial molar enthalpies $\bar{h}_1$ and $\bar{h}_2$ for a point at composition x_1. These values are found at the intersection of the tangent to the curve h–x, at the given composition x_1, with the y-axis at $x_1=1$ and $x_1=0$, respectively.

In fact, by analogy to Equation 6.17, we have:

$$\left(\frac{\partial h}{\partial x_1}\right)_{T,P} = \frac{\bar{h}_1 - h}{1-x_1} = \frac{h-\bar{h}_2}{x_1} \tag{23.5}$$

Thus, the tangent to the curve at a point x_1 fixes the position of both partial molar enthalpies at this composition.

Heats of mixing also provide the experimental information necessary to evaluate the temperature dependence of the parameters used in models for the excess Gibbs energy of binary systems. These models use the Lewis convention of ideality, and in this case, as shown by Equation 23.3, the heat of mixing Δh_m is equal to the excess enthalpy h_E. Adding Equation 16.13b over all compounds, we write:

$$\Delta h_m = -RT^2\left(\frac{\partial\left(g^E/RT\right)}{\partial T}\right)_{P,n} \tag{23.6}$$

Conversely, knowledge of (g^E/RT) as a function of temperature makes it possible to estimate values of Δh_m.

ENTHALPY–CONCENTRATION DIAGRAMS

The measurement of heats of mixing is performed in calorimeters. With this experimental information available and the help of Equation 23.3, it is possible to construct diagrams or tables of enthalpy versus composition (concentration) for a mixture of interest. Although for most cases of practical interest this information is handled directly by computer programs, for the treatment of new situations it is important to understand the basic principles used for the construction of such diagrams or tables. For this purpose, Equation 23.3 is better written as:

$$h = \Delta h_m + \sum_i x_i h_i^0 \tag{23.3a}$$

This form of Equation 23.3 shows that while Δh_m is experimentally measurable, the value of h depends on the arbitrarily chosen reference states used to set the enthalpy to zero for the pure compounds. For water, to be in agreement with the reference state used in the steam tables, the enthalpy is set at zero for saturated liquid water at 273.15 K or 32.018°F. In fact, the pure compound as saturated liquid at 273.15 K is a common reference state for compounds that are liquids at this temperature. For solutes such as NaCl and NaOH, it is conventional to consider the partial molar enthalpy of the solute at infinite dilution in the solvent, water, for example, as zero at 273.15 K. This choice, which may seem odd at first, has its basis on Equation 23.3 written as:

$$\Delta h_m = \sum_i^c x_i\left(\bar{h}_i - h_i^0\right) \tag{23.3b}$$

For a binary system then,

$$\Delta h_m = x_1\left(\bar{h}_1 - h_1^0\right) + x_2\left(\bar{h}_2 - h_2^0\right)$$

The heat of mixing per mole of solute is given by:

$$\tilde{\Delta h}_S\left[\frac{J}{\text{mol of solute}}\right] \equiv \frac{\Delta h_m\left[\dfrac{J}{\text{mol of solution}}\right]}{x_2\left[\dfrac{\text{mol of solute}}{\text{mol of solution}}\right]}$$

For solid solutes, it is convenient to express the concentration of solute as molality, m. Then,

$$\tilde{\Delta h}_S\left[\frac{J}{\text{mol of solute}}\right] \equiv \frac{\Delta h_m\left[\dfrac{J}{\text{kg of solvent}}\right]}{m_2\left[\dfrac{\text{mol of solute}}{\text{kg of solvent}}\right]}$$

At high dilution of a solute 2 in a solvent 1, x_2 is small and x_1 is virtually unity. In this case, $\bar{h}_1 = h_1^0$ and the first term of the right-hand side of Equation 23.3b cancels out. In addition, $\bar{h}_2 = h_2^\infty$. Thus, arbitrarily setting $\bar{h}_2^\infty = 0$, we write:

$$\Delta\tilde{h}_S^\infty = \lim_{x_2 \to 0} \frac{\Delta h_m}{x_2} = -h_2^0 \tag{23.7}$$

The left-hand side of Equation 23.7 is the heat of solution, that is, the heat of mixing per mole of solute at high solute dilution. This is a measurable quantity. Hence, the convention of arbitrarily setting $\bar{h}_2^\infty = 0$ fixes the value of the enthalpy of the pure solute at the temperature of the system, indicating once more that the enthalpy values are relative to the zero fixed for their scale. As an example of practical tables for enthalpy, Tables 2–182 and 2–183 (pages 2–203 to 2–206) of Perry's handbook [1] give heats of solution per mole of solute for inorganic and organic compounds. These values are reported as heat evolved, and thus they are negative values of the heats of solution. The values of Table 2–182 of Perry's handbook [1] are reported either at infinite dilution, indicating number of moles of water (e.g., 400 and 1,000) necessary to dissolve 1 mole of solute, or at "aq.," indicating an aqueous solution of unspecified high dilution. This is justified because the heat of solution per mole of solute is quite constant from infinite dilution up to 20 moles of solvent per mole of solute. A practical consequence of this fact is that, as also suggested by Equation 23.7, for diluted solutions we can use the approximation:

$$\Delta h_m^{dil} \approx \left(\Delta\tilde{h}_S^\infty\right) x_2 \tag{23.8}$$

At higher concentrations of solute, errors can be large and experimental values are required. In some cases, when heats of mixing or heats of solution are not available, it is possible to find information on heats of formation of aqueous solutions. These refer to the hypothetical formation of an aqueous mixture starting from the elements forming the solute and a certain number of molecules of water. As an example, Table 2–178 of Perry's handbook, page 2–203 [1], provides these data. The use of this information is better illustrated with Example 23.2.

Example 23.2: Calculation of a Heat of Dilution from Heats of Formation

For $BaBr_2$, Table 2–182 of Perry's handbook [1], page 2–203, reports that the heat evolved when forming an infinitely dilute solution in water is +5.3 kcal/mol of solute. Considering that Table 2–182 gives heat evolved, we write $\Delta h_s^\infty = -5.3(\text{kcal/mol})$. Verify this value using information on heats of formation.

Solution

Table 2–178 of Perry's handbook [1], page 2–186, reports the heats of formation of $BaBr_2$(c) and of $BaBr_2$(aq.,400) as –180.38 and –185.67 kcal/mol, respectively. These values refer to the following formation equations:

1. $Ba+Br_2+(400)\ H_2O=BaBr_2\ (aq,400)$	$\Delta h_f=-185.67$ kcal/mol
2. $Ba+Br_2=BaBr_2\ (c)$	$\Delta h_f=-180.38$ kcal/mol

Subtracting Equation 2 from Equation 1, we get

$$BaBr_2\,(c)+(400)H_2O = BaBr_2\,(aq,400) \quad \Delta h = (-185.67)-(-180.38)$$

$$\Delta h = -5.29\ (\text{kcal/mol})$$

This value compares well with $\Delta h_s^\infty = -5.3\ (\text{kcal/mol})$ obtained from Table 2–182 of Perry's handbook [1].

Thus, either from experimental measurements or from estimates, we can have values of Δh_m as a function of composition. Subsequently, using Equation 23.2, we can prepare diagrams or tables of enthalpy versus composition.

Typical examples are the enthalpy–concentration diagrams for aqueous solutions of NaOH and H_2SO_4 presented in Figures 2–29 and 2–31 of Perry's handbook [1], pages 2–403 and 2–409. Similarly, as mentioned in Example 23.2, Table 2–182, page 2–203, Perry's handbook [1] gives values of the heat evolved per mole of solute when forming dilute solutions. The conceptually interesting point is the relation between the enthalpy–concentration diagrams and the information given in this table. Again, in this case, the relation between these two pieces of information is better illustrated by an example.

Example 23.3: Calculation of the Enthalpy of a Solution

Figure 2–29, page 2–403, of Perry's handbook [1] contains the information presented in Table 23.2.
Compare these data with values obtained using Equation 23.3a:

$$h = \Delta h_m + \sum_i x_i h_i^0$$

Solution at 18°C

Finding the Enthalpies of Pure Compounds, h_i^0

For NaOH: Temperatures of 64°F and 248°F correspond to 18°C and 120°C. From Table 2–182, page 2–203, of Perry's handbook [1], the heat evolved per mole of solute when forming an

TABLE 23.2
Enthalpy Data for Aqueous Solution of NaOH

Temperature t (°F)	Enthalpy of the Solution, h (Btu/lb of Solution)	
	$X_{NaOH}=0.05$	$X_{NaOH}=0.10$
64	35	32
248	209	198

Source: Green, D. W., and Perry, R. H., *Perry's Chemical Engineers' Handbook*, 8th ed., McGraw-Hill Professional Publishing, Blacklick, OH, 2007, Figure 2-29, p. 2-403.

Note: The resolution of the plot allows for enthalpy readings within ±3 units on the last place. X_{NaOH} is the weight fraction of NaOH in the solution.

infinitely dilute solution of NaOH in water at 18°C is 10.18 (kcal/mol). Thus, the heat of solution per mole of solute at infinite dilution is the negative of this value:

$$\Delta\tilde{h}_s^{\infty} = -10.18\ (\text{kcal/mol of solute})$$

Therefore, according to Equation 23.7, the enthalpy of NaOH (solid) at 18°C is arbitrarily fixed at $h_2^0 = 10.18\ (\text{kcal/mol})$ of NaOH or (using the factor 1800 [(Btu/lb-mol)/(kcal/mol)]):

$$h_2^0 = 10.18\ (\text{kcal/mol}) \cdot 1800\left[\frac{\text{Btu}/\text{lbmol}}{\text{kcal}/\text{mol}}\right] = 18{,}324\left(\text{Btu}/\text{lbmol}\right)\text{of NaOH}$$

or

$$h_2^0 = 18{,}324\left(\text{Btu}/\text{lbmol}\right)\frac{1(\text{lbmol})}{40(\text{lb})} = 458.1\left(\text{Btu}/\text{lb}\right)\text{of NaOH}$$

For H_2O: For liquid water, Table 2–305 of Perry's handbook, page 2–413 [1], gives the information presented below in kilojoules per mole. As 18°C is equal to 291.15 K, we need to perform the interpolation to get the value of enthalpy at the required temperature (see Table 23.3). Thus,

$$h_1^0 = 1.3609\ (\text{kJ/mol}) \cdot 429.92\left[\frac{\text{Btu}/\text{lbmol}}{\text{kJ}/\text{mol}}\right] = 585\left(\text{Btu}/\text{lbmol}\right)\text{or } h_1^0 = 32.5\left(\text{Btu}/\text{lb}\right)\text{of water}$$

Finding Δh_m
From Equation 23.8,

$$\Delta h_m^{dil} \approx \left(\Delta\tilde{h}_S^{\infty}\right)x_2$$

This approximation is valid when there are more than 20 moles of solvent per mole of solute.

For a more concentrated solution, at a weight fraction $X_2=0.10$, in 100 lb of mixture there are 10 lb of NaOH and 90 lb of H_2O. Remembering that $M_2=40.0$ and $M_1=18.0$, this is equivalent to 0.25 lb-mol of NaOH and 5.0 lb-mol of H_2O. Thus, the ratio of moles of water per mole of solute is 20, which falls in a safe region for application of Equation 23.8. In addition, from Equation 23.7,

$$\Delta\tilde{h}_S^{\infty} = -h_2^0$$

TABLE 23.3
Enthalpy of Liquid Water at Different Temperatures

Temperature (K)	h^{sL} (kJ/mol)
290	1.2742
300	2.0279
Interpolating	
291.15	1.3609

Source: Green, D. W., and Perry, R. H., *Perry's Chemical Engineers' Handbook*, 8th ed., McGraw-Hill Professional Publishing, Blacklick, OH, 2007, Table 2-305, p. 2-413.

At a weight fraction $X_2=0.10$,

$$\Delta h_m = (-18{,}324)\cdot\left(0.25/5.25\right) = -872.6\left(\frac{\text{Btu}}{\text{lbmol mixture}}\right)$$

$$\Delta h_m = (-872.6)\cdot\left(5.25/100\right) = -45.8\left(\frac{\text{Btu}}{\text{lb mixture}}\right)$$

From Equation 23.3a,

$h=(-45.8)+(0.90)(32.5)+(0.10)(458.1)$

$h=29$ (Btu/lb)

This value should be compared with $h=32$ (Btu/lb) as reported in Table 23.2.
At a weight fraction $X_2=0.05$, similar calculations give:

$$\Delta h_m = -22.9\left(\frac{\text{Btu}}{\text{lb mixture}}\right)$$

and

$h=31$ (Btu/lb)

which compare well with $h=35$ (Btu/lb) presented in Table 23.2.

Solution at 120°C
In order to estimate h at 120°C, we assume that Δh_m does not change with temperature, but that the enthalpies of the pure compounds need to be corrected for temperature.

For H_2O: Table 2–305 of Perry's handbook, page 2–413 [1], gives for liquid water the information presented below in kilojoules per mole. As 120°C is equal to 393.15 K, we need to perform the interpolation to get the value of enthalpy at the required temperature (see Table 23.4). Thus,

$$h_1^0 = 3903\left(\text{Btu}/\text{lb mol}\right) \text{ or } h_1^0 = 216.84\left(\text{Btu}/\text{lb}\right) \text{ of water}$$

For NaOH: For NaOH, no values of the heat of solution per mole of solute at infinite dilution are reported at other temperatures. On the other hand, values for C_p for NaOH are reported in [2]. The trend line in the temperature range of 298 K–400 K has the form:

TABLE 23.4
Enthalpies of Saturated Liquid Water at Different Temperatures

Temperature (K)	h^{sL}(kJ/mol)
390	8.8354
400	9.6013
Interpolating	
393.15	9.0767

Source: Green, D. W., and Perry, R. H., *Perry's Chemical Engineers' Handbook*, 8th ed., McGraw-Hill Professional Publishing, Blacklick, OH, 2007, Table 2-305, p. 2-413.

$$C_P\left(\text{cal}/_{\text{deg}\cdot\text{mol}}\right) = 6\cdot 10^{-5}t^2 + 0.0041t + 14.122 \text{ with } R^2 = 1$$

where temperature t is in degrees Celsius. As the precision of the information available does not justify an integration, considering that at 18°C, C_p=0.355 cal/(g deg)=0.640 Btu/(lb·deg), and at 120°C, C_p=0.387 cal/(g deg)=0.697 Btu/(lb·deg), we assume $\bar{C}_p \approx 0.668\,\text{Btu/(lb}\cdot\text{deg)}$ as an average value.

$$\Delta h_2^0 = C_p\Delta T = 0.668\cdot(120-18) = 68.1\,\text{Btu/lbNaOH}$$

Then, at 120°C,

$$h_2^0 = 458.1 + 68.1 = 526.2\left(\text{Btu/lbNaOH}\right)$$

Thus, using Equation 23.3a,

At a weight fraction X_2=0.05: h=206 Btu/lb versus h=209 Btu/lb (Table 23.2)

At a weight fraction X_2=0.10: h=195 Btu/lb versus h=198 Btu/lb (Table 23.2)

These results compare well with those given in Table 23.2 that were obtained from the enthalpy–concentration diagram and show the consistency of different sources of information.

USE OF ENTHALPY–CONCENTRATION INFORMATION

Enthalpy–concentration diagrams or tables are useful tools in chemical engineering calculations. Here, we discuss two important cases: (1) mixing two solutions of different concentrations and at different temperatures and (2) concentrating a solution of a nonvolatile solute by evaporation of the solvent.

Mixing of Two Solutions

The general scheme of the mixing process is shown in Figure 23.4. This case also includes the dilution of a mixture by the addition of pure solvent, that is, X_s=0. The symbols F, S, and P (feed, solvent, and product) denote the total amounts (during a given time interval) of the streams added; X_F, X_S, and X_P denote the mole fractions or weight fractions of the solute in each stream; and t_F, t_S, and t_P are the corresponding temperatures. We can write two equations for the material balances and one equation for the energy balance (neglecting the work of stirring and heat losses).

The material balances (see Appendix A) are:

$$P = F + S \tag{23.9}$$

$$PX_P = FX_F + SX_S \tag{23.10}$$

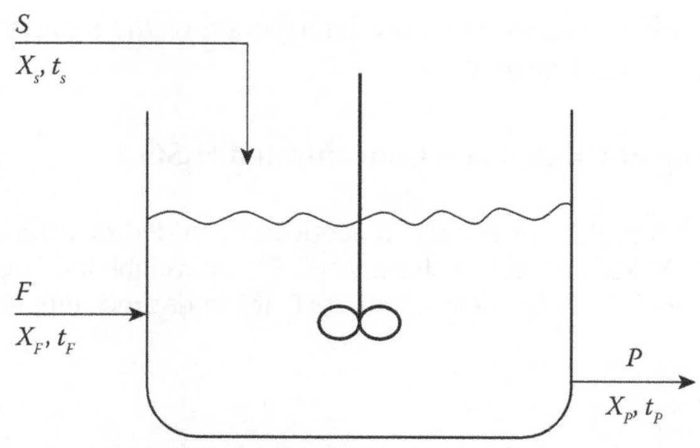

FIGURE 23.4 General scheme of the mixing process.

TABLE 23.5
Different Cases for Solving the Material and Energy Balances of Mixing

Known	To Be Evaluated	
F, X_F, S, X_S	$P=F+S$	$X_P=(FX_F+SX_S)/P$
F, X_F, P, X_P	$S=P-F$	$X_s=(PX_P-FX_F)/S$
F, X_F, X_S, X_P	$P=F(X_F-X_S)/(X_P-X_S)$	$S=P-F$
X_F, X_s, X_P, P	$F=P(X_P-X_S)/(X_F-X_S)$	$S=P-F$

and the energy balance is:

$$Q = Ph_p - (Fh_F + Sh_s) \tag{23.11}$$

In most practical cases, the material balances can be solved separately, and the results are then used to solve the energy balance. Table 23.5 shows some common situations. Other cases can be similarly solved.

From the point of view of the energy balance, two cases are of special interest. The first is the case of an adiabatic mixing process for which we know all the variables of the material balances, t_F and t_S, and we want to obtain t_p^{ad}. With (t_F, X_F) and (t_S, X_S), we determine h_F and h_S, respectively, from an enthalpy–concentration diagram. From Equation 23.11, with $Q^{ad}=0$, we obtain:

$$h_p^{ad} = (F\ h_F + S\ h_s)/P \tag{23.12}$$

Then, with h_p^{ad} and X_p we obtain t_p^{ad} from the enthalpy–concentration diagram.

The second case of interest is one in which we want to determine the heat Q that needs to be exchanged with the surroundings for a desired value of t_p. Here, we assume that first we form the mixture adiabatically, and then we exchange heat to change the temperature of the mixture from t_p^{ad} to t_p. Thus, with (t_p, X_p) we read h_p from the enthalpy–concentration diagram and

$$Q = P\left(h_p - h_p^{ad}\right) \tag{23.13}$$

In fact, by substituting Equation 23.12 into Equation 23.13, we recover Equation 23.11. Eliminating F, S, and P from Equations 23.9, 23.10, and 23.12, it is possible to show that h_p^{ad} can be directly read in an enthalpy–concentration diagram from the intersection of the straight line joining the points (X_F, t_F), (X_S, t_S) with the vertical line at X_p.

Example 23.4: Mixing of Diluted and Concentrated H_2SO_4

Calculate the heat, in kilojoules per hour, that needs to be added or withdrawn from a continuous mixer receiving 4500 kg/h of dilute sulfuric acid, 5% by weight and 200°F, and 8000 kg/h of concentrated sulfuric acid, 90% by weight and 70°F, if the desired temperature of the resulting mixture is 100°F.

Solution

Let us identify the dilute acid with the stream *S*, $S=4500$ kg/h, $X_S=0.05$, $t_s=200$°F, and the concentrated acid with the stream *F*, $F=8000$ (kg/h), $X_F=0.90$, $t_F=70$°F. For the product *P*, $t_P=100$°F. From the material balances,

$$P = F + S = 12{,}500\ (kg/h)$$

and

$$X_P = \frac{FX_F + SX_S}{P} = \frac{8000 \cdot 0.90 + 4500 \cdot 0.05}{12{,}500} = 0.59$$

From Figure 2–31, page 2–409, of Perry's handbook [1],

$$X_F=0.9,\ t_F=70°F;\ h_F=-60 (Btu/lb)=-140\ (kJ/kg)$$

$$X_S=0.05,\ t_S=200°F;\ h_S=+145\ (Btu/lb)=+337\ (kJ/kg)$$

$$X_P=0.59,\ t_P=100°F;\ h_P=-100\ (Btu/lb)=-233\ (kJ/kg)$$

Then, from Equation 23.11,

$$Q=-3309\ 10^3\ (kJ/h)$$

Thus, heat needs to be withdrawn from the system. Note the fact that the values of *X* are on a weight basis and including the vapor in Figure 2–31 of Perry's handbook [1]. This is of no concern below the line of boiling points at 1 atm.

CONCENTRATION OF A SOLUTION OF A NONVOLATILE SOLUTE BY EVAPORATION OF SOLVENT

The general scheme of the evaporation process is shown in Figure 23.5.

In this case, the stream *S* containing pure solvent ($X_S=0$) leaves the system. Usually, the temperatures of the streams *S* and *P* are the same, $t_S=t_P$. Since the boiling temperature of the concentrated solution *P* is higher than the boiling point of the solvent at the same pressure, the assumption of thermal equilibrium, $t_S=t_P$, means that the solvent leaves the system as superheated vapor. In this case, the material balances take the form:

$$F = S + P \tag{23.14}$$

$$F X_F = P X_P \tag{23.15}$$

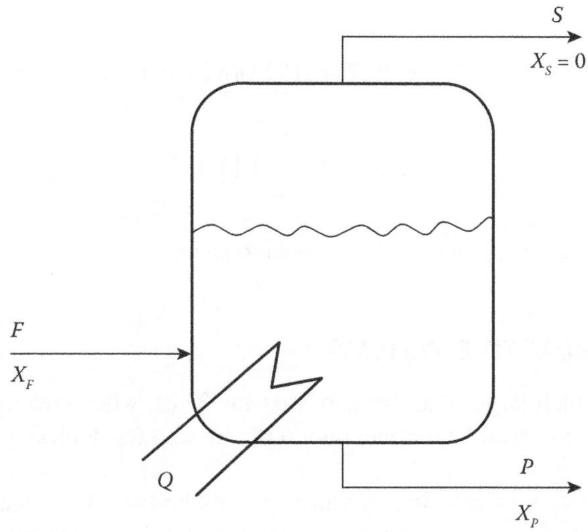

FIGURE 23.5 General scheme of the evaporation process.

and the energy balance can be written as:

$$Q = (Sh_S + Ph_P) - Fh_F \tag{23.16}$$

Example 23.5: Concentration of a Dilute NaOH Solution

Calculate the heat required to concentrate 12,000 kg/h of a dilute solution of sodium hydroxide from 5% to 65% by weight of solute. The diluted solution enters the process at 70°F, and the concentrated solution leaves the evaporator at 260°F. The evaporator operates at atmospheric pressure.

Solution

In this case, $F=12{,}000$ kg/h, $X_F=0.05$, $t_F=70$°F; $t_P=t_S=260$°F, $X_P=0.65$.

From the material balances,

$$P = FX_F/X_P = (12{,}000)(0.05)/(0.65) = 923\ (\text{kg/h})$$

$$S = F - P = 12{,}000 - 923 = 11{,}077\ (\text{kg / h})$$

From Figure 2–29, page 2–403, of Perry's handbook [1],

$$X_F = 0.05, t_F = 70°F; h_F = 30(\text{Btu/lb}) = 70\ (\text{kJ/kg})$$

$$X_P = 0.65, t_P = 260°F; h_P = 355(\text{Btu/lb}) = 826\ (\text{kJ/kg})$$

For superheated steam (1 atm, 260°F) from [10], h_s= (1173.56 Btu/lb)= 2729.70 (kJ/kg).

From Equation 23.16,

$$Q = \left[(11{,}077)(2729.70) + (923)(826)\right] - (12{,}000)(70)$$

$$Q = 30{,}159\ 10^3\left(\text{kJ/h}\right)$$

Thus, as expected, it is necessary to add heat to the system.

HEAT EFFECTS IN REACTIVE SYSTEMS

Heats of reaction are much larger than heats of mixing. Thus, when chemical reactions occur, the small contribution to the overall heat effect due to the mixing (or demixing) of compounds can be totally ignored.

To calculate heat effects in reacting mixtures, we make use of the standard heat of reaction discussed in Chapter 21. This standard heat of reaction at any temperature *T*, ΔH_T^θ can be calculated from Equations 21.11 and 21.15 or 21.17. We recall that this value of ΔH_T^θ corresponds to the enthalpy change of the standard reaction; that is, when the reactants enter in stoichiometric proportions, all at the same temperature *T*, the reaction is complete and the stoichiometric amounts of products leave at the same temperature *T*. In practical cases, the reactants are rarely added in stoichiometric proportions or added all at the same temperature. Frequently, "inert" compounds, that is, compounds that do not participate in the reactions, are present in one or more streams. The reaction itself is usually not complete or at equilibrium. Finally, there can be more than one reaction in the system. To fix ideas, Figure 23.6 illustrates the case of a reactor in which an excess of reactant *A* is fed at a temperature t_A to react with an expensive reactant *B*, diluted with an inert compound *I*, at a temperature t_B. The products *C* and *D*, plus unreacted amounts of *A* and *B* and the inert *I*, leave the

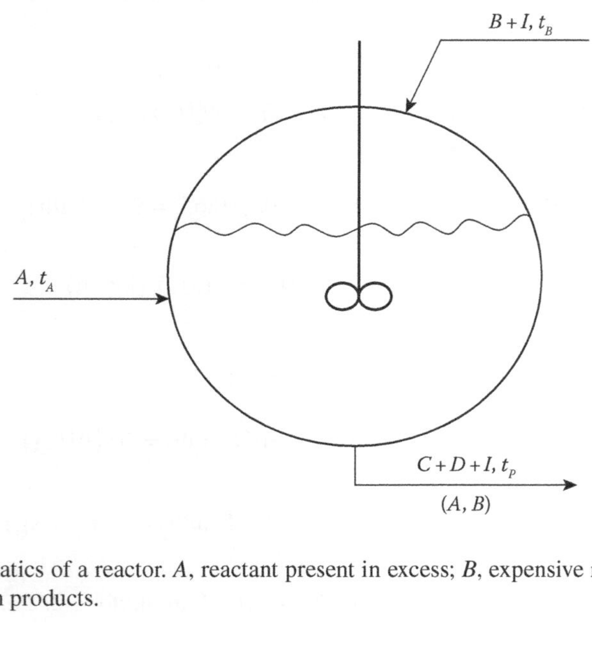

FIGURE 23.6 Schematics of a reactor. *A*, reactant present in excess; *B*, expensive reactant; *I*, inert dilutant for *B*; *C* and *D*, reaction products.

reactor at a temperature t_P. The study of heat effects in reacting systems always starts by writing the equations for the material balances and the equation for the energy balance and solving first, if at all possible, the material balances. In some situations, it may be necessary to solve simultaneously the material and energy balances. In the discussion that follows, we concentrate on how to solve the energy balance equation assuming the extent of reaction to be known.

For the energy balance, we start from Equation 23.1, written either for a certain lapse of time or for a certain amount of a given stream. The value of Δh is usually calculated assuming four separate steps:

1. The amounts of materials in the feed that actually react are taken from their temperature at the entrance conditions to a common temperature T conveniently chosen.
2. The reactions are carried out at the temperature T, to the extent that they actually occur in the system.
3. The products of reaction are taken from T to the temperature t_P of the product stream.
4. The unreacted reactants plus inerts are taken from their temperature at the feed to t_P. Each of these steps is described in detail below.

For simplicity, the discussion is presented in terms of flows, F (mol/h). Other bases may also be used.

Step 1: We first arbitrarily select a temperature at which the reactions will be carried out. In most cases, required data are tabulated at 25°C; thus, by choosing the reaction temperature as $T=T_0=298.15$ K, the calculations for step 2 are greatly simplified. In any case, designating by T the chosen temperature, the enthalpy change for step 1 is:

$$\Delta h_1 = \sum_i F_i^R \overline{C_{P_i}} (T - T_i) \tag{23.17}$$

where the sum is taken over all the reactants in the amounts F_i^R that actually react, each at its temperature T_i.

Step 2: In this step, we calculate the standard heat of reaction for each reaction that has taken place, being careful to express the result of each reaction per mole of a particular compound k, arbitrarily chosen but clearly specified. This standard heat of reaction is calculated with Equation 21.11 if $T=T_0$ or with Equation 21.15 if $T \neq T_0$. In any case, for step 2,

$$\Delta h_2 = \sum_r \left[F_k \Delta H_T^{\theta} \left(\text{per mol of } k\right)\right]_r \tag{23.18}$$

where F_k denotes the moles of the compound k, chosen as reference, that are actually participating in the particular reaction r and the sum is carried out over all reactions. The base for F_k (per hour, per 100 lbs of product, etc.) must be the same as the base used for F_i in Equation 23.17.

Step 3: In this step, all the products of the reactions, designated generically by the index j, are taken from the temperature T to the temperature T_P of the product stream:

$$\Delta h_3 = \sum_j F_j \overline{C_{P_j}} (T_P - T) \tag{23.19}$$

Step 4: Finally, all the unreacted reactants plus inerts entering into the feed, in the amounts F_i^I, are taken from their respective temperatures at entrance T_i to T_P:

$$\Delta h_4 = \sum_i F_i^I \overline{C_{P_i}} \left(T_P - T_i\right) \tag{23.20}$$

Note that the value of C_{Pi} for step 1 is the mean heat capacity in the temperature range $(T_i - T)$, while in step 4 it is its value in the range $(T_P - T_i)$. These two values are not necessarily the same.

Finally, Equation 23.1 can be rewritten as:

$$q = \Delta h_1 + \Delta h_2 + \Delta h_3 + \Delta h_4 \tag{23.21}$$

One type of reaction for which the heat effects are of special importance is the combustion of organic compounds. These reactions are so important that their standard heats of reaction at $T_0 = 298.15$ K are available in tabulations. For example, values for the combustion of carbon, carbon monoxide, and other inorganic compounds are reported in Table 2–178, pp. 2–186 to 2–194, of Perry's handbook [1]. Table 2–179, pp. 2–195 to 2–200, contains the same information for hydrogen and organic compounds. It is important to observe that the values reported in this table are the negative values of the heat of combustion, that is, $\Delta H_{c,T_0}$, and that it is necessary to decide if the product of combustion is liquid water or gaseous water. In general, for an organic compound of the form $C_nH_{m(g)}$, the heat of combustion to produce liquid water refers to the reaction:

$$\text{A. } C_nH_{m(g)} + \left(n + \frac{m}{4}\right)O_{2(g)} \rightarrow nCO_{2(g)} + \frac{m}{2}H_2O_{(l)} \quad \Delta H_{c,C_nH_{m(g)},T_0}$$

The combustion of $C_{(s)}$ or $H_{2(g)}$ are just special cases of this reaction and correspond to the reactions of formation of $CO_{2(g)}$ or $H_2O_{(l)}$, respectively:

$$\text{B. } C_{(s)} + O_{2(g)} \rightarrow CO_{2(g)} \quad \Delta H^0_{f,CO_{2(g)},T_0} = \Delta H^0_{c,C_{(s)},T_0}$$

$$\text{C. } H_{2(s)} + \frac{1}{2}O_{2(g)} \rightarrow H_2O_{(l)} \quad \Delta H^0_{f,H_2O_{(l)},T_0} = \Delta H^0_{c,H_{2(s)},T_0}$$

We recall that, by convention, the heats of formation of the elements in their elemental standard state at $T_0 = 298.15$ K are considered to be equal to zero.

Adding reaction B multiplied by n to reaction C multiplied by $(m/2)$ and subtracting reaction A, we obtain:

$$nC_{(s)} + \frac{m}{2}H_{2(g)} \rightarrow C_nH_{m(g)} \quad \Delta H^0_{f,C_nH_{m(g)}}$$

Thus,

$$\Delta H^0_{f,C_nH_{m(g)},T_0} = n\Delta H^0_{f,CO_{2(g)},T_0} + \frac{m}{2}\Delta H^0_{f,H_2O_{(l)},T_0} - \Delta H^0_{c,C_nH_{m(g)},T_0} \tag{23.22}$$

For the case of the standard heat of formation of a liquid compound $C_nH_{m(l)}$, the last term of the right-hand side of Equation 23.22 must be $\Delta H^0_{c,C_nH_{m(l)},T_0}$. Similarly, when one chooses to use the value of the heat of combustion of the compound C_nH_m, either liquid or gas, referred to the formation of water gas instead of liquid water, the value of $\Delta H^0_{f,H_2O_{(l)},T_0}$ in Equation 23.22 must be replaced by $\Delta H^0_{f,H_2O_{(g)},T_0}$.

Table 2–179 of Perry's handbook [1] contains values of ideal-gas enthalpy of formation and the heats of combustions for organic compounds with gaseous water as one of the products. Both of these values can be used if one takes into account the enthalpies of vaporization.

If the compound of interest is not included in either of these two tables, it is possible in most cases to use a group contribution method to obtain the value of its heat of formation. Values of the group contributions for the method of Domalski–Hearing are given in Table 2–343, pages 2–479 to 2–484, of Perry's handbook [1]. Other group methods are also available.

As a final remark, it is worth mentioning here that for reactions involving only organic compounds for which the heats of combustion are available, it is possible to skip the step of calculating the heat of formation of each compound in order to evaluate the standard heat of reaction ΔH^{θ}_{T0}. According to Equation 21.11, the standard heat of reaction or standard enthalpy change of reaction is evaluated as:

$$\Delta H^{\theta}_{T0} \equiv \sum_{k}^{c} v_k \Delta H^{0}_{f,k,T0} \tag{21.11}$$

We recall that in Equation 21.11, v_k is positive for products and negative for reactants. The standard heats of combustion refer to the breaking down of an organic compound into CO_2 and H_2O by the action of oxygen, while the standard heat of formation refers to the formation of the compound starting from the elements in their standard states. In addition, as the atoms of each element forming the organic molecules on both sides of the stoichiometric equation are balanced, the terms including the heat of formation of CO_2 and of H_2O cancel out. Thus,

$$\Delta H^{\theta}_{T0} \equiv -\sum_{k}^{c} v_k \Delta H^{0}_{c,k,T0} \tag{23.23}$$

The sum on the right-hand side of Equation 23.23 includes the values of the standard heat of combustion of all compounds participating in the reaction. Thus, if $H_{2(g)}$ or $C_{(s)}$, as such, participates in the reaction, their heats of combustion must also be included. In addition, the heats of combustion must all refer to the formation of either liquid water or water gas; in no case should the information be mixed.

Example 23.6: Calculation of the Heat of Formation

Calculate the value of the heat of formation of propylene by the following methods: (a) using heats of combustion with $H_2O_{(g)}$ as a product, (b) using heats of combustion with $H_2O_{(l)}$ as a product, and (c) using Domalski–Hearing's group method.

Solution

1. From Table 2–179, Perry's handbook [1], the reported heat of combustion of propylene is with $H_2O_{(g)}$ as a product. Thus,

$$\Delta H^{0}_{c,C_3H_{6(g)}} = -1926.2\,\text{kJ/mol}$$

$$\Delta H^{0}_{c,H_{2(g)}} = \Delta H_{f,H_2O_{(g)}} = -241.826\,\text{kJ/mol}$$

$$\Delta H_{c,C_{(s)}} = \Delta H_{f,CO_{2(g)}} = -393.514\,\text{kJ/mol}$$

Using Equation 23.22 with $n=3$, $m=6$,

$$\Delta H_{f,C_3H_6(g)} = (3)(-393.514)+\left(\frac{6}{2}\right)(-241.826)-(-1926.2)$$

$$= 20.18\,\text{kJ/mol}$$

The calculated value compares well to the one reported in Table 2–179 [1] of 20.23 kJ/mol.

2. From [3], we obtain a value of heat of combustion with $H_2O_{(l)}$ as a product:

$$\Delta H^0_{c,C_3H_6(g)} = -2057.8\,\text{kJ/mol}$$

$$\Delta H^0_{c,H_2(g)} = \Delta H_{f,H_2O(l)} = -285.84\,\text{kJ/mol}$$

$$\Delta H^0_{f,C_3H_6(g)} = (3)(-393.514)+\left(\frac{6}{2}\right)(-285.84)-(-2057.8)$$

$$= 19.74\,\text{kJ/mol}$$

3. Domalski and Hearing group contribution method [4].

To apply this method, it is necessary to understand its nomenclature. The groups are denoted by the atom other than hydrogen joined with a line to the other atoms to which it is joined in the molecule. For example, C–(3H)(C) means "atom of carbon joined to another atom of carbon and to three hydrogen atoms," that is, the group $-CH_3$. When an atom of carbon has a double bond to another atom of carbon, it is denoted by C_d and only the other two connections are specified. For example, C_d–(H)(C) means that the other two connections are carbon atoms (joined to others by single bonds) and an atom of hydrogen. On the other hand, C_d–(2H) means that a carbon atom is connected to a double bond and two hydrogen atoms. Table 23.6 reports the appropriate values

TABLE 23.6
Calculation of ΔH_f^0 Propene from the Domalski and Hearing Group Contribution Method

Groups	ΔH_f^0 (KJ/mol)
C_d–(2H)	26.32*
C_d–(H)(C)	36.32
C–(3H)(C)	–42.26
$CH_2=CH-CH_3$	20.38

Note: Observe (and correct) error of the sign in Perry's Table 2-343 [1]. The correct sign can be found in the original source [4]. In the sixth edition of Perry's Table 3-335, for the group contribution method of Verma and Doraiswamy, there is an error of sign for the $-CH_3$ group.

from Table 2–343 of Perry's handbook [1]. For propylene, the structure is $CH_{2=CH}$–CH_3. The value obtained is 20.38 kJ/mol.

Example 23.7: Calculation of Standard Heats of Reaction

Obtain the values of the standard heat of reaction at 298.15 K for the following reactions:

1. (Propylene)+(Chlorine)=(Allylchloride)+(Hydrochloric acid)

$$CH_2 = CH - CH_3 + Cl_2 \rightarrow CH_2 = CH - CH_2Cl + HCl$$

2. (Propylene)+(Chlorine)=1,2-Dichloropropane

$$CH_2 = CH - CH_3 + Cl_2 \rightarrow CH_2Cl - CHCl - CH_3$$

Solution

In order to use Equation 21.11 or 23.22, we need the values of the heats of formation or combustion for allyl chloride and 1,2-dichloropropane, and these values are not available in Table 2–179 of Perry's handbook [1].

In order to use the Domalski–Hearing group contribution method, we need the values of the contributions of all groups, and in Table 2–343 [1], the contribution for the group Cl is not included. Benson's method, as reported in Table 3.4 in [5], contains enough information to estimate the standard heat of formation for 1,2-dichloropropane and allyl chloride (see Tables 23.7 and 23.8, respectively) in the gaseous state:

$$\Delta H^0_{f,gas}(1,2-\text{dichloropropane}) = -73.21\,\text{kJ/mol}$$

TABLE 23.7
Calculation of ΔH_f^0 of 1,2-Dichloropropane (CH_2Cl–CHCl–CH_3) from the Benson Group Contribution Method

Groups	ΔH_f^0 (KJ/mol)
CH_2Cl–C	–69.07
CHCl–(2C)	–61.95
CH_3–(C)	–42.19
CH_2Cl–CHCl–CH_3	–173.21

TABLE 23.8
Calculation of ΔH_f^0 of Allyl Chloride (CH_2Cl–CH =CH2) from the Benson Group Contribution Method

Groups	ΔH_f^0 (KJ/mol)
CH_2Cl–(=C)	–68.65
=CH–(C)	35.96
=CH2	26.20
CH_2Cl–CHCl–CH_3	–6.49

This is a first approximation to the result because at 25°C, 1,2-dichloropropane is in a liquid state. Benson's method, on the other hand, provides estimates for ideal-gas properties. Therefore, it is necessary to take into account the heat of vaporization ΔH_{vap} [6]. Thus,

$$\Delta H^0_{f,liq.}(1,2-\text{dichloropropane}) = -173.21 - 35.26 = -208.47\,\text{kJ/mol}$$

$$\Delta H^0_{f,gas}(\text{allyl chloride}) = -6.49\ \text{kJ/mol}$$

As above, at 25°C allyl chloride is also a liquid. Thus, an analogous correction has to be made:

$$\Delta H^0_f(\text{allyl chloride}) = -6.49 - 29.04 = -35.53\ \text{kJ/mol}$$

From Table 2-178 of Perry's handbook, for HCl,

$$\Delta H^0_{f,HCl} = -22.063\,\text{kcal/mol} = -92.31\ \text{kJ/mol}$$

and for propylene, from Example 23.6,

$$\Delta H_{f,C_3H_6(g)} = 20.23\ \text{kJ/mol}$$

For reaction 1, from Equation 21.11, in the gaseous state approximation, we get:

$$\Delta H^\theta_{T0,gas} = (-92.31) - 6.49 - 20.23 = -119.37\ \text{kJ/mol}$$

and

$$\Delta H^\theta_{T0,liq.} = (-92.31) - 35.53 - 20.23 = -148.07\ \text{kJ/mol}$$

taking into account the liquid state.

For reaction 2, correspondingly we obtain the following results:

$$\Delta H^\theta_{T0,gas} = (-173.21) - 20.23 = -193.44\ \text{kJ/mol}$$

$$\Delta H^\theta_{T0,liq.} = (-208.47) - 20.23 = -228.70\ \text{kJ/mol}$$

Example 23.8: Heat Effects in Reacting Mixtures

A gas mixture containing 80 mol% propane and 20 mol% butane is burned with 50% excess air. Assuming that the combustion is complete, calculate the heat per kilogram of gas that is transferred to a boiler if the initial gas mixture and the air enter the burner at 20°C and the products of combustion enter the stack at 180°C. The boiler heating chamber is at atmospheric pressure.

Solution

Material Balance

The line of reasoning followed below is quite standard in combustion problems, and for this reason, it is explained in detail. In this case, we take as a basis of calculation 1 kg of gas entering the burner. The molecular weights of propane and butane are, respectively, 44.09 and 58.12. One mole of mixture weighs [(0.8) (44.09)+(0.2)(58.12)]=46.90 g, and thus in 1000g (l kg) we have 21.32 moles of mixture, that is, 17.06 moles of propane and 4.26 moles of butane. The combustion reactions are:

$$C_3H_8 + 5O_2 = 3CO_{2(g)} + 4H_2O_{(g)}$$

$$C_4H_{10} + \frac{13}{2}O_2 = 4CO_{2(g)} + 5H_2O_{(g)}$$

The complete burning of 17.06 moles of propane consumes 85.30 moles of O_2 and produces 51.18 moles of CO_2 and 68.24 moles of H_2O. Similarly, for 4.26 moles of butane, the respective amounts are 27.72, 17.06, and 21.32 moles. Thus, the total production of CO_2 and H_2O is 68.24 and 89.56 moles, respectively. The stoichiometric requirement of O_2 is 113.02 moles. Using 50% excess air means feeding 169.52 moles of O_2, from which 113.02 react and 56.51 pass the burner unreacted. The amount of nitrogen fed to the system with the air is (0.79/0.21) (169.52)=637.76 moles. These results are presented in Table 23.9.

Energy Balance

We choose to carry out the reaction at T_0=298.15 K and follow the four steps given by Equations 23.16–23.20. For steps 1, 3, and 4, we need average values of the specific heats of the compounds. Hence, with the help of information from Section C, page A-35, of reference [5], Table 23.10 can be prepared. On the right-hand side of Table 23.10, we have included the mean values of $c_P{}^*$ at an average temperature for steps 1, 3, and 4.

TABLE 23.9
Data for Example 23.8

(mol/kg Fuel Gas)

Compound	Feed Stream	Consumed	Produced	Unreacted	Product Stream
C_3H_8(g)	17.06	17.06	0	0	0
C_4H_{10}(g)	4.26	4.26	0	0	0
O_2(g)	169.52	113.02	0	56.51	56.51
N_2(g)	637.76	0	0	637.76	637.76
CO_2(g)	0	0	68.24	0	68.24
H_2O(g)	0	0	89.56	0	89.56

TABLE 23.10
Data Used in the Energy Balance of Example 23.8

$c_P^*/R = a + bT + cT^2 + dT^3 + eT^4, T\,(K)$

						(cal/mol)		
Compound	a	$b \cdot 10^3$	$c \cdot 10^5$	$d \cdot 10^8$	$e \cdot 10^{11}$	$\left(\bar{c}_P^*\right)_1$ 20°C–25°C	$\left(\bar{c}_P^*\right)_3$ 25°C–180°C	$\left(\bar{c}_P^*\right)_4$ 20°C–80°C
C_3H_8	3.847	5.131	6.011	−7.893	3.079	17.50		
C_4H_{10}	5.547	5.536	8.057	−10.571	4.134	23.47		
O_2	3.63	−1.794	0.658	−0.601	0.179	6.96		
N_2	3.539	−0.261	0.007	0.157	−0.099			7.17
CO_2	3.259	1.356	1.502	−2.374	1.056		9.58	7.00
H_2O	4.395	−4.186	1.405	−1.564	0.632		8.17	

Step 1: Preheating Reactants from 20°C to 25°C

$$\Delta H_1 = [(17.06)(17.50)+(4.26)(23.33)+(112.99)(7.01)](298.15-293.15) = +6.0\,\text{kcal/kg fuel}$$

Step 2: Combustion
From Table 2–179 of Perry's handbook [1], with $H_2O_{(g)}$ as a product,

$$\Delta H^0_{c,C_3H_8} = -2.04311\cdot 10^9\,\text{J / kmol} = -488.315(\text{kcal/molC}_3\text{H}_8)$$

$$\Delta H^0_{c,C_4H_{10}} = -2.65732\cdot 10^9\,\text{J / kmol} = -635.115(\text{kcal/molC}_3\text{H}_8)$$

$$\Delta H_2 = (17.06)(-488.315)+(4.26)(-635.115) = -11{,}039\ \text{kcal/kg fuel}$$

Step 3: Heating Products from 25°C to 180°C

$$\Delta H_3 = [(68.24)(9.48)+(89.56)(8.19)](453.15-298.15) = +214.0\ \text{kcal/kg fuel}$$

Step 4: Heating Nonreacting Gases from 20°C to 180°C

$$\Delta H_4 = [(56.51)(7.17)+(637.73)(7.00)](453-293) = +778.9\ \text{kcal/kg fuel}$$

Thus, the total energy produced per kilogram of fuel, using Equation 23.20, is:

$$Q = 11{,}039+6+214+779 = 10{,}254\ \text{kcal/kg fuel}$$

Example 23.9: Production of Allyl Chloride in an Adiabatic Continuous Stirred Tank Reactor

For the production of allyl chloride from propylene and chlorine, J. M. Smith [7] has shown that the best yields are obtained with an adiabatic continuous stirred tank reactor. In a given run of a pilot plant reactor with a feed of 0.170(lb-mol/h) of Cl_2 and 0.680 (lb-mol/h) of propylene, 70% of the Cl_2 feed is converted to allyl chloride and 14% is converted to the undesired by-product 1,2-dichloropropane. The small conversion to other by-products can be ignored. The reactants are initially at 20°C, and after being preheated separately, they enter the adiabatic reactor at 200°C. Calculate (1) the total heat effect of the process, in (kcal/h) and (2) the temperature of the gas products leaving the reactor.

Solution

Material Balance

The reactions are those given in Example 23.7. In 1 h of operation, 0.170 lb-moles of Cl_2 and 0.680 lb-moles of propylene, C_3H_8, are fed to the reactor. Of the chlorine feed, (0.70)(0.170)=0.119 lb-moles react to produce 0.119 lb-mol of allyl chloride and 0.119 lb-mol of hydrochloric acid, and (0.14)(0.170)=0.024 lb-mol react to produce 0.024 lb-mol of 1,2-dichloropropane. The unreacted chlorine, 0.027 lb-mol, leaves with the product stream. For each lb-mol of chlorine reacted, in either reaction, 1 lb-mol of propylene is consumed. Thus, (0.119+0.024)=0.143 lb-moles of propylene react and (0.680 – 0.143)=0.537 lb-moles go to the product stream unreacted. These results are summarized in Table 23.11.

TABLE 23.11
Data for Material Balance for Example 23.9

Compound	Feed Stream	Consumed	Produced	Inerts and Unreacted	Product Stream
			F (lb-mol/h)		
C_3H_6	0.680	0.143	0	0.537	0.537
Cl_2	0.170	0.143	0	0.027	0.027
C_3H_5Cl	0	0	0.119	0	0.199
HCl	0	0	0.119	0	0.199
$C_3H_6Cl_2$	0	0	0.024	0	0.024

TABLE 23.12
Data Used in the Energy Balance of Example 23.9

	$c_P^*/R = a + bT + cT^2 + dT^3 + eT^4$ [T(K)]					(c_P^*) (cal/mol)		
Compound	***a***	***b*·10[3]**	***c*·10[5]**	***d*·10[8]**	**e·10[11]**	**20°C**	**25°C**	**200°C**
C_3H_6	3.834	3.893	4.688	–6.0130	2.283	15.217	15.398	21.752
Cl_2	3.056	5.3708	–0.8098	0.5693	–0.15256	8.081	8.100	8.567
C_3H_5Cl[a]	4.365	9.895	5.366	–7.708	3.12	20.201	20.446	28.732
HCl	3.827	–2.936	0.879	–1.031	0.439	6.944	6.944	7.022
$C_3H_6Cl_2$	1.697	40.582	–2.247	–0.038	0.377	23.212	23.486	31.828

Note: The [a]indicates approximated as 1-chloropropane.

Energy Balance

As most of the process happens at fairly high temperatures, we start with the approximation of the gaseous state.

In the preheating stage, the reactants are taken from 293.2 K to 473.2 K. In the adiabatic reactor, we carry out the reaction at 298.2 K, and then we take all products of the reaction from T_0=298.2 K to T_p (unknown), and also the unreacted reactants from 473 K to T_p. Thus, for the preheating stage, we need c_p for C_3H_6 and Cl_2 for the range of 293–473 K. For the same compounds, we need c_p for the range of 298.2–473 K and for the range 473 K to T_p. For the products of reaction, we need c_p for the range 298.2 K to T_p. From Example 23.7, we know that both reactions are exothermic, so the temperature T_p is higher than 473 K. We can guess the temperature and use approximated values of c_p. Subsequently, using any optimizing technique, Solver in Excel, for example, we calculate the required value of T_p. Say that we assume T_p=800 K. From Section C, page A-35 of [5], we get the information needed to prepare Table 23.12.

Preliminary Step: Preheating Reactants from 20°C to 200°C

Converting the pound-moles to moles, 453.6 (mol/lb-mol), and using the average values of $\left(\bar{C}_P^*\right)$,

$$\Delta H_0 = \Delta H_{preheating} = \left[(0.680)(18.48)+(0.170)(8.32)\right](473-293)(0.4536) = 1141(\text{kcal/h})$$

Since the reactor is adiabatic, the total heat effect of the process is Q=1,142 (kcal/h); this heat enters the system (Q>0).

Step 1: Cooling the Amounts of Reactants That Reacted from 200°C to 25°C

From Equations 23.17 through 23.19,

$$\Delta H_1 = \Delta H_{cooling} = \left[(0.143)(18.58)+(0.143)(8.33)\right](298-473)(0.4536) = -373.2\ (\text{kcal/h})$$

Step 2: Reactions a and b at 25°C
For ΔH_2, we use the calculated heats of reactions from Example 23.7, noting that the amounts in the table of material balances are in pound-moles.

$$\Delta H_2 = \Delta H_{reaction} = \left[(0.119)(-35.39)+(0.024)(-54.66)\right](453.6) = -2500.4 \text{ (kcal/h)}$$

Step 3: Heating the Products from 25°C to T_p

$$\Delta H_3 = \left[(0.119)(30.0)+(0.119)(7.1)+(0.024)(32.74)\right](0.4536)(T_P - 298) = 2.36\ (T_P - 298) \text{ (kcal/h)}$$

Step 4: Heating Unreacted Reactants from 200°C to T_p

$$\Delta H_4 = \left[(0.537)(26.3)+(0.027)(8.7)\right](0.4536)(T_P - 473) = 6.50\ (T_P - 473) \text{ (kcal/h)}$$

For the final energy balance, since $Q=0$,

$$O = 1141.8 - 373.2 - 2034.8 + 2.36(T_P - 298) + 6.50(T_P - 473)$$

$$8.86\ T_P = 5044$$

$$T_P = 570 \text{ K}$$

Using $T_p=570$ K, the values of c_p that depend on T_P can be revaluated. They are presented in Table 23.13.
Thus,

$$\Delta H_3 = \left[(0.119)(26.6)+(0.119)(7.0)+(0.024)(29.5)\right](0.4536)(T_P - 298) = 2.135\ (T_P - 298) \text{ (kcal/h)}$$

$$\Delta H_4 = \left[(0.537)(23.4)+(0.027)(8.6)\right](0.4536)(T_P - 473) = 5.805\ (T_P - 473) \text{ (kcal/h)}$$

$$O = 1141.8 - 373.2 - 2034.8 + 2.135\ (T_P - 298) + 5.805\ (T_P - 473)$$

TABLE 23.13
Average Heat Capacity of Compounds from Example 23.9 as a Function of Temperature T_p and the Approximation Used for Allyl Chloride Data

	$(\bar{c}_P^*)$ (cal/mol)					
	T_p=570 K		T_p=661 K		T_p=678 K	
Compound	**Step 3**	**Step 4**	**Step 3**	**Step 4**	**Step 3**	**Step 4**
C_3H_6	—	23.35		24.67		24.89
Cl_2	—	8.63		8.68		8.68
C_3H_5Cl	26.58[a]	—	28.16[a]		24.29[b]	
HCl	7.01	—	7.04		7.04	
$C_3H_6Cl_2$	29.48	—	30.94		31.18	

[a] Approximated as 1-chloropropane.
[b] Benson's group method approximation.

$$7.94\ T_P = 4648.2$$

$$T_P = 585.4\ \text{K}$$

This process can be repeated. The convergence is slow, as the direct optimization method gives the value of T_p=661 K.

When Benson's approximation for the $\left(c_P^*\right)$ value of allyl chloride is used, the result is T_p=678 K. Table 23.13 compares different final values of $\left(c_P^*\right)$.

Releasing the first assumption of products being in the gaseous state changes the final result to 647 K. This difference can be understood when we realize that the heats of vaporization of products that have to be subtracted from their respective heats of formation to take into account their liquid state at 25°C in step 2 need to be added again in step 3 when the products are heated above their boiling points. Thus, the only difference between the approaches is heating of liquid products from 25°C to their boiling temperatures. All information required for these calculations can be found in [8,9].

REFERENCES

1. Green, D. W., and Perry, R. H. 2007. *Perry's Chemical Engineers' Handbook*, 8th ed. Blacklick, OH: McGraw-Hill Professional Publishing.
2. Douglas, T. B., and Dever, J. L. 1954. Thermal conductivity of molten materials. II. The heat capacity of anhydrous sodium hydroxide from 0°C to 700°C. *J. Res. Natl. Bur. Stand.* 53(2): Research Paper 2519.
3. Wiberg, K. B., and Fenoglio, R. A. 1968. Heats of formation of C4H6 hydrocarbons. *J. Am. Chem. Soc.* 90: 3395–3397.
4. Domalski, E. S., and Hearing, E. D. 1988. Estimation of the thermodynamic properties of hydrocarbons at 298.15 K. *J. Phys. Chem. Ref. Data* (4): 1637–1678.
5. Poling, B. E., Prausnitz, J. M., and O'Connell, J. P. 2001. *The Properties of Gases and Liquids*, 5th ed. New York: McGraw-Hill.
6. Varushchenko, R. M., Druzhinina, A. I., Kuramshina, G. M., and Dorofeeva, O. V. 2007. Thermodynamics of vaporization of some freons and halogenated ethanes and propanes. *Fluid Phase Equil.* 256(1–2): 112–122.
7. Smith, J. M., 1981. *Chemical Engineering Kinetics*, 3rd ed. New York: McGraw-Hill.
8. National Center for Biotechnology Information. PubChem compound database. CID=7850. Available at https://pubchem.ncbi.nlm.nih.gov/compound/7850 (accessed November 16, 2015).
9. National Center for Biotechnology Information. PubChem compound database. CID=6564. Available at https://pubchem.ncbi.nlm.nih.gov/compound/6564 (accessed November 16, 2015).
10. https://che.k-state.edu/docs/imported/SteamTable.pdf (accessed August 23, 2024)

† One may ask how it is possible to have one mole of solute at infinite dilution and how to measure the heat effect in this case. A plot of the heat of solution against the concentration of the salt shows a curve that starts from the point (0, 0). This curve is closely a straight line in the diluted region, and its slope is the value of the heat of solute per mole of solute at infinite dilution. The heat of solution should not be confused with the heat of dilution that is the heat effect when more solvent is added to an existing solution.

PROBLEMS

23.1 Construction of a pressure–enthalpy diagram

Use the PRSV equation of state (Appendix C) to draw a pressure–enthalpy diagram for ammonia, NH_3. Make it similar to Figure 23.1, that is, include the saturation envelope and one isotherm at T = 285.5 K. Obtain the PRSV parameters from Table C.3. The triple-point temperature is [1,2] T_{tp} = 195.49 K. The isobaric ideal-gas heat capacity is given by [1]

$$\frac{c_P^*}{R} = c_0 + \sum_{k=1}^{3} v_k \left(\frac{u_k}{T}\right)^2 \frac{\exp(u_k/T)}{\left[\exp(u_k/T)-1\right]^2}$$

with $c_0 = 4, v_1 = 2.224, v_2 = 3.148, v_3 = 0.9579,\ u_1 = 1646$ K, $u_2 = 3965$ K, $u_3 = 7231$ K, and T in K. Set the reference state as $h^{*0} = 28990.8025$ kJ/kmol at 300 K, to allow comparison with values in [2].

23.2 **Heat of solution of binary systems**

A frequent way of reporting heats of mixing of binary systems, especially in older references, is in the form of *integral heat of solution*, Δh^s, which is the enthalpy change upon mixing one (mass or mol) unit of one component (the "solute") and r units of the other (the "solvent").

a. Sketch the general shape of the Δh^s versus r curve for an exothermic mixing process. Explain the physical meaning of the slope of this curve and find its value at $r = 0$ and $r \to \infty$.
b. Table 23.14 gives integral heats of solution (kJ/kg NH_3) measured when 1 kg NH_3 is mixed with r kg H_2O at 0°C and 1 atm. Convert these values to excess enthalpies of mixing (kJ/kg of mixture) and plot them in an enthalpy–composition diagram versus mass fraction of NH3.
c. An 80% by weight aqueous NH_3 solution is to be brought to 50% by weight by diluting it with pure water. How much heat must be supplied or removed per kg of the final mixture if the process takes place at 0°C and 1 atm?

TABLE 23.14
Integral Heat of Solution of Aqueous Ammonia (Computed from Data in Ref. [3])

r (kg H_2O/kg NH_3)	Δh^s (kJ/kg NH_3)
0	0.00
0.111	−37.45
0.250	−79.97
0.428	−127.8
0.667	−179.0
1.00	−230.8
1.50	−284.9
2.33	−325.2
4.00	−346.6
9.00	−362.4
∞	−377.7

23.3 **Dilution of an aqueous NaOH solution**

A saturated aqueous solution of NaOH at 300 K is slowly added to a stirred storage tank that contains 1,000 kg of pure water, also at 300 K, until the concentration reaches 30% NaOH by weight. The process lasts one hour, over which time the temperature in the tank is kept at 300 K by means of a cooling system that extracts heat at a fixed rate, that is, a constant amount per unit time. Find the total heat removed, and the feed rate of the saturated solution as a function of time. Use the enthalpy–concentration diagram of Figure 23.7.

23.4 **Evaporation of a sugar solution**

10,000 kg/h of a 10% by weight aqueous solution of pure sucrose (molar mass 342.3 g/mol) are to be concentrated to 50% by evaporation. The dilute solution enters the process at 27°C, and the heating medium is saturated steam at 200 kPa. Calculate the steam rate in kg/h required if the process takes place in:

a. A single evaporator that operates at 1 atm.

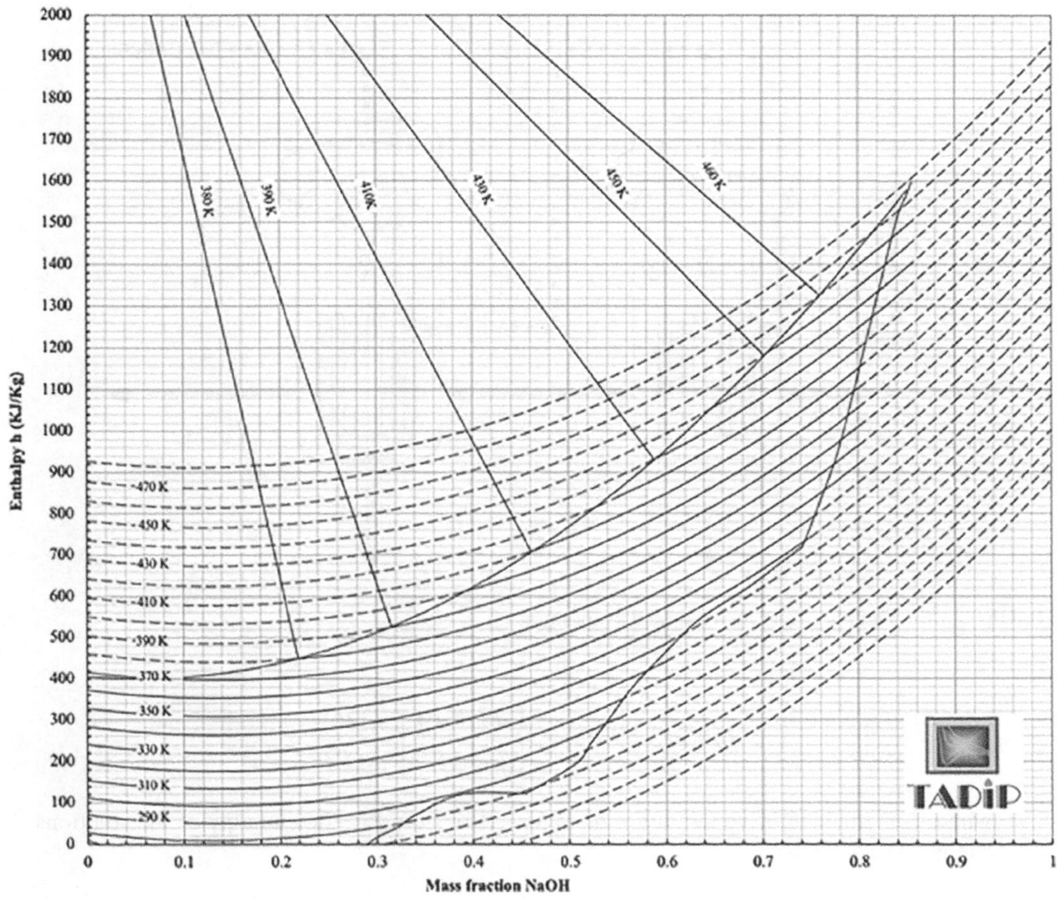

FIGURE 23.7 Enthalpy–concentration diagram in SI units for aqueous NaOH solutions. Reference states: Pure liquid water at 273.15 K; infinitely dilute NaOH at 298.15 K. Updated from the original version by Fernández et al. [4].

b. A single evaporator that operates at 3.17 kPa.
c. Two evaporator bodies ("effects") in a co-current flow pattern ("forward feed"): the dilute solution F enters the first effect, where it is evaporated at 1 atm by heating with the live steam S_0; the partially concentrated liquor L_1 then enters the second effect, which operates at 3.17 kPa and is heated by the water vapor S_1 evolved in the first effect (see Figure 23.8).

An online calculator of steam and water properties is available at checalc.com/calc/steam.html. The boiling point rise ("BPR") of a solution of solute mass fraction X is the difference between the boiling temperatures $T(P,X)$ of the solution and $T^S_{H_2O}(P)$ of pure water at the same pressure P. Values of the BPR of sucrose solutions at 1 atm ($T^S_{H_2O} = 100°C$) found in [5] can be correlated by

$$BPR(P_0 = 1 \text{ atm}, X) = X\left(0.9141 + 2.485X + 5.620X^2\right)$$

The *BPR* at any other pressure is given by the thermodynamic formula known as Tishchenko's equation [6],

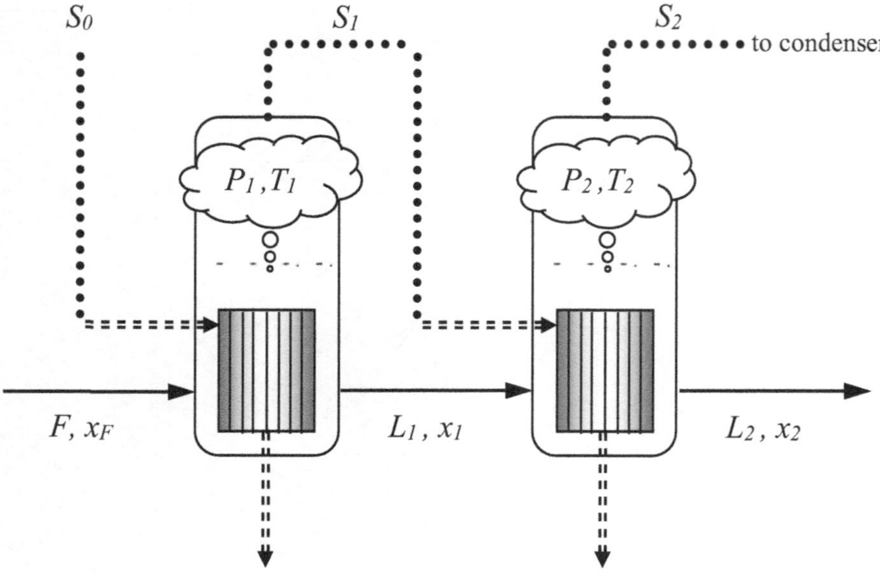

FIGURE 23.8 Double-effect forward feed evaporator.

$$BPR(P) = BPR(P_0)\left[\frac{T^S_{H_2O}(P)}{T^S_{H_2O}(P_0)}\right]^2 \frac{\lambda(P_0)}{\lambda(P)}$$

where λ is the heat of vaporization of water. Also, from [5], the enthalpy of the solutions can be expressed as:

$$h\ (\text{kJ/kg}) = (4.187 - 2.51X + 0.00375Xt)t$$

where t is the temperature in °C.

23.5 **Enthalpy balance in a nonadiabatic reaction**

A gas mixture composed of 9% SO_2, 10% O_2 and 81% N_2 at 450°C enters a reactor where 80% of the SO_2 is oxidized to SO_3 according to the reaction 2 $SO_{2(g)}+O_{2(g)} \rightarrow 2$ $SO_{3(g)}$. The reaction products exit at 625°C. The process takes place at 0.1 MPa. Estimate the heat losses from the reactor.

Enthalpies of formation $\Delta h^0_{f,298}$ [kJ/kmol] at 298.15 K and 1 bar: SO_2=–296,831, SO_3=–394,952

	Mean isobaric heat capacities [kJ/(kmol·K)]			
Temperature range, K	**SO_2**	**SO_3**	**N_2**	**O_2**
298.15–723.15	48.98	70.58	29.77	31.22
298.15–898.15	50.23	73.30	30.26	31.96

REFERENCES FOR PROBLEMS

1. Gao, K., Wu, J., Bell, I. H., Harvey, A. H., and Lemmon, E. W. 2023. A reference equation of state with an associating term for the thermodynamic properties of ammonia. *J. Phys. Chem. Ref. Data* 52: 013102.

2. Lemmon, E. W., Bell, I. H., Huber, M. L., and McLinden, M. O., Thermophysical properties of fluid systems. In *NIST Chemistry WebBook, NIST Standard Reference Database Number 69*, eds. P. J. Linstrom and W. G. Mallard. Gaithersburg, MD: National Institute of Standards and Technology. Available at https://doi.org/10.18434/T4D303 (accessed November 10, 2023).
3. Balzhiser, R. E., Samuels, M. R., and Eliassen, J. D., 1972. *Chemical Engineering Thermodynamics*. Englewood Cliffs, NJ: Prentice-Hall.
4. Fernández, C. V., Marcano, J. R., and Olivera-Fuentes, C. G. 1998. Enthalpy-concentration diagram in SI units for the sodium hydroxide – water system (in Spanish). *Proceedings, Tenth Scientific and Technical Engineering Meeting*, vol. II, QUI-20–QUI-27: Maracaibo, Venezuela. Available at https://www.researchgate.net/publication/259090205.
5. Asadi, M. 2007. *Beet-Sugar Handbook*. Hoboken, NJ: Wiley-Interscience.
6. Bates, F. J. 1942. *Polarimetry, Saccharimetry, and the Sugars*, NBS Circular 440. Washington, DC: Government Printing Office.

24 Adsorption of Gases on Solids

Abstract

This chapter presents a clear departure from the idea of a "simple system", that is, a system that is homogeneous and isotropic, introduced in Chapter 1. In gas adsorbed in a solid, the density of the gas close to the solid surface is higher than the density of the gas far from the surface. The chapter starts with a presentation of classical models of physical and chemical adsorption isotherms. Gibbs adsorption treatment, presented here, considers the difference in properties between the adsorbed system and an ideal nonadsorbed system. The adsorption of gas mixtures and thermal effects in adsorption are presented.

INTRODUCTION

There are many ways of studying adsorption [1–6]. In this chapter, we first give a molecular view of the mechanisms by which a compound in a gas phase gets adsorbed on a solid and then give a detailed method to handle adsorption data for practical purposes.

There are two distinct mechanisms by which gases adsorb on solid surfaces, physisorption and chemisorption. Physisorption is exothermic and the forces of attraction involving interactions between dipoles and induced dipoles are physical in nature, similar to the van der Waals forces between molecules in the fluid phase. As activation energy is required for a chemical reaction to happen, chemisorption is endothermic and the binding forces, being chemical in nature, are strong. Thus, chemisorption is rarely reversible. In this chapter, we focus on physisorption and call it simply adsorption from here on.

A MOLECULAR VIEW OF ADSORPTION IN GAS–SOLID SYSTEMS

The basic piece of information needed for the study of adsorption is the amount of gas adsorbed per mass unit of the adsorbent as a function of pressure at constant temperature. Curves presenting this information are known as adsorption isotherms and in general it is not possible to foresee their shape without first collecting experimental data for the system of interest. As thermodynamics deals only with equilibrium processes, it can only consider cases in which the adsorption and desorption isotherms coincide. Generally, this is true for the case of solids with pores smaller than 2 nm in diameter. When the adsorption and desorption curves do not coincide, there is no equilibrium, and it is said that there is hysteresis. This happens when the diameter of the pores is large enough for condensation of the adsorbate to occur [1].

An IUPAC technical report [7] distinguishes the six basic types of physical adsorption of gases on solids. The report also presents five main types of capillary hysteresis loops. Numerous equations have been proposed for the numerical representation of adsorption isotherms. The three most frequently used are discussed below.

Freundlich Adsorption Isotherm

Freundlich adsorption isotherm [8] is an empirical relationship between the amount of an adsorbed gas on a solid surface, called loading, and the gas pressure:

$$\text{Loading of adsorbate onto the solid} = c\left[\frac{\text{mass, mole or volume of adsorbate}}{\text{mass, mole or volume of solid}}\right]$$

DOI: 10.1201/9781003431985-27

It is expressed as:

$$c = KP^{1/n} \quad or \quad \log c = \log K + \frac{1}{n}\log P \tag{24.1}$$

where

c – loading of adsorbate,
P – equilibrium pressure of the adsorbed gas,
K – distribution coefficient
n – correction factor
K and n are constants characteristic of the system studied. They can be obtained from the linear range of the plot of $\log c$ versus $\log P$. Freundlich equation works well at low pressure and fails for higher pressures.

Langmuir Adsorption Isotherm [9]

This model is based on the following simplifying assumptions:

1. The adsorbing surface is flat and homogenous. All active sites are identical.
2. An active site of an adsorbing surface can accept only one molecule of the adsorbed gas.
3. There is no interaction between the molecules of the adsorbed gas.

From the above, it follows that only a monolayer can be formed.

The kinetic derivation of the Langmuir equation considers two reversible processes: adsorption (*ads*) and desorption (*des*) taking place on a mass unit of an adsorbent. Assuming the θ_{ads} is fraction of occupied adsorption sites:

$$\theta_{ads} = \frac{N_{ads}(\text{number of occupied sites})}{N_{total}(\text{total number of sites})}$$

the fraction of sites free for adsorption is $(1-\theta_{ad})$. The rate of adsorption, v_{ads} of gas molecules is proportional to the fraction of unoccupied adsorption sites and the gas pressure:

$$v_{ads} = k_{ads}P(1-\theta_{ads})$$

For pressures when the nonideality of gas has to be taken into account, the fugacity of the gas should be used instead of the pressure P.

The rate of desorption, v_{des}, is proportional to the fraction of occupied sites:

$$v_{des} = k_{des}\theta_{ads}$$

At equilibrium $v_{ads} = v_{des}$, thus,

$$k_{ads}P(1-\theta_{ads}) = k_{des}\theta_{ads}$$

The equilibrium constant of the adsorption process is defined as:

$$K_{ads} = \frac{k_{ads}}{k_{des}}.$$

Thus,

$$K_{ads} = \frac{\theta_{ads}}{P\,(1-\theta_{ads})}.$$

Rearranging gives the Langmuir equation:

$$\theta_{ads} = \frac{K_{ads}P}{1+K_{ads}P} \tag{24.2}$$

Taking the reciprocal of Equation 24.2, we obtain:

$$\frac{P}{\theta_{ads}} = \frac{1}{K_{ads}} + P$$

Remembering that:

$$\theta_{ads} = \frac{N_{ads}}{N_{total}} = \frac{N_{ads}}{N_m}$$

$$\frac{P}{N_{ads}} = \frac{1}{N_m K_{ads}} + \frac{P}{N_m} \tag{24.3}$$

where N_{total} is the total number of sites available for adsorption, that is, the monolayer capacity, N_m, per unit of mass of the absorbent. Any loading of the adsorbed gas, c, like its mass adsorbed per g of adsorbent, its volume, or the surface it occupies, is directly proportional to N_{ads} and N_m. Thus,

$$\frac{P}{c} = \frac{1}{c_m K_{ads}} + \frac{P}{c_m} \tag{24.4}$$

If the Langmuir model applies, plotting $\frac{P}{c}$ versus the equilibrium pressure P (or vs. fugacity), should give a straight line with the slope $\frac{1}{c_m}$ and the intercept $\frac{1}{c_m K_{ads}}$.

The Langmuir adsorption model can also be applied to competitive and dissociative adsorption. In addition, many extensions and empirical modifications exist [10].

Brunauer–Emmett–Teller (BET) Theory [11]

Langmuir's theory of monolayer adsorption is valid only at low pressures and high temperatures. When the pressure is increased or the temperature is lowered, additional layers are formed. The BET theory is an extension of the Langmuir theory applied to each layer of a multilayer system. The number of layers is not limited. The molecules of gas interact with the solid adsorbent and with adjacent adsorbed layers. The heat of adsorption for the first layer is greater than for the subsequent layers. The BET equation has the form:

$$\theta = \frac{K_{BET}P}{\left(1-P/P_0\right)\left[P_0 + P(K_{BET}-1)\right]} \tag{24.5}$$

Where the surface coverage, θ, is defined as $\theta = c_{ads}/c_m$ with c_{ads} defined as amount of the adsorbed gas, and c_m is monolayer equivalent, that is, the amount of gas necessary to cover the surface with a single layer; K_{BET} is the BET constant; P equilibrium pressure and P_0 is the saturation pressure at the adsorption temperature. At an adsorption temperature above the critical, where there is no condensation and P_0 does not exist, one uses the equilibrium pressure P only.

If the amount of gas, x, is expressed in volume units, and $\varphi = \frac{P}{P_0}$, Equation 24.5 is transformed to give:

$$\frac{\varphi}{v[1-\varphi]} = \frac{K_{BET}-1}{v_m K_{BET}}\varphi + \frac{1}{v_m K_{BET}} \tag{24.6}$$

With

$$K_{BET} = \exp\left(\frac{E_1 - E_{2+}}{RT}\right)$$

E_1 is the heat of adsorption of the first layer

E_{2+} is the heat of adsorption of the second and higher layers which is equal in its absolute value to the heat of liquefaction

Plotting $\frac{\varphi}{v[1-\varphi]}$ on the y-axis versus φ on the x-axis gives a straight line ("the BET plot") with the slope$=\frac{K_{BET}-1}{v_m K_{BET}}$ and the intercept$=\frac{1}{v_m K_{BET}}$. From there the parameters of the BET equation can be determined:

$$v_m = \frac{1}{\text{slope} + \text{intercept}} \quad \text{and} \quad K_{BET} = \frac{\text{slope}}{\text{intercept}} + 1 \tag{24.7}$$

Equation 24.6 is usually linear for type II isotherms in the range of $0.05 < \varphi < 0.35$, which leads to discarding around 70% of data. In practice, it is difficult to choose the experimental conditions and the linear region of data which would produce parameters with some physical meaning. A procedure described in [7] helps to avoid this subjectivity.

Modified Virial Equation

Thermodynamic properties can be calculated from experimental data fitted with an analytical equation. Most supercritical isotherms can be reproduced well with the modified virial equation [12,13], which although it expresses P as a function of moles of adsorbed gas per unit of mass of adsorbing solid (n), according to [12,13] fits the data better than expressing the loading, n, as a function of pressure:

$$P(n) = \frac{n}{K}\left[\frac{m}{m-n}\right]\exp\left[\beta_1 n + \beta_2 n^2 + \beta_3 n^2 + \ldots\right] \quad (n < m) \tag{24.8}$$

where

P – external pressure in the gas phase
n – moles of adsorbed gas per unit of mass of adsorbing solid
m – saturation capacity in moles per unit of mass of adsorbing solid
$K = \left(\frac{dn}{dP}\right)_{P\to 0}$ is Henry's constant
β_i – virial coefficients; three virial coefficients are usually sufficient.

If the virial coefficients are zero Equation 24.8 reduces to the Langmuir Equation 24.4.

To use Equation 24.8, it is necessary to determine Henry's constant by plotting the $ln\left(\frac{P}{n}\right)$. The intercept gives the value of *1/K*. It is also possible to determine Henry's constant experimentally.

Interpolation of adsorption isotherms in terms of temperature is based on Equation 24.9:

$$\bar{h} = R\left[\frac{\partial \ln P}{\partial\left(\frac{1}{T}\right)}\right]_n \tag{24.9}$$

where $\bar{h}$ is isosteric differential heat of adsorption. As the physical adsorption process is exothermic, its value is negative. For high pressures P should be replaced by the fugacity of the gas.

Equation 24.9 can be approximated as:

$$\bar{h}(n) = D_0 + D_1 n + D_2 n^2 + D_3 n^3 + (\ldots)$$

The constants D_i are temperature-independent.

Using the integrated form of Equation 24.9, we get the final expression:

$$P(n,T) = \frac{n}{K}\left[\frac{m}{m-n}\right]\exp\left[\frac{\bar{h}(n)}{R}\left(\frac{1}{T}-\frac{1}{T^*}\right)\right]\exp\left[\beta_1 n + \beta_2 n^2 + \beta_3 n^3 + \ldots\right] \tag{24.10}$$

All constants refer to temperature T^*. Equation 24.10 shows that for a system of one gas and one adsorbing solid, combining one isotherm at temperature T^* with the knowledge of the differential heat of adsorption contains complete information of this system. All other thermodynamic functions can be calculated based on Equation 24.10 and thermodynamics relations presented in the next sections of this chapter.

A TWO-DIMENSIONAL THERMODYNAMIC APPROACH

As is usual in thermodynamics, the idea is to systematize the information that can be obtained from the study of the phenomena for pure compounds with the purpose of using it for interpolation and extrapolation, as well as for extension to the treatment of mixtures.

Here we follow Chao and Greenkorn [14] presentation of the Gibbs treatment which is a simple way of studying the adsorption of a gas phase on a solid surface. The solid surface could be that of a porous solid or that of a granular adsorbent material. The treatment is applicable, *mutatis mutandis*, to the adsorption of a solute from a liquid mixture.

At atmospheric pressure, or at pressures not higher than 5 atm, gases can be safely considered to behave as an ideal gas. For an ideal gas at pressure, P, and temperature, T, the molar density, d_g, is given by:

$$d_g = \frac{n_g}{V_g} = \frac{P}{RT}$$

Consider a vessel in which n_g moles of gas are charged together with a solid adsorbent material. The gas occupies a volume V_g, and the solid occupies a volume V_s. At equilibrium, the gas phase is at a pressure P and temperature T of the system. Because of adsorption, there is a gradual increase in the density of the gas in the proximity of the solid surface. Thus, it is not a simple matter to decide what is the region of adsorbed gas and what is the region of bulk nonadsorbed gas. The Gibbs treatment of adsorption solves this problem elegantly by defining an ideal nonadsorbing system, indicated by a superscript *id*. In the ideal nonadsorbing system, the same gas volume V_g, at the pressure P and temperature T of the actual system, has the same molar density d_g as that of the bulk gas, that is, the gas that is far from the adsorbing surface, in the actual system. Thus, the number of moles of gas, n_g^{id}, that would be present in the ideal nonadsorbing system is:

$$n_g^{id} = V_g d_g \tag{24.11}$$

As the actual adsorbed system has a higher gas density in the proximity of the solid surface, the number of moles of gas in the actual adsorbed system will be larger than the number of moles of gas in the ideal nonadsorbing system.

$$n_g > n_g^{id}$$

The moles of gas adsorbed, n_g^{ads}, are then given by:

$$n_g^{ads} = n_g - n_g^{id} \tag{24.12}$$

For its use in further derivations, this equation is better written as:

$$n_g^{id} = n_g - n_g^{ads} \tag{24.12a}$$

From this form of the equation, it is clear that the number of moles of gas in the bulk gas phase, far from the effects of the solid surface, is in fact equal to the number of moles of gas in the ideal nonadsorbing system:

$$n_g^{bulk} = n_g^{id} \tag{24.13}$$

In the adsorbed system there are three phases present: (1) the solid phase, (2) the adsorbed gas phase, and (3) the bulk gas phase. For each phase at internal equilibrium, the Gibbs–Duhem equation, Equation 6.9, applies:

$$\sum_i n_i d\mu_i = VdP - SdT \tag{24.14}$$

Adding these equations for the three phases present at equilibrium, we obtain:

$$n_s d\mu_s + n_g^{ads} d\mu_g^{ads} + n_g^{bulk} d\mu_g^{bulk} = \left(V_s + V_g\right)dP - \left(n_s s_s + n_g^{ads} s_g^{ads} + n_g^{bulk} s_g^{bulk}\right)dT$$

The left-hand side of this equation can be simplified considering that at equilibrium the chemical potential of the gas adsorbed is equal to the chemical potential of the bulk gas. Thus,

$$d\mu_g^{ads} = d\mu_g^{bulk} = d\mu_g$$

In addition, the total number of moles of gas is equal to the sum of the moles of gas adsorbed and gas in the bulk. On the right-hand side, we identify the number of moles of gas in the bulk phase with the number of moles in the ideal nonadsorbing system and recall that the entropy of the gas in the bulk phase is equal to the entropy of the gas at pressure P and temperature T of the system. Therefore, we rewrite this equation as:

$$n_s d\mu_s + n_g d\mu_g = \left(V_s + V_g\right)dP - \left(n_s s_s + n_g^{ads} s_g^{ads} + n_g^{id} s_g\right)dT$$

For the ideal nonadsorbing system, the Gibbs treatment considers that although the entropy of the solid does not change by the superficial load of adsorbate, the enthalpy of the solid surface with adsorbate is different from the enthalpy of the solid surface without adsorbate. This difference is evident from the thermal effects of adsorption. The chemical potential of the solid is the difference between its molar enthalpy and the product of the absolute temperature times its molar entropy. Hence, to differentiate the chemical potential of the nonadsorbing solid surface from the chemical potential of the solid with adsorbate existing in the real adsorption system, the former is marked with a superscript 0. Repeating the exercise of adding the Gibbs–Duhem equations for the solid and the gas phase present in the ideal nonadsorbing system, we write:

$$n_s d\mu_s^0 + n_g^{id} d\mu_g = \left(V_s + V_g\right) dP - \left(n_s s_s + n_g^{id} s_g\right) dT$$

Subtracting this equation for the ideal nonadsorbing system from the equation describing the actual adsorbing system gives:

$$n_s\left(d\mu_s - d\mu_s^0\right) + \left(n_g - n_g^{id}\right) d\mu_g = -n_g^{ads} s_g^{ads} dT$$

Or, using Equation 24.12 and dividing by the number of moles of gas adsorbed, we get:

$$\frac{n_s d\left(\mu_s - \mu_s^0\right)}{n_g^{ads}} + d\mu_g = -s_g^{ads} dT \qquad (24.15)$$

The Gibbs treatment introduces here three useful definitions. Considering that the total active area of the adsorbing solid is α, the number of moles of gas adsorbed per unit of surface area of the adsorbent solid is defined as the *surface excess concentration* by:

$$\Gamma \equiv \frac{n_g^{ads}}{\alpha}\left[\frac{\text{moles of adsorbate}}{\text{solid surface area}}\right] \qquad (24.16)$$

The reciprocal value of the surface excess concentration, that is, the area of solid surface per mole of adsorbate, is designated by:

$$\bar{\alpha} \equiv \frac{\alpha}{n_g^{ads}} = \frac{1}{\Gamma} \qquad (24.17)$$

Finally, and most importantly, the "spreading pressure," or the energy per unit of surface area of solid that tends to spread the adsorbate over the surface, is defined as:

$$\pi \equiv -\frac{n_s\left(\mu_s - \mu_s^0\right)}{\alpha} \qquad (24.18)$$

In differential form, this definition is written as:

$$-\alpha\, d\pi = n_s\, d\left(\mu_s - \mu_s^0\right)$$

Dividing this expression by the number of moles of gas adsorbed, using Equation 24.17, and replacing in Equation 24.15, we obtain:

$$-\bar{\alpha}\, d\pi + d\mu_g = -s_g^{ads} dT \qquad (24.15a)$$

So, remembering that at equilibrium the chemical potential of the bulk gas is equal to the chemical potential of the adsorbed gas, we rewrite Equation 24.15a as:

$$d\mu_g^{ads} = \bar{\alpha}\, d\pi - s_g^{ads} dT \tag{24.15b}$$

All the terms in Equation 24.15b correspond to terms of the adsorbed phase. From this equation, we obtain:

$$\left(\frac{\partial \mu_g^{ads}}{\partial T}\right)_\pi = -s_g^{ads} \tag{24.19}$$

and

$$\left(\frac{\partial \mu_g^{ads}}{\partial \pi}\right)_T = \bar{\alpha}\left[\frac{\text{solid surface area}}{\text{moles of adsorbate}}\right] \tag{24.20}$$

It is illustrative to compare Equations 24.15b, 24.19, and 24.20 with the corresponding equations for a pure compound in "volumetric" thermodynamics. In volumetric thermodynamics, from Equation 7.8, for a pure compound we have:

$$d\mu = \mathrm{v}dP - sdT$$

$$\left(\frac{\partial \mu}{\partial T}\right)_P = -s$$

and

$$\left(\frac{\partial \mu}{\partial P}\right)_T = \mathrm{v}\left[\frac{\text{volume of container}}{\text{moles of gas}}\right]$$

These analogies suggest the identification of π with a spreading pressure for the case of the adsorbed gas in the two-dimensional system.

It is important to note that the above treatment is also valid for the case of adsorption of a gas in the pores of a porous solid. In this case the volume of the gas is usually considered equal to the void volume of the solid pores.

GIBBS TREATMENT FOR THE ISOTHERMAL ADSORPTION OF A PURE COMPOUND GAS AT A SOLID SURFACE

For an isothermal process of adsorption, Equation 24.15a can be written as:

$$d\pi = \frac{1}{\alpha}\left(d\mu_g\right)_T = \Gamma\left(d\mu_g\right)_T$$

At constant temperature, for a change in the chemical potential of the gas in the bulk phase in terms of its fugacity, f_g, we write:

$$\left(d\mu_g\right)_T = RTd\ln f_{g,T}$$

so,

$$d\pi = RT\Gamma d \ln f_{g,T}$$

Integrating between $\pi = 0$ at $f_g = 0$, $\mu_s = \mu_s^0$, and the value π at $f_g = \phi_g P$, $\mu_s \neq \mu_s^0$, we obtain:

$$\pi = RT \int_0^{f_g} \Gamma d \ln f_{g,T} \tag{24.21}$$

At this point, it is necessary to introduce a variable of practical use in adsorption studies.

$$\text{Specific surface area of adsorbing solid} = a\left[\frac{\text{solid surface area}}{\text{mass, mole or volume of solid}}\right]$$

And remember that:

$$\text{Loading of adsorbate onto the solid} = c\left[\frac{\text{mass, mole or volume of adsorbate}}{\text{mass, mole or volume of solid}}\right]$$

Both variables, a and c, must be referred to the same units for the amount of solid. For example,

$$c\left[\frac{\text{std cm}^3 \text{of adsorbate}}{g \text{ of solid}}\right], a\left[\frac{\text{solid surface area}}{g \text{ of solid}}\right]$$

In this case,

$$\Gamma = \frac{c}{a\text{v}_0}\left[\frac{\text{moles of adsorbate}}{\text{solid surface area}}\right] \tag{24.22}$$

with the standard molar volume of gas at a pressure of 1 atm and at a temperature of 0°C,

$$\text{v}_0 = 22,400\left[\frac{\text{std.cm}^3}{\text{gmol of gas}}\right]$$

or alternatively, use

$$c'\left[\frac{g \text{ of adsorbate}}{g \text{ of solid}}\right], a'\left[\frac{\text{solid surface area}}{g \text{ of solid}}\right]$$

In this case,

$$\Gamma = \frac{c'}{a'M_w}\left[\frac{\text{moles of adsorbate}}{\text{solid surface area}}\right] \tag{24.23}$$

where M_w is the molecular weight of the gas.

Using Equation 24.22, we write Equation 24.21 as:

$$\pi = \frac{RT}{a\text{v}_0} \int_0^{f_g} c \, d \ln f_{g,T} \tag{24.21a}$$

For convenience in the calculations, it is useful to separate the constant terms from the variables in the adsorption process and define:

$$\Psi \equiv T\int_0^{f_g} c\, d\ln f_{g,T}\left[\frac{\text{std. cm}^3\text{of adsorbate}\cdot\text{Kelvins}}{g\text{ of solid}}\right] \tag{24.24}$$

The units of Ψ depend on the choice of units for c. For the case considered here, the relation between this new function and the spreading pressure is:

$$\pi = \frac{R}{a\nu_0}\Psi \tag{24.21b}$$

For the treatment of the adsorption of gas mixtures with the approach presented here, it is necessary to have plots of Ψ as a function of pressure for the different compounds forming the gas mixture. The first step for this purpose is to have constant temperature data, as a function of pressure, of the loading of the compounds adsorbing in the solid surface. Data of this type are either collected experimentally for different temperatures close to the temperature of interest [1–4] or obtained from computer simulations [5,6]. Figure 24.1 depicts a schematic plot of two *adsorption isotherms*, for adsorption of CO_2 on charcoal. At constant temperature, the loading increases with an increase in pressure. As the temperature increases, the load of adsorbate decreases at constant pressure. From these constant temperature data, smoothed using one of many model equations [1–4], it is possible to evaluate Ψ as a function of pressure. Thus, the graphical representation presented here can be replaced by a numerical approach.

The integration of the right-hand side of Equation 24.24 requires some discussion. The logarithm of the fugacity of the gas goes to minus infinity as the fugacity goes to zero at zero pressure. Thus, a transformation is necessary, and we write Equation 24.24 as:

$$\Psi \equiv T\int_0^{f_g} \frac{c}{f_g}\, df_{g,T} \tag{24.24a}$$

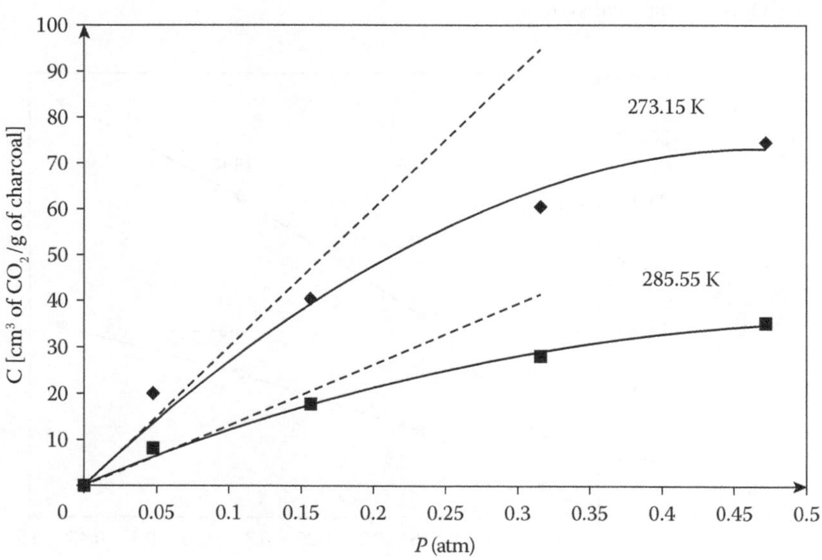

FIGURE 24.1 Example of adsorption isotherms of CO_2 in charcoal. Experimental data acquired in the Physical Chemistry Undergraduate Laboratory.

As the pressure goes to zero, both c and f_g go to zero, so the limiting value of the ratio can be obtained by the L'Hôpital rule:

$$\lim_{P\to 0}\frac{c}{f_g}=\left(\frac{dc}{dP}\right)_{P\to 0}$$

As shown in Figure 24.1, the value of the lower limit of the integrand is the slope at the origin of the adsorption isotherm. A graphical representation of the integrand at two temperatures is presented in Figure 24.2. The result of this type of integration, at two temperatures, is represented schematically in Figure 24.3. As the carbon dioxide gas used in these measurements behaves as an ideal gas at the conditions of the experiments, it was assumed that $\phi_g=1$, that is, $f_g=P$.

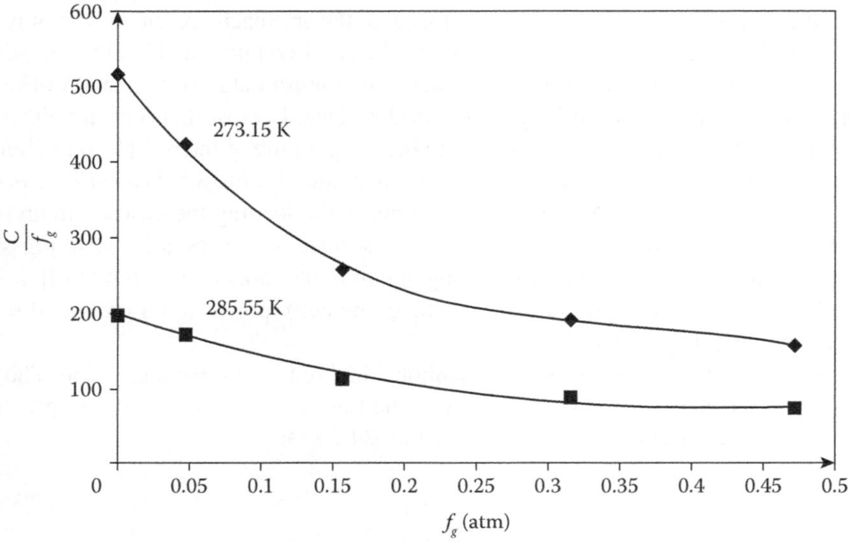

FIGURE 24.2 Plot of the ratio of loading of adsorbate gas over its fugacity as a function of fugacity and temperature in CO_2 in the charcoal system.

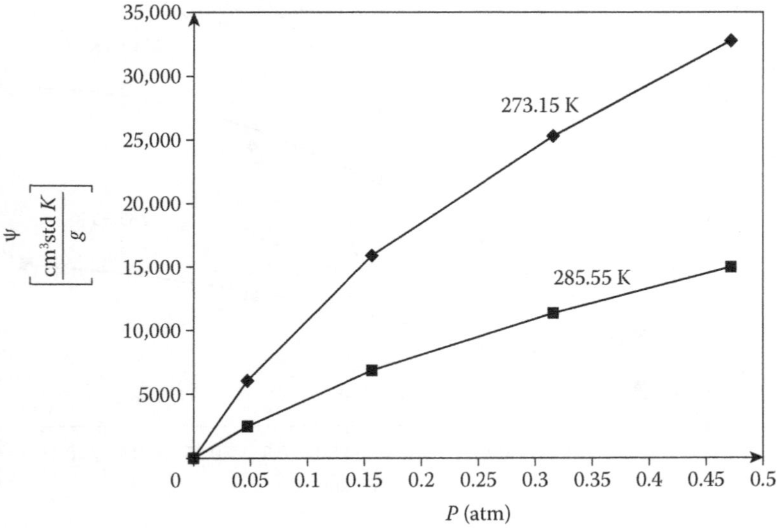

FIGURE 24.3 Pressure and temperature dependence of function ψ, Equation 24.24.

ISOTHERMAL ADSORPTION OF A GAS MIXTURE

For a total active area of the adsorbing solid α, the number of moles of gas i adsorbed per unit of surface area of adsorbent solid is:

$$\Gamma_i \equiv \frac{n_i^{ads}}{\alpha} = \frac{n_i - n_i^{id}}{\alpha}\left[\frac{\text{moles of } i \text{ adsorbate}}{\text{solid surface area}}\right] \tag{24.25}$$

The superscript *id* is used here for the ideal nonadsorbing case, and the subscript i refers to compound i. In the case of a mixture of gases with a mole fraction y_i of compound i in the bulk gas phase, we have:

$$n_i^{id} = V_g d_i = V_g y_i d_g$$

Here, we repeat exactly the treatment leading to Equation 24.15. The only difference is that in the Gibbs–Duhem equation, instead of the differential of the chemical potential of a single gas, we have now the sum of the chemical potential changes over all compounds. There is no difference in the term referring to the solid surface area. As in the discussion of the pure compound adsorption, we consider the isothermal case here and write:

$$n_s\left(d\mu_s - d\mu_s^0\right) + \sum_k \left(n_k - n_k^{id}\right) d\mu_k = 0\big|_T$$

or

$$n_s\left(d\mu_s - d\mu_s^0\right) + \sum_k n_k^{ads} d\mu_k = 0\big|_T$$

Dividing by the total number of moles of gas adsorbed,

$$n_g^{ads} = \sum_j n_j^{ads}$$

we obtain:

$$-\bar{\alpha}\, d\pi + \sum_k x_k d\mu_k = 0\big|_T \tag{24.26}$$

In Equation 24.26, we have used the same definition of terms given by Equations 24.16–24.18. In addition, we have defined the mole fraction of compound k in the adsorbed phase as:

$$x_k \equiv \frac{n_k^{ads}}{n_g^{ads}}$$

Equation 24.26 can be rewritten as:

$$\sum_k x_k \left(\frac{\partial \mu_k}{\partial \pi}\right)_{T,n} = \bar{\alpha}\big|_T \tag{24.26a}$$

This form suggests the definition:

$$\bar{\alpha}_k \equiv \left(\frac{\partial \mu_k}{\partial \pi}\right)_{T,n} \tag{24.27}$$

Thus, Equation 24.26a can be written as:

$$\sum_k x_k \overline{\alpha}_k = \overline{\alpha}\big|_T \tag{24.26b}$$

Comparison of Equations 24.26b and 24.27 with the following equations used in volumetric thermodynamics:

$$\overline{v}_k \equiv \left(\frac{\partial \mu_k}{\partial P}\right)_{T,n}$$

and

$$\sum_k x_k \overline{v}_k = v\big|_T$$

suggests the interpretation of $\overline{\alpha}_k$ as a partial molar surface area occupied by the adsorbed compound k at temperature T and spreading pressure π.

Finally, we observe that from the definitions given by Equations 24.16 and 24.25, for the mole fraction of compound i in the adsorbed phase, we get:

$$x_i = \frac{\Gamma_i}{\Gamma} \tag{24.28}$$

EXAMPLE OF THE GIBBS TREATMENT FOR THE ADSORPTION OF A GAS MIXTURE

We consider here the case of an ideal adsorption, that occurs without heat effects due to mixing of the adsorbed gases and having the property that the surface covered by the adsorbed mixture corresponds to the sum of the surface areas covered by the pure compounds at the same temperature and spreading pressure. This corresponds exactly with the analogy of forming an ideal mixture in volumetric thermodynamics, which happens without heat effects and without change in volume. In the case of forming an ideal mixture in volumetric thermodynamics, the partial molar volume of compound i in the mixture is equal to the pure compound molar volume at the same temperature and pressure. Thus, in surface thermodynamics, for ideal mixing in an adsorbed phase, we write:

$$\overline{\alpha}^{ads,im}_{i(T,\pi)} = \overline{\alpha}^{*}_{i(T,\pi)} \tag{24.29}$$

where the superscript indicates the surface area of solid covered by the pure compound i at the same temperature and spreading pressure of the mixture. The ideal mixture considered here is indicated by a superscript *im*, and it should not be confused with Gibbs's ideal case of a nonadsorbing system. Following the treatment of an ideal mixture in volumetric thermodynamics, for an ideal adsorbed phase we have:

$$f_i^{ads,im} = x_i f^{\theta}_{i(T,\pi)} \tag{24.30}$$

where, using the normal convention, the superscript θ indicates the standard state of pure adsorbed compound i at temperature T and spreading pressure π given by Equation 24.18. Experimental studies [12–14] have shown that the assumption of ideal mixture gives satisfactory results for a wide variety of systems. For the treatment of a nonideal adsorbed phase, for which experimental information may be available, the introduction of activity coefficients may be necessary. Assuming ideal mixing in the adsorbed phase, from Equation 24.26b, we write:

$$\overline{\alpha}^{im}_{(T,\pi)} = \sum_k x_k \overline{\alpha}_{k(T,\pi)} \tag{24.31}$$

Example

Assuming ideal adsorption of a gas mixture onto a solid surface, obtain the loading of each adsorbate at a temperature T and a pressure P when the composition of the gas mixture is given in terms of the mole fraction y_i of the components. Adsorption isotherms for pure compounds are available.

Solution
At equilibrium, assuming an ideal adsorbate phase,

$$f_i^{gas\ mixture} = f_i^{ads,im} = x_i f_{i(T,\pi)}^{\theta}$$

or

$$y_i \phi_i P = x_i f_{i(T,\pi)}^{\theta}$$

At low pressure, the fugacity coefficient ϕ_i can be set to unity. The standard state for the compound i in the adsorbed phase is conveniently set as equal to the pressure (or fugacity, if necessary) $P_{i(T,\pi)}^{*}$ required to give a spreading pressure π, that is, the corresponding value of ψ, when adsorbed as a pure compound at temperature T. At equilibrium, the spreading pressure is defined by Equation 24.18, and thus it is a property of the solid surface. This means it has the same value for all compounds adsorbed:

$$y_i P = x_i P_{i(T,\pi)}^{*}$$

or

$$x_i = \frac{y_i P}{P_{i(T,\pi)}^{*}} \tag{24.32}$$

Thus,

$$\sum \frac{y_i P}{P_{i(T,\pi)}^{*}} = 1 \tag{24.33}$$

Strategy for the Solution of the Problem
The pressure, P, the temperature, T, and the mole fractions of all compounds in the bulk gas phase, y_i, are known. Assume a value for ψ and from the isotherm corresponding to the temperature of the system (in a figure like Figure 24.3), evaluate the pressure P_i^* for each compound. Test the result obtained from Equation 24.33 and continue by trial and error until this equation is satisfied. If experimental data are smoothed by an analytical expression, the value of ψ can be obtained numerically by solving Equation 24.24a. Subsequently, using Equation 24.32, calculate the values of x_i for each compound in the adsorbed phase. At this point, the mole fractions of all adsorbates are known, but not their loadings c_i. From Equations 24.22 and 24.28, we write:

$$c_i = x_i c$$

The value of the total loading c is not known. From Equations 24.17 and 24.22, we write:

$$\bar{\alpha}_{(T,\pi)} = \frac{a v_0}{c_{(T,\pi)}} \tag{24.34a}$$

and for a pure compound,

$$\bar{\alpha}_{i(T,\pi)} = \frac{a v_0}{c_{i(T,\pi)}} \tag{24.34b}$$

For an ideal adsorption, according to Equation 24.26b,

$$\alpha^{im}_{(T,\pi)} = \sum_k x_k \bar{\alpha}^*_{k(T,\pi)}$$

Replacing Equations 24.34a and 24.34b in this expression, we obtain:

$$\frac{1}{c_{(T,\pi)}} = \sum \frac{x_k}{c^*_{k(T,\pi)}}$$

The values of $c^*_{i(T,\pi)}$ for all pure compounds are obtained from Figure 24.2. With these values using the above equation, the values of the loadings c_i are obtained:

$$c_i = x_i c$$

THERMAL EFFECTS OF ADSORPTION

Physical adsorption is exothermic. For a pure compound gas adsorbed onto a solid surface, at constant temperature and pressure,

$$\Delta h^{ads} = \left(h_g^{ads} - h_g\right) + \left(h_s^{ads} - h_s^0\right) \tag{24.35}$$

In Equation 24.35, all terms are expressed per mole of gas adsorbed.

For each phase, we know:

$$\mu = g = h - Ts$$

or

$$h = \mu + Ts$$

For the adsorbed gas phase, per mole of gas adsorbed,

$$h_g^{ads} = \mu_g^{ads} + Ts_g^{ads}$$

In an ideal nonadsorbed system,

$$h_g = \mu_g + Ts_g$$

The contribution of the adsorbed phase per mole of adsorbate is:

$$\left(h_g^{ads} - h_g\right) = T\left(s_g^{ads} - s_g\right) \tag{24.36}$$

For n_s moles of solid, remembering that adsorption does not change the entropy of the solid phase, we get:

$$H_s^{ads} = n_s\mu_s + TS_s$$

$$H_s^0 = n_s\mu_s^0 + TS_s$$

So, subtracting and dividing by n_g^{ads}, we obtain:

$$h_s^{ads} - h_s^0 = \frac{n_s\left(\mu_s - \mu_s^0\right)}{n_g^{ads}} = -\bar{\alpha}\,\pi \tag{24.37}$$

where

$$\bar{\alpha} = \frac{\alpha}{n_g^{ads}} = \frac{1}{\Gamma}$$

and

$$\pi = \frac{-n_s\left(\mu_s - \mu_s^0\right)}{\alpha}$$

Substituting Equations 24.36 and 24.37 into Equation 24.35, we obtain:

$$\Delta h^{ads} = T\left(s_g^{ads} - s_g\right) - \bar{\alpha}\pi$$

This equation is normally written as:

$$-\Delta h^{ads} = T\left(s_g - s_g^{ads}\right) + \bar{\alpha}\pi \tag{24.38}$$

Equation 24.38 can be transformed to a more practical form by eliminating the entropy term.
For the gas phase, we have:

$$d\mu_g = -s_g dT + \mathrm{v}_g dP$$

Taking the temperature derivative of both sides at constant spreading pressure π (or, equivalently, at constant Ψ) gives:

$$\left(\frac{\partial \mu_g}{\partial T}\right)_\pi = -s_g + \mathrm{v}_g\left(\frac{\partial P}{\partial T}\right)_\pi$$

But we know that:

$$\left(\frac{\partial \mu_g^{ads}}{\partial T}\right)_\pi = -s_g^{ads} \text{ and } \mu_g = \mu_g^{ads}$$

So,

$$\left(\frac{\partial \mu_g^{ads}}{\partial T}\right)_\pi = -s_g^{ads} = -s_g + \mathrm{v}_g\left(\frac{\partial P}{\partial T}\right)_\pi$$

Considering

$$\mathrm{v}_g = \frac{zRT}{P}$$

we write:

$$s_g - s_g^{ads} = \frac{zRT}{P}\left(\frac{\partial P}{\partial T}\right)_\pi = zRT\left(\frac{\partial \ln P}{\partial T}\right)_\pi$$

So, Equation 24.38 takes the form:

$$-\Delta h^{ads} = zRT^2\left(\frac{\partial \ln P}{\partial T}\right)_\pi + \bar{\alpha}\pi \tag{24.39}$$

This equation is sometimes further transformed using Equations 24.21b and 24.34a to obtain:

$$-\Delta h^{ads} = zRT^2\left(\frac{\partial \ln P}{\partial T}\right)_\pi + \frac{R\Psi}{c} \tag{24.40}$$

This important equation contains only measurable variables.

From Figure 24.3, we get the temperature–pressure dependence under constant spreading pressure, as shown in Figure 24.4a.

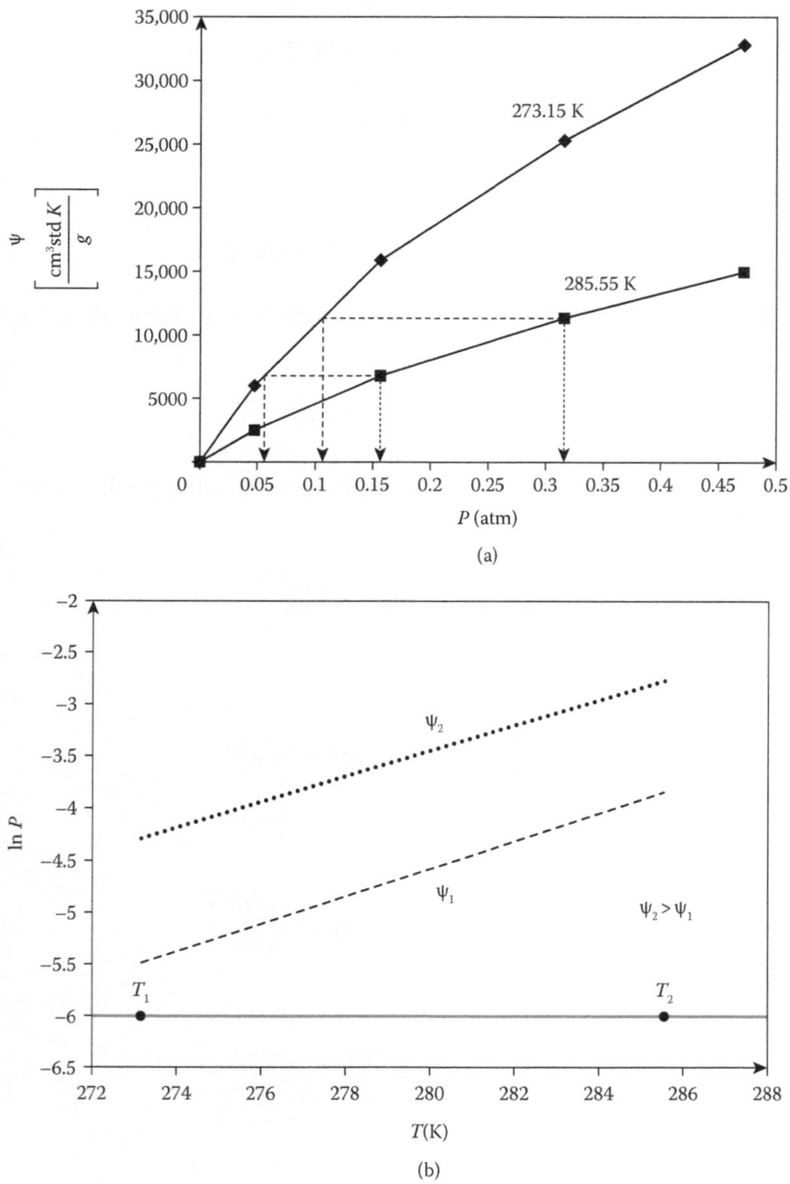

FIGURE 24.4 (a) Determination of P vs. T dependence at constant spreading pressure. (b) Determination of the enthalpy of adsorption.

This allows for the determination of the derivative of Equation 24.40. If only two isotherms are available, we evaluate $\left(\frac{\partial \ln P}{\partial T}\right)_\pi$ as shown in Figure 24.4b. From Figure 24.1, we get c. Thus, $-\Delta h_{ads}$ is evaluated.

If many isotherms are available, Equation 24.40 should be transformed to give:

$$-\Delta h^{ads} = zR\left(\frac{\partial \ln P}{\partial\left(\frac{1}{T}\right)}\right)_\pi + \frac{R\Psi}{c} \tag{24.40a}$$

and the slope of the plot of ln P versus T^{-1} should be used.

HEAT EFFECTS IN IDEAL ADSORPTION OF A MIXTURE OF GASES

We defined ideal adsorption as the case in which there is no change in the surface area of the solid covered by the effect of mixing. As in volumetric thermodynamics, the definition of ideal adsorption is completed by stating that there are no heat effects by mixing the adsorbed pure compounds at the same temperature T and spreading pressure π. Thus, for the ideal adsorption of a mixture of gases,

$$\left(-\Delta h^{ads}\right) = \sum x_i\left[h^*_{i(g)} - \left(h^E_i\right)^*_{(T,\pi)}\right] + \sum x_i\left[\bar{h}_{i(g)} - h^*_{i(g)}\right]$$

The first term on the right-hand side of the equation is the sum of the individual heats of adsorption of the pure compounds at T and π, which are evaluated as shown previously.

The second term is a minor correction due to variation of pressure from P for $\bar{h}_{i(g)}$ in the bulk gas phase to P_i^* for $h^*_{i(g)}$ in the adsorbed gas phase, both values at the temperature T of the system. For ideal gases, this second term is null, as the enthalpy of an ideal gas is only a function of temperature. For real gases, it can be obtained from the equation of state or generalized plots. In both terms, we observe that the mole fractions used refer to the adsorbed phase x_i and not to the gas phase mole fraction y_i.

CONCLUSIONS

There is no single adsorption isotherm that can handle all the complex phenomena of physical adsorption of pure or mixed gases on a porous surface. It is therefore recommended to consult the literature listed below for a more in-depth knowledge of these phenomena.

REFERENCES

1. Myers, A. L. 2004. Thermodynamics of adsorption. In *Chemical Thermodynamics for Industry*, ed. T. M. Letcher. Letchworth, UK: Royal Society for Chemistry, pp. 243–253.
2. Keller, J. U., and Staudt, R. 2005. *Gas Adsorption Equilibria: Experimental Methods and Adsorptive Isotherms*. Boston, MA: Springer.
3. Tien, C. 1994. *Adsorption Calculations and Modeling*. Boston, MA: Butterworth-Heinemann.
4. Ruthven, D. M. 1984. *Principles of Adsorption and Adsorption Processes*. New York: John Wiley & Sons.
5. Pikunic, J., Lastoskie, C. M., and Gubbins, K. E. 2002. Molecular modeling of adsorption from the gas phase. In *Handbook of Porous Solids*, eds. F. Schuth, K. S. W. Sing, and J. Weitkamp. Weinheim: Wiley-VCH, pp. 182–236.
6. Palmer, J. C., Moore, J. D., Roussel, T. J., Brennan, J. K., and Gubbins, K. E. 2011. Adsorptive behavior of CO2, CH4 and their mixtures in carbon nanospace. A molecular simulation study. *Phys. Chem. Chem. Phys.* 13(9): 3985–3996.

7. Thommes, M., Kaneko, K., Neimark, A. V., Olivier, J. P., Rodriguez-Reinoso, F., Rouquerol, J., and Sing, K. S. W. 2015. Physisorption of gases, with special reference to the evaluation of surface area and pore size distribution (IUPAC Technical Report). *Pure Appl. Chem.* 87(9–10): 1051–1069. https://doi.org/10.1515/pac-2014-1117.
8. Freundlich, H. 1909. Kapillarchemie, eine Darstellung der Chemie der Kolloide und verwandter Gebiete. *Akademische Verlagsgesellschaft.* viii, 591.
9. Langmuir, I. 1918. The adsorption of gases on plane surface of glass, mica and platinum. *J. Am. Chem. Soc.* 40(9): 1361–1402. https://doi.org/10.1021/ja02242a004.
10. Langmuir adsorption model – Wikipedia https://en.wikipedia.org/wiki/Langmuir_adsorption_model (accessed February 6, 2024).
11. Brunauer, S., Emmett, P. H., and Teller, E. (1938). Adsorption of gases in multimolecular layers. *J. Am. Chem. Soc.* 60(2): 309–319. Bibcode:1938JAChS..60..309B. https://doi.org/10.1021/ja01269a023. ISSN 0002-7863.
12. Siperstein, F. R., and Myers, A. L. 2001. Mixed-gas adsorption. *AIChE J.* 47: 1141–1159.
13. Myers, A. L., and Monson, P. A. 2002. Adsorption in porous materials at high pressure: Theory and experiment. *Langmuir* 18: 10261–10273.
14. Chao, K. C., and Greenkorn, R. A. 1975. *Thermodynamics of Fluids: An Introduction to Equilibrium Theory.* Series: Chemical Processing and Engineering, Volume 4. New York: M. Dekker.

PROBLEMS

The following experimental data of CO_2 adsorption on charcoal were acquired in an undergraduate physical chemistry laboratory at McGill University. As the goal of this chapter is to present basic concepts of the physical adsorption of gases on solids, the following problems simplify inherent intricacies of research quality data acquisition and the treatment. For readers interested in a more advanced approach, see [1–3].

24.1 **Freundlich Isotherm**

Using data from Table 24.1 plot the logarithmic version of Equation 24.1 and determine the characteristic constants K and n.

24.2 **Langmuir Isotherm**

Using data from Table 24.1 plot Equation 24.4 and determine the monolayer capacity, x_m, and the adsorption equilibrium constant, K_{ads}.

24.3 **Surface Area**

Using results from Problem 24.2, calculate the surface area of 1 g of the charcoal adsorbate. The density of CO_2 in the liquid state is 1101 kg/m^3 [2].

TABLE 24.1
Weights w_{co2} of adsorbed CO_2 in mg on degassed charcoal as a function of temperature and pressure

Pressure CO_2 (torr)	w_{co2} (mg) at given temperatures, (°C)			
	0°C	12.4°C	25°C	35°C
0	0	0	0	0
36	19.81	8.38	5.55	4.28
119	39.92	18.24	12.69	10.08
240	59.76	28.88	20.01	16.38
358.6	73.71	36.16	25.58	21.54
500	87.64	45.17	31.27	26.86
Weight of charcoal (g), $w_{charcoal}$	0.5030	0.5225	0.5089	0.4985

24.4 **Analytical Expressions for the function Ψ**

At pressures for which the fugacity of the gas can be replaced by the total pressure of the system, derive analytical expressions for the function Ψ defined by Equation 24.24 for an adsorption system obeying

A. Freundlich Isotherm
B. Langmuir Isotherm.

24.5 **Isosteric Heat of Adsorption**

Using Freundlich equation calculate the isosteric heat of adsorption of CO_2 on charcoal. The saturation pressure for this system is assumed to be 2.505 MPa [1].

REFERENCES FOR PROBLEMS

1. Cazorla-Amorós, D., Alcañiz-Monge, J., de la Casa-Lillo, M. A., and Linares-Solano, A. 1998. CO_2 as an adsorptive to characterize carbon molecular sieves and activated carbons. *Langmuir* 14: 4589–4596.
2. https://en.wikipedia.org/wiki/Liquid_carbon_dioxide , accessed February 14, 2024.

Section IV

Special Topics

If you do not make mistakes, you're not working on hard enough problems. And that's a big mistake.*

NOTE

* The unsolicited advice that Frank Wilczek, recipient of the Nobel Prize in Physics (2004), gives to his graduate students. From Wilczek, F., and Devine, B. 1987. *Longing for the Harmonies: Themes and Variations from Modern Physics*, W. W. Norton & Company, New York.

DOI: 10.1201/9781003431985-28

Section IV

Special Topics

25 Thermodynamics of Flow of Compressible Fluids

ABSTRACT

The chapter emphasizes the difference between the flow of incompressible fluids and compressible fluids. For the flow of compressible fluids, an important parameter is the speed of sound in the conditions of flow. At low speeds compared with the speed of sound and relatively short ducts, Equation 3.16 is applicable. With some simplifications, it is shown that the maximum entropy of the gas under the conditions of flow corresponds to the point when the velocity of the gas reaches the speed of sound. Beyond this point the flow becomes unstable.

For completeness, in this chapter, we present a brief introduction to the thermodynamics of flow of compressible fluids. When a gas flows at steady state in a horizontal adiabatic duct of constant cross-section A, the initial velocity at the entrance of the duct determines the kind of flow that the gas will develop. The initial velocity can be subsonic, sonic, or supersonic, depending on whether it is lower than, equal to, or higher than the velocity of propagation of sound in the gas at the prevailing temperature. From basic physics, the velocity of sound is given by:

$$\dot{\upsilon}_s = \sqrt{\frac{\gamma RT}{M}} \tag{25.1}$$

where γ is the ratio of the constant pressure heat capacity to the constant volume heat capacity of the gas, R is the universal gas constant, T is the absolute temperature, and M is the molecular weight of the gas. The Mach number, Ma, is defined as the ratio of the linear velocity of the gas in the duct to the velocity of sound:

$$Ma = \frac{\dot{\upsilon}}{\dot{\upsilon}_s}$$

At steady state, the mass flow rate $\dot{m}$ is constant, and it is related to the linear velocity $\dot{\upsilon}$ and the specific volume v of the gas by:

$$\dot{m} = \frac{\dot{\upsilon}A}{\mathrm{v}} \tag{25.2}$$

SUBSONIC, SONIC, AND SUPERSONIC FLOW

We observe here that Equation 25.2 uses the specific volume and not the molar volume. This is due to the need to combine the thermodynamic equations with fluid mechanics. Experiments show that if the Mach number at the inlet of the duct is smaller than unity, there is a pressure drop down the duct, the gas expands, and its linear velocity increases. If the duct is long enough, eventually the gas reaches sonic velocity and the velocity at the inlet cannot be maintained. If, on the contrary, the Mach number at the inlet is larger than unity, the pressure increases down the duct, the gas is compressed, and the linear velocity decreases as it flows. If the duct is long enough, eventually the sonic velocity is reached and a shock wave is produced. The longer the duct extends beyond the critical length, the

DOI: 10.1201/9781003431985-29

further back the shock wave moves until it eventually reaches the source feeding the supersonic gas. Thermodynamics helps us to understand this phenomenon, but to determine the value of the critical length, it is necessary to resort to fluid mechanics, which is beyond the scope of this chapter. For the case of subsonic flow at the entrance of a duct and ducts not excessively long, all equations of Chapter 3 are applicable to the flow of compressible fluids. The only difference between both cases is that for steady-state flow of an incompressible fluid in a constant cross-section duct, the linear velocity is constant while for a compressible fluid it increases. This is so because as the pressure drops the fluid expands and to keep the mass flow rate constant the linear velocity must increase.

THERMODYNAMIC UNDERSTANDING OF THE LIMITING SPEED

If the horizontal duct is well insulated, there is no heat transfer with the surroundings, but as there is friction in the duct, the irreversibility of the process increases the entropy of the gas as it flows. The fact that in both cases, of subsonic or supersonic entrance velocities, there is a critical length of the duct beyond which it is not possible to keep a stable flow suggests that at that length, the entropy has reached a maximum value. Any further increase in the length of the flow path would require a decrease in entropy that, thermodynamics tells us, is not allowed to happen. Hence, it is of interest to calculate the linear velocity that the gas would have at a maximum value of its entropy.

Assuming ideal gas behavior, according to Equations 5.26 and 5.28b, for a maximum value of the entropy, we need to have:

$$ds^* = \frac{c_P^* \, dT}{T} - \left(\frac{\partial \mathrm{v}^*}{\partial T}\right)_P dP = \frac{dh^*}{T} - \left(\frac{\partial \mathrm{v}^*}{\partial T}\right)_P dP = 0$$

or

$$dh^*_{s\ \max} = T\left(\frac{\partial \mathrm{v}^*}{\partial T}\right)_P dP$$

Using the ideal gas equation of state for the specific volume,

$$\mathrm{v}^* = \frac{RT}{MP} \tag{25.3}$$

we write:

$$dh^*_{s\ \max} = T\left(\frac{\partial \mathrm{v}^*}{\partial T}\right)_P dP = \mathrm{v}\ dP \tag{25.4}$$

For simplicity, from here we drop the asterisk with the understanding that we are working with an ideal gas.

An incremental enthalpy change is related to the velocity of the gas by the differential form of the energy balance, which, for the horizontal adiabatic duct from Equations 3.15 and 3.16b, takes the form:

$$dh + \frac{1}{2} d\dot{\upsilon}^2 = 0 \tag{25.5}$$

Hence, from Equation 25.2,

$$d\dot{\upsilon}^2 = 2\left(\frac{\dot{m}}{A}\right)^2 \mathrm{v}\ d\mathrm{v}$$

Thus, by replacing this expression in Equation 25.5 and rearranging, we get:

$$dh = -\left(\frac{\dot{m}}{A}\right)^2 \mathrm{v}\, d\mathrm{v} \tag{25.6}$$

Replacing this expression in Equation 25.4 and rearranging with the help of Equation 25.2, we write:

$$\left(\frac{dP}{d\mathrm{v}}\right)_{s\ \max} = -\left(\frac{\dot{m}}{A}\right)^2 = -\left(\frac{\dot{\upsilon}}{\mathrm{v}}\right)^2$$

or

$$\dot{\upsilon}_{s\ \max} = \sqrt{-\mathrm{v}^2\left(\frac{dP}{d\mathrm{v}}\right)_{s\ \max}} \tag{25.7}$$

Although, admittedly, this is a crude approximation for an irreversible process, for the adiabatic expansion of the gas, from Equation 5.34,

$$P\mathrm{v}^{\gamma} = C$$

Differentiating and rearranging, assuming γ to be constant,

$$-\mathrm{v}^2\left(\frac{dP}{d\mathrm{v}}\right) = \gamma\frac{C}{\mathrm{v}^{\gamma}}\mathrm{v} = \gamma P\mathrm{v} = \frac{\gamma RT}{M}$$

or

$$\dot{\upsilon}_{s\ \max} = \sqrt{\frac{\gamma RT}{M}} \tag{25.8}$$

Comparison of this result with Equation 25.1 confirms that the flow becomes unstable once the speed of sound is reached and that this speed corresponds to the maximum of the gas entropy. Any further change would cause a decrease in the entropy violating the second law.

Considering that when the gas enters the duct at subsonic velocity there is a pressure drop and the gas expands, its specific enthalpy decreases as the gas flows while its entropy increases.

When the gas enters at supersonic velocity, there is a compression of the gas and its specific enthalpy increases as the gas flows. A plot of enthalpy versus entropy must show that the curves for supersonic and subsonic flow for the same mass flow rate meet at the maximum of the entropy, that is, when the velocity of the gas is equal to the velocity of sound.

ENTHALPY AND ENTROPY CHANGES IN COMPRESSIBLE FLOW

In publications dealing with the flow of compressible fluids, the differential equations for the enthalpy and for the entropy are sometimes written in a form that is difficult to recognize. For the enthalpy change for an ideal gas, the simple form of Equation 5.28b,

$$dh = c_P dT$$

is transformed using the ideal gas relation for specific heats, Equation 5.29,

$$c_P - c_\mathrm{v} = \frac{R}{M}$$

and the definition of γ, Equation 5.32,

$$\gamma \equiv \frac{c_P}{c_v}$$

to write:

$$c_P = \frac{(c_P - c_v)c_P}{(c_P - c_v)} = \frac{(c_P - c_v)\frac{c_P}{c_v}}{\left(\frac{c_P}{c_v} - 1\right)} = \frac{R\gamma}{M(\gamma - 1)}$$

and

$$dh = \frac{R\gamma}{M(\gamma - 1)} dT \tag{25.9}$$

or using the ideal gas equation of state, Equation 5.27, we write:

$$dh = \frac{\gamma}{(\gamma - 1)} d(Pv) \tag{25.10}$$

Similarly, for the entropy differential,

$$ds = \frac{R\gamma}{M(\gamma - 1)} \frac{dT}{T} - \frac{R}{M} \frac{dP}{P} \tag{25.11}$$

With the help of the equation of state, it is possible to integrate Equations 25.10 and 25.11 between the conditions at the entrance of the duct and subsequent states, so the curves for h versus s can be calculated.

RELATION OF FLUID PROPERTIES TO MACH NUMBER AT ENTRANCE CONDITION

It is also useful to obtain a relation between the fluid properties down the duct and the properties at the entrance of the duct. Combining Equations 25.6 and 25.10 and rearranging, we get

$$d(Pv) + \left(\frac{\gamma - 1}{\gamma}\right)\left(\frac{\dot{m}}{A}\right)^2 v\, dv = 0$$

Integrating this expression,

$$Pv + \frac{1}{2}\left(\frac{\gamma - 1}{\gamma}\right)\left(\frac{\dot{m}}{A}\right)^2 v^2 = \text{constant}$$

Applying this relation for any two states 1 and 2, and rearranging,

$$\frac{P_2 v_2}{P_1 v_1} = \frac{T_2}{T_1} = 1 + \left(\frac{\gamma - 1}{\gamma}\right) \frac{\left(\frac{\dot{m}}{A}\right)^2 v_1^2}{2P_1 v_1} \left(1 - \frac{v_2^2}{v_1^2}\right)$$

This expression can be conveniently put in terms of the Mach number at the entrance of the duct:

$$Ma^2 = \frac{\dot{\upsilon}_1^2}{\gamma (R/M) T} = \frac{\left(\frac{\dot{m}}{A}\right)^2 v_1^2}{\gamma P_1 v_1}$$

Thus,

$$\frac{P_2 v_2}{P_1 v_1} = \frac{T_2}{T_1} = 1 + \left(\frac{\gamma - 1}{2}\right) Ma^2 \left[1 - \left(\frac{v_2}{v_1}\right)^2\right] \quad (25.12)$$

DETERMINATION OF CRITICAL LENGTH

The determination of the critical length of the duct where the sonic velocity is attained requires using the mechanical energy balance, which in its differential form for compressible fluids flowing at a steady state in a horizontal adiabatic duct can be written as:

$$v\, dP + \left(\frac{\dot{m}}{A}\right) v\, dv + \delta l_w = 0 \quad (25.13)$$

The determination of the lost work by friction belongs to the field of fluid mechanics, and it is beyond the scope of this chapter. More complete treatments of the flow of compressible fluids in horizontal insulated ducts presented in [1–4] or other texts can now be followed, after this introduction linking them to the equations discussed in the text.

REFERENCES

1. Cengel, Y. A., and Cimbala, J. M. 2014. *Fluid Mechanics: Fundamentals and Applications*, 3rd ed. New York: McGraw-Hill.
2. Oosthuizen, P. H., and Carscallen, W. E. 2013. *Introduction to Compressible Fluid Flow*, 2nd ed. Boca Raton, FL: CRC Press/Taylor and Francis.
3. Anderson Jr., J. D. 2003 [1982]. *Modern Compressible Flow*, 3rd ed. New York: McGraw-Hill.
4. Hougen, O. A., Watson, K. M., and Ragatz, R. A. 1959. *Chemical Process Principles. Part II. Thermodynamics*, 2nd ed. New York: John Wiley & Sons.

PROBLEMS

25.1 The Bernoulli equation

Consider the energy balance, Equation 3.7, in the case of steady but not necessarily incompressible flow of single stream. Derive the corresponding expression for the mechanical energy balance that applies instead of Equation 3.16a. In what special cases does this reduce to *Bernoulli's equation*, given below?

$$\frac{P_1}{\rho} + \frac{\dot{\upsilon}_1^2}{2} + gZ_1 = \frac{P_2}{\rho} + \frac{\dot{\upsilon}_2^2}{2} + gZ_2$$

25.2 Joule–Thomson expansion of a nonideal gas

Refrigerant R-12 (dichlorodifluoromethane, CCl_2F_2) at 700 kPa and 310 K flowing steadily in a horizontal pipe passes through a thermally insulated throttling valve from which it exits at 140 kPa. Find (a) the outlet temperature, and (b) the entropy change of the fluid. Use the PRSV equation of state with parameters from Table C.3. The ideal gas heat capacity of R-12 is given by the fourth-degree polynomial [1]

$$\frac{c_P}{R} = \sum_{k=0}^{4} a_k T^k$$

a_0	a_1	a_2	a_3	a_4
2.185	3.1251×10^{-2}	-3.724×10^{-5}	1.930×10^{-8}	-3.23×10^{-12}

25.3 The Pitot-static tube

The Pitot-static tube is an instrument that measures flow velocity by applying the Bernoulli equation (Problem 25.1). As illustrated very sketchily for pipe flow in Figure 25.1, it consists of an inner tube with a small opening that faces straight into the flow and measures the *stagnation* or *dynamic pressure* P_2, when the fluid is brought to a stop and all its kinetic energy is converted into pressure, and an outer tube that has openings or "ports" tangential to the flow direction and measures the *static pressure* P_1 of the fluid unperturbed by the instrument. The difference $P_2 - P_1$ is measured by connecting the outer ends of the tubes to a manometer or other type of differential pressure metering device. For a well-designed tube, frictional effects can be neglected [2]. Pitot tubes are the standard instruments that measure the speed of airplanes in flight.

A Pitot-static tube installed in an airplane flying at an altitude of 10,000 m records a pressure difference $P_2 - P_1 = 32.4$ kPa. What is the velocity, and the Mach number? According to the International Standard Atmosphere model, the dependence of air pressure and temperature on altitude Z in the troposphere (altitude less than 18,000 m) is given by [3]:

$$T = T_0 - LZ; \quad P = P_0\left(1 - \frac{LZ}{T_0}\right)^{\frac{gM}{RL}}$$

where $T_0 = 288.15$ K and $P_0 = 101{,}325$ Pa are reference values at sea level, $L = 0.0065$ K/m is the *temperature lapse rate*, $g = 9.80665$ m/s^2 is the gravitational acceleration,

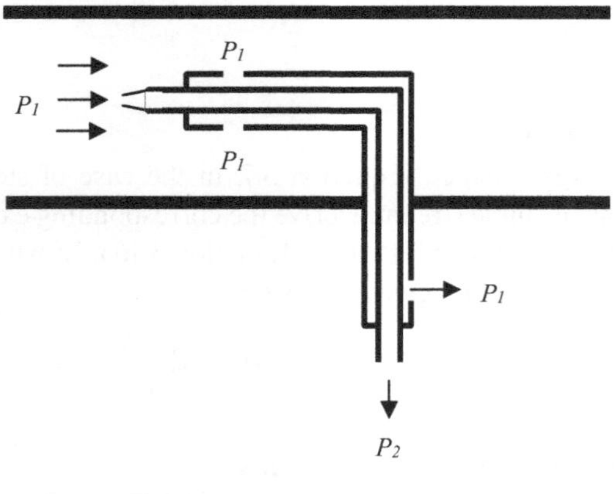

FIGURE 25.1 Sketch of a Pitot-static tube for fluid velocity measurement in a duct.

$M = 0.0289652$ is the molar mass of standard air, and $R = 8.31446$ J/(mol.K) is the gas constant. The specific heat ratio of air in the troposphere is fairly constant at $\gamma = 1.402$ [4].

25.4 **The orifice plate**

Another useful flow-measuring device in ducts is the orifice plate or orifice meter, which consists of a flat plate placed at right angles to the flow direction and blocking the entire cross-section except for a circular hole at the center. A sketch of the flow arrangement is shown in Figure 25.2.

a. Using the Bernoulli equation for frictionless, incompressible steady flow, derive an expression for the fluid velocity upstream of the orifice plate in terms of the pressures P_1 before the plate and P_2 immediately after the plate, if the cross areas are A_1 for the pipe and A_2 for the orifice.
b. Find the velocity of water flowing at 300 K in a pipe where the pressure drop measured by an orifice plate with $A_2/A_1 = 0.20$ is 20 kPa. Note that, as depicted in Figure 25.2, the downstream pressure is not measured as the fluid issues from the orifice, but at a nearby location down the pipe.

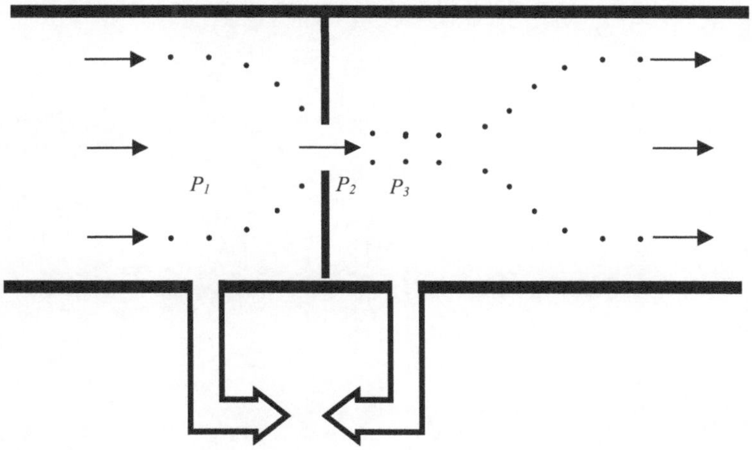

FIGURE 25.2 Sketch of the orifice plate for fluid velocity measurement in a duct

25.5 **The Fanno line**

As mentioned in this chapter, with the help of the equation of state of a gas it is possible to integrate the differential equations for the enthalpy and the entropy, and thus calculate the curve for s versus h along a duct. Plotted in enthalpy versus entropy coordinates, this is called the *Fanno line*.

a. Write the continuity and energy (not Bernoulli!) equations that apply to the steady, adiabatic flow of an ideal gas of constant heat capacity in a horizontal duct of constant area. From these and the properties of ideal gases, derive a relation between the entropy and the enthalpy of the gas as it flows downstream. (Hint: Use $T_0 = 0$ K as the reference temperature for the gas enthalpies).
b. Air at 700 kPa and 310 K, moving at 150 m/s, enters a thermally insulated horizontal duct of constant cross-section. Compute and plot the Fanno line. Assume ideal gas behavior with constant $\gamma = c_P/c_v = 1.40$.

REFERENCES FOR PROBLEMS

1. Poling, B. E., Prausnitz, J. M., and O'Connell, J. P. 2001. *The Properties of Gases and Liquids*, 5th ed. New York: McGraw-Hill.
2. de Nevers, N. 2005. *Fluid Mechanics for Chemical Engineers*, 3rd ed. New York: McGraw-Hill Education.
3. Wikipedia. *Density of Air.* Available at https://en.wikipedia.org/wiki/Density_of_air (accessed March 04, 2024).
4. National Bureau of Standards. 1955. *Tables of Thermal Properties of Gases*, NBS Circular 564. Washington, DC: U.S. Govt. Printing Office.

26 Elements of Statistical Thermodynamics

ABSTRACT

The canonical, or (N, V, T) ensemble, is a large number of systems having the same values of numbers of molecules N, volume V, and temperature T. The Gibbs ensemble has in common N, P, and T. The chemical potentials and the equation of state are obtained as a function of the partition function, which in the (N, V, T) ensemble is defined as the sum of the exponentials of the dimensionless internal energies of the very many different possible quantum states compatible with the canonical variables. Boltzmann equation relating the entropy of the system to the probabilities of finding the different quantum states is derived and the Third Law of Thermodynamics is stated. The treatment is extended to cover the Gibbs (N, P, T) ensemble.

Statistical thermodynamics is a branch of knowledge that has its own postulates and techniques. We do not attempt to give here even an introduction to the field. In this chapter, we elaborate on the ideas presented by T. H. Hill [1] with the direct purpose of giving a molecular justification to equations of state and excess Gibbs energy models used in practical work.

CANONICAL ENSEMBLE AND PROBABILITY

For this introduction, we consider a closed system having rigid and diathermal walls, immersed in a thermostatic bath at a temperature T. The system has a fixed number of molecules inside of the same or of different kinds, and a fixed volume, and it can only exchange energy in the form of heat with its surroundings. The molecules in the system are moving around, hitting each other, and hitting the rigid walls. A detailed description of the internal energy of this system would require the specification of the position and momentum coordinates, that is, the external degrees of freedom, of each molecule at any given instant. In addition, it would be necessary to specify the rotational, vibrational, and electronic coordinates, that is, the internal degrees of freedom, of each molecule at any given instant. Because per each mole inside the system there are 10^{23} molecules, this detailed description becomes an impossible task. On the other hand, macroscopically the system is characterized by three parameters: the temperature T, the total volume V, and the number of moles n or, more properly, the number of molecules N that it contains. Each set of values of the internal and external coordinates of the molecules compatible with the values of the macroscopic parameters T, V, and N represents a possible quantum state of the system. There are very many such quantum states, each with a value E_i for the internal energy of the macroscopic system. The internal energy U of the system is then a macroscopic average of the internal energies of these very many possible quantum states. The probability $\mathcal{P}_i$ of having a particular quantum state i with an internal energy E_i will be equal to the ratio between the number of quantum states $\mathcal{N}(E_i)$ having that particular value E_i of the internal energy and the total number of quantum states compatible with the macroscopic parameters T, V, and N:

$$\mathcal{P}_i = \frac{\mathcal{N}(E_i)}{\sum_j \mathcal{N}(E_i)} \tag{26.1}$$

DOI: 10.1201/9781003431985-30

with

$$\sum_j \mathcal{P}_j = 1 \tag{26.2}$$

INTERNAL ENERGY AND THE PARTITION FUNCTION

Hence, according to Equation 26.1, as the internal energy $U(T, V, N)$ of the system is a weighted average of the internal energy E_i of these quantum states, it takes the form:

$$U = \sum_j \mathcal{P}_j E_j \tag{26.3}$$

Considering the large number of possible quantum states, the probability of finding one particular state with internal energy E_i is well represented by a normalized Boltzmann distribution function, namely,

$$\mathcal{P}_i = \frac{e^{-\frac{E_i}{kT}}}{\sum_j e^{-\frac{E_j}{kT}}} \tag{26.4}$$

where k is the Boltzmann constant, given by the ratio of the gas constant over Avogadro's number. In this equation, it is conventional to define the partition function Q as:

$$\mathcal{Q} \equiv \sum_j e^{\frac{-E_j}{kT}} \tag{26.5}$$

With the purpose of connecting these expressions with classical thermodynamics, we differentiate Equation 26.3 and write:

$$dU = \sum_j \mathcal{P}_j dE_j + \sum_j E_j d\mathcal{P}_j \tag{26.6}$$

The next step is to express the right-hand side of this equation in terms of the macroscopic thermodynamic variables. In Classical Thermodynamics, we have:

$$\left(\frac{\partial U}{\partial V}\right)_{S,n} = -P$$

So, by analogy, we define:

$$\left(\frac{\partial E_j}{\partial V}\right)_{T,n} = -P_j$$

or, at constant T and n,

$$dE_j = -p_j dV$$

Multiplying this equation by $\mathcal{P}_j$ and adding over all j, we have:

$$\sum_j \mathcal{P}_j dE_j = -\left(\sum_j \mathcal{P}_j p_j\right) dV = -PdV \tag{26.7}$$

In this step, we have identified the pressure of the system with the weighted average of values of p_j.

For the second term of the right-hand side of Equation 26.6, from Equation 26.4 and the definition of the partition function, Equation 26.5, we write for a quantum state j:

$$E_j = -kT \ln \mathcal{P}_j - kT \ln \mathcal{Q} \tag{26.8}$$

Multiplying this expression by $d\,\mathcal{P}_j$ and adding over all j,

$$\sum_j E_j d\mathcal{P}_j = -kT \sum_j \ln \mathcal{P}_j d\mathcal{P}_j - kT \ln \mathcal{Q} \sum_j d\mathcal{P}_j$$

But according to Equation 26.2,

$$\sum_j d\mathcal{P}_j = d \sum_j \mathcal{P}_j = 0$$

so the second term on the left-hand side vanishes, and we write:

$$\sum_j E_j d\mathcal{P}_j = -kT \sum_j \ln \mathcal{P}_j d\mathcal{P}_j$$

This expression can be transformed further considering that:

$$d \sum_j \mathcal{P}_j \ln \mathcal{P}_j = \sum_j \mathcal{P}_j d \ln \mathcal{P}_j + \sum_j \ln \mathcal{P}_j d\mathcal{P}_j$$

but

$$\sum_j \mathcal{P}_j d \ln \mathcal{P}_j = \sum_j \mathcal{P}_j \frac{1}{\mathcal{P}_j} d\mathcal{P}_j = \sum_j d\mathcal{P}_j = 0$$

so

$$d \sum_j \mathcal{P}_j \ln \mathcal{P}_j = \sum_j \ln \mathcal{P}_j d\mathcal{P}_j$$

Thus,

$$\sum_j E_j d\mathcal{P}_j = -kTd \sum_j \mathcal{P}_j \ln \mathcal{P}_j = -Td\left(k \sum_j \mathcal{P}_j \ln \mathcal{P}_j\right) \tag{26.9}$$

and replacing Equations 26.7 and 26.9 in Equation 26.6, we obtain:

$$dU = -P\ dV + Td\left(-k \sum_j \mathcal{P}_j \ln \mathcal{P}_j\right) \tag{26.10}$$

ENTROPY, PROBABILITY, AND THE THIRD LAW OF THERMODYNAMICS

From the combination of the first and second laws of thermodynamics, we know that:

$$dU = TdS - P\,dV$$

So, by comparison with Equation 26.10, we conclude that the entropy is related to the probability by the Boltzmann equation,

$$S = -k\sum_j \mathcal{P}_j \ln \mathcal{P}_j \tag{26.11}$$

Equation 26.11 is of importance in information theory.

A perfect crystal at 0 K is in its ground state. Thus, for this single quantum state the value of $\mathcal{P}_j$ is unity and $S=0$. This is a statement of the Third Law of Thermodynamics.

EQUATION OF STATE, CHEMICAL POTENTIAL, AND PARTITION FUNCTION

After obtaining this first bridge between statistical and classical thermodynamics, we multiply Equation 26.8 by $\mathcal{P}_j$, and add over all quantum states to obtain:

$$\sum_j \mathcal{P}_j E_j = T\left(-k\sum_j \mathcal{P}_j \ln \mathcal{P}_j\right) - kT\ln Q\sum_j \mathcal{P}_j$$

Thus, replacing in this expression Equations 26.2, 26.3, and 26.11,

$$U = TS - kT\ln Q$$

or

$$U - TS = -kT\ln Q$$

But from thermodynamics we know that the Helmholtz function is defined by Equation 4.1 as:

$$A = U - TS$$

Thus, we conclude that:

$$A = -kT\ln Q \tag{26.12}$$

Equation 26.12 is one of the most important links between statistical thermodynamics and classical thermodynamics, as it permits us to obtain expressions for the equation of state and for the chemical potential based on the exact relations.

From Equation 4.15,

$$P = -\left(\frac{\partial A}{\partial V}\right)_{T,n} = kT\left(\frac{\partial \ln Q}{\partial V}\right)_{T,n} \tag{26.13}$$

And from Equation 4.16,

$$\mu_i = \left(\frac{\partial A}{\partial n_i}\right)_{T,V,n_{j\neq i}} = -kT\left(\frac{\partial \ln Q}{\partial n_i}\right)_{T,V,n_{j\neq i}} \tag{26.14}$$

The next step, then, is to say something about the partition function for gases and liquids. This is discussed in the chapters that follow.

MICROCANONICAL, GRAND CANONICAL, AND GIBBS ENSEMBLES

The *canonical* ensemble discussed above corresponds to actual systems at constant volume (V), temperature (T), and numbers of molecules (N). The internal energy of each subsystem fluctuates around a value U, corresponding to the internal energy of the system of interest. The probability of finding a subsystem of the ensemble with a value of its internal energy E_i was estimated to be well represented by a Boltzmann normalized distribution function, Equation 26.4. In a similar way, we can consider ensembles having in common the properties (U, V, N), (V, T, μ) and (P, T, N).

The ensemble (U, V, N), known as *microcanonical* ensemble, is rather useless as the actual system must be isolated and, as such, it does not have any interaction with its environment. The ensemble (V, T, μ), sometimes called *macrocanonical* or *grand canonical* ensemble, is useful for the study of open isothermal systems.

For computer simulations, the *Gibbs* or (P, T, N) ensemble, defined by the values of pressure P, temperature T, and numbers of molecules N, all the same or of different kinds, is of much practical interest. In this ensemble, the volumes of the closed isothermal-isobaric subsystems fluctuate around the value of the volume V of the system of interest.

In the case of the canonical ensemble, the key step to make some progress was to use the normalized Boltzmann distribution function, Equation 26.4 for the probability of finding a particular state with the value of an internal energy E_i and the values of V and N of the system of interest. In the case of the Gibbs or (P, T, N) ensemble, it is also assumed that the probability of finding a subsystem with a volume V_i and the values of T and N of the actual system is given by a normalized Boltzmann distribution function. Now, the probability of finding a subsystem with a volume V_i and the same values of T and N of the actual system, is equivalent to finding the subsystem with a canonical partition function Q (V_i, T, N), and thus,

$$\mathcal{P}_{(V_i)} = \frac{Q(V_i,T,N)e^{\frac{-PV_i}{RT}}}{\sum_j Q(V_j,T,N)e^{\frac{-PV_j}{RT}}} \tag{26.15}$$

Note that in order to get a more direct connection between classical and statistical thermodynamics, in this equation the total volume is written in molar terms and the gas constant is used instead of the Boltzmann constant. In addition, we have introduced a change in the nomenclature with respect to the one used in the discussion of the canonical ensemble. Due to the need of distinguishing between the partition functions of the different ensembles, we have kept the symbol Q for "partition function" and noted in brackets the variables considered in each case. The partition function Q (T, V, N) was simply denoted by Q in the presentation of the canonical ensemble. For the partition function of the Gibbs ensemble, we use here Q (T, P, N). Other authors, instead of using Q, and indicating the variables (T, P, N), use a different symbol for the Gibbs ensemble. One symbol commonly used is capital delta (Δ).

In the nomenclature used here, the partition function Q (T, P, N) of the Gibbs ensemble is then:

$$Q(T,P,N) = \sum_j Q(V_j,T,N)\, e^{\frac{-PV_j}{RT}} \tag{26.16}$$

As it is normally done in statistical thermodynamics, this summation is replaced by its maximum term and we write:

$$Q\,(T,P,N) = Q(T,V,N)\; e^{\frac{-PV}{RT}} \tag{26.17a}$$

or

$$-RT \ln Q\,(T,P,N) = -RT\ \ln Q(T,V,N) + PV \tag{26.17b}$$

Replacing Equation 26.12, and recalling that from classical thermodynamics $G=A+PV$, similar to Equation 26.12, the link between the classical and statistical thermodynamics in this case takes the form:

$$G = -RT \ln Q\,(T,P,N) \tag{26.18}$$

REFERENCE

1. Hill, T. L. 1960. *An Introduction to Statistical Thermodynamics.* Reading, MA: Addison-Wesley.

PROBLEMS

26.1 **Thermodynamic properties from the partition function**

Starting from Equation 26.12, derive equations for the entropy S, internal energy U, enthalpy H, and Gibbs function G in terms of the partition function Q.

26.2 **Classical approximation to the quantum partition function**

According to quantum theory, the vibrational energy levels of a diatomic molecule are

$$E_{vib} = \left(\nu + \frac{1}{2}\right)\hbar f, \qquad \nu = 0,1,2,\ldots$$

where ν are vibrational quantum numbers, $\hbar$ is Planck's constant, and f is a substance-dependent characteristic frequency. Find the partition function Q_{vib}

a. By summation as defined in Equation 26.5;

b. By assuming that the energy levels are so closely spaced that the summation can be replaced by an integral (the *classical approximation*).

In both cases, find the vibrational contribution to the internal energy U and to the isochoric heat capacity C_v. What are the limiting values of these contributions at very low and at very high temperatures?

26.3 **Boltzmann's equation for entropy**

The best-known version of Boltzmann' s seminal equation for entropy in statistical terms is

$$S(N,V,U) = k \ln \Omega \tag{26.15a}$$

where Ω is the number of configurations (microstates) accessible to a system given its total number of particles N, volume V, and energy U, whereas Equation 26.11 has been derived for known N, V and temperature T. Prove that Equation 26.15a follows from Equation 26.11 by constraining a system of given (N, V, T) to only those microstates that have a particular energy E, just one of the possible values at such conditions.

26.4 **Negative absolute temperatures**

Consider a system containing N particles that have only two possible energy levels: a ground state with $E_1 = 0$, and an excited state with $E_2 = \varepsilon$. Find the energy and entropy of the system as a function of the number m of particles in the excited state. Show that the system's absolute temperature can become negative at the higher energies and interpret this in physical terms.

27 Statistical Thermodynamics Basis of Equations of State

Abstract

It is shown that the equation of state (EOS) for the ideal gas is obtained from the partition function when there are no intermolecular interactions and the molecules are just points in space; the virial EOS is obtained when the compressibility factor is expanded in Taylor series around the ideal gas limit at zero pressure and the cubic EOS are obtained with simple assumptions for the intermolecular potential and the reduction in free volume caused by the bulk of the molecules. Explicit relations are given for the second virial coefficient of pure compounds and binary mixtures in terms of intermolecular potentials.

The starting point for the statistical thermodynamics' justification of equations of state (EOSs) for pure compounds is Equation 26.13, namely,

$$P = -\left(\frac{\partial A}{\partial V}\right)_{T,N} = kT\left(\frac{\partial \ln \mathcal{Q}}{\partial V}\right)_{T,N} \tag{26.13}$$

with

$$\mathcal{Q} \equiv \sum_j e^{\frac{-E_j}{kT}} \tag{26.5}$$

The next step is to say something about the partition function Q in terms of the temperature T, the total volume V, and the number of identical molecules N of a pure compound in the gas phase. Considering that the number of molecules present is of the order of 10^{23}, it is legitimate to replace quantum statistics by classic statistics and transform the sum over the quantum states with energy E_j into an integral over the range of parameters characterizing the different modes of motion. Thus, Equation 26.5 takes the form:

$$\mathcal{Q} = \frac{1}{N!}\int\limits_I\int\limits_M\int\limits_r e^{-[E_I + E_M + E(r)]/kT} dI_N dM_N dr_N$$

where the integrations are taken over the I internal degrees of motion (rotational, vibrational, and electronic), M momentum coordinates describing the external state of motion (momentum) of the molecules, and r position coordinates describing the translational modes of motion for the N molecules in the three-dimensional space. The inverse factorial term in front of the integrals is included to avoid double counting of the identical molecules. As an approximation, which is acceptable for the gas phase, the different modes of motion are considered to be independent of each other. Thus, the integrals are separable, and we write:

$$\mathcal{Q} = \frac{1}{N!}\int\limits_I e^{-E_I/kT} dI_N \int\limits_M e^{-E_M/kT} dM_N \int\limits_r e^{-E(r)/kT} d\overline{r_1}\ldots d\overline{r_N}$$

where the notation $d\overline{r_i}$ implies the three coordinates of position of molecule i.

DOI: 10.1201/9781003431985-31

In a shorthand nomenclature, designating each integral by the symbol Q with the corresponding subscript, we have:

$$\mathcal{Q} = \frac{1}{N!}\, \mathcal{Q}_I\, \mathcal{Q}_M\, \mathcal{Q}_r$$

$$\ln \mathcal{Q} = \ln \mathcal{Q}_I + \ln \mathcal{Q}_M + \ln \mathcal{Q}_r - \ln(N!)$$

As indicated by Equation 26.13, for the EOS we are only interested in the dependence of the partition function on volume. Thus, considering that the internal and external modes of motion of the molecules are only temperature dependent, Equation 26.13 takes the form:

$$P = -\left(\frac{\partial A}{\partial V}\right)_{T,N} = kT\left(\frac{\partial \ln \mathcal{Q}_r}{\partial V}\right)_{T,N} \tag{27.1}$$

Although it is only a change of formalism sometimes, due to its importance, Q_r is given the special symbol Z_N and called the classical configurational integral,

$$Z_N \equiv \mathcal{Q}_r \equiv \sum_r e^{-\frac{E(r)}{kT}}\, d\bar{r}_1 \ldots d\bar{r}_N \tag{27.2}$$

so, we write:

$$P = -\left(\frac{\partial A}{\partial V}\right)_{T,N} = kT\left(\frac{\partial \ln Z_N}{\partial V}\right)_{T,N} \tag{27.1a}$$

In practical work, we define the compressibility factor z by the general relation:

$$z \equiv \frac{Pv}{RT} = \frac{P}{\rho kT} \tag{27.3}$$

where $\rho = N/V$ is the number density of the fluid. The compressibility factor z should not be confused with the classical configurational integral Z_N.

EQUATION OF STATE FOR THE IDEAL GAS

As a first example of application of this treatment, we consider that there are no intermolecular interactions, that is, $E(r)=0$, and that the molecules are just points in space and do not occupy any volume, in which case there is no contribution from the relative orientation of the molecules, hence only three external coordinates fully describe their configuration in space. Thus, the integral for each of the N molecules is just the total volume of the system,

$$Z_N^* = \int_r dr_1 \ldots dr_{3N} = V^N \tag{27.4}$$

and from Equation 27.1a,

$$P^* = \frac{NkT}{V} = \frac{\frac{N}{N_A}(N_A k)T}{V} = \frac{nRT}{V} = \frac{RT}{v} \tag{27.5}$$

where N_A is Avogadro's number and $N_A k = R$ is the gas constant. Thus, according to Equation 27.5, the compressibility factor for the ideal gas is equal to unity:

$$z^* \equiv \frac{P^* \mathrm{v}}{RT} = 1 \tag{27.5a}$$

VIRIAL EQUATIONS OF STATE

As an improvement toward more realistic systems, we may expand the compressibility factor z, as a Taylor series around the real gas limit at pressure tending to zero, that is, $P \rightarrow 0$; $\mathrm{v} \rightarrow \infty$; $\rho \rightarrow 0$:

$$z = z_{\rho=0} + \left(\frac{\partial z}{\partial \rho}\right)_{\rho=0} (\rho - 0) + \frac{1}{2!}\left(\frac{\partial^2 z}{\partial \rho^2}\right)_{\rho=0} (\rho - 0)^2 + \ldots$$

The first term of the series is given by Equation 27.5, and for convenience, a symbol is given to the value of the coefficients of the density powers:

$$B = \left(\frac{\partial z}{\partial \rho}\right)_{\rho=0}; \qquad C = \frac{1}{2!}\left(\frac{\partial^2 z}{\partial \rho^2}\right)_{\rho=0} \ldots$$

These coefficients are called the second, third, and so forth, virial coefficients. They are evaluated at the limit of the pressure tending to zero for each compound and are a function of temperature only. Thus, we write:

$$z = 1 + B\,\rho + C\,\rho^2 + \ldots \tag{27.6}$$

Comparison of Equations 27.5a and 27.6 shows that the virial terms are perturbations to the ideal gas case. The term *virial* means force, and it reflects the fact that these perturbations are due to molecular interactions.

It is also possible to expand the compressibility factor as a Taylor pressure series around the real gas limit,

$$z = z_{P=0} + \left(\frac{\partial z}{\partial P}\right)_{P=0} (P - 0) + \frac{1}{2!}\left(\frac{\partial^2 z}{\partial P^2}\right)_{P=0} (P - 0)^2 + \ldots$$

In this case, we write:

$$z = 1 + B'P + C'P^2 + \ldots \tag{27.7}$$

The relationship between the virial coefficients of the two series is obtained considering that by combining Equations 27.3 and 27.6, we obtain:

$$P = RT\left[\rho + B\rho^2 + C\rho^3 + \ldots\right]$$

Replacing this expression in Equation 27.7, rearranging and comparing the coefficients up to the second power of density with those of Equation 27.6, we obtain:

$$B' = \frac{B}{RT}; \quad C' = \frac{C - B^2}{(RT)^2}, \text{ etc.}$$

The second and third virial coefficients of expansion in terms of density, B and C, are known, B more accurately than C, for many substances. Higher coefficients are seldom used in practice. Hence, the truncated pressure series of practical use takes the form:

$$z = 1 + \frac{B}{RT}P + \frac{C - B^2}{(RT)^2}P^2 + \ldots \tag{27.7a}$$

Equation 27.7a, truncated after the second term, is particularly useful for calculating fugacity coefficients of pure gases at low pressure using Equation 13.7a. The virial EOS is further discussed in Appendix B.

CUBIC EQUATIONS OF STATE

In these equations, the pressure is a function of the third power of the molar volume. This dependence makes the volume roots of these equations easily solved for particular values of the pressure and temperature. The practical details for working with these equations are presented in Appendix C. The need to use more complex expressions for the EOS appears because as the pressure increases, at fixed volume and temperature, the interactions between molecules become stronger and the intermolecular potential cannot be ignored, nor can the volume occupied by the molecules. In this case, it is convenient to define a mean intermolecular potential φ by the expression:

$$e^{-\frac{\varphi}{kT}} \equiv \frac{\int_r e^{-\frac{E(r)}{kT}} dr_1 \ldots dr_{3N}}{\int_r dr_1 \ldots dr_{3N}} \tag{27.8}$$

For the case of the ideal gas, the integral appearing in the denominator of this expression is identified with the total volume of the system to the N-th power because for each molecule, the integral runs in the three directions of space to cover the total volume of the system. If the molecules themselves occupy part of the volume, their translational movements can only happen in the free volume, V_f, left unoccupied by the molecules:

$$\int_r dr_1 \ldots dr_{3N} = V_f^N$$

Introducing this expression in Equation 27.8 and combining with Equation 27.2, we write:

$$Z_N = \int_r e^{-\frac{E(r)}{kT}} dr_1 \ldots dr_{3N} = V_f^N e^{-\frac{\varphi}{kT}} \tag{27.9}$$

Thus, from Equation 27.1a,

$$P = NkT\left(\frac{\partial \ln V_f}{\partial V}\right)_{T,N} - \left(\frac{\partial \varphi}{\partial V}\right)_{T,N} = nRT\left(\frac{\partial \ln V_f}{\partial V}\right)_{T,n} - \left(\frac{\partial \varphi}{\partial V}\right)_{T,n} \tag{27.10}$$

To continue, it is necessary to introduce some assumptions for the free volume and the intermolecular potential. Designating by b the volume occupied by the molecules per mole of gas, as a first crude approximation we write:

$$V_f = V - nb$$

The van der Waals model is a simple treatment retaining the most basic elements causing important effects shown by real systems. In this model, the repulsive part of the intermolecular potential is represented by the decrease in free volume caused by the bulkiness of the molecules as they get closer and closer together, represented by the parameter b in the term $(V - nb)$.

The attractive part of the intermolecular potential, which by convention is negative, is proportional to the number of moles present, and it tends to vanish as the volume increases, thus as a first approximation it is inversely proportional to the molar volume. Hence,

$$\varphi = -\frac{na_c}{\text{v}} = -\frac{n^2 a_c}{V}$$

where a_c is the proportionality constant. Introducing these two assumptions in Equation 27.10, we obtain:

$$P = \frac{nRT}{V - nb} - \frac{n^2 a_c}{V^2} = \frac{RT}{\text{v} - b} - \frac{a_c}{\text{v}^2} \tag{27.11}$$

This is the van der Waals EOS [1].

Next, we can assume that the intermolecular potential is a more complex function of volume and temperature, say:

$$\varphi = -n^2 a_c F(V)\,\alpha(T)$$

Thus, the EOS would take the form:

$$P = \frac{RT}{\text{v} - b} + n^2 a_c\,\alpha(T)\left(\frac{\partial F(V)}{\partial V}\right)_{T,n}$$

For this form, there are the two most successful cubic EOSs:

Redlich–Kwong equation [2]

$$P = \frac{RT}{\text{v} - b} - \frac{a_c\,\alpha_{RK}(T)}{\text{v}(\text{v} + b)}$$

Peng–Robinson equation [3]

$$P = \frac{RT}{\text{v} - b} - \frac{a_c\,\alpha_{PR}(T)}{\text{v}^2 + 2b\text{v} - b^2}$$

Each of these has a different function of temperature $\alpha(T)$, which at this point we leave unspecified and use $a = a_c\alpha(T)$. The use of a cubic EOS is further discussed in Appendix C. For each of these EOSs, it is possible to back-calculate the form of the function $F(V)$. Although, at our present degree of knowledge, there is not much understanding to gain with this further step for completeness, we include these functions below:

$$F_{RK}(V) = \frac{1}{nb}\ln\left(\frac{V + nb}{V}\right)$$

$$F_{PR}(V) = \frac{1}{2nb\sqrt{2}}\ln\left(\frac{V + nb - nb\sqrt{2}}{V + nb + nb\sqrt{2}}\right)$$

RELATION OF INTERMOLECULAR POTENTIALS TO THE SECOND VIRIAL COEFFICIENT

To make some progress, we return to statistical thermodynamics treatment and relate the virial coefficients to intermolecular potentials. For this, we add and subtract one dimensionless unit to the integrand of the configurational integral and write Equation 27.2 as:

$$Z_N = \int_r e^{-\frac{E(r)}{kT}} dr_1 \ldots dr_{3N} = V^N \left[1 + \frac{1}{V^N} \int_r \left[e^{-\frac{E(r)}{kT}} - 1 \right] d\overline{r}_1 \ldots d\overline{r}_N \right] \tag{27.12}$$

where, in the low-pressure region of applicability of the virial equation, we have used the approximation given by Equation 27.4. To continue, let us concentrate attention on the integral appearing in Equation 27.12, namely,

$$\mathrm{I} \equiv \frac{1}{V^N} \int_r \left[e^{-\frac{E(r)}{kT}} - 1 \right] d\overline{r}_1 \ldots d\overline{r}_N$$

Equation 27.12 can be written as:

$$Z_N = V^N [1 + I] \tag{27.12a}$$

or

$$\ln Z_N = N \ln V + \ln[1 + I]$$

Considering that the definition of I includes a denominator V^N, its value is very small, $I \lll 1$, so

$$\ln[I + 1] \approx I$$

and

$$\ln Z_N = N \ln V + I \tag{27.13}$$

So, from Equation 27.1a,

$$P = \frac{NkT}{V} + kT \left(\frac{\partial I}{\partial V} \right)_{T,N} \tag{27.14}$$

At low pressure, interactions of three or more molecules are extremely rare. Thus, considering only pair interactions, the number of possible arrays is:

$$\frac{N(N-1)}{2} \approx \frac{N^2}{2}$$

For each array, only the coordinates of the two interacting molecules $\overline{r}_1$ and $\overline{r}_2$ are important for the intermolecular potential $E(\overline{r}_1, \overline{r}_2)$, while the rest of the noninteracting molecules will produce a term of the form:

$$\int_r d\overline{r}_3 \ldots\ldots d\overline{r}_N = V^{N-2}$$

Thus,

$$\mathrm{I} = \frac{N^2 V^{N-2}}{2V^N} \int_r \left[e^{-\frac{E(\bar{r}_1,\bar{r}_2)}{kT}} - 1 \right] d\bar{r}_1 d\bar{r}_2 = \frac{N^2}{2V^2} \int_r \left[e^{-\frac{E(\bar{r}_1,\bar{r}_2)}{kT}} - 1 \right] d\bar{r}_1 d\bar{r}_2$$

In addition, as the interaction potential between two molecules is a function of the distance r between the two molecules, considering a central molecule 1, we may consider the element of volume in which there is interaction with molecule 2 as $4\pi r^2 dr$ and integrate over $d\bar{r}_1$ to cover the whole volume V:

$$I = \frac{N^2}{2V^2} \left\{ \int_r \left[e^{-\frac{E(r)}{kT}} - 1 \right] 4\pi r^2 dr \right\} V = \frac{2\pi N^2}{V} \left\{ \int_r \left[e^{-\frac{E(r)}{kT}} - 1 \right] r^2 dr \right\}$$

In order to simplify the equation, let us define:

$$B \equiv -2\pi N_A \int_r \left[e^{-\frac{E(r)}{kT}} - 1 \right] r^2 dr \tag{27.15}$$

or

$$I = -\frac{\left(N^2 / N_A \right)}{V} B$$

And, having been integrated over the entire fluid volume, B is a function of temperature only:

$$\left(\frac{\partial I}{\partial V} \right)_{T.N} = \frac{\left(N^2 / N_A \right)}{V^2} B$$

Hence, Equation 27.14 takes the form:

$$P = \frac{NkT}{V} + kT \frac{\left(N^2 / N_A \right) B}{V^2} = \frac{(N/N_A)(N_A k)T}{V} + (N_A k)T \frac{(N/N_A)^2 B}{V^2} \tag{27.14a}$$

$$P = \frac{RT}{\mathrm{v}} + \frac{RT}{\mathrm{v}^2} B$$

So,

$$z = 1 + \frac{B}{\mathrm{v}} = 1 + B\rho$$

The comparison of this result with Equation 27.6 shows that Equation 27.15 gives the relation of the second virial coefficient with the two-body intermolecular potential. In practice, Equation 27.15 is used to study potential functions starting from experimentally measured second virial coefficients.

SECOND VIRIAL COEFFICIENTS OF BINARY MIXTURES

For a binary mixture of compounds i and j, the configurational integral takes the form:

$$Z_N = V^N \left[1 + \frac{1}{V^N} \sum_i \sum_j \int_r \left[e^{-\frac{E_{ij}(r)}{kT}} - 1 \right] d\bar{r}_i d\bar{r}_j \right] \tag{27.16}$$

where again we have considered only pair interactions and integrated over r for the $(N-2)$ noninteracting molecules. As above, we may consider again a central molecule i, integrate over $d\overline{r_i}$ to obtain V and express $d\overline{r_j}$ as $4\pi r^2 dr$ and integrate over dr to obtain:

$$Z_N = V^N\left[1+\frac{4\pi}{V}\sum_i\sum_j\int_r\left[e^{-\frac{E_{ij}(r)}{kT}}-1\right]r^2dr\right] \tag{27.17}$$

In the sums, we distinguish three types of interactions: i–i interactions, repeated $N_i^2/2$ times; i–j interactions, repeated N_iN_j times; and j–j interactions, repeated $N_j^2/2$ times. Thus, we write:

$$Z_N = V^N\left[1+\frac{2\pi N_i^2}{V}\int_r\left[e^{-\frac{E_{ii}(r)}{kT}}-1\right]r^2dr+\frac{4\pi N_iN_j}{V}\int_r\left[e^{-\frac{E_{ij}(r)}{kT}}-1\right]r^2dr+\frac{2\pi N_j^2}{V}\int_r\left[e^{-\frac{E_{jj}(r)}{kT}}-1\right]r^2dr\right]$$

In order to identify the second virial coefficient for the binary mixture, there is no need to derive the EOS. In short notation, we write:

$$Z_N = V^N\left[1+\frac{N^2B_m}{N_AV}\right]$$

According to Equation 27.15,

$$B_{ii} \equiv -2\pi N_A\int_r\left[e^{-\frac{E_{ii}(r)}{kT}}-1\right]r^2dr$$

$$B_{ij} \equiv -2\pi N_A\int_r\left[e^{-\frac{E_{ij}(r)}{kT}}-1\right]r^2dr$$

$$B_{jj} \equiv -2\pi N_A\int_r\left[e^{-\frac{E_{jj}(r)}{kT}}-1\right]r^2dr$$

Thus,

$$B_m = x_i^2B_{ii}+2x_ix_jB_{ij}+x_j^2B_{jj} \tag{27.18}$$

with $x_i=N_i/N$.

The term B_{ij} is called the cross second virial coefficient of compounds i and j.

MIXING RULE FOR CUBIC EQUATIONS OF STATE

Expanding Equation 27.11, or any of the cubic EOSs of the van der Waals family, into a series in terms of density gives

$$z = 1+\frac{B}{\mathrm{v}}+\ldots = 1+B\rho+\ldots \tag{27.19}$$

with

$$B = b-\frac{a}{RT} \tag{27.20}$$

and

$$a = a_C\alpha$$

Comparison of this result with Equation 27.18 suggests the use of the following mixing rules for the extension of cubic equations to mixtures:

$$b_m = x_i^2 b_{ii} + 2x_i x_j b_{ij} + x_j^2 b_{jj} \tag{27.21}$$

and

$$a_m = x_i^2 a_{ii} + 2x_i x_j a_{ij} + x_j^2 a_{jj} \tag{27.22}$$

For parameter b, which represents the bulkiness of the molecules, it seems logical to eliminate the double subindex and take the cross-term as an arithmetic average of the values for the pure compounds. Thus,

$$b_m = x_i^2 b_i + 2x_i x_j \frac{b_i + b_j}{2} + x_j^2 b_j = x_i\left(x_i + x_j\right)b_i + x_j\left(x_i + x_j\right)b_j$$

or

$$b_m = x_i b_i + x_j b_j \tag{27.23}$$

For the parameter a, which is related to the intermolecular potentials, the form of Equation 27.22 is kept and the cross-term a_{ij} is approximated by:

$$a_{ij} = \left(a_{ii}a_{jj}\right)^{0.5}\left(1 - \delta_{ij}\right) \tag{27.24}$$

where δ_{ij} is the Prausnitz empirical adjustable parameter.

NONCUBIC EQUATIONS OF STATE OF THEORETICAL INTEREST

There is a very large number of cubic and noncubic EOSs proposed in the literature. Normally, the variables fixed for the calculation of fugacities in phase equilibrium problems are the pressure and the temperature, so due to the additional iterations necessary to find the roots of volume, noncubic EOSs are not of common use in practice. In this chapter, we only give a general discussion of the subject based on our original work.

The *free volume* term in van der Waals' generalized model, presented above, represents the repulsive forces arising from the interactions of the hard core of the molecules, while the intermolecular potential represents the attractive forces between molecules. Van der Waals's model considers that the pressure is determined by the difference of these two contributions. In fact, any deficiency of the van der Waals free volume term representing hard-core effects tends to be compensated by the term representing attraction effects.

Another equally valid assumption is that of Dieterici, in which the potential directly perturbs the hard-core contribution. The Dieterici equation [4] of state has the form:

$$P = \frac{RT}{\mathrm{v} - b}\exp\left(-\frac{a}{RT\mathrm{v}}\right) \tag{27.25}$$

As discussed elsewhere by Polishuk and Vera [5,6], both potentials are not mutually exclusive, and they can be combined in a single expression. Van der Waals potential represents the cohesive energy keeping the molecules together, while the Dieterici potential considers that only those molecules with enough kinetic energy to escape the cohesive field contribute to the pressure exerted by the fluid over the walls of the system. Perhaps the most serious theoretical shortcoming of both the van der Waals and Dieterici equations is that their repulsive term fails to reproduce well-known results for hard-sphere fluids. Stated shortly, in a close-packed system of hard spheres of equal diameter, it is impossible to fill all the volume of the system. In round numbers, the hard spheres fill 0.74 of the total volume and the rest is empty space between the close-packed hard spheres. Thus, the compressibility factor goes to infinity when the hard spheres are close-packed but not all the volume of the system is filled. Wang et al. [7] proposed a solution to this problem using the scaled particle theory with a lattice formed by Kelvin's tetrakaidecahedron cells [8,9]. The following two basic forms were studied:

$$z = \frac{\sum_{i=1} C_i \xi^{i-1}}{(1-\xi)^n} \tag{27.26}$$

and

$$z = \frac{1}{\xi} \sum_{i=1} C_i \left(\frac{\xi}{1-\xi} \right)^i \tag{27.27}$$

with

$$\xi = \frac{V_0}{V} \tag{27.28}$$

where V_0 is the close-packed volume of the hard spheres.

The incorporation of an intermolecular potential for a system of hard spheres of equal diameter (pure compounds) was studied by Khoshkbarchi and Vera [10]. EOSs for mixtures of hard spheres of different diameters were studied by Khoshkbarchi and Vera [11], with corrections that were noted by Taghikhani et al. [12]. Further studies were presented by Ghotbi and Vera [13–15]. Although we consider the details of this material too specialized to be included here, we quote the above references as a guide for those interested in pursuing further the study of this subject.

REFERENCES

1. van der Waals, J. D. 1873. *Over de Continuiteit van den Gas- en Vloeistoftoestand*. Leiden: A. W. Sijthoff.. English translation: *On the Continuity of the Gaseous and Liquid States*. J. D. van der Waals, ed. Rowlinson, J. S. 1988. Mineola, NY: Dover
2. Redlich, O., and Kwong, J. N. S. 1949. On the thermodynamics of solutions. V. An equation of state. Fugacities of gaseous solutions. *Chem. Rev.* 44(1): 233–244.
3. Peng, D. Y., and Robinson, D. B. 1976. A new two-constant equation of state. *Ind. Eng. Chem. Fundam.* 15(1): 50–64.
4. Dieterici, C. 1899. Über den kritischen Zustand. *Ann. Phys. Chem.* 69: 685–705.
5. Polishuk, I., and Vera, J. H. 2005. A novel EOS that combines the van der Waals and the Dieterici potentials. *AIChE J.* 51: 2077–2088.
6. Polishuk, I., and Vera, J. H. 2005. A novel equation of state for the prediction of thermodynamic properties of fluids. *J. Phys. Chem. B* 109: 5977–5984.
7. Wang, W., Khoshkbarchi, M. K., and Vera, J. H. 1996. A new volume dependence for the equations of state of hard spheres. *Fluid Phase Equilib.* 115: 25–38.
8. Kelvin, L. 1894. On homogeneous division of space. *Proc. R. Soc. (Lond.)* 55: 1–16.
9. Smith, V. J. 1982. *Geometrical and Structural Crystallography*. New York: Wiley & Sons.

10. Khoshkbarchi, M. K., and Vera, J. H. 1997. A simplified hard-sphere equation of state meeting the high- and low-density limits. *Fluid Phase Equilib.* 130: 189–194.
11. Khoshkbarchi, M. K., and Vera, J. H. 1998. A generalized mixing rule for hard-sphere equations of state of Percus-Yevick type. *Fluid Phase Equilib.* 142: 131–147.
12. Taghikhani, V., Khoshkbarchi, M. K., and Vera, J. H. 1999. On the expression for the chemical potential in mixtures of hard spheres. *Fluid Phase Equilib.* 165: 141-146, 279.
13. Ghotbi, C., and Vera, J. H. 2001. Performance of three mixing rules using different equations of state for hard-spheres. *Fluid Phase Equilib.* 187-188: 321–336.
14. Ghotbi, C., and Vera, J. H. 2001. Extension to mixtures of two robust hard-sphere equations of state satisfying the ordered close-packed limit. *Can. J. Chem. Eng.* 79: 678–686.
15. Ghotbi, C., and Vera, J. H. 2002. A general expression for the ordered-packed volume fraction of hard spheres of different diameters. *Ind. Eng. Chem. Res.* 41(5): 1122–1128.

PROBLEMS

27.1 **Virial expansion of cubic equations of state of van der Waals type**

Cubic EOSs of van der Waals' type (the so-called "2P1T" cubic EOS) have the general form

$$P = \frac{RT}{\mathrm{v} - b} - \frac{a(T)}{(\mathrm{v} + mb)(\mathrm{v} + nb)}$$

where m, n are numerical constants, for example, $m = 0, n = 1$ for the Redlich–Kwong EOS, and $m = 1 - \sqrt{2}, n = 1 + \sqrt{2}$ for the Peng–Robinson equation. Prove Equation 27.20 for this cubic model and find an analogous formula for the third virial coefficient C.

27.2 **Configurational integral of a hard-sphere fluid**

The Carnahan–Starling [1] EOS for a fluid composed of nonattracting rigid spherical molecules is

$$z = \frac{1 + \eta + \eta^2 - \eta^3}{(1 - \eta)^3}$$

where $\eta = (\pi\sigma^3 / 6)\rho$ is a scaled (dimensionless) density, σ being the molecular diameter. Find the configurational integral Z_N corresponding to this fluid.

27.3 **Second virial coefficient from the square-well intermolecular potential**

The square-well intermolecular potential combines hard-core repulsion with finite-range attraction, and is defined as

$$E_r(r) = \begin{cases} \infty & r \le \sigma \\ -\varepsilon & \sigma < r \le R\sigma \\ 0 & r > R\sigma \end{cases}$$

where σ is the molecular diameter, ε is the well depth, and $(R-1)\sigma$ is the well width (see Figure 27.1). Derive an expression for the second virial coefficient of a fluid that obeys this potential.

27.4 **Alternative integral for the second virial coefficient**

Prove that Equation 27.15 can be rewritten as [2]

$$B = -\frac{2\pi N_A}{3kT} \int_r r^3 \frac{dE(r)}{dr} e^{-\frac{E(r)}{kT}} dr$$

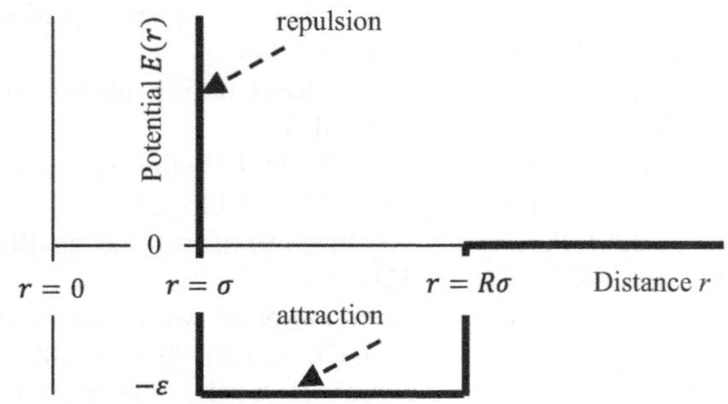

FIGURE 27.1 The square-well intermolecular potential.

27.5 **The principle of equipartition of energy**

In addition to the translational kinetic energy, which depends on the three velocity components $\upsilon_x^2, \upsilon_y^2, \upsilon_z^2$, polyatomic ideal gases may possess intramolecular energies arising from rotational motion, dependent on the angular velocity ω^2, and from vibrational motion, dependent on the displacement δ^2 (Hooke's law) and on the speed of displacement $\dot{\delta}^2$. The common feature of all these contributions is their dependence on the square of the respective variable. Assume that an ideal gas has F such degrees of freedom, of the general quadratic form $E_i(x_i) = a_i x_i^2,\ i = 1,\ldots,F$. Prove that the isochoric molar heat capacity is $C_v = F(R/2)$, where R is the ideal gas constant.

27.6 **Configurational and residual properties**

In classical thermodynamics, the deviation of real fluid behavior from that of a hypothetical ideal gas at the same temperature and density is represented by *residual* properties, defined, for example, for the Helmholtz function as

$$A^r(T,V,n) \equiv A(T,V,n) - A^*(T,V,n)$$

In statistical thermodynamics, on the other hand, the deviation arises from the configurational integral that depends on the intermolecular potentials (which ideal gases do not have), cf. Equation 27.2. The *configurational* Helmholtz energy is therefore given by Equation 26.12 as

$$A^{conf}(T,V,N) = -kT \ln Z_N$$

What is the connection between A^r and A^{conf}? Also find the relation between residual and configurational internal energy and pressure.

REFERENCES FOR PROBLEMS

1. Carnahan, N. F., and Starling, K. E. 1969. Equation of state for nonattracting rigid spheres. *J. Chem. Phys.* 51(2): 635–636.
2. Hirschfelder, J. O., Curtiss, C. F., and Bird, R. B. 1954. *Molecular Theory of Gases and Liquids*. New York: Wiley.

28 Statistical Thermodynamics Justification of Some Commonly Used Expressions for the Excess Gibbs Energy

Abstract

Van Laar model and its variation regular solution theory together with Margules Equation, Redlich–Kister Polynomial, Flory–Huggins, and Guggenheim–Staverman models, are all put in context. Guggenheim's combinatorial term normalized with Guggenheim's and with Wang's nonrandom factors, leading to the quasi-chemical theory and to the universal quasi-chemical (UNIQUAC) and Wilson's equations, respectively, are presented. The group approaches, universal functional activity coefficient (UNIFAC) and analytical solution of groups (ASOG), arising from these treatments, are discussed.

Models of mixtures are analytical expressions that, based on molecular concepts, aim to represent the thermodynamic behavior of solutions. As discussed in Chapter 17, the thermodynamic behavior of organic mixtures depends strongly on the nature of the functional groups present and on the length of the hydrocarbon chains of the molecules. The thermodynamic behavior of aqueous mixtures depends on the nature of the solute. A preliminary classification of the nature of organic mixtures helps in deciding what kind of modeling would be appropriate to describe their behavior. Considering Equation 16.12a,

$$g_\theta^E = h_\theta^E - Ts_\theta^E \tag{16.12a}$$

we may distinguish two extreme cases.

In the first one, there are no strong interactions between the functional groups forming the molecules and the predominant effect is due to difference in the hydrocarbon length of the compounds. In this case, the mixing will be almost athermal; that is, it will not be necessary to have heat exchanged to keep the temperature constant while mixing compounds at the same temperature. As the change in enthalpy is related to heat exchange, we associate this athermal process with an entropic effect. Thus, for athermal mixing, as an approximation, we write:

$$g_\theta^E \approx -Ts_\theta^E$$

$$h_\theta^E = \sum x_i \bar{h}_i^E \approx 0$$

So, $\bar{h}_i^E \approx 0$, and from Equation 16.17,

$$\bar{h}_i^E = -RT^2\left(\frac{\partial \ln \gamma_{i,\theta}}{\partial T}\right) \approx 0 \tag{28.1}$$

DOI: 10.1201/9781003431985-32

Hence, $\gamma_{i,\theta}$ is independent of temperature.

In the second case, the heat effect of mixing is present, and the process is called nonathermal, which means that it is thermal. This situation, when the enthalpic contribution dominates, is the most encountered in practice,

$$g_\theta^E \approx h_\theta^E$$

$$s_\theta^E = \sum x_i \overline{s}_i^E \approx 0$$

So, $\overline{s}_i^E \approx 0$, and from Equation 16.18,

$$\overline{s}_i^E = -RT\left(\frac{\partial \ln \gamma_{i,\theta}}{\partial T}\right) - R\ln \gamma_{i,\theta\theta} = -\left(\frac{\partial RT\ln \gamma_{i,\theta}}{\partial T}\right) \approx 0 \qquad (28.2)$$

Thus, in this case, $(RT \ln\gamma_{i,\theta})$ is independent of temperature.

In actual mixtures, the two extreme situations considered above are only approximations of reality. In any case, the relation between the excess enthalpy and the excess Gibbs energy is given by the Gibbs–Helmholtz equation, Equation 16.13a:

$$h_\theta^E = -RT^2\left(\frac{\partial\left(g_\theta^E / RT\right)}{\partial T}\right)_{P,n} \qquad (16.13a)$$

Therefore, any model that would represent well the excess Gibbs energy as a function of temperature could be used to obtain values for the excess enthalpy, that is, the heats of mixing. Reciprocally, information on heats of mixing can be used with advantage to fit parameters for the excess Gibbs energy as a function of temperature.

VAN LAAR MODEL

One of the first successful models of solutions was proposed by van Laar [1,2], who, being a student of van der Waals, based his work on the van der Waals equation of state. As this equation can be linked to basic statistical thermodynamics, it is only fair to start this chapter with a presentation of the van Laar model that is useful for nonathermal, that is, enthalpic, mixtures.

In the van Laar model, it is assumed that the mixing in the liquid phase occurs without change in volume; thus,

$$g_0^E \approx h_0^E = u_0^E + Pv_0^E \approx u_0^E = u - u^{id.0} = \Delta u$$

Subsequently, the two pure compounds are decompressed isothermally from the liquid phase to the ideal gas state, the two ideal gases are mixed at a constant temperature, and the mixture is isothermally compressed from the ideal gas state to the liquid phase. As discussed in Chapter 11, forming an ideal gas mixture happens without change in internal energy. Therefore, only the first and third steps contribute to the change in internal energy per mole of mixture,

$$g_0^E = \sum x_i \Delta u_i^{vaporization} - \Delta u_{mixture}^{vaporization}$$

From Equation 5.21, for an isothermal change assuming the validity of the van der Waals equation (Equation 25.11),

$$\left(\frac{\partial u}{\partial \text{v}}\right)_T = \frac{a}{\text{v}^2}$$

Thus, integrating between the liquid volume and an infinite volume for the ideal gas state, we get:

$$\Delta u^{vaporization} = \frac{a}{\text{v}^L} \tag{28.3}$$

and

$$g_0^E = \left(x_1 \frac{a_1}{\text{v}_1^L} + x_2 \frac{a_2}{\text{v}_2^L}\right) - \frac{a_m}{\text{v}_m^L} \tag{28.4}$$

As van Laar assumed no volume change, for a binary mixture, we can write:

$$\text{v}_m^L = x_1 \text{v}_1^L + x_2 \text{v}_2^L$$

For a_m, he used the mixing rule given by Equation 25.20:

$$a_m = x_1^2 a_1 + 2x_1 x_2 \sqrt{a_1 a_2} + x_2^2 a_2$$

Replacing and rearranging gives:

$$g_0^E = \frac{\left(x_1 \text{v}_1^L\right)\left(x_2 \text{v}_2^L\right)}{x_1 \text{v}_1^L + x_2 \text{v}_2^L}\left(\frac{\sqrt{a_1}}{\text{v}_1^L} - \frac{\sqrt{a_2}}{\text{v}_2^L}\right)^2 = \left(x_1 \text{v}_1^L + x_2 \text{v}_2^L\right)\Phi_1 \Phi_2 \left(\frac{\sqrt{a_1}}{\text{v}_1^L} - \frac{\sqrt{a_2}}{\text{v}_2^L}\right)^2 \tag{28.4a}$$

with

$$\Phi_i = \frac{x_i \text{v}_i^L}{x_1 \text{v}_1^L + x_2 \text{v}_2^L} \tag{28.5}$$

The activity coefficients obtained from this expression for the excess Gibbs energy using Equation 16.20 are:

$$\ln \gamma_1 = \frac{A}{\left(1 + \frac{A}{B}\frac{x_1}{x_2}\right)^2} \tag{28.6}$$

$$\ln \gamma_2 = \frac{B}{\left(1 + \frac{B}{A}\frac{x_2}{x_1}\right)^2} = \frac{B}{\left(\frac{B}{A}\frac{x_2}{x_1}\right)^2 \left(1 + \frac{A}{B}\frac{x_1}{x_2}\right)^2} \tag{28.7}$$

with

$$A = \frac{\text{v}_1^L}{RT}\left(\frac{\sqrt{a_1}}{\text{v}_1^L} - \frac{\sqrt{a_2}}{\text{v}_2^L}\right)^2 \tag{28.8}$$

and

$$B = \frac{\text{v}_2^L}{RT}\left(\frac{\sqrt{a_1}}{\text{v}_1^L} - \frac{\sqrt{a_2}}{\text{v}_2^L}\right)^2 \tag{28.9}$$

van Laar replaced the liquid volume by the van der Waals parameter b. For all practical purposes, this is immaterial. Independently of the values of the parameters, according to Equation 28.4a this model always gives positive values of the excess Gibbs energy; so, it can only predict positive deviations from ideality. In addition, being a thermal model, the activity coefficients satisfy Equation 28.2. In practice, the van Laar equation is used with adjustable parameters A and B. Thus, Equations 28.8 and 28.9 are irrelevant and the van Laar equation can represent positive or negative deviations from ideality. The van Laar equation is particularly useful for back-of-the-envelope calculations as the two parameters A and B can be directly evaluated from the knowledge of the coordinates of a single vapor–liquid equilibrium point using Equation 16.2 or Equation 16.4. For example, if the binary system has an azeotropic point, it is enough to know the coordinates of this point to calculate both parameters. For this direct evaluation of the parameters, it is useful to observe that from Equations 28.6 and 28.7,

$$\frac{\ln \gamma_1}{\ln \gamma_2} = \frac{B}{A}\left(\frac{x_2}{x_1}\right)^2 \tag{28.10}$$

REGULAR SOLUTION THEORY

The regular solution theory, independently developed by Hildebrand [3] and Scatchard [4], follows closely the van Laar equation. This theory defines the solubility parameter δ_i as:

$$\delta_i \equiv \sqrt{\frac{\Delta u_i^{vap}}{\mathrm{v}_i^L}} \tag{28.11}$$

Then from Equation 28.3,

$$\delta_i = \frac{\sqrt{a_i}}{\mathrm{v}_i^L}$$

Direct replacement into Equation 28.4 gives:

$$g_0^E = \left(x_1\mathrm{v}_1^L + x_2\mathrm{v}_2^L\right)\Phi_1\Phi_2\left(\delta_1 - \delta_2\right)^2 \tag{28.12}$$

and then

$$\ln \gamma_1 = \frac{\mathrm{v}_1^L\Phi_2^2}{RT}\left(\delta_1 - \delta_2\right)^2 \tag{28.13}$$

and

$$\ln \gamma_2 = \frac{\mathrm{v}_2^L\Phi_1^2}{RT}\left(\delta_1 - \delta_2\right)^2 \tag{28.14}$$

The form of Equations 28.12 through 28.14 explains why the parameter δ was called the solubility parameter. If δ has the same value for both compounds, the solution is ideal and the two compounds are totally soluble one in the other. If a solvent has a value of the solubility parameter close to that of a solute, it is a good solvent for that solute. Solubility parameters for different compounds are available in the literature [5,6].

The extension to mixtures of the regular solution theory follows directly from a generalization of Equation 28.4 with the van der Waals mixing rules:

$$g_0^E = \left(\sum_i x_i \frac{a_i}{v_i^L}\right) - \frac{a_m}{\sum_k x_k v_k^L} \tag{28.15}$$

with

$$a_m = \sum_j \sum_k x_j x_k a_{jk}$$

The activity coefficient for a component *i* in a regular solution mixture is given by:

$$\ln \gamma_i = \frac{v_i^L}{RT}\left(\delta_i - \bar{\delta}\right)^2 \tag{28.16}$$

with

$$\bar{\delta} = \sum_j \Phi_j \delta_j \tag{28.17}$$

There are two facts worth observing from these results. There is a major simplification in the equations caused by the introduction of the definition of the solubility parameter. Although both the van Laar and the regular solution treatments appear to be the same, the latter provides a clearer connection of the parameters with intermolecular forces through Equation 28.11. The regular solution theory interprets the ratio of the molar internal energy change for the total vaporization over the liquid molar volume as a cohesive energy density, that is, the energy that holds together the molecules in the liquid phase. In fact, as discussed in Chapter 27, the relation of the parameter *a* with intermolecular forces was already made in the derivation of the van der Waals equation of state.

TWO-SUFFIX MARGULES EQUATION AND THE REDLICH–KISTER POLYNOMIAL

As a corollary to this comparison between the van Laar and regular solution treatments, it is interesting to see their connection with empirical forms, which were used in the first half of the twentieth century. Considering the molar volumes of compounds 1 and 2 to be the same, say v^L, and defining a new parameter A_0 by

$$A_0 \equiv v^L\left(\delta_1 - \delta_2\right)^2 = v^L\left(\frac{\sqrt{a_1}}{v^L} - \frac{\sqrt{a_2}}{v^L}\right)^2 \tag{28.18}$$

Equation 28.4a can be written as:

$$g_0^E = A_0 x_1 x_2 \tag{28.19}$$

and Equations 28.6a and 28.7a become:

$$\ln \gamma_1 = \frac{A_0}{RT} x_2^2 \tag{28.20}$$

and

$$\ln \gamma_2 = \frac{A_0}{RT} x_1^2 \tag{28.21}$$

Considering the definition of A_0 by Equation 28.18, this is again a thermal model. Nevertheless, Equation 28.19 was an empirical proposal independent of the van Laar model or the regular solution theory. Thus, parameter A_0 was adjusted freely, and it could be positive or negative depending on the deviation of the solution from ideality. One problem with Equation 28.19, known as Margules two-suffix form [7], is that it gives a symmetric excess Gibbs energy. In order to account for asymmetry, extra terms were introduced in what was called a Redlich–Kister [8] expansion of the form,

$$g_0^E = x_1x_2\left[A_0 + B_0\left(x_1 - x_2\right) + C_0\left(x_1 - x_2\right)^2\right] \tag{28.22}$$

These empirical forms were used until the early fifties.

FLORY–HUGGINS AND GUGGENHEIM–STAVERMAN MODELS FOR ATHERMAL MIXTURES

In many cases of solutions of polymers in organic solvents, functional group interactions are either not present or weak. In these cases, we expect the entropic contribution to the excess Gibbs energy to be predominant. Huggins [9] and Flory [10] recognized this fact and, almost simultaneously, proposed a combinatorial expression based on the array of segments of the molecules in a two-dimensional lattice. Without entering into the details of the derivations, we give here the results for the excess entropy that, in this case, directly relates to the excess Gibbs energy. For the Flory–Huggins model,

$$\left(\frac{g_0^E}{RT}\right)_{FH} = -\frac{s_0^E}{R} = \sum_j x_j \ln\frac{\Phi_j}{x_j} \tag{28.23}$$

with

$$\Phi_j = \frac{x_jr_j}{\sum_k x_kr_k} \tag{28.24}$$

We observe that Equation 28.24 is just a generalization of Equation 28.5, considering $r_j = v_j^L / v_0$, where v_0 is the volume of a standard segment and r_j is then the number of segments of the molecule j. The activity coefficients obtained from the Flory–Huggins model are of the form:

$$\left(\ln\gamma_i\right)_{FH} = 1 - \frac{\Phi_i}{x_i} + \ln\frac{\Phi_i}{x_i} \tag{28.25}$$ [1]

For the Guggenheim–Staverman model,

$$\left(\frac{g_0^E}{RT}\right)_{GS} = -\frac{s_0^E}{R} = \sum_j x_j \ln\frac{\Phi_j}{x_j} + \sum_j x_jq_j \ln\frac{\theta_j}{\Phi_j} \tag{28.26}$$

with

$$\theta_j = \frac{x_jq_j}{\sum_k x_kq_k} \tag{28.27}$$

In Equation 28.27, q_i is a surface area parameter of molecule i. As detailed in a review by Sayegh and Vera [11], Guggenheim [12] considered that the surface area of the segments was important for the combinatorial array in a lattice, and Staverman [13] demonstrated that Guggenheim's combinatorial expression was applicable not only to open-chain r-mers but also to bulky molecules.

When a new theory is presented, it is not easy to get rid of older nomenclature and symbols that are unnecessary in the new approach. In his combinatorial analysis, Guggenheim retained the nomenclature coming from lattice models. Thus, in Equations 28.26 and 28.27, we have taken the liberty of modifying Guggenheim's nomenclature and used the symbol q_j for what in Guggenheim's treatment appears as $zq_j/2$, with z being the coordination number, that is, the number of closest neighbors to a site in a lattice. This change is only formal, as it does not change the concepts, or the algebra involved. The activity coefficients obtained from the Guggenheim–Staverman model are of the form:

$$\left(\ln\gamma_i\right)_{GS} = \left(1 - \frac{\Phi_i}{x_i} + \ln\frac{\Phi_i}{x_i}\right) - q_i\left(1 - \frac{\Phi_i}{\theta_i} + \ln\frac{\Phi_i}{\theta_i}\right) \tag{28.28}$$

If q_i is directly proportional to r_i, Equation 28.27 reduces to Equation 28.24 and Equation 28.28 reduces to Equation 28.25.

COMPLETE REGULAR SOLUTIONS AND FLORY–HUGGINS EXPRESSIONS

It did not take long for researchers to realize that both the regular solution theory and the Flory–Huggins theory were using only half of Equation 16.12a:

$$g_\theta^E = h_\theta^E - Ts_\theta^E \tag{16.12a}$$

For a binary polymer–solvent mixture, the Flory–Huggins complete form is:

$$\frac{g_0^E}{RT} = \left(x_s \ln\frac{\Phi_s}{x_s} + x_p \ln\frac{\Phi_p}{x_p}\right) + v_s^L\left(x_s + x_p r_p\right)\Phi_s\Phi_p\chi_F \tag{28.29}$$

and for a binary mixture, the regular solution theory complete form is:

$$\frac{g_0^E}{RT} = \left(x_1 \ln\frac{\Phi_1}{x_1} + x_2 \ln\frac{\Phi_2}{x_2}\right) + \left(x_1 v_1^L + x_2 v_2^L\right)\Phi_1\Phi_2\frac{\left(\delta_1 - \delta_2\right)^2}{RT} \tag{28.30}$$

or, defining $r_2 = v_2^L / v_1^L$,

$$\frac{g_0^E}{RT} = \left(x_1 \ln\frac{\Phi_1}{x_1} + x_2 \ln\frac{\Phi_2}{x_2}\right) + v_1^L\left(x_1 + x_2 r_2\right)\Phi_1\Phi_2\frac{\left(\delta_1 - \delta_2\right)^2}{RT} \tag{28.30a}$$

Comparison of Equation 28.29 and Equation 28.30a shows that the Flory parameter is directly related to the solubility parameters. However, in practice, the Flory parameter is used as an adjustable parameter and sometimes is considered to be a function of temperature. The activity coefficient of a compound i obtained from the complete form of both theories is just the sum of the two contributions. For the complete regular solution theory, for example, from Equations 28.16 and 28.25,

$$\ln\gamma_i = \left(1 - \frac{\Phi_i}{x_i} + \ln\frac{\Phi_i}{x_i}\right) + \frac{v_i^L}{RT}\left(\delta_i - \bar{\delta}\right)^2 \tag{28.31}$$

MODERN TIMES

The second half of the twentieth century was a period of intense search for models of the excess Gibbs energy or, equivalently, of activity coefficients for the calculation of phase equilibria involving nonideal liquid mixtures. One stream followed the quasi-chemical theory of Guggenheim [14–18].

Another stream developed models based either on the two-fluid approach [19] or on the local composition concept proposed by Wilson [20] and elaborated further by Prausnitz and his collaborators [21,22], among others. These later models were and are extremely successful, and with the advent of computers, they were easy to use. Both kinds of treatments eventually evolved into approaches based on the functional groups forming the molecules. The quasi-chemical theory evolved into the form proposed by Kehiaian et al. [15], while the local composition models into the analytical solution of groups (ASOG) [23–25] and universal functional activity coefficient (UNIFAC) [26]. We present here a brief description of the quasi-chemical theory and the two-fluid and local composition models.

QUASI-CHEMICAL THEORY

In this presentation, we distinguish between the Guggenheim combinatorial partition function and the Guggenheim quasi-chemical partition function, and we also get rid of the coordination number that was included in this treatment as a leftover from old lattice approaches. Following the work of Kehiaian et al. [15], we adopt from the beginning the group approach, which reduces to a molecular approach when one considers the molecules of different compounds as individual groups. The Guggenheim partition function is based on the idea that in a liquid mixture, what is important is the way in which different molecules or the segments of the molecules contact each other. The segments are characteristic functional groups, so the whole liquid mixture can be considered a mixture of groups. The importance of the functional groups forming the molecules has been discussed in Chapter 17. The interaction between functional groups is determinant in the thermodynamic behavior of the mixture. Thus, Guggenheim considered that the mixture behavior is characterized by how the contacts between groups arrange in the liquid phase. Based on the combinatorial array for the number of contacts of each type, Guggenheim's combinatorial partition function is written as:

$$Q_{CG} = \frac{\left(\sum_{k=1}^{G} \eta_{kk} + \sum_{k=1}^{G} \sum_{l \neq k}^{G} \frac{1}{2} \eta_{kl}\right)!}{\prod_{k=1}^{G} (\eta_{kk})! \prod_{k \neq l}^{G} \left(\frac{1}{2} \eta_{kl}\right)!} \tag{28.32}$$

In this expression, η_{kl} represents the "number of contacts" between groups k and l, considered to be proportional to the surface area of group k available for contacts with group l. This availability for contacts will certainly depend on the energetic compatibility of the groups. Interactions between groups are always attractive, but the interactions between groups of different kinds can be stronger or weaker than the interactions between groups of the same kind. Stronger interactions will tend to produce negative deviations from ideality, and weaker interactions will tend to produce positive deviations from ideality. If the interactions are similar, the array will be random and the mixture will approach the behavior of an ideal mixture. The summation of the number of contacts will be the same for the random and nonrandom arrays, as the surface areas of the groups do not change. Only the way in which the contacts are arranged changes. Hence,

$$2\eta_{kk} + \sum_{l \neq k}^{G} \eta_{kl} = \alpha_k \tag{28.33a}$$

and

$$\left(\sum_{k=1}^{G} \eta_{kk} + \sum_{k=1}^{G} \sum_{l \neq k}^{G} \frac{1}{2} \eta_{kl}\right) = \left(\sum_{k=1}^{G} \eta_{kk}^{0} + \sum_{k=1}^{G} \sum_{l \neq k}^{G} \frac{1}{2} \eta_{kl}^{0}\right) = \sum_{k=l}^{G} \frac{\alpha_k}{2} \tag{28.33b}$$

where η_{kl}^{0} represents the number of contacts of type k-l in a random array and α_k is the total surface area of groups k available for external contacts. For compatibility with the counting of the number of contacts, α_k is better interpreted as the total number of contact points of group k, each point representing a unit of surface area. This total number is proportional to the surface area available for external contacts in group k. The factor 1/2, which is immaterial, is introduced to avoid double counting of contacts. Hence, Equation 28.32 for a random array can be written as:

$$Q_{CG}^{0} = \frac{\left(\sum_{k=1}^{G} \frac{\alpha_k}{2}\right)!}{\prod_{k=1}^{G} \left(n_{kk}^{0}\right)! \prod_{k \neq l}^{G} \left(n_{kl}^{0}\right)!} \tag{28.32a}$$

In his quasi-chemical approach, Guggenheim normalized the combinatorial partition function to modify the total random array of contacts and multiplied the combinatorial partition function by a factor g_R of the form:

$$g_R = \frac{\prod_{k=1}^{G} \left(n_{kk}^{0}\right)! \prod_{k \neq l}^{G} \left(\frac{1}{2} n_{kl}^{0}\right)!}{\prod_{k=1}^{G} \left(n_{kk}\right)! \prod_{k \neq l}^{G} \left(\frac{1}{2} n_{kl}\right)!} \tag{28.34}$$

So, for the complete quasi-chemical partition function, he obtained:

$$Q_{QC} = Q_{CG}^{0} g_R \exp\left[\frac{-E}{RT}\right] \tag{28.35}$$

with

$$-E = \sum_{k=1}^{G} \eta_{kk} \varepsilon_{kk} + \sum_{k=1}^{G-1} \sum_{l \triangleright k}^{G} \eta_{kl} \varepsilon_{kl} \tag{28.36}$$

where ε_{kl} is the interaction energy between groups k and l. By minimization of the Helmholtz function for the equilibrium condition, Guggenheim obtained the following relation:

$$\frac{4\eta_{kk}\eta_{ll}}{\eta_{kl}^{2}} = \exp\left[\frac{\Delta\varepsilon_{kl}}{RT}\right] \tag{28.37}$$

Due to the similitude of this expression with the form of a chemical equilibrium constant, Guggenheim called this treatment the quasi-chemical theory. A complete derivation of the thermodynamic functions generated from this theory was presented by Wilczek-Vera and Vera [18]. Panayiotou and Vera [27] identified the rigorous use of the concept of local composition in the quasi-chemical theory.

The quasi-chemical theory of Guggenheim, attractive as it seems from a theoretical point of view, has some serious drawbacks. On the theoretical side, Prigogine et al. [28] and also Kemeny et al. [29] have argued that Guggenheim's normalization of the partition function satisfying the condition for total randomness of the contacts between groups makes this theory valid only for cases close to total randomness and not for cases of very specific interactions between groups. On the practical side, the quasi-chemical theory has two major drawbacks. First, by Equation 28.37, it ties the interactions between different groups kl to the interactions of the individual groups kk and ll. Thus, it limits the treatment to a single binary parameter when it is well known that the thermodynamic treatment of binary mixtures needs two parameters due to the different behaviors of a mixture in the two regions of high dilution. The second drawback is that for more than two groups, the quasi-chemical relation, Equation 28.37, is far from simple to solve. A general solution method was developed by Abusleme and Vera [30], but it is certainly not easy to apply. Notably, both problems presented by the quasi-chemical treatment arise from the form of the normalization function g_R used by Guggenheim, Equation 28.34, which is not part of the combinatorial expression, Equation 28.32a.

WILSON EQUATION FOR THE EXCESS GIBBS ENERGY

The structure of the equation for the excess Gibbs energy proposed by Wilson [20] is best seen in the context of Flory treatment, Equations 28.23–28.25. Wilson argued that due to specific interactions in a solution of components i and j, the local mole fraction of compound j around a central molecule i, x_{ij}, would be different from the overall mole fraction of compound j, x_j, in the mixture. For this local composition, he proposed:

$$\frac{x_{ij}}{x_{ii}} = \frac{x_j}{x_i}\exp\left(\frac{-(\lambda_{ij}-\lambda_{ii})}{RT}\right) = \frac{x_j}{x_i}\tau_{ij} \tag{28.38}$$

So, instead of Equation 28.24, Wilson wrote a "local" volume, or segment, fraction as:

$$\Phi_j^w = \frac{x_{jj}r_j}{\sum_k x_{jk}r_k} = \frac{x_j r_j}{\sum_k x_k r_k \tau_{jk}} = \frac{\Phi_j}{\sum_k \Phi_k \tau_{jk}} \tag{28.39}$$

with Φ_j given by Equation 28.5. Replacing this expression in Equation 28.23,

$$\left(\frac{g_0^E}{RT}\right)_w = \sum_j x_j \ln\frac{\Phi_j^w}{x_j} = \sum_j x_j \ln\frac{\Phi_j}{x_j\sum_k \Phi_k\tau_{jk}} = \sum_j x_j \ln\frac{\Phi_j}{x_j} - \sum_j x_j \ln\sum_k \Phi_k\tau_{jk} \tag{28.40a}$$

On the right-hand side of Equation 28.40a, we recognize a combinatorial contribution and the introduction of an enthalpic term. The Wilson equation is usually written in a more compact form as:

$$\left(\frac{g_0^E}{RT}\right)_w = -\sum_j x_j \ln\sum_k x_k\Lambda_{jk} \tag{28.40b}$$

with

$$\Lambda_{jk} = \frac{r_k}{r_j}\tau_{jk} \tag{28.41}$$

We observe here that by their definition in Equations 28.38 and 28.41, both τ_{jk} and Λ_{jk} are considered to be temperature dependent.

The expression for the activity coefficient obtained from Equation 28.40a is:

$$\ln\gamma_i = \left(1-\frac{\Phi_i}{x_i}+\ln\frac{\Phi_i}{x_i}\right)+\left(\frac{\Phi_i}{x_i}-\ln\sum_k \Phi_k\tau_{ik}-\frac{\Phi_i}{x_i}\sum_k \frac{x_k\tau_{ki}}{\Sigma_j\Phi_j\tau_{kj}}\right) \tag{28.42a}$$

And the activity coefficient obtained from Equation 28.40b is:

$$\ln\gamma_i = 1-\ln\sum_k x_k\Lambda_{ik} - \sum_k \frac{x_k\Lambda_{ki}}{\Sigma_j x_j\Lambda_{kj}} \tag{28.42b}$$

Equations 28.42a and 28.42b are just a transformation of one another. The form shown in Equation 28.42a explicitly shows the combinatorial contribution and the contribution due to interactions between molecules. Realizing that Equation 28.40 (a or b) was unable to describe the case of instability of a liquid mixture, that is, liquid–liquid phase separation, Wilson [20] added a third parameter, C_W, which should be greater than unity for the representation of mixtures splitting into two liquid phases:

$$\left(\frac{g_0^E}{RT}\right)_w = -C_W \sum_j x_j \ln \sum_k x_k \Lambda_{jk} \tag{28.43}$$

This three-parameter form of the Wilson equation satisfies the condition for instability.

Flemr [31] and also McDermott and Ashton [32] questioned the theoretical basis of Wilson's local composition expression. A redefinition of local composition and local surface area fractions compatible with the quasi-chemical theory was later presented by Panayiotou and Vera [27]. A new approach to this problem is discussed at the end of this chapter.

BIRTH OF THE SOLUTION OF GROUPS METHOD TO CALCULATE ACTIVITY COEFFICIENTS IN LIQUID MIXTURES

Superior minds sometimes can foresee relations that normal intellects will take time to justify. This is the case with the proposal of the Wilson equation, Wilson and Deal's proposal of a group-contribution method to evaluate the activity coefficients in organic mixtures, the formulation of ASOG by Derr, and the proposal of universal quasi-chemical (UNIQUAC). We briefly review these ground-breaking advances that followed Wilson's equation. We use our own nomenclature.

Wilson and Deal [23], based on some previous studies on heats of mixing [33,34], and obviously aware of the two contributions to the excess Gibbs energy shown by Equation 16.12a, proposed that the activity coefficient of a compound in an organic mixture be expressed as the sum of a Flory combinatorial term and an "enthalpic term" arising from the interactions between the groups:

$$\ln \gamma_i = (\ln \gamma_i)_{combinatorial} + (\ln \gamma_i)_{interactions}$$

$$\ln \gamma_i = \left(1 - \frac{r_i}{\sum_k x_k r_k} + \ln \frac{r_i}{\sum_k x_k r_k}\right) + \sum_k \nu_{ki}\left[\ln \Gamma_k - \ln \Gamma_k^{(i)}\right] \tag{28.44}$$

The presentation of this expression, so familiar to us now, is so impressive that it is necessary to say something about each of the two terms. For the Flory combinatorial term, Wilson and Deal, combining Equations 28.24 and 28.25, explicitly showed the size parameter r_i of compound i, which they proposed to identify with the number of atoms different from hydrogen in molecule i. This way, they clearly were counting the functional groups present in compound i. In the second term, ν_{ki} is the number of groups of type k in molecule i, Γ_k is the activity coefficient of a group of type k with respect to an arbitrary but well-defined standard state, and $\Gamma_k^{(i)}$ is the activity coefficient of a group of type k in pure compound i, with respect to the same standard state as Γ_k.

Here, we observe that Γ_k is a function of the (group) composition of the mixture, while $\Gamma_k^{(i)}$ is a function of the group composition in pure compound i. The (mole) fraction of group k is given by:

$$X_k = \frac{\sum_j x_j \nu_{kj}}{\sum_l \sum_m x_m \nu_{lm}} \tag{28.45a}$$

and in pure compound i,

$$X_k^{(i)} = \frac{\nu_{ki}}{\sum_l \nu_{li}} \quad (28.45b)$$

Thus, according to Equation 28.44, the activity coefficient $\gamma_{i,0}$ is normalized to unity at the condition of a pure compound i. The group activity coefficient Γ_k is normalized to unity in a system formed only by group k. This is the case of the group CH_2 in alkanes, as no distinction is made between the methyl (CH_3–) and methylene (–CH_2–) groups. This normalization was assumed to also be valid for group –OH in pure water. Water in mixtures was considered to be 1.5 (–OH) groups. To test the model, they studied several binary and ternary systems: alcohol–alkane, alcohol–alcohol, water–alcohol, and so forth. With the assumption that the group activity coefficient is only a function of the group composition of the mixture, and it does not depend on the way that the groups are attached in the molecules, data from one system could be used for a different system at the same group composition. Activity coefficient data of water in alcohol–water systems, from methanol to decyl alcohol, provided the information necessary to calculate Γ_{OH} at different X_{OH} values in CH_2/OH systems. Similarly, from activity coefficient data of the hydrocarbon in hydrocarbon–alcohol systems, it is possible to calculate Γ_{CH_2} at different X_{CH_2} values in CH_2/OH systems. The method was direct and simple. In the evaluation of the group activity coefficients, the authors did not use any model, other than Equation 28.44, which does not contain any adjustable parameter. The combinatorial contribution in Equation 28.44 is calculated and subtracted from the logarithm of the activity coefficient γ. Plots of group activity coefficients Γ_k against the group k mole fraction, coming from very different systems at the same temperature, all fell over a single curve for a given group. In addition, it was observed that the curves of group activity coefficients against group fraction presented a remarkable symmetry, with curves of compound activity coefficients versus mole fraction of the compound. Figure 28.1 depicts the values of the activity coefficient of group OH in the system methanol–hexane at 60°C. Figure 28.2 presents the values of the activity coefficient of methanol in the same system.

Studies of other alkane–alcohol systems gave values of the activity coefficient of the OH group overlapping and extending the curve shown in Figure 28.1. These results are so impressive that they should be shown in any lecture dealing with group methods. Further results were presented

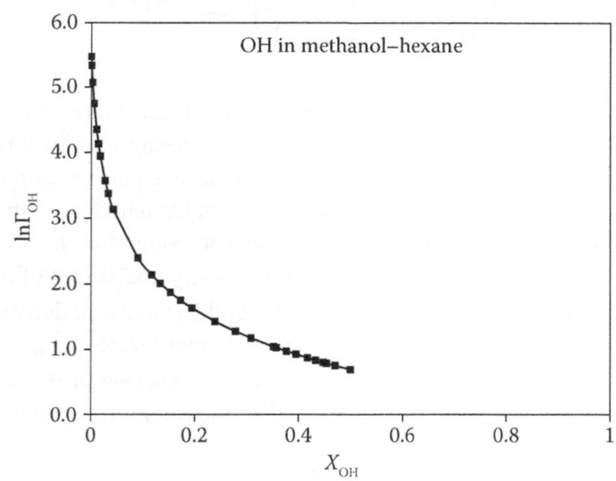

FIGURE 28.1 Activity coefficient of group OH in the system methanol–hexane at 60°C. (Data from Von Schellehr, M., Schubert, Koennecke, H. G., *J. Prakt. Chem.*, 311, 974–982, 1969.)

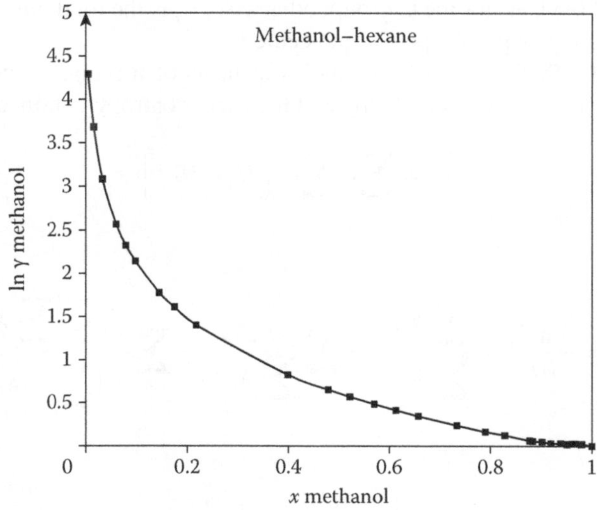

FIGURE 28.2 Activity coefficient of methanol in the system methanol–hexane at 60°C. (Data from Von Schellehr, M., Schubert, Koennecke, H. G., *J. Prakt. Chem.*, 311, 974–982, 1969.)

by Deal and Derr [24], who announced that Derr had developed ASOG, to be published later. The publication of Derr, based on the use of the Wilson equation for the excess Gibbs energy, came a year later [25].

ANALYTICAL SOLUTION OF GROUPS (ASOG)

After obtaining plots of $\ln \Gamma_k$ versus X_k, without using a single model and observing that the curves looked like curves of $\ln \gamma_i$ versus x_i, Derr and Deal [25] used the Wilson equation for the correlation of the experimental values of $\ln \Gamma_k$. From Equation 28.42b, in terms of group fractions, in its more general form, we write:

$$\ln \Gamma_k = 1 - \ln \sum\nolimits_l X_l A_{kl} - \sum\nolimits_l \frac{X_l A_{lk}}{\sum_m X_m A_{lm}} \tag{28.46a}$$

and

$$\ln \Gamma_k^{(i)} = 1 - \ln \sum_l X_l^{(i)} A_{kl} - \sum_l \frac{X_l^{(i)} A_{lk}}{\sum_m X_m^{(i)} A_{lm}} \tag{28.46b}$$

Values for the group interaction parameters were published much later by Kojima and Tochigi [35a,b], using the form:

$$A_{kl} = \exp\left(m_{kl} + \frac{n_{kl}}{T} \right) \tag{28.47}$$

Group parameters m_{kl} and n_{kl} for this version of ASOG, usually designated as ASOG-KT, are available in the literature [35a,b], and they are reproduced with permission in Appendix D. Other applications of ASOG-KT for high-pressure or polymer solutions have also been developed [36,37]. For the counting of groups (ν_{kj}), the ASOG-KT method recommends using 1.6 OH for water, 0.8 CH_2

for the CH group, and 0.5 CH_2 for the C group; otherwise, ν_{kj} is the total number of atoms different from hydrogen in all the groups of type k in molecule i.

As the proposal of ASOG started from studies of heats of mixing, for completeness, we give below the expressions for the excess enthalpy and the excess entropy obtained from ASOG [38]:

$$h^E = \sum_j x_j \sum_k \nu_{kj} \left[H_k - H_k^{(j)} \right]$$

with

$$\frac{H_k}{RT^2} = \frac{\partial \ln \Gamma_k}{\partial T} = \frac{\sum_l X_l B_{kl}}{\sum_j X_j A_{kj}} + \sum_j \frac{X_j B_{jk}}{\sum_l X_l A_{jl}} - \sum_j \frac{X_j A_{jk} \left(\sum_l X_l B_{jl} \right)}{\left(\sum_l X_l A_{jl} \right)^2}$$

where

$$B_{kl} = \frac{\partial A_{kl}}{\partial T} = -A_{kl} \frac{n_{kl}}{T^2}$$

and

$$s^E = \sum_j x_j \sum_k \nu_{kj} \left[S_k - S_k^{(j)} \right]$$

with

$$\frac{S_k}{R} = \frac{H_k}{RT} - \ln \Gamma_k$$

NONRANDOM TWO-LIQUID EQUATION FOR THE EXCESS GIBBS ENERGY

Based on Scott's two-fluid model [39], coupled with Wilson's local composition concept [20], Renon and Prausnitz [40] proposed a three-parameter expression for the excess Gibbs energy of the form:

$$\left(\frac{g_0^E}{RT} \right)_{NRTL} = \sum_k x_k \frac{\sum_j x_j G_{jk} \tau_{jk}}{\sum_l x_l G_{lk}} \tag{28.48}$$

with

$$G_{jk} = \exp\left(-\alpha_{jk} \tau_{jk} \right) \tag{28.49}$$

and

$$\tau_{jk} = \frac{\Delta g_{jk}}{RT} \tag{28.50}$$

In these expressions, the adjustable parameters in a binary system are Δg_{12}, Δg_{21}, and $\alpha_{12}=\alpha_{21}$. No additional parameters are needed for multicomponent systems. The difference between the nonrandom two-liquid equation (NRTL) and the three-parameter Wilson equation is that in the latter one, once the third parameter is fixed, it has the same value for all the binary systems in the multicomponent system under study. As it was observed by Renon and Prausnitz [19], following a suggestion by R. L. Scott, had they made their two-fluid derivation for the excess enthalpy instead of the excess Gibbs energy, they would have obtained Wilson's three-parameter equation.

The expressions for the activity coefficients obtained from Equation 28.48 take the form:

$$\ln\gamma_i = \frac{\sum_j x_j G_{ji}\tau_{ji}}{\sum_l x_l G_{ji}} + \sum_j \frac{x_j G_{ij}}{\sum_l x_l G_{ll}}\left[\tau_{ij} - \frac{\sum_r x_r G_{rj}\tau_{rj}}{\sum_l x_l G_{ll}}\right] \tag{28.51a}$$

For a binary system, the expressions for the activity coefficients take a friendlier form,

$$\ln\gamma_1 = x_2^2\left[\tau_{21}\left(\frac{G_{21}}{x_1 + x_2 G_{21}}\right)^2 + \frac{\tau_{12}G_{12}}{(x_2 + x_1 G_{12})^2}\right] \tag{28.51b}$$

and

$$\ln\gamma_2 = x_1^2\left[\tau_{12}\left(\frac{G_{12}}{x_2 + x_1 G_{12}}\right)^2 + \frac{\tau_{21}G_{21}}{(x_1 + x_2 G_{21})^2}\right] \tag{28.51c}$$

Although the theoretical basis of both the local composition concept, as formulated by Wilson, and the two-fluid approach have been questioned [41], the NRTL equation has been found to be extremely successful for the correlation and prediction of vapor–liquid and liquid–liquid equilibria.

UNIQUAC EQUATION

As stated by Prausnitz et al. [42a], the UNIQUAC equation proposed by Abrams and Prausnitz [21] was derived by Maurer and Prausnitz [22] using phenomenological arguments based on the local composition concept and the two-liquid approach, in a similar way as these concepts were used in the derivation of NRTL. These authors conclude that UNIQUAC, being a very useful two-parameter equation, "cannot claim as much theoretical foundation as was hoped originally" [22, p.96]. Looking in retrospect, the name UNIQUAC was a misnomer. The main advantage of UNIQUAC is that it is unrelated to the quasi-chemical theory, so it does not have any of its limitations. An alternative derivation of UNIQUAC relying only on the local composition concept was proposed by Vera et al. [43]. This derivation does not need the two-liquid approach. In addition, it can be related with a renormalization of Guggenheim's partition function proposed by Wang and Vera [44]. The main point distinguishing Guggenheim's from Flory–Huggins's combinatorial term was that the important factor for the interaction of the molecules (or segments of the molecules) was the surface area available for contact. Thus, the key innovation introduced by UNIQUAC was the incorporation of the surface area fraction, as defined by Equation 28.27. Here, we follow the simpler derivation suggested by Vera et al. [43], and following Wilson, we define a local surface area by the equivalent to Equation 28.42:

$$\xi_j^{LS} = \frac{x_{jj}q_j}{\sum_k x_{jk}q_k} = \frac{x_j q_j}{\sum_k x_k q_k \tau_{jk}} = \frac{\theta_j}{\sum_k \theta_k \tau_{jk}} \tag{28.52}$$

Replacing this expression in Equation 28.26 gives:

$$\left(\frac{g_0^E}{RT}\right)^{LS} = \left(\sum_j x_j \ln\frac{\Phi_j}{x_j} + \sum_j x_j q_j \ln\frac{\theta_j}{\Phi_j}\right) - \sum_j x_j q_j \ln\sum_k \theta_k \tau_{jk} \tag{28.53}$$

This equation, called the local surface Guggenheim (LSG) form, only differs from UNIQUAC [42b] by a factor ($z/2$) that was left in the combinatorial part of the latter from its historical lattice roots. The two first terms of the right-hand side of Equation 28.53 correspond to the combinatorial expression coming from Equation 28.26. Models of solutions keep changing with time, so when using parameters reported in the literature, it is mandatory to use the exact analytical form that was used for the evaluation of the parameters.

The expression for the activity coefficient derived from LSG, Equation 28.53, has the form:

$$(\ln \gamma_i) = \left(1 - \frac{\Phi_i}{x_i} + \ln \frac{\Phi_i}{x_i}\right) - q_i\left(1 - \frac{\Phi_i}{\theta_i} + \ln \frac{\Phi_i}{\theta_i}\right) + q_i\left[1 - \ln \sum \theta_k \tau_{ik} - \sum_k \frac{\theta_k \tau_{ki}}{\sum_j \theta_j \tau_{kj}}\right] \tag{28.54a}$$

The first two brackets in Equation 28.54a correspond to the combinatorial contribution given by Equation 28.28. Equation 28.54a can be written in a more compact form as:

$$(\ln \gamma_i) = \left(1 - \frac{\Phi_i}{x_i} + \ln \frac{\Phi_i}{x_i}\right) + q_i\left[\frac{\Phi_i}{\theta_i} - \ln\left(\frac{\Phi_i}{\theta_i}\sum_k \theta_k \tau_{ik}\right) - \sum_k \frac{\theta_k \tau_{ki}}{\sum_j \theta_j \tau_{kj}}\right] \tag{28.54b}$$

For UNIQUAC, the activity coefficients are given by:

$$(\ln \gamma_i) = \left(1 - \frac{\Phi_i}{x_i} + \ln \frac{\Phi_i}{x_i}\right) - \frac{z q_i}{2}\left(1 - \frac{\Phi_i}{\theta_i} + \ln \frac{\Phi_i}{\theta_i}\right) + q_i\left[1 - \ln \sum_k \theta_k \tau_{ik} - \sum_k \frac{\theta_k \tau_{ki}}{\sum_j \theta_j \tau_{kj}}\right] \tag{28.55}$$

with $z=10$. Notably, if for all compounds q_i is directly proportional to r_i, then $\theta_i=\Phi_i$ and both Equation 28.54a or Equation 28.54b and Equation 28.55 reduce to:

$$\ln \gamma_i = \left(1 - \frac{\Phi_i}{x_i} + \ln \frac{\Phi_i}{x_i}\right) + q_i\left[1 - \ln \sum_k \Phi_i \tau_{ik} - \sum_k \frac{\Phi_i \tau_{ki}}{\sum_j \Phi_i \tau_{kj}}\right]$$

This form is similar but not identical to Wilson's equation, Equation 28.42a. As discussed by Vera et al. [43], Wilson's equation is only recovered when both Equations 28.39 and 28.52 are combined with Guggenheim's combinatorial form and for all compounds q_i is directly proportional to r_i.

UNIFAC

There are many versions of UNIFAC. We give here the original form only. Basically, UNIFAC follows the scheme of ASOG, and it considers the activity coefficient of a compound given by a combinatorial contribution and the contribution of the group interactions. For the combinatorial term, UNIFAC uses Guggenheim's form, and for the group contribution, UNIFAC follows the suggestion of Wilson and Deal [23] and writes:

$$\ln\gamma_i = \left(1-\frac{\Phi_i}{x_i}+\ln\frac{\Phi_i}{x_i}\right)-\frac{zq_i}{2}\left(1-\frac{\Phi_i}{\theta_i}+\ln\frac{\Phi_i}{\theta_i}\right)+\sum_k \nu_{ki}\left[\ln\Gamma_k-\ln\Gamma_k^{(i)}\right] \tag{28.56}$$

with ϕ_i and θ_i given by Equations 28.24 and 28.27, respectively, but using

$$r_i = \sum_k \nu_{ki}R_k \tag{28.57}$$

and

$$q_i = \sum_k \nu_{ki}Q_k \tag{28.58}$$

where R_k and Q_k are the size and surface area parameters of group k, respectively. These values are tabulated for the different groups.

For the activity coefficients of the groups, UNIFAC uses the same form as the term for interactions in UNIQUAC but written in terms of groups, namely,

$$\ln\Gamma_k = Q_k\left[1-\ln\sum_l \Theta_l a_{kl}-\sum_l\frac{\Theta_l a_{lk}}{\sum_m \Theta_m a_{lm}}\right] \tag{28.59}$$

with

$$\Theta_k = \frac{\sum_j x_j\nu_{kj}Q_k}{\sum_l\sum_m x_m\nu_{lm}Q_m} \tag{28.60}$$

and in pure compound i,

$$\Theta_k^{(i)} = \frac{\nu_{ki}Q_k}{\sum_l \nu_{li}Q_l} \tag{28.60a}$$

The values of the group interaction parameters are a function of temperature,

$$a_{kl} = \exp\left[\frac{B_{kl}}{T}\right] \tag{28.61}$$

Comparisons between ASOG and UNIFAC are scarce and depend on the systems chosen for the comparison. See, for example, a publication by Gmehling et al. [45].

WANG'S RENORMALIZATION OF GUGGENHEIM'S PARTITION FUNCTION

As stated in the presentation of the quasi-chemical theory, its major drawbacks come from the normalization factor introduced by Guggenheim. Thus, Wang and Vera [44] proposed to normalize Guggenheim's combinatorial term with a different function meeting the limits of total randomness, indicated with a superscript 0, and total order, indicated with a superscript *, and wrote:

$$Q = Q_{CG}F_\alpha \tag{28.62}$$

For total disorder, Guggenheim's combinatorial term gives the right limiting behavior, that is, $Q = Q_{CG}^0$, so

$$F_\alpha^0 = 1 \tag{28.63}$$

For total order, indicated with an asterisk, $Q^* = 1$, because when all contacts are of type k-k, and all $\eta_{kl}^* = 0$ for $l \neq k$, there is one single array for the system. Thus, for complete order,

$$F_\alpha^* = \frac{1}{Q_{CG}^*} = \frac{\prod_{k=1}^{G} \left(\eta_{kk}^*\right)!}{\left(\sum_{k=1}^{G} \frac{\alpha_k}{2}\right)!}$$

From Equation 28.33, when all $\eta_{kl}^* = 0$ for $l \neq k$, $2\eta_{kk}^* = \alpha_k$, and we write:

$$F_\alpha^* = \frac{\prod_{k=1}^{G} \left(\frac{\alpha_k}{2}\right)!}{\left(\sum_{k=1}^{G} \frac{\alpha_k}{2}\right)!}$$

Using Stirling's approximation for the factorials, we obtain:

$$F_\alpha^* = \prod_{k=1}^{G} \left[\frac{\frac{\alpha_k}{2}}{\sum_{k=1}^{G} \frac{\alpha_k}{2}} \right]^{\frac{\alpha_k}{2}} \tag{28.64}$$

The function satisfying both limits proposed by Wang is:

$$F_\alpha = \prod_{k=1}^{G} \left[\frac{\sum_{l=1}^{G} \frac{\alpha_k}{2} \tau_{lk}}{\sum_{k=1}^{G} \frac{\alpha_k}{2}} \right]^{\frac{\alpha_k}{2}} \tag{28.65}$$

with $\tau_{kk} = \tau_{ll} = 1$ and, in general, $\tau_{kl} \neq \tau_{lk}$. For totally disordered systems, $\tau_{kl}^0 = \tau_{lk}^0 = 1$, and for totally ordered systems, $\tau_{kl}^* = \tau_{lk}^* = 0$. The values of the parameters τ_{kl} and τ_{lk} characterize the degree of order in the mixture and are not related to any local composition.

We observe that Wang transferred the temperature dependence of the partition function from the exponential term used in Guggenheim's formulation, Equation 28.35, to the temperature dependence of the interaction parameters. Guggenheim's formulation needed the exponential term; otherwise, Equation 28.37 would not have had any adjustable parameter to represent energetic interactions. In Wang's formulation, these energy interactions are represented by the parameters characterizing the interactions of groups. According to Equation 24.14, the chemical potential is obtained by differentiation of the logarithmic form of Equation 28.62 with respect to n_i at a constant volume and constant number of moles of all other compounds different from i, so we write the logarithm of Equation 28.65 as:

$$\ln F_\alpha = \sum_{k=1}^{G} \frac{\alpha_k}{2} \ln \sum_{l=1}^{G} \frac{\alpha_l}{2} \tau_{lk} - \sum_{k=1}^{G} \frac{\alpha_k}{2} \ln \sum_{l=1}^{G} \frac{\alpha_l}{2} \tag{28.66}$$

Because the only reason to have the factor 1/2 correcting the surface area of interaction was to avoid the double counting of contacts, we eliminate it here and write:

$$\frac{\alpha_k}{2} = \sum_{j=1}^{N} n_j \nu_{kj} Q_k \tag{28.67}$$

where n_j is the number of moles of compound j in the mixture, ν_{kj} is the number of groups of type k in molecule j, and Q_k is the surface parameter of group k, already introduced in Equation 28.58. Following the standard thermodynamic path to obtain activity coefficients,

$$\frac{\mu_i}{RT} = -\left(\frac{\partial \ln Q_{CG}}{\partial n_i}\right)_{T,V,n_{j\neq 1}} - \left(\frac{\partial \ln F_\alpha}{\partial n_i}\right)_{T,V,n_{j\neq 1}}$$

$$\ln(x_i\gamma_i) = \frac{\mu_i}{RT} - \frac{\mu_i^{(i)}}{RT}$$

and using Equation 28.28 for the combinatorial term, for the activity coefficient of compound i in terms of groups, we write:

$$\ln\gamma_i = \left(1 - \frac{\Phi_i}{x_i} + \ln\frac{\Phi_i}{x_i}\right) - q_i\left(1 - \frac{\Phi_i}{\theta_i} + \ln\frac{\Phi_i}{\theta_i}\right) + \sum_k \nu_{ki}\left[\ln\Gamma_k - \ln\Gamma_k^{(i)}\right] \tag{28.68}$$

$$\ln\Gamma_k = Q_k\left[1 - \ln\sum_l \Theta_l a_{kl} - \sum_l \frac{\Theta_l a_{lk}}{\sum_m \Theta_m a_{lm}}\right] \tag{28.69}$$

Working in terms of molecules, the result is:

$$\ln\gamma_i = \left(1 - \frac{\Phi_i}{x_i} + \ln\frac{\Phi_i}{x_i}\right) - q_i\left(1 - \frac{\Phi_i}{\theta_i} + \ln\frac{\Phi_i}{\theta_i}\right) + q_i\left[1 - \ln\sum_k \theta_k\tau_{ik} - \sum_k \frac{\theta_k\tau_{ki}}{\sum_j \theta_j\tau_{kj}}\right] \tag{28.70}$$

For the equations for activity coefficients in terms of molecules, it is interesting to compare Equation 28.70 with Equation 28.54a for LSG and with Equation 28.55 for UNIQUAC. If for all k, $r_k = q_k = 1$, Equation 28.70 reduces to Wilson equation (28.42b). For the equations in terms of groups, it is worth comparing Equations 28.68 and 28.69 with Equations 28.56 and 28.59 for UNIFAC. The important point to notice is that Equations 28.68–28.70 come from a proper normalization of Guggenheim's combinatorial partition function and are unrelated to the concept of local composition or the concept of local surface area fraction.

SAFT

The statistical associating fluid theory (SAFT) [46] is perhaps one of the most active international efforts in our days aiming at the understanding of the behavior of fluids. As Prausnitz et al. [42c] state, "The literature on SAFT is complex and confusing." Basically, SAFT is a successful computer modeling of fluid behavior. It considers that molecules are formed by groups, that is, by repulsive hard-core segments that are affected by attractive potentials. These potentials arise from chain attachments and from interactions with other segments. These interactions can be only physical or also include association. By adding parameters to account for the different effects, SAFT has been very successful in reproducing and predicting the behavior of pure compounds and even complex mixtures. The whole simulation of the behavior of pure compounds and mixtures is computer-based, and it is discussed in detail in Chapters 32 and 33. A good overall view of SAFT has been presented by Prausnitz et al. [42d], as well as by Müller and Gubbins [47].

CONCLUSIONS

Models come and go. There are several versions of UNIFAC and ASOG in the literature – too many to mention. In our view, what is important is to have a perspective of the historical path of the development of models of solutions so when new ideas are generated, they can be seen in context and the experience of past developments is not lost.

Summary of Some Useful Equations for the Correlation of Activity Coefficients

Van Laar Model

$$\ln \gamma_1 = \frac{A}{\left(1+\frac{A}{B}\frac{x_1}{x_2}\right)^2} \tag{28.6}$$

$$\ln \gamma_2 = \frac{B}{\left(1+\frac{B}{A}\frac{x_2}{x_1}\right)^2} = \frac{B}{\left(\frac{B}{A}\frac{x_2}{x_1}\right)^2\left(1+\frac{A}{B}\frac{x_1}{x_2}\right)^2} \tag{28.7}$$

where A and B are adjustable parameters.

REGULAR SOLUTION THEORY

$$\ln \gamma_1 = \frac{v_1^L \Phi_2^2}{RT}(\delta_1 - \delta_2)^2 \tag{28.13}$$

and

$$\ln \gamma_2 = \frac{v_2^L \Phi_1^2}{RT}(\delta_1 - \delta_2)^2 \tag{28.14}$$

$$\ln \gamma_i = \frac{v_i^L}{RT}\left(\delta_i - \bar{\delta}\right)^2 \tag{28.16}$$

$$\bar{\delta} = \sum_j \Phi_j \delta_j \tag{28.17}$$

with

$$\Phi_j = \frac{x_i r_i}{\sum_k x_k r_k} \tag{28.24}$$

The regular solution theory uses $r_i = v_i^L$. Solubility parameters δ_j are tabulated for many compounds [5].

WILSON EQUATION

$$\ln \gamma_i = 1 - \ln \sum_k x_k \Lambda_{ik} - \sum_k \frac{x_k \Lambda_{ki}}{\sum_j x_j \Lambda_{kj}} \tag{28.42b}$$

where Λ_{ik} are adjustable parameters obtained from binary systems data (Λ_{12} and Λ_{21}).

NRTL EQUATION

$$\ln \gamma_i = \frac{\sum_j x_j G_{ji} \tau_{ji}}{\sum_l x_l G_{ji}} + \sum_j \frac{x_j G_{ij}}{\sum_l x_l G_{ll}} \left[\tau_{ij} - \frac{\sum_r x_r G_{rj} \tau_{rj}}{\sum_l x_l G_{ll}} \right] \tag{28.51a}$$

For a binary system,

$$\ln \gamma_1 = x_2^2 \left[\tau_{21} \left(\frac{G_{21}}{x_1 + x_2 G_{21}} \right)^2 + \frac{\tau_{12} G_{12}}{(x_2 + x_1 G_{12})^2} \right] \tag{28.51b}$$

and

$$\ln \gamma_2 = x_1^2 \left[\tau_{12} \left(\frac{G_{12}}{x_2 + x_1 G_{12}} \right)^2 + \frac{\tau_{21} G_{21}}{(x_1 + x_2 G_{21})^2} \right] \tag{28.51c}$$

with

$$G_{jk} = \exp(-\alpha_{jk} \tau_{jk}) \tag{28.49}$$

and

$$\tau_{jk} = \frac{\Delta g_{jk}}{RT} \tag{28.50}$$

The adjustable parameters in a binary system are Δg_{12}, Δg_{21} and $\alpha_{12} = \alpha_{21}$.

UNIQUAC Equation

$$\ln y_i = \left(1 - \frac{\Phi_i}{x_i} + \ln \frac{\Phi_i}{x_i} \right) - \frac{z q_i}{2} \left(1 - \frac{\Phi_i}{\theta_i} + \ln \frac{\Phi_i}{\theta_i} \right)$$
$$+ q_i \left[1 - \ln \sum_k \theta_k \tau_{ik} - \sum_k \frac{\theta_k \tau_{ki}}{\sum_j \theta_j \tau_{kj}} \right] \tag{28.55}$$

with

$$\theta_i = \frac{x_i q_i}{\sum_k x_k q_k} \tag{28.27}$$

and Φ_i given by Equation 28.5 above. The adjustable binary parameters are τ_{12} and τ_{21}. The values of r_i and q_i are tabulated, $z = 10$.

NOTE

1 In the study of polymer solutions, Equation 28.25 is written in an almost unrecognizable way. It is easy to show that for a solvent s having in solution a polymer of r_p segments, $\frac{\Phi_s}{x_s} = \Phi_s + \frac{1}{r_p}\Phi_p = 1 - \left(1 - \frac{1}{r_p}\right)\Phi_p$. Thus, Equation 28.25 can be written as $\ln\gamma_s = \left(1 - \frac{1}{r_p}\right)\Phi_p + \ln\left[1 - \left(1 - \frac{1}{r_p}\right)\Phi_p\right]$ and $\ln a_s \equiv \ln(x_s\gamma_s) = \left(1 - \frac{1}{r_p}\right)\Phi_p + \ln(1 - \Phi_p)$.

REFERENCES

1. van Laar, J. J. 1910. Über Dampfspannungen von binären Gemischen. *Z. Phys. Chem.* 72: 723–751.
2. van Laar, J. J. 1929. Über die Verschiebung des heterogenen Gleichgewichts von zwei geschmolzenen Metallen mit ihren Salzen durch einen dritten, indifferenten Stoff. *Z. anorgan. allg. Chem.* 185: 35–48.
3. Hildebrand, J. H. 1947. The entropy of solution of molecules of different sizes. *J. Chem. Phys.* 15: 225–228.
4. Scatchard, G. 1949. Equilibrium in nonelectrolyte mixtures. *Chem. Rev.* 44: 7–35.
5. Barton, A. F. M. 1991. *Handbook of Solubility Parameters and Other Cohesion Parameters*, 2nd ed. Boca Raton, FL: CRC Press.
6. Hansen, Ch. M. 2007. *Hansen Solubility Parameters: A User's Handbook*, 2nd ed. Boca Raton, FL: CRC Press, Taylor & Francis Group.
7. Margules, M. 1895. Über die Zusammensetzung der gesättigten Dämpfe von Misschungen. *Sitzungs berichte der Kaiserliche Akademie der Wissenschaften Wien Mathematisch-Naturwissenschaftliche Klasse II* 104: 1243–1278. Available at https://archive.org/details/sitzungsbericht10wiengoog.
8. Redlich, O., and Kister, A. T. 1948. Algebraic representation of thermodynamic properties and the classification of solutions. *Ind. Eng. Chem.* 40(2): 345–348.
9. Huggins, M. L. 1942. Thermodynamic properties of solutions of long-chain compounds. *Ann. N.Y. Acad. Sci.* 43: 1–32.
10. Flory, P. J. 1942. Thermodynamics of high polymer solutions. *J. Chem. Phys.* 10: 51–61.
11. Sayegh, S. G., and Vera, J. H. 1980. Lattice-model expressions for the combinatorial entropy of liquid mixtures: A critical discussion. *Chem. Eng. J.* 19: 1–10.
12. Guggenheim, E. A. 1944. Statistical thermodynamics of mixtures with zero energies of mixing. *Proc. R. Soc. Lond. A* 183: 203–212.
13. Staverman, A. J. 1950. The entropy of polymer solutions. *Rec. Trav. Chim. Pays-Bas.* 69: 163–174.
14. Guggenheim, E. A. 1944. Statistical thermodynamics of mixtures with nonzero energies of mixing. *Proc. R. Soc. Lond. A* 183: 213–227.
15. Kehiaian, H. V., Grolier, J.-P. E., and Benson, G. C. 1978. Thermodynamics of organic mixtures—Generalized quasi-chemical theory in terms of group surface interactions. *J. Chim. Phys. Phys. Chim. Biol.* 75(11–1): 1031–1048.
16. Panayiotou, C., and Vera, J. H. 1980. The quasi-chemical approach for non-randomness in liquid mixtures. Expressions for local surfaces and local compositions with an application to polymer solutions. *Fluid Phase Equilib.* 5(1–2): 55–80.
17. Abusleme, J. A., and Vera, J. H. 1985. The quasi-chemical group solution theory for organic mixtures. *Fluid Phase Equilib.* 22(2): 123–138.

18. Wilczek-Vera, G., and Vera, J. H. 1990. A general derivation of thermodynamic functions from the quasichemical theory and their use in terms of non-random factors. *Fluid Phase Equilib.* 59(1): 15–30.
19. Renon, H., and Prausnitz, J. M. 1969. Derivation of the three-parameter Wilson equation for the excess Gibbs energy of liquid mixtures. *AIChE J.* 15(5): 785–787.
20. Wilson, G. M. 1964. Vapor-liquid equilibrium. XI. A new expression for the excess free energy of mixing. *J. Am. Chem. Soc.* 86(2): 127–130.
21. Abrams, D. S., and Prausnitz, J. M. 1975. Statistical thermodynamics of liquid mixtures. New expression for the excess Gibbs energy of partly or completely miscible systems. *AIChE J.* 21(1): 116–128.
22. Maurer, G., and Prausnitz, J. M. 1978. On the derivation and extension of the UNIQUAC equation. *Fluid Phase Equilib.* 2(2): 91–99.
23. Wilson, G. M., and Deal, C. H. 1962. Activity coefficients and molecular structure. *Ind. Eng. Chem. Fundam.* 1: 20–23.
24. Deal, C. H., and Derr, E. L. 1968. Group contributions in mixtures. *Ind. Eng. Chem.* 60(4): 28–38.
25. Derr, E. L., and Deal Jr., C. H. 1969. Analytical solutions of groups. Correlation of activity coefficients through structural group parameters. *Proc. Int. Symp. Distill.* 3: 40–51.
26. Fredenslund, A., Jones, R. L., and Prausnitz, J. M. 1975. Group contribution estimation of activity coefficients in nonideal solutions. *AIChE J.* 21: 1086–1099.
27. Panayiotou, C., and Vera, J. H. 1981. Local compositions and local surface area fractions: A theoretical discussion. *Can. J. Chem. Eng.* 59(4): 501–505.
28. Prigogine, I., Mathot-Sarolea, L., and van Hove, L. 1952. Combinatory factor in regular assemblies. *Trans. Faraday Soc.* 48: 485–492.
29. Keméný, S., Balog, Gy., Radnai, Gy., Sawinsky, J., and Rezessy, G. 1990. An improved quasilattice expression for liquid phase order-disorder. *Fluid Phase Equilib.* 54: 247–275.
30. Abusleme, J. A., and Vera, J. H. 1985. A generalized solution method for the quasichemical local composition equations. *Can. J. Chem. Eng.* 63: 845–849.
31. Flemr, V. 1976. A note on excess Gibbs energy equations based on local composition concept. *Coll. Czech. Chem. Commun.* 41: 3347–3349.
32. McDermott, C., and Ashton, N. 1977. Note on the definition of local composition. *Fluid Phase Equilib.* 1: 33–35.
33. Redlich, O., Derr, E. L., and Pierotti, G. J. 1959. Group interaction. I. A model for interaction in solutions. *J. Am. Chem. Soc.* 81(10): 2283–2285.
34. Papadopoulos, M. N., and Derr, E. L. 1959. Group interaction. II. A test of the group model on binary solutions of hydrocarbons. *J. Am. Chem. Soc.* 81(10): 2285–2289.
35. (a) Kojima, K., and Tochigi, K. 1979. *Prediction of Vapor-Liquid Equilibria by ASOG Method.* Tokyo: Kodasha-Elsevier. (b) Tochigi, K., Tiegs, D., Gmeling, J., and Kojima, K. 1990. Determination of new ASOG parameters. *J. Chem. Eng. J. Jpn.* 23: 453–463.
36. Tochigi, K. 1995. Prediction of high-pressure vapor-liquid equilibria using ASOG. *Fluid Phase Equilib.* 104: 253–260.
37. Tochigi, K. 1998. Prediction of vapor–liquid equilibria in non-polymer and polymer solutions using an ASOG-based equation of state (PRASOG). *Fluid Phase Equilib.* 144: 59–68.
38. Ashraf, F. A., and Vera, J. H. 1980. A simplified group method analysis. *Fluid Phase Equilib.* 4: 211–228.
39. Scott, R. L. 1956. Corresponding states treatment of nonelectrolyte solutions. *J. Chem. Phys.* 25: 193–205.
40. Renon, H., and Prausnitz, J. M. 1968. Local composition in thermodynamic excess functions for liquid mixtures. *AIChE J.* 14: 135–144.
41. Vera, J. H. 1982. On the two-fluid local composition expressions. *Fluid Phase Equilib.* 8(3): 315–318.
42. Prausnitz, J. M., Lichtenthaler, R. N., and Gomes de Azevedo, E. 1999. *Molecular Thermodynamics of Fluid Phase Equilibria*, 3rd ed. Upper Saddle River, NJ: Prentice Hall, (a) p. 346, (b) p. 290, (c) p. 368, (d) pp. 390–400.
43. Vera, J. H., Sayegh, S. G., and Ratcliff, G. A. 1977. A quasi lattice-local composition model for the excess Gibbs free energy of liquid mixtures. *Fluid Phase Equilib.* 1: 113–135.
44. Wang, W., and Vera, J. H. 1993. Statistical thermodynamics of disordered and ordered systems. A properly normalized local order theory. *Fluid Phase Equilib.* 85: 1–18.
45. Gmehling, J., Tiegs, D., and Knipp, U. 1990. A comparison of the predictive capability of different group contribution methods. *Fluid Phase Equilib.* 54: 147–165.
46. Chapman, W. G., Jackson, G., and Gubbins, K. E. 1988. Phase equilibria of associating fluids: Chain molecules with multiple bonding sites. *Mol. Phys.* 65: 1057–1079.
47. Müller, E. A., and Gubbins, K. E. 2001. Molecular-based equations of state for associating fluids: A review of SAFT and related approaches. *IEC Res.* 40: 2193–2211.

PROBLEMS

28.1 Van Laar revisited

Actually, Equation 28.4 was derived by van Laar for the excess *enthalpy* of mixing and, as far as can be ascertained from his intricate presentation [1], simply *assumed* to apply also to the excess Gibbs energy. The concepts of ideal solution, partial properties, and fugacity, to be introduced by G. N. Lewis, were still decades in the future. Nowadays we know that excess Gibbs energy and fugacity coefficients are related by Equation 16.24.

a. Using van der Waals' equation of state, as van Laar did, derive an exact expression for the excess Gibbs energy of mixing of a multicomponent solution. Identify the terms appearing in the equation and obtain the limiting form at *infinite pressure*.
b. Van Laar stated that (translated from the original German) "Anyone who has read the main work [*] of van der Waals: '*Die Continuität*, etc.' will know that liquid mixtures only behave as ideal if the two components do not influence each other, that is, if:

$$\frac{a_m}{b_m} = \frac{x_1^2 a_1 + 2x_1x_2a_{12} + x_2^2 a_2}{x_1b_1 + x_2b_2} = x_1\frac{a_1}{b_1} + x_2\frac{a_2}{b_2}$$

And of course this will not be the case." Yet in a highly authoritative modern textbook on molecular thermodynamics, we read that "had he [van Laar] assumed that

$$a_m = x_1a_1 + x_2a_2$$

he would have obtained that $\gamma_1 = \gamma_2 = 1$ [i.e., that solutions are ideal] for all x". Which of these two conflicting statements is right, and what does this imply for the mixing rules of cubic equations of state of van der Waals type?

[*] Van Laar was not citing van der Waals' original (1873) Doctoral thesis, in which the author did not apply his equation of state to mixtures. The "main work" referred to must be the second German edition published in two volumes (1899, 1900), the second of which incorporated van der Waals' paper on binary mixtures [2]. This paper is included in the English translation of the thesis, edited by Rowlinson [3].

28.2 Parameters for g_0^E models from infinite-dilution activity coefficients

A convenient feature of two-parameter excess Gibbs energy models is that parameters for a binary system can be estimated from a single experimental point, or alternatively from the infinite-dilution activity coefficients [IDACs] of the two components. As an example of the latter, consider the system pentane (1)-dichloromethane (2) at 298.19 K, for which IDACs and pure-component saturation pressures are reported as [4]

$$\gamma_1^\infty = 3.7830, \gamma_2^\infty = 2.7314, P_1^s = 68.412\,\text{kPa}, P_2^s = 58.358\,\text{kPa}$$

Predict the vapor–liquid equilibrium of the system at the given temperature using (a) the van Laar model, (b) the Wilson model. Since the equilibrium pressures are low, use the modified Raoult's law, that is, neglect the vapor phase fugacity coefficients and the Poynting correction factors.

28.3 Entropy change on mixing two pure-fluid streams

Two streams of a pure fluid, $\dot{m}_1$ mol/s at temperature T_1 and $\dot{m}_2$ mol/s at temperature T_2, are mixed in a steady-state flow process to obtain $\dot{m}$ mol/s of output stream at temperature T. The process is adiabatic and there is no pressure loss. The molar heat capacity of the fluid, c_P can be assumed constant. Express the entropy change per mol of mixture as a function of the ratios $x_i = \dot{m}_i / \dot{m}$ and $\beta = T_1 / T_2$. What does the result remind you of? How is this problem related to the subject of this chapter?

28.4 **Wilson parameters from azeotropic data**

Consider again the system n-pentane (1)-dichloromethane (2) of Problem 28.2. Vapor–liquid equilibrium data of this system are reported in [4] also at 348.17 K (and at 398.16 K). The azeotropic coordinates are not given, but can be estimated analytically from a polynomial fit of the extensive experimental data. They are given in the following table, together with the vapor pressures and molar volumes of the pure liquids taken from the same reference:

T (K)	P_1^s (kPa)	P_2^s (kPa)	v_1^L (mL/mol)	v_2^L (mL/mol)	$x_1^{az} = y_1^{az}$	P^{az} (kPa)
298.19	68.412	58.358	116.19	64.49	0.5482	82.527
348.17	323.905	308.143	127.12	69.32	0.4853	385.243

a. From the azeotropic data, obtain the Wilson parameters of the system and use them to construct the *P*-*x*-y diagram at each temperature. Assume valid the modified Raoult's law. Compare your results with those obtained in Problem 28.2.
b. Is the temperature dependence of the Wilson parameters in agreement with Equations 28.38 and 28.41?

28.5 **Prediction of vapor–liquid equilibrium using a group-contribution method**

Group-contribution models provide a purely predictive approach to estimate the phase equilibrium of a system of interest for which experimental data are unavailable, fragmentary or unreliable. Use the ASOG-KT method (Appendix D) to construct the $T - x - y$ diagram of the binary system ethyl acetate (1)-benzene (2) at 1 atm. It has already been found in Problem 14.3 that vapor phase nonidealities can be neglected. Antoine equations for the pure-component vapor pressures are [5]:

$$\log_{10}\left(P_1^s / \text{mmHg}\right) = 7.30692 - \frac{1359.48}{t / °\text{C} + 230}$$

$$\log_{10}\left(P_2^s / \text{mmHg}\right) = 6.89745 - \frac{1206.35}{t / °\text{C} + 220.24}$$

REFERENCES FOR PROBLEMS

1. van Laar, J. J. 1910. Über Dampfspannungen von binären Gemischen. *Z. Phys. Chem.* 72: 723–751.
2. van der Waals, J. D. 1890. Molekulartheorie eines Körpers, der aus zwei verschiedenen Stoffe besteht. *Z. Phys. Chem.* 5U(1): 133–173.
3. van der Waals, J. D. 1988. *On the Continuity of the Gaseous and Liquid States.* Mineola, NY: Dover.
4. Khurma, J. G., Muthu, O., Munjal, S., and Smith, B. D. 1983. Total-pressure vapor–liquid equilibrium data for binary systems of dichloromethane with pentane, acetone, ethyl acetate, methanol, and acetonitrile. *J. Chem. Eng. Data* 28(4): 412–419.
5. Miller, R. N. 1959. *An Improved Vapor-liquid Equilibrium Still. M. S. Thesis in Chemical Engineering.* Corvallis, OR: Oregon State College.

29 Statistical Thermodynamics Approach to the Generalized van der Waals Theory

Toward Understanding of the Molecular Behavior of Fluids

ABSTRACT

This chapter discusses, first, intermolecular potentials, the radial distribution function, and structure factors and recalls the essentials of the Carnahan–Starling equation of state (EOS), the generalized van der Waals theory of fluids, the Flory–Patterson–Prigogine theory, the Lattice-Fluid theory, the Perturbed Hard-Chain Theory, and the PACT, COMPACT, and APACT EOS theories.

Molecules in a fluid phase interact at distance through intermolecular potentials which determine, to a large extent, the thermodynamic properties of the system. This chapter gives a rather simplified view of the present understanding of these effects. For this purpose, some common intermolecular potentials and related basic concepts are reviewed first.

INTERMOLECULAR POTENTIALS

Common models of intermolecular potentials for spherical molecules are the *hard-sphere* potential, the *square-well* potential and the *Lennard-Jones* potential [1,2]. The hard-sphere potential, shown graphically in Figure 29.1, has a steep repulsive term which considers that two interacting molecules do not overlap, or inter-penetrate at a distance r below the sphere diameter, σ, thus:

$$u(r)=\begin{bmatrix} \infty & r\le\sigma \\ 0 & r>\sigma \end{bmatrix} \quad (\text{hard-sphere}) \tag{29.1}$$

Thus, the second virial coefficient, $B_2(T)$, of hard-sphere molecules is given by (cf. Equation 27.15):

$$B_2(T)=2\pi\int_0^\infty\left[1-\exp\left(-\frac{u(r)}{kT}\right)\right]r^2dr=2\pi\int_0^\sigma r^2dr+2\pi\int_\sigma^\infty(1-1)r^2dr=\frac{2\pi\sigma^3}{3} \tag{29.2}$$

Note that, to be in agreement with the literature on the subject, here we change the convention of symbols for the virial coefficients. The square-well potential, shown graphically in Figure 29.1, has, both, repulsive and attractive terms given by:

$$u(r)=\begin{bmatrix} \infty & r\le\sigma \\ -\varepsilon & \sigma<r<\lambda\sigma \\ 0 & r>\lambda\sigma \end{bmatrix} \quad (\text{square-well}) \tag{29.3}$$

DOI: 10.1201/9781003431985-33

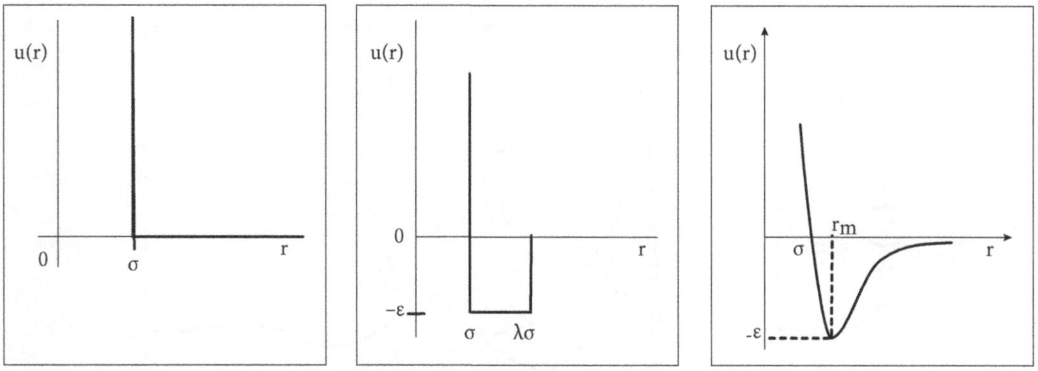

FIGURE 29.1 The hard-sphere potential (left), square-well potential (center), and Lennard-Jones 12–6 potential (right).

The Lennard-Jones potential for nonpolar spherical molecules has also repulsive and attractive terms with a smooth continuous function, shown graphically in Figure 29.1, given by:

$$u(r) = 4\varepsilon\left[\left(\frac{\sigma}{r}\right)^{12} - \left(\frac{\sigma}{r}\right)^{6}\right] \quad \text{(Lennard-Jones)} \tag{29.4}$$

As shown in Figure 29.1, the Lennard-Jones potential exhibits a minimum at $r = r_m = 2^{1/6}\sigma$, where the potential energy is equal to $-\varepsilon$. At $r = \sigma$, the potential energy is zero.

RADIAL DISTRIBUTION FUNCTION

Further progress in modeling the behavior of fluids requires an estimate of the average distance between the interacting molecules. The structure of a fluid around a central molecule is described by the *pair-correlation function* or *radial distribution function* (RDF), $g(r)$, defined as the ratio of the number density, $\rho_N(r)$ of molecules, at distance r from the center (of mass) of the central molecule, over the average or bulk number density, $\rho_N(\infty) = N/V$, of the fluid. A typical plot of $g(r)$ versus r is shown in Figure 29.2. If the molecules are considered to be hard spheres of diameter, σ, no centers of other molecules would be found at distances $r < \sigma$. At infinite distance, the central molecule has no effect on the molecular distribution and $g(r) = 1$. At closer distances, however, the central molecule influences this distribution due to the intermolecular potentials. Thus, the number of molecular centers in the differential volume of the shell from r to $r + dr$ is:

$$\left(4\pi r^2 dr\right)\rho_N(r) = \left(4\pi r^2 dr\right)\left(N/V\right)g(r)$$

The RDF is a measure of the *nonrandomness* of the molecular distribution in the volume of the fluid. Equivalent to RDF is the *pair-correlation function*, $h(r) = g(r) - 1$.

For simple isotropic fluids at low densities, the RDF and the interaction potential between two molecules are related by:

$$g(r) = \exp\left(-\frac{u(r)}{kT}\right) \quad \text{(dilute gas)} \tag{29.5}$$

Thus, if $u(r)$ is given by the Lennard-Jones potential, the first peaks in Figure 29.2 would correspond to the distance, r_m, of the lowest interaction energy.

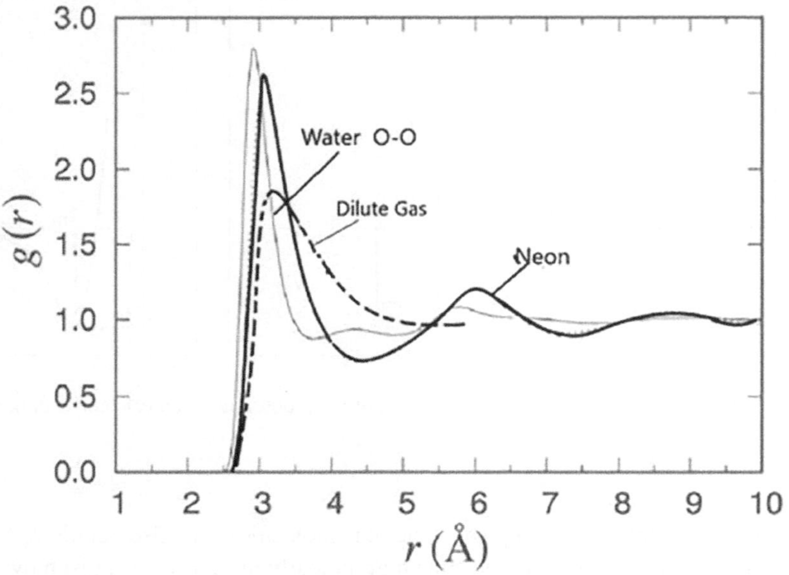

FIGURE 29.2 The radial distribution function (RDF) of Neon – a representative simple monatomic fluid (solid curve). For comparison, the RDF of a dilute gas (dashed curve) and of the oxygen in liquid Water (thin curve) are also shown. Adapted from Ref. [3] (fig. II-1.2) with permission.

In crystals, these peaks would correspond to first neighbors, second neighbors, etc. In liquids, the peaks do not have the regularity of the peaks of a crystal but they would show that the liquid has some degree of short-range order around each central molecule, reminiscent of the long-range order in the corresponding crystal.

It is interesting to examine the interconnection of RDF with the second virial coefficient, with the concept of *excluded volume*, that is, the volume of a molecule which is impenetrable by the center of another molecule, and with the *free volume*, V_f, discussed in Chapter 27 in relation to the second virial coefficient, $B_2(T)$ (cf. Equation 27.6):

$$\frac{P}{\rho kT} = 1 + B_2(T)\rho + \ldots \tag{29.6}$$

For $B_2(T)$,

$$B_2(T) = -2\pi\int_0^\infty \left[\exp\left(-\frac{u(r)}{kT}\right) - 1\right] r^2 dr = -2\pi\int_0^\infty \left[g(r) - 1\right] r^2 dr \tag{29.7a}$$

and

$$B_2(T) = -2\pi\int_0^\infty \left[\exp\left(-\frac{u(r)}{kT}\right) - 1\right] r^2 dr = -2\pi\int_0^\infty f(r)\, r^2 dr = -\frac{1}{2}\beta_1 \tag{29.7b}$$

In Equation 29.7b the Mayer *cluster function*, $f(r)$, (or, rather $f_{ij}(r_{ij})$, between the interacting pair of molecules *i–j*) [2], was introduced. The first *irreducible integral*, β_1, takes the form

$$\beta_1 = 4\pi\int_0^\infty f_{ij}(r_{ij})\, r_{ij}^2 dr_{ij}$$

and conventionally it is denoted graphically as . The filled circles in this graph represent the two molecules i and j, whose positions are integrated over the system volume. The connecting line indicates interaction between these two molecules. This is the starting point of the *Graph Theory* or Graph representation of Statistical Mechanics. The full representation of the configurational integral of an assembly of molecules involves a number of successively complicated irreducible integrals and corresponding graphs. The second irreducible integral, β_2, involves three-body interactions of a cyclic form between three molecules, 1, 2, and 3:

$$\beta_2 = \frac{1}{2V}\int_V\int_V\int_V f_{12}\, f_{23}\, f_{31}\, d\bar{r}_1\, d\bar{r}_2\, d\bar{r}_3$$

which is denoted graphically as 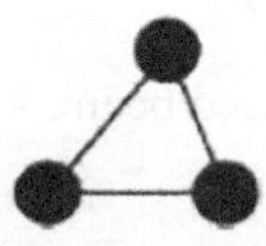 ($d\bar{r}_i$ is volume element around vector $\bar{r}_i$ position). The third virial coefficient is given as $B_3(T)=-2\beta_2/3$. A full account of the Graph Theory, however, exceeds the scope of this book and will not be discussed further here.

Assuming that the molecules are hard spheres of diameter, σ, the closest distance of their centers is equal to σ. Thus, each of these two spheres *excludes* a volume, equal to $4\pi\sigma^3/3$, from the reach of the other sphere. Assuming, then, that the fluid density is sufficiently low, so that each molecule has, on the average, at most one neighbor, the free volume of the fluid is given by:

$$V_f = V - \frac{N}{2}\left(\frac{4}{3}\pi\sigma^3\right) = V - \frac{2\pi N}{3}\sigma^3 = V - 4N\frac{4\pi}{3}\left(\frac{\sigma}{2}\right)^3 = V - 4Nb_0 = V - Nb \qquad (29.8)$$

where b_0 is the volume of the molecular hard sphere and b is the excluded volume per molecule.

The thermodynamic description of liquids, however, is not as simple as presented above. We do not discuss here contributions from orientations of multiatom, multisite, or anisotropic molecules, where RDF must be expressed in polar coordinates, $g(r, \varphi, \theta)$. In longer flexible molecules, internal rotations may lead to different conformers of the very same molecule, each of which is attributed to its own RDF in polar coordinates.

Perturbation-type approximations, analogous to those discussed in Chapter 27, may be used for the RDF of liquids. Equation 29.5 is, typically, extended by polynomials of fluid density:

$$g(r) = \exp\left(-\frac{u(r)}{kT}\right)\left(1 + \sum_n \rho^n y_n(r)\right) \qquad (29.9)$$

whose coefficients, $y_n(r)$, are determined from experimental or computer-simulation data. On the other hand, at higher densities, the close neighbors of a central molecule are more than one and the excluded volumes overlap, as shown on the right-hand side of Figure 29.3. Thus, at higher densities $V_f > V - Nb$.

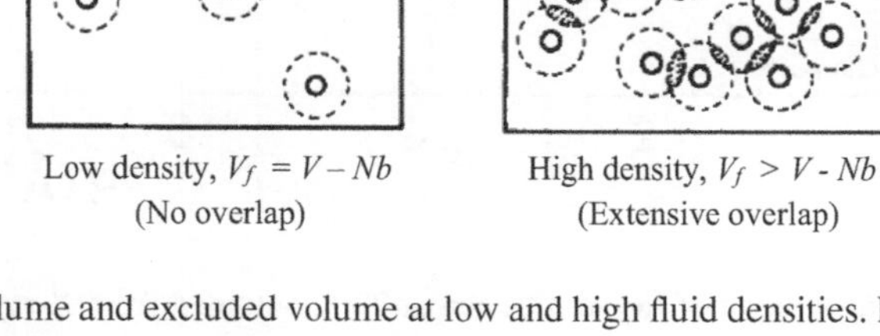

Low density, $V_f = V - Nb$ (No overlap)   High density, $V_f > V - Nb$ (Extensive overlap)

FIGURE 29.3 Free volume and excluded volume at low and high fluid densities. Reprinted with permission from Ref. [4].

RADIAL DISTRIBUTION FUNCTION FROM EXPERIMENT AND FROM MOLECULAR SIMULATIONS

The usefulness of RDF resides on the fact that it is experimentally measurable, and it can be obtained from Monte Carlo (MC) or Molecular Dynamics (MD) simulations. In scattering (small-angle neutron scattering, SANS, or X-ray diffraction) experiments, the quantity of central importance is the *structure factor*, $S(q)$. In a typical static light scattering experiment from an assembly of molecules, if θ is the angle between the incident beam of wavelength, λ, and the scattered beam, for nearly elastically scattered photons, the *scattering wave vector*, q. is given by:

$$q = \frac{4\pi}{\lambda}\sin\left(\frac{\theta}{2}\right) \tag{29.10}$$

From measurements of the intensity, $I(\theta)$, of the scattered beam, it is possible to evaluate the structure factor, $S(q)$, defined by:

$$S(q) = 1 + \rho g(q) = 1 + \rho\int g(r)\exp\left(i\bar{q}\cdot\bar{r}\right)dr = 1 + \frac{4\pi\rho}{q}\int_0^\infty rg(r)\sin(qr)\,dr \tag{29.11}$$

where, $g(q)$ is the Fourier transform of the RDF, $g(r)$. This is the most important equation interconnecting RDF with the measurable structure factor, $S(q)$. The experimental structure factor of liquid nitrogen is shown in Figure 29.4. As seen, the shape of the structure factor is similar to the corresponding RDF. Since Fourier transforms are unique, RDF may be obtained by inversion of $S(q)$. Thus, for isotropic fluids, we have:

$$g(r) = \frac{1}{\rho}\int_0^\infty q^2\left(S(q) - 1\right)\frac{\sin(qr)}{qr}\,dq \tag{29.11a}$$

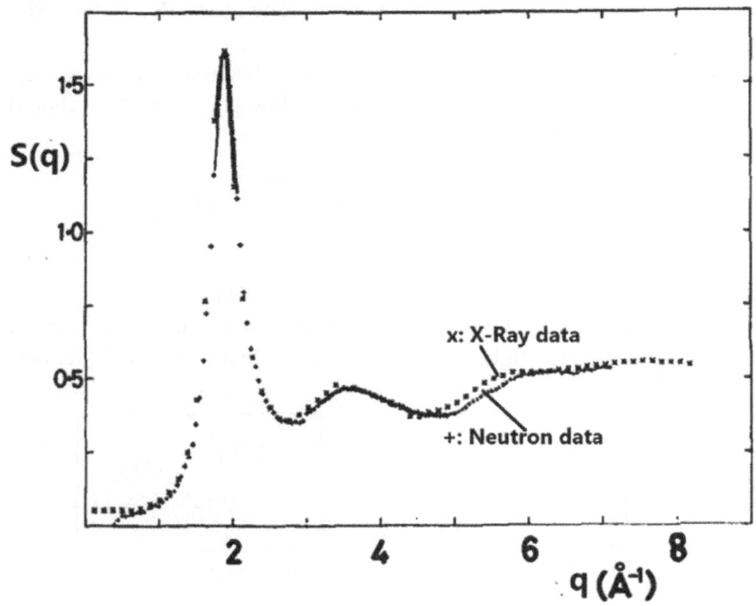

FIGURE 29.4 The structure factor of liquid nitrogen at 77 K obtained from X-ray and neutron diffraction measurements. Adapted from Ref. [5, fig. 5] with permission.

This practice is particularly useful in the study of dense liquids or of amorphous materials, such as polymers.

In polyatomic molecules, which is the vast majority of fluids of practical interest, the scattering experiments bring up contributions from all scattering separations, intramolecular (scattering atoms of the same molecule) and intermolecular (scattering atoms from different molecules). If ρ_N is the number density of the fluid, the intermolecular contributions arise from atoms j at distance r from central atom i of another molecule, whose density is $\rho_N g_{ij}(r)$. The intramolecular contributions arise from scattering separations of atoms j at distance r from a central atom i of the same molecule, whose probability distribution is $s_{ij}(r)$. From scattering experiments in Fourier-transform space, a linear combination is obtained:

$$G_{ij}(r) = \rho_N g_{ij}(r) + s_{ij}(r) \tag{29.12}$$

This pair distribution function for liquid nitrogen, G_{NN}, is shown schematically in Figure 29.5.

The interatomic distance (bond length) N–N in the molecule of N_2 is 1.1 Å and this is what the first peak in Figure 29.5 represents. In dense liquids the molecules may be considered touching each other. The second peak, then, in Figure 29.5 indicates that a van der Waals diameter of 3.3 Å may be attributed to liquid N_2. Since atoms are attaching atoms of another molecule, a peak is expected at separations 3.3+1.1=4.4 Å, and this is what the shoulder in the second broader peak represents. Integration over this broader peak (the first coordination shell) gives:

$$n = 4\pi \int_{3Å}^{5.6Å} G_{NN}(r)\, r^2 dr \approx 12 \tag{29.13}$$

This is a useful result, since n may be considered as the *coordination number* of the fluid. In lattice models, the coordination number, z, varies typically between 6 and 14 with a commonly used average, $z=10$. In a sense, Equation 29.13 is interconnecting the two diverse approaches to the structure of liquids, namely, the RDF approach and the quasi-lattice or combinatorial approach, discussed in other chapters of this book.

In addition to scattering experiments, computer simulations may also be used for extracting information for RDF. In contrast to the complexities of theoretical RDF calculations, especially with more complicated molecular systems, computer simulations do not require any theory or reference system information. The two most common simulation techniques are MC and MD [6].

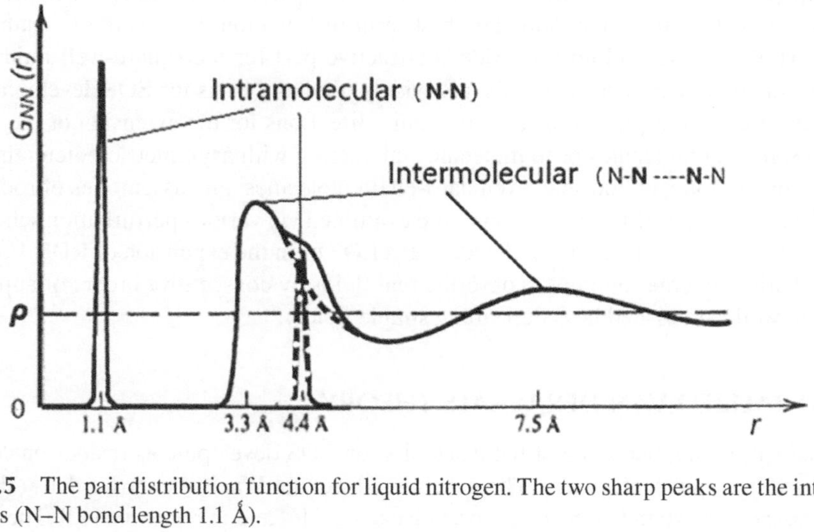

FIGURE 29.5 The pair distribution function for liquid nitrogen. The two sharp peaks are the intramolecular contributions (N–N bond length 1.1 Å).

The behavior of a collection of molecules of any complexity is directly simulated in one of these two ways. In MC simulations of a stochastic character, random-number generators are used for molecular displacements and the generation of new states according to appropriate Boltzmann distributions. The average values of properties are determined over such probable states. In MD simulation, the dynamical equations of motions of the molecules (dictated by the adopted *force-field* or set of intermolecular potentials) are solved and the time evolution of the system is followed. Equilibrium properties are obtained in MC simulations while, both, equilibrium and nonequilibrium properties are obtained in MD simulations. Inverse techniques are applied to these simulations for obtaining the structure factors and, as above, the RDF.

RDF AND EQUATION-OF-STATE MODELS

The above simulation techniques give numerical values for the properties in specified states for the adopted intermolecular potentials/force-fields. These numerical values are treated as "experimental" data and used for determining various model parameters. Model parameters determined in this way are not considered adjustable and are often referred to as *universal* parameters. Equation-of-state (EOS) models with such universal parameters often exhibit improved versatility in the description of real experimental data. Experimental data from neutron scattering or X-ray diffraction and data from computer simulations (MC or MD) lead to the prevailing idea that the RDF of real fluids is dictated, mainly, by the repulsive part of the intermolecular potential. The attractive forces on a central molecule from its surrounding molecules, nearly cancel out or have rather minor effect on the average structure of the liquid although they are essential for the field experienced by the moving molecule. Typically, then, the RDF of nearly spherical molecules, at liquid densities, are expressed by the corresponding RDF of hard-sphere fluids and, most often, by the expression derived from the Carnahan–Starling EOS [7], which describes well the free-volume increase shown in Figure 29.3. In fact, the Carnahan–Starling repulsive form has been the basis for the development of numerous *perturbed hard-sphere* EOS models.

Of much interest is the equation of Alder et al. [8] for the square-well fluid of well-depth ε and hard-sphere diameter σ:

$$P = \frac{RT}{V}\left(\frac{1+\eta+\eta^2-\eta^3}{(1-\eta)^3}\right) - \sum_i \sum_j A_{ij}\left(\frac{\varepsilon}{kT}\right)^i \left(\rho\sigma^3\right)^j \qquad (29.14)$$

where $\eta = \pi\rho\sigma^3/6$, is the reduced density of the fluid and A_{ij} are 36 universal constants determined from computer (MD) simulation data. The first term in Equation 29.14 is the Carnahan–Starling repulsive part and the second term is Alder's attractive part for the square-well fluid. These two terms, either in combination or separately, are widely used as a basis for EOS developments.

Barker and Henderson [9] also discuss the main directions for the extension of the above treatment to nonspherical molecules or to molecules interacting with asymmetric potentials. Hard convex, dumbbells, discotic, cylindrical, even hard-chain molecules, and assemblies of rods in parallel nematic-like liquid-crystalline orientations were examined via various perturbation schemes [9]. As in the virial expansion departure from the ideal-gas EOS or in the expansion of RDF, Equation 29.9, these perturbation schemes attempt to describe real fluids by consecutive orders of approximation, starting from well-established laws describing simple fluids.

THE GENERALIZED VAN DER WAALS THEORY

It may sound surprising, but much of the discussion on EOS developments resides on concepts and ideas proposed over one century ago. The generalized van der Waals theory for dense fluids [1] was further discussed and expanded in an interpretative review [4] and in reference [10]. The ideas of van der Waals keep inspiring developers of contemporary equations of state even in unexpected applications as in the dense stellar matter [11]. The central equation in the discussion of the generalized van der Waals theory is the canonical partition function of the real fluid (equation A-5 of Ref. [4]):

$$Q = \frac{1}{N!}\left[V_f\right]^N \left[\exp\left(\frac{-\varphi}{2kT}\right)\right]^N \left[\frac{q_{r,v,e}}{\Lambda^3}\right]^N \tag{29.15}$$

Λ in Equation 29.15 is the de Broglie wavelength ($\Lambda = h(2\pi mkT)^{-1/2}$, h is Plank's constant, m is the molecular mass), $q_{r,v,e}$ is the partition function for the internal (rotational, vibrational, and electronic) degrees of freedom, and φ the potential energy of interaction between any one molecule and all others in the system. The division by 2 in the denominator corrects for double counting of contacts. The potential energy, φ, is given by (equation A-6 of Ref. [4]):

$$\varphi = \frac{N}{V}\int_0^\infty \Gamma(r)g(r)4\pi r^2\,dr \tag{29.16}$$

where $\Gamma(r)$ is the pair potential function (only two-body interactions are considered). For simple Argon-like molecules, $q_{r,v,e}$ does not depend on density and does not contribute to the EOS. For chain-like molecules, however, in general, this simplification does not apply. In the original van der Waals theory, φ was considered to be inversely proportional to volume, V:

$$\varphi = -2\left(\frac{N}{N_{Av}}\right)\frac{a}{V} \tag{29.17}$$

where, a is the van der Waals attractive constant.

As discussed in Chapter 26, all basic thermodynamic quantities can be obtained from the form for the partition function. In particular, the equation for the Helmholtz free energy, A, is obtained from Equation 26.12 and the EOS from Equation 26.13:

$$A = -kT\ln Q \tag{26.12}$$

and

$$P = kT\left(\partial \ln Q/\partial V\right)_{T,N} \tag{26.13}$$

THE FLORY–PATTERSON–PRIGOGINE (FPP) EQUATION-OF-STATE THEORY

The EOS developments for high polymers followed a substantially different direction. Significant progress was achieved by the introduction of the Flory–Patterson–Prigogine (FPP) EOS theory [12–16], which was the first combined and systematic effort for the exploration of the central role of free volume on the LCST behavior of polymer systems.

Prigogine [12] introduced the *Cell Model* of liquids as a modification of the classical rigid-lattice model through a different description of intermolecular interactions. Each segment of an r-mer molecule was considered to occupy a lattice site but was allowed to be displaced from its central position, due to thermal fluctuations. This displacement, however, is restricted, due to the repulsive interactions of the neighboring segments, which form the confining "cell" within which the segment may fluctuate. In this "cell" model, rotational and vibrational modes of motion are no longer volume independent and, thus, the number of external degrees of freedom are not just the three translational modes but include some of the rotational and vibrational modes of motion, as well. In the FPP model, the equivalent external motions are set equal to $3c$, leaving $3r-3c$ motions to be internal to the molecules. Equation 29.15 may, then, be written as (equation F-2 of Ref. [4]):

$$Q=\frac{1}{N!}\left[\frac{V_f^{1/3}}{\Lambda}\right]^{3Nc}\left[\exp\left(\frac{-\varphi}{2kT}\right)\right]^N[f(T)]^N \tag{29.18}$$

or, in a more general form, as:

$$Q=\frac{1}{N!}[\psi(T,V)]^{cN}\left[\exp\left(\frac{-\varphi}{2kT}\right)\right]^N[f(T)]^N \tag{29.19}$$

The FPP model for high polymers expresses φ in Equation 29.19, as in the original van der Waals attractive term, to be inversely proportional to V. Prigogine had adopted the Lennard-Jones intermolecular potential and defined free volume (cf. Equation 29.18) as:

$$V_f=(N/N_{Av})rv^*g\left(\tilde{v}^{1/3}-2^{-1/6}\right)^3 \tag{29.20}$$

where $\tilde{v}=\frac{(VN_{Av}/N)}{rv^*}=\frac{V_m}{V^*}$. v* is the hard-core volume of each segment and g a geometric constant. The potential energy, U, that is, the energy of placing the N molecules in the cells, may be expressed in van der Waals terms, as (Equation F-6 of Ref. [4]):

$$U=\frac{N\varphi}{2}=-\left(\frac{N}{N_{Av}}\right)^2\frac{u^*rv^*}{V}=-\left(\frac{N}{N_{Av}}\right)\frac{u^*}{\tilde{v}} \tag{29.21}$$

where u^* is a characteristic molar interaction energy. With these key assumptions, Prigogine developed a full EOS model of liquid mixtures [12].

Patterson et al. [13,14] combined ideas from the Krigbaum–Flory model, discussed in Chapter 19, and explored Prigogine's cell model with some modifications for the description of the LCST behavior of nonpolar polymer solutions, not only as a Θ temperature- but also as a free-volume-driven process, thus, explaining the common occurrence of LCST above UCST in these polymer solutions (cf. Chapter 19).

Flory et al. [15,16] also simplified Prigogine's cell model, attributed $3c$ external degrees of freedom to each segment, and adopted for ψ, cf. Equation 29.19, the cell partition function term of Eyring and Hirschfelder [17]:

$$\psi(V)=(\text{const.})Nrv^*\left(\tilde{v}^{1/3}-1\right)^3 \tag{29.22a}$$

or, equivalently, for the free volume, the equation:

$$V_f=Nrv^*g\left(\tilde{v}^{1/3}-1\right)^3 \tag{29.22b}$$

Flory adopted a van der Waals-type volume-dependent molar potential energy given by:

$$U_m=-r\frac{N}{N_{Av}}\frac{s\varepsilon}{2\tilde{v}}=-rn\frac{\varepsilon^*}{\tilde{v}} \tag{29.23}$$

where ε is the segment-segment interaction energy, and s is the mean number of external contacts per segment.

With these assumptions Flory developed a full equation-of-state model for polymeric liquids, which is the most widely used variation of the cell models. The EOS in this model takes the following form:

$$\frac{\tilde{P}\tilde{v}}{\tilde{T}} = \frac{\tilde{v}^{1/3}}{\tilde{v}^{1/3}-1} - \frac{1}{\tilde{v}\tilde{T}} \tag{29.24}$$

$\tilde{P}$ and $\tilde{T}$ are the reduced pressure and temperature, respectively, and are defined as follows:

$$\tilde{P} = \frac{P}{P^*} = \frac{P\left(v^*\right)^2}{\varepsilon^*}, \qquad \tilde{T} = \frac{T}{T^*} = \frac{cRTv^*}{\varepsilon^*} \tag{29.25}$$

The parameters ε^*, v^*, and c or, alternatively, the parameters T^*, P^*, and v^*, are the three *scaling constants* or EOS parameters of each liquid. The model was extended to mixtures with the appropriate mixing and combining rules [18].

Equation 29.24 does not reduce to the ideal-gas EOS in the limit of zero densities. Thus, FPP is a model of the liquid state only. To remove this drawback, most of the EOS developments after this pioneering FPP work, focused on models that could apply to the gas state as well. Two major families of models in this direction are the Generalized Lattice-Fluid (GLF) models and the tangent-sphere or perturbation-theory models, which include the family of Perturbed Hard-Chain Theory (PHCT) models and the family of Statistical Associating Fluid Theory (SAFT) models. Chapter 30 is devoted to GLF models, and in subsequent chapters, the large SAFT family of models are discussed. Here, the simple Lattice-Fluid (LF) model and the essentials of the PHCT models are discussed in some detail.

THE LATTICE-FLUID EQUATION-OF-STATE APPROACH

As the name implies, the LF approach [19–23] is also based on the incompressible lattice model but the lattice loses its rigidity by letting some of its sites be unoccupied or empty. This is close to the actual physical picture of the liquid state, where there is some empty space or free volume between molecules. In fact, the LF model may be derived, entirely and equivalently as an off-lattice model [20,23].

With the lattice picture in mind, Sanchez and Lacombe [19,20] considered the molecules divided into r segments occupying one lattice site each. If N r-mer molecules are arranged on the quasi-lattice, leaving N_0 empty sites, the total lattice sites are:

$$N_r = rN + N_0 \tag{29.26}$$

If v^* is the volume of each lattice site, the total volume is:

$$V = \left(rN + N_0\right)v^* = V^* + N_0 v^* \tag{29.27}$$

V^* is the volume occupied by the r-mers and $N_0 v^*$ is the unoccupied or free volume. The reduced volume $\tilde{v}$ and the reduced density $\tilde{\rho}$ are defined by:

$$\tilde{v} = \frac{V}{V^*} = \frac{N_r}{rN} = \frac{1}{\tilde{\rho}} \tag{29.28}$$

The quasi-lattice energy is considered to depend only on the nearest-neighbor binary interactions and the interaction energy with an empty site is set equal to zero. If ε is the mer–mer interaction energy in a pure fluid, the total energy is the product of ε with the number of mer–mer contacts. Assuming random arrangement of molecules and empty sites in the lattice, the total number of mer–mer contacts is just the number of mers multiplied by the probability of finding a mer for contact. From its definition, $\tilde{\rho}$ is the probability of finding a segment or mer in the volume of the system, Thus, the total configurational potential energy is:

$$-U = rN\tilde{\rho}\, s\varepsilon/2 = rN\tilde{\rho}\varepsilon^* \tag{29.29}$$

where s is the average number of external contacts per segment, an analog of the lattice coordination number. Equation 29.29 is entirely analogous to Equation 29.23 of the FPP model.

However, as mentioned above, it is not necessary to adopt the above lattice picture and consider ε* as a nearest-neighbor binary interaction energy. If $\bar{u}$ is the average interaction energy of a mer with all other mers in the system, quite generally the configurational potential energy can be written as:

$$U = rN\bar{u}/2 \tag{29.30}$$

with

$$\bar{u} = 4\pi\rho \int_0^\infty u(L)g(L)L^2 dL \tag{29.31}$$

$u(L)$ and $g(L)$ in Equation 29.31 are the intermolecular potential and the RDF, respectively, and ρ is the number density of mers. (L is here the distance and r is the number of segments per molecule.) Assuming a Sutherland-type intermolecular potential, $u(L)$ (hard-core or $\sigma = (v^*)^{1/3}$ plus attractive tail,

$$\begin{aligned} u(L) &= \infty \quad \text{for} \quad L < \sigma \\ u(L) &= -\varepsilon_0 \left(\frac{\sigma}{L}\right)^n \quad \text{for} \quad L > \sigma \end{aligned} \tag{29.32}$$

in the mean-field approximation, the hard-core intermolecular potential gives for RDF:

$$\begin{aligned} g(L) &= 0 \quad \text{for} \quad L < \sigma \\ g(L) &= 1 \quad \text{for} \quad L > \sigma \end{aligned} \tag{29.33}$$

Substitution of Equation 29.32 and 29.33 in Equation 29.30 and 29.31 gives:

$$-U = rN\tilde{\rho}\frac{2\pi\varepsilon_0}{n-3} = rN\tilde{\rho}\varepsilon^* \tag{29.34}$$

Hence, Equation 29.29 is recovered but now the interaction energy ε* is proportional to the depth of the potential energy well, ε_0. With the usual value of $n=6$, $\varepsilon^* = 2\pi\varepsilon_0/3$. Thus, for sufficiently short-range intermolecular potential (n>3), the fluid potential energy in the mean-field approximation is of the van der Waals type, that is, it is inversely proportional to the volume. In essence, then, the lattice and the off-lattice formulations are equivalent and interchangeable.

In their original work [19,20], Sanchez and Lacombe adopted the Flory–Huggins combinatorial term of the partition function (cf. Chapter 30) for the most probable arrangements of molecules and vacant sites on the quasi-lattice. They also adopted Equation 29.29, or 29.34, for the potential energy in the maximum term approximation [1,2]. In this way, they obtained the following equation for the Gibbs energy for pure fluids:

$$G = N\mu = rN\varepsilon^*\left[-\tilde{\rho}+\tilde{P}\tilde{v}+\tilde{T}\tilde{v}\left[(1-\tilde{\rho})\ln(1-\tilde{\rho})+\frac{\tilde{\rho}}{r}\ln\tilde{\rho}\right]\right] \tag{29.35}$$

where μ is the chemical potential and the reduced temperature and pressure are defined by the following equations:

$$\tilde{T}=\frac{T}{T^*}=\frac{RT}{\varepsilon^*},\quad \tilde{P}=\frac{P}{P^*}=\frac{Pv^*}{\varepsilon^*} \tag{29.36}$$

From Equation 29.35 for G, the equations for the various thermodynamic quantities can be obtained [1, 2].

The number of vacant sites or the reduced volume at equilibrium is obtained by minimization of G with respect to $\tilde{v}$. Setting the derivative with respect to $\tilde{v}$ equal to zero, the following equation of state is obtained:

$$\tilde{P}+\tilde{\rho}^2+\tilde{T}\left[\ln(1-\tilde{\rho})+(1-1/r)\tilde{\rho}\right]=0 \tag{29.37}$$

The important point is that this equation meets the ideal-gas limit and, thus, LF model is also applicable to the gas state. For high polymers this equation takes the following simple form:

$$\tilde{P}+\tilde{\rho}^2+\tilde{T}\left[\ln(1-\tilde{\rho})+\tilde{\rho}\right]=0 \quad \text{(high polymers)} \tag{29.38}$$

This equation suggests a corresponding states behavior of high polymers. Indeed, by plotting $\tilde{\rho}$ versus $\tilde{T}$, the experimental data for various polymers fall on lines of the same reduced pressure [20]. Such plots may be used for a quick estimation of densities of polymeric liquids at various temperatures and pressures.

Thermal expansion coefficient, α, and isothermal compressibility, κ_T, are given by:

$$\alpha T=\frac{1+\tilde{P}\tilde{v}^2}{\tilde{T}\tilde{v}\left[1/(\tilde{v}-1)+1/r\right]-2} \tag{29.39}$$

and

$$\kappa_T P=\frac{\tilde{P}\tilde{v}^2}{\tilde{T}\tilde{v}\left[1/(\tilde{v}-1)+1/r\right]-2} \tag{29.40}$$

The extension of the LF model to mixtures is straightforward if appropriate mixing and combining rules are adopted. Various options for these rules were applied in the literature. In the original Sanchez–Lacombe model [19,20], the number of segments r_k of component k in a mixture varied with mixture composition. For coherence of presentation, we adopt here the more generalized mixing rules with constant r_k [21,22], which is also used in subsequent chapters.

Let us consider a multicomponent mixture of t types of molecules at temperature, T, and pressure, P. Each of the N_k molecule of type k consists of r_k segments of volume v_k^* each and their average per-segment interaction energy is $\varepsilon_k^* = s_k\,\varepsilon_{kk}/2$, where s_k is the average number of external contacts per segment. The total number of segments in the mixture is:

$$\sum_{k=1}^{t} r_k N_k=\left(\sum_{m=1}^{t} N_m\right)\left(\sum_{k=1}^{t} r_k x_k\right)=N\sum_{k=1}^{t} r_k x_k=rN \tag{29.41}$$

and the volume occupied by all molecular segments is:

$$V^* = \sum_{k=1}^{t} r_k v_k^* N_k = rN\left(\sum_{k=1}^{t} \phi_k v_k^*\right) = rNv^* \tag{29.42}$$

where, the volume fraction or, rather, the segment fraction, ϕ_k, is given by:

$$\phi_k = \frac{r_k N_k}{rN} = \frac{r_k x_k}{r} \tag{29.43}$$

If there are N_0 vacant lattice sites in the system, the total number of lattice sites is:

$$N_r = rN + N_0 \tag{29.44}$$

and the total volume is:

$$V = rNv^* + N_0 v^* = N_r v^* \tag{29.45}$$

Segment fractions, ϕ_k, should be distinguished from the site fractions, f_k, defined by:

$$f_k = \frac{r_k N_k}{N_r} = \frac{r_k N_k}{rN}\frac{rN}{N_r} = \phi_k \tilde{\rho} \tag{29.46}$$

and

$$f_0 = \frac{N_0}{N_r} = \frac{N_r - rN}{N_r} = 1 - \tilde{\rho} \tag{29.47}$$

In a similar manner, the contact surface area fraction or, simply, the contact fraction is defined by:

$$\theta_k = \frac{s_k r_k N_k}{\sum_{m=1}^{t} s_m r_m N_m} = \frac{s_k r_k N_k}{srN} = \frac{\phi_k s_k}{s} \tag{29.48}$$

With these definitions, the average per-segment interaction energy of the mixture is given by the following mixing and combining rules:

$$\varepsilon^* = \frac{s}{2}\sum_{i=1}^{t}\sum_{j=1}^{t} \theta_i \theta_j \varepsilon_{ij} \tag{29.49}$$

and

$$\varepsilon_{ij} = \zeta_{ij}\sqrt{\varepsilon_{ii}\varepsilon_{jj}} \tag{29.50}$$

ζ_{ij} is a binary correction parameter expected to be close to one. The potential energy of the mixture is then given by an equation identical in form to Equation 29.29 but with ε^* given by Equation 29.49. The Gibbs energy of the mixture is given by:

$$\frac{G}{RT} = rN\left\{-\frac{\tilde{\rho}}{\tilde{T}} + \frac{\tilde{P}\tilde{v}}{\tilde{T}} + (\tilde{v} - 1)\ln(1 - \tilde{\rho}) + \frac{1}{r}\ln\tilde{\rho} + \sum_{k=1}^{t}\frac{\phi_k}{r_k}\ln\frac{\phi_k}{\omega_k}\right\} \tag{29.51}$$

and the EOS for the mixture obtained is also identical in form to Equation 29.37, but the scaling quantities ε*, v*, and r, are now given by Equations 29.49, 29.42, and 29.41, respectively. The ω_k parameter accounts for chain flexibility and symmetry of component k and cancels out in all applications of interest here. The chemical potential, obtained from Equation 29.51, is given by:

$$\frac{\mu_k}{RT}=\left(\frac{\partial G}{\partial N_k}\right)_{T,P,\tilde{\rho},N_{m\neq k}}=\ln\frac{\phi_k}{\omega_k}+1-\frac{r_k}{r}+r_k\left\{\frac{\tilde{P}\tilde{v}D_k-\tilde{\rho}}{\tilde{T}}+(\tilde{v}-1)\ln(1-\tilde{\rho})+\frac{1}{r_k}\ln\tilde{\rho}\right\}$$
$$+r_k\tilde{\rho}\left\{\sum_{j\prec k}\theta_j\frac{s_k}{s_j}X_{jk}+\sum_{j\succ k}\theta_jX_{kj}-\sum_{i\prec j}\sum_{j}\theta_i\theta_j\frac{s_k}{s_i}X_{ij}\right\} \tag{29.52}$$

where

$$X_{ij}=\frac{\varepsilon_i^*+(s_i/s_j)\varepsilon_j^*-2\sqrt{(s_i/s_j)}\varepsilon_{ij}^*}{RT} \tag{29.53}$$

and

$$\varepsilon_{ij}^*=\zeta_{ij}\sqrt{\varepsilon_i^*\varepsilon_j^*} \tag{29.54}$$

The factor D_k depends on the mixing rule for v*. In the usual case of common v* (= 9.75 cm^3/mol) $D_k=1$.

Heats of mixing and volumes of mixing are given by Equations 29.55 and 29.56, respectively:

$$H^E=U^E+PV^E=rN\left[\sum_k\tilde{\rho}_k\phi_k\varepsilon_k^*-\tilde{\rho}\varepsilon^*\right]+PV^E \tag{29.55}$$

where

$$V^E=rN\left[\tilde{v}v^*-\sum_k\tilde{v}_k\phi_k v_k^*\right] \tag{29.56}$$

Note that, as discussed in Chapter 15, ideal mixtures form without heat or volumetric effects, that is with null enthalpy or volume change. Thus, the excess properties for enthalpy and volume are identical to the change of these properties by mixing. To be in agreement with the literature on the subject, the symbol for excess property is used here for enthalpy and volume instead of the delta normally used for property change of mixing.

The above constitute a complete set of equations for the basic thermodynamic quantities of multicomponent fluid mixtures. The surface-to-volume ratios s_i/s_j are literature values – not adjustable parameters, or are set equal to unity as in the original Sanchez–Lacombe model [19,20]. In this case, $\phi_i=\theta_i$ and the above formalism simplifies significantly.

THE PERTURBED HARD-CHAIN THEORY

Beret and Prausnitz [24] started from Equations 29.15 and 29.16 and made some key assumptions and modifications which led to the PHCT EOS model, valid for both the liquid and the gas states for molecules of any size. They considered chain molecules consisting of tangentially connected hard-sphere segments, with $3c$ external degrees of freedom. The EOS was considered to consist of a repulsive part given by the Carnahan–Starling EOS [7], and an attractive part given by an

Alder-type term [8] by adopting a square-well interaction potential of well width equal to half the diameter σ of the molecule, with 24 A_{ij} constants, cf. Equation 29.14.

Alder's results affect the potential field, which in PHCT model takes the form:

$$\varphi = \varepsilon q W\left(\tilde{T}, \tilde{v}\right) = \varepsilon q \sum_{i=1}^{4} \sum_{j=1}^{J} \left(\frac{A_{ij}}{\tilde{v}^{j}} \right) \left(\frac{1}{\tilde{T}^{i-1}} \right) \tag{29.57}$$

J in Equation 29.57 varies depending on the value of i (for i equal to 1, 2, 3 or 4, J is 6, 9, 5 or 4, respectively). The characteristic interaction energy per molecule, εq, is the product of the energy per unit surface area, ε, and the parameter q which is proportional to the external surface area of the molecule. The reducing temperature, T^*, pressure, P^*, and volume, v^*, are now:

$$T^* = \frac{\varepsilon q}{cR}, \qquad P^* = \frac{\varepsilon q}{r v^*} \qquad \text{and} \qquad V^* = rNv^* = V/\tilde{v}$$

where, r is the number of segments per molecule and v^* is the close-packed hard-core volume per segment.

As in Equation 29.18, the partition function for the rotational and vibrational modes of motion is split in a product of two terms, one external, $q_{ext}(T, V)$ and one internal, $q_{int}(T)$. In order to make the model applicable to both, gas and liquid states, the following expression is adopted for q_{ext}:

$$q_{ext} = \left(\frac{V_f}{V} \right)^{c-1} \tag{29.58}$$

With the above assumptions, the final EOS of the PHCT model is:

$$\frac{\tilde{P}\tilde{v}}{\tilde{T}} = \frac{1}{c} + \frac{4(\tau/\tilde{v}) - 2(\tau/\tilde{v})^2}{(1-\tau/\tilde{v})^3} - \frac{\tilde{v}}{\tilde{T}} \left(\frac{\partial W}{\partial \tilde{v}} \right)_{\tilde{T}} \tag{29.59}$$

where $\tau = 0.7405$. Equation 29.59 was tested with good results against experimental volumetric data for several amorphous polymers [24]. The PHCT model, with some minor modifications, was further tested with good results by Donohue and Prausnitz [25] for pure fluids of low molar mass and their mixtures.

PACT, COMPACT, AND APACT EQUATION-OF-STATE MODELS

The PHCT model was further modified by Vimalchand and Donohue [26] and extended to include multipolar anisotropic interactions, leading to the PACT (Perturbed Anisotropic Chain Theory) EOS model. Anisotropic forces due to dipoles and quadrupoles are accounted for with the theory of Gubbins and Twu [27]. A second-order perturbation of the Lennard-Jones intermolecular potential was adopted. The Helmholtz free energy was expanded as a power series in reciprocal temperature as follows:

$$A = A^{IG} + A^{rep} + \frac{A_1^{iso}}{T} + \frac{A_2^{iso}}{T^2} + \ldots + \frac{A_2^{ani}}{T^2} + \frac{A_3^{ani}}{T^3} + \ldots \tag{29.60}$$

where the superscripts *IG*, *rep*, *iso*, and *ani* indicate ideal-gas, repulsive, isotropic, and anisotropic interaction contributions, respectively. The size of molecular segments is described by two parameters, the hard-sphere diameter, d, and the soft-sphere diameter, σ. Their ratio d/σ is given by a fourth-order polynomial in reduced temperature. Two similar Pade approximants are used to account for contributions due to the higher-order terms in the perturbation expansion, Equation 29.60, one for the isotropic and the other for the anisotropic interactions.

PACT is applicable to linear molecules of any size. Mixture properties of quadrupolar fluids are well predicted. Experimental data over a range of external conditions are well correlated with small deviations values of the adjustable binary interaction parameters [26].

Ikonomou and Donohue [28] extended PACT to the association APACT (Associated Perturbed Anisotropic Chain Theory) EOS model by incorporating in PACT the Chemical Theory of Association, which is discussed in the chapter on hydrogen bonding, Chapter 32. As shown by extensive applications [28], APACT is a versatile and successful EOS model for associated fluids.

PHCT and PACT are rather complex mathematically. Thus, an effort was made to simplify them, and this led to the SPHCT (Simplified PHCT) [29]. SPHCT consists of an ideal-gas term, a repulsive term taken from the Carnahan–Starling EOS [7] and an attractive term derived from the generalized van der Waals partition function using the local composition model of Lee et al. [30]. An interesting aspect of the Lee et al. model [30] is the estimation of the variation of the coordination number with temperature and density on a sound theoretical basis. SPHCT was also combined with the Chemical Theory of Association and led to the COMPACT EOS for associated fluids and mixtures.

All the above models are, in a sense, extensions of the generalized van der Waals theory [1, 4], which since 1975 received a renewed interest. Barker and Henderson [9] presented a comprehensive review of the literature on this subject up to 1975. Apparently, van der Waals' ideas will keep inspiring researchers and developers of EOSs fluids.

Association models are discussed in Chapter 32 on Hydrogen Bonding. The family of SAFT models, which combine various concepts discussed in this chapter are also discussed in Chapters 32 and 33.

REFERENCES

1. Hill, T. L. 1960. *An Introduction to Statistical Thermodynamics*. Reading, MA: Addison-Wesley.
2. Chandler, D. 1987. *Introduction to Modern Statistical Mechanics*. Oxford: Oxford University Press.
3. Hirata, F. 2020. *Exploring Life Phenomena with Statistical Mechanics of Molecular Liquids*. Boca Raton, FL, CRC Press.
4. Vera, J. H., and Prausnitz, J. M. 1972. Generalized van der Waals Theory for Dense Fluids. *Chem. Eng. J.* 3: 1–13.
5. Dore, J.C., Walford, G., and Page, D. I. 1975. The structure factor for liquid nitrogen and liquid oxygen from neutron scattering studies. *Mol. Phys.* 29:2, 565–580
6. Frenkel, D., and Smit, B. 2002. *Understanding Molecular Simulation: From Algorithms to Applications*. San Diego, CA: Academic Press.
7. Carnahan, N. F., and Starling, K. E. 1972. Intermolecular repulsions and the equation of state for fluids. *AIChE J.* 18: 1184–1188.
8. Alder, B. J., Young, D. A., and Mark, M. A. 1972. Studies in molecular dynamics. X. Corrections to the augmented van der Waals theory for square-well fluid. *J. Chem. Phys.* 56: 3013–3029.
9. Barker, J. A., and Henderson, D. J. 1976. What is "liquid"? Understanding the states of matter. *Rev. Modern Phys.* 45(4): 587–671.
10. Vera, J. H., and Prausnitz, J. M. 1971. A semi-theoretical van der Waals equation for calculation of vapor-liquid equilibria at high pressures. *Canad. J. Chem.* 49: 2037–2043.
11. Andrew, K., Steinfelds, E. V., and Andrew, K. A. 2023. The van der Waals Hexaquark chemical potential in dense stellar matter. *Particles* 6: 556–567.
12. Prigogine, I. (with the collaboration of Bellemans, A. and Mathot, V.). *The Molecular Theory of Solutions*, ch. 16 and 17. Amsterdam: North Holland,
13. Delmas, G., Patterson, D., and Somcynsky, T. 1962. Thermodynamics of polyisobutylene – n-alkane systems. *J. Polym. Sci.* 57: 79–98.
14. Patterson, D., Delmas, G., and Somcynsky, T. 1967. A comparison of lower critical solution temperatures of some polymer solutions. *Polymer* 8: 503–516.
15. Flory, P. J., Orwoll, R. A., and Vrij, A. 1964. Statistical thermodynamics of chain molecule liquids. II. Liquid mixtures of normal paraffin hydrocarbons. *J. Am. Chem. Soc.* 86: 3515–3520.
16. Flory, P. J. 1970. Thermodynamics of polymer solutions. Discuss. *Faraday Soc.* 49: 7–29.
17. Eyring, H., and Hirschfelder, J. 1937. The theory of the liquid state. *J. Phys. Chem.* 41: 249–257.

18. Panayiotou, C. 1984. Statistical thermodynamics of polymer solutions. *J. Chem. Soc. Faraday Trans.* 2(80): 1435–1449.
19. Sanchez, I. C., and Lacombe, R. H. 1976. An elementary molecular theory of classical fluids. Pure fluids. *J. Phys. Chem.* 80: 2352–2362.
20. Sanchez, I. C., and Lacombe, R. H. 1978. Statistical thermodynamics of polymer solutions. *Macromolecules* 11: 1145–1156.
21. Panayiotou, C. 1987. Lattice-fluid theory of polymer solutions. *Macromolecules* 20: 861–871.
22. Panayiotou, C., and Sanchez, I. C. 1991. Hydrogen bonding in fluids: An equation of-state approach. *J. Phys. Chem.* 95: 10090–10097.
23. Sanchez, I. C., and Panayiotou, C. 1994. Equations of state thermodynamics of polymer and related solutions. In *Models for Thermodynamic and Phase Equilibria Calculations*, ed. S. Sandler. New York: Marcel Dekker Inc, pp. 187–286.
24. Beret, S., and Prausnitz, J. M. 1975. A generalized van der Waals equation for polymers and other fluids. *Macromolecules* 8: 878–882.
25. Donohue, M. D., and Prausnitz, J. M. 1978. Perturbed hard chain theory for fluid mixtures: Thermodynamic properties for mixtures in natural gas and petroleum technology. *AIChE J.* 24: 849–860.
26. Vimalchand, P., and Donohue, M. D. 1985. Thermodynamics of quadrupolar molecules: The perturbed-anisotropic-chain theory. *Ind. Eng. Chem. Fundam.* 24: 246–257.
27. Gubbins, K. E., and Twu, C. H. 1979. Thermodynamics of polyatomic fluid mixtures. *Chem. Eng. Sci.* 34: 951–958.
28. Ikonomou, G. D., and Donohue, M. D. 1988. Extension of the associated perturbed anisotropic chain theory to mixtures with more than one associating component. *Fluid Phase Equilb.* 39: 129–159.
29. Ikonomou, G. D., and Donohue, M. D. 1987. COMPACT: A simplified equation of state for associated molecules. *Fluid Phase Equilib.* 33: 61–90.
30. Lee, K.-H., Lombardo, M., and Sandler, S. I. 1985. The generalized van der Waals partition function II. Application to the Square-Well fluid. *Fluid Phase Equilib.* 21: 177–196.

30 Generalized Lattice-Fluid Model

Abstract

This chapter presents the general formalism of the lattice-fluid model derived from a generalized form of the partition function via statistical thermodynamics. Three alternative combinatorial terms are examined. The quasi-chemical theory is used to account for the nonrandom distribution of molecular species and free volume. Both the molecular and the group-contribution formalism are presented.

The capacity of the lattice-fluid (LF) model [1–5] to describe the thermodynamic behavior of polymer systems was presented in Chapter 29. This relatively simple equation-of-state (EOS) model is used in a broad variety of applications ranging from vapors and supercritical fluids up to polymer solutions, glassy polymers, gels/rubber networks, ionic polymers, bulk phases, and interfaces. Some representative applications were discussed in Chapter 20. This model, which considers a more realistic picture than rigid-lattice models, can account for a temperature- and pressure-dependent free volume of the system. As discussed in Chapter 29, the LF model may also be derived as an off-lattice model by adopting a Sutherland-type intermolecular potential (a repulsive hard-core plus an attractive tail) and the radial distribution function formalism [2,5].

The LF model assumes a random distribution of molecules and of free volume and it adopts the plain Flory–Huggins combinatorial term [6] in the partition-function formulation. However, strongly polar systems, anisotropic molecules of varying size/shape and flexibility and, especially, molecules interacting with strong specific forces, such as hydrogen bonding (HB), require additional refinements in the LF model for their adequate description. One successful refinement is made via a generalized form of the partition function (PF) and the refined LF model is referred to as the *generalized lattice-fluid* (GLF) model.

The general form of the partition function for a multicomponent system of fluids in the canonical ensemble and in its maximum-term approximation [7,8], may be written in the form of Equation 30.1. Following the nomenclature used in the literature on this topic, we have used the symbol $\{N\}$ instead of N to denote the fixed numbers of moles of all species N_1, N_2, …, in the system:

$$Q(T,V,\{N\}) = g_{comb}\exp\left(\frac{-U}{RT}\right) = g_{NR}g^{R}_{comb}\exp\left(\frac{-U}{RT}\right) = g_{HB}g_{NHB}g^{R}_{comb}\exp\left(\frac{-U}{RT}\right) \quad (30.1)$$

What changes in the above alternative form of PF is the pre-exponential combinatorial factor, g_{comb}. In the first variation, the combinatorial term is split into two terms. The combinatorial term g^{R}_{comb} disregards intermolecular interactions and considers the molecules distributed perfectly randomly in the volume of the system. The correction term g_{NR} accounts for the molecular ordering in the system due to the intermolecular interactions. This correction is meaningful in the case of strong interaction energies which exceed thermal energy, RT. An approximation for g_{NR} is Guggenheim's quasi-chemical term [8] (cf. Equation 28.34):

$$g_{QC} = \frac{\prod_{i=1}^{t} N^{0}_{ii}!\prod_{i\neq j}^{t}\left[\left(\frac{N^{0}_{ij}}{2}\right)!\right]^2}{\prod_{i=1}^{t} N_{ii}!\prod_{i\neq j}^{t}\left[\left(\frac{N_{ij}}{2}\right)!\right]^2} \quad (30.2)$$

DOI: 10.1201/9781003431985-34

In this equation, N_{ij} is the number of external contacts between molecular segments of type i and j. The superscript 0 refers to the case of randomly distributed segments. Thus, g_{QC} becomes equal to unity in the case of random distribution. It is to be expected that g_{QC} will not be the right correction in the case of strong hydrogen-bonding interactions. Thus, in the second variation of the combinatorial term in Equation 30.1, the nonrandomness factor, g_{NR}, is split into two factors. The first factor, g_{HB}, accounts for the ordering in the system due to the presence of strong specific forces or HB interactions. The second factor, g_{QC}, accounts for the nonrandomness due to all other interactions in the system, except for the HB ones. This second term is set either equal to unity, when the non-HB interactions are considered weak with negligible effect on the overall nonrandomness or is set equal to the quasi-chemical term of Equation 30.2, when the non-HB interactions are considered sufficiently strong to cause nonrandomness in the molecular arrangements.

A more versatile expression for g_{HB} is presented in Chapter 32. The rest of this chapter focuses on the other terms of Equation 30.1, except for the hydrogen-bonding term.

CLASSICAL COMBINATORIAL TERMS

For convenience, the various rigid-lattice expressions for the combinatorial term, g_{comb}^R [6–9]. discussed in Chapter 28 are summarized below:

- The Flory–Huggins combinatorial term [6,7] for a binary solvent (1) – polymer (2) mixture with $N_r = N_1 + r_2N_2$,

$$g_{comb}^R = \frac{r_2^{N_2 r_2}}{N_2!}\left\{\frac{(N_r/r_2)!}{(N_1/r_2)!}\right\}^r \left[\frac{z-1}{N_r}\right]^{N_2(r_2-1)} \quad \text{(Flory–Huggins)} \tag{30.3}$$

- The Guggenheim–Orr–Miller combinatorial term [7,8] with $N_r = \sum_i r_i N_i$, $N_q = \sum_i q_i N_i$

$$g_{comb}^R = \prod_i \delta_i^{N_i} \frac{N_r!}{\prod_i N_i!}\left(\frac{N_q!}{N_r!}\right)^{z/2} \quad \text{(Guggenheim)} \tag{30.4}$$

- The Guggenheim–Staverman combinatorial term [9]

$$g_{comb}^R = \prod_i \delta_i^{N_i} \frac{N_r!\prod_i N_r^{l_i N_i}}{\prod_i N_i!}\left(\frac{N_q!}{N_r!}\right)^{z/2} \quad \text{(Guggenheim–Staverman)} \tag{30.5}$$

where

$$l_i = \frac{z}{2}(r_i - q_i) - (r_i - 1) \tag{30.6}$$

and z is the lattice coordination number. The rest of the symbols used here are the same as in Chapter 29. There is a variety of combinations of the pre-exponential terms in Equation 30.1, which lead to a variety of models discussed below.

THE PARTITION FUNCTION OF LF MODEL

The plain LF model that was discussed extensively in Chapter 29 may be derived by assuming perfect randomness in the molecular distribution, or $g_{NR}=1$, and by adopting Equation 30.3 for the basic combinatorial term. In the LF approach, the rigid-lattice Equations 30.3–30.5 are modified to account for the free volume in the system. This is done by discretizing the free volume into N_0 empty sites dispersed in the volume of the system as an additional inert molecular species ($r_0 = 1$, $\varepsilon_{0i}=0$). For convenience, Equation 30.1 is often written in the Gibbs (N, P, T) ensemble. Thus, for the plain LF model [1–5], Equation 30.1 takes the form:

$$Q(T,P,\{N\})=\left(\frac{1}{f_0}\right)^{N_0}\left(\frac{\omega_1}{f_1}\right)^{N_1}\left(\frac{\omega_2}{f_2}\right)^{N_2}\exp\left[-\frac{U+PV}{RT}\right] \tag{30.7}$$

where

$$f_0=\frac{N_0}{N_r},\quad f_1=\frac{r_1N_1}{N_r},\quad f_2=\frac{r_2N_2}{N_r} \tag{30.8}$$

$$\omega_i=\frac{\delta_i r_i}{\sigma_i\exp(r_i-1)}\qquad i=1,2 \tag{30.9}$$

and

$$N_r=N_0+\sum_i r_iN_i \tag{30.10}$$

Both partition functions are related by:

$$\ln\, Q(T,P,\{N\})=\ln\, Q(T,V,\{N\})-PV/RT \tag{30.11a}$$

so that, in the same way that the Helmholtz energy A follows from $Q(T, V,\{N\})$, cf. Equation 26.12, the Gibbs energy $G=A+PV$ may be obtained directly from Equation 30.7 as:

$$G(T,P,\{N\})=-RT\ln Q(T,P,\{N\}) \tag{30.11b}$$

All basic thermodynamic quantities are then derived from Equation 30.11b as explained in Chapter 29. As an example, the reduced volume and reduced density are obtained by:

$$\tilde{v}=\frac{V}{V^*}=\frac{N_r v^*}{(r_1N_1+r_2N_2)v^*}=\frac{N_r}{rN}=\frac{1}{\tilde{\rho}} \tag{30.12}$$

where v* is the hard-core volume per segment in the system (cf. Chapter 29).

THE PV EQUATION-OF-STATE MODEL

The PV EOS model [10] accounts for nonrandomness and adopts Equation 30.2 for g_{NR}, and Equation 30.4 for the basic combinatorial term. The free volume is introduced as in LF, and Equations 30.10 and 30.12 hold true, in this case, as well. We stress the point that subscript i in Equation 30.4 runs from $i=0$ in this case. Each r-mer fluid is considered to have zq external intermolecular contacts given by:

$$zq = (z-2)r + 2 \tag{30.13}$$

In the PV model (and in nearly all GLF models), v* is considered a universal constant equal to 9.75 cm³/mol [10].

The quasi-chemical conditions, Equation 28.37, apply here as well. For the derivation of the PV formalism, the following nonrandomness factors, Γ_{ij} for the i–j contacts have been introduced [10]:

$$\Gamma_{ij} = \frac{N_{ij}}{N_{ij}^0} \tag{30.14}$$

Replacing in Equation 28.37, we obtain the following alternative form of this contact conservation/balance equation:

$$\frac{4N_{ii}N_{jj}}{N_{ij}^2} = \frac{\Gamma_{ii}\Gamma_{jj}}{\Gamma_{ij}^2} = \frac{\tau_{ii}\tau_{jj}}{\tau_{ij}^2} \tag{30.15}$$

where

$$\tau_{ij} = \exp\left(-\frac{\varepsilon_{ij}}{RT}\right) \tag{30.16}$$

with $\varepsilon_{0j} = 0$.

For a pure fluid the nonrandomness factors Γ_{11}, $\Gamma_{10,}$ and Γ_{00} are given by simple analytical equations [10]. For multicomponent multigroup systems, however, they are obtained numerically. For this purpose, the nonrandomness factors in Equation 30.15 may be split into individual molecule-, or group-, or segment-specific nonrandomness factors, Γ_n. But this cannot be done arbitrarily and the quasi-chemical condition, Equation 30.15, should not be violated. The consistent split may be done, then, by the following equation:

$$\Gamma_{mn} = \Gamma_m \Gamma_n \tau_{mn} \tag{30.17}$$

For the number of contacts in the case of random distribution, we have:

$$\begin{aligned} N_{ii}^0 &= \frac{z}{2} q_i N_i \frac{q_i N_i}{N_q} = \frac{z}{2} q_i N_i \theta_i \theta_r \\ N_{ij}^0 &= z q_i N_i \frac{q_j N_j}{N_q} = z q_i N_i \theta_j \theta_r = z q_j N_j \theta_i \theta_r \quad i \neq j \end{aligned} \tag{30.18}$$

where

$$\theta_r = \frac{qN}{N_q} = 1 - \theta_0 = \frac{q/r}{q/r + \tilde{v} - 1} \tag{30.19}$$

and

$$N_{00}^0 = \frac{1}{2} N_0 z \frac{N_0}{N_q} = \frac{z}{2} N_0 \theta_0 \tag{30.20}$$

The contact balance equations are [8,10]:

$$2N_{ii} + \sum_j N_{ji} = z q_i N_i \quad \text{for all } i \tag{30.21}$$

By combining Equations 30.14–30.20, we obtain from Equation 30.21:

$$\theta_i\Gamma_{ii} + \sum_{j\neq i}\theta_j\Gamma_{ji} = \sum_{j=0}^{t}\theta_j\Gamma_{ji} = \Gamma_i\sum_{j=0}^{t}\theta_j\Gamma_j\tau_{ij} = 1 \tag{30.22}$$

or

$$\Gamma_i = 1/\sum_{j}\theta_j\Gamma_j\tau_{ij} \quad \text{for all } i,\ j \tag{30.23}$$

In logarithmic terms, Equation 30.23 may be rewritten as follows:

$$\ln\Gamma_j = -\ln\sum_{i}\theta_i\Gamma_i\tau_{ij} \quad \text{for all } j \tag{30.24}$$

Equation 30.24, or rather the system of Equation 30.23 or 30.24, is a recursive one, useful for the development of an iterative numerical solution algorithm for obtaining Γ_j s and Γ_{ij} s. It is in fact a very efficient and robust algorithm converging rapidly from an initial guess of $\Gamma_j = 1$ for all molecules or groups *j*.

The EOS for the PV model is obtained by minimizing the free energy, *G* (cf. Equation 30.11), with respect to the reduced density, $\tilde{\rho}$, and is:

$$\tilde{P} + \tilde{T}\left\{\ln(1-\tilde{\rho}) - \frac{z}{2}\left(1+\frac{q}{r}\tilde{\rho}-\tilde{\rho}\right) + \frac{z}{2}\ln\Gamma_{00}\right\} = 0 \tag{30.25}$$

In the original PV model [10], (where there was no distinction of strong specific interactions) and to simplify the calculation of the nonrandomness factors, it was assumed that only the molecular species are distributed nonrandomly, while the free volume was distributed randomly throughout the volume of the system. In this case, the EOS takes the following form (for both pure component and mixture):

$$\tilde{P} + \tilde{T}\left\{\ln(1-\tilde{\rho}) - \frac{z}{2}\left(1+\frac{q}{r}\tilde{\rho}-\tilde{\rho}\right) + \frac{\theta_r^2}{\tilde{T}}\right\} = 0 \quad \left(\text{random free - volume distribution}\right) \tag{30.26}$$

and the equation for the activity coefficient is given by:

$$\ln\gamma_1 x_1 = \ln\varphi_1 + \ln\frac{\tilde{\rho}}{\tilde{\rho}_1} + q_1\ln\left(\frac{1-\tilde{\rho}_1}{1-\tilde{\rho}}\right) + q_1\left(\frac{2\theta_{r,1}-\theta_r}{\tilde{T}_1} - \frac{\theta_r}{\tilde{T}}\right) + \frac{zq_1}{2}\ln\dot{\Gamma}_{11} \tag{30.27}$$

In the last equation, $\theta_{r,1}$ is the surface area fraction θ_r of pure component 1, and $\dot{\Gamma}_{11}$ is obtained from:

$$\theta_1\dot{\Gamma}_{11} + \theta_2\dot{\Gamma}_{12} = 1, \quad \dot{\Gamma}_{12} = \frac{2}{1+\sqrt{1-4\theta_1\theta_2(1-A_{12})}} \quad \text{and} \quad A_{12} = \exp\left[\theta_r\frac{\varepsilon_{11}+\varepsilon_{22}-2\varepsilon_{12}}{RT}\right]$$

In another variation of the PV model [11], the bulk of the nonrandom distribution of molecular species was accounted for by the hydrogen-bonding factor, g_{HB}, and g_{QC} was used to account for the nonrandom distribution of the free volume by making no distinction of the neighboring molecular segments. This model was applied to volumetric properties and vapor–liquid equilibria and was extended to polydisperse polymers and to inhomogeneous systems for the calculation of interfacial properties, surface tension, or composition profiles across interfaces in hydrogen-bonded mixtures.

EXTENSION OF THE PV MODEL TO A GROUP-CONTRIBUTION EQUATION-OF-STATE MODEL

High and Danner [12,13] have extended the PV model and developed a group-contribution scheme for the estimation of the two scaling constants, ε_{ii} and v*, of the model. In this way, the model can predict polymer-solvent equilibria given only the structure of the polymer and solvent molecules.

As discussed in Chapter 28, the group-contribution scheme requires the definition of the surface area fractions of the various groups in the molecules they belong to. Thus, if group k has a surface area Q_k and there are v_k^i such groups in molecule i, then, the surface area fraction of group k in molecule i, θ_k^i, is given by:

$$\theta_k^i = \frac{v_k^i Q_k}{\sum_m v_m^i Q_m} \tag{30.28}$$

Besides Q_k, each group is characterized by a hard-core or reference volume, R_k. The hard-core volume of molecule i at temperature, T, may be calculated, then, by:

$$V_{i,T}^* = a_T + \sum_k v_k^i R_k \tag{30.29}$$

A Berthelot-type mixing rule is adopted for the cross-interaction energy, ε_{ij} at temperature, T, giving:

$$\varepsilon_{ij,T} = \sum_k \sum_l \theta_k^i \theta_l^j \sqrt{\varepsilon_{kk,T}\varepsilon_{ll,T}} \tag{30.30}$$

Interaction energies, ε_{kk}, surface areas, Q_k and reference volumes, R_k, for groups in alkanes, alkenes, cycloalkanes, aromatics, esters, alcohols, ethers, water, ketones, aromatic ketones, amines, siloxanes, and monochloroalkanes, have been reported for 300 and 400 K [12,13]. A simple linear interpolation scheme is proposed for values at intermediate temperatures. With these parameters, one may use the equations of the PV model and perform phase equilibrium calculations in polymer-solvent systems.

Experimental activity coefficient data in polymer solutions are more often reported as weight fractions activity coefficients of the solvent, Ω_1, or

$$\ln\Omega_1 = \ln\frac{\gamma_1 x_1}{w_1} = \ln\frac{a_1}{w_1} \tag{30.31}$$

and Equation 30.27 is used for the activity $a_1 = \gamma_1 x_1$.

A detailed procedure for the use of this group-contribution method is described in Ref. [13], along with example calculations. The calculations are facilitated significantly, and the method may be extended to multicomponent systems by using the numerical solution algorithm of the system of Equation 30.24 for the nonrandomness factors. High and Danner [12,13] have used a more tedious procedure for estimating nonrandomness factors, which had posed some unnecessary restrictions on the extensive application of the method.

NEWER REFINEMENTS OF THE GLF MODEL

In the original PV model [10], Equation 30.4 was used for the basic combinatorial term, g^R_{comb}. More versatile and popular is, however, the Guggenheim–Staverman (GS) term of Equation 30.5. The GS combinatorial term has been adopted in a refined GLF model [14,15]. This GLF model is referred to in the literature as the NRHB (nonrandomness and hydrogen-bonding) since it incorporates also a g_{HB} – pre-exponential factor in the partition function (cf. Equation 30.1) that accounts for hydrogen-bonding interactions in the system. This g_{HB} factor will be discussed extensively in Chapter 32 on hydrogen bonding. In this section we will confine ourselves to the presentation of the non-hydrogen-bonding part of the model. A broader overview of developments and applications of the NRHB model can be found in Ref. [16].

The adoption of the GS combinatorial term implies some changes in the GLF formalism. The EOS for the mixture takes the following form:

$$\tilde{P}+\tilde{T}\left[\ln(1-\tilde{\rho})-\tilde{\rho}\sum_{i=1}^{t}\phi_i\frac{l_i}{r_i}-\frac{z}{2}\ln\left[1-\tilde{\rho}+\frac{q}{r}\tilde{\rho}\right]+\frac{z}{2}\ln\Gamma_{00}\right]=0 \tag{30.32}$$

Since $\Gamma_{00}=\Gamma_0^2$, the last term in Equation 30.32 may be written, also, as $z\ln\Gamma_0$.

As in the LF model, each fluid is characterized by three scaling constants, namely, an interaction energy, ε^*, a hard-core volume per segment, v^*, and a number of segments, r, per molecule or, equivalently, a specific volume, v^*_{sp}. The reducing quantities for the scaling constants are the same as in the LF model discussed extensively in Chapter 29. Thus, $\varepsilon_i^*=\frac{z}{2}\varepsilon_{ii}$ and, for the mixture:

$$\varepsilon^*=\sum_{i=1}^{t}\sum_{j=1}^{t}\theta_i\theta_j\varepsilon_{ij}^*=RT^*=P^*v^* \tag{30.33}$$

with

$$\varepsilon_{ij}^*=\sqrt{\varepsilon_i^*\varepsilon_j^*}\left(1-k_{ij}\right) \tag{30.34}$$

while the segment volume, v*, is treated as a universal constant equal to 9.75 cm^3/mol [10]. Equations 30.24 and 30.32 are coupled equations and must be solved simultaneously for the reduced density and the nonrandomness factors.

The chemical potential of each component in the mixture is obtained from the Gibbs energy derivation:

$$\mu_i=\left(\frac{\partial G}{\partial N_i}\right)_{T,P,N_{j,j\neq i},N_{10},\ldots,N_{t0},\tilde{\rho}} \tag{30.35}$$

which gives for the "physical" contribution to the chemical potential:

$$\begin{aligned}\frac{\mu_{i,PH}}{RT}&=\ln\frac{\varphi_i}{\delta_i r_i}-r_i\sum_j\frac{\varphi_j l_j}{r_j}+\ln\tilde{\rho}+r_i(\tilde{v}-1)\ln(1-\tilde{\rho})\\&-\frac{z}{2}r_i\left[\tilde{v}-1+\frac{q_i}{r_i}\right]\ln\left[1-\tilde{\rho}+\frac{q}{r}\tilde{\rho}\right]+\frac{zq_i}{2}\left[\ln\Gamma_{ii}+\frac{r_i}{q_i}(\tilde{v}-1)\ln\Gamma_{00}\right]-\frac{q_i}{\tilde{T}_i}+r_i\frac{\tilde{P}\tilde{v}}{\tilde{T}}\frac{v_i^*}{v^*}\end{aligned} \tag{30.36}$$

The equation for the chemical potential of pure component 1 is obtained from Equation 30.36 by setting $\varphi_i=\theta_i=1$. With the PV practice of a universal segment volume, $v_{i*}=v^*=9.75$ cm^3/mol [10], one may substitute the last term in Equation 30.36 from the EOS, Equation 30.32, and obtain:

$$\begin{aligned}\frac{\mu_{i,PH}}{RT}&=\ln\frac{\varphi_i}{\delta_i r_i}+\ln\tilde{\rho}-r_i\ln(1-\tilde{\rho})+\frac{z}{2}r_i\left[1-\frac{q_i}{r_i}\right]\ln\left[1-\tilde{\rho}+\frac{q}{r}\tilde{\rho}\right]-\frac{q_i}{\tilde{T}_i}\\&+\frac{zq_i}{2}\ln\Gamma_{ii}-\frac{zr_i}{2}\ln\Gamma_{00}\end{aligned} \tag{30.36a}$$

Replacing from Equation 30.17, we obtain:

$$\frac{\mu_{i,PH}}{RT} = \ln\frac{\varphi_i}{\delta_i r_i} + \ln\tilde{\rho} - r_i \ln(1-\tilde{\rho}) + \frac{z}{2}[r_i - q_i]\ln\left[1-\tilde{\rho}+\frac{q}{r}\tilde{\rho}\right] + zq_i \ln\Gamma_i - zr_i \ln\Gamma_0 \tag{30.36b}$$

The volume of mixing V^E is obtained as follows:

$$V_{PH}^E = V - \sum_i V_i^0 = rN\tilde{v}\,v^* - \sum_i r_i N_i \tilde{v}_i v_i^* = rN\left[\tilde{v}v^* - \sum_i \varphi_i \tilde{v}_i v_i^*\right] \tag{30.37}$$

and with a universal v*, as follows:

$$V_{PH}^E = V - \sum_i V_i^0 = rNv^*\left[\tilde{v} - \sum_i \varphi_i \tilde{v}_i\right] \tag{30.37a}$$

The heats of mixing are obtained in a similar manner:

$$H_{PH}^E = H - \sum_i H_i^0 = -\sum_{j\geq i}^{t} N_{ij}\varepsilon_{ij} + \sum_{i=1}^{t} N_{ii}^0 \varepsilon_{ii} + PV_{PH}^E \tag{30.38}$$

When a universal hard-core volume per segment, v* = 9.75 cm³/mol is adopted [10], the mean interaction energy is assumed to depend on temperature as follows:

$$\varepsilon^* = \varepsilon_h^* + (T - 298.15)\varepsilon_s^* \tag{30.39}$$

and the characteristic specific volume as follows:

$$v_{sp}^* = v_{sp,0}^* + (T - 298.15)v_{sp,s}^* \tag{30.40}$$

Sources for the scaling constants are reviewed in Ref. [16].

The above formalism accounts for the contribution of the "physical" interactions in the system, that is, all interactions except for the strong specific interactions such as the hydrogen-bonding ones. The full expressions for the corresponding thermodynamic quantities require also the corresponding hydrogen-bonding contributions, discussed in Chapter 32.

As mentioned already, the above general formalism for the physical contributions has been simplified in two directions. In the first direction, only the nonrandom distribution of free volume is considered and the distribution of all other molecular contacts (except for the hydrogen-bonding ones, of course) is considered random [11]. In the second direction, the opposite is considered, namely, the molecular contacts are distributed nonrandomly and the free volume randomly, as in the original PV model [10]. These simplifications affect the above formalism. The essentials of the first direction are presented in the next subsection.

THE GLF MODEL WITH NONRANDOM FREE VOLUME DISTRIBUTION

The GLF model is simplified significantly when the bulk of the nonrandom distribution of molecular species is accounted for by the hydrogen-bonding factor, g_{HB}, and g_{QC} is used to account for the nonrandom distribution of the free volume by making no distinction of the neighboring molecular segments. There is, then, only one type of molecule – empty site interaction which will be indicated as $r-0$ interaction. In this case, the factor, $g_{NR}=g_{QC}$ for the mixture becomes:

$$g_{NR} = \frac{N_{rr}^0!N_{00}^0!\left[\left(\frac{N_{r0}^0}{2}\right)!\right]^2}{N_{rr}!N_{00}!\left[\left(\frac{N_{r0}}{2}\right)!\right]^2} \tag{30.41}$$

N_{rr} is the number of external intermolecular contacts, N_{00} is the number of contacts between the empty sites, and N_{r0} is the number of contacts between a molecular segment and an empty site. In the random case N_{rr}, N_{00}, and N_{r0} take the form (cf. Equations 30.18–30.20):

$$N_{rr}^0 = \sum_i N_{ii}^0 + \sum_i \sum_{j>i} N_{ij}^0 = \frac{1}{2} zqN \frac{qN}{N_0+qN} = \frac{z}{2} qN\theta_r \tag{30.42}$$

$$N_{00}^0 = \frac{1}{2} N_0 z \frac{N_0}{N_q} = \frac{z}{2} N_0\theta_0 \tag{30.43}$$

$$N_{r0}^0 = zqN\frac{N_0}{N_q} = zN_0\frac{qN}{N_q} = zqN\theta_0 = zN_0\theta_0 \tag{30.44}$$

where θ_r and θ_0 are given by Equations 30.19–30.20.

The material balance equations are now:

$$\begin{aligned} 2N_{00} + N_{0r} = zN_0 \quad &\text{or} \quad \theta_0\Gamma_{00} + \theta_r\Gamma_{r0} = 1 \\ 2N_{rr} + N_{0r} = zqN \quad &\text{or} \quad \theta_r\Gamma_{rr} + \theta_0\Gamma_{r0} = 1 \end{aligned} \tag{30.45}$$

The potential energy U in this case is given by:

$$-U = \sum_i N_{ii}\varepsilon_{ii} + \sum_i \sum_{j>i} N_{ij}\varepsilon_{ij} = \Gamma_{rr} qN\theta_r\varepsilon^* \tag{30.46}$$

while the EOS is identical in form to Equation 30.32. There is only one quasi-chemical condition in this case, given by:

$$\frac{4N_{rr}N_{00}}{N_{r0}^2} = \frac{\Gamma_{rr}\Gamma_{00}}{\Gamma_{r0}^2} = \exp\left(\frac{2\varepsilon^*/z}{RT}\right) = A \tag{30.47}$$

Replacing from Equation 30.45 into Equation 30.47, we obtain a quadratic equation for Γ_{r0}, with the physically meaningful solution:

$$\Gamma_{r0} = \frac{2}{1+[1-4\theta_0\theta_r(1-A)]^{1/2}} \tag{30.48}$$

Γ_{rr} and Γ_{00} are, then, obtained from Equation 30.45.

The chemical potential of component i is given by:

$$\frac{\mu_i}{RT} = \ln\frac{\varphi_i}{\delta_i r_i} - r_i\sum_j \frac{\varphi_j l_j}{r_j} + \ln\tilde{\rho} + r_i(\tilde{v}-1)\ln(1-\tilde{\rho}) - 2q_i\theta_r\Gamma_{rr}\frac{\sum_k \Theta_k\varepsilon_{ik}^* - \varepsilon^*}{RT}$$
$$-\frac{z}{2}r_i\left[\tilde{v}-1+\frac{q_i}{r_i}\right]\ln\left[1-\tilde{\rho}+\frac{q}{r}\tilde{\rho}\right] + \frac{zq_i}{2}\left[\ln\Gamma_{rr} + \frac{r_i}{q_i}(\tilde{v}-1)\ln\Gamma_{00}\right] - \frac{q_i}{\tilde{T}} + r_i\frac{\tilde{P}\tilde{v}}{\tilde{T}}\frac{v_i^*}{v^*} \quad (30.49)$$

In the case of a binary mixture, Equation 30.49 reduces to:

$$\frac{\mu_1}{RT} = \ln\frac{\varphi_1}{\delta_1 r_1} - r_1\frac{l}{r} + \ln\tilde{\rho} + r_1(\tilde{v}-1)\ln(1-\tilde{\rho}) - \frac{z}{2}r_1\left[\tilde{v}-1+\frac{q_1}{r_1}\right]\ln\left[1-\tilde{\rho}+\frac{q}{r}\tilde{\rho}\right]$$
$$+\frac{zq_1}{2}\left[\ln\Gamma_{rr} + \frac{r_1}{q_1}(\tilde{v}-1)\ln\Gamma_{00}\right] - \frac{q_1}{\tilde{T}} - 2q_1\theta_r\Gamma_{rr}\Theta_2\frac{\varepsilon_1^* - \varepsilon_{12}^* - \Theta_2 RTX_{12}}{RT} + r_1\frac{\tilde{P}\tilde{v}}{\tilde{T}}\frac{v_1^*}{v^*} \quad (30.50)$$

where

$$\Theta_i = \frac{N_i q_i}{qN} = \frac{N_i q_i}{N_q}\frac{N_q}{qN} = \frac{\theta_i}{\theta_r}$$

Again, the above formalism accounts for the contribution of the "physical" interactions in the system. The full expressions for the corresponding thermodynamic quantities require also the corresponding hydrogen-bonding contributions, discussed in Chapter 32.

GLF, LATTICE-CLUSTER, AND THE BGY EQUATION-OF-STATE MODELS

Two advanced lattice models, primarily, for polymer systems, are the LCT (lattice-cluster theory) model of Freed et al. [17–20] and the BGY (Born–Green–Yvon) model of Lipson et al. [21–23]. Both models use GLF as a reference model for the liquid structure, and in their current development are rather comprehensive EOS models covering a wide range of applications.

The LCT model makes it possible to describe the influence of macromolecules' architecture (short and long branching, pendant rings, etc.) on the free energy of their mixtures. The molecular structure-dependent partition function, Q, is developed as a double series expansion in the inverse coordination number $1/z$ and the dimensionless interaction energy $\tilde{\varepsilon}$. Q was developed to second order in both of these parameters and for both incompressible and compressible systems. The LCT incompressible variant has been tested rigorously against Monte-Carlo simulations, and rather successfully to solutions and melts of hyper-branched polymers with relatively weak interactions. Incorporation of strong or specific interactions leads to rather complex formalism and has been done for polymers with end groups exhibiting specific interactions [20]. Graph theory is extensively used in LCT developments and also provides expressions for the structure factor (cf. Chapter 29) in polymer mixtures, which is particularly valuable in scattering (small-angle neutron scattering, X-ray diffraction) experiments with polymer systems.

The zero-order series expansion of the incompressible LCT variant yields the Flory–Huggins model and the zero-order series expansion of the compressible LCT EOS model yields the LF model.

Lipson et al. [21–24] have used the integral-equation approach of Born–Green–Yvon [25] and derived analytical expressions for the basic thermodynamic quantities of both incompressible and compressible-lattice mixtures. They also derived expressions for the structure factor and the interpretation of SANS experimental data for polymer mixtures. For the compressible mixture, as in LCT

EOS model, they have adopted the LF practice for the free volume consisting of empty lattice sites dispersed throughout the volume of the system. The compressible-lattice BGY EOS for pure fluid is:

$$\tilde{P}+\tilde{T}\left[\ln(1-\tilde{\rho})-(z/2)\ln\left[(1-\tilde{\rho})/(1-\theta)\right]+(1-1/r)\tilde{\rho}\delta_{q,r}+(z/2)J\theta^2\right] \tag{30.51}$$

where J is given by:

$$J=\left[\exp\left(-2/z\tilde{T}\right)-1\right]/\left[\theta\exp\left(-2/z\tilde{T}\right)+(1-\theta)\right] \tag{30.52}$$

The symbols are as in the above LF model and $\delta_{q,r}$ is the Kronecker delta.

The PV EOS [10] is obtained by expanding J up to $O(\varepsilon/kT)$. On the other hand, by further setting $q=r$, expanding J and keeping terms up to $O(\varepsilon/kT)$, Equation 30.51 becomes identical in form to the LF EOS. As PV, the BGY model accounts also for nonrandom mixing. Thus, the BGY model could be considered an extended GLF EOS model. In Ref. [23], the BGY model was tested against experimental *PVT* data for various polymers and normal alkanes from C_6 to C_{30}. The performance of the BGY EOS was the best. In most cases, the performance of PV and BGY were nearly equivalent. The performance of the plain LF was somewhat inferior.

A quantum-mechanics-based model, which will be discussed in the next chapter, is at present a rigid-lattice model. However, as we will see, its very structure facilitates the extension of its formalism into a GLF formalism, in close analogy with the LCT and especially the BGY and PVEOS models.

REFERENCES

1. Sanchez, I. C., and Lacombe, R. 1976. An elementary molecular theory of classical fluids. Pure fluids. *J. Phys. Chem.* 80: 2352–2362
2. Sanchez, I. C., and Lacombe, R. 1978. Statistical thermodynamics of polymer solutions. *Macromolecules* 11: 1145–1156.
3. Panayiotou, C. 1987. Lattice-fluid theory of polymer solutions. *Macromolecules* 20: 861–871.
4. Panayiotou, C., and Sanchez, I. C. 1991. Hydrogen bonding in fluids: An equation of-state approach. *J. Phys. Chem.* 95: 10090–10097.
5. Sanchez, I. C., and Panayiotou, C. 1994. Equations of state thermodynamics of polymer and related solutions. In *Models for Thermodynamic and Phase Equilibria Calculations*, ed. S. Sandler. New York.: Marcel Dekker Inc, pp. 187–286.
6. Flory, P. J. 1953. *Principles of Polymer Chemistry*. Ithaca, NY: Cornell University Press.
7. Hill, T. L. 1986. *An Introduction to Statistical Thermodynamics*. New York: Dover Publications Inc.
8. Guggenheim, E. A. 1952. *Mixtures*. Oxford: Oxford University Press.
9. Staverman, A. J. 1950. The entropy of high polymer solutions. Generalization of formulae. *Rec. Trav. Chim. Pays-Bas.* 69: 163–174.
10. Panayiotou, C., and Vera, J. H. 1982. Statistical thermodynamics of r-mer fluids and their mixtures. *Polymer J.* 14: 681–694.
11. Panayiotou, C. 2003. The QCHB model of fluids and their mixtures. *J. Chem. Thermodyn.* 35: 349–381.
12. High, M. S., and Danner, R. P. 1990. Application of the group contribution lattice-fluid EOS to polymer solutions, *AIChE J.* 36: 1625–1632.
13. High, M. S., and Danner, R. P. 1993. *Handbook of Polymer Solution Thermodynamics*. New York: AIChE, DIPPR.
14. Panayiotou, C., Stefanis, E., Tsivintzelis, I., Pantoula, M., and Economou, I. 2004. Nonrandom hydrogen-bonding model of fluids and their mixtures 1. Pure fluids. *Ind. Eng. Chem. Res.* 43: 6592–6606
15. Panayiotou, C., Tsivintzelis, I., and Economou, I. G. 2007. Nonrandom hydrogen-bonding model of fluids and their mixtures. 2. Multicomponent mixtures *Ind. Eng. Chem. Res.* 46: 2628–2636.
16. Mensitieri, G., Scherillo, G., Panayiotou, C., and Musto, P. 2020. Towards a predictive thermodynamic description of sorption processes in polymers: The synergy between theoretical EOS models and vibrational spectroscopy, *Mater. Sci. Eng. R.* 140: 100525.

17. Freed, K. F., and Bawendi, M. G. 1989. Lattice theories of polymeric fluids. *J. Phys. Chem.* 93: 2194–2203.
18. Dudowicz, J., and Freed, K. F. 1991. Effect of monomer structure and compressibility on the properties of multicomponent polymer blends and solutions. 3. Application to deuterated polystyrene [PS(D)] poly(vinyl methyl ether) (PVME) blends *Macromolecules* 24: 5112–5123.
19. Buta, D., Freed, K. F., and Szleifer, I. 2000. Thermodynamic properties of lattice polymers: Monte Carlo simulations and mean-field theories. *J. Chem. Phys.* 112: 6040–6048.
20. Xu, W.-S., and Freed, K. F. 2014. Lattice cluster theory for polymer melts with specific interactions. *J. Chem. Phys.* 141: 044909.
21. Lipson, J. E. G. 1991. Polymer blends and polymer solutions: a Born-Green-Yvon integral equation treatment. *Macromolecules* 24: 1334–1340.
22. Lipson, J. E. G. 1992. A Born-Green-Yvon integral equation treatment of incompressible lattice mixtures. *J. Chem. Phys.* 96: 1418–1425.
23. Lipson, J. E. G., and Andrews, S. S. 1992. A Born-Green-Yvon integral equation treatment of a compressible fluid. *J. Chem. Phys.* 96: 1426–1434
24. White, R. P., Lipson, J. E. G., and Higgins, J. S. 2012. How pure components control polymer blend miscibility. *Macromolecules* 45: 8861–8871.
25. (a) Born, M., and Green, H. S. 1946. A general kinetic theory of liquids. I. The molecular distribution functions. *Proc. R. Soc. Lond. A* 188: 10–18; (b) Yvon, J. 1935. *La Théorie Statistique des Fluides et l'Équation d'État.* Paris: Hermann et Cie.

31 COSMO-RS-Type Models

Abstract

This chapter examines, first, the quantum chemical and the statistical thermodynamic aspects of the COnductor-like Screening MOdel for Real Solvents (COSMO-RS) a-priori predictive model. Subsequently, the central working equations of the model are derived in a consistent manner from Guggenheim's quasi-chemical theory. COSMO-RS is, then, extended into an equation-of-state model, and the perspectives are discussed.

INTRODUCTION

This chapter reviews the advances in solution thermodynamics initiated by Andreas Klamt with the development of a novel a-priori predictive COSMO-RS model [1–4]. COSMO-RS stands for COnductor-like Screening MOdel for Real Solvents. The meaning of these terms, coming from quantum mechanics, will be understood with the presentation of this topic below. It combines quantum chemical calculations with statistical thermodynamics and is rooted in the large family of dielectric continuum solvation models (CSM) [5–7] of quantum mechanics. COSMO modification of the dielectric CSMs [7] replaced the dielectric boundary conditions by a much simpler scaled-conductor boundary condition. In a second crucial step, Klamt made the COSMO solvation approach into a powerful thermodynamic model by bringing quantum chemical information into statistical thermodynamics. With the assumption of perfect charge screening with a conductor of infinite dielectric constant, it made possible thermodynamic treatment of the electrostatic interactions as binary intermolecular interactions. In this way, COSMO-RS model became usable even by nonexperts in quantum mechanics.

After the pioneering work by Klamt, various important versions of COSMO-RS model have appeared in the literature [8–10], most notably the COSMO-SAC model by Sandler and co-workers [8], where SAC stands for Segment Activity Coefficient model. This further step made the COSMO-RS model easily recognizable by those familiar with group activity coefficient models (cf., Chapter 28 and Appendix D).

COSMO-RS has two main components, one quantum chemical and one statistical thermodynamic. In what follows we examine briefly these two components. As this COSMO treatment is still in the stage of development, we depart from the style of presenting standard thermodynamic methods used in this book and, in an interpretative way, follow the different stages of its development until the time of this writing.

QUANTUM CHEMICAL ASPECTS OF COSMO-RS

In COSMO-RS, quantum chemical calculations are, first, done from a start geometry, that is, a chosen molecular structure with the corresponding molecular surface area, in order to obtain the surface electron density and the generated electrostatic potential, which is mapped on the molecular surface as a surface charge distribution. The solvation of the solute molecules by the solvent-conductor may be visualized as the molecules residing in their molecule-shaped cavities and embedded in a conductor of infinite dielectric constant. In this state, the electric field from the charge distribution of each molecule polarizes the embedding medium and produces a new charge distribution on the surface of the molecular cavity. This, in turn, counterbalances and screens the charge distribution of the molecule. These conductor-like screening charges are included as additional external charges

DOI: 10.1201/9781003431985-35

in the next iteration of quantum chemical calculation. The iterations continue until self-consistency. In each repetition, the molecular structure is optimized from local energy minimization and the process continues until no perceptible change.

The charge distribution from the polarization of the medium is easily generated by quantum chemical calculations at a medium level of density functional theory/basis-sets combinations. Without entering into the details of quantum chemical calculations, let us state that when self-consistency is attained, the results give a detailed charge distribution over the surface of the molecular cavity. Due to the COSMO method, out of which the molecule is brought into the perfectly screened state, the results are obtained in a reasonable computing time. In this perfectly screened state, the molecules move freely and independently in the medium, without any intermolecular electrostatic interactions.

The molecular surface may be divided into surface segments by selecting a standard surface segment area, α_{eff}, and, with the appropriate averaging calculations, a surface charge density, σ, which is the ratio of the average charge over the segment area, may be attributed to each surface segment. Since there may be many surface segments with the same charge density, a desirable information is the surface charge distribution profile, or *sigma profile*, $p_i(\sigma_m)$, which is the probability of finding a segment with charge density σ_m on the surface of the molecular cavity of molecule i. If ν_{mi} is the number of segments with charge density σ_m on the surface of molecule i, with total surface area α_i, the probability $p_i(\sigma_m)$ is given by:

$$p_i(\sigma_m) = \frac{\nu_{mi}\alpha_{eff}}{\sum_j \nu_{ji}\alpha_{eff}} = \frac{\alpha_{mi}}{\sum_j \alpha_{ji}} = \frac{\alpha_{mi}}{\alpha_i} \tag{31.1}$$

where α_{mi} is the surface area of molecule i with charge density σ_m. This sigma profile is the most important piece of information obtained from quantum chemical calculations and is utilized in the statistical thermodynamic developments of the model.

However, in real solutions, the molecules are surrounded by other molecules and not by a perfect conductor. Thus, the next step is to bring two molecules into direct contact by removing the medium from the contacted surface segments. In this case, the medium screens only the rest of the molecular surfaces, that is, the molecular surface of a complex or dimer. This may lead to an energy change depending on the charge densities, σ_1 and σ_2 of the two contacted surfaces. If they are exactly opposite, there will be overall zero screening charge, However, $\sigma_1 + \sigma_2$ will not always be zero and this residual charge density, σ_{misf}, will lead to an energy change, called *misfit* energy, considered to be proportional to $\alpha_{eff}(\sigma_1 + \sigma_2)^2$.

The above process is continued until all surface segments of a molecule are only in direct contact with other molecules, and the conductor medium is entirely removed. The result is a new picture of the liquid state in which all molecules are in direct contact with the surrounding molecules without any voids in between, as shown in Figure 31.1.

The first consequence of this rather rigid structure is the attribution of relatively large *cosmo* molecular volumes, v_{cosm}, and *cosmo* molecular surface areas, a_{cosm}, to the molecules or to the molecular cavities. This is done by an appropriate increase of atom radii by about 17% compared to the classical Bondi's radii [3] of widely used models such as UNIQUAC. Thus, the molecular cavities are as rigid as the van der Waals volume of a molecule calculated from Bondi's atom radii. The second consequence is the need for revision of the realism or the justification of this representation. One justification for the direct contacts of the molecules, with no void space in between, is that it does not reflect the classical notion of molecules in their van der Waals volume, but the *solvated* solute molecules by the surrounding solvent. This has led to a change in the name of the model, and currently RS in COSMO-RS stands for *Realistic Solvation* [11]. On the other hand, the misfit energy does not seem sufficient to describe realistically the full range of polar interactions, especially when

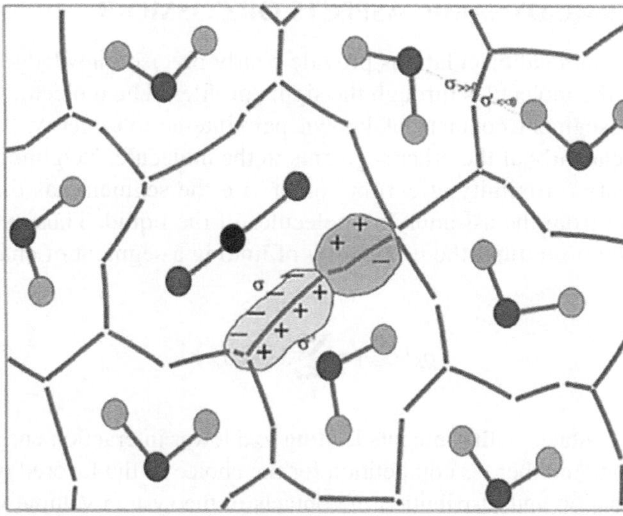

FIGURE 31.1 Schematic illustration of contacting molecular cavities and segment–segment contact interactions. Reproduced with permission from Ref. [3, fig. 1].

the molecules interact with strong specific forces, such as in strong hydrogen bonding. In order to account for hydrogen bonding, the misfit energy is augmented with an optional term, which is nonzero when the interacting opposite surface charge densities exceed in absolute terms a cutoff value, σ_{hb}. Thus, the final form of the pair interaction energy between two contacted surface segments of charge densities σ_i and σ_j is given by:

$$\varepsilon_{pair}(\sigma_i,\sigma_j)=(\alpha'\alpha_{eff}/2)(\sigma_i+\sigma_j)^2+c_{hb}\max[0,\sigma_{acc}-\sigma_{hb}]\min[0,\sigma_{don}+\sigma_{hb}] \quad (31.2)$$

The first term on the right-hand side is the misfit energy with α' being a universal misfit constant. The second term is the hydrogen-bonding energy, in which the positive charge density is labeled with the subscript *acc* (acceptor) and the negative one with the subscript *don* (donor), and c_{hb} is the universal hydrogen-bonding constant. In fact, c_{hb} is temperature-dependent and this is taken into account with:

$$c_{hb}(T)=c_{hb}\max\left[0,1-c_{hb}^{T}\left(1-\frac{298.15K}{T}\right)\right] \quad (31.3)$$

c_{hb}^{T} here is a universal constant for this temperature dependence.

The weak nonelectrostatic interactions (mainly dispersion or London-type interactions) are treated separately in several different ways. These, as well as a number of other subtleties of the model, such as the method of smoothing along the crevices of intersecting atom spheres in the cavity construction, or the correction for outlying charge errors, do not add much to this brief presentation and are not discussed here. The essential point is that the model relaxes some of its initial assumptions in order to adapt itself to a more realistic picture of the liquid state. Nevertheless, it is worth noticing that there is an important assumption in the way that the quantum chemical calculations are done for the extraction of the sigma profiles of the molecules: All segments in the liquid are able to make contacts independent of one another as if the segments were cut loose from the rigid molecular surfaces. This assumption facilitates the combination of quantum mechanics with statistical thermodynamics. The important point of Figure 31.1 is that illustrates the way of handling intermolecular interactions as binary interactions, that is of two interacting sites only.

STATISTICAL THERMODYNAMIC ASPECTS OF COSMO-RS

The above quantum chemical calculations provide nearly precise knowledge of the surface segment composition of the molecules through the sigma profiles. The molecules in a liquid interact by making segment–segment contacts of known pair interaction energy. The segments make these contacts independently of the other segments in the molecule. Neighboring segments influence the charge density (proximity effect) but otherwise the segment has the freedom to select its partner for contact from the assembly of molecules of the liquid. Thus, when the system S is a multicomponent liquid mixture, the probability of finding a segment of charge density σ_m may be obtained from:

$$p_S(\sigma_m) = \sum_i x_i p_i(\sigma_m) \tag{31.4}$$

Each segment favors contacts with segments leading to a lower interaction energy but all segments must find their partners and there is competition for the choice of the favored ones. Ultimately, the prevailing accommodation and distribution of contacts in the system volume will be the one with the lowest system-free energy.

This competition in the segment–segment pairing implies that the contacts of two segments are not done randomly or in agreement with their overall concentration in the system, as given by Equation 31.4. As Klamt sees it, each contact must be weighted by a Boltzmann-type factor with the exchange energy of the contact given by Equation 31.2 and this dictates the distribution in the contact pair population. Statistical thermodynamics needs to be used in order to obtain a consistent solution for the nonrandom distribution of binary (segment–segment) contacts in the liquid. At the early stages of developing the above treatment, Klamt did not relate it to Guggenheim's quasi-chemical approach, which had given an answer to this problem. Thus, he proposed to solve the problem with the following recursive equation:

$$\mu_S(\sigma) = -RT \ln\left[\int p_S(\sigma')\exp\left\{\frac{\mu_S(\sigma') - \varepsilon_{pair}(\sigma,\sigma')}{RT}\right\}d\sigma'\right] \tag{31.5}$$

In the first attempts, there was a second term on the right-hand side, added to represent the combinatorial contribution to the chemical potential $\mu_S(\sigma)$ of a segment of σ charge density in the system S. This term was dropped later. The combinatorial term may be meaningful when it concerns molecules, but not when it concerns a surface segment. Much effort was made to attribute a meaning to $\mu_S(\sigma)$. Initially it was often interpreted as pseudochemical potential for the solvation of the segment in the Ben-Naim sense [12]. At some stage, $\mu_S(\sigma)/RT$ was set equal to $\ln\gamma_S(\sigma)$, and Equation 31.5 was rewritten in a form that it "may look more familiar to chemical engineers" [3]:

$$\ln\gamma_S(\sigma) = -\ln\left[\int p_S(\sigma')\gamma_S(\sigma')\exp\left\{\frac{-\varepsilon_{pair}(\sigma,\sigma')}{RT}\right\}d\sigma'\right] \tag{31.5a}$$

In this equation, $\gamma_S(\sigma)$ was called the *activity factor* of the segment. For a better understanding of Equation (31.5), we recall some essentials from the quasi-chemical approach relevant to the COSMO-RS model.

A RELEVANT NOTE ON THE QUASI-CHEMICAL APPROACH

The nonrandomness factors, Γ_{ij}, were introduced [13] in the quasi-chemical theory as the ratio of N_{ij} contacts in the real system over the corresponding number of contacts, N_{ij}^0, in a system of perfect randomness, or:

$$\Gamma_{ij} = \frac{N_{ij}}{N_{ij}^0}, \quad \Gamma_{ii} = \frac{N_{ii}}{N_{ii}^0} \tag{31.6}$$

With these factors, the quasi-chemical condition, Equation 28.37, takes the form (cf. Equation 30.15):

$$\frac{\Gamma_{ij}^2}{\Gamma_{ii}\Gamma_{jj}} = \frac{\tau_{ij}^2}{\tau_{ii}\tau_{jj}}, \quad \tau_{ij} = \exp\left(-\frac{\varepsilon_{ij}}{kT}\right) \tag{31.7}$$

Implementation of these nonrandomness factors into Guggenheim's quasi-chemical theory gives the following equation for the activity coefficient of component i in the mixture [13]:

$$\ln \gamma_i = (\ln \gamma_i)_{comb} + \sum_k \nu_{ki} \frac{Q_k}{2}\left[\ln \Gamma_{kk} - \ln \Gamma_{kk}^{(i)}\right] \tag{31.8}$$

Superscript (i) in this equation indicates pure component i property and Q_k is the number of external contacts (or contact surface area) of group k. As discussed in Chapter 30, the nonrandomness factor Γ_{mn} for the distribution of a segment m around a central segment n can be split into segment-specific nonrandomness factors, Γ_i, as:

$$\Gamma_{mn} = \Gamma_m \Gamma_n \tau_{mn} \tag{31.9}$$

Equation 31.9 satisfies the quasi-chemical condition. The nonrandomness factors satisfy the contact balance or conservation equation [13] (cf. Equation 30.22):

$$\sum_n \Theta_n \Gamma_{mn} = 1 \tag{31.10}$$

where Θ_n is the surface area fraction of group or segment n in the system. Following Klamt, the development in this section is in terms of groups or segments, thus the appropriate surface area fraction is Θ_m for the group m and not θ_i for the molecule i, as it was the case in Chapter 30. Replacing Equation 31.9 in Equation 31.10, the conservation equation takes the form:

$$\sum_n \Theta_n \Gamma_m \Gamma_n \tau_{mn} = \Gamma_m \sum_n \Theta_n \Gamma_n \tau_{mn} = 1 \tag{31.11}$$

or, in logarithmic form:

$$\ln \Gamma_m = -\ln \sum_n \Theta_n \Gamma_n \tau_{mn} \tag{31.12}$$

A QUASI-CHEMICAL COSMO-RS MODEL

With the above elements from the quasi-chemical approach, replacing the exponential term in Equation 31.5a by $\tau(\sigma,\sigma')$, it takes the form:

$$\ln \gamma_S(\sigma) = -\ln\left[\int p_S(\sigma')\gamma_S(\sigma')\tau(\sigma,\sigma')d\sigma'\right] \tag{31.5b}$$

The sigma profiles are normally reported in a two-column table with a varying number of rows, equal to the number of surface segments of the molecule. For all practical purposes, the integral in Equation 31.5 or 31.5a can be replaced by a sum, that is:

$$\ln\gamma_S(\sigma_m) = -\ln\left[\sum_n p_S(\sigma_n)\gamma_S(\sigma_n)\tau(\sigma_m,\sigma_n)\right] \tag{31.5c}$$

In addition, as seen in Equation 31.1, $p_i(\sigma_m)$ is equal to the surface fraction of segments with charge density σ_m in molecule i, that is, it is equal to Θ_{mi}. Replacing in Equation 31.4 gives:

$$p_S(\sigma_m) = \sum_i x_i\Theta_{mi} = \Theta_m \tag{31.4a}$$

So, Equation 31.5c takes the form:

$$\ln\gamma_S(\sigma_m) = -\ln\left[\sum_n \Theta_m\gamma_S(\sigma_n)\tau(\sigma_m,\sigma_n)\right] \tag{31.5d}$$

Equation 31.5d is identical to Equation 31.12 with γ replacing Γ. If it is considered that the "activity factor" $\gamma_S(\sigma_m)$ is just the segment-specific nonrandomness factor Γ_m, then the Klamt model appears as a reformulation of the quasi-chemical theory!

In the quasi-chemical theory, the thermodynamically consistent equation for the activity coefficient of a molecular compound i is given by Equation 31.8; thus, it is essential to see how this equation was obtained [13]. For this, taking the derivative with respect to n_i of the partition function, ln Z given by Equation 28.35, keeping all other numbers $n_{j\neq i}$ constant, gives [13]:

$$\frac{\mu_i}{RT} = -\left(\frac{\partial\ln Z}{\partial n_i}\right)_{n_{j\neq i}} = \frac{\mu_i^R}{RT} + \sum_k \nu_{ki}\frac{Q_k}{2}\ln\Gamma_{kk} \tag{31.13}$$

where the superscript R indicates the reference state of randomness. This derivation is very much facilitated by the fact that ln Z was minimized with respect to Γ_{kl} when obtaining the quasi-chemical condition, Equation 31.7 (cf. Ref. [13] and Chapter 30). The activity coefficient is then obtained by subtracting the chemical potential of the pure compound:

$$\ln\gamma_i = \frac{\mu_i - \mu_{i,\text{pure}}}{RT} - \ln x_i \tag{31.14}$$

Substituting from Equation 31.13 gives:

$$\ln\gamma_i = \left(\frac{\mu_i^R - \mu_{i,\text{pure}}^R}{RT} - \ln x_i\right) + \sum_k \nu_{ki}\frac{Q_k}{2}\left[\ln\Gamma_{kk} - \ln\Gamma_{kk}^{(i)}\right] \tag{31.15}$$

The term in parenthesis is the combinatorial term, and thus, Equation 31.8 is obtained. Replacing Equation 31.9 in Equation 31.8 gives:

$$\begin{aligned}\ln\gamma_i &= (\ln\gamma_i)_{comb} + \sum_k \nu_{ki}\frac{Q_k}{2}\left[\ln\left(\Gamma_k^2\tau_{kk}\right) - \ln\left(\Gamma_{kk}^2\tau_{kk}\right)^{(i)}\right] \\ &= (\ln\gamma_i)_{comb} + \sum_k \nu_{ki}Q_k\left[\ln\Gamma_k - \ln\Gamma_k^{(i)}\right]\end{aligned} \tag{31.16}$$

It should be stressed that Γ_k in this equation is the group-specific or surface-segment-specific *nonrandomness factor* and $\nu_{ki}Q_k$ is the contact surface area of all the groups k in the surface area of molecule i. In terms of the COSMO-RS symbols:

$$\frac{\nu_{mi}Q_m}{\alpha_i/\alpha_{eff}} = \frac{\alpha_i(\sigma_m)}{\alpha_i} = p_i(\sigma_m) \tag{31.17}$$

Substituting in Equation 31.16 gives:

$$\ln\gamma_i = (\ln\gamma_i)_{comb} + \frac{\alpha_i}{\alpha_{eff}}\sum_k p_i(\sigma_k)\left[\ln\Gamma_k - \ln\Gamma_k^{(i)}\right] \tag{31.18}$$

which is precisely the COSMO-RS Equation for the activity coefficient!

The interesting question is how Klamt reached this equation starting from Equation 31.5 or 31.5a. In 2002 a new variation was presented under the name COSMOSPACE [14]. In COSMOSPACE, Γ_k is defined as the *activity coefficient* of segment of type k through the equation:

$$kT\ln\Gamma_k = \mu^k - \mu^{kk} \tag{31.19}$$

where μ^k is the pseudochemical potential of segment k in the mixture and μ^{kk} is the pseudochemical potential of segment k in a liquid phase consisting of segments k only. In spite of the rigidity implied by Figure 31.1, here the segments are treated as independent and accessible by any other segment for contact.

It is rather hard to reconcile the quasi-chemical theory with Equation 31.19 and its ad-hoc assumptions. Although the equations for the activity coefficient in quasi-chemical theory and COSMO-RS are identical in form, they were obtained by different routes. It is clear that the starting point of COSMO-RS, Equation 31.5, is essentially identical to the recursive Equation 31.12 which is just an algorithm for the numerical solution of a system of equations, as we have seen in Chapter 30, and not a self-consistent theory of the liquid state. Even if one has these Γ's, the problem is how they will be used consistently in a thermodynamic model of liquid mixtures.

CURRENT STATUS AND PERSPECTIVES OF COSMO-RS MODEL

COSMO-RS can be considered as a group-contribution model, in which the interaction energies are obtained from manageable quantum chemical calculations. Experimental data are not needed except for the parameterization of the model, that is, the determination of some of its universal constants. Once these few parameters are fixed, the model can be used for the prediction of phase equilibria and related properties of complex multicomponent systems. It may run calculations even with compounds not yet synthesized and is thus a valuable tool for the design of new materials and solvent screening applications. There are various ways of extending COSMO-RS. An interesting example is an extension to electrolyte systems, COSMO-RS-ES [15] in which Equation 31.2 is modified in order to account for ion interactions. Vapor-liquid equilibria are predicted satisfactorily by the model. Hsieh et al. [16] have added a temperature term in the pair interaction term of COSMO-SAC and they further differentiated the hydrogen-bonding term depending on the acceptor atom, leading to significantly improved prediction of both vapor-liquid and liquid-liquid equilibria.

One of the shortcomings of COSMO-RS and its variations is its inability to account for volume changes and pressure effects. This arises from the rigid structure shown in Figure 31.1. In a sense, the model was parameterized by disregarding free volume and with the molar volume corresponding to the molar *cosmo volume*, V_{cosm}. The question of whether this structure could become more realistic without violating quantum chemical calculations has been recently addressed [17]. It was shown that COSMO-RS may be extended to a COSMO-RS equation-of-state (COSMO-RS-EOS) model with statistical thermodynamics and without violating any principles or quantum chemical calculations. The same approach may also be used for the derivation of a consistent equation for the solvation-free energy. Quantum mechanics-based calculations of solvation quantities opens interesting perspectives in the molecular thermodynamics of mixtures. The essentials of this extension will be summarized in the next section.

A COSMO-RS EQUATION-OF-STATE MODEL

As seen above, the COSMO-RS equations may be derived in a straightforward manner from quasi-chemical theory with no recourse to any fictitious (pseudo)chemical potentials and related ad-hoc assumptions. This, then, makes it possible to use COSMO-RS as a vehicle and bring quantum chemical calculations into EOS frameworks such as the generalized lattice fluid (GLF) approach discussed in Chapter 30. With this in mind, there are two immediate issues to be addressed: (1) In what way could the free volume be incorporated in the above rigid structure of COSMO-RS and produce an EOS meeting the ideal-gas limit? And (2) Should σ-profiles be changed for use in such a COSMO-RS-EOS model?

The answer to the first question is facilitated by the very structure of the GLF approach. In fact, the introduction of free volume in the system through a number of empty (lattice) sites may be handled consistently with the quasi-chemical theory in the very same way used above for the derivation of the rigid COSMO-RS model [17,18]. The answer to the second question, however, needs some more care.

As already mentioned, the current size of the COSMO-RS molecular cavities is significantly larger than the conventional van der Waals volumes calculated from Bondi's radii. The presence of free volume will force these molecular cavities to shrink down to approximately the van der Waals volumes. By shrinking the molecular volume, the molecular surface area is also shrinking and, as a consequence, the surface charge density, σ, is increasing inversely proportionately. However, as seen in Equation 31.2, what counts in the intermolecular interactions is the actual charges that interact with each other, that is, the product of σ with the corresponding surface area. Thus, the required quantity, $\sigma \times \alpha$, for the interaction energy $\varepsilon_{pair}(\sigma, \sigma')$, is an invariant in going from the rigid COSMO-RS structure to a fluid structure with rigid molecular volumes and free volume in between.

The above transition from the rigid to the fluid structure poses an algebraic problem: Could Equation 31.2 be applicable to the gas state? In the ideal-gas state all intermolecular interactions should be null. The misfit interaction energy, however, of a molecular segment of σ charge density with the surrounding empty sites of zero charge density is $(\alpha'\alpha_{eff}/2)(\sigma+0)^2$, which is not equal to zero. Thus, the first thing to be done is to replace the (always positive) misfit interaction by an interaction compatible with the gas state. As discussed in Ref. [17], this can be done by adopting the classical expression for electrostatic interactions, or

$$\varepsilon_{pair}(\sigma_i,\sigma_j)=\alpha'(\alpha_i\sigma_i)(\alpha_j\sigma_j)+c_{hb}\alpha_{acc}\alpha_{don}\max[0,\sigma_{acc}-\sigma_{hb}]\min[0,\sigma_{don}+\sigma_{hb}] \quad (31.20)$$

Equation 31.20 preserves the above-mentioned invariant (interacting surface charges). As discussed in Ref. [17], Equation 31.20 must be augmented by a "dispersion" term for the weak London-type dispersive interactions. There are numerous more or less empirical options for the dispersion term in the open literature ranging from simple ones of the general van der Waals form $-\tilde{\rho}\, D_{ij}$, up to involved equations such as the Alder-type polynomials [19] discussed in Chapter 29. $\tilde{\rho}$ is the usual (GLF) reduced density of the system and in Ref. [17] D_{ij} was considered proportional to the square root of the product of the interacting molecular surfaces. The discussion on dispersive interactions is not repeated here. Suffice it to say that, with a dispersion term added, the first term on the right-hand side of Equation 31.20 may be considered to represent all polar interactions, except for the strong specific ones accounted for by the last term of the equation. At this point it is possible to bring COSMO-RS into the GLF framework.

In the GLF framework, the free volume is discretized into voids, empty sites, or inert molecular entities of zero surface charge density. It is considered that all molecular surfaces are in contact with either other molecules or with some voids. Any contact with a void, or between voids, is characterized by zero interaction energy. The relevant GLF nomenclature of Chapter 30 is applicable here. Thus, if the system consists of N r-mers and N_0 empty sites, the total number of sites, N_r, is equal to $(rN+N_0)$. If v* is the volume of each site, the total volume V of the system is equal to $N_r v^*$ and the free volume is $V_0=N_0 v^*$. The overall van der Waals volume, squeezed COSMO volume, or hard-core volume V^* of the system is equal to rNv^*. With these values, the reduced volume and reduced density are given by:

$$\tilde{v} = \frac{V}{V^*} = \frac{V}{V-V_0} = \frac{1}{\tilde{\rho}} \tag{31.21}$$

The essential point here is that the free volume has been introduced in the system as an additional mono-segmental component indicated by subscript 0.

The COSMO-RS-EOS model for a multicomponent multigroup system may easily be obtained even accounting for the nonrandom distribution of the free volume throughout the system. In this case, one more segment type (subscript 0) should be added in the system of the recursive Equations 31.10–31.12. Thus, the corresponding additional equations for the voids in the system would read:

$$\sum_{n=0}^{t} \Theta_n \Gamma_{0n} = \sum_{n=0}^{t} \Theta_n \Gamma_0 \Gamma_n = \Gamma_0 \sum_{n=0}^{t} \Theta_n \Gamma_n = 1 \tag{31.22}$$

or

$$\ln \Gamma_0 = -\ln \sum_{n=0}^{t} \Theta_n \Gamma_n \tag{31.23}$$

All other Equations 31.10–31.12 will keep their form, but the summation will run from 0 instead of 1. The rest of the formalism comes straightforwardly via the GLF approach.

In this case, the EOS is already available since 1982 and it reads [17,18,20,21]:

$$\frac{PV}{NRT} = -\ln(1-\tilde{\rho}) + \frac{z}{2}\ln\left(1-\tilde{\rho}+\frac{q}{r}\tilde{\rho}\right) - z\ln\Gamma_0 \tag{31.24}$$

COSMO-RS already uses r for the number of segments and q/r as a surface-to-volume ratio for the molecular shape. The same way is used here for obtaining the r and q of the mixture from the corresponding r_i and q_i of pure components i. and the volume fractions φ_i. The coordination number, z, is a remnant from the quasi-lattice picture that was used initially for deriving the working equations but it may be just replaced by the number of external contacts per void entity, or by the universal surface area ratio α_0/α_{eff}. The overall surface area fraction, θ, of molecular segments in the real system, containing N_0 empty sites, is given by:

$$\theta = \frac{qN}{qN+N_0} = \frac{\frac{q}{r}\tilde{\rho}}{\frac{q}{r}\tilde{\rho}+1-\tilde{\rho}} \tag{31.25}$$

COSMO-RS typically uses the Guggenheim-Staverman combinatorial term [1–6]. In this case, Equation 31.24 becomes [17]:

$$\frac{PV}{NRT} = -\ln(1-\tilde{\rho}) + \tilde{\rho}\frac{l}{r} + \frac{z}{2}\ln\left(1-\tilde{\rho}+\frac{q}{r}\tilde{\rho}\right) - z\ln\Gamma_0 \tag{31.26}$$

where $l=(z/2)\times(r-q)-(r-1)$ is a measure of the departure from linearity of the molecule. For linear molecules, $l=0$ and Equation 31.24 is recovered.

The equation for the chemical potential has the form [17]:

$$\frac{\mu_i}{RT} = \ln\frac{\varphi_i}{\omega_i r_i} + \ln\tilde{\rho} - r_i\ln(1-\tilde{\rho}) + \frac{z}{2}[r_i - q_i]\ln\left[1-\tilde{\rho}+\frac{q}{r}\tilde{\rho}\right] - zr_i\ln\Gamma_0 + zq_i\left[\sum_m \Theta_{mi}\ln\Gamma_m\right] \tag{31.27}$$

ω_i is a characteristic quantity for each fluid that takes into account the flexibility, conformational changes, and the symmetry of the molecule. The corresponding equation for pure component is:

$$\frac{\mu_i^0}{RT} = \ln\tilde{\rho}_i - r_i\ln(1-\tilde{\rho}_i) + \frac{z}{2}[r_i - q_i]\ln\left[1-\tilde{\rho}_i+\frac{q_i}{r_i}\tilde{\rho}_i\right] - zr_i\ln\Gamma_0 + zq_i\left[\sum_m \Theta_m^i\ln\Gamma_m\right] \tag{31.28}$$

Vapor-liquid and liquid-liquid equilibria are described by the equality of chemical potentials, Equation 31.27, in all phases at equilibrium. Activity coefficients are obtained by subtraction of Equation 31.28 from Equation 31.27.

From Equations 31.27 and 31.28, the equation for the solvation-free energy is obtained [17]:

$$\frac{\Delta G_i^{*S}}{RT} = \frac{\mu_i - \mu_i^{IG}}{RT} + \ln Z = \ln\frac{\omega_i^{IG}}{\omega_i} - r_i\ln(1-\tilde{\rho}) + \frac{z}{2}[r_i - q_i]\ln\left[1-\tilde{\rho}+\frac{q}{r}\tilde{\rho}\right] - zr_i\ln\Gamma_0 + zq_i\sum_{m=1}^{k}\Theta_m^i\ln\Gamma_m \tag{31.29}$$

Z being the compressibility factor. In the limit of infinite dilution of solute 1 in solvent 2, Equation 31.29 becomes:

$$\frac{\Delta G_{1/2}^{*S}}{RT} = \ln\frac{\omega_1^{IG}}{\omega_1} - r_1\ln(1-\tilde{\rho}_2) + \frac{z}{2}[r_1 - q_1]\ln\left[1-\tilde{\rho}_2+\frac{q_2}{r_2}\tilde{\rho}_2\right] - zr_1\ln\Gamma_0^{(2)} + zq_1\sum_{m=1}^{k}\Theta_m^1\ln\Gamma_m \tag{31.30}$$

The first three terms in Equation 31.30 constitute the *cavitation/restructuring* term:

$$\frac{\Delta G_{1/2,cav}^{*S}}{RT} = \ln\frac{\omega_1^{IG}}{\omega_1} - r_1\ln(1-\tilde{\rho}_2) + \frac{z}{2}[r_1 - q_1]\ln\left[1-\tilde{\rho}_2+\frac{q_2}{r_2}\tilde{\rho}_2\right] \tag{31.31}$$

This term corresponds to Abraham's L descriptor [17] when the solvent, component 2, is n-hexadecane.

The first term in Equation 31.31 is the conformational/flexibility contribution to the cavitation term. This term contains information on the change in conformer population of solute in going from the ideal-gas state to the solvent. In addition, it contains information on the effect of standard segment volume that was used to define r_1 and r_2. Since the admission of free volume requires small standard segment volumes, this may have an effect on the combinatorial contribution, especially for highly asymmetrical solute/solvent pairs. The correction to this would be proportional to the ratio of athermal combinatorial factors in the partition function, $g_{comb}(R, Q)/g_{comb}(r, q)$, where the pair R, Q signifies COSMO-RS values while the pair r, q signifies COSMO-RS-EOS values.

Heats of solvation are obtained from the classical defining equation:

$$\Delta H_{1/2}^{*S} = -T^2 \frac{\partial\left(\Delta G_{1/2}^{*S}/T\right)}{\partial T} \tag{31.32}$$

and the volume changes on solvation from the derivative of solvation-free energy with respect to the external pressure, P.

Thus, all needed working equations are available for the application of COSMO-RS-EOS to real systems. If carefully parameterized, it may become a valuable predictive tool in molecular thermodynamics. Equation 31.20 is the key equation for the determination of the binary GLF parameter $-\varepsilon_{ij} = \varepsilon_{pair}(\sigma_i, \sigma_j)$. The hard-core molar volume, V^*, varies linearly with the COSMO volume [22] while with the widely used GLF practice for the corresponding site or segment volume, $v^* = 9.75$ cm^3/mol the required segment number is obtained as $r_i = V_i^*/9.75$. All the above equations may be re-expressed with the corresponding COSMO-RS nomenclature (replace Θ_{mi} by $p_i(\sigma_m)$, etc.). The guidelines for this exchange of nomenclature are also available in the literature [23].

The above extensions bring into COSMO-RS a more realistic picture of the liquid state and turn it into an interesting EOS model with a-priori predictive capacity. They even restore the initial name and RS would again mean <u>R</u>eal <u>S</u>olvents! There is great potential in COSMO-RS and, hopefully, the above EOS dimension will stir the deserved interest in the open literature for more open developments of this remarkable model.

REFERENCES

1. Klamt, A. 1995. Conductor-like screening model for real solvents: A new approach to the quantitative calculation of solvation phenomena. *J. Phys. Chem.* 99(7): 2224–2235.
2. Klamt, A., Jonas, V., Burger, T., and Lohrenz, J. 1998. Refinement and parameterization of COSMO-RS. *J. Phys. Chem.* 102(26): 5074–5085.
3. Klamt, A., and Eckert, F. 2000. COSMO-RS: A novel and efficient algorithm for a-priori prediction of thermophysical data of liquids. *Fluid Phase Equilib.* 172(1): 43–72.
4. Klamt, A. 2005. *COSMO-RS from Quantum Chemistry to Fluid Phase Thermodynamics and Drug Design*. Amsterdam: Elsevier.
5. Tomassi, J., and Persico, M. 1994. Molecular interactions in solution: An overview of methods based on continuous distributions of the solvent. *Chem. Rev.* 94(7): 2027–2094.
6. Bottcher, C. J. F., and Bordewijk, P. 1978. *Theory of Electric Polarization*. Amsterdam: Elsevier.
7. Klamt, A., and Schüürmann, G. 1993. COSMO: A new approach to dielectric screening in solvents with explicit expressions for the screening energy and its gradient. *J. Chem. Soc., Perkin Trans.* 2: 799–805.
8. Lin, S. T., and Sandler, S. I. 2002. A priori phase equilibrium prediction from a segment contribution solvation model. *Ind. Eng. Chem. Res.* 41: 899–913.
9. Grensemann, H., and Gmehling, J. 2005. Performance of a conductor-like screening model for real solvents model in comparison to classical group contribution methods. *Ind. Eng. Chem. Res.* 44: 1610–1624.
10. Pye, C. C., Ziegler, T., van Lenthe, E., and Louwen, J. N. 2009. An implementation of the conductor-like screening model of solvation within the Amsterdam density functional package, Part II. COSMO for real solvents. *Can. J. Chem.* 87: 790–797.

11. Klamt, A., Eckert, F., and Arlt, W. 2010. COSMO-RS: An alternative to simulation for calculating thermodynamic properties of liquid mixtures. *Annu. Rev. Chem. Biomol. Eng.* 1: 101–122.
12. Ben-Naim, A. 1978. *Solvation Thermodynamics*. New York: Plenum Press.
13. Panayiotou, C., and Vera, J. H. 1980. The quasi-chemical approach for non-randomness in liquid mixtures. Expressions for local surfaces and local compositions with an application to polymer solutions. *Fluid Phase Equilib.* 5: 55–80.
14. Klamt, A., Krooshof, G. J. P., and Taylor, R. 2002. COSMOSPACE: Alternative to conventional activity-coefficient models. *AIChE J.* 48(10): 2332–2349.
15. Gerlach, T., Müller, S., and Smirnova, I. 2017. Development of a COSMO-RS based model for the calculation of phase equilibria in electrolyte systems. *AIChE J.* 64: 272–285.
16. Hsieh, C.-M., Sandler, S. I., and Lin, S.-T. 2010. Improvements of COSMO-SAC for vapor-liquid and liquid-liquid equilibrium predictions. *Fluid Phase Equilib.* 297: 90–97.
17. Acree, W. E., Panayiotou, C., and Zuburtikudis, I. 2023. COSMO-RS and LSER models of solution thermodynamics. Towards a COSMO-LSER equation-of-state model of fluids. *J. Mol. Liq.* 390 A: 122992.
18. Panayiotou, C. P., and Vera, J. H. 1982. Statistical thermodynamics of r-mer fluids and their mixtures. *Polymer J.* 14: 681–694.
19. Alder, B. J., Young, D. A., and Mark, M. A. 1972. Studies in molecular dynamics. X. Corrections to the augmented van der Waals theory for square-well fluid. *J. Chem. Phys.* 56: 3013–3029.
20. High, M. S., and Danner, R. P. 1989. A group contribution equation of state for polymer solutions. *Fluid Phase Equilib.* 53: 323–330.
21. High, M. S., and Danner, R. P. 1990, Application of the group contribution lattice-fluid EOS to polymer solutions. *AIChE J.* 36: 1625–1632.
22. Panayiotou, C., Mastrogeorgopoulos, S., Ataman, M., Hadadi, N., and Hatzimanikatis, V. 2016. Molecular thermodynamics of metabolism: Hydration quantities and the equation-of-state approach. *Phys. Chem. Chem. Phys.* 18: 3257–32592.
23. Panayiotou, C., Tsivintzelis, I., Aslanidou, D., and Hatzimanikatis, V. 2015. Solvation quantities from a COSMO-RS equation of state. *J. Chem. Thermodyn.* 90: 294–309.

32 Thermodynamics of Hydrogen Bonding

ABSTRACT

This chapter summarizes, first, the essentials of the chemical theory of hydrogen bonding. Subsequently, three representative hydrogen-bonding approaches of current use in the literature are examined, namely, the Statistical Associating Fluid Theory, the COnductor like Screening MOdel for Real Solvents (COSMO-RS), and the Lattice Fluid with Hydrogen-Bonding approaches. The variant Nonrandomness and Hydrogen Bonding is also presented, and representative applications are reported.

INTRODUCTION

As discussed in Chapter 17, Berg's classification of liquid mixtures recognized the importance of the behavior of liquid mixtures of hydrogen bonding, and the presence of electron donor and electron acceptor compounds. Water and other chemical substances such as alcohols, amines, carboxylic acids, and glycols possess strong hydrogen bonding. Thus, one of the central features of modern molecular thermodynamic models is their capacity to handle efficiently strong specific interactions, such as hydrogen bonding and Lewis's acid–base interactions. Modern equation-of-state models are also able to handle polymer systems forming hydrogen bonds.

The literature on hydrogen-bonding modeling is vast [1–7]. In the present chapter, we first review some basic concepts and summarize the essentials of chemical theories of hydrogen bonding [3,5–7]. Subsequently, we discuss three approaches, prevalent in equation-of-state or activity-coefficient models.

BASIC CONCEPTS

Hydrogen bonding is considered to occur under the scheme, A–H – –B, that is, when hydrogen covalently bonded to an electronegative atom, A, interacts intermolecularly with an electronegative atom, B. Atom A can be, O, N, halogen and, also S and C in aromatic, fluoro- and chloro-compounds, such as chloroform or fluoroform. Atom B often is O or N located at an appropriate range of distances and angles with respect to the bond A–H. For this reason, hydrogen bonding (HB) is referred to as *directional* interaction.

As an intermolecular interaction, HB exhibits strengths higher than the dispersion or other weaker polar interactions but lower than covalent bonds. HB strength depends on the specific pair of the interacting molecular moieties (A–H and B) and, quite often, it has values of the order of 30 kJ/mol. This is significantly higher than the ordinary van der Waals interactions with average strengths of around 3–4 kJ/mol. For this reason, HB is often referred to as *strong specific* intermolecular interaction.

Traditionally, HB has been treated as an electrostatic interaction. Thus, the range of HB interaction ($\sim 1/r^2$) is not as short as the ordinary dispersion interactions ($\sim 1/r^6$). This picture has been partly abandoned in recent years, as many more types of weaker interactions are being considered to be HB interactions [4].

HB may be *intermolecular* (between different molecules) or *intramolecular* (within a single molecule). Alkoxyalkanols and glycols are typical examples of molecules exhibiting intramolecular HB interactions, as shown in the scheme below. On the left-hand side, three alkoxy ethanol molecules are connected.

DOI: 10.1201/9781003431985-36

with two *intermolecular* hydrogen bonds. On the right-hand side, the alkoxy ethanol forms a five-member ring with an *intramolecular* hydrogen bond. Typically, intramolecular bonds are weaker than the intermolecular bonds but in cases of significant dilution, they may be predominant, simply because their intermolecular contacts are not statistically favored.

Quite often hydrogen bonds form cooperatively stabilized ring structures. The phenomenon is called *HB cooperativity* and is particularly important in biology. A simple example of HB cooperativity is carboxylic acid dimerization. Once the first intermolecular hydrogen bond between carboxyls is formed, the second one is formed more easily, not only because the carboxyl stereochemistry facilitates it, but also because the entropy reduction in the second hydrogen bond is lower than in the first. Water structure consists of three-dimensional cooperative HB substructures. Polymer chains, such as polyamides (nylons) are organized in their crystalline state via an extended HB (N–H – – O=C) cooperativity.

It should be kept in mind that HB is a strong interaction between two sites of *complementary* character, one positively charged or *proton donor* or (Lewis) *acidic* and another negatively charged, *proton-acceptor* or (Lewis) *basic*. It is this complementary character that differentiates HB from all other interactions of nonionic substances when issues of solubility or miscibility come into play. The famous dictum "*similia similibus solvuntur*" applies to the weaker interactions but *not* to HB.

Hydrogen bonds are not permanent. They are created and disrupted continuously. However, the probability of finding a site B in the immediate neighborhood of site A–H is significantly higher than what would correspond to the overall concentration. In the lower pure alkanols, as an example, the probability of engagement of their hydroxyls in HB or *the degree of hydrogen bonding* of OH groups, is well above 90%, reflecting a significant *ordering* in the system.

With respect to *ordering in solutions*, we stress the point that this molecular distribution is one of the most challenging problems in molecular thermodynamics. The ordering or the nonrandomness in the distribution of molecular species in a mixture can be treated by two alternative approaches of statistical thermodynamics: the *radial distribution function* (RDF) approach (cf. Chapter 29) and the statistical or *combinatorial* approach (cf. Chapter 30). In later years, there have been remarkable advances in both directions, and thus, both are now particularly attractive.

CHEMICAL OR ASSOCIATION THEORIES OF HYDROGEN BONDING

The oldest known thermodynamic approach to HB is that of Dolezalek [8], which postulates the existence of distinct molecular species or *associates* in solution. The existence of these associates, although a matter of controversy in the literature, is the central assumption in the so-called *chemical* or *association* theories of HB [1–3,5–17]. These associates are oligomers with a repeating unit of the associated molecule.

As an example, in the case of 1-alkanols, R–OH, the associates are of the form shown on the upper part of the scheme below.

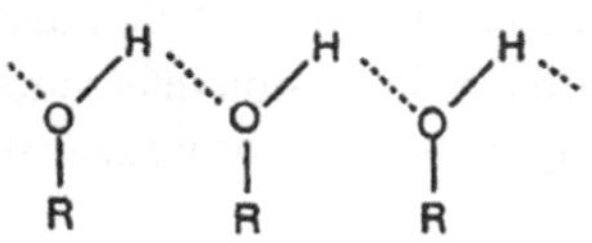

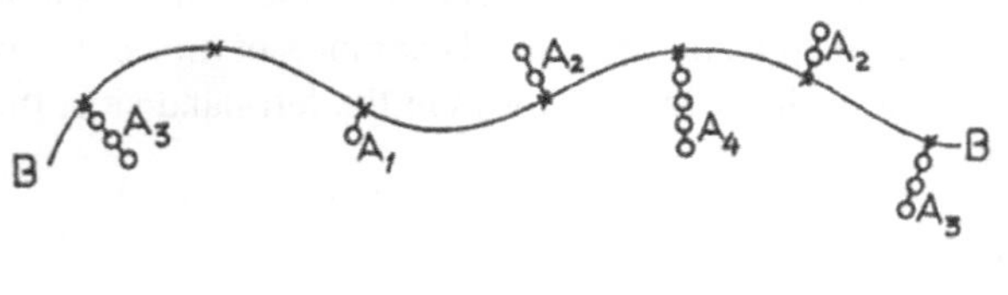

Thus, in pure alkanol or in a mixture with an inert nonassociated solvent, the chemical theory postulates the existence of a "polydisperse polymer" solution, to which it even attributes an average degree of "polymerization" or association. Analogous associates are postulated in the case of mixtures of self-associated and cross-associated compounds. Besides linear chain-like oligomers, branched structures may be also postulated. On the lower part of the above scheme is shown a branched structure consisting of a backbone polymeric chain with proton-acceptor or basic sites, such as poly(ethylene oxide), and of oligomeric branches of a self-associating solvent, such as ethanol [13]. Heidemann and Prausnitz [14] seem to have been the first to implement the chemical theory in a cubic equation of state and their idea of combining an empirical or theoretical equation of state with association keeps inspiring researchers till today.

To understand the thermodynamic arguments of the chemical or association approach to HB, let us consider a self-associated compound, A, in its pure state or in a mixture with an inert compound. In the system, we have all kinds of consecutive quasi-reactions: $A_1+A_1=A_2$, …, $A_{i-1}+A_1=A_i$, …, $A_m+A_n=A_{m+n}$. For each of these quasi-reactions, we may write corresponding equations with chemical potentials, such as, $\mu_{A_{i-1}}+\mu_{A_1}=\mu_{A_i}$, or attribute an equilibrium constant, K_{i-1}, in each formation quasi-reaction $A_{i-1}+A_1=A_i$. For simplicity, let us consider all K_is equal, that is, $K_1=K_2=\ldots=K_i=K$. As with ordinary reactions, we may attribute to K a free-energy change, ΔG^H, upon formation of each hydrogen bond, or

$$-RT\ln K=\Delta G^H=\Delta H^H-T\Delta S^H=\Delta U^H+P\Delta V^H-T\Delta S^H \tag{32.1}$$

where ΔH^H, ΔS^H, ΔU^H, and ΔV^H are the corresponding enthalpy, entropy, energy, and volume changes upon hydrogen-bond formation.

The population of associates in this system is obtained from simple material balance equations. If N_A is the number of molecules of compound A, which are distributed in the various associates, the following material balance holds:

$$\sum_i iN_{A_i}=N_A \tag{32.2}$$

For the total number, N_T, of the true molecular species in the system, we have:

$$\sum_i N_{A_i}=N_T \tag{32.3}$$

Their ratio gives the average degree of association, or

$$\bar{i}=\frac{\sum_i iN_{A_i}}{\sum_i N_{A_i}}=\sum_i iX_{A_i}=\frac{N_A}{N_T} \tag{32.4}$$

where X_{A_i} is the apparent mole fraction of the associate A_i.

If r_A is the number of segments per molecule A, we may define the segment fraction of associate A_i in the polydisperse mixture, as it is usual in polymer contexts (cf. Chapters 20 and 29):

$$\phi_{A_i}=\frac{ir_AN_{A_i}}{\sum_j jr_AN_{A_j}}=\frac{iN_{A_i}}{N_A} \tag{32.5}$$

As presented in Chapters 19 and 20, an elegant thermodynamic treatment of this system was done by Flory [18] who considered it as an athermal polydisperse polymer mixture arranged on a quasi-lattice. Flory defined the equilibrium constant for each polymerization step as:

$$K^F = \frac{\phi_{A_{i+j}}}{\phi_{A_i}\phi_{A_j}}\frac{ij}{i+j} = \frac{N_{A_{i+j}}}{N_{A_i}N_{A_j}}N_A \tag{32.6}$$

In a real system, however, with free volume, K^F is not a constant. If as discussed in Chapters 20 and 29 the Lattice Fluid (LF) model is adopted to account for free volume, the true equilibrium constant, K, takes the form [12]:

$$K = \frac{r_A}{\tilde{\rho}_A}\frac{\phi_{A_{i+j}}}{\phi_{A_i}\phi_{A_j}}\frac{ij}{i+j} = \frac{r_A}{\tilde{\rho}}K^F \tag{32.7}$$

By applying Equation 32.6 consecutively for the series of associate numbers, N_{A_i}, upon summation ($K^F\phi_{A_1} < 1$), the following important equation is obtained [12]:

$$\sum_i iN_{A_i} = \frac{N_{A_1}}{\left(1-K^F\phi_{A_1}\right)^2} = N_A \quad \text{or} \quad \frac{\phi_{A_1}}{\left(1-K^F\phi_{A_1}\right)^2} = 1 \tag{32.8}$$

From this equation, the segment fraction of the free monomer is given by:

$$\phi_{A_1} = \frac{1+2K^F - \sqrt{1+4K^F}}{2\left(K^F\right)^2} \tag{32.9}$$

From classical thermodynamic considerations [19], the chemical potential, μ_A, of compound A is equal to the chemical potential of its free monomer, A_1. Following the procedure entirely analogous to the one used in Chapter 20 or 29, an expression for the chemical potential in terms of ϕ_{A_1} and K^F for the associated compound, A, in a mixture is obtained. Thus, in a mixture of the self-associated compound, A, with an inert compound, B, the chemical potential of A is given by [12]:

$$\frac{\mu_A}{RT} = \frac{\mu_{A_1}}{RT} = \ln\phi_{A_1} + \phi_A K^F\phi_{A_1} + \left(1-\frac{r_A}{r_B}\right)\phi_B + r_A\tilde{\rho}X_{AB}\theta_B^2 + r_A(\tilde{v}-1)\ln(1-\tilde{\rho}) + \\ + \ln\frac{\tilde{\rho}}{\omega_{A_1}} - \frac{r_A\tilde{\rho}}{\tilde{T}_A} + r_A\frac{\tilde{P}\tilde{v}}{\tilde{T}}D_{AB} \tag{32.10}$$

Equation 32.10 is identical in form to Equation 20.14 except for the first two terms with:

$$\phi_{A_1} = \frac{1+2\phi_A K^F - \sqrt{1+4\phi_A K^F}}{2\phi_A\left(K^F\right)^2} \tag{32.11}$$

The Flory constant K^F is again given by Equation 32.6 except for N_A which is replaced by rN/r_A. The chemical potential for component B is given by:

$$\frac{\mu_B}{RT} = \phi_A\frac{r_B}{r_A}K^F\phi_{A_1} + \ln\phi_B + \left(1-\frac{r_B}{r_A}\right)\phi_A + r_B\tilde{\rho}X_{BA}\theta_A^2 + r_B(\tilde{v}-1)\ln(1-\tilde{\rho}) + \\ + \ln\frac{\tilde{\rho}}{\omega_B} - \frac{r_B\tilde{\rho}}{\tilde{T}_B} + r_B\frac{\tilde{P}\tilde{v}}{\tilde{T}}D_{BA} \tag{32.12}$$

The equation of state is identical in form to Equation 29.37 except for the term $1/r$, which is replaced by the term:

$$\frac{1}{\bar{r}} = \frac{1}{r} - \frac{\phi_A}{r_A}K^F\phi_{A_1} \tag{32.13}$$

Phase-equilibrium calculations can be done by using Equations 32.10 and 32.12. Volumes of mixing are calculated with Equation 32.14:

$$\frac{V^E}{N} = r\left(\tilde{v}v^* - \phi_A \tilde{v}_A v_A^* - \phi_B \tilde{v}_B v_B^*\right) + x_A \Delta V^H \left[K^F \phi_{A_1} - \left(K^F \phi_{A_1}\right)_{pure\,A}\right] \quad (32.14)$$

and the heats of mixing with Equation 32.15:

$$\frac{H^E}{N} = -r\left(\tilde{\rho}\varepsilon^* - \phi_A \tilde{\rho}_A \varepsilon_A^* - \phi_B \tilde{\rho}_B \varepsilon_B^*\right) + x_A \Delta U^H \left[K^F \phi_{A_1} - \left(K^F \phi_{A_1}\right)_{pure\,A}\right] + P\frac{V^E}{N} \quad (32.15)$$

This model, referred to as the Lattice Fluid with Association (LFAS) model, has been extended to mixtures of a self-associated compound A with a polymer B having proton-acceptor sites only [13]. LFAS was tested against a variety of experimental data and properties [12,13,21–23]. It is worth pointing out that all key Equations 32.10, 32.12, 32.14, and 32.15, used to calculate measurable quantities, may be separated into a "physical" and a "chemical" contribution. The first is described by the plain LF equations of Chapters 20 and 29 and the second by the HB terms. This splitting of contributions to basic thermodynamic quantities is common to most Association Models (AM).

The above rationale of the chemical approach to HB and linear consecutive association, expressed with Equations 32.1–32.9, is common to numerous equation-of-state or activity-coefficient models [1–3,5–20]. A typical example is the combination of the perturbed anisotropic-chain theory (PACT) equation of state with association, resulting in APACT (associated PACT) model [16], discussed in Chapter 29. In APACT the compressibility factor Z is given by the sum of a "chemical" and a "physical" contribution split further into three subcontributions as follows:

$$Z = \frac{PV}{N_A RT} = \frac{N_T}{N_A} + Z^{phys} = \frac{N_T}{N_A} + Z^{rep} + Z^{is} + Z^{anis} \quad (32.16)$$

Superscripts *rep*, *is*, and *anis* refer, respectively, to repulsive, isotropic, and anisotropic contributions to the compressibility factor. The chemical contribution N_T/N_A is calculated as above, cf. Equation 32.4.

Economou and Donohue [3,17] have reviewed Association Models and approaches to HB and concluded that the key equations of the type of Equation 32.9, or Equation 32.11, are very similar in the various approaches and the corresponding models are nearly equivalent in common areas of application. Their conclusions hold till today. The crucial differentiating point is their areas of application.

All models of the chemical approach apply to systems where HB clusters or associates may be represented as linear chain-like structures or tree-like branched structures with linear branches. Intramolecular association, ring structures, cooperative closed structures, or three-dimensional interconnected networks of hydrogen bonds are rather outside the realm of the chemical approach. Even its applicability to water and aqueous systems is questioned since water association does not lead to linear associates only. Levine and Perram [24] and Huyskens et al. [25], among others, had already indicated the inadequacy of the chemical theory models for the treatment of systems forming three-dimensional HB networks. As more adequate, they proposed either a *combinatorial* approach to account efficiently for the ordering due to HB or opt for an *off-lattice* picture of associated liquids. It should be noticed that a manageable treatment of polymers with the RDF approach was rather inconceivable at their time. We now proceed to discuss the newer approaches to HB. These could be divided into two main classes, those that model molecular distribution via RDF and those that use refined combinatorial approaches to model molecular or segmental distribution.

THE RDF APPROACH TO HYDROGEN BONDING: WERTHEIM'S PERTURBATION THEORY AND SAFT

SAFT stands for Statistical Associating Fluid Theory. The essentials of SAFT are presented in Chapter 33. Here we present the SAFT approach to HB, which is based on Wertheim's first-order perturbation approximation for association in fluids with highly directional attractive forces [31–36]. In SAFT, each molecule consists of m spherical segments bound or associated tangentially and forming a chain. HB is a distinct association and occurs at specific sites in some of the chain segments. Typically, a Lennard-Jones or a Mie potential is used for the interaction of spherical segments. A square-well potential is used for the HB interaction [37–40]. The Helmholtz free energy in SAFT is split into additive terms for the contributions of the various types of repulsive and attractive intermolecular interactions, one of which is the "association" term [37–39]. This association term, A^{assoc}, captures the contribution of HB between molecules of a fluid and is given by:

$$\frac{A^{assoc}}{RT} = \sum_{A=1}^{M}\left[\ln X^A - \frac{X^A}{2}\right] + \frac{M}{2} \tag{32.17}$$

where M is the number of association sites per molecule and X^A is the fraction of free (nonassociated) molecules not bonded at site A. The fraction X^A is given by the recursive equation:

$$X^A = \frac{1}{\left(1 + \sum_{B=1}^{M} \rho X^B \Delta^{AB}\right)} \tag{32.18}$$

where ρ is the density of the system. In the case of mixtures, these equations take the following forms:

$$\frac{A^{assoc}}{RT} = \sum_i x_i \left\{\sum_{A_i}\left[\ln X^{A_i} - \frac{X^{A_i}}{2}\right] + \frac{M_i}{2}\right\} \tag{32.19}$$

and

$$X^{A_i} = \frac{1}{\left(1 + \sum_j \sum_{B_j}^{M} \rho_j X^{B_j} \Delta^{A_i B_j}\right)} \tag{32.20}$$

where M_i is the number of association sites on molecule i and X^{A_i} is the fraction of molecules i not bonded at site A.

For the "association strength," $\Delta^{A_iB_j}$, a square-well potential is adopted with well depth being the HB interaction energy, $\varepsilon^{A_iB_j}$ and well width, $\kappa^{A_iB_j}$. The expression for $\Delta^{A_iB_j}$of the bond $A_i - B_j$ between a site A of molecule i and a site B of a different molecule j, is then obtained from the first-order approximation of Wertheim's thermodynamic perturbation theory (TPT1) [31–36], given by:

$$\Delta^{A_iB_j} = d_{ij}^3 g_{ij}\left(d_{ij}\right)^{seg}\left[\exp\left(\varepsilon^{A_iB_j}/kT\right) - 1\right]\kappa^{A_iB_j} \tag{32.21}$$

where $d_{ij} = (d_{ii} + d_{jj})/2$ is the temperature-dependent average segment diameter of interacting segments i and j, and $g_{ij}\left(d_{ij}\right)^{seg}$ is the RDF of hard-sphere segments. In the various versions of SAFT, different expressions are used for RDF, for the temperature dependence of the diameter d, or for the volume v^0 (= $\pi d^3/6$) of the spherical segment. A simple form for the close-packed *hard-core* volume, v^0, is the following [39,40]:

$$v^0 = v^{00}\left[1 - 0.12\exp\left(-3u^0/kT\right)\right]^3 \text{ (Huang and Radosz)} \tag{32.22}$$

The segment volume v^{00} (= $\pi\sigma^3/6$) of the fluid is independent of temperature as are the size parameter, σ, and the energy parameter, u^0, of the segmental interaction potential.

Typically, the segmental RDF is approximated by an expression for mixtures of hard spheres derived from the Carnahan–Starling equation of state [41]:

$$g_{ij}\left(d_{ij}\right)^{seg} \approx \frac{1}{1-\zeta_3} + \frac{d_i d_j}{d_i + d_j}\frac{3\zeta_2}{\left(1-\zeta_3\right)^3} + \left(\frac{d_i d_j}{d_i + d_j}\right)^2 \frac{2\zeta_2^2}{\left(1-\zeta_3\right)^3} \tag{32.23}$$

where

$$\zeta_k = \frac{\pi}{6} N_{Av}\rho \sum_i x_i m_i d_{ii}^k \tag{32.24}$$

Quite often, however, a simpler expression is used [39]. This expression, which is also used in the variants PC-SAFT [42] and CPA (Cubic plus Association) [43], has the form:

$$g^{hs} = \frac{1-0.5\eta}{(1-\eta)^3} \tag{32.25}$$

where the reduced density, η, is given by:

$$\eta = 0.74048\,\rho m v^0 \tag{32.26}$$

With Equation 32.25 for RDF, the association strength may be expressed by:

$$\Delta^{AB} = \sqrt{2}v^{00}\frac{1-0.5\eta}{(1-\eta)^3}\left[\exp\left(\varepsilon^{AB}/kT\right) - 1\right]\kappa^{AB} \text{ (Huang and Radosz)} \tag{32.27}$$

or by a similar equation, with the v^{00} replaced by v^0 (Chapman et al. [37,38]).

Thus, m, v^{00}, and u^0/k are the three adjustable SAFT parameters for each non-hydrogen-bonded fluid. In hydrogen-bonded pure fluids, each type of A–B HB interaction requires the knowledge of the association energy, ε^{AB}, and the association volume, κ^{AB}, both of which are temperature independent and fitted to experimental vapor pressure and density data or, less frequently, to spectroscopic data.

In the general case, Equation 32.18 or 32.20 must be solved numerically. In simple cases of hydrogen bonds of the same Δ's, these recursive equations lead to analytic explicit equations [39].

As an example, in the association scheme called 3B, the molecule has three association sites, A, B, and C, but nonzero HB occurs only between sites A and C and between sites B and C ($\Delta^{AC} = \Delta^{BC} \neq 0$), the relations between nonbonded fractions are: $X^A = X^B$ and $X^C = 2X^A - 1$, and the fraction X^A is given by:

$$X^A = \frac{-1 + \rho\Delta^{AC} + \sqrt{\left(1+\rho\Delta^{AC}\right)^2 + 4\rho\Delta^{AC}}}{4\rho\Delta^{AC}} \text{ [3B scheme]} \tag{32.28}$$

In the scheme called 4C, the molecules have four association sites, A, B, C, and D, with nonzero HB between A and C, A and D, B and C, and B and D, with all Δ's equal and fractions X equal, the equal fractions are given by:

$$X^A = \frac{-1+\sqrt{1+16\rho\Delta^{AC}}}{8\rho\Delta^{AC}} \quad [\text{4C scheme}] \tag{32.29}$$

SAFT uses both above schemes to describe HB in water. For alkanols, the scheme $3B$ is used and also a simpler two-site scheme $2B$ is used, where

$$X^A = X^B = \frac{-1+\sqrt{1+4\rho\Delta^{AB}}}{2\rho\Delta^{AB}} \quad [\text{2B scheme}] \tag{32.30}$$

In SAFT, molecules, even with one single association site, may interact with HB. The scheme is named "1" and the nonbonded fraction is given by:

$$X^A = \frac{-1+\sqrt{1+4\rho\Delta^{AA}}}{2\rho\Delta^{AA}} \quad [\text{scheme 1}] \tag{32.31}$$

This scheme is often used to describe HB in dimerizable carboxylic acids.

The use of Equation 32.28 for water and the use of Equation 32.31 for carboxylic acids are examples indicating that SAFT conceives HB interactions in its own way and, thus, HB SAFT parameters should be used with care. As an example, SAFT HB interaction energies, ε/k, for water, reported in the literature [2], vary from 825.9 K up to 2,286 K. Even when using the same scheme, 4C, the reported values range from 1,195 K up to 1,804 K. Similarly, ε/k values for methanol, from SAFT variants, vary from 1,470 K up to 2,964 K [2]. Similar is the variation of the size parameter, κ, which is considered to reflect the entropy change upon hydrogen-bond formation.

The nonbonded fraction, X^A, is a spectroscopically measurable quantity and is used for comparison with other association theories. An example of such a comparison is shown in Figure 32.1 [44]. Beyond monomer fractions, the SAFT HB contributions are not easily separable from the rest of contributions to the property studied and comparative studies are not easy to make neither they are conclusive. Nevertheless, the prevailing opinion is that the HB approaches of chemical theory [3], SAFT, and Lattice Fluid with Hydrogen Bonding (LFHB) [44] are, essentially, equivalent in the common areas of applications [17]. The LFHB and Nonrandom Hydrogen Bonding (NRHB) are

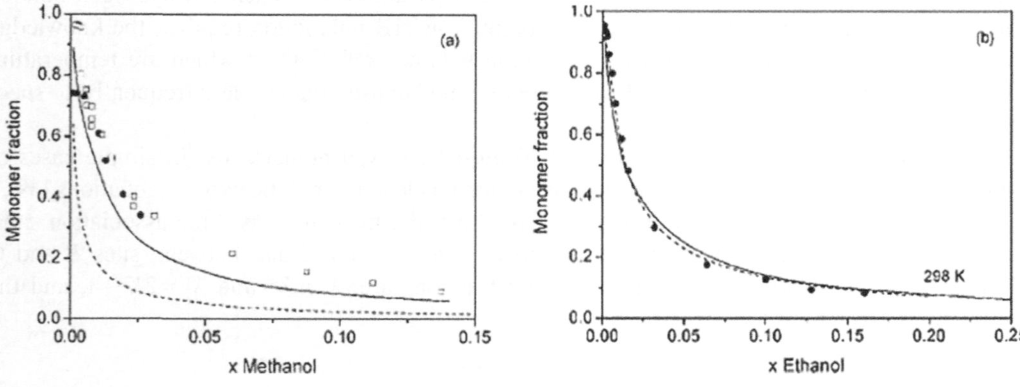

FIGURE 32.1 Fraction of non-hydrogen-bonded molecules in n-hexane – 1-alkanol mixtures. Experimental data (symbols) [45,46]. NRHB predictions (solid line), sPC-SAFT predictions (dashed line) in mixtures of n-hexane with (a) methanol and (b) ethanol. Reprinted from Ref. [44, fig. 9] with permission.

examined below. It is worth mentioning here that Gupta and Johnson [47] have shown that the SAFT and the LFHB approaches are essentially equivalent and interchangeable.

The original first-order Wertheim approach to HB cannot handle intramolecular HB and this is inherited in SAFT. Attempts have been made to remove this shortcoming [48]. In a recent effort [49] to account for ring formation from intramolecular and intermolecular HB, the SAFT HB formalism has been restructured based on a second-order Wertheim approximation. The model was tested successfully against computer simulation data. No comparison was made with experimental data of real systems. These SAFT derivations are rather complex, and no attempt is made to reproduce them here. More details are found in the original publications.

Besides intramolecular HB, basic SAFT does not seem to be able to handle hydrogen-bonding cooperativity or handle systems of self-associated polymers, like polyamides (nylons), poly(vinyl alcohol), poly(vinyl phenol), cellulose, starch, etc., in a straightforward manner. Also, it does not seem to be able to handle easily the effect of HB on the glass transition of polymers.

THE QUANTUM CHEMISTRY–BASED COMBINATORIAL APPROACH TO HYDROGEN BONDING

Combinatorial contributions are traditionally used to account for entropy changes in a variety of phenomena. Thus, entropy reduction from molecular ordering due to HB is treated by refined combinatorial approaches. Two elegant refined combinatorial approaches are discussed below. The broader statistical mechanical framework of the partition-function (PF) formulation [50,51] was discussed in Chapter 30.

The quantum chemistry–based approach was discussed in Chapter 31 along with the COnductor like Screening MOdel for Real Solvents (COSMO-RS model [26–30]. In COSMO-RS, the pre-exponential factor of Equation 30.1 is given by the product of the quasi-chemical g_{NR} term, Equation 30.2, and the Guggenheim–Staverman combinatorial term [52], Equation 30.5. As discussed in Chapter 31, COSMO-RS sets a threshold on the absolute molecular surface charge distribution, σ, which should be exceeded for the site to be qualified as an HB one. Depending on the charge sign, positive or negative, the site is qualified as an HB proton donor or acceptor site, respectively. HB occurs when two complementary sites, one HB donor and one HB acceptor, interact.

As discussed in Chapter 31, the HB part of the interaction energy is given by:

$$\varepsilon_{pair}^{HB}(\sigma_i, \sigma_j) = c_{hb} \max[0, \sigma_{acc} - \sigma_{hb}] \min[0, \sigma_{don} + \sigma_{hb}] \tag{32.32}$$

where

$$c_{hb}(T) = c_{hb} \max\left[0, \left(1 - c_{hb}^{T} \frac{T - 298.15K}{T}\right)\right] \tag{32.33}$$

and c_{hb}^{T} is a universal constant, as is also the c_{hb} constant. Some comments regarding Equations 32.32 and 32.33 are necessary.

The pair interaction energy in Equation 32.32 is always negative (attractive) and contributes to the nonrandomness correction term, g_{NR}, of the partition function as a weighing Boltzmann factor, $\exp(-\varepsilon_{pair}^{HB}/kT)$ of the corresponding number of interaction contacts. Thus, the ordering in the system due to HB is accounted for via the quasi-chemical approach and Equation 30.2. It is essential to point out here that it is the introduction of Equation 32.32 which leads to an efficient account of strong specific interactions via the quasi-chemical approach. Equation 32.33 is a correction to the HB interaction energy, when the temperature changes. Alternatively, it may be considered as an entropic correction to the HB strength. In its present form, COSMO-RS does not account for volume changes and, thus, it does not account for the effect of pressure on the HB strength. An extension of COSMO-RS to an equation-of-state model was discussed in Chapter 31.

The key quantities in the quasi-chemical approach are the nonrandomness factors, $\Gamma_{ij} = \Gamma_i \Gamma_j \tau_{ij}$, where $\tau_{ij} = \exp\left(-\varepsilon_{pair,ij}/RT\right)$.

In COSMO-RS, where there are numerous interacting sites, the formalism is of the group-contribution type rather than the molecular type and Γs are obtained numerically from the system of recursive equations (cf. Equation 30.23):

$$\Gamma_m = 1/\sum_n \Theta_n \Gamma_n \tau_{mn} \quad \text{for all } m, n \tag{32.34}$$

where Θ_n is the surface area fraction of site / group n.

Intramolecular HB is accounted for by considering the corresponding molecular structure as an alternative conformer of the compound for which quantum chemical calculations are conducted, starting from the conformer geometry [26,27]. However, it is not clear to what extent this practice accounts for intramolecular HB. Any number of HB types may be handled with COSMO-RS. Self-associated molecules with any number of HB sites may be handled easily. Subtle effects, such as group-proximity effects, may be considered from medium to high level quantum chemical calculations and the *sigma* profiles discussed in Chapter 31. However, stereochemical factors may not be taken properly into account. Quantum chemical calculations with molecules of large numbers of non-hydrogen atoms require excessively long computation times. This poses limitations in use of COSMO-RS for the study of hydrogen-bonded polymers.

VEYTSMAN STATISTICS, LFHB, AND NRHB MODELS

The approach to HB, on which we focus here, also belongs to the broader combinatorial approach to strong specific interactions [24,25,53–55]. Combinatorial models still consider HB as a quasi-reaction with a free energy of bond formation but focus on the number of donor–acceptor contacts and seek combinatorial expressions for the number of ways of forming hydrogen bonds in systems of known populations of proton-donor and proton-acceptor groups. These models imply that the molecules tend to be distributed in the system nonrandomly for more efficient HB interaction. We have recalled above the essentials of the chemical theory to HB [12,13,21–23] and seen that the extension of the chemical association models to polymer solutions of self-associated polymers, or to any systems forming rings or network-like associates, presents challenging problems. The goal in what follows is to consider how to overcome these shortcomings of the chemical theory models.

In 1990, Veytsman [54] published a two-page letter entitled "Are Lattice Models Valid for Fluids with Hydrogen Bonds?", with which he was responding to the criticism of Huyskens et al. [25] for the suitability of lattice models to handle water and aqueous systems. Veytsman considered systems of molecules with one type of proton-donor group and one type of proton-acceptor group and proposed an approximate combinatorial expression for the number of ways of forming hydrogen bonds in the system which, in the limit, were leading to equations identical to those of Huyskens et al. [25]. The value of the ideas presented in that letter was immediately recognized. They were generalized and properly modified to be incorporated in an equation-of-state framework, leading to the development of the LFHB model [55]. In LFHB, Veytsman's combinatorial expression has been extended to systems of molecules with any number and type of HB groups and applied successfully to a variety of real hydrogen-bonded systems. In this section, we review the essentials of this generalized combinatorial formalism to HB [55].

For convenience, the partition-function formulation in the Gibbs ensemble, Q (T, P, N), and in its maximum-term approximation [50], is adopted. As in Equation 30.1, Q for a mixture is split into HB (subscript HB) and non-HB (subscript NHB) factors as follows:

$$Q\left(T,P,\{N\}\right) = g_{HB} g_{NHB} g^{R}_{comb} \exp\left(-\frac{U+PV}{RT}\right) \tag{32.35}$$

g_{HB} is the HB contribution to nonrandomness factor, on which we focus here. The pre-exponential factors g_{NHB}, g^R_{comb} are those defined after Equation 30.1. U is the internal energy and the symbol $\{N\}$ indicates the fixed number of molecules, N_1, N_2, …, of each species present in the system under consideration. According to Equation 26.17 for the Gibbs ensemble, the Gibbs energy G of the system is obtained from Q [50,55] as:

$$G = -RT \ln Q(T, P, \{N\}) \tag{32.36}$$

and from it all basic thermodynamic quantities may be derived. However, we consider that the HB contribution is separable, implying that Equation 32.35 becomes:

$$Q = \left[g_{HB} \exp\left(-\frac{U_{HB} + PV_{HB}}{RT} \right) \right] \left[g_{NHB} g^R_{comb} \exp\left(-\frac{U_{NHB} + PV_{NHB}}{RT} \right) \right] = Q_{HB} Q_P \tag{32.37}$$

Thus, Q is split into a "chemical" or HB factor, Q_{HB}, and a "physical" factor, Q_P, encompassing contributions from the rest of the non-HB interactions. In Q_P (or rather, in $-RTlnQ_P$), we may use any appropriate model that does not account explicitly for HB, such as UNIQUAC, NRTL, FPP model of Chapter 29 or even SAFT model by disregarding its own HB term. Appropriate terms from Chapter 30 for Q_P may also be used. Thus, now we focus on the HB factor, Q_{HB}.

Generalized Veytsman Statistics

The HB factor Q_{HB} consists of an exponential and a pre-exponential factor, g_{HB} (cf. Equation 32.37). To derive an equation for g_{HB}, let us consider a multicomponent system of t types of molecules, in which there are m different kinds of HB donors and n kinds of HB acceptors. Let d_i^k be the number of hydrogen-bond donors of type i ($i = 1,m$) in each molecule of type k ($k = 1,t$) and α_j^k the number of hydrogen-bond acceptors of type j ($j = 1,n$) in each molecule of type k. The total number N_d^i of hydrogen-bond donors i in the system is given by:

$$N_d^i = \sum_k^t d_i^k N_k \tag{32.38}$$

and the total number N_a^j of hydrogen-bond acceptors j by:

$$N_a^j = \sum_k^t \alpha_j^k N_k \tag{32.39}$$

Equation 32.37 implies that the HB potential energy of the system is *in excess* of that due to physical interactions. The total HB energy U_{HB} of the system is given by:

$$U_{HB} = \sum_i^m \sum_j^n N_{ij} U_{ij}^H \tag{32.40}$$

and the total number of hydrogen bonds is:

$$N_H = \sum_i^m \sum_j^n N_{ij} \tag{32.41}$$

where N_{ij} is the number of hydrogen bonds between HB donors of type i and HB acceptors of type j, and U_{ij}^H the corresponding HB energy of the i–j interaction contact.

The key step now is to find the number, Ω, of distinguishable ways the N_{ij} bonds may be distributed among the functional groups of the system. Let us start with the simple case of a system of molecules with one proton-donor and one proton-acceptor group, which self-associate, such as the systems with hydroxyl groups. The number of the different iso-energetic configurations of the system are then found following the steps below:

a. Find the number of different ways of selecting the associated donor sites out of the donor population.
b. Find the number of different ways of selecting the associated acceptor sites out of the acceptor population.
c. Find the number of different ways of making hydrogen bonds between the selected donor and acceptor sites.

The number of system configurations is the product of the numbers of these three terms (a through c).

To make it more concrete, let us have a system with N molecules, each having one donor site (type 1) and one acceptor site (type 1) with N_{11} hydrogen bonds among them. The number of ways of selecting the N_{11} associated donors, out of the donor population N, is just the binomial coefficient $N!/(N–N_{11})!N_{11}!$. Similarly, the number of ways of selecting the N_{11} associated acceptors, out of the acceptor population, N, is again the binomial coefficient $N!/(N–N_{11})!N_{11}!$. The free donor and acceptor groups are $N–N_{11} = N_{10}=N_{01}$, each, in this particular system. One specific donor can hydrogen bond with any of the N_{11} acceptors, a second donor can hydrogen bond with any of the remaining $N_{11} - 1$ acceptors, and so on. The number of ways that N_{11} bonds can be formed between N_{11} donors and N_{11} acceptors is just $N_{11} \times (N_{11} - 1) \times (N_{11} - 2) \times \ldots \times 1 = N_{11}!$ Thus, the final number of ways the N_{11} bonds can be formed between N donors and N acceptors is the product of the above three terms, or

$$\Omega = \frac{N!}{(N-N_{11})!N_{11}!}\frac{N!}{(N-N_{11})!N_{11}!}N_{11}! = \frac{N!}{N_{10}!}\frac{N!}{N_{10}!N_{11}!} \tag{32.42}$$

These arguments, when generalized for the case of multisite molecules, lead to Equation 32.43 [55]:

$$\begin{aligned}\Omega &= \prod_i^m \frac{N_d^i!}{N_{i0}!N_{i1}!\ldots N_{in}!}\prod_j^n \frac{N_a^j!}{N_{0j}!N_{1j}!\ldots N_{mj}!}\prod_i^m\prod_j^n N_{ij}! \\ &= \prod_i^m \frac{N_d^i!}{N_{i0}!}\prod_j^n \frac{N_a^j!}{N_{0j}!}\prod_i^m\prod_j^n \frac{1}{N_{ij}!}\end{aligned} \tag{32.43}$$

where N_{i0} is the number of free (non-hydrogen-bonded) donor sites of type i and N_{0j} the number of free acceptor groups of type j:

$$N_{i0} = N_d^i - \sum_j^n N_{ij} \tag{32.44}$$

$$N_{0j} = N_a^j - \sum_i^m N_{ij} \tag{32.45}$$

Density Effects on Hydrogen-Bonding

However, to form a hydrogen bond between two specific interacting sites, the sites must be in direct contact. The mean-field probability P_{ij} for this proximity is inversely proportional to the system volume, or $P_{ij} \sim 1/V$. Even spatial proximity does not guarantee that a bond is formed. Bond formation requires that donor and acceptor be properly oriented with respect to each other (*directional* interaction). Formation of the bond is also accompanied by a loss of rotational degrees of freedom. Steric considerations also come into play in bond formation. In general, this probability is approximated as [55]:

$$P_{ij} = e^{S_{ij}^H/R}\frac{\tilde{\rho}}{rN} \tag{32.46}$$

where S_{ij}^H is the entropy loss due to hydrogen-bond i–j formation. $\tilde{\rho}$ is the reduced density, and rN is the total number of segments, here estimated from the GLF model of the physical term, Q_P (cf. Chapter 30).

With these definitions, the HB factor, Q_{HB}, takes the form:

$$Q_{HB}\left(T,N_0,\{N_k\},\{N_{ij}\}\right)=\left(\frac{\tilde{\rho}}{rN}\right)^{N_H}\prod_i^m\frac{N_d^i!}{N_{i0}!}\prod_j^n\frac{N_a^j!}{N_{0j}!}\prod_i^m\prod_j^n\frac{\exp\left(-G_{ij}^H/RT\right)}{N_{ij}!} \tag{32.47}$$

where

$$G_{ij}^H = U_{ij}^H + PV_{ij}^H - TS_{ij}^H \tag{32.48}$$

For strong HB interactions, Equation 32.47 may be used as is. For relatively weak HB interactions Equation 32.47 should be corrected in the way the quasi-chemical factor, g_{QC}, was corrected to become equal to unity in the absence of interactions. In direct analogy, Equation 32.47 may be rewritten as follows [51,56]:

$$Q_{HB}\left(T,N_0,\{N_k\},\{N_{ij}\}\right)=\left(\frac{\tilde{\rho}}{rN}\right)^{N_H-N_H^0}\prod_i^m\frac{N_{i0}^0!}{N_{i0}!}\prod_j^n\frac{N_{0j}^0!}{N_{0j}!}\prod_i^m\prod_j^n\frac{N_{ij}^0!}{N_{ij}!}\exp\left(-\frac{G_{ij}^H}{RT}\right) \tag{32.49}$$

where superscript 0 indicates quantities in the limit of $G_{ij}^H = 0$.

The LFHB Model

Having the partition function, Equation 32.47 or 32.49, one may derive equations for the HB contribution to all basic thermodynamic quantities. However, these contributions have to be combined with the contributions from the rest of intermolecular interactions. As mentioned above, this can be done with various options for Q_P, such as, UNIQUAC, SAFT, LF etc. To fix ideas, let us use LF model for the Gibbs energy, G_P, and Equation 29.51 of Chapter 29, or

$$\frac{G_P}{RT}=\frac{G_{LF}}{RT}=rN\left\{-\frac{\tilde{\rho}}{\tilde{T}}+\frac{\tilde{P}\tilde{v}}{\tilde{T}}+(\tilde{v}-1)\ln(1-\tilde{\rho})+\frac{1}{r}\ln\tilde{\rho}+\sum_{k=1}^{t}\frac{\phi_k}{r_k}\ln\frac{\phi_k}{\omega_k}\right\} \tag{29.51}$$

The symbols are the usual LF ones of Chapters 20 or 29. In a similar manner, the HB contribution, G_{HB}, to the system Gibbs energy, obtained from Equation 32.47, is given by:

$$\frac{G_{HB}}{RT}=rN\left\{\sum_i^m\sum_j^n\nu_{ij}\left[1+\frac{G_{ij}^H}{RT}+\ln\left(\frac{\tilde{\upsilon}\nu_{ij}}{\nu_{i0}\nu_{0j}}\right)\right]+\sum_i^m\nu_d^i\ln\frac{\nu_{i0}}{\nu_d^i}+\sum_i^m\nu_a^j\ln\frac{\nu_{0j}}{\nu_a^j}\right\} \tag{32.50}$$

where

$$\nu_{ij}\equiv\frac{N_{ij}}{rN}\qquad \nu_{i0}\equiv\frac{N_{i0}}{rN}\qquad \nu_d^i\equiv\frac{N_d^i}{rN}\qquad\text{etc.} \tag{32.51}$$

Equation 32.50 is the same regardless of the selected G_P term, When the LF Equation 29.51 is combined with the HB Equation 32.50, the model is referred to as LFHB.

The total volume, V, of the system, with the LFHB model, is given by:

$$V = V_P + V_{HB} = rN\tilde{v}v^* + \sum_i^m \sum_j^n N_{ij} V_{ij}^H \tag{32.52}$$

where V_{ij}^H is the volume change upon formation of an i–j hydrogen bond.

The numbers of hydrogen bonds, N_{ij}, are obtained from the following Gibbs energy minimization conditions:

$$\left(\frac{\partial G}{\partial N_{ij}}\right)_{T,P,\{N_k\},\{N_{rs\neq ij}\}} = 0 \tag{32.53}$$

which leads to the following equations:

$$\frac{\nu_{ij}}{\nu_{i0}\nu_{0j}} = \tilde{\rho}\exp\left(-\frac{G_{ij}^H}{RT}\right) \quad \text{for all } i, j \tag{32.54}$$

or

$$\nu_{ij} = \left[\nu_d^i - \sum_k^n \nu_{ik}\right]\left[\nu_a^j - \sum_k^m \nu_{kj}\right]\tilde{\rho}\exp\left(-\frac{G_{ij}^H}{RT}\right) \tag{32.55}$$

This is a system of $(m \times n)$ quadratic equations for v_{ij}. These fractions can be calculated from the following $(m \times n)$ equations:

$$r\nu_{ij} = \frac{1}{2}\left\{P_{ij} - \sqrt{\left(P_{ij}^2 - 4Q_{ij}\right)}\right\} \quad \text{for all } i, j \text{ pairs} \tag{A}$$

where

$$P_{ij} = A_{ij} + \sum_k d_i^k x_k + \sum_k a_j^k x_k - \sum_{m\neq j} r\nu_{im} - \sum_{n\neq i} r\nu_{nj} \tag{B}$$

and

$$Q_{ij} = \left(\sum_k d_i^k x_k - \sum_{m\neq j} r\nu_{im}\right)\left(\sum_k a_j^k x_k - \sum_{n\neq i} r\nu_{nj}\right) \tag{C}$$

A fast and robust algorithm for solving numerically these equations is as follows: Assume some initial values for the v_{ij}s and calculate the P_{ij}s and Q_{ij}s, which are substituted into Equation A for a first estimation of v_{ij}s. These v_{ij}s are substituted into Equations B and C and the new P_{ij}s and Q_{ij}s are substituted into A for a second estimation of v_{ij}s. This procedure is repeated until two consecutive calculations of the v_{ij}s differ by less than a preset tolerance.

However, $\tilde{\rho}$ is also needed, which is obtained from the equation of state:

$$\tilde{P} + \tilde{\rho}^2 + \tilde{T}\left[\ln(1-\tilde{\rho}) + \tilde{\rho}\left(1 - \frac{1}{\bar{r}}\right)\right] = 0 \text{ (LFHB)} \tag{32.56}$$

Equation 32.56 is identical in form to the LF Equation 29.37 except for the term $1/r$ which is, now, replaced by:

$$\frac{1}{\bar{r}} = \frac{1}{r} - \sum_{i}^{m}\sum_{j}^{n} v_{ij} = \frac{1}{r} - v_H \tag{32.57}$$

v_H is the reduced total number of hydrogen bonds in the system, $\nu_H = N_H / rN$. It is worth comparing this equation with Equation 32.13. Both equations indicate that HB implies an increased average molecular size in the equation of state.

The above system of Equations 32.55 must be solved in combination with the equation of state, which means that there is a system of $(m \times n + 1)$ coupled nonlinear equations for $\tilde{\rho}$ and v_{ij}. This system is solved by the above iterative process.

The HB contribution to the chemical potential, obtained from Equation 32.50, is:

$$\frac{\mu_{k,HB}}{RT} = \frac{1}{RT}\left(\frac{\partial G_{HB}}{\partial N_k}\right)_{T,P,N_j,\tilde{\upsilon},\{N_{ij}\}} = r_k \nu_H - \sum_{i}^{m} d_i^k \ln\frac{\nu_d^i}{\nu_{i0}} - \sum_{j}^{n} a_j^k \ln\frac{\nu_a^j}{\nu_{0j}} \tag{32.58a}$$

If Equation 32.49 is used, instead of Equation 32.47, the following equation is obtained:

$$\frac{\mu_{k,HB}}{RT} = r_k\left(\nu_H - \nu_H^0\right) - \sum_{i}^{m} d_i^k \ln\frac{\nu_{i0}^0}{\nu_{i0}} - \sum_{j}^{n} a_j^k \ln\frac{\nu_{0j}^0}{\nu_{0j}} \tag{32.58b}$$

Heats of mixing and volumes of mixing are also split into a physical (LF) and a chemical/hydrogen-bonding (HB) contribution. Thus, the volume of mixing, with LFHB model, is:

$$V^E = V_{LF}^E + V_{HB}^E = \left\{ rN\left[\tilde{v}v^* - \sum_{k}\tilde{v}_k\phi_k v_k^*\right]\right\} + \left\{ rN\sum_{i}^{m}\sum_{j}^{n} V_{ij}^H\left[\nu_{ij} - \sum_{k}^{t}\nu_{ij}^k\right]\right\} \tag{32.59}$$

and the heat of mixing is:

$$H^E = rN\left[\sum_{k}\tilde{\rho}_k\phi_k\varepsilon_k^* - \tilde{\rho}\varepsilon^*\right] + rN\sum_{i}^{m}\sum_{j}^{n} U_{ij}^H\left[\nu_{ij} - \sum_{k}^{t}\nu_{ij}^k\right] + PV^E \tag{32.60}$$

The LFHB model [55,57] has been applied to numerous systems for VLE, LLE, orthobaric densities, heats of mixing, or volumes of mixing in ordinary liquids as well as systems of high polymers, and to the supercritical and the glassy state and for the study of the effect of HB in the glass-transition behavior of polymer mixtures [58].

The QCHB and NRHB Models

In an entirely similar manner, the G_P term can be tailor-made by the appropriate selection of the pre-exponential terms of the physical part of Equation 32.37. By selecting the quasi-chemical, g_{QC}, and Guggenheim's combinatorial term, Equation 30.4, the model is referred to as *QCHB* (Quasi-Chemical with Hydrogen Bonding) and has been presented in Ref. [59]. *QCHB* has been applied to phase equilibria and volumetric properties and has been extended to polydisperse polymer systems and to inhomogeneous systems and interfaces [59].

By selecting g_{QC} for the nonrandomness term and the Guggenheim–Staverman combinatorial term (Equation 30.5), the model is referred to as NRHB (Nonrandomness and Hydrogen Bonding) and was presented in Refs. [60,61]. The broad range of applications of NRHB in bulk phases, interfaces, glassy polymers, network swelling, and vibrational spectroscopy are presented in an extensive recent review [62]. The HB part in NRHB and QCHB is identical to that of LFHB above. The physical part varies, depending on whether the molecular species and/or the free volume (empty sites) are distributed nonrandomly in the system. Again, the system of Equations 32.55 must be solved along with the corresponding equation of state.

An extensive comparison was made between NRHB and SAFT for pure fluids and mixtures [44]. Analysis of the results has shown that the NRHB model is more accurate for the correlation of saturated liquid densities and vapor pressures of most of the pure fluids. Also, NRHB is clearly more accurate than SAFT for the correlation of phase equilibria of binary mixtures that contain one self-associating fluid, while the accuracy of the two theories is comparable for the case of nonassociating mixtures.

Another extensive comparison was made between NRHB and a successful SAFT variant, the sPC-SAFT [44]. Both models yielded satisfactory and often very similar results for the phase behavior of the investigated mixtures. sPC-SAFT yielded more accurate predictions, while NRHB yielded more accurate correlations, in mixtures of water with n-alkanes and cycloalkanes. In aqueous aromatic hydrocarbon solutions, satisfactory correlations were obtained when solvation was accounted for. Both models resulted in satisfactory correlations for all other mixtures, while for specific mixtures, one model may perform better than the other. Finally, despite being based on different approaches for the HB treatment, both models yielded similar predictions for the fraction of non-hydrogen-bonded molecules (monomer fraction) in pure 1-alkanols and in 1-alkanol-n-hexane mixtures [63].

INTRAMOLECULAR HYDROGEN BONDING

In molecules possessing both proton-donor and proton-acceptor groups, *intra*molecular association may occur if the location of the association groups and stereochemical factors do not prohibit such an interaction. Intramolecular HB is often an important contribution to the overall HB in fluid systems, especially in systems of biological interest. Of interest is the case of very dilute systems in inert solvents where intermolecular HB is absent and the only observed hydrogen bonds are the intramolecular ones.

In this section, we extend the above HB formalism to account for intramolecular association. For simplicity, we confine the treatment to the case of molecules, like polyethoxyalcohols (PEA) with one donor group (-––OH) and with w equivalent proton-acceptor sites (-–O–) per molecule. The formalism can easily be extended to more complex cases. The Q_{HB} term of the partition function may be written, in this case, as:

$$Q_{HB} = \left(\frac{\tilde{\rho}}{rN}\right)^{N_H} g_{intra} \exp\left(-\frac{\sum N_{ij} G_{ij}^H}{RT}\right) \tag{32.61}$$

The focus now is on the factor g_{intra}.

Let us consider a system with N_1 PEA molecules and N_2 molecules of an inert solvent ($N = N_1 + N_2$). The number of proton donors of type 1 (–OH) is N_1, of proton acceptors of type 1 (–OH) is N_1, and of proton acceptors of type 2 (–O–) is wN_1. Let there be N_{11} hydrogen bonds OH- - -OH, N_{12} intermolecular bonds OH- - -O–, and B intramolecular bonds OH- - -O– in the system. The number of free proton donors, N_{10}, is:

$$N_{10} = N_1 - N_{11} - N_{12} - B \tag{32.62}$$

g_{intra} can be found by applying the above three-step rationale to find the different number of iso-energetic configurations of the system. We select, first, the N_{11}, N_{12}, B, and N_{10} donors out of the N_1 donor population. This can be done in $N_{1!}/[B!N11!N12!N10!]$ ways. Then, we select the N_{11} acceptors-1 out of the N_1 acceptor population. This can be done in $N_{1!}/[N_{11!}(N_1-N_{11})!]$ ways. In a third step we select the B acceptors-2 out of the wN_I acceptor population. However, once we have selected the B proton donors that participate in intramolecular bonds, we have also selected the molecules with the acceptor-2 sites that participate in the B intramolecular bonds. Assuming that all w acceptor sites are equivalent for intramolecular HB in these B molecules, we must now select one acceptor-2 site, in each of the B molecules, for the intramolecular bond out of the w acceptor-2 population. This can be done in $w!/[1!(w-1)!]$ ways and, for all B molecules, it can be done in $\{w!/[1!(w-1)!]\}^B = w^B$ ways. Having selected the B acceptor-2 sites we must now select, out of the remaining $(wN_1 - B)$ acceptor-2 population, the N_{12} which participate in the intermolecular bonds. This can be done in $(wN_1 - B)!/[(wN_1-B-N_{12})!N12!]$ ways. The N_{11} and N_{12} bonds can be done in $N_{11}!N_{12}!$ ways while the B bonds in only one way after we have selected, both, the donor and the acceptor site in each molecule. Thus, the number of configurations in the HB system is:

$$g_{intra} = \frac{N_1!}{B!N_{11}!N_{12}!N_{10}!}\frac{N_1!}{N_{11}!(N_1-N_{11})!}\left(\frac{w!}{(w-1)!}\right)^B \frac{(wN_1-B)!}{(wN_1-B-N_{11})!N_{12}!}N_{11}!N_{12}!$$
$$= \frac{w^B(N_1!)^2(wN_1-B)!}{B!N_{11}!N_{12}!N_{10}!(N_1-N_{11})!(wN_1-B-N_{12})!} \tag{32.63}$$

In the system studied, there are three types of hydrogen bonds: N_{11} 1-1 bonds with free energy of bond formation $G_{11}{}^H$, N_{12} 1-2 intermolecular bonds with free energy of bond formation $G_{12}{}^H$, B 1-2 intramolecular bonds with free energy of bond formation $G_B{}^H$. Thus, the HB term Q_{HB} of the partition function can be written as follows:

$$Q_{HB} = \left(\frac{\tilde{\rho}}{rN}\right)^{N_{11}+N_{12}} c^B \frac{w^B(N_1!)^2(wN_1-B)!}{B!N_{11}!N_{12}!N_{10}!(N_1-N_{11})!(wN_1-B-N_{12})!}$$
$$\exp\left(-\frac{N_{11}G_{11}^H+N_{12}G_{12}^H+N_BG_B^H}{RT}\right) \tag{32.64}$$

By minimizing the Gibbs energy, $G_{HB}=-RT\ln Q_{HB}$, with respect to the unknowns N_{11}, N_{12}, and B, we obtain the following coupled equations:

$$\frac{B(wN_1-B)}{(wN_1-B-N_{12})N_{10}w} = c\exp\left(-\frac{G_B^H}{RT}\right) = K_B \tag{32.65}$$

$$\frac{N_{11}}{(N_1-N_{11})N_{10}} = \frac{\tilde{\rho}}{rN}\exp\left(-\frac{G_{11}^H}{RT}\right) = \frac{K_{11}}{N} \tag{32.66}$$

$$\frac{N_{12}}{(wN_1-B-N_{12})N_{10}} = \frac{\tilde{\rho}}{rN}\exp\left(-\frac{G_{12}^H}{RT}\right) = \frac{K_{12}}{N} \tag{32.67}$$

After some algebra, the above three equations lead to the following set of equations:

$$N_{12} = \frac{K_{12}}{K_B N\,w} B(wN_1 - B) \tag{32.68}$$

$$N_{11} = \frac{K_{11}}{K_B N\,w + B(K_{11} - K_{12})} BN_1 \tag{32.69}$$

$$B = \frac{(NwK_B - K_{12}B)}{N} \left[N_1 - B - B\frac{K_{12}}{wK_B}\frac{wN_1 - B}{N} - N_1 \frac{K_{11}B}{wN\,K_B + B(K_{11} - K_{12})} \right] \tag{32.70}$$

The last equation contains only the unknown B and it may be solved numerically by successive substitutions. The solution for B may then be replaced in Equations 32.68 and 32.69 to obtain N_{12} and N_{11}, respectively.

The equation for the heat of mixing of the binary system becomes:

$$\begin{aligned} H^E &= x_1 r_1 \tilde{\rho}_1 \varepsilon_1^* + x_2 r_2 \tilde{\rho}_2 \varepsilon_2^* - r\tilde{\rho}\varepsilon^* + \\ &N_{11}U_{11}^H + N_{12}U_{12}^H + BU_B^H - x_1\left(N_{11}^0 U_{11}^H + N_{12}^0 U_{12}^H + B^0 U_B^H\right) \end{aligned} \tag{32.71}$$

The equation for the volume of mixing is obtained similarly.

The model has been tested extensively against spectroscopic, calorimetric, volumetric, and phase-equilibrium experimental data [64–66]. Of much interest is the capacity of the model to describe the degree of inter- and intramolecular HB over the full composition range and to differentiate the various contributions to the heats of mixing, as shown in Figure 32.2. The model calculations explain the exothermic character of the intramolecular HB contribution in contrast to the endothermic character of the intermolecular one [64,66].

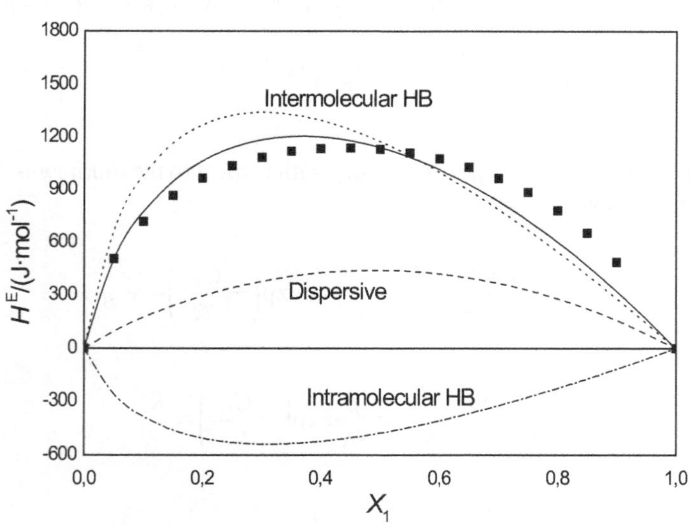

FIGURE 32.2 Experimental [64], ■, and calculated, —, heats of mixing for the system 2-ethoxyethanol(1)+n-octane(2) at 25°C. The contributions from dispersive interactions, - - -, intermolecular HB, ···, and intramolecular HB, –·–·–·, are shown. Reprinted from Ref. [64, fig. 7] with permission.

CARBOXYLIC ACID DIMERIZATION

Carboxylic acids are fluids with peculiar HB behavior arising from the formation of two hydrogen bonds between two acid molecules in a ring structure. Once the first bond is formed, the formation of the second is very much facilitated (hydrogen-bond cooperativity). This cooperativity results in the formation of stable dimers and, thus, the formation of linear oligomers, typical in other association schemes, is not favored.

Accurate modeling of such fluid mixtures is important for many chemical and biochemical processes. This peculiar HB is modeled quite easily by the above LFHB or NRHB model. Below we briefly present the essentials of the modeling of a mixture of N_1 molecules of carboxylic acid (1) with N_2 molecules of an inert solvent (2). Details as well as applications to real systems may be found in Ref. [67].

Let N_{dm} be the number of dimers formed at equilibrium. First it is necessary to select the $2N_{dm}$ acid molecules which form these dimers and then, find the number of ways the selected molecules may form the N_{dm} dimers. The number of ways this selection of the $2N_{dm}$ acid molecules can be done is:

$$\Omega_1 = \frac{N_1!}{(2N_{dm})!(N_1 - 2N_{dm})!} \tag{32.72}$$

For the number of ways the dimers may be formed, we could think as follows: Once the first member of the dimer has been selected, the second one may be any of the remaining $2N_{dm}-1$. For the second dimer, once the first member of the dimer was selected, the second member may be any of the remaining $2N_{dm}-3$. Similarly, all other pairs are selected until ending up to the last pair, which can be selected in just one way. If we set $M=2N_{dm}-1$, we may write for the double factorial, $M!!$, the following equation:

$$\Omega_2 = M!! = (2N_{dm}-1)(2N_{dm}-3)...(3)(1) = \frac{(2N_{dm})!}{2^{N_{dm}} N_{dm}!} \tag{32.73}$$

Thus, the total number of dimer formation, $g_{dm}=\Omega_1\Omega_2$, is:

$$g_{dm} = \frac{N_1!}{(2N_{dm})!(N_1 - 2N_{dm})!} \frac{(2N_{dm})!}{2^{N_{dm}} N_{dm}!} = \frac{N_1!}{2^{N_{dm}} N_{dm}!(N_1 - 2N_{dm})!} \tag{32.74}$$

As previously, cf. Equation 32.47, the HB partition function is now:

$$Q_{HB} = \frac{N_1!}{2^{N_{dm}} N_{dm}!(N_1 - 2N_{dm})!} \left(\frac{\tilde{\rho}}{rN}\right)^{N_{dm}} \exp\left(-\frac{N_{dm}G^{dm}}{RT}\right) \tag{32.75}$$

where the free-energy change upon dimer formation is:

$$G^{dm} = U^{dm} + PV^{dm} - TS^{dm} = H^{dm} - TS^{dm} \tag{32.76}$$

From Equation 32.75, we obtain the HB contribution, G_H, to the Gibbs energy and, by minimization, the number of dimers in the system:

$$\left(\frac{\partial G_H}{\partial N_{dm}}\right)_{T,P,N_i} = -RT\left(\frac{\partial \ln Q_{HB}}{\partial N_{dm}}\right)_{T,P,N_i} = 0 \text{ , which gives:}$$

$$\nu_{dm} = \frac{N_{dm}}{rN} = \frac{2x_1 + \frac{1}{K_{dm}} - \sqrt{\frac{1}{K_{dm}^2} + \frac{4x_1}{K_{dm}}}}{4r} \tag{32.77}$$

$$\text{where } K_{dm} = \frac{\tilde{\rho}}{r}\exp\left(-\frac{G^{dm}}{RT}\right) \quad (32.78)$$

The HB contribution to the chemical potential of the acid (component 1) is:

$$\frac{\mu_1^H}{RT} = \frac{1}{RT}\left(\frac{\partial G_H}{\partial N_1}\right) = r_1\nu_{dm} - \ln\frac{x_1}{x_1 - 2r\nu_{dm}} \quad (32.79)$$

and the corresponding contribution of the inert component 2 is:

$$\frac{\mu_2^H}{RT} = \frac{1}{RT}\left(\frac{\partial G_H}{\partial N_2}\right) = r_2\nu_{dm} \quad (32.80)$$

These HB contributions may be added to the corresponding physical contributions (LFHB or NRHB) to obtain the full expressions for the chemical potential of each component.

HYDROGEN BONDING IN GLASSY POLYMERS

Chapter 20 discussed the role of the entropy equation in accounting for the glass transition and the thermodynamics of the glassy state. This equation can now be enhanced by the HB contribution in hydrogen-bonded polymer systems. As with the other thermodynamic quantities, the entropy equation consists of two parts, the physical, S_P, and the HB contribution, S_H. In Chapter 20, the plain LF equation was used for S_P [58,62]. In the case of LFHB model, the physical contribution is given by [58]:

$$\frac{S_P}{rNR} = \frac{S_{LF}}{rNR} = \frac{1}{r}\ln\omega + (1-\tilde{\nu})\ln(1-\tilde{\rho}) + \frac{\ln(r\,\tilde{\nu})}{r} + \frac{1}{r} - 1 - \frac{x_1\ln x_1 + x_2\ln x_2}{r} + \\ + \frac{\phi_1}{r_1}\ln\delta_1 + \frac{\phi_2}{r_2}\ln\delta_2 \quad (32.81)$$

In the case of the NRHB, this contribution is:

$$\frac{S_P}{rNR} = \frac{1}{r}\ln\omega + (1-\tilde{\nu})\ln(1-\tilde{\rho}) + \frac{l + \ln(r\tilde{\nu})}{r} + \frac{z}{2}\left[\tilde{\nu} - 1 + \frac{q}{r}\right]\ln\left[1 - \tilde{\rho} + \frac{q}{r}\tilde{\rho}\right] \\ + \frac{zq}{2r}\left[(1-\theta_r)\Gamma_{r0}\frac{2}{z\tilde{T}} - \ln\Gamma_{rr} - (\tilde{\nu}-1)\ln\Gamma_{00}\right] \quad (32.82)$$

and the HB contribution for both cases is given by:

$$\frac{S_H}{RrN} = -\nu_H - \sum_i^m \nu_d^i \ln\frac{\nu_{i0}}{\nu_d^i} - \sum_j^n \nu_a^j \ln\frac{\nu_{0j}}{\nu_a^j} \quad (32.83)$$

Otherwise, the model is used in the way described in Chapter 20 and in the frame of the Gibbs–DiMarzio approach to glass transition [68].

As mentioned above, a comprehensive account of the broad range of applications of the LFHB and NRHB models may be found in the recent review article [62]. Apart from the ordinary phase equilibria and related properties, these models have been extensively used in inhomogeneous systems for the study of polymer foaming and the design of nano/microcellular polymeric materials, in the extension of the NELF model, discussed in Chapter 20, in sol–gel transition, gel swelling and collapse, in sorption of gases by glassy polymers and the retrograde vitrification, in the solubility of pharmaceuticals in supercritical fluids, in copolymer systems with tunable LCST, in systems of ionic liquids, etc. [62].

REFERENCES

1. Gilli, G., and Gilli, P. 2009. *The Nature of the Hydrogen Bond*. New York: Oxford University Press.
2. Kontogeorgis, G. M, and Folas, G. K. 2010. *Thermodynamic Models for Industrial Applications. From Classical and Advanced Mixing Rules to Association Theories*. Chichester, UK: Wiley.
3. Economou, I. G., and Donohue, M. D. 1991. Chemical, quasi-chemical, and perturbation theories for associating fluids. *AIChE J.* 37(12): 1875–1894.
4. Scheiner, S. 2020. The hydrogen bond: A hundred years and counting. *J. Indian Inst. Sci.* 100: 61–76.
5. Joesten, M. D., and Schaad, L. J. 1974. *Hydrogen Bonding*. New York: Marcel Dekker.
6. Acree, W. E. 1984. *Thermodynamic Properties of Nonelectrolyte Solutions*. Orlando, FL: Academic Press.
7. Prausnitz, J. M., Lichtenthaler, R. N., and Gomes de Azevedo, E. 1986. *Molecular Thermodynamics of Fluid-Phase Equilibria*, 2nd ed. Englewood Cliffs, NJ: Prentice-Hall.
8. Dolezalek, F., 1908. Zur Theorie der binären Gemische und konzentrierten Lösungen. *Z. Physik. Chem.* 64U(1): 727–747.
9. Kempter, H., and Mecke, R. 1940. Spektroskopische Bestimmung von Assoziationsgleichgewichten. *Z. Physik. Chem.* 46B(1): 229–241.
10. Renon, H., and Prausnitz, J. M. 1967. On the thermodynamics of alcohol–hydrocarbon solutions. *Chem. Eng. Sci.* 22(3): 299–307.
11. Nagata, I., and Tamura, K. 1982. Thermodynamics of solutions of methanol and solvating components. *Thermochim. Acta* 57(3): 331–349.
12. Panayiotou, C. G. 1988. Thermodynamics of alkanol-alkane mixtures. *J. Phys. Chem.* 92(10): 2960–2969.
13. Panayiotou, C., and Sanchez, I. C. 1991. Statistical thermodynamics of associated polymer solutions. *Macromolecules* 24(23): 6231–6237.
14. Heidemann, R. A., and Prausnitz, J. M. 1976. A van der Waals-type equation of state for fluids with associating molecules. *Proc. Natl. Acad. Sci. USA* 73(6):1773–1776.
15. Elliott, J. R., Suresh, S. J., and Donohue, M. D. 1990. A simple equation of state for nonspherical and associating molecules. *Ind. Eng. Chem. Res.* 29(7): 1476–1485.
16. Economou, G. D., and Donohue, M. D. 1988. Extension of the associated perturbed anisotropic chain theory to mixtures with more than one associating component. *Fluid Phase Equilib.* 3(2): 129–159.
17. Economou, I. G., and Donohue, M.D., 1996, Equations of state for hydrogen bonding systems. *Fluid Phase Equilib.* 116(1–2): 518–529.
18. Flory, P. J. 1944. Thermodynamics of heterogeneous polymers and their solutions. *J. Chem. Phys.* 12(11): 425–438.
19. Prigogine, I., and Defay, R. 1965. *Chemical Thermodynamics*. New York: Wiley.
20. Anderko, A. 1989. Extension of the AEOS model to systems containing any number of associating and inert components. *Fluid Phase Equilib.* 50(1–2): 21–52.
21. Panayiotou, C. G. 1989. Applications of the LFAS theory to VLE and volumetric properties in 1-alkanol+n-alkane mixtures. *Pure Appl. Chem.* 61: 1453–1460.
22. Panayiotou, C. G. 1990. Thermodynamics of associated solutions. Mixtures of 1-alkanols. *Fluid Phase Equilib.* 56: 171–188.
23. Panayiotou, C. G. 1991. Thermodynamics of mixtures of amines with n-alkanes and 1-alkanols. *J Solution Chem.* 20(1): 97–114.
24. Levine, S., and Perram, J. W. 1968. A statistical mechanical treatment of hydrogen-bonding in water. In *Hydrogen-Bonded Solvent Systems*, eds. A. K. Covington, and P. Jones. London: Taylor and Francis.
25. Huyskens, P. L., Haulait-Pirson, M. C., Siegel, G. G., and Kapuku, F. 1988. Gibbs energy of some binary systems containing hydrogen bonds. *J. Phys. Chem.* 92(23): 6841–6847.
26. Klamt, A. 1995. Conductor-like screening model for real solvents: A new approach to the quantitative calculation of solvation phenomena. *J. Phys. Chem.* 99(7): 2224–2235.
27. Klamt, A. 2005. *COSMO-RS from Quantum Chemistry to Fluid Phase Thermodynamics and Drug Design*. Amsterdam: Elsevier.
28. Lin, S. T., and Sandler, S. I. 2002. A priori phase equilibrium prediction from a segment contribution solvation model. *Ind. Eng. Chem. Res.* 41(5): 899–913.
29. Grensemann, H., and Gmehling, J. 2005. Performance of a conductor-like screening model for real solvents model in comparison to classical group contribution methods. *Ind. Eng. Chem. Res.* 44(5): 1610–1624.

30. Pye, C. C., Ziegler, T., van Lenthe, E., and Louwen, J. N. 2009. An implementation of the conductor-like screening model of solvation within the Amsterdam density functional package, Part II. COSMO for real solvents. *Can. J. Chem.* 87(7): 790–797.
31. Wertheim, M. S. 1984. Fluids with highly directional attractive forces. I. Statistical thermodynamics. *J. Stat. Phys.* 35(1/2): 19–34.
32. Wertheim, M. S. 1984. Fluids with highly directional attractive forces. II. Thermodynamic perturbation theory and integral equations. *J. Stat. Phys.* 35(1/2): 35–47.
33. Wertheim, M. S. 1986. Fluids with highly directional attractive forces. III. Multiple attraction sites. *J. Stat. Phys.* 42(3/4): 459–476.
34. Wertheim, M. S. 1986. Fluids with highly directional attractive forces. IV. Equilibrium polymerization. *J. Stat. Phys.* 42(3/4): 477–492.
35. Wertheim, M. S. 1986. Fluids of dimerizing hard spheres, and fluid mixtures of hard spheres and dispheres. *J. Chem. Phys.* 85(5): 2929–2936.
36. Wertheim, M. S. 1987. Thermodynamic perturbation theory of polymerization. *J. Chem. Phys.* 87(12): 7323–7331.
37. Chapman, W. G., Gubbins, K. E., Jackson, G., and Radosz, M. 1989. SAFT: Equation-of-state solution model for associating fluids. *Fluid Phase Equilib.* 52: 31–38.
38. Chapman, W. G., Gubbins, K. E., Jackson, G., and Radosz, M. 1990. New reference equation of state for associating liquids. *Ind. Eng. Chem. Res.* 29(8): 1709–1721.
39. Huang, S. H., and M. Radosz, M. 1990. Equation of state for small, large, polydisperse, and associating molecules. *Ind. Eng. Chem. Res.* 29(11): 2284–2294.
40. Müller, E. A., and Gubbins, K. E. 2001. Molecular-based equations of state for associating fluids: A review of SAFT and related approaches. *Ind. Eng. Chem. Res.* 40(10): 2193–2211.
41. Carnahan, N. F., and Starling, K. E. 1970. Thermodynamic properties of a rigid-sphere fluid. *J. Chem. Phys.* 53(2): 600–603.
42. Gross, J., and Sadowski, G. 2001. Perturbed-chain SAFT: An equation of state based on a perturbation theory for chain molecules. *Ind. Eng. Chem. Res.* 40(4): 1244–1260.
43. Kontogeorgis, G. M., Voutsas, E. C., Yakoumis, I. V., and Tassios, D. P. 1996. An equation of state for associating fluids. *Ind. Eng. Chem. Res.* 35(11): 4310–4318.
44. Tsivintzelis, I., Grenner, A., Economou, I. G., and Kontogeorgis, G. M. 2008. Evaluation of the non-random hydrogen bonding (NRHB) theory and the simplified perturbed-chain-statistical associating fluid theory (sPC-SAFT). 2. Liquid-liquid equilibria and prediction of monomer fraction in hydrogen bonding systems. *Ind. Eng. Chem. Res.* 47(15): 5651–5659.
45. Asprion, N., Hasse, H., and Maurer, G. 2001. FT-IR spectroscopic investigations of hydrogen bonding in alcohol-hydrocarbon solutions. *Fluid Phase Equilib.* 186(1–2): 1–25.
46. Martínez, S. 1986. Methanol/n-hexane system-I. Infrared studies. *Spectrochim. Acta* 42A(4): 531–536.
47. Gupta, R., and Johnston, K. 1994. Lattice fluid hydrogen bonding model with a local segment density. *Fluid Phase Equilib.* 99: 135–151.
48. Avlund, A. S., Kontogeorgis, G. M., and Chapman, W. G. 2011. Intramolecular association within the SAFT framework. *Mol. Phys.* 109(14): 1759–1769.
49. Febra, S. A., Aasen, A., Adjiman, C. S., Jackson, G., and Galindo, A. 2019. Intramolecular bonding in a statistical associating fluid theory of ring aggregates. *Mol. Phys.* 117(23–24): 3884–3912.
50. Hill, T. L. 1986. *An Introduction to Statistical Thermodynamics*. New York: Dover.
51. Guggenheim, E. A. 1952. *Mixtures*. London: Oxford University Press.
52. Staverman, A. J. 1950. The entropy of high polymer solutions. Generalization of formulae. *Rec. Trav. Chim. Pays-Bas.* 69(2): 163–174.
53. Luck, W. A. P. 1980. Modellbetrachtung von Flüssigkeiten mit Wasserstoffbrücken. *Angew. Chem.* 92(1): 29–42.
54. Veytsman, B. A. 1990. Are lattice models valid for fluids with hydrogen bonds? *J. Phys. Chem.* 94(23): 8499–8500.
55. Panayiotou, C., and Sanchez, I.C. 1991. Hydrogen bonding in fluids: An equation-of-state approach. *J. Phys. Chem.* 95(24): 10090–10097.
56. Shin, M. S., and Kim, H. 2007. A multi-fluid nonrandom associating lattice-fluid model. *Fluid Phase Equilb.* 253: 29–35.
57. Sanchez, I. C., and Panayiotou, C. G. 1994. Equations of state thermodynamics of polymer and related solutions. In *Models for Thermodynamic and Phase Equilibria Calculations*, ed. S. I. Sandler. New York: Marcel Dekker.

58. Prinos, J., and Panayiotou, C. 1995. Glass transition temperature in hydrogen-bonded polymer mixtures. *Polymer* 36(6): 1223–1227.
59. Panayiotou, C. 2003. The QCHB model of fluids and their mixtures. *J. Chem. Thermodyn.* 35(2): 349–381.
60. Panayiotou, C., Stefanis, E., Tsivintzelis, I., Pantoula, M., and Economou, I. 2004. Nonrandom hydrogen-bonding model of fluids and their mixtures. 1. Pure fluids. *Ind. Eng. Chem. Res.* 43(20): 6592–6606.
61. Panayiotou, C., Tsivintzelis, I., and Economou, I. G. 2007. Nonrandom hydrogen-bonding model of fluids and their mixtures. 2. Multicomponent mixtures. *Ind. Eng. Chem. Res.* 46(8): 2628–2636.
62. Mensitieri, G., Scherillo, G., Panayiotou, C., and Musto, P. 2020. Towards a predictive thermodynamic description of sorption processes in polymers: The synergy between theoretical EoS models and vibrational spectroscopy. *Mater. Sci. Eng. R Rep.* 140: 100525.
63. Tsivintzelis, I., Spyriouni, T., and Economou, I. G. 2007. Modeling of fluid phase equilibria with two thermodynamic theories: Non-random hydrogen bonding (NRHB) and statistical associating fluid theory (SAFT). *Fluid Phase Equilib.* 253(1): 19–28 .
64. Missopolinou, D., Ioannou, K., Prinos, I., and Panayiotou, C. 2002. Thermodynamics of alkoxyethanol+alkane mixtures. *Z. Phys. Chem.* 216(7): 905–918, 1–14.
65. Missopolinou, D., Tsivintzelis, I., and Panayiotou, C. 2005. Phase compositions and saturated densities for the binary system of carbon dioxide with 2-ethoxyethanol. *Fluid Phase Equilb.* 238(2): 204–209.
66. Missopolinou, D., Tsivintzelis, I., and Panayiotou, C. 2006. Excess enthalpies of binary mixtures of 2-ethoxyethanol with four hydrocarbons at 298.15, 308.15, and 318.15K: An experimental and theoretical study. *Fluid Phase Equilb.* 245(2): 89–101.
67. Tsivintzelis, I., Kontogeorgis, G. M., and Panayiotou, C. 2017. Dimerization of carboxylic acids: An equation of state approach. *J. Phys. Chem. B* 121(9): 2153–2163.
68. Gibbs, J. H., and DiMarzio, E. A. 1958. Nature of the glass transition and the glassy state. *J. Chem. Phys.* 28(3): 373–383.

33 The Family of SAFT Equation-of-State Models

Abstract

This chapter summarizes, first, the essentials of the statistical associating fluid theory (SAFT) equation-of-state model. Subsequently, it briefly discusses the development of coarse-grained SAFT force fields for use in molecular simulations. Two examples of the use of Wertheim's association term in other equations of state (the Cubic-plus-Association and Elliot–Suresh–Donohue models) are presented.

INTRODUCTION

The ultimate dream of molecular thermodynamics would be the capacity to account for the contribution of every structural and interactional molecular detail to the macroscopic-thermodynamic behavior of matter. The required information for this goal refers not only to the atomic constitution, size, and shape of the molecules but also to the structure of their molecular assemblies, the distances, and relative orientations in their arrangements in the macroscopic fluid or solid matter. Statistical thermodynamics reduces drastically the enormous amount of required information and strengthens the hope that, by improving our understanding of the structural molecular details, the capacity to describe and even predict reliably the macroscopic behavior will keep improving as well.

As discussed in Chapter 29, statistical mechanics has reached a level of nearly perfect description of the radial distribution function and the thermodynamic behavior of simple fluids considered as hard spheres or soft spheres interacting with spherically symmetric interaction potentials. Significant progress has been made also in the case of nonspherical molecules interacting with relatively weak intermolecular forces. However, when these intermolecular forces are sufficiently strong, the above description becomes very difficult even in the case of structurally simple molecules such as water.

As discussed in the previous chapter, water is an example of molecules (though an exceptional one) interacting with strong directional specific forces, named hydrogen bonding (HB) or association interactions. Fluids of molecules exhibiting association interactions are named *associating fluids*. The systems of much technological interest and of the main interest of modern molecular thermodynamics, very often consist of such associating fluids leading to a highly nonideal system behavior. Asymmetric or chain molecules, oligomers or polymers of high technological interest further enhance this nonideal behavior, as seen in Chapters 19, 20, 29, and 30. There is, thus, a persistent shift of priorities in addressing the challenging real-life problems even though the provenance of the observed thermodynamic behavior remains largely elusive.

Some realities, however, cannot be disregarded nowadays. The unprecedented and ever-increasing computing speed and power available today, even at the desktop level, influences drastically the way the above thermodynamic challenging issues are addressed. In addition, experimental measurements of thermodynamic properties are becoming rather scarce as they are considered tedious and often even dangerous. Molecular simulations, on the other hand, have reached a level of approximation that also strengthens the hope for a reliable and safe substitution of laboratory experiments with computer-based ones. Protein folding and their tertiary structures are predicted successfully today in such computer experiments with artificial intelligence tools. Thermodynamicists, especially chemical engineers, cannot disregard these realities when designing processes or new products.

 DOI: 10.1201/9781003431985-37

In the early 1980s, Michael Wertheim developed his pioneering work on the thermodynamic perturbation theory (TPT) [1–6] and has shown that the properties of an associating or multi-mer chain fluid may be obtained from the knowledge of the Helmholtz free energy and structure of the monomer fluid. The significance of this cannot be overemphasized. Since the behavior of simple monomer fluids is already described very well or may be very reliably obtained from molecular simulations with realistically detailed interaction potentials or atomistic-level force fields, TPT may be the dreamed means for the description or even prediction of the behavior of real-life systems of fluids. Of central importance in these TPT and subsequent developments is the inclusion in Helmholtz free energy of fluids of an *association* term, A^{Assoc}, which accounts explicitly for highly directional, short-range, strong specific attractive forces, such as HB, Lewis acid–base and other association interactions. This term, based on Wertheim's first-order thermodynamic perturbation theory (TPT1) [1–6], was discussed extensively in the previous chapter on HB and, as mentioned there, it is widely used in a variety of thermodynamic models, in empirical equations of state, in lattice-based models, in UNIQUAC, etc. in order to account for the association interactions.

Chapman, Jackson, Gubbins, and Radosz in the late 1980s [7–8] extended Wertheim's theory to mixtures of spheres and chain molecules, by considering the association between monomers in its limit (complete association) and replacing the association interactions with covalent, chain-forming bonds. This permitted a straightforward extension to nonspherical molecules up to high polymers. These developments were implemented in an equation-of-state framework for which they have coined the term *statistical associating fluid theory* – SAFT. The SAFT approach has proven to be very successful for modeling associating chain-like fluids, including polymers. Today, it is considered one of the most versatile and successful methods for modeling the thermodynamic behavior of systems of complex fluids. In fact, under the name SAFT resides today a large and ever-growing family of models of fluids and their mixtures, such as SAFT-HS, SAFT-HR, SAFT-VR, soft-SAFT, SAFT-VR Mie 2006, SAFT-VR Mie 2013, SAFT-γ, SAFT-γ Mie, PC-SAFT, eSAFT, iSAFT, and NE-SAFT. The family of SAFT models has created a large family of SAFT model developers and users worldwide and this is probably one of the most interesting examples of common international efforts in mixture thermodynamics for modeling systems of challenging complexity, such as aqueous solutions, polymers, surfactants, phase inhomogeneities, electrolytes, ionic liquids, and supercritical fluids.

A comprehensive overview of the SAFT model variants is not easy and their comparison is even more difficult. As Prausnitz et al. [9] state, "The literature on SAFT is complex and confusing." However, the interested reader is referred to the first main SAFT papers [7,8] and to reviews by Müller and Gubbins [10], Economou [11], McCabe and Galindo [12], and Vega and Llovell [13]. Wertheim's TPT was developed by making extensive use of graph theory, as implemented in statistical mechanics, for the graphical summation of interactions between different species and the interested reader is referred to the pedagogical review by Zmpitas and Gross [14].

Thus, in what follows we will confine ourselves to a short presentation of the SAFT essentials as an equation of state and as a tool for force field developments in Monte-Carlo (MC) and molecular dynamics (MD) simulations. In addition, two examples of other than SAFT models using the same association term will be given.

THE ESSENTIALS OF SAFT EQUATION OF STATE

To understand SAFT, it is useful to recall, first, some very basic elements of perturbation theory. In discussing the virial equation of state of real gases we have structured it as an expansion around a well-known reference term perfectly valid for a perfectly well-described reference fluid – the ideal gas. A similar approach was used for the virial equation of dilute polymer solutions. These are examples of perturbation expansions and the underlying idea of TPT. When the goal is an equation of state for associating chain molecules, one has to select, first, properly a reference system, whose Helmholtz free energy, A^{ref}, or compressibility factor, Z^{ref}, are well known. Obviously, the closer the reference system to the studied fluid, the better its choice for the construction of the expansion or perturbation part of the final equation of state. A good reference system should reflect, as an example, to a sufficient extent the structure of the real system in the dual sense mentioned above, namely, the structure of the molecule and the molecular arrangements, the radial distribution functions or pair-correlation functions. In agreement with this structure, the interaction potential should be split properly into a reference part, pertinent to the reference system, and a perturbation part for the correction of the inadequacies of the reference term. This, in turn, will result in the expansion of the radial distribution function into a reference part, g^{ref}, and a perturbation correction part. The desired expansion here is the one with as negligible as possible perturbation part. Although this sounds difficult, as mentioned in Chapter 29, extensive experimental measurements coupled with theoretical calculations have shown that most often the structure of simple dense fluids is dictated by the short-range interactions, primarily, the repulsive ones.

In addition to the above, Wertheim's TPT offers two important elements for the construction of the perturbation part. The first is the already mentioned association term, A^{Ass}, and the second is the term A^{xhain} that accounts for the contribution of the bonding of monomers for the formation of the chain molecules. With the above in mind, let us examine now the way real fluid molecules are perceived and treated in SAFT.

In contrast to the COSMO-RS quantum-chemical picture of the molecule with details at the atomic level, discussed in Chapter 31, SAFT replaces this physical picture with a much simpler one, wherein molecules are built up of spherical units bound tangentially at their binding or sticky sites. One would expect an integer number of such units per molecule but, typically, carbon dioxide consists of 1.42 such units, n-decane of 7.53, n-eicosane of 13.94, benzene of 3.75, and ethanol of 2.46 units [15]. These average noninteger numbers may not reflect the quantum-chemical picture of the real molecule but the selection of the repeating unit leading to this division of the molecule and its intermolecular interaction features are of primordial importance in the development of SAFT models. This division of molecules into fictitious segments is not an exclusive SAFT feature but, rather, the standard practice in chemical engineering thermodynamics.

Spherical units are relatively easily handled with the radial distribution function (RDF) statistical mechanical formulation. Through Wertheim's TPT approach, much of the information on RDF and intermolecular interaction of these autonomous units is retained when bound tangentially to form multi-segment or chain molecules. Thus, the focus in SAFT is on the rigorous algebraic description of the properties of these rather fictitious multi-segment molecules. These fictitious molecular entities are easily studied in MC and MD simulations and, as already mentioned, SAFT developments are intimately connected to computer simulations from Wertheim's time. Molecular simulation data over a range of conditions are extensively used for parameterizing the model and even for expressing model parameters as universal equations with universal constants by adopting a corresponding-states approach.

The bulk of SAFT development is heavily based on the first-order approximation of Wertheim's TPT (TPT1) in which the reference fluid molecules, typically spherical molecules, may have one or more short-range association sites to mediate intermolecular association (HB). As mentioned above, when the association interactions are taken in the limit of covalent bonding, the spherical molecules are associated/bonded tangentially and lead to chain molecules. On this basis, SAFT considers that the Helmholtz free energy, A, of the system may be split into additive terms, expressing different contributions to it, as shown by Equation 33.1:

$$A = A^{ideal} + A^{Mono} + A^{chain} + A^{assoc} = A^{ideal} + \left(A^{hs} + \mathrm{A}^{disp}\right) + A^{chain} + A^{assoc} \tag{33.1}$$

The first term in Equation 33.1, A^{ideal}, captures all kinetic contributions of molecular particles and reduces to the ideal-gas equation of state, in the limit. It is given by:

$$\frac{A^{ideal}}{RT} = \ln\left(\rho_N \Lambda(T)^3\right) - 1 \tag{33.2}$$

$\Lambda(T)$ is the temperature-dependent de Broglie wavelength of the particle and ρ_N the number density. In this way, this ideal-gas term is taken as the reference and the sum of the rest of the terms is the residual Helmholtz free energy, A^{res}.

The second term, A^{Mono}, is split into a hard-sphere term, A^{hs}, and a dispersion term, A^{disp}. The A^{hs} term reflects the hard-sphere repulsive interactions and is given by the classical Carnahan–Starling equation [16] (cf. Chapter 29):

$$\frac{A^{hs}}{RT} = m\frac{4\eta - 3\eta^2}{(1-\eta)^2} \tag{33.3}$$

In the SAFT terminology, m is the number of segments per molecule (the corresponding of r in Flory–Huggins and LF models) and η is the reduced density (the corresponding $\tilde{\rho}$ of Flory–Huggins and LF models), given by the equation:

$$\eta = 0.74048\rho m v^0 \tag{33.4}$$

where ρ is the density of the system and v^0 is the close-packed hard-core molar volume calculated from the close-packed segment volume v^{00} of the fluid at T=0 by two alternative equations, the simpler one being [15]

$$v^0 = v^{00}\left[1 - C\exp\left(-3u^0 / kT\right)\right]^3 \quad \text{(Huang and Radosz)} \tag{33.5}$$

The constant C is equal to 0.120 (except for hydrogen for which it is equal to 0.241) and u^0 is the dispersion energy per segment. Thus, m, v^{00}, and u^0/k are the three adjustable SAFT parameters for each non-hydrogen-bonded fluid. Typically, they are obtained by fitting experimental vapor pressure, density, etc. data over a range of temperature and pressure. An explicit expression for the dispersion term is also given (see below) although more often the monomer term, A^{Mono}, is kept without splitting and is written, alternatively, as:

$$\frac{A^{Mono}}{RT} = m\frac{a^{Mono}}{RT}$$

with a^{Mono} being the contribution per segment unit.

The third term, A^{chain}, in Equation 33.1 is the contribution of the $m-1$ covalent bonds forming the chain molecule and is obtained from Wertheim's TPT1 approximation as:

$$\frac{A^{chain}}{RT} = -(m-1)\ln y^{Mono}(\sigma) = -(m-1)\ln g^{Mono}(\sigma) \tag{33.6}$$

where, $y^{Mono}(\sigma)$ is the cavity distribution function related to the RDF $g^{Mono}(\sigma)$ by:

$$y^{Mono}(\sigma) = g^{Mono}(\sigma)\exp\left(\frac{u^{Mono}(r)}{kT}\right) \tag{33.7}$$

where $u^{Mono}(r)$ is the pair potential between tangentially bonded monomers with σ the monomer segment diameter. For hard-sphere monomers Equation 33.6 becomes:

$$\frac{A^{chain}}{RT} = -(m-1)\ln g = -(m-1)\ln\frac{1-0.5\eta}{(1-\eta)^3} \tag{33.8}$$

These equations are written in tractable analytical form when using the first-order TPT1 approximation with the simple two-body RDF, g. Higher-order TPT with many-body RDF that capture bond angles and dihedrals cannot be easily written in tractable analytical forms.

The association term, A^{assoc}, was already discussed in Chapter 32 on HB. It captures the contribution of strong specific directional intermolecular interactions and is given by:

$$\frac{A^{assoc}}{RT} = \sum_{A=1}^{M}\left[\ln X^A - \frac{X^A}{2}\right] + \frac{M}{2} \tag{33.9}$$

M is the number of association sites per molecule and X^A is the fraction of free (nonassociated) molecules not bonded at site A. The sum is over the M association sites of the molecule and the fraction X^A is given by the recursive equation:

$$X^A = \frac{1}{\left(1+\sum_{B=1}^{M}\rho X^B \Delta^{AB}\right)} \tag{33.10}$$

In a simplified version, the association strength, Δ^{AB}, for the bond $A-B$ is given by the equation [15]:

$$\Delta^{AB} = gf^{AB}K^{AB} = \left[\frac{1-0.5\eta}{(1-\eta)^3}\right] f^{AB}\left[\sqrt{2}v^{00}\kappa^{AB}\right] \quad \text{(Huang and Radosz)} \tag{33.11}$$

or by a similar equation with the v^{00} replaced by v^0 [8–9,17]. $f^{AB} = \left[\exp\left(\varepsilon^{AB}/kT\right)-1\right]$ in Equation 33.11 is the Mayer function (cf. Chapter 29). Thus, each type of AB HB interaction in SAFT requires the knowledge of two parameters, the association energy, ε^{AB}, and the association volume, κ^{AB}, both considered to be temperature independent and fitted, in practice, to experimental vapor pressure and density data.

All other intermolecular interactions were named *dispersion* interactions in the early SAFT developments and are captured by the A^{disp} term for which Chapman et al. [7,8] have proposed the following equation, initially proposed by Cotterman et al. [18]:

$$\frac{A^{disp}}{RT} = m\frac{u^0}{kT}\left(a_1^{disp} + \frac{u^0}{kT}a_2^{disp}\right) \tag{33.12}$$

where a_1^{disp} and a_2^{disp} are given by two fourth-order polynomials of the reduced density η. Huang and Radosz [15] proposed an alternative Alder's type [19] polynomial equation for A^{disp} based on MD simulation data for the square-well fluid (cf. Chapter 29):

$$\frac{A^{disp}}{RT} = m\sum\sum D_{ij}\left(\frac{u}{kT}\right)^i\left(\frac{\eta}{0.74048}\right)^j \tag{33.13}$$

where $u/k = \left(u^0/k\right)(1+10/T)$ for all compounds with few exceptions and D_{ij} are universal constants.

The extension to mixtures is rather straightforward. The Mansoori equation for hard spheres [20] is used for the A^{hs} term, while the A^{chain} and A^{assoc} terms are obtained rigorously from the Wertheim's TPT1 approximation. Thus, mixing rules are only needed for the dispersion term, A^{disp} in the one-fluid approach and are:

$$m=\sum_i x_i m \tag{33.14}$$

$$\frac{u}{k}=\frac{\sum_i\sum_j x_i x_j m_i m_j \frac{u_{ij}}{k}\mathrm{v}_{ij}^0}{\sum_i\sum_j x_i x_j m_i m_j \mathrm{v}_{ij}^0}\quad \text{(Huang and Radosz)} \tag{33.15}$$

$$\mathrm{v}_{ij}^0=\left[\frac{\left(\mathrm{v}_{ii}^0\right)^{1/3}+\left(\mathrm{v}_{jj}^0\right)^{1/3}}{2}\right]^3 \tag{33.16}$$

$$\frac{u_{ij}}{k}=\left(\frac{u_{ii}}{k}\frac{u_{jj}}{k}\right)^{1/2}\left(1-k_{ij}\right)\quad \text{(Huang and Radosz)} \tag{33.17}$$

In these equations x_i is the mole fraction of component i and k_{ij} is an adjustable parameter for the pair i–j interaction. Mixing rules for v^0, u^9 and u_{ij}^0 were also proposed by Chapman et al. [7,8,17].

This is the main SAFT equation-of-state model. Parameters for using this SAFT equation are given, as an example, by Huang and Radosz [15]. Equations for the chemical potential or the pressure are obtained from Helmholtz free energy by the classical derivations.

There are numerous extensions and simplifications of the main SAFT. Of much interest is the transferability of the HB (association) term, as we have mentioned in Chapter 32 on HB. It may be combined with a cubic equation of state giving the **c**ubic-**p**lus-**a**ssociation (CPA) equation of state [21]. It may also be replaced by analogous HB terms, like the LFHB term from Veytsman statistics [22,23], which was discussed extensively in Chapter 32. Such a replacement was done by Gupta and Johnston [24] with improved performance of the model. The HB terms of SAFT and its contemporary LFHB may also replace each other in activity coefficient models as well, as shown by Gupta et al. [25] with the UNIQUAC-HB.

The way of splitting the Helmholtz free energy in equation 1 implies that all nonassociation interactions, that is, London-type dispersion interactions and all polar or other electrostatic interactions have been combined in the monomer term and treated via a mean-field approach [15]. Thus, this SAFT version and its contemporary LFHB are mean-field models. The key difference is that the dispersion terms in SAFT, Equations 33.12–33.13, are polynomials of the reduced density, η, and, thus, composition-dependent This means that consistent equations for the chemical potential or the activity coefficient may not be very tractable analytical expressions and would be rather handled numerically even in nonassociating systems. In polar-SAFT variants, in addition to the above segment/dispersion term, there are separate polarity terms, A^{Polar}, of Helmholtz energy in Equation 33.1.

The chain term is kept separate in most SAFT versions and is obtained consistently from Wertheim's TPT. However, Gross and Sadowski [26] have proposed a most interesting version of SAFT, the **p**erturbed-**c**hain SAFT (PC-SAFT), by applying the perturbation approximation directly to preformed chain molecules, thus, dropping out the chain term in Equation 33.1, which may be written, now, as follows:

$$\frac{A}{RT}=\frac{A^{Ideal}}{RT}+\frac{A^{HC}}{RT}+\frac{A^{PC}}{RT}+\frac{A^{Assoc}}{RT}\quad \text{(PC-SAFT)} \tag{33.18}$$

The second term on the right-hand side of this equation corresponds to the free energy of a reference *hard-sphere-chain* fluid, given by Equation 33.6, that is, the monomer fluid is, now, a chain fluid of hard spheres. The term, A^{PC}, corresponds to the *chain perturbation* contribution, which is based on an earlier theory for square-well chain molecules and is determined from a Taylor series expansion fitted to experimental vapor-liquid equilibrium data of pure alkanes. This is a crucial step in PC-SAFT development. The fitting to experimental data greatly enhances the accuracy of the model for the description of experimental systems, but at the same time it implies that the direct link to the reference term is lost, and it is no longer straightforward to define the underlying intermolecular potential of the model. Thus, comparison against computer simulations cannot be used in the assessment of further theoretical developments. This fitting for the dispersion term is preceded by an intermediate step assuming a Lennard-Jones perturbing potential for the calculation of two integral expressions previously derived for square-well chains and by using average interchain segment-segment RDF.

As regards polymer systems, PC-SAFT performs better than the main SAFT [8,9] and this has opened the way for SAFT to enter into new technologically important applications with polymer systems. Polymer systems are also well described by the variable-range SAFT-VR [27–29].

An interesting alternative PC-SAFT version, the PC-SAFT-D(imer), was proposed by Chapman et al. [30] based on *chain-of-dimers* reference and was demonstrated to work better for polymers than the *chain-of-monomers* PC-SAFT variant [26]. The chain in PC-SAFT-D is formed through an intermediate step of di-sphere formation. The advantage of a dimer reference is that it incorporates the structural information on the dimer fluid; thus, it takes into account not just the neighboring segments in forming the bonds but also the next neighbor effects. This dimer version utilizes the hard-disphere RDF, g^{hd}, and the hard-sphere RDF, g^{hs}, to calculate the chain term as follows:

$$A^{chain} = -\frac{m}{2}\ln g^{hs}(\sigma) - \left(\frac{m}{2} - 1\right)\ln g^{hd}(\sigma) \quad \text{(PC-SAFT-D)} \tag{33.19}$$

with the number of segments, m, larger than 2.

There are numerous other SAFT extensions and applications in solutions of electrolytes, liquid crystals, ionic liquids, amphiphiles, etc., some of which are reviewed in Refs. [12] and [31]. Explicit polarity and polarizability interaction terms have been added also in Equation 33.1 or in Equation 33.18. New parameters are added in this way which further increase the versatility of the SAFT equation of state. Accounting, however, for the fact that the TPT itself may not be sufficiently accurate, some care must be exercised in the physical interpretation of these SAFT parameters.

A most interesting and significant feature of SAFT is its capacity for direct connection and exchange of information with MD or MC molecular simulation approaches, especially when experimental data are scarce or nonexistent. This capacity is touched upon very briefly in the next section. For more in-depth information, the interested reader is referred to the comprehensive review by Müller and Jackson [32].

SAFT FORCE FIELDS AND MOLECULAR SIMULATIONS

Wertheim has prepared the ground with his TPT developments [1–6] and, today, SAFT developers take advantage of the current computational speeds, use realistic intermolecular potentials, and develop versatile SAFT force fields for MD simulations in a variety of important applications.

The motion of particles simulating real molecules in an MD box is dictated by the attractive and repulsive forces or the force field exercised by the other particles in the box. The moving particles strive to minimize the overall potential of the system and reach equilibrium. Current atomistic-level MD simulations use highly advanced universal force fields that capture intermolecular interactions in much detail at the atomic level, the covalent bonds and bond angles. When the description is at this atomic level, the force field is named *all-atom* force field. The use of such a force field is, however, computationally expensive with lengthy computation times. In a compromising approach, the force field could become more affordable by uniting atoms or lumping carbon atoms with their hydrogens into one independent interaction center (one bead per heavy atom) and capture intermolecular interactions at the level of these beads. This force field is named *united-atom* force field. The next level of approximation in this process is to treat the segment or monomer as one bead and capture interactions at the level of segments with more than one heavy atoms, reminiscent of the SAFT division of molecules in segments. This force field is referred to as *coarse-grained* (CG) force field. CG force fields lead to significantly faster computer simulations due to the much-reduced number of interactions to determine and to faster diffusion of particles. From the practical point of view, it is also important that the CG-level interaction potentials are usually softer, thus allowing for increased time steps in MD simulations, access to larger length scales as well as to longer time scales, which are essential in demanding applications.

It is at this coarse-grained level that SAFT has become a very useful tool for the parameterization and development of efficient force fields for MD simulations. SAFT force fields are parameterized with the so-called *top-down* approach, in which models of lower resolution or macroscopic experimental data are typically used and correlated for obtaining the CG force field parameters. SAFT force fields are obtained via a two-step process. First, the form of the intermolecular potential is selected and all SAFT analytical equations for the macroscopic properties are expressed in terms of the intermolecular potential parameters. In the second step, these analytical equations are used to correlate the experimental data over a range of external conditions. In this second step, the potential parameters are determined via a classical minimization of an objective function with the deviations of calculations from experimental data. These parameters are, then, transferred to the CG force field (SAFT force field) for running MD simulations. The above is done in a way guaranteeing a one-to-one relation between the CG force field parameters and the corresponding macroscopic properties. With these MD simulations other thermophysical properties or fluid structures may be obtained, which could not be easily obtained directly from SAFT analytical equations.

Apart from robustness, the criteria for good CG descriptions are the *representability* of the reference experimental set or the reference model and the *transferability* to new conditions and new models, such as the transfer from a single bead to an m-bead molecule or to combinations of different beads in different ways in a molecule. All these are done successfully in the SAFT CG process [27–29,32]. Representative properties such as vapor pressures and saturated liquid densities, are calculated much faster with analytical SAFT equations than by MD simulations and the parameters of the potential are optimized over a broader range of states for improving transferability. These important transferability features of SAFT force fields are directly related to the very structure of the model based on Wertheim's TPT developments. But it took quite some time until SAFT was reformulated in terms of a sufficiently versatile continuous intermolecular potential and a realistic SAFT force field could be obtained. With continuous potential, both MC and MD simulations can be run.

In 1997, Gil-Villegas et al. [27] developed the SAFT-VR version with **v**ariable-**r**ange (VR) potentials. The Mie potential was mainly used in the subsequent SAFT-VR-Mie versions. In 2006, Lafitte et al. [28] demonstrated the better performance of the SAFT-VR-Mie model over previous SAFT versions (PC-SAFT, SAFT-VR-SW, etc.) for the description of vapor-liquid equilibria and derivative properties (compressibility, thermal expansivity, speed of sound). This version was further expanded and improved in 2013 by Lafitte et al. [29] and is referred to as SAFT-VR-Mie 2013 version. In parallel, the SAFT-γ Mie version was developed in a group-contribution formulation to treat heteronuclear molecules formed from segments of differing chemical functionality [33–35]. These two SAFT versions are common today for the development of SAFT force fields.

These SAFT CG force fields are based on the Mie intermolecular potential for spherical isotropic beads of the following general form:

$$u(r) = C\varepsilon\left[\left(\frac{\sigma}{r}\right)^{\lambda_r} - \left(\frac{\sigma}{r}\right)^{\lambda_a}\right] \tag{33.20}$$

where

$$C = \left(\frac{\lambda_r}{\lambda_r - \lambda_a}\right)\left(\frac{\lambda_r}{\lambda_a}\right)^{\lambda_a/(\lambda_r-\lambda_a)} \tag{33.21}$$

In this equation, r is the center-to-center distance between two interacting spherical beads, σ is the length scaling parameter, ε is the minimum depth of the potential well, λ_r is the parameter that controls the steepness of the repulsion and λ_a is the parameter that controls the shape and range of the attractive part of the potential. Individual beads may be connected in linear chains of m beads. As seen, the Mie potential is a generalization of the classical Lennard-Jones (12–6) potential (cf. Chapter 29). The possibility to change λ_r and λ_a independently makes Mie potential a versatile one and the same holds true for the CG SAFT force fields based on it. Appropriate mixing rules apply in the case of mixtures leading to transferable cross-interaction parameters valid over a broad range of external conditions. In this way, a broad spectrum of challenging applications is reached by SAFT, which would be rather difficult for a plain equation-of-state model to reach [35].

THE CUBIC-PLUS-ASSOCIATION (CPA) AND ESD EQUATIONS OF STATE

As mentioned above, the heart of the SAFT models is the association term, A^{ass}, given by Wertheim's TPT1 [1–6]. This term has also been used in equation-of-state models which do not bear the SAFT name. In this section we will briefly discuss two such examples [21,36,37]. Details may be found in the original works.

At the outset the developments of equation-of-state molecular models, especially in the chemical engineering context, are almost invariably dictated by the need to reconcile molecular theory and simplicity. In numerous product- and process-design applications, the quantity of primordial interest is fugacity whose evaluation must be repeated over and over and should never fail. There are excellent theoretical developments around but rather too complex to make their way into routine engineering calculations. In this spirit, equation-of-state models seeking to improve their capacity in handling complex systems of associating fluids, were combined with Wertheim's association approach, which has added power without affecting severely their simplicity.

The first example is the Elliot–Suresh–Donohue (ESD) model developed in the early 1990s [36–37]. As in SAFT, ESD uses expressions mimicking molecular physics results with sound physical interpretation and yet simple enough for engineering applications. The starting point is the division of the compressibility factor into associating and nonassociating terms:

$$Z = Z^{nass} + Z^{ass} \tag{33.22}$$

The nonassociating term may be split into the ideal term and two terms corresponding to the repulsive and the attractive interactions by adapting a spherical square-well potential for the interactions between nonspherical molecules:

$$Z^{nass} = 1 + Z^{rep} + Z^{att} = 1 + \frac{4c\eta}{1-1.9\eta} + \frac{z_m q\eta Y}{1+k_1\eta Y} \quad \text{(ESD)} \tag{33.23}$$

where η is the reduced density, c and q are shape factors, $Y=\exp(\varepsilon/kT)-k_2$, ε is the square-well depth, and z_m, k_1, and k_2 are model constants.

The association term, obtained from Wertheim's TPT, reads:

$$Z^{ass} = -\frac{1-X^A}{1-1.9\eta} \text{ (TPT1)} \tag{33.24}$$

where for a two-site associating molecule, the extent of HB (cf. Chapter 32) is given by:

$$X^A = \frac{N_T}{N_0} = \frac{2}{1+\sqrt{1+4\rho_0\Delta_{AB}}} \tag{33.25}$$

and

$$\rho_0\Delta_{AB} = \rho_0 g^{(\sigma)} f_{AB}\kappa_{AB} = \frac{\eta_0}{1-1.9\eta_0}\frac{\kappa_{AB}}{v^*}\left[\exp\left(\frac{\varepsilon_{HB}}{RT}\right)-1\right] \tag{33.26}$$

N_T is the true number of moles, N_0 is the superficial (no-association) number in the case of the two-site model, v* is the volume parameter of the model, ε_{HB} is the HB square-well depth, $g^{(\sigma)}$ the pair-correlation function at contact, f_{AB} the Mayer function (cf. Chapter 29) and κ_{AB} is a measure of bonding volume available to the molecule, as in SAFT. The subscript 0 in the last two equations implies that all properties of these terms can be evaluated as if association is ignored.

The nonassociating part of this equation is cubic. Alternative expressions for the association term have been derived for fluids with three or four-site association interactions. In contrast to purely empirical equations of state, ESD possesses one degree of sound theoretical basis. The model has been tested successfully with phase equilibrium data of a variety of associating fluid systems.

The next example is the CPA model [21,38]. CPA also combines association and nonassociation terms. For the nonassociation part, simple cubic equations of state, mainly, the Soave–Redlich–Kwong (SRK) or the Peng–Robinson (PR) ones, are used. The association term is taken from Wertheim's TPT1 [1–6]. Following the cubic-equation practice, CPA is expressed in terms of pressure as follows:

$$P = P^{nass} + P^{ass} \tag{33.27}$$

$$P^{nass} = \frac{RT}{V-b} - \frac{a(T)}{V(V+b)} \tag{33.28}$$

and

$$P^{ass} = -\frac{RT}{2V}\left[1+\frac{1}{V}\frac{\partial \ln g}{\partial(1/V)}\right]\sum_i x_i \sum_{A_i}\left(1-X^{Ai}\right) \tag{33.29}$$

As usual in cubic equations, V, a, and b are the molar volume, the interaction energy parameter and the co-volume, respectively. X^{Ai} is the mole fraction of component i not bonded at site A_i. and is obtained from Equation 33.10 (cf. Chapter 32 for different association types). Initially, g was taken as the Carnahan–Starling hard-sphere RDF but later [38], the following reduced hard-sphere RDF was adopted:

$$g = \left[1-\frac{1.9b}{4V}\right]^{-1} \tag{33.30}$$

The addition of the association term makes the equation noncubic but still easily solvable. In addition to the three cubic-equation constants, two more parameters per hydrogen-bond type are needed as above in SAFT.

This relatively simple association model has been tested extensively for a variety of systems over a broad range of external conditions and, very often, with gratifying results. Of much interest is an extensive comparison of CPA with eight versions of the representative and most commonly used PC-SAFT and SAFT-VR models for their capacity to describe water in the entire liquid region and at supercritical conditions and for densities and second-order properties (response functions) over a very broad range of temperature and pressure [39]. The findings did not permit to draw conclusions on the accuracy and general performance of these equations because different equations reproduce reasonably well different properties and at different thermodynamic conditions but fail in other instances. An association-dependent PC-SAFT version, especially designed and parameterized for water exhibits somewhat better performance but still cannot be considered as overall reliable.

An in-depth assessment of TPT and SAFT may be found in Ref. [40].

REFERENCES

1. Wertheim, M. S. 1984. Fluids with highly directional attractive forces. I. Statistical thermodynamics. *J. Stat. Phys.* 35(1/2): 19–34.
2. Wertheim, M. S. 1984. Fluids with highly directional attractive forces. II. Thermodynamic perturbation theory and integral equations. *J. Stat. Phys.* 35(1/2): 35–47.
3. Wertheim, M. S. 1986. Fluids with highly directional attractive forces. III. Multiple attraction sites. *J. Stat. Phys.* 42(3/4): 459–476.
4. Wertheim, M. S. 1986. Fluids with highly directional attractive forces. IV. Equilibrium polymerization. *J. Stat. Phys.* 42(3/4): 477–492.
5. Wertheim, M. S. 1986. Fluids of dimerizing hard spheres, and fluid mixtures of hard spheres and dispheres. *J. Chem. Phys.* 85(5): 2929–2936.
6. Wertheim, M. S. 1987. Thermodynamic perturbation theory of polymerization. *J. Chem. Phys.* 87(12): 7323–7331.
7. Chapman, W. G., Gubbins, K. E., Jackson, G., and Radosz, M. 1989. SAFT: Equation-of-state solution model for associating fluids. *Fluid Phase Equilib.* 52: 31–38.
8. Chapman, W. G., Gubbins, K. E., Jackson, G., and Radosz, M. 1990. New reference equation of state for associating liquids. *Ind. Eng. Chem. Res.* 29(8): 1709–1721.
9. Prausnitz, J. M., Lichtenthaler, R. N., and Gomes de Azevedo, E. 1999. *Molecular Thermodynamics of Fluid Phase Equilibria*, 3rd ed. Upper Saddle River, NJ: Prentice Hall, (a) p. 346, (b) p. 290, (c) p. 368, (d) pp. 390–400.
10. Müller, E. A., and Gubbins, K. E. 2001. Molecular-based equations of state for associating fluids: A review of SAFT and related approaches. *Ind. Eng. Chem. Res.* 40(10): 2193–2211.
11. Economou, I. G. 2002. Statistical associating fluid theory: A successful model for the calculation of thermodynamic and phase equilibrium properties of complex fluid mixtures. *Ind. Eng. Chem. Res.* 41(5): 953–962.
12. McCabe, C., and Galindo, A. 2010. SAFT associating fluids and fluid mixtures. In *Applied Thermodynamics of Fluids*, eds. A. R. H. Goodwin, J. V. Sengers, and C. J. Peters. Cambridge, UK: The Royal Society of Chemistry, pp. 215–279.
13. Vega, L. F., and Llovell, F. 2016. Review and new insights into the application of molecular-based equations of state to water and aqueous solutions. *Fluid Phase Equilib.* 416:150–173.
14. Zmpitas, W., and Gross, J. 2016. Detailed pedagogical review and analysis of Wertheim's thermodynamic perturbation theory. *Fluid Phase Equilib.* 428: 121–152.
15. Huang, S. H., and Radosz, M. 1990. Equation of state for small, large, polydisperse, and associating molecules, *Ind. Eng. Chem. Res.* 29(11): 2284–2294.
16. Carnahan, N. F., and Starling, K. E. 1972, Intermolecular repulsions and the equation of state for fluids. *AIChE J.* 18(6): 1184–1188
17. Chapman, W. G., Jackson, G., and Gubbins, K. E. 1988. Phase equilibria of associating fluids. Chain molecules with multiple bonding sites. *Mol. Phys.* 65(5): 1057–1079.
18. Cotterman, R. L., Schwarz, B. J., and Prausnitz, J. M. 1986. Molecular thermodynamics for fluids at low and high densities. Part I: Pure fluids containing small or large molecules. *AIChE J.* 32(11): 1787–1798.

19. Alder, B. J., Young, D. A., and Mark, M. A. 1972. Studies in molecular dynamics. X. Corrections to the augmented van der Waals theory for the square-well fluid. *J. Chem. Phys.* 56(6): 3013–3029.
20. Mansoori, G. A., Carnahan, N. F., Starling, K. E., and Leland, T. W. 1971. Equilibrium thermodynamic properties of the mixture of hard spheres. *J. Chem. Phys.* 54(4): 1523–1525.
21. Kontogeorgis, G. M., Voutsas, E. C., Yakoumis, I. V., and Tassios, D. P. 1996. An equation of state for associating fluids. *Ind. Eng. Chem. Res.* 35(11): 4310–4318.
22. Veytsman, B. A. 1990. Are lattice models valid for fluids with hydrogen bonds? *J. Phys. Chem.* 94(23):8499–8500.
23. Panayiotou, C., and Sanchez, I. C. 1991. Hydrogen bonding in fluids: An equation-of-state approach. *J. Phys. Chem.* 95(24):10090–10097.
24. Gupta, R. B., and Johnston, K. P. 1994. Lattice fluid hydrogen bonding model with a local segment density. *Fluid Phase Equilib.* 99: 135–151.
25. Gupta, R. B., Kumar, R., and Betageri, G. V. 1997. Phase behavior of mixtures containing antibiotics. Chloramphenicol partitioning. *Ind. Eng. Chem. Res.* 36(9): 3954–3959.
26. Gross, J., and Sadowski, G. 2001. Perturbed-chain SAFT: An equation of state based on a perturbation theory for chain molecules. *Ind. Eng. Chem. Res.* 40(4): 1244–1260.
27. Gil-Villegas, A., Galindo, A., Whitehead, P. J., Mills, S. J., Jackson, G., and Burgess, A. N. 1997. Statistical associating fluid theory for chain molecules with attractive potentials of variable range. *J. Chem. Phys.* 106(10): 4168–4186.
28. Lafitte, T., Bessieres, D., Piñeiro, M. M., and Daridon, J.-L. 2006. Simultaneous estimation of phase behavior and second-derivative properties using the statistical associating fluid theory with variable range approach. *J. Chem. Phys.* 124(2): 024509.
29. Lafitte, T., Apostolakou, A., Avendaño, C., Galindo, A., Adjiman, C. S., Müller, E. A., and Jackson, G. 2013. Accurate statistical associating fluid theory for chain molecules formed from Mie segments. *J. Chem. Phys.* 139(15): 154504.
30. Dominik, A., Jain, S., and Chapman, W. G. 2007. New equation of state for polymer solutions based on the statistical associating fluid theory (SAFT)-Dimer equation for hard-chain molecules. *Ind. Eng. Chem. Res.* 46(17): 5766–5774.
31. Tan, S. P., Adidharma, H., and Radosz, M. 2008. Recent advances and applications of statistical associating fluid theory. *Ind. Eng. Chem. Res.* 47(21): 8063–8082.
32. Müller, E. A., and Jackson, G. 2014. Force-field parameters from the SAFT-γ equation of state for use in coarse-grained molecular simulations. *Annu. Rev. Chem. Biomol. Eng.* 5: 405–427.
33. Lymperiadis, A., Adjiman, C. S., Galindo, A., and Jackson, G. 2007. A group contribution method for associating chain molecules based on the statistical associating fluid theory (SAFT-γ). *J. Chem. Phys.* 127(23): 234903.
34. Lymperiadis, A., Adjiman, C. S., Jackson, G., and Galindo, A. 2008. A generalization of the SAFT-γ group contribution method for groups comprising multiple spherical segments. *Fluid Phase Equilib.* 274 (1–2): 85–104.
35. Fayaz-Torshizi, M., and Müller, E. A., 2021. Coarse-grained molecular dynamics study of the self-assembly of polyphilic bolaamphiphiles using the SAFT-γ Mie force field. *Mol. Syst. Des. Eng.* 6(8): 594–608.
36. Elliott Jr., J. R., Suresh, S. J., and Donohue, M. D. 1990. A simple equation of state for nonspherical and associating molecules. *Ind. Eng.Chem. Res.* 29(7): 1476–1485.
37. Suresh, S. J., and Elliott Jr., J. R. 1992. Multiphase equilibrium analysis via a generalized equation of state for associating mixtures. *Ind. Eng. Chem. Res.* 31(12): 2783–2794.
38. Kontogeorgis, G. M., Yakoumis, I. V., Meijer, H., Hendriks, E., and Moorwood, T. 1999. Multicomponent phase equilibrium calculations for water-methanol-alkane mixtures. *Fluid Phase Equilib.* 158–160: 201–209.
39. Nezbeda, I., Klajmon, M., and Hrubý, J. 2022. Thermodynamic properties of water from SAFT and CPA equations of state: A comprehensive assessment. *J. Mol. Liq.* 362: 119769.
40. Nezbeda, I. 2020. On molecular-based equations of state: Perturbation theories, simple models, and SAFT modeling. *Front. Phys.* 8: 287.

34 The Activity of Individual Ions—Past, Present and Future *In Memory of Dr. Dan Fraenkel, April 1, 1946–March 8, 2023*

Abstract

Values of the single-ion activity coefficients (SIAC) in aqueous solutions obtained measuring the potential difference between an ion-selective electrode and a reference electrode, immersed in the aqueous electrolyte solution, are presented. The contact between the solution in the reference electrode and the sample solution creates a "junction potential" of rather unknown magnitude which it has been said that it may not allow the determination of the SIACs. The whole chapter is dedicated to demonstrating that with proper calibration, even with changing of the liquid junction, the SIACs are measurable and can be interpreted by physically meaningful theories.

In keeping with the structure of this book, the presentation of the activity of individual ions, which is presently receiving a new impulse in its state of development, is presented here, separate from material belonging to the permanent structure of classical thermodynamics and together with other subjects in progress. For a long period, the measurability of the activity of individual ions was not considered to be settled and some researchers maintained that these measurements were impossible to achieve in concentrated solutions above 0.01 m. A historical perspective of roots of that point of view is given elsewhere [1]. In addition, two discussions available in the literature illustrate the extent of the disagreement that existed on this matter [2–10]. In both cases, the editor of the respective journal closed the debate once the informed reader could judge how much each of the two sides knew about the subject. At the time when this subject was still unsettled, arguments in favor of the measurability of the activity of individual ions were repeatedly presented in the literature [1,2,6,11–30]. In this chapter, we first repeat the material included in the first edition of this book and then present a short overview of recent theoretical advances in the field.

CRUX OF THE PROBLEM

Chapter 18 discussed the direct electrochemical measurement of the mean ionic activity coefficient of an electrolyte in an aqueous solution by measuring the potential difference between the responses of an ion-selective electrode (ISE) sensitive to the cation and an ISE sensitive to the anion. For clarity, some of the information given in Chapter 18 is repeated here. It is an accepted fact that the voltage response $E_{i,k}$ of an ISE sensitive to ion i at a molality $m_{i,k}$ in an aqueous solution, is given by the Nernst equation:

$$E_{i,k} = E_{i,0} + \frac{RT}{Z_i F}\ln\left(a_{i,k}\right) \tag{18.23}$$

DOI: 10.1201/9781003431985-38

The symbols used in this equation, and in the equations that follow, are detailed in Chapter 18. When an ISE for the cation and an ISE for the anion of an electrolyte are both immersed in the same solution at the same time and their responses are measured directly one against the other in a voltmeter, the voltage difference is:

$$\Delta E_k = \Delta E_0 + \frac{RT}{F}\ln\left[a_{+,k}^{\frac{1}{Z_+}}\, a_{-,k}^{\frac{1}{|Z_-|}}\right] \tag{18.24a}$$

Using Equations 18.1b, 18.1c, and 18.7, Equation 18.4a, at a dimensionless value $\tilde{m}_k$ of the solution molality, can be written as:

$$\Delta E_k = \Delta E_0 + \frac{RT}{(\nu_+ Z_+)F}\ln\left[\tilde{m}_k^{\nu}\gamma_\pm^{\nu}\left(\nu_+^{\nu_+}\nu_-^{\nu_-}\right)\right] \tag{18.24b}$$

The term ΔE_0 is a constant if measurements are done in a continuous run by adding electrolyte into the sample solution over a short time. Thus, ΔE_0 can be evaluated by calibration, measuring several points in the dilute region where the mean ionic activity coefficient is given by the type of the extended Debye–Hückel equation:

$$\ln\gamma_\pm^{dil} = -\frac{A_D|Z_+Z_-|I^{1/2}}{1+I^{1/2}}$$

It has been repeatedly proved that the mean ionic activity coefficient values for different electrolytes obtained by these direct electrochemical measurements agree with the values obtained by independent measurements of osmotic coefficients, also presented in Chapter 18.

An interesting result is obtained if instead of connecting directly both ISEs to the voltmeter and reading one response against the other, each of the two voltage responses is separately measured against a half-cell single-junction Ag/AgCl reference electrode, as shown in Figure 18.4. The half-cell reference electrode is immersed in an internal standard solution that leaks into the sample solution. A 4 M KCl solution is commonly, but not necessarily, used as internal standard. A liquid junction potential $E_{J,k}$ is created at the tip of the reference electrode, where the reference solution is in contact with the sample solution. This contact of the two solutions is necessary for closing the electric circuit.

In this case, the voltage response of an ISE takes the form:

$$E_{i,k}^R = E_{i,0}^R + \frac{RT}{Z_i F}\ln\left(a_{i,k}\right) + E_{J,k} \tag{34.1}$$

So for each of the electrodes,

$$E_{+,k}^R = E_{+,0}^R + \frac{RT}{Z_+ F}\ln\left(a_{+,k}\right) + E_{J,k}$$

$$E_{-,k}^R = E_{-,0}^R - \frac{RT}{|Z_-|F}\ln\left(a_{-,k}\right) + E_{J,k}$$

The superscript R indicates that in this case, the voltage response is measured against a reference electrode. The junction potential has the same value for both responses in simultaneous measurements. Taking the difference of both independent readings, we obtain:

$$\Delta E_k^R = \Delta E_0^R + \frac{RT}{F}\ln\left[a_{+,k}^{1/Z_+}\; a_{-,k}^{1/|Z_-|}\right] \tag{34.2}$$

Equation 34.2 is formally identical to Equation 18.24a. By independent calibration of the values of $E^R_{+,0}$ and $E^R_{-,0}$, the value of ΔE^R_0 is obtained, so values for the mean ionic activity coefficients of the electrolyte can be calculated from the independent measurements of the two ISEs. These values of the mean ionic activity coefficient agree with the values obtained by the direct electrochemical procedure and with the values obtained from measurements of the osmotic coefficient. This result is not surprising at all, since it is clear that the new element $E_{J,k}$ introduced by the presence of a junction potential cancels out in this exercise. Although the value of $E_{J,k}$ changes from one experimental point to the next because the sample solution changes composition, in simultaneous measurements its value is the same for both ISEs. Even accepting this fact, one cannot avoid the feeling that the expression for the individual responses of the cation and the anion ISEs, written as:

$$E^R_{i,k} = E^R_{i,0} + \frac{RT}{Z_i F}\ln\left(\tilde{m}_{i,k}\gamma_{i,k}\right) + E_{J,k} \tag{34.1a}$$

contains the information of the individual activity of the ions $\tilde{m}_{i,k}\ \gamma_{i,k}$, or more specifically, about the individual activity coefficient of the ions $\gamma_{i,k}$. Almost a century ago, Harned [31] called attention to the main problem here, stating at the beginning of his manuscript, "Since the difficulty in estimating liquid junction potentials *is the only cause* of our not being able to estimate exactly the individual activity of ions, an analysis of the problem of liquid junction potentials will be made" (emphasis ours). Admittedly, the agreement of the results of the independent measurement of the two ISEs with the known values of the mean ionic activity coefficients of the electrolyte is a necessary condition for accepting the goodness of the measurements, but it is not a sufficient condition for extracting from that information the value of the activity coefficients of the individual ions. Harned's difficulty is resolved only if it can be demonstrated that the value of the junction potential present in the experimental measurements of the individual ISE, against a half-cell single-junction reference electrode, does not significantly affect the results obtained. This is an important point to keep in mind. Before addressing this point, however, it is necessary to dispel a wrong concept that seems to be behind the objection to the measurability of the activity of the individual ions.

The chemical potential is defined as the partial derivative of the Gibbs energy with respect to the number of moles of species i at constant temperature and pressure, and the number of all other species in the system:

$$\mu_i = \left(\frac{\partial G}{\partial n_i}\right)_{T,\,P,n_{j\neq i}} \tag{7.1}$$

On the other hand, for an ion i, the chemical potential is related to its activity by:

$$\mu_i = \mu_i^{\theta} + RT\ln\left(a_{i,\theta}\right) \tag{12.12a}$$

In an aqueous solution of an electrolyte, due to the condition of electroneutrality, it is not possible to change the number of moles of one ion and keep constant the moles of all other ions. Thus, one may argue that it is not possible to measure the activity of an individual ion. This is a seemingly convincing argument, indeed.

The inconsistency of this argument is better understood considering the case of the measurement of the activity coefficient of species i in a mixture of nonelectrolytes. As discussed in Chapter 16, this is done by measuring the equilibrium pressure and the compositions of the vapor and liquid phases of a binary system at constant temperature. Equation 16.2 can be written as:

$$\ln\left(x_i\gamma_{i,0}\right) = \ln\left(\frac{y_i P}{P_i^s}\right) - \left[\ln\left(\frac{\varphi_i^{0,s}}{\varphi_i^V}\right) + \frac{v_i^{0,L}\left(P - P_i^s\right)}{RT}\right] \tag{16.2b}$$

In a more detailed consideration, however, in vapor–liquid equilibrium of a binary mixture, according to the phase rule, there are two degrees of freedom. Thus, only two variables can be fixed or changed independently. At constant temperature, the pressure cannot be kept constant if the number of moles of one of the species is changed and the number of moles of the second species is kept constant. Despite this fact, nobody would argue that because one cannot change the number of moles of one compound and keep the pressure and temperature constant, as required by the definition of the chemical potential by Equation 7.1, the activity of a compound cannot be obtained from Equation 16.2b. Even more, one observes that in the use of Equation 16.2b, there is no change of composition required to obtain the activity of species *i*. At each experimental point, the measurement is carried out without a change in composition. In complete analogy, in the case of the measurement of the activity of species *i* in an aqueous solution of an electrolyte, Equation 34.1a can be written as:

$$\ln\left(\tilde{m}_{i,k}\gamma_{i,k}\right)=\frac{Z_iF}{RT}\left(E_{i,k}^R-E_{i,0}^R\right)-\left[\frac{Z_iF}{RT}E_{J,k}\right] \tag{34.1b}$$

Again, in this case, there is no change of composition required to obtain the activity of species *i*. In both Equations 16.2b and 34.1b, the first term of the right-hand side contains the measured variables and gives the main contribution to the activity of species *i*, while the second term represents a small correction, and it is usually evaluated using appropriate correlations. The truth is that this later statement is accepted without discussion for the case of nonelectrolytes, and it was precisely the matter under discussion for the case of electrolytes. In addition, as discussed in the next section, changing the nature of the liquid junction in different measurements demonstrates the negligible effect of the junction potential on the experimental values of the individual ionic activities.

Results presented from here on are those obtained by members of the McGill University and Universidad de Satiago de Compostela research groups under the direction of the senior author of this text. These results are compared with literature values from other researchers in the later part of this chapter. Measurements were performed in continuous isothermal runs carried out in short time spans. It is important to realize that ISEs are transient devices. They change with time and so does their response. The same ISE immersed in the same electrolyte aqueous solution at the same temperature will give different responses on different days.

METHODS USED TO REDUCE THE EXPERIMENTAL DATA

Equation 34.1a can be written as:

$$\xi_{i,k}\equiv\left(E_{i,k}^R-E_{J,k}\right)=E_{i,0}^R+S_i\ln\left(\tilde{m}_{i,k}\gamma_{i,k}\right) \tag{34.1c}$$

For a Nernstian ISE, the slope is $S_i=RT/Z_iF$, and for a set of measurements done in a reasonably short time, $E_{i,0}^R$ is a constant value. If an approximate value for $E_{J,k}$ is entered, the value of $\xi_{i,k}$ can be calculated for each experimental point from the known measured values of $E_{i,k}^R$. Equations for the junction potential are discussed in the last section of this chapter. The first studies [11] used an approximation to the junction potential proposed by Henderson [32]. For the purposes of this presentation, at this point, the analytical form of this approximation is immaterial. At first, the following indirect approach was used.

Indirect Electrochemical Approach

As a way of avoiding the evaluation of $E_{i,0}^R$ and imposing the Nernstian value for S_i, two arbitrary points of the same experimental run, at molalities $\tilde{m}_{i,p}$ and $\tilde{m}_{i,q}$ were selected and Equation 34.1c was used to eliminate these two terms, thus obtaining:

$$\zeta_{i,k} = \frac{\xi_{i,k} - \xi_{i,p}}{\xi_{i,q} - \xi_{i,p}} = \alpha_i + \beta_i \ln\left(\tilde{m}_{i,k}\gamma_{i,k}\right) \quad (34.3)$$

with

$$\alpha_i = \frac{-\ln\left(\tilde{m}_{i,p}\gamma_{i,p}\right)}{\ln\left(\tilde{m}_{i,p}\gamma_{i,p} / \tilde{m}_{i,q}\gamma_{i,q}\right)}$$

and

$$\beta_i = \frac{1}{\ln\left(\tilde{m}_{i,p}\gamma_{i,p} / \tilde{m}_{i,q}\gamma_{i,q}\right)}$$

For each experimental run, the values of α_i and β_i, although unknown, are constant. These two values were then obtained by calibration of the response of the electrode in the region below 0.01 m. In this very diluted region, data of ionic activities exist [33], and it has long been recognized that these values are well represented by the Guntelberg equation [34]. For monovalent ions, this equation gives:

$$\ln \gamma_{i,k}^{(\tilde{m}<0.01)} = \frac{-1.1762\tilde{m}^{0.5}}{1+\tilde{m}^{0.5}}$$

The important point here is that, as shown later in this chapter, the values obtained for the activity of the individual ions in concentrated solutions are repeatable in different experimental runs and independent of the two points selected in each run to eliminate $E_{i,0}^R$ and S_i.

Direct Electrochemical Approach

Realizing that the above approach to reduce the data assumes that $E_{i,0}^R$ and S_i are constant values for a particular run, Equation 34.1c was used directly to evaluate $E_{i,0}^R$ and S_i for each run of experiments. Measurements were repeated [14], and instead of eliminating these two constants, their values were obtained by calibration of the response of the electrode for compositions below 0.01 m. Within the accuracy suggested by the first indirect approach, the results were in total agreement. An agreement in the activity coefficient around ±0.02 is normally obtained.

As both methods handle differently the contribution of the junction potential, the close agreement of the results obtained with the two methods for reducing the data was considered an indication of the negligible influence of the approximation used for the junction potential.

EFFECT OF AN ERROR IN THE SIGN OF THE JUNCTION POTENTIAL

However, it may be argued that although both methods handle the data in different ways, both still rely on the same approximation to the junction potential. At this point, one fortuitous event provided an additional element for estimating the interference introduced by the approximation used to calculate the junction potential. Armin C. Schneider [35], working independently at Institut fur Energie und Umweltverfahrenstechnik, Universitat Duisburg, Germany, realized that the approximation for the junction potential used in the previous calculations was written in the "potentiometry convention," while the ISE and the reference electrode were connected in the "pH convention" [2]. That is, the correction for the junction potential had been used with the wrong sign. This caused a recalculation of all measured values [2]. The most notable observation coming out of this exercise was that the error of sign in the approximation to the junction potential did not change the results up to concentrations of 2 m for the case of the ions potassium and chloride in solutions of KCl. For the ions sodium and chloride in NaCl solutions, on the other hand, errors were only slightly above the expected accuracy up to concentrations of 0.5 m and increased thereafter. This unplanned event had three positive outcomes. In the first place, it confirmed previous theoretical predictions [20] that an error in the estimate of the junction potential would affect the results more in NaCl solutions than in KCl solutions. It also suggested new ways of carrying out the experimental measurements, as well as promoted the development of better approximations for the junction potential. These three aspects are addressed in what follows.

The measurement of the activity of the individual ions in KCl aqueous solutions using a half-cell single-junction reference electrode filled with a 4 M KCl standard solution is a very special case of a homoionic system. The largest junction potential happens when the concentrated standard KCl solution leaks into the most dilute sample solutions in the calibration range of the ISE. As the activity coefficient is set to its theoretical value in this range of composition, this error gets absorbed in the value of $E_{i,0}^{R}$ [20]. As the sample solution gets more concentrated in KCl, the value of the junction potential decreases, and it would be null at a sample solution concentration of 4 M.

ADDITIONAL EXPERIMENTAL STUDIES

The previous reasoning suggested the idea of measuring the activity of different ions using not only 4 M KCl but also 2 M NH_4Cl as standard solution of the reference electrode [22]. Additional studies reported results of measurements of the activity of sodium and chloride ions in NaCl solutions using NaCl at different concentrations, 0.5, 1, 1.5, 2, and 3 m, as standard solution of the reference electrode [24]. Finally, in this series of experimental studies, measurements of the activities of potassium and chloride ions in KCl solutions using KCl, NaCl, and CsCl in the reference electrode and of sodium and chloride in NaCl solutions using CsCl in the reference electrode [6] confirmed the fact that the method of calibration of the ISEs absorbs the effect introduced by the approximation made for the junction potential. The negligible effect of the junction potential in the results for potassium and chloride ions in KCl solutions is shown in Figure 34.1, and for the ions sodium and chloride in NaCl solutions is shown in Figure 34.2.

On the side of theory, a study of the activities of potassium and chloride ions in KCl solutions [29] showed that the use of the theoretically sound approximation to the junction potential described below does not significantly change the value of the activity of the individual ions reported previously.

DERIVATION OF EQUATIONS TO CALCULATE THE JUNCTION POTENTIAL

Those only interested in the practical use of the activity of individual ions can go directly to the next section. The treatment that follows is reproduced [2] with permission from John Wiley & Sons. Equations to calculate the junction potential between two solutions of different concentrations are obtained considering that no net electric current is transported by the ions through the junction. As the current would be carried by ions in solution, we consider first the molar flux of ions i, J_i, through a unit surface area A. This flux is proportional to the molar concentration of ions i, C_i, and to the negative of the gradient of the electrochemical potential of species i, μ_i, over the path l, that is, the thermodynamic force acting over the ions i. Thus, we write:

$$J_i = \frac{D_i}{RT} C_i \left(-\frac{\partial \mu_i}{\partial l} \right) \tag{34.4}$$

In Equation 34.4, D_i is the diffusion coefficient of i, R is the gas constant, and T is the absolute temperature. The electrochemical potential of species i has the form:

$$\mu_i = \mu_i^o + RT \ln a_i + Z_i F \Psi \tag{34.5}$$

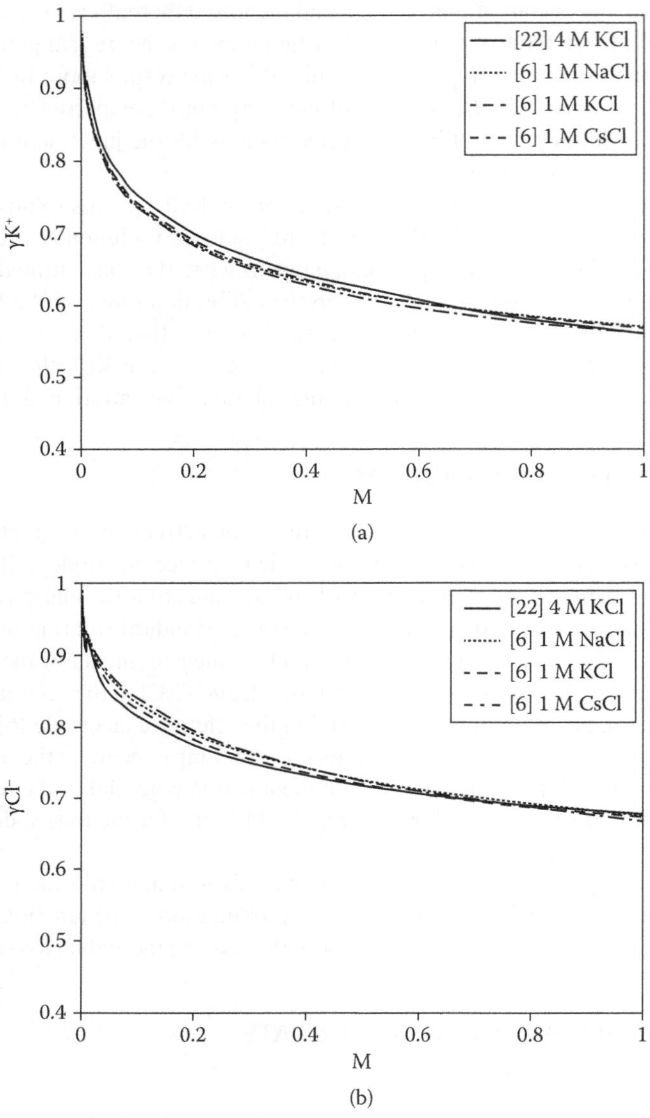

FIGURE 34.1 Effect of different filling solutions on the measurements of the individual activity of ions in an aqueous KCl system at 298.15 K. Square brackets indicate the number of the reference from which data were taken. (a) Activity coefficients of K^+ ion. (b) Activity coefficients of Cl^- ion. Reproduced with permission from Ref. [41].

where a_i and Z_i are the activity and the charge of ion i, respectively; F is the Faraday constant; and Ψ is the electric potential of the liquid phase. Thus, the gradient of the electrochemical potential takes the form:

$$\frac{\partial \mu_i}{\partial l} = RT\frac{\partial \ln a_i}{\partial l} + Z_i F\frac{\partial \Psi}{\partial l} \tag{34.6}$$

To continue, it is necessary to express the diffusion coefficient in terms of the molar ionic conductivity of ions moving under the effect of an electric potential only. For this, it is useful to define the

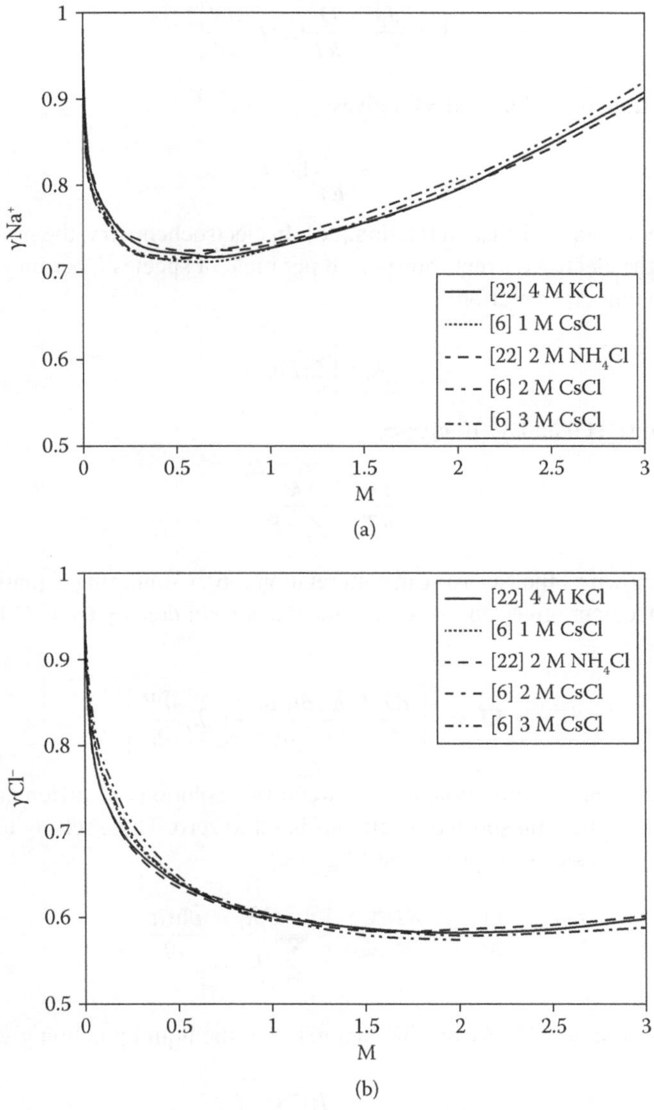

FIGURE 34.2 Effect of different filling solutions on the measurements of the individual activity of ions in an aqueous NaCl system at 298.15 K. Square brackets indicate the number of the reference from which data were taken. (a) Activity coefficients of Na^+ ion. (b) Activity coefficients of Cl^- ion. Reproduced with permission from Ref. [41].

drift speed s_i attained by ion i when the drag forces balance the forces acting over the ion. This drift speed is proportional to the gradient of the applied electrical field; thus,

$$s_i = \frac{|J_i|}{C_i} = u_i \left| \frac{\partial \Psi}{\partial l} \right| \tag{34.7}$$

The proportionality constant u_i, known as the electrical mobility of ion i, represents the limiting steady-state velocity attained by ion i under a unit of potential gradient. From Equations 34.4 and 34.6, for an ion moving under the effect of an electric field only, in the absence of a concentration gradient in the solution, the drift speed takes the form:

$$s_i = \frac{|J_i|}{C_i} = \frac{D_i}{RT}|Z_i|F\left|\frac{\partial \Psi}{\partial l}\right| \tag{34.8}$$

Hence, combining Equations 34.7 and 34.8 gives:

$$u_i = \frac{D_i}{RT}|Z_i|F \tag{34.9}$$

Equation 34.9 is known as the Einstein relation [36]. In electrochemistry, the molar ionic conductivity λ_i is defined as the electric current transported per mole of species i, per unit electrical potential gradient, through a unit cross section.

Thus,

$$\lambda_i = |Z_i|Fu_i \tag{34.10}$$

Combining Equations 34.9 and 34.10 gives:

$$\frac{D_i}{RT} = \frac{\lambda_i}{|Z_i|^2 F^2} \tag{34.11}$$

Equation 34.11 is known as the Nernst–Einstein relation [36]. Combining Equations 34.4, 34.6, and 34.11, after rearrangement, gives an expression for the current density ($Z_i\, F\, J_i$) transported by ions of type i:

$$z_i F J_i = -\left[\frac{RT}{F}\frac{C_i\lambda_i}{Z_i}\frac{\partial \ln a_i}{\partial l} + C_i\lambda_i\frac{\partial \Psi}{\partial l}\right] \tag{34.12}$$

For the calculation of the junction potential between two solutions of different compositions, the sum of the current densities transported by all ions is set to zero. Thus, setting to zero the sum over all ions of the right-hand side of Equation 34.12 gives:

$$\frac{\partial \Psi}{\partial l} = -\frac{RT}{F}\sum_i \frac{1}{Z_i}\frac{C_i\lambda_i}{\sum_j C_j\lambda_j}\frac{\partial \ln a_i}{\partial l} \tag{34.13}$$

Finally, integrating Equation 34.13 over the thickness l of the liquid junction gives:

$$E_J = \Psi_2 - \Psi_1 = -\frac{RT}{F}\sum_i \int_1^2 \frac{t_i}{Z_i} d\ln a_i \tag{34.14}$$

where, to follow the conventional nomenclature, we have used the symbol E_J to indicate the potential difference $\Delta\Psi = \Psi_2 - \Psi_1$, across the junction, and introduced the symbol t_i for the transference number of ion i, defined as:

$$t_i = \frac{C_i\lambda_i}{\sum_j C_j\lambda_j} = \frac{\tilde{C}_i\tilde{\lambda}_i}{\sum_j \tilde{C}_j\tilde{\lambda}_j} = \frac{|Z_i|C_i\tilde{\lambda}_i}{\sum_j |Z_j|C_j\tilde{\lambda}_j} \tag{34.15}$$

It is interesting to note that the transference number t_i appears naturally in Equation 34.13. This term, which has caused some confusion in the literature, can be written in alternative ways. In Equation 34.15, the equivalent conductivity $\tilde{\lambda}_i$ and the concentration of ions i, $\tilde{C}_i$, in equivalent per liter, are related to the corresponding molar quantities, λ_i and C_i, by

$$\lambda_i = |Z_i|\tilde{\lambda}_i \tag{34.16}$$

and

$$C_i = \frac{\tilde{C}_i}{|Z_i|} \tag{34.17}$$

An erroneous interpretation of these terms found in the literature is discussed in the section dealing with the Henderson approximation.

The next step is to obtain an integrated form of the equation for the junction potential. For the purposes of this work, we follow the potentiometric convention. In this convention, the ISE electrode is written at the right and the reference electrode at the left. In agreement with the Stockholm convention, the junction potential, in this case, corresponds to the potential of the sample solution (*k*) minus the potential of the solution at the reference electrode (*r*). Thus, the lower limit of the integral in Equation 34.14 is the state of the ion at the reference solution and the upper limit is the state of the ion in the sample solution. Hence, Equation 34.14 takes the form:

$$E_{J,k} = \Psi_k - \Psi_r = -\frac{RT}{F}\sum_i \int_r^k \frac{1}{Z_i} \frac{C_i\lambda_i}{\sum_j C_j\lambda_j} \frac{da_i}{a_i} \tag{34.18}$$

For simplicity we rewrite Equation 34.18 as:

$$E_{J,k} = +\frac{RT}{F}\sum_i \tau_i \tag{34.19}$$

with τ_i defined as:

$$\tau_i = \frac{1}{Z_i}\int_k^r \frac{C_i\lambda_i}{\sum_j C_j\lambda_j} \frac{da_i}{a_i} \tag{34.20}$$

The key problem to determine the activity of ions is the evaluation of τ_i. This topic is discussed below.

Henderson's Approximation

Henderson's approximation [32] is the best-known equation for the junction potential. Thus, we discuss here its relation with the equations presented above. To avoid the problem of the undefined terms in the case that ion *i* is not present at one of the two solutions, Henderson introduced the auxiliary variable α and considered a linear variation of the concentration of the ions with α across the junction:

$$C_i = C_{i,k} + \alpha\Delta C_i \tag{34.21}$$

with

$$\Delta C_i = C_{i,r} - C_{i,k} \tag{34.22}$$

In addition, Henderson introduced two stringent simplifications. He assumed that the activity of ion *i* was equal to its molar concentration and that the molar conductivity was independent of concentration and equal to the molar conductivity of the ion at infinite dilution, λ_i^∞. Thus, setting $a_i = C_i$ and also $\lambda_{i,k} = \lambda_{i,r} = \lambda_i^\infty$ in Equation 34.20 gives:

$$\tau_i^H = \frac{\lambda_i^\infty \Delta C_i}{Z_i\left[\sum_j C_{j,r}\lambda_j^\infty - \sum_j C_{j,k}\lambda_j^\infty\right]} \ln \frac{\sum_j C_{j,r}\lambda_j^\infty}{\sum_j C_{j,k}\lambda_j^\infty} \tag{34.23}$$

Hence, for the schematics of the electrochemical cell written in the potentiometric convention, as it is considered in the case of an ISE, we write:

$$E_{J,K}^H = \frac{RT}{F}\left[\frac{\sum_i \frac{1}{Z_i} C_{i,r}\lambda_i^\infty - \sum_i \frac{1}{Z_i} C_{i,k}\lambda_i^\infty}{\sum_i C_{i,r}\lambda_i^\infty - \sum_i C_{i,k}\lambda_i^\infty}\right] \ln \frac{\sum_i C_{i,r}\lambda_i^\infty}{\sum_i C_{i,k}\lambda_i^\infty} \tag{34.24}$$

The usual form for the Henderson equation found in the literature [37a,b] is written for the pH convention, which considers the ISE to be written at the left and the reference electrode at the right. In addition, it is normally written in terms of the equivalent limiting conductivity, $\tilde{\lambda}_i$, of the ion instead of the molar limiting conductivity, λ_i These two limiting conductivities are related by Equation 34.16. Hence, the form of the Henderson equation found in the literature is as follows:

$$\left(E_{J,K}^H\right)^* = -E_{J,K}^H = \frac{RT}{F}\left[\frac{\sum_i \frac{|Z_i|}{Z_i} C_{i,k}\tilde{\lambda}_i^\infty - \sum_i \frac{|Z_i|}{Z_i} C_{i,r}\tilde{\lambda}_i^\infty}{\sum_i |Z_i| C_{i,k}\tilde{\lambda}_i^\infty - \sum_i |Z_i| C_{i,r}\tilde{\lambda}_i^\infty}\right] \ln \frac{\sum_i |Z_i| C_{i,k}\tilde{\lambda}_i^\infty}{\sum_i |Z_i| C_{i,r}\tilde{\lambda}_i^\infty} \tag{34.25}$$

The ratio of the absolute value of the charge over the value of the charge of ion i, appearing in the numerator inside the brackets of the right-hand side of Equation 34.25, is +1 for cations and −1 for anions. Hence, this term is usually not explicitly written and the sum over all ions is written as the sum over the cations minus the sum over the anions:

$$\left(E_{J,K}^H\right)^* = \frac{RT}{F}\,\frac{\left[\sum C_+\tilde{\lambda}_+^\infty - \sum C_-\tilde{\lambda}_-^\infty\right]_k - Y_1}{\left[\sum C_+\tilde{\lambda}_+^\infty |Z_+| + \sum C_-\tilde{\lambda}_-^\infty |Z_-|\right]_k - Y_2} \times \ln \frac{\left[\sum C_+\tilde{\lambda}_+^\infty |Z_+| + \sum C_-\tilde{\lambda}_-^\infty |Z_-|\right]_k}{Y_2} \tag{34.26}$$

In Equation 34.26, the symbols Y_1 and Y_2 represent numerical values that depend on the solution r used in the reference electrode. In Henderson's approximation, these constants are obtained as:

$$Y_1 = C_{+,r}\tilde{\lambda}_{+,r}^\infty - C_{-,r}\tilde{\lambda}_{-,r}^\infty \tag{34.27}$$

and

$$Y_2 = C_{+,r}\tilde{\lambda}_{+,r}^\infty |Z_+|_r + C_{-,r}\tilde{\lambda}_{-,r}^\infty |Z_-|_r \tag{34.28}$$

At 298.15 K, for a 4 M KCl reference electrode solution, $Y_1 = -11.6$ and $Y_2 = 623$. At this temperature, the value of the ratio (RT/F) is 25.693 (mV). However, more important is to note that the terms in each sum contain the product of the molar concentration times the equivalent limiting conductivity. Standard references in the literature are in error here. In some places, it is indicated that in Equation 34.26 both the concentration and the conductivity are in molar units [37a], while in others it is stated that both the concentration and the conductivity are in terms of equivalents [37b]. Both of these statements are wrong. Although these errors do not affect the value of the junction potential for 1:1 electrolytes, they give wrong results for 1:2 and 2:1 electrolytes. Values tabulated in the literature for these latter cases are in error.

AN IMPROVED EQUATION TO CALCULATE LIQUID JUNCTION POTENTIALS

Two early attempts were made to improve Henderson's equation for the calculation of the liquid junction potential. In 1926, Harned [31] proposed a modification that corrects for the variation of the ionic conductivity with concentration, and in 1985, Harper [38] included the correction for nonideality of the solutions and for the variation of ionic conductivity. While the method of Harned resulted in an analytical solution that does not correct for activities, the method of Harper produced complex expressions that required numerical solutions. In more recent times, a new analytical equation to calculate the value of the liquid junction potential that includes corrections for nonideality and for the variation of conductivity with concentration was proposed [2]. The same mathematical artifice used by Henderson was employed for the integration of Equation 34.20. The integration path followed a linear variation of the product $(C\lambda)_i$ with α. Thus,

$$C_i\lambda_i = (C\lambda)_{i,k} + \alpha\Delta(C\lambda)_i \tag{34.29}$$

with

$$\Delta(C\lambda)_i = (C\lambda)_{i,r} - (C\lambda)_{i,k} \tag{34.30}$$

As the integral in Equation 34.20 depends only on the two limiting electrolyte solutions, the following form, meeting the conditions of being zero at the electrolyte solution k and unity at the reference solution r, was used for the dummy variable α:

$$\alpha = \frac{a_i - a_{i,k}}{\Delta a_i} \tag{34.31}$$

with

$$\Delta a_i = a_{i,r} - a_{i,k} \tag{34.32}$$

With these definitions, Equation 34.20 can be written as:

$$\tau_i = \frac{\Delta a_i}{Z_i} \int_0^1 \frac{\left[(C\lambda)_{i,k} + \alpha\Delta(C\lambda)_i\right] d\alpha}{\left[a_{i,k} + \alpha\Delta a_i\right]\left[\beta_k + \alpha\Delta\beta\right]} \tag{34.33}$$

with

$$\beta_k = \sum_j C_{j,k}\lambda_{j,k} \tag{34.34}$$

and

$$\Delta\beta = \beta_r - \beta_k \tag{34.35}$$

where

$$\beta_r = \sum_j C_{j,r}\lambda_{j,r} \tag{34.36}$$

The integration of Equation 34.33 is analytical, and after rearrangement, it produces the following general expression for τ_i for the case when ion i is present in both solutions k and r:

$$\tau_i = \frac{1}{Z_i}\left[\frac{\left[a_{i,k}(C\lambda)_{i,r} - a_{i,r}(C\lambda)_{i,k}\right]}{\left[\alpha_{i,k}\beta_r - \alpha_{i,r}\beta_k\right]}\ln\frac{a_{i,r}}{a_{i,k}} + \frac{\Delta a_i}{\Delta\beta}\frac{\left[(C\lambda)_{i,k}\beta_r - (C\lambda)_{i,r}\beta_k\right]}{\left[a_{i,k}\beta_r - a_{i,r}\beta_k\right]}\ln\frac{\beta_r}{\beta_k}\right] \tag{34.37}$$

As expected, the integrated form of the term τ_i is only a function of the compositions of the end solutions, and it is independent of the integration path. If an ion i is not present in one of the two solutions, the first term inside the brackets of the right-hand side of Equation 34.37 vanishes. For the case of ion i present only in the reference electrode solution r, its composition and activity are equal to zero in solution k and Equation 34.37 simplifies to:

$$\tau_{i,r} = +\frac{1}{Z_i}\frac{C_{i,r}\lambda_{i,r}}{\Delta\beta}\ln\frac{\beta_r}{\beta_k} \tag{34.38}$$

Similarly, for the case of ion i present only in the sample solution k, Equation 34.37 simplifies to:

$$\tau_{i,k} = -\frac{1}{Z_i}\frac{C_{i,k}\lambda_{i,k}}{\Delta\beta}\ln\frac{\beta_r}{\beta_k} \tag{34.39}$$

Notably, these two expressions are independent of the activity of the ions. Equations 34.19 and 34.37 suffice for the calculation of the liquid junction potential to be used in the measurement of the activity of individual ions. All that is required is to have experimentally determined values of the molar conductivity λ_i of the ions and good initial estimates of the activity of the individual ions present in both solutions forming the liquid junction. There is a clear advantage in using homoionic systems for measuring the activity of ions.

The interesting point to observe here is that to measure the activity of individual ions, it is necessary to know these activities to calculate the junction potential. This is not a "vicious circle," as stated by Guggenheim [39], but is just an iteration loop in the calculations. In fact, recent work [29] used exactly this iteration scheme. The first estimate of the activity coefficient of the individual ions was obtained by reducing the data assuming Henderson's approximation for the junction potential. Another alternative is to follow Harper's suggestion [38] and obtain a first crude estimate of the activity coefficients of the individual ions from the mean ionic activity coefficient of the electrolyte at the concentration of the solution. By direct application of the Debye–Hückel equation,

$$\ln\gamma_{i,k} = \frac{Z_i^2}{|Z_+Z_-|}\ln\gamma_{\pm,k} \tag{34.40}$$

USE OF THE ACTIVITY OF INDIVIDUAL IONS IN MULTI-ION AQUEOUS SOLUTIONS

It is an established fact that in concentrated electrolyte solution, the activity of an individual ion depends on the nature of the counterion present in the solution. Thus, the key problem when dealing with a multiion concentrated aqueous solution is how to combine information gathered by studying the behavior of each ion in systems with a single counterion present. Lin and Lee [40], in an impressively insightful derivation, elegantly solved this problem. Although their derivation was coupled with a new equation the authors were proposing, their final result is totally general. Lin and Lee considered the ionic composition of the multiion aqueous solution to be known by an independent chemical analysis. In their mathematical treatment, the solvent-free mixture of ions is brought from infinite dilution in the solvent to the ionic strength of the multiion solution, as defined by Equation 18.17. Throughout this process, the composition of the solvent-free mix of ions is constant. Lin and Lee clearly presented the algebra in detail, so it is not necessary to repeat it here. The Lin and Lee equation for the osmotic coefficient of a multiion aqueous solution at an ionic strength I_k takes the form:

$$\varphi_k = 1 + \sum_i Y_{i,k}\left[\ln\gamma_{i,k} - \frac{1}{I_k}\int_0^{I_k}(\ln\gamma_i)\,dI\right] \tag{34.41}$$

with

$$Y_{i,k} = \frac{m_{i,k}}{\sum_j m_{j,k}} \tag{34.42}$$

The activity of water is then obtained from Equation 18.11 for a multiion solution, written as:

$$\ln a_{w,k} = -\frac{M_w \sum_j \tilde{m}_{j,k}}{1000}\varphi_k$$

A particularly useful correlation for the activity coefficient of individual ions proposed in the literature [2] has the form:

$$\ln\gamma_{i,k} = \frac{-AZ_i^2 I_k^{0.5}}{1+I_k^{0.5}} + \frac{b_i I_k}{(1+1.5I_k)^2} + d_i I_k \tag{34.43}$$

For aqueous solutions at 298.15 K, $A = 1.1762$ and b_i and d_i are adjustable parameters, the values of which are given in the publication referenced above [2]. The first term of the right-hand side of this equation represents the contribution of long-range interactions, and the second term represents the contribution of the short-range (specific) interactions. Further study using the Lin and Lee equation [26] showed that a particularly simple expression for the osmotic coefficient is obtained with a minor modification of Equation 34.43, namely,

$$\ln\gamma_{i,k} = -AZ_i^2\left[\frac{I_k^{0.5}}{1+I_k^{0.5}} + 2\ln\left(1+I_k^{0.5}\right)\right] + \frac{b_i I_k}{(1+1.5I_k)^2} + d_i I_k \tag{34.44}$$

The second term for the long-range interactions finds justification in the expression for the mean ionic activity coefficients obtained from the radial distribution function,[1] and it has little effect on the fitting of the activities of the individual ions. If necessary, the parameters b_i and d_i can be refitted from those reported in the literature [2]. The expression for the osmotic coefficient obtained using Equation 34.43 and the Lin and Lee equation is:

$$\varphi_k = 1 + \sum_i Y_{i,k}\left\{\begin{array}{c} -AZ_i^2\left[1 - 2I_k^{0.5} + \dfrac{2}{I_k}\ln\left(1+I_k^{0.5}\right)\right] \\ +\bar{b}_{i,k}\left[\dfrac{3I_k+1}{1.5(1+1.5I_k)^2} - \dfrac{1}{1.5^2 I}\ln(1+1.5I_k)\right] + \dfrac{\bar{d}_{i,k}}{2}I_k \end{array}\right\} \tag{34.45}$$

while using Equation 34.44, it is:

$$\varphi_k = 1 + \sum_i Y_{i,k} \left\{ \begin{array}{c} -AZ_i^2 \left[\dfrac{I_k^{0.5}}{1+I_k^{0.5}} \right] \\ +\bar{b}_{i,k} \left[\dfrac{3I_k+1}{1.5(1+1.5I_k)^2} - \dfrac{1}{1.5^2 I} \ln(1+1.5I_k) \right] + \dfrac{\bar{d}_{i,k}}{2} I_k \end{array} \right\} \tag{34.46}$$

For the parameters $\bar{b}_{i,k}$ and $\bar{d}_{i,k}$ of the term representing short-range interactions in a multiion aqueous solution, the following mixing rules are used:

$$\bar{b}_{i,k} = \sum_j X_{j,k} b_{i/j}^0 \tag{34.47}$$

$$\bar{d}_{i,k} = \sum_j X_{j,k} b_{i/j}^0 \tag{34.48}$$

where for an ion i the sums run over all the counterions present, with the counterion charge mole fractions defined as:

$$X_{j,k} = \frac{m_{j,k}|Z_j|}{\sum_l m_{l,k}|Z_l|} \tag{34.49}$$

In Equations 34.47 and 34.48, $b_{i/j}^0$ and $d_{i/j}^0$ are the values of the parameters in the binary i/j ion–counterion system. These parameters are independent of composition. Figure 34.3 presents the results

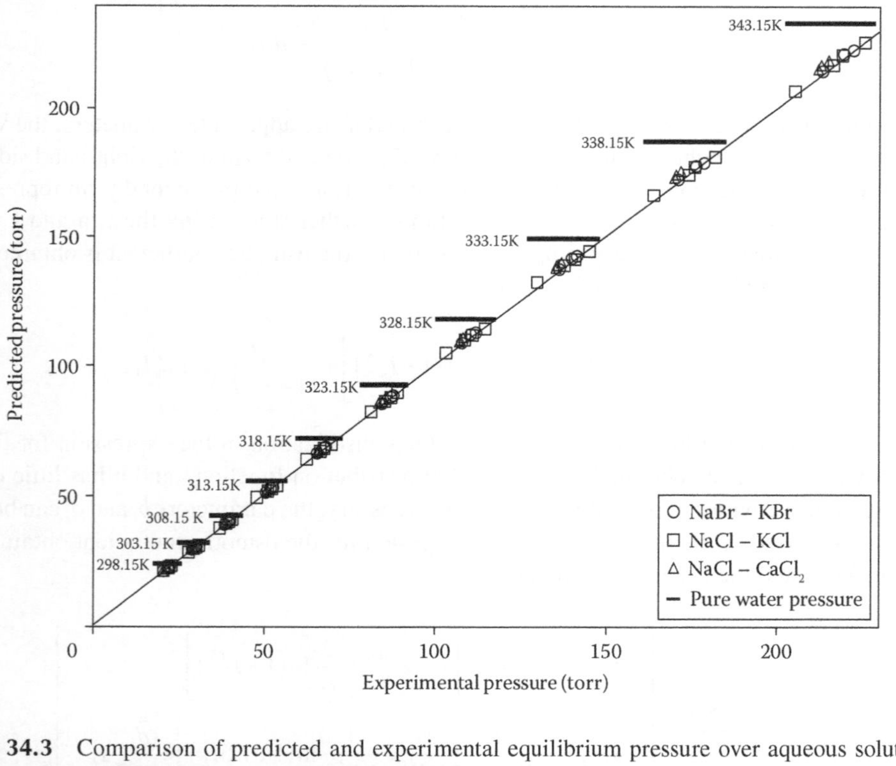

FIGURE 34.3 Comparison of predicted and experimental equilibrium pressure over aqueous solutions of mixed electrolytes with a common ion. The horizontal lines cutting the 45° line indicate the vapor pressure of pure water at each experimental temperature. Reproduced with permission from Ref. [26].

of the prediction of the equilibrium pressure over aqueous solutions of mixed electrolytes with a common ion in comparison with experimental data available in the literature [26]. It is important to point out that the parameters used for the prediction were evaluated at 298.15 K only.

The importance of this new thermodynamic approach for the treatment of electrolyte solutions will be emphasized after demonstrating the consistency of the experimental information.

COMPARISON OF EXPERIMENTAL RESULTS FOR THE ACTIVITY OF INDIVIDUAL IONS OBTAINED BY DIFFERENT RESEARCHERS

Experimental[2] values of the activity of individual ions in concentrated solutions had been measured by several researchers [41]. These experimental results were questioned by some, arguing that such measurements are "impossible" to make. The negative attitude against the mere concept of measuring ionic activities was so strong that previous researchers publishing these experimental results were forced to pay lip service to the prevailing opinion and admit, as the paradigm required, that theory did not support the concept of individual activity of ions [42–44].

First, we consider the results for aqueous solutions of sodium salts. As the measurements reported were performed at different concentrations, and any smoothing equation would introduce its own bias, a graphical comparison is probably the most direct way of proceeding. Figure 34.4 presents the comparison of experimentally measured individual activity coefficients of Na^+ and Cl^- in the aqueous NaCl system determined in five different laboratories with significantly different experimental techniques (see Table 34.1). In addition to our original results from 2004 [2], we also include later measurements done with the improved experimental setups [22] and the results obtained with a Ag/AgCl reference electrode filled with 2 M NH_4Cl instead of a saturated KCl solution, as was done before.

Figure 34.5a depicts comparison of our data with results reported by Hurlen and Breivik [45] for aqueous solutions of NaBr. We used Na ISE and Br ISE. Our measurements were performed at different times and with different inner electrolytes in the single-junction Ag/AgCl reference electrode. Hurlen's measurements of anion data were done with an Ag/AgBr electrode, while the activity coefficients for the Na^+ ions were calculated from the mean ionic activity coefficient data. Figure 34.5b presents data for the system of aqueous Na_2SO_4. In our study [2], we used the Na^+-sensitive glass electrode to measure the activity of the sodium ion. The activities of the sulfate ion were calculated from the mean ionic activity coefficient. Hurlen and Breivik [45], on the other hand, measured only the anion using an Hg/Hg_2SO_4 electrode, and the activity of cation was calculated.

Figures 34.6 and 34.7a depict a comparison of results for 1:1 chlorides LiCl, CsCl, and NH_4Cl, while Figure 34.7b presents the results for the LiBr system. In all these cases, the measurements were done for anions only, while the missing information for cations was calculated from the mean ionic activity coefficient data. Due to the use of Ag/AgCl and Ag/AgBr electrodes, Hurlen [44,45] could extend his measurements up to 4 m, while the use of a Cl ISE and Br ISE limited the range of measurements to 3 m.

Figures 34.8 and 34.9 show comparisons for the aqueous system's alkaline–earth chlorides. As before, only the activity coefficients for anions were determined experimentally. The corresponding information for cations was calculated from the mean ionic activity coefficient data. It is interesting to observe that in the systems of aqueous $MgCl_2$ and $CaCl_2$ at lower concentrations, the activity coefficients of cations are smaller than the activity coefficients of anions, but at higher concentrations this trend is reversed. As the activity coefficients for the $BaCl_2$ were determined below 2 m only, it is difficult to say if the same behavior would be observed in this system as well.

Due to the main assumption that $\gamma_- = \gamma_\pm$ for saturated KCl aqueous solution used by Hurlen [44–47] to calibrate the results of his method, we can expect a discrepancy between his and our data. The comparison presented in Figure 34.10 confirms this prediction. On the other hand, the agreement of our data with data of Schneider et al. [35] is quite satisfactory.

In Figure 34.10, we have included our oldest data from 2004 [2], as well as newer data [22] in which two different inner solutions were used in the single-junction Ag/AgCl reference electrode. It is clear that with the increasing experience of the experimenters, the quality of our data improved.

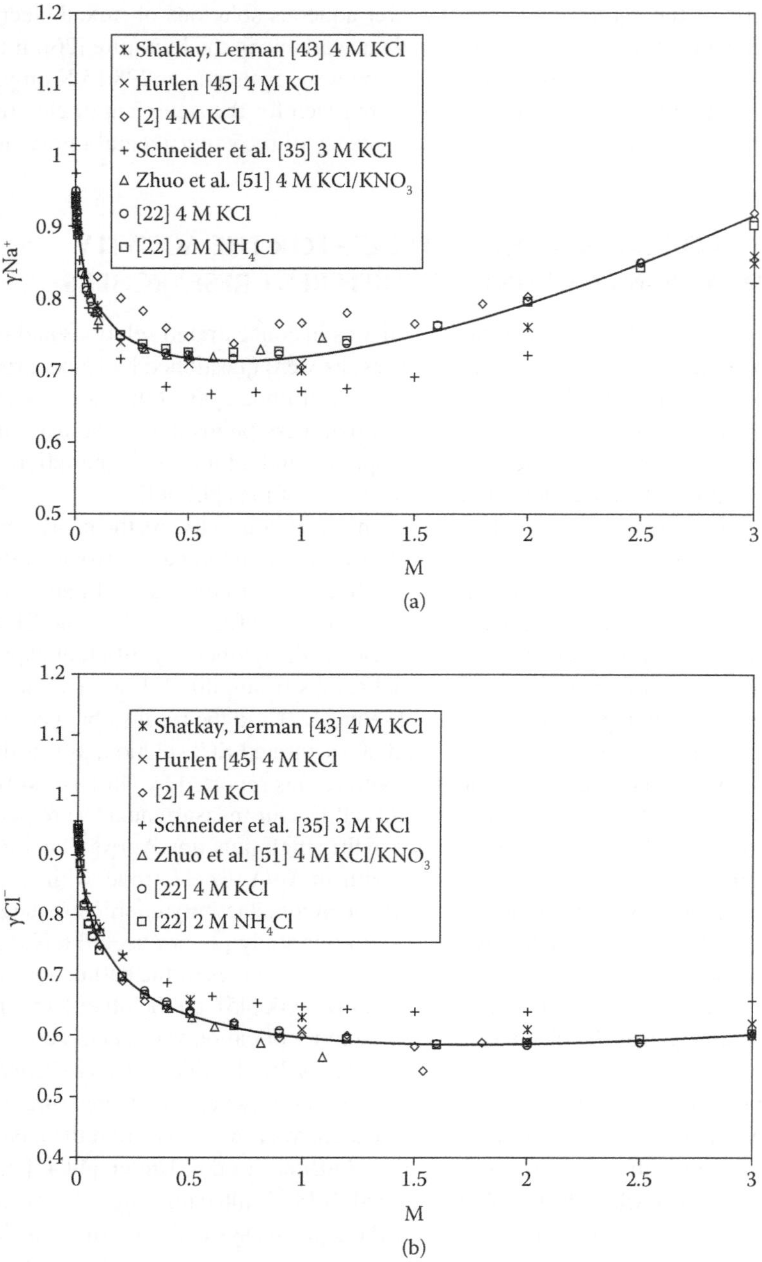

FIGURE 34.4 Comparison of different experimental results obtained for the individual ionic activity coefficients in the aqueous solutions of NaCl. (a) Individual activity coefficients of Na^+. (b) Individual activity coefficients of Cl^-. (Inner solutions of reference electrodes are listed in the corresponding reference.) Reproduced with permission from Ref. [41].

Figure 34.11 presents comparison of data measured by different authors in the years 1935–2011. The lines on the graph correspond to our smoothed data. It is clear that there are big discrepancies of different data sets. The data of Sakaida and Kakiuchi [48] plotted together with their error limits and the data of Schneider et al. [49] seemed to be in mutual agreement. The data of Szabo [50], on the other hand, show a reverse relation compared with all other data between the activity coefficients of H^+ and those of Cl^-.

TABLE 34.1
Summary of Different Techniques Used for the Measurements of Ionic Activity Coefficients

ISE Cation	ISE Anion	Reference Electrode	Runs	Calibration
		Szabo, 1935 [50]		
Pt/H_2, HCl	Hg/HgCl, HCl	Liquid chains of cells of identical electrodes were measured	Discontinuous	At low concentration range, junction potential evaluated through diffusion potentials measured in the same study
		Shatkay and Lerman, 1969 [42,43]		
Glass sodium electrode	Ag/AgCl	Standard calomel electrode (SCE), KCl bridge	Discontinuous	By taking $\gamma_{\pm}=0.903$ for 0.01 m NaCl; E_J Henderson [32]
		Hurlen, 1979–1983 [44–47]		
Calculated	Ag/AgCl electrode [44,46] Ag/AgBr electrode [45] Hg/Hg_2SO_4 electrode [47]	SCE, SJ, inner solution sat. KCl	Discontinuous	$\gamma_{-}=\gamma_{\pm}$ for sat. KCl aq.; E_J Henderson [32]
		Khoshkbarchi and Vera, 1996 [11]		
Na ISE K ISE	Cl ISE, Br ISE	Ag/AgCl, SJ, inner solution sat. KCl	Continuous	Based on 2 points at intermediate concentrations; E_J Henderson [32]; wrong sign
		Schneider et al., 2003–2004 [35,49]		
Na ISE, K ISE Ca ISE	Cl ISE	Ag/AgCl, inner solution: 3 M KCl [35] 3 M KCl [49]	Continuous	Extended Debye–Hückel equation used up to $I_m=0.01$ m; (1) E_o and S fitted; (2) S Nernstian, E_o fitted; E_J Henderson [32]
		Vera et al., 1999–2009 [2,6,22,24]		
K ISE, Na ISE	Br ISE, Cl ISE, F ISE, NO_3 ISE	Ag/AgCl, SJ, inner solutions: KCl, NaCl, CsCl, or NH_4Cl	Continuous	Based directly on least-squares fit of several points in the dilute region
		Zhuo et al., 2008 [51]		
Na glass ISE	Br ISE, Cl ISE, F ISE	Ag/AgCl, DJ, inner solution: KCl, outer solution: 10% KNO_3	Continuous	Based on one intermediate point and a model equation with two parameters; uses all points of the run for the fit; assumes E_J constant
		Sakaida and Kakiuchi, 2011 [48]		
$Pt/H_2,H^+$	Ag/AgCl	Ionic liquid salt bridge	Discontinuous	Use of DHL at 0.01 M HCl aq.

Note: All measurements were performed at 25°C. SCE, standard calomel electrode; SJ, single junction; DJ, double junction; DHL, Debye–Hückel limiting law; aq., aqueous; sat., saturated.

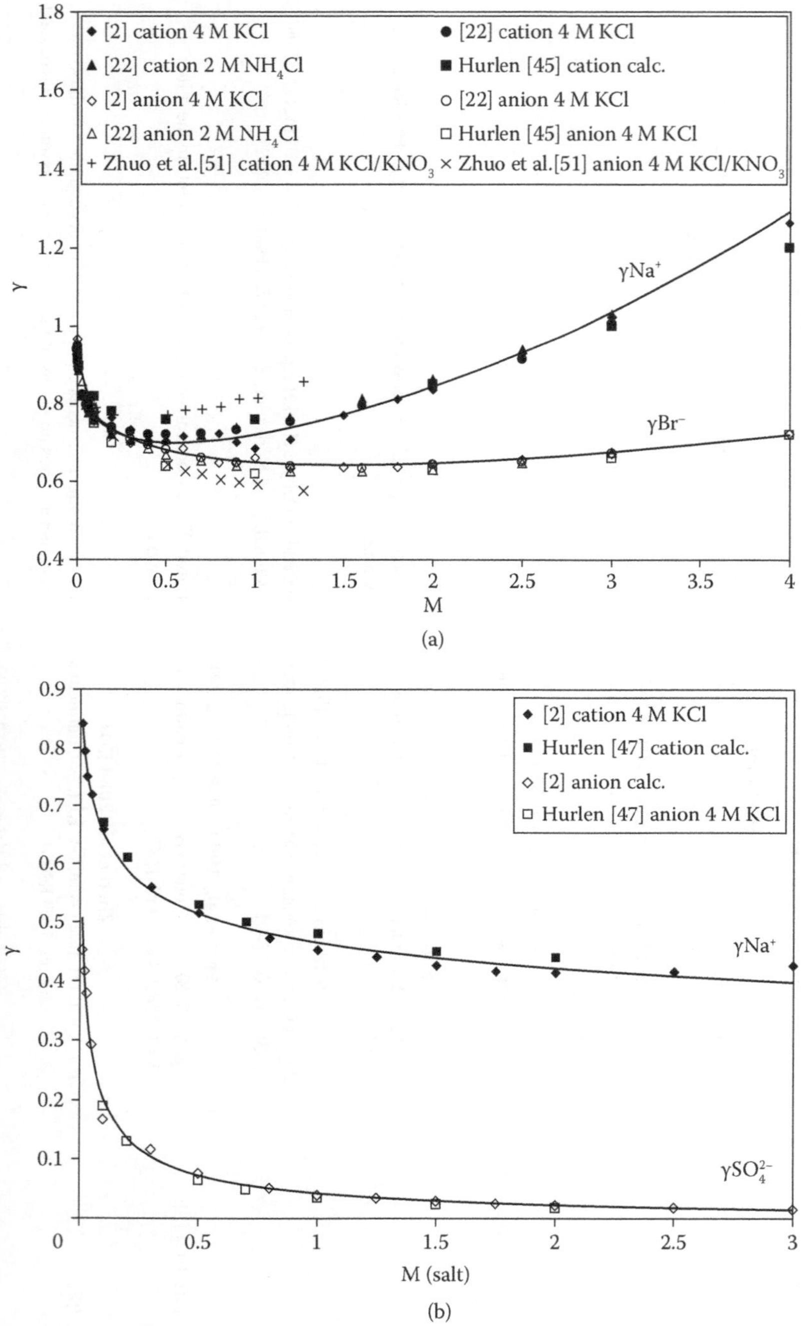

FIGURE 34.5 Comparison of different experimental ionic activity coefficients in aqueous solutions of (a) NaBr and (b) Na_2SO_4. (Inner solutions of reference electrodes are listed in the corresponding reference.) Reproduced with permission from Ref. [41].

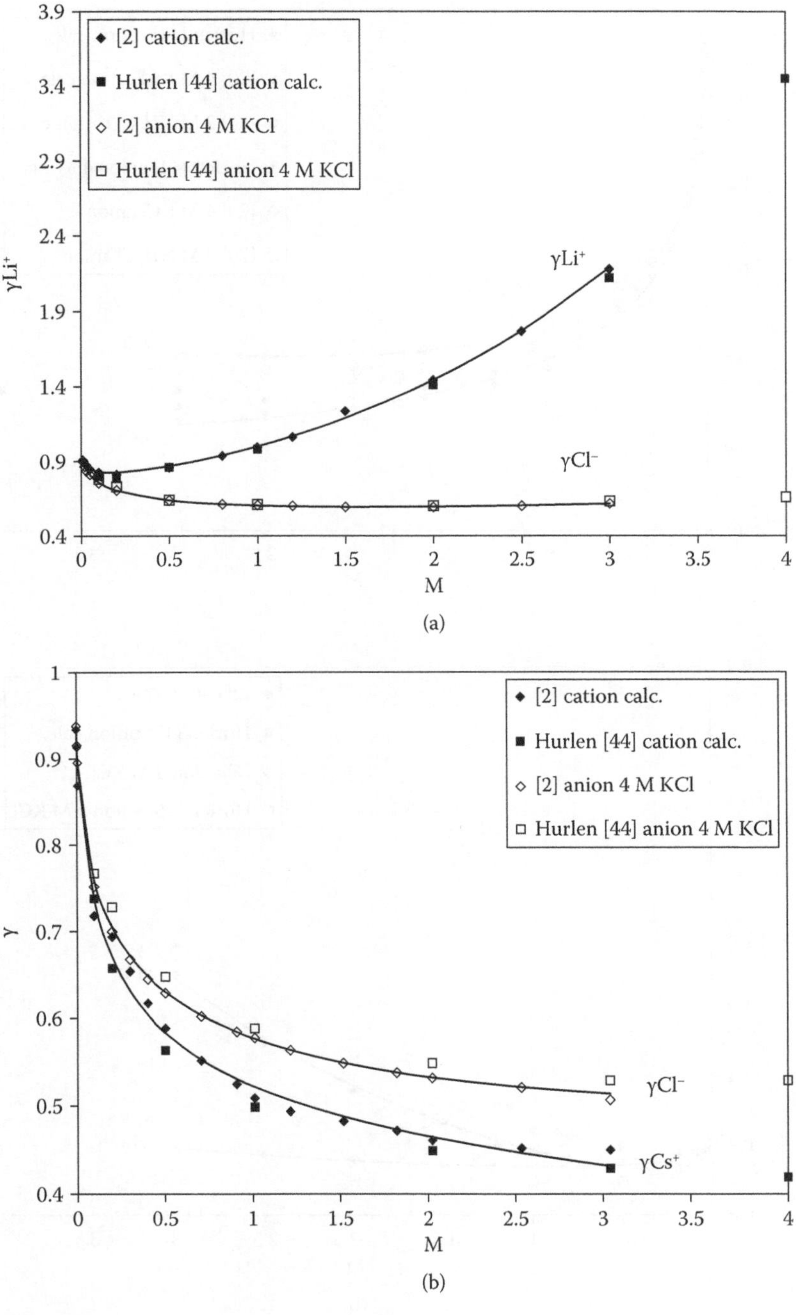

FIGURE 34.6 Comparison of ionic activity coefficients in aqueous solutions of (a) LiCl and (b) CsCl. Note that the activity coefficients of cations were calculated. Reproduced with permission from Ref. [41].

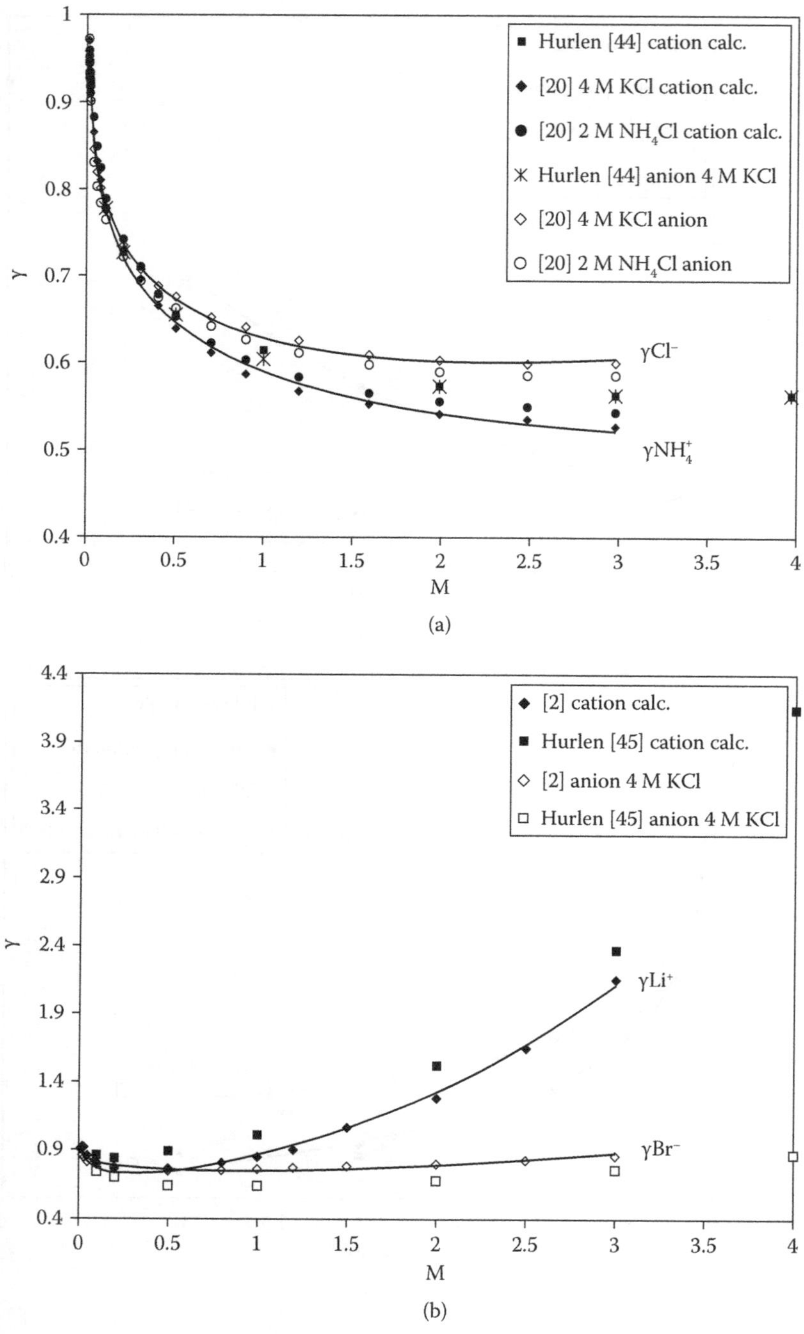

FIGURE 34.7 Comparison of ionic activity coefficients in aqueous solutions of (a) NH_4Cl and (b) LiBr. Note that the activity coefficients of cations were calculated. Reproduced with permission from Ref. [41].

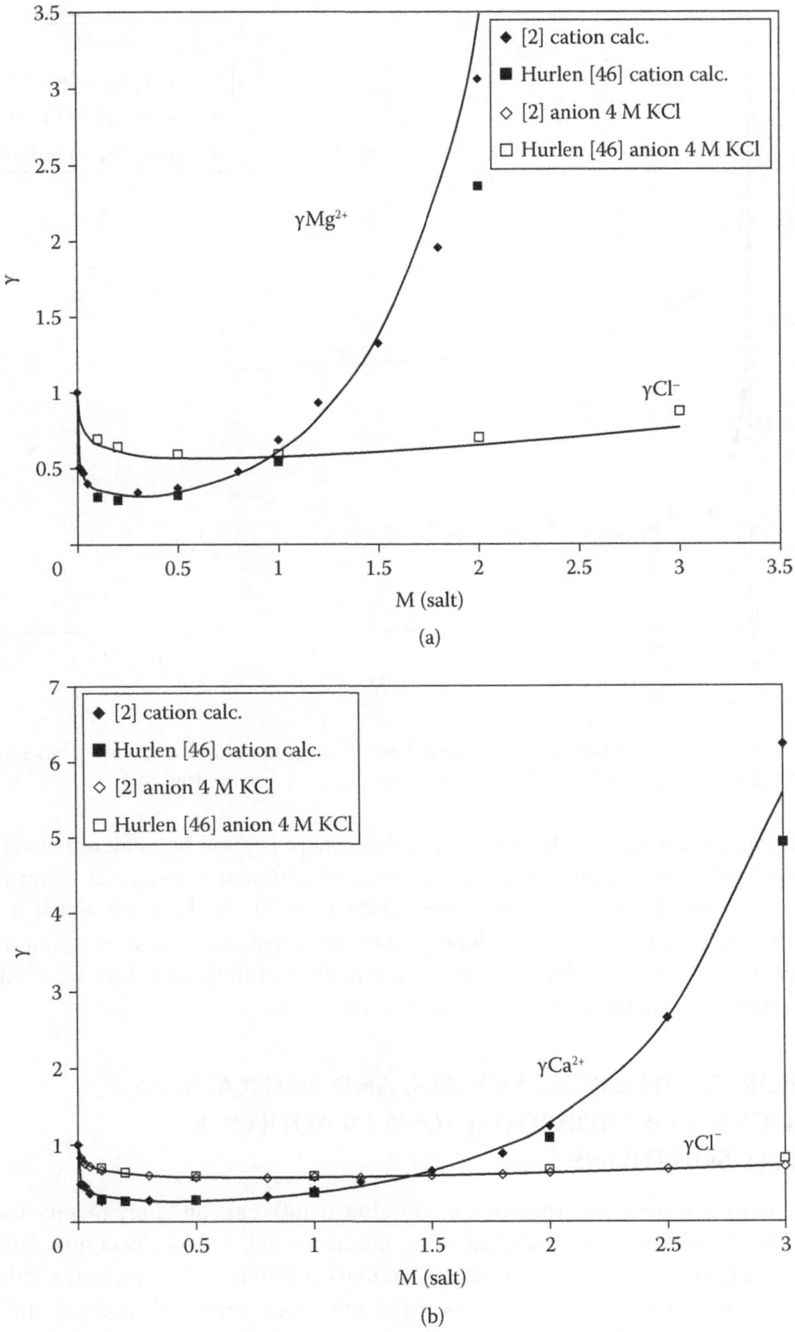

FIGURE 34.8 Comparison of ionic activity coefficients in aqueous solutions of (a) $MgCl_2$ and (b) $CaCl_2$. Note that the activity coefficients of cations were calculated. Reproduced with permission from Ref. [41].

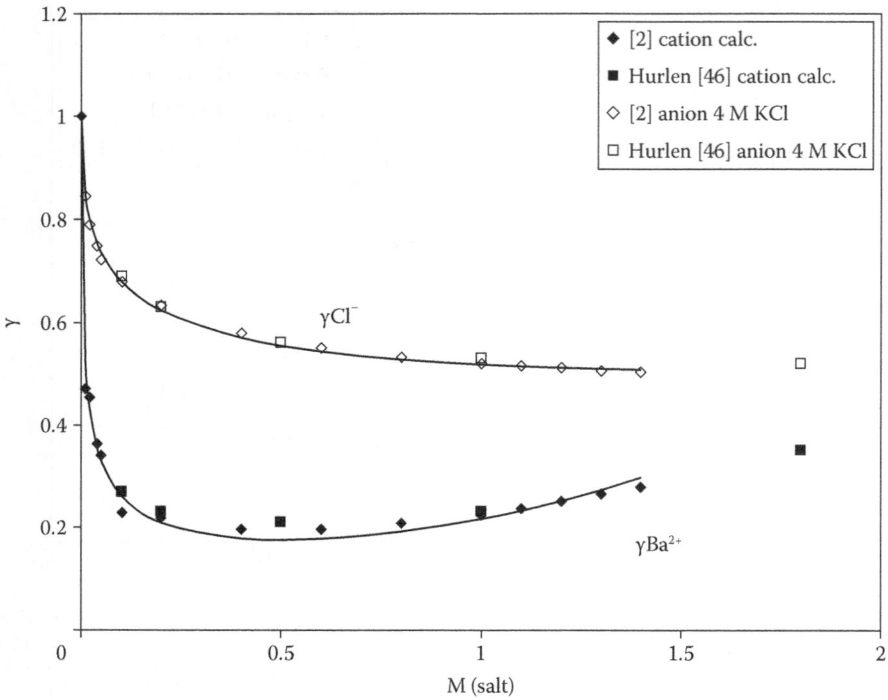

FIGURE 34.9 Comparison of ionic activity coefficients in aqueous solutions of $BaCl_2$. Note that the activity coefficients of cations were calculated. Reproduced with permission from Ref. [41].

Summarizing, except for data for the HCl system, there is good agreement between the values of the activity coefficients of individual ions reported by different laboratories using very different methods for collecting the data. In some cases, such as NaCl and KCl, for which several reports coincide, while some measurements are clearly out of agreement. In no case attention should be paid to experimental or calculated values of activity coefficients of individual ions which do not satisfy the known values of the mean ionic activity coefficients.

CONTEMPORARY THEORIES, MODELS, AND SIMULATIONS OF THE ACTIVITY OF INDIVIDUAL IONS IN AQUEOUS ELECTROLYTE SOLUTIONS

The field of electrolyte solutions was stagnant for almost 100 years and only received new attention after the advent of biotechnology. It was the norm in science that a theory becomes validated once it is confirmed by experimental results. Hence, it is remarkable that in the case of the individual activity of ions, the situation has been the inverse. In recent times, several theoretical approaches have been presented in support of the veracity of the experimentally measured activity of the individual ions. Without attempting to give all the theoretical background behind each of these treatments, we review the most important trends below.

EARLY THEORIES

Three early theories supporting the experimentally measured activity of individual ions, which used well-defined, physically meaningful parameters, closely reproduced the experimentally measured ionic activities in aqueous solutions were those of Fraenkel [53], of Liu and Eisenberg [54] and of Valisko and Boda [55]. A brief description of these theories is given below.

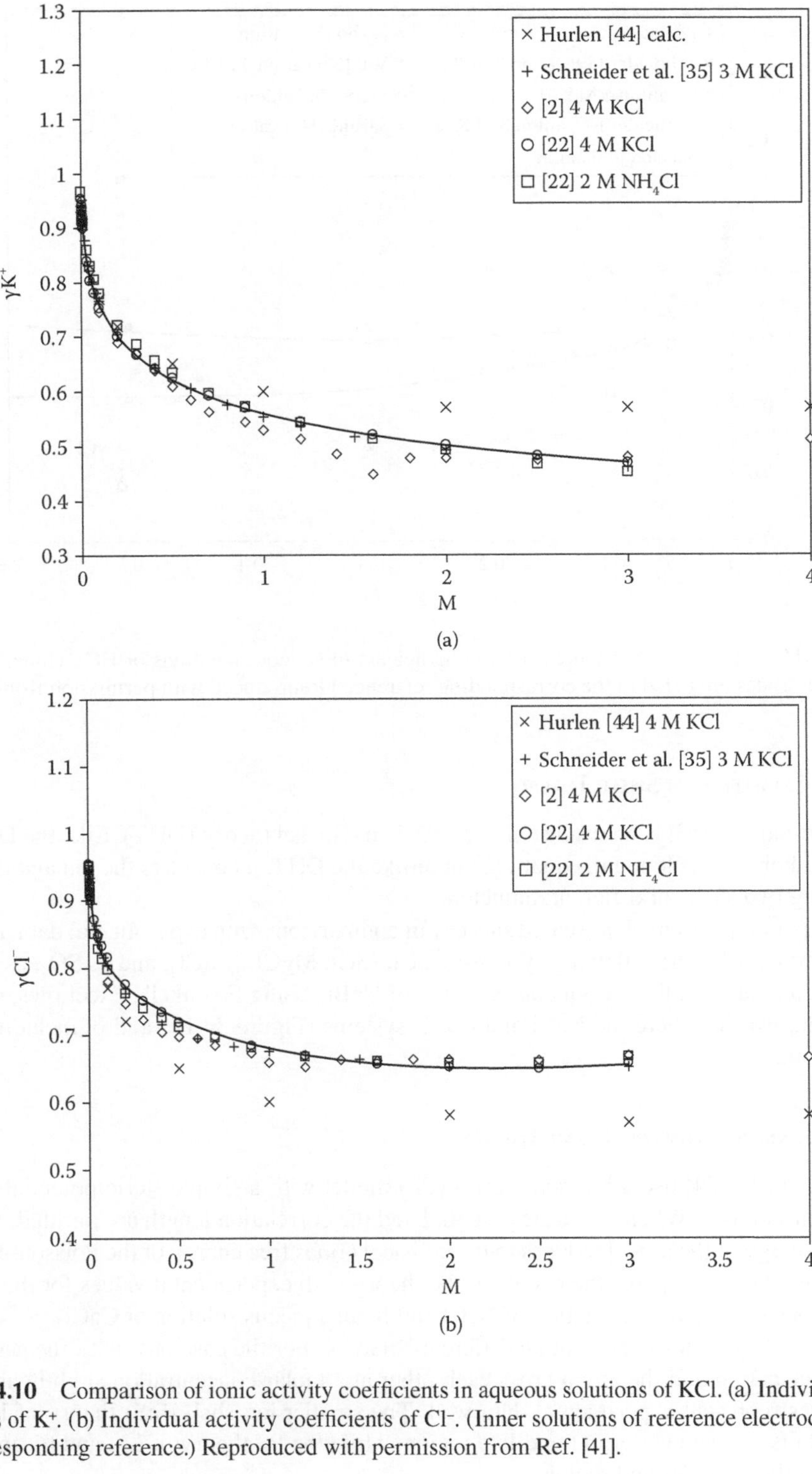

FIGURE 34.10 Comparison of ionic activity coefficients in aqueous solutions of KCl. (a) Individual activity coefficients of K^+. (b) Individual activity coefficients of Cl^-. (Inner solutions of reference electrodes are listed in the corresponding reference.) Reproduced with permission from Ref. [41].

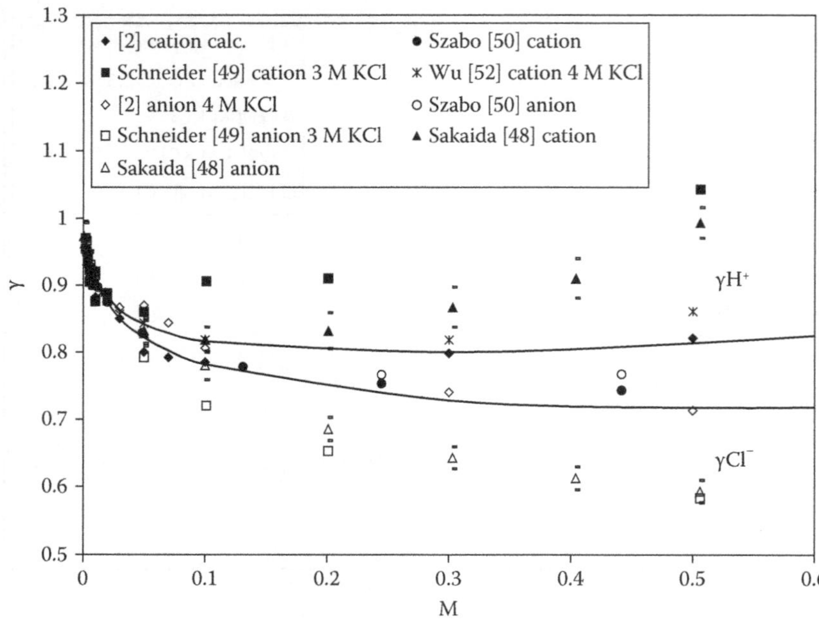

FIGURE 34.11 Comparison of ionic activity coefficients in aqueous solutions of HCl. (Inner solutions of reference electrodes are listed in the corresponding reference.) Reproduced with permission from Ref. [41].

Fraenkel's Smaller-Ion Shell Theory

Fraenkel's treatment [53] is an extension of the Debye–Hückel theory (DHT). Like the DHT, it uses a linearized Poisson–Boltzmann equation, but unlike the DHT, it considers the ion and the counter-ion as having two shells of different diameters.

In his paper [53], Fraenkel presented successful comparisons with experimental data measured at McGill University for the following systems: NaCl, LiCl, $MgCl_2$, $CaCl_2$, and K_2SO_4, as well as for the Zhuo et al. data [51] for the aqueous solution of NaBr. Using Fraenkel's Excel files, sent kindly to us by him, we show here the NaCl and $CaCl_2$ systems (Figure 34.12a and b) including also the K_2SO_4 system.

Liu and Eisenberg Poisson–Fermi Theory

Liu and Eisenberg [54] use a Poisson–Fermi (PF) model with a simple steric potential and a correlation length of ions. When the steric potential and the correlation length are omitted, the Gibbs–Fermi free energy of the model reduces to the classical Gibbs free energy of the Poisson–Boltzmann model. Figure 34.13 compares the results of the theory with experimental values for the activity of individual ions in an aqueous solution of NaCl and in an aqueous solution of $CaCl_2$.

These two electrolytes present quite different behavior. For the case of $CaCl_2$, the activity coefficients of the cation and the anion cross each other at a molar concentration slightly above unity. Both theories represent experimental data well. The smaller-ion shell (SiS) theory of Fraenkel is simpler and fits the data better in the dilute region, while the PF theory of Liu and Eisenberg gives a better fit of the concentrated region.

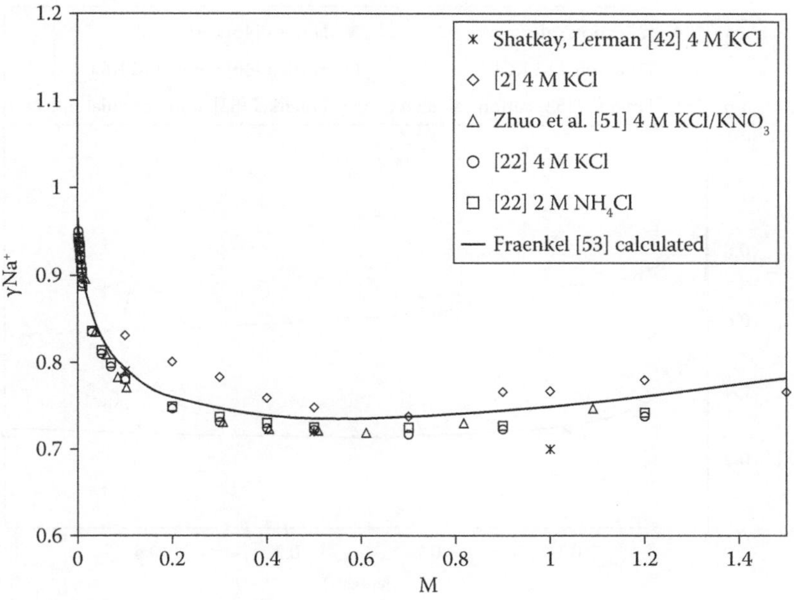
Shatkay, Lerman [42] 4 M KCl
[2] 4 M KCl
Zhuo et al. [51] 4 M KCl/KNO3
[22] 4 M KCl
[22] 2 M NH4Cl
Fraenkel [53] calculated
γNa+
M
1.2
1.1
1
0.9
0.8
0.7
0.6
0
0.2
0.4
0.6
0.8
1
1.2
1.4

(a)

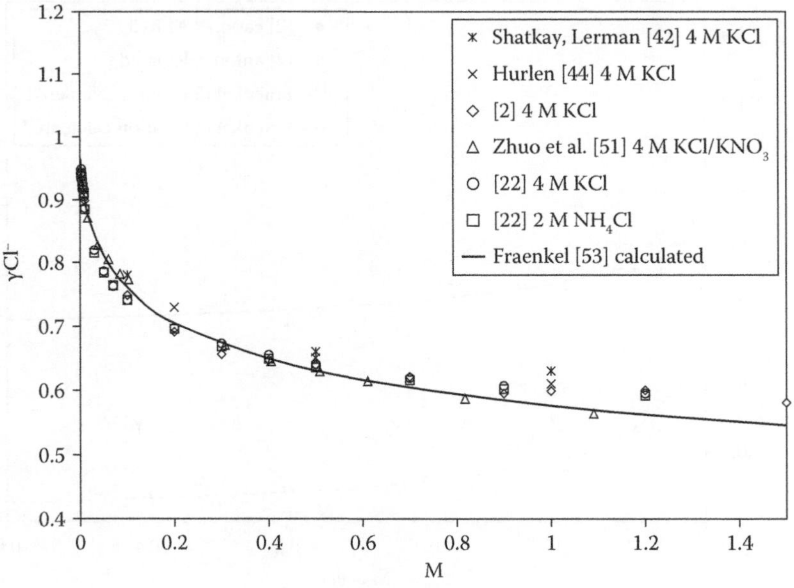
Shatkay, Lerman [42] 4 M KCl
Hurlen [44] 4 M KCl
[2] 4 M KCl
Zhuo et al. [51] 4 M KCl/KNO3
[22] 4 M KCl
[22] 2 M NH4Cl
Fraenkel [53] calculated
γCl−
M
1.2
1.1
1
0.9
0.8
0.7
0.6
0.5
0.4
0
0.2
0.4
0.6
0.8
1
1.2
1.4

(b)

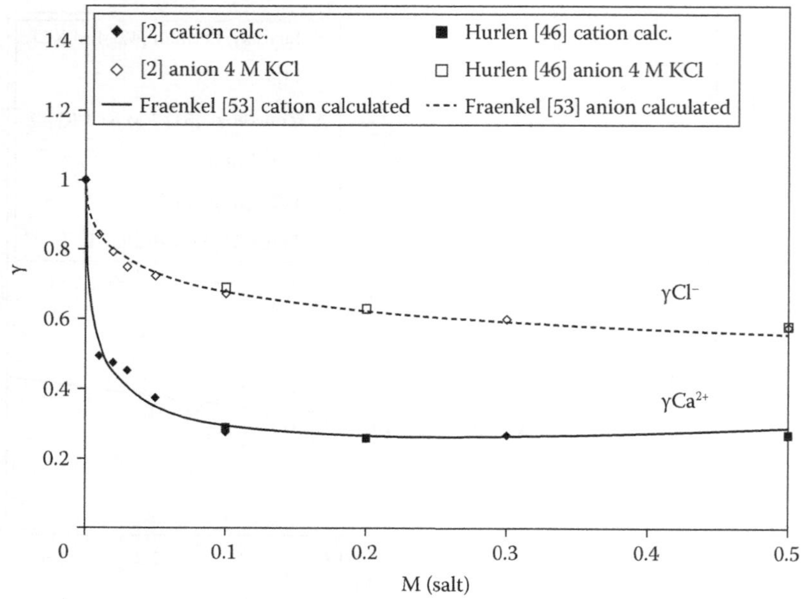

(c)

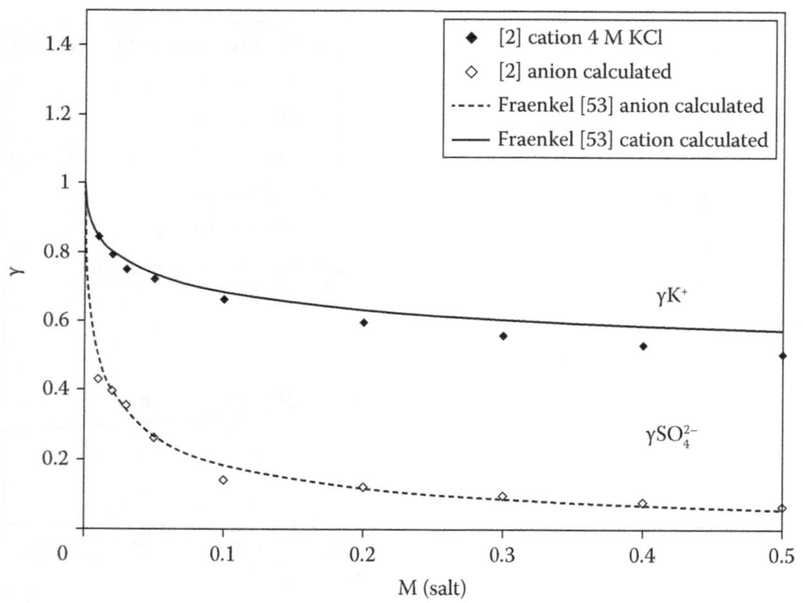

(d)

FIGURE 34.12 Comparison of experimental and calculated ionic activity coefficients in aqueous solutions of NaCl, $CaCl_2$, and K_2SO_4. (a) Individual activity coefficients of Na^+ in NaCl. (b) Individual activity coefficients of Cl^- in NaCl. (c) $CaCl_2$. (d) K_2SO_4. (Inner solutions of reference electrodes are listed in [53]. Calculated data for the systems were obtained from Dr. Dan Fraenkel through personal communication.) Reproduced with permission from Ref. [41].

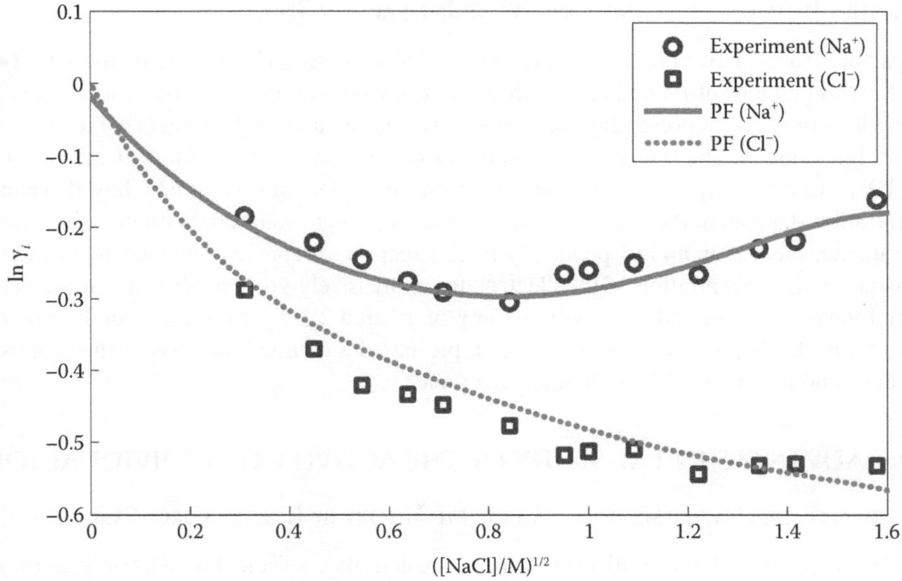

(a)

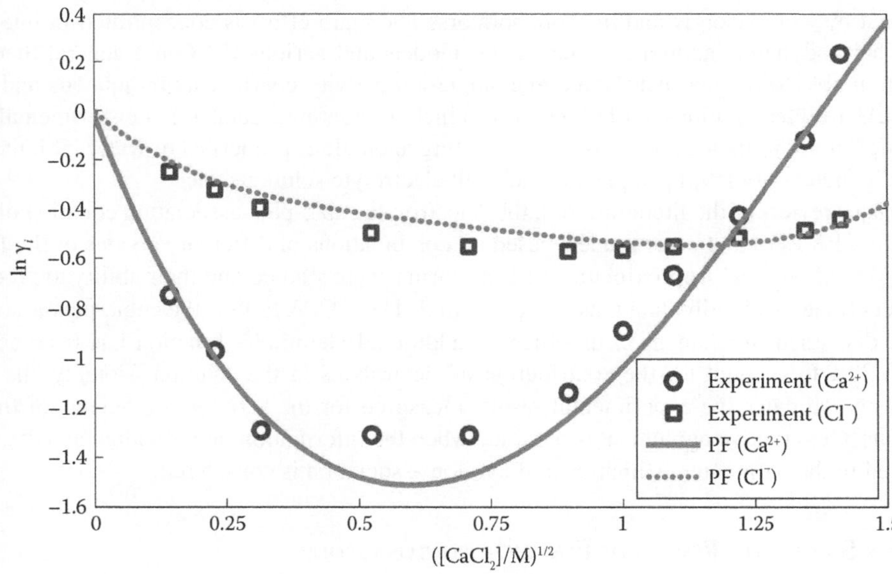

(b)

FIGURE 34.13 Comparison of experimental and calculated ionic activity coefficients in aqueous solutions of NaCl and $CaCl_2$. (a) NaCl. (b) $CaCl_2$. (Reprinted from *Chemical Physics Letters*, 637, Liu, J.-L., and Eisenberg, B., Poisson–Fermi model of single ion activities in aqueous solutions, 1–6, Copyright 2015, with permission from Elsevier.) Reproduced with permission from Ref. [41].

Valisko and Boda Ion–Ion and Ion–Water Interaction Theory

In a departure from the two previous treatments, Valisko and Boda [55] modeled the activity coefficients of individual ions of 1:1 and 2:1 electrolytes in aqueous solutions splitting the chemical potential of the ions in a term representing the ion–ion (II) interactions and a term (IW) representing the ion–water interactions. The II term was computed from "grand canonical Monte Carlo simulations" and the IW term was computed "on the basis of Born's treatment of solvation." They also considered a concentration-dependent dielectric constant of the electrolyte solution. Without using any adjustable parameter, other than an independently fitted function to represent the variation of the dielectric constant with concentration of the electrolyte, impressively good predictions of the activity of individual ions were obtained for a wide variety of 1:1 and 2:1 electrolytes. Rounding their work, Valisko and Boda [56], quoting 282 references, presented a detailed analysis of the proposed (II)/(IW) theory and its relationship with other treatments.

RECENT ADVANCES IN THE STUDY OF THE ACTIVITY OF INDIVIDUAL IONS

Work toward the Development of a Complete Theory by Kontogeorgis' Group

In some sense, the three theories discussed above are activity coefficient models only, as they do not consider physical interactions such as dispersion and association.

A wide range and systematic research toward the understanding of electrolytes systems is being carried out by Kontogeorgis and his collaborators. The main effort is concentrated in integrating diverse thermodynamic approaches such as G^E models and various EOS on a general theoretical formulation able to predict mean ionic and single-ion activity coefficients in aqueous and mixed solvents electrolytes solutions. Part of their effort include creating a database of experimental values of densities of alkali halide solutions [57], modeling anomalous properties of water [58], modeling electrical conductivity [59] in single and multisalt electrolyte solutions, etc.

In a major review of the literature [60], the "electrolyte cubic-plus-association equation of state," in short, e-CPA EOS and other models based on combinations of different versions of the Debye–Hückel, Mean Spherical Approximation or Born terms, were studied and their ability to predict the activity coefficients of individual ions was compared. The e-CPA EOS is the cubic-plus-association equation discussed in Chapter 32 in which an additional Helmholtz function has been added to Equation 32.1 to account for the contribution of electrolytes in the solution. Notably, the theory qualitatively validates the experimental results measured for the activity coefficients of the individual ions. Quantitative agreement is obtained when the information on individual activity of ions is included in the parameter estimation and ion–ion association is considered.

Computer Simulation Results of Panagiotopoulos Group

We close this literature review quoting a computer simulation study by Saravi and Panagiotopoulos [61] which provides a deep inside view of a particular case of the activity of individual ions that has been puzzling some researchers in the field. As it happens, at the same concentration of the electrolyte in an aqueous solution, the individual activity coefficient of the sodium ion is larger than the activity coefficient of the chloride ion. As an example, Figure 34.4 shows that at 2M concentration of NaCl, the activity coefficient of sodium is around 0.75, while the activity coefficient of chloride is around 0.6. In Figure 34.10, at the same concentration of 2M KCl, the activity coefficient of potassium is 0.5, while the activity coefficient of chloride is 0.65. In principle, there is nothing that could cause alarm with these values, because it is known that the ions potassium and sodium affect differently other molecules in solution. One good example is the effect of these ions on the solubility of amino acids in aqueous solutions, to the extreme that one of these ions can cause salting in effects while the other causes salting out effect [62]. The reason for this concern is that there is a publication [63] showing that the activity of the potassium ion is also higher than the activity of the chloride ion at the same concentration. Close observation of the data presented in this publication quoted in two important studies [61,64] shows that the values reported do not satisfy the values of the mean ionic activity coefficient of KCl, a *sine qua non* condition to accept experimental results. Notably, Saravi and Panagiotopoulos not only obtained the right trends without any adjustment of interaction parameters in their molecular dynamic simulation, but explained why this is so. Observing the shapes of the pair correlation function $g(r)$ of both cations in aqueous solution they deduced that the sodium ion has tendency toward hydration while the potassium ion is "more drawn towards its counterion than water." This conclusion alone proves the value of well-designed simulation studies to understand the molecular behavior of mixtures.

Unveiling the Invisible Term in Pitzer's Equations

As stated elsewhere [65], "in the second half of last century, few people devoted as much effort as Kenneth S. Pitzer to advance the understanding of the thermodynamic behavior of aqueous electrolyte solutions." An extended list of Pitzer's publications on this subject can be found in this reference. The final form resulting from Pitzer's derivation can be conveniently written as:

$$\ln \gamma_+ = Z_+^2 \, F_I + \frac{\nu}{2\nu_+} \ln \gamma_\pm \; + Z_+ \, \Phi \tag{34.50}$$

and for the anion,

$$\ln \gamma_- = Z_-^2 \, F_I + \frac{\nu}{2\nu_-} \ln \gamma_\pm \; - |Z_-| \, \Phi \tag{34.51}$$

where F_I is a shorthand notation for the Debye–Hückel contribution and all other symbols are those defined in Chapter 18. When replaced into Equation 18.8, these equations regenerate the expression for the mean ionic activity coefficient as the last term of the right-hand side of both cancels out due to the condition of electroneutrality, Equation 18.1a. Thus, from information of the mean ionic activity coefficient only, the value of Φ cannot be determined. Foreseeing this situation, Pitzer canceled the last term of both equations, and what are known now as "Pitzer Equations" are the truncated forms of these expressions. The availability of values of the individual ionic activities allowed to determine the value of Φ for 1:1, 1:2 and 2:1 electrolyte. Notably, the values of Φ, correlate well with a simple three-term polynomial in terms of the square root of the ionic strength [65].

A PREVIEW OF THE FUTURE

Judging by the progress made in recent years, it is clear that, on the experimental side, measurement of the activity coefficient of individual ions in solvents other than water and in mixed solvents will require developing new approaches to gather the data. On the theoretical side, if one extrapolates from the present state of knowledge, to include mixed solvents, the work will follow the lines of two recent publications [66,67] on equations of state models for electrolyte in aqueous and non-aqueous solutions.

CONCLUSIONS

Perhaps the most convincing proof of the veracity of the measured activity of individual ions is given by a detailed consideration of the results for Na_2SO_4 presented in Figure 34.5b. Once we have made a case for supporting the measurability of ionic activities in concentrated aqueous solutions, we briefly address here both the evolution of thoughts on this matter and the potential use of the new thermodynamics of electrolyte solutions [26] that this process has generated.

Gilbert N. Lewis [68–72] introduced the concepts of activity and activity coefficients of individual ions and obtained values for these properties in dilute aqueous solutions of electrolytes from solubility and free energy of dilution data [68]. He also established the connection between the activity of individual ions and the electromotive force in an electrochemical cell [69]. In the first edition of their book *Thermodynamics*, G. N. Lewis and M. Randall [73] stated that "it would be of much theoretical interest if we could determine the actual activity of an ion in a solution of any concentration," and added, "This indeed might be accomplished if we had any general method of calculating the potential at a liquid junction." In the second edition of the same text, revised by K. S. Pitzer and L. Brewer [74], these new authors wrote, "In order to measure the activity of a single ion in solution, it is necessary to measure experimentally some process which transfers that single charged species into or out of the solution. The enormous magnitude of the space-charge energies prevents such measurements by conventional methods which require transfer of macroscopic amounts of materials." However, they added, "Nevertheless, single-ion properties are potentially measurable." In the third edition of the same text [75], this time edited by K. S. Pitzer alone, the activity coefficients of the individual ions appear only as analytical expressions derived from differentiation of the excess Gibbs energy based on mean ionic activity coefficients in what is called the Pitzer theory. Although in this latter text the possibility of measuring the activity of individual ions in concentrated solutions is ignored, the following statement in Pitzer's text [75] clearly justifies the new thermodynamics of electrolyte solutions [26] based on the actual measurement of these properties: "Indeed, for many applications to complex mixtures, it is simpler to use the single-ion expressions." These applications are well illustrated in another text by the same author [76].

EPILOGUE

In some respect, the long delay in the acceptance of the possibility of ever measuring the activity of individual ions brings to mind the following children's story:

> The Grand Vizier of a remote kingdom died. To choose his replacement the King called the best astrologer, the best wizard and the most successful farmer of the land. He had them entering a tower, closed the door and said to them: "The first of you to come out of the tower will be my Grand Vizier". Then, through a small window, he threw inside a hermetic box containing the key to the door.
>
> While the farmer was sitting quietly in one corner of the dark room, the astrologer consulted all his maps showing the positions of the stars and the wizard played all his tricks around the box trying to figure out how to open it.
>
> Suddenly, the room was full of light and the farmer was standing outside the open door.
>
> "How did you open the door?", the wizard and the astrologer asked him in unison. The farmer calmly answered: "The King never said that the door was locked."
>
> Harned [31] saw clearly where the problem of measuring the activity of the individual ions was when he wrote: "The difficulty in estimating liquid junction potentials is the only cause of our not being able to estimate exactly the individual activity coefficient of the ions." He just was working too early in this problem to know that the door was not locked.

NOTES

1 See Equation 16.42 from [40], "pressure equation," for the mean ionic activity coefficient written as $\ln\gamma_\pm = -\frac{b}{3a}\left[\frac{Ka}{1+Ka} + 2\ln(1+Ka)\right]$ where according to Equation 16.17 of [41], $Ka \approx I_k^{0.5}$.

2 Reprinted and partially modified from *The Journal of Chemical Thermodynamics*, 99, Wilczek-Vera, G., and Vera J. H., How much do we know about the activity of individual ions?, 65–69, Copyright 2016, with permission from Elsevier.

REFERENCES

1. Wilczek-Vera, G., and Vera, J. H. 2005. On the measurement of individual ion activities. *Fluid Phase Equilib.* 236: 96–110.
2. Wilczek-Vera, G., Rodil, E., and Vera, J. H. 2004. On the activity of ions and the junction potential. Revised values for all data. *AIChE J.* 50: 445–462.
3. Malatesta, F. 2005. On a recent purported determination of individual ion activity coefficients. *AIChE J.* 52: 785–791.
4. Sandler, S. I. 2006. Preface to the letters to the editor. *AIChE J.* 52: 2306.
5. Vera, J. H. 2006. Letter to the editor. *AIChE J.* 52: 2306.
6. Lladosa, E., Arce, A., Wilczek-Vera, G., and Vera, J. H. 2010. Effect of the reference solution in the measurement of ion activity coefficients using cells with transference at T=298.15 K. *J. Chem. Thermodyn.* 42: 244–250.
7. Zarubin, D. P. 2011. The nature of single-ion activity coefficients calculated from potentiometric measurements on cells with liquid junctions. *J. Chem. Thermodyn.* 43: 1135–1152.
8. Vera, J. H., and Wilczek-Vera, G. 2012. Comment on “The nature of single-ion activity coefficients calculated from potentiometric measurements on cells with liquid junctions” by Dmitri P. Zarubin, *J. Chem. Thermodyn.* 43, 47 (2011): 442–444, 1135–1152.
9. Zarubin, D. P. 2012. The nature of single-ion activity coefficients calculated from potentiometric measurements on cell with liquid-junction. *J. Chem. Thermodyn.* 47: 445–448.
10. Wilczek-Vera, G., and Vera, J. H. 2012. Short answer to the reply from D.P. Zarubin to our comment on “The nature of single-ion activity coefficients calculated from potentiometric measurements on cell with liquid-junction.” *J. Chem. Thermodyn.* 47: 449–450.
11. Khoshkbarchi, M. K., and Vera, J. H. 1996. Measurement and correlation of ion activity in aqueous single electrolyte solutions. *AIChE J.* 42(1): 249–258.
12. Khoshkbarchi, M. K., and Vera, J. H. 1996. Measurement and correlation of ion activity coefficients in aqueous solutions of mixed electrolyte with a common ion. *Fluid Phase Equilib.* 121(1–2): 253–265.
13. del Mar Marcos-Arroyo, M., Khoshkbarchi, M. K., and Vera, J. H. 1996. Activity coefficients of sodium, potassium, and nitrate ions in aqueous solutions of NaNO3, KNO3, and NaNO3+KNO3 at 25°C. *J. Solution Chem.* 25(10): 983–1000.
14. Rabie, H. R., Wilczek-Vera, G., and Vera, J. H. 1999. Activities of individual ions. From infinite dilution to saturated solutions. *J. Solution Chem.* 28: 885–914.
15. Taghikhani, V., Modarress, H., and Vera, J. H. 1999. Individual anionic activity coefficients in aqueous electrolyte solutions of LiCl and LiBr. *Fluid Phase Equilib.* 166(1): 67–77.
16. Taghikhani, V., Modarress, H., and Vera, J. H. 2000. Measurement and correlation of the individual ionic activity coefficients of aqueous electrolyte solutions of KF, NaF and KBr. *Can. J. Chem. Eng.* 78(1): 175–181.
17. Rodil, E., and Vera, J. H. 2001. Individual activity coefficients of chloride ions in aqueous solutions of MgCl2, CaCl2 and BaCl2 at 298.2 K. *Fluid Phase Equilib.* 187–188: 15–27.
18. Rodil, E., and Vera, J. H. 2001. Measurement and correlation of the activity coefficients of individual ions in aqueous electrolyte solutions of Na2SO4 and K2SO4. *Can. J. Chem. Eng.* 79(5): 771–776.
19. Rodil, E., Persson, K., Wilczek-Vera, G., and Vera, J. H. 2001. Determination of the activity of H+ ions within and beyond the pH meter range. *AIChE J.* 47(12): 2807–2817.
20. Rodil, E., and Vera, J. H. 2003. The activity of ions: Analysis of the theory and data for aqueous solutions of MgBr2, CaBr2 and BaBr2 at 298.2 K. *Fluid Phase Equilib.* 205(1): 115–132. Erratum: *Fluid Phase Equilib.* 211(2): 289, 2003.
21. Wilczek-Vera, G., and Vera, J. H. 2003. Peculiarities of the thermodynamics of electrolyte solutions: A critical discussion. *Can. J. Chem. Eng.* 81: 70–79.
22. Wilczek-Vera, G., Rodil, E., and Vera, J. H. 2006. Towards accurate values of individual ion activities. Additional data for NaCl, NaBr and KCl, and new data for NH4Cl. *Fluid Phase Equilib.* 241: 59–69.
23. Wilczek-Vera, G., Rodil, E., and Vera, J. H. 2006. A complete discussion of the rationale supporting the experimental determination of individual ionic activities. *Fluid Phase Equilib.* 244: 33–45.
24. Arce, A., Wilczek-Vera, G., and Vera, J. H. 2007. Activities of aqueous Na+ and Cl− ions from homoionic measurements with null junction potentials at different concentrations. *Chem. Eng. Sci.* 52: 3849–3857.
25. Rodil, E., Arce, A., Wilczek-Vera, G., and Vera, J. H. 2009. Measurement of ion activity coefficients in aqueous solutions of mixed electrolyte with a common ion: NaNO3+KNO3, NaCl+KCl and NaBr+NaCl. *J. Chem. Eng. Data* 54: 345–350.
26. Wilczek-Vera, G., and Vera, J. H. 2009. On the predictive ability of the new thermodynamics of electrolyte solutions. *Ind. Eng. Chem. Res.* 48(13): 6436–6440.

27. Wilczek-Vera, G., and Vera, J. H. 2009. Answer to the comments by Francesco Malatesta on J. *Chem. Eng. Data*, 2009, 54, 345–350 by Rodil, E., Arce, A., Wilczek-Vera, G., Vera, J.H. *J. Chem. Eng. Data*: 2979.
28. Wilczek-Vera, G., Arce, A., and Vera, J. H. 2010. Answer to "Comment on individual ion activities of Na+ and Cl– by Arce, Wilczek-Vera and Vera" by F. Malatesta. *Chem. Eng. Sci.* 65: 2263–2264.
29. Wilczek-Vera, G., and Vera, J. H. 2011. On the measurement of the real values of individual ionic activities: A chemical engineering perspective. *Chem. Eng. Sci.* 66: 3782–3791.
30. Wilczek-Vera, G., and Vera, J. H. 2011. The activity of individual ions. A conceptual discussion of the relation between the theory and the experimentally measured values. *Fluid Phase Equilib.* 312: 79–84. Corrigendum to "The activity of individual ions. A conceptual discussion of the relation between the theory and the experimentally measured values" [*Fluid Phase Equilib.* 312 (2011) 79–84]. *Fluid Phase Equilib.* 314: 203, 2012.
31. Harned, H. S. 1926. Individual thermodynamic behaviors of ions in concentrated solutions including discussion of the thermodynamic method of computing liquid junction potentials. *J. Phys. Chem.* 30: 433–456.
32. Henderson, P. 1907. Zur thermodynamik der flussigkeitsketten. *Z. Phys. Chem.* 59: 118–127.
33. Lewis, G. N., and Randall, M. 1921. The activity coefficient of strong electrolytes. *J. Am. Chem. Soc.* 43: 1112–1159.
34. Guntelberg, E. 1926. Interaction of ions. *Z. Phys. Chem.* 123: 199–247. Quoted from Guggenheim, E. A. 1950. *Thermodynamics: An Advanced Treatise for Chemists and Physicists*, 2nd ed. New York: Interscience Publishers.
35. Schneider, A. C., Pasel, C., Luckas, M., Schmidt, K. G., and Herbell, J.-D. 2003. Bestimmung von Ionenaktivitatskoeffizienten in wassrigen Losungen mit Hilfe ionenselektiver Elektroden. *Chem. Ing. Tech.* 75: 244–249.
36. Atkins, P. W. 1986. *Physical Chemistry*, 3rd ed. New York: Freeman, p. 676.
37. (a) Bates, R. G. 1964. *Determination of pH, Theory and Practice*. New York: John Wiley. (b) 2nd ed., 1973.
38. Harper, H. W. 1985. Calculation of liquid junction potentials. *J. Phys. Chem.* 89(9): 1659–1664.
39. Guggenheim, E. A. 1929. The conceptions of electrical potential difference between two phases and the individual activity of ions. *J. Phys. Chem.* 33: 842–849.
40. Lin, C.-L., and Lee, L.-S. 2003. A two-ionic-parameter approach for ion activity coefficients of aqueous electrolyte solutions. *Fluid Phase Equilib.* 205: 69–88.
41. Wilczek-Vera, G., and Vera, J. H. 2016. How much do we know about the activity of individual ions? *J. Chem. Thermodyn.* 99: 65–69.
42. Shatkay, A., and Lerman, A. 1969. Individual activities of sodium and chloride in aqueous solutions of sodium chloride. *Anal. Chem.* 41: 514–517.
43. Shatkay, A. 1970. Further study of individual ion activities in pure aqueous sodium chloride solution. *Talanta* 17: 381–390.
44. Hurlen, T. 1979. Convenient single-ion activities. *Acta Chem. Scand. A* 33: 631–635.
45. Hurlen, T., and Breivik, T. R. 1981. Ion activities of alkali metal bromides in aqueous solution. *Acta Chem. Scand. A* 35: 415–418.
46. Hurlen, T. 1979. Ion activities of alkaline-earth chlorides in aqueous solution. *Acta Chem. Scand. A* 33: 637–640.
47. Hurlen, T. 1983. Single-ion activities of sodium sulfate in aqueous solution. *Acta Chem. Scand. A* 37: 739–742.
48. Sakaida, H., and Kakiuchi, T. 2011. Determination of single-ion activities of H+ and Cl– in aqueous hydrochloric acid solutions by use of an ionic liquid salt bridge. *J. Phys. Chem. B* 115(45): 13222–13226.
49. Schneider, A. C., Pasel, C., Luckas, M., Schmidt, K. G., and Herbell, J.-D. 2004. Determination of hydrogen single ion activity co efficients in aqueous HCl solutions at 25°C. *J. Solution Chem.* 33(3): 257–273.
50. Szabo, Z. 1935. Uber die Ionenaktivitaten der Salzsaure. *Z. Phys. Chem. A* 174: 22–32.
51. Zhuo, K., Dong, W., Wang, W., and Wang, J. 2008. Activity coefficients of individual ions in aqueous solutions of sodium halides at 298.15 K. *Fluid Phase Equilib.* 274: 80–84.
52. Wu, Y. Ch., Feng, D., and Koch, W. F. 1989. Evaluation of liquid junction potentials and determination of pH values of strong acids at moderate ionic strengths. *J. Solution Chem.* 18(7): 641–649.
53. Fraenkel, D. 2012. Single-ion activity: Experiment versus theory. *J. Phys. Chem. B* 116: 3603–3612.
54. Liu, J.-L., and Eisenberg, B. 2015. Poisson-Fermi model of single ion activities in aqueous solutions. *Chem. Phys. Lett.* 637: 1–6.

55. Valisko, M., and Boda, D. 2015. Unraveling the behavior of the individual ionic activity coefficients on the basis of the balance of ion–ion and ion–water interactions *J. Phys. Chem. B* 119(4): 1546–1557.
56. Valisko, M., and Boda, D. 2023. Resurrection of Hückel's idea: decoupling ion-ion and ion-water terms in activity coefficients via state-dependent dielectric constant. *Fluid Phase Equilib.* 572 (15): 113826–113847.
57. Yang, F., Qu, J., Kontogeorgis, G. M., and de Hemptinne, J.- Ch. 2022. Reference Density database for 20 aqueous alkali halide solutions. *J. Phys. Chem. Ref. Data* 51: 043104. https://doi.org/10.1063/5.0124173.
58. Novak, N., Liang, X., and Kontogeorgis, G. M. 2024. Prediction of water anomalous properties by introducing the two-state theory in SAFT. *J. Chem. Phys.*, 160(10):104505. https://doi.org/10.1063/5.0186752.
59. Boroujeni, S. N., Maribo-Mogensen, B., Liang, X., and Kontogeorgis, G. M. 2024. Novel model for predicting the electrical conductivity of multisalt electrolyte solutions. *J. Phys. Chem. B* 128(2): 536–550. https://doi.org/10.1021/acs.jpcb.3c05718.
60. Li, S. Liang, X., von Solms, N., and Kontogeorgis, G. M. 2020. Analysis of some electrolyte models including their ability to predict the activity coefficients of individual ions. *Ind. Eng. Chem. Res.* 59: 11790–11809.
61. Saravi, S. H., and Panagiotopoulos, Z. 2021. Individual ion activity coefficients in aqueous electrolytes from explicit-water molecular dynamics simulations. *J. Phys.Chem. B.* 125(30): 8511–8521. https://doi.org/10.1021/acs.jpcb.1c04019.
62. Vera, J. H., and Wilczek-Vera, G. (Eds). 2018. *Ionic Surfactants and Aqueous Solutions. Biomolecules, Metals and Nanoparticles*. Berlin/Boston, MA: De Gruyter Chapter 3.
63. Dash, D., Kumar, S., Mallika, C., and Kamachi Mudali, U. 2012. New data on activity coefficients of potassium, nitrate, and chloride ions in aqueous solutions of KNO3 and KCl by Ion Selective Electrodes. *ISRN Chem. Eng.* 2012: Article ID 730154. https://doi.org/10.5402/2012/730154.
64. Silva, G. M., Liang, X., and Kontogeorgis, G. M. 2022. On the derivation of the Debye-Hückel equations. *Mol. Phys.* 120(10): Article e2064353. https://doi.org/10.1080/00268976.2022.2064353.
65. Manaure, E., Olivera-Fuentes, C., Wilczek-Vera, G., and Vera, J. H., 2021. Pitzer Equations and a model-free versions of the ion interaction approach for the activity of ions. *Chem. Eng. Sci.* 241: 116619. https://doi.org/10.1016/j.ces.2021.116619.
66. Olsen, M. D., Kontogeorgis, G. M., de Hemptinne, J.-C., Liang, X., and von Helms, N., 2023. Comparison of equation of state models for electrolytes: e-CPA and e-PC-SAFT. *Fluid Phase Equilib.* 571: 113804 (33 pages).
67. Novak, N., Kontogeorgis, G. M., Castier, M., and Economou, I. G. 2023. Extension of the e-SAFT-VR Mie equation of state from aqueous to non-aqueous electrolyte solutions. *Fluid Phase Equilib.* 565: 113618. https://doi.org/10.1016/j.fluid.2022.113618.
68. Lewis, G. N. 1907. Outlines of a new system of thermodynamic chemistry. *Proc. Am. Acad. Arts Sci.* 43: 259–296.
69. Lewis, G. N. 1907. Umriss eines neuen Systems der chemischen Thermodynamik. *Z. Phys. Chem.* 61: 129–165.
70. Lewis, G. N. 1909. The use and abuse of the ionic theory. *Z. Phys. Chem.* 70: 212–219.
71. Lewis, G. N. 1912. The activity of the ions and the degree of dissociation of strong electrolytes. *J. Am. Chem. Soc.* 34(12): 1631–1644.
72. Lewis, G. N. 1913. The free energy of chemical substances. Introduction. *J. Am. Chem. Soc.* 35: 1–30.
73. Lewis, G. N., and Randall, M. 1923. *Thermodynamics*, 1st ed. New York: McGraw-Hill.
74. Lewis, G. N., and Randall, M. 1961. *Thermodynamics*, 2nd ed., rev. K. S. Pitzer and L. Brewer. New York: McGraw-Hill.
75. Pitzer, K. S. 1995. *Thermodynamics*, 3rd ed. New York: McGraw-Hill.
76. Pitzer, K. S. 1991. *Activity Coefficients in Electrolyte Solutions*, 2nd ed. Boca Raton, FL: CRC Press.

Section V

Appendices

ASOG-KT + PRSV = Design

DOI: 10.1201/9781003431985-39

Appendix A
Material Balances

ABSTRACT

Equations are given to deal with open systems for which streams of matter enter and/or leave, both in an unsteady state, when the amount of matter in the system changes with time, and in steady state. It is explained how to proceed when there are chemical reactions in the system.

MATERIAL BALANCES FOR SYSTEMS WITHOUT CHEMICAL REACTIONS

The basic idea of a material balance is that the mass of a closed system does not change unless there is interconversion between mass and energy. Thus, for processes in which the integrity of atoms is unaffected, the mass of the system changes only if mass enters or leaves a system. For this case, in a process occurring in an open system during a time interval *dt*, the differential change of mass of the system is given by:

$$[\delta m]_{in} - [\delta m]_{out} = [dm]_{system} \tag{A.1}$$

The quantities $[\delta m]$ (*in* or *out*) represent the amount of material that enters and leaves the system during the time interval, *dt*, and $[dm]$ is the change in the mass of the system during that same time interval. Note that the differential symbol, *d*, is not used for the mass entering or leaving the system, as these are not exact differentials. Dividing all the terms in Equation A.1 by *dt*, we obtain:

$$\dot{m}_{in} - \dot{m}_{out} = \frac{dm}{dt} \tag{A.2}$$

In Equation A.2, the symbol *m* with an upper dot is used for mass flow rate (lbm/h, kg/s, or else), and *dm*/*dt* is the rate of change of the mass, that is, the accumulation of mass in the system. This accumulation can be positive or negative. For the special case of "steady-state operation," the values of the properties of the system at any particular point do not change with time. Hence, for steady state, there is no accumulation of material in the system and

$$\dot{m}_{in} = \dot{m}_{out} \tag{A.3}$$

In the case of a fluid of density ρ (kg/m^3) flowing inside a duct or a pipe of cross-sectional area A (m^2) at a linear velocity $\dot{\upsilon}$ (m/s), the mass flow rate is given by:

$$\dot{m} = \rho\dot{\upsilon}A \tag{A.4}$$

Hence, combining Equations A.3 and A.4 for steady flow in a duct, we write:

$$(\rho\dot{\upsilon}A)_{in} = (\rho\dot{\upsilon}A)_{out} \tag{A.5}$$

For the case of the flow of a liquid, that is, an incompressible fluid, the density is constant and the linear velocity is inversely proportional to the cross-sectional area of the flow.

DOI: 10.1201/9781003431985-40

MATERIAL BALANCES FOR SYSTEMS WITH CHEMICAL REACTIONS

When there are chemical reactions in a system at steady state, the mass entering the system per unit time is still equal to the mass leaving the system per unit time. However, the number of moles of different compounds entering the system per unit time can be different from the number of moles leaving the system per unit time. If there are no chemical reactions, the number of moles of each compound is conserved. If there *are* chemical reactions occurring in the system, neither the mass nor the number of moles of the individual compounds participating in the reactions is conserved. However, the total mass is conserved and the number of moles of each *chemical element* is conserved. Thus, in the case of chemical reactions, it is easier to perform material balances on each of the chemical elements rather than on the chemical compounds. If it is desired to perform a material balance on a *chemical compound i*, in terms of moles, Equation A.1 for steady state takes the form:

$$[\text{mols of } i]_{in} + [\text{mols of } i]_{produced} - [\text{mols of } i]_{consumed} = [\text{mols of } i]_{out} \tag{A.6}$$

At *unsteady* state, the difference between the right-hand side and the left-hand side of Equation A.6 gives the accumulation of moles of compound i in the system.

Appendix B
Working with the Virial EOS

ABSTRACT

Correlations of the second and third virial coefficients, in terms of the acentric factor and reduced temperature, for pure compounds and mixtures are presented. The equations for the fugacity coefficients are made available for these cases.

VIRIAL EOS TRUNCATED AFTER THE THIRD TERM

Due to its direct connection with intermolecular forces, the virial equation of state (EOS) has been of permanent theoretical interest. An excellent presentation and compilation of information on this subject can be found in the text by Prausnitz et al. [1]. From a practical point of view, its use is limited to the calculation of fugacities in the gas phase at low pressures. The compressibility factor expansion in terms of density is:

$$z = 1 + B\rho + C\rho^2 + \ldots \tag{27.6}$$

and in terms of pressure,

$$z = 1 + \frac{B}{RT}P + \frac{C - B^2}{(RT)^2}P^2 + \ldots. \tag{27.7a}$$

The second and third virial coefficients of expansion in terms of density, B and C, are known with accuracy for many substances. Higher coefficients are seldom used in practice.

There are many correlations available for the calculation of values of the second and third virial coefficients of pure compounds. We reproduce here two simple ones that give satisfactory approximations.

For the second virial coefficient of nonpolar compounds, Tsonopoulos [2] recommended the form:

$$B = \frac{RT_c}{P_c}\left(B^{(0)} + \omega B^{(1)}\right) \tag{B.1}$$

where T_c and P_c are the critical temperature and critical pressure of the compound, both in absolute scales. The two contributions to the second virial coefficient for nonpolar compounds are given by:

$$B^{(0)} = 0.1445 - \frac{0.330}{T_r} - \frac{0.1385}{T_r^2} - \frac{0.0121}{T_r^3} - \frac{0.000607}{T_r^8} \tag{B.2}$$

$$B^{(1)} = 0.0637 + \frac{0.331}{T_r^2} - \frac{0.423}{T_r^3} - \frac{0.008}{T_r^8} \tag{B.3}$$

DOI: 10.1201/9781003431985-41

where $T_r = T/T_c$ is the reduced temperature and the acentric factor is defined as:

$$\omega = -\log_{10}\frac{P^s_{\text{at } T_r=0.7}}{P_c} - 1 \tag{B.4}$$

For the third virial coefficient of nonpolar compounds, Orbey and Vera [3] proposed the correlation:

$$C = \left(\frac{RT_c}{P_c}\right)^2\left(C^{(0)} + \omega C^{(1)}\right) \tag{B.5}$$

with

$$C^{(0)} = 0.01407 + \frac{0.02432}{T_r^{2.8}} - \frac{0.00313}{T_r^{10.5}} \tag{B.6}$$

$$C^{(1)} = -0.02676 + \frac{0.01770}{T_r^{2.8}} + \frac{0.040}{T_r^{3.0}} - \frac{0.003}{T_r^{6.0}} - \frac{0.00228}{T_r^{10.5}} \tag{B.7}$$

FUGACITY COEFFICIENT OF A PURE COMPOUND

From Equations 13.7a and 27.7a truncated after the third term, the fugacity coefficient of a pure nonpolar compound is given by:

$$\ln\phi_i^0 = \frac{BP}{RT} + \frac{\left(C - B^2\right)P^2}{2(RT)^2} + \ldots \tag{B.8}$$

For polar compounds, Abusleme and Vera [4] developed a group contribution method that reduces to Tsonopoulos's correlation for nonpolar compounds. For polar compounds, the correlation uses Tsonopoulos's correlation with a modified acentric factor that, following the work of Hayden and O'Connell [5], is calculated using the mean radius of gyration of the molecules obtained from a program available without cost from the supplemental material of the Hayden–O'Connell publication. In addition, the correlation uses an energetic contribution calculated in terms of group interaction parameters presented by Abusleme and Vera [4] in tabular form.

VIRIAL EOS FOR MIXTURES

As discussed in Chapter 25, the virial coefficients can be related to intermolecular potentials, and the second virial coefficient of a mixture of nonpolar compounds is related to the respective pure compound virial coefficients by:

$$B_m = \sum_j\sum_k y_j y_k B_{jk} \tag{B.9}$$

By a similar argument, it can be shown that:

$$C_m = \sum_j\sum_k\sum_l y_j y_k y_l C_{jkl} \tag{B.10}$$

Hence, for a binary mixture of nonpolar compounds,

$$B_m = y_1^2 B_{11} + 2y_1 y_2 B_{12} + y_2^2 B_{22} \tag{B.9a}$$

and

$$C_m = y_1^3 C_{111} + 3y_1^2 y_2 C_{112} + 3y_1 y_2^2 C_{122} + y_2^3 C_{222} \tag{B.10a}$$

The problem here is to obtain the values of the cross-virial coefficients B_{12}, C_{112}, and C_{122}. For this, empirical mixing rules need to be introduced and the treatment loses its elegance. This explains why the use of the virial EOS for mixtures is restricted to the pressure range up to which the second virial coefficient suffices to represent the volumetric behavior of the mixture.

FUGACITY COEFFICIENT OF A COMPOUND IN A MIXTURE

For the range of pressure, where the second virial term suffices, the fugacity coefficient of component i is given by:

$$\ln \phi_i = \left(2\sum_k y_k B_{ik} - B_m \right) \frac{P}{RT} \tag{B.11}$$

In the optimum situation, if the second virial coefficient of the binary mixture is known from the experiment as a function of composition, the cross second virial coefficient can be evaluated by fitting the data with Equation B.9a. If this is not the case, crude mixing rules can be used. As suggested by Equations 27.19 and 27.20, the second virial coefficient is the result of two contributions: a hard-core contribution that is independent of temperature, $B_{ii,0}$, and a temperature-dependent contribution arising from the intermolecular potentials, $B_{ii,T}$,

$$B_{ii} = B_{ii,0} - B_{ii,T} \tag{B.12}$$

Judging by the success of the mixing rules used for cubic EOSs, Equations 27.21 and 27.22, although not tested, a logical approximation to obtain the cross second virial coefficient is

$$B_{jk,0} = \frac{B_{jj,0} + B_{kk,0}}{2} \tag{B.13}$$

$$B_{jk,T} = \sqrt{B_{jj,T} B_{kk,T}} \left(1 - \delta_{jk}\right) \tag{B.14}$$

$$B_{jk} = B_{jk,0} - B_{jk,T} \tag{B.15}$$

where δ_{jk} is a binary parameter. More complex mixing rules are available in the literature [1].

For the second virial coefficient of polar, associating, and quantum compounds, several correlations proposed in the literature are reviewed by Prausnitz et al. [1]. A compilation of values of virial coefficients is available in the literature [6], [7].

REFERENCES

1. Prausnitz, J. M., Lichtenthaler, R. N., and Gomes de Azevedo, E. 1999. *Molecular Thermodynamics of Fluid Phase Equilibria*, 3rd ed. Upper Saddle River, NJ: Prentice Hall.
2. Tsonopoulos, C. 1974. An empirical correlation of second virial coefficients. *AIChE J.* 20: 263–272.
3. Orbey, H., and Vera, J. H. 1983. Correlation for the third virial coefficient using *Tc*, *Pc* and ω as parameters. *AIChE J.* 29: 107–113.

4. Abusleme, J. A., and Vera, J. H. 1989. A group contribution method for second virial coefficients. *AIChE J.* 35: 481–489.
5. Hayden, J. G., and O'Connell, J. P. 1975. A generalized method for predicting second virial coefficients. *Ind. Eng. Chem. Process Des. Dev.* 14: 209–216.
6. Dymond,, J. H., and Smith, E. B. 1979. *The Virial Coefficients of Pure Gases and Mixtures.* Oxford: Clarendon.
7. Dymond, J. H., Marsh, K. N., Wilhoit, R. C., and Wong, K. C. 2002. *Virial Coefficients of Pure Gases and Mixtures*, ed. Frenkel, M., and Marsh, K. N. Group IV, Volume 21, Subvolumes A and B of Landolt-Börnstein Numerical Data and Functional Relationships in Science and Technology, New Series. Berlin, Heidelberg: Springer-Verlag.

Appendix C
Working with the PRSV EOS

ABSTRACT

The use of the PRSV EOS is described to compute vapor and liquid volumes, saturation pressures, enthalpies, heats of vaporization, and fugacities of pure compounds, in dimensional (T, P, v) and dimensionless (T_r, P_r, z) forms. Examples show how to calculate vapor pressures given the PRSV parameters, and conversely, how to obtain the PRSV parameters from experimental vapor pressures. The application to multicomponent systems by introducing mixing rules and binary interaction parameters is presented, and equations for the fugacity coefficients are derived for gas mixtures and for more complex, strongly nonideal solutions. The combination of the EOS with excess Gibbs energy models is briefly described. Table C.3 presents PRSV parameters for over 800 compounds classified by families, and Tables C.1 and C.2 give additional parameters for further improvement of predicted liquid volumes.

THE PRSV EOS

The PRSV EOS [1–3] is a cubic equation of state (EOS) specially designed to give reliable values of the vapor pressure and to assure equality of the fugacities of saturated phases for nonpolar, polar, and associating pure compounds. Its extension to calculate the fugacity of the components of a mixture is done using appropriate mixing rules.

The PRSV EOS uses the framework of the Peng–Robinson (PR) EOS [4], namely,

$$P = \frac{RT}{(\mathrm{v} - b)} - \frac{a_c\,\alpha}{\mathrm{v}^2 + 2b\mathrm{v} - b^2} \tag{C.1}$$

with

$$a_c = A_c\frac{(RT_c)^2}{P_c} \quad \text{with} \quad A_c = \frac{8}{27}(1+2B_c)^3 = 0.45723553\ldots \tag{C.2}$$

$$b = B_c\frac{RT_c}{P_c} \quad \text{with} \quad B_c = \frac{1}{\left[\left(2-\sqrt{2}\right)^{1/3}+\left(2+\sqrt{2}\right)^{1/3}\right]^3} = 0.07779607\ldots \tag{C.3}$$

The numerical values of the constants A_c and B_c of Equations C.2 and C.3 are usually approximated to 0.45724 and 0.07780.

The PR EOS, following previous work by Soave [5] on improving the Redlich–Kwong EOS [6], gives the temperature dependence of α as:

$$\alpha = [1+\kappa(1-T_r^{0.5})]^2 \tag{C.4}$$

with κ, considered to be a constant for each compound, given as a function of the acentric factor ω defined by Equation 5.38. At the critical temperature, from Equation C.4, $\alpha = 1$.

Stryjek [1,2] introduced two important modifications to the PR EOS by considering that κ is not only a function of the acentric factor but it also depends on temperature, with the form,

$$\kappa = \kappa_0 + \kappa_1(1+T_r^{0.5})(0.7-T_r) \tag{C.5}$$

DOI: 10.1201/9781003431985-42

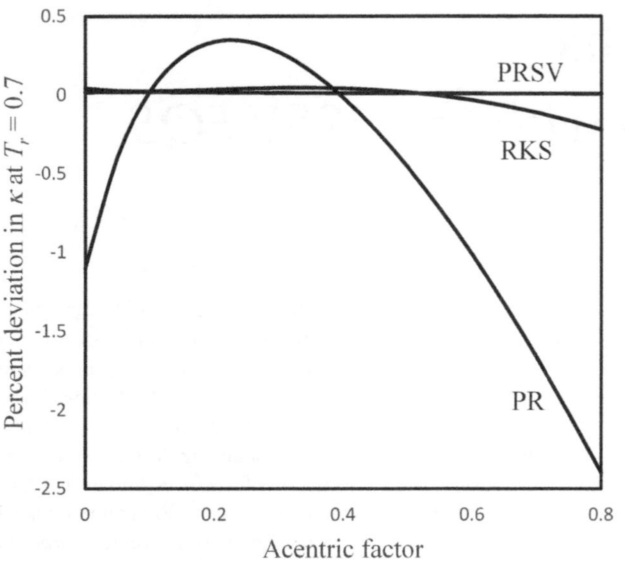

FIGURE C.1 Relative deviation in the value of κ at $T_r = 0.7$ by the Redlich–Kwong–Soave (1972) [5], Peng–Robinson (1976) [4], and PRSV (Stryjek and Vera 1986) [1–3] EOSs.

Only at $T_r = 0.7$, $\kappa = \kappa_0$. In addition, fitting vapor pressure data of a large number of nonpolar, polar and associating compounds at $T_r = 0.7$, Stryjek [1,3] proposed a function for κ_0 in terms of ω different from the correlation for κ used in the original PR EOS [4],

$$\kappa_0 = 0.37889 + 1.48971\omega - 0.17132\omega^2 + 0.01965\omega^3 \tag{C.6}$$

Figure C.1, redrawn after reference [7], compares the relative deviations in the value of κ at T_r = 0.7 obtained with Stryjek's correlation, Equation C.6, with the relative deviations obtained with the correlations used in the Redlich–Kwong–Soave [5] and Peng–Robinson [4] equations of state. Clearly, the function used in the PR EOS is only satisfactory for medium-weight hydrocarbons, say from pentane to heptane, which are of interest in the petroleum industry. The relative deviations obtained with Equation C.6 are minimal even for compounds with acentric factors as high as 0.8. Values of κ_1 were then determined for each individual compound over a wide temperature range. The temperature dependence introduced in Equation C.5 generates a term in Equation C.4 which can be shown to correspond to the form of a function previously proposed by Mathias [8] for a different cubic EOS. With these changes, the PRSV EOS [1,2] extended the use of cubic equations of state to alcohols, acids, water, and all other kinds of functional compounds.

PRSV2 [3] is a modification of PRSV including two extra pure compound parameters in the temperature dependence of the function α. In PRSV2, Equation C.5 is replaced by Equation C.5a:

$$\kappa = \kappa_0 + \left[\kappa_1 + \kappa_2(\kappa_3 - T_r)\left(1 - T_r^{0.5}\right)\right]\left(1 + T_r^{0.5}\right)(0.7 - T_r) \tag{C.5a}$$

Values of κ_1, κ_2, and κ_3 are reported in the original publications [1–3].

The extensive lists of parameters κ_1, reported in the literature [1–3,9] together with ω and the critical constants T_c and P_c necessary for calculations with the PRSV EOS, became obsolete after Olivera-Fuentes [10] introduced a major simplification by reporting directly the values of a_c, b, κ_0, and κ_1 required by Equations C.1, C.4, and C.5. The values of these parameters for over 800 compounds, based on the work of Figueira et al. [11] using data from the DIPPR compilation [12], are reproduced in Table C.3 at the end of this Appendix.

The value of the critical temperature necessary to calculate the reduced temperature in Equations C.4 and C.5, can then be obtained from the combination of Equations C.2 and C.3 as:

$$T_c / \mathrm{K} = 0.020463647 \frac{a_c}{b} \tag{C.7}$$

where we have used $R = 8.314472$ [kPa m^3 kmol^{-1} K^{-1}] and the numerical factor of the right-hand side has units of 1/R.

If required, the critical pressure can be obtained from Equation C.3:

$$P_c / \mathrm{kPa} = 0.013236568 \frac{a_c}{b^2} \tag{C.8}$$

Values of κ_0 reported in Table C.3 were correlated [11] with the acentric factor of the compounds with a regression coefficient of 0.99766 by Equation C.9:

$$\kappa_0 = 0.379368 + 1.459994\ \omega - 0.125563\ \omega^2 \tag{C.9}$$

Notably Equation C.9, which can be used to calculate the value of the acentric factor if needed, is quite close although not identical to Equation C.6, because of the different dataset and optimization procedure used in its development (see Example C.2 below). For most cases of practical interest, PRSV gives accurate enough values of vapor pressure and the equality of fugacities, as well. If a better precision is required, PRSV2 can be used.

USES OF THE PRSV EOS

Cubic EOS are useful tools for the calculation of thermodynamic properties. The complexity of their behavior has been discussed elsewhere [13]. In this appendix, we first give two cubic forms of the PRSV EOS as a function of temperature and pressure: one with dimensional terms and the other with terms in dimensionless form. We then give the analytic method of solution of a cubic equation and also an alternative iterative method of solution of the cubic equation which has advantages for the solution of the EOS in dimensionless form.

Once the values of the liquid and vapor molar volumes are obtained at the pressure and temperature of the system, the values of the fugacity or the fugacity coefficients can be calculated by the equations presented in Chapter 13. As a step for obtaining the equilibrium vapor pressure of a pure compound, we give a useful adaptation of the Antoine equation proposed by Kolasinska [14] and refer also to an accurate approximation formula developed by Soave [15]. The extension of PRSV to calculate liquid densities, enthalpies, and heats of vaporization of pure compounds [16] follows, and the appendix is completed with a consideration of mixing rules and the expressions to calculate fugacity coefficients in mixtures [17–31], and the table of recommended PRSV parameters for a large number of fluids.

CUBIC FORM OF THE PRSV EOS

The PRSV EOS cubic expansion is:

$$v^3 + [b - \frac{RT}{P}]v^2 + [\frac{a_c\,\alpha(T_r)}{P} - 2\frac{RT}{P}b - 3b^2]v + [b^2 + \frac{RT}{P}b - \frac{a_c\,\alpha(T_r)}{P}]b = 0 \tag{C.10}$$

which corresponds to the general cubic form:

$$v^3 + C_1 v^2 + C_2 v + C_3 = 0 \tag{C.11}$$

with

$$C_1 = b - \frac{RT}{P} \tag{C.12}$$

$$C_2 = \frac{a_c\,\alpha(T_r)}{P} - 2\frac{RT}{P}b - 3b^2 \tag{C.13}$$

$$C_3 = [b^2 + \frac{RT}{P}b - \frac{a_c\,\alpha(T_r)}{P}]b \tag{C.14}$$

Equation C.10 can also be written in dimensionless terms as:

$$z^3 + (B_P - 1)z^2 + \left(A_P - 2B_P - 3B_P^2\right)z + \left(B_P^2 + B_P - A_P\right)B_P = 0 \tag{C.15}$$

which corresponds to the general form of Equation C.11 with

$$C_1 = B_P - 1 \tag{C.16}$$

$$C_2 = A_P - 2B_P - 3B_P^2 \tag{C.17}$$

$$C_3 = \left(B_P^2 + B_P - A_P\right)B_P \tag{C.18}$$

where

$$A_P = \frac{a_c\alpha P}{(RT)^2} = A_c\alpha\frac{P_r}{T_r^2} \tag{C.19}$$

$$B_P = \frac{bP}{RT} = B_c\frac{P_r}{T_r} \tag{C.20}$$

Analytic Solution of a Cubic Equation

The exact Cardano method [32] to evaluate the roots of a cubic equation, given in Perry's Handbook [33] and presented in a simplified way elsewhere [13], is reproduced here. Consider the general form of a cubic equation:

$$v^3 + C_1v^2 + C_2v + C_3 = 0 \tag{C.21}$$

where C_1, C_2, and C_3 are numeric coefficients. For simplicity, define

$$C_4 = C_1/3 \tag{C.22}$$

$$D = \frac{1}{2}(C_2\,C_4 - C_3) - C_4^{\,3} \tag{C.23}$$

$$E = C_4^{\,2} - (C_2/3) \tag{C.24}$$

and

$$\Delta = D^2 - E^3 \tag{C.25}$$

If $\Delta = 0$, say $|\Delta| < 10^{-3}$, there are three real roots with at least two equal roots:

$$v_1 = 2\sqrt[3]{D} - C_4 \tag{C.26}$$

$$v_2 = v_3 = -\sqrt[3]{D} - C_4 \tag{C.27}$$

If $\Delta > 0$, there is one real root and two complex conjugate roots. In this case, define

$$F = \sqrt[3]{D + \sqrt{\Delta}} \tag{C.28}$$

and

$$G = \sqrt[3]{D - \sqrt{\Delta}} \tag{C.29}$$

$$H = -\left[\frac{1}{2}(F + G) + C_4\right] \tag{C.30}$$

$$I = \frac{\sqrt{3}}{2}(F - G) \tag{C.31}$$

$$v_1 = F + G - C_4 \tag{C.32}$$

$$v_2 = H + I\,i \tag{C.33}$$

$$v_3 = H - I\,i \tag{C.34}$$

If $\Delta < 0$, there are three real and unequal roots. In this case, define

$$\vartheta\,(\text{radians}) = \text{arc}\cos\left(\frac{D}{\sqrt{E^3}}\right) \tag{C.35}$$

$$v_1 = 2\sqrt{E}\cos(\vartheta / 3) - C_4 \tag{C.36}$$

$$v_2 = 2\sqrt{E}\cos(\vartheta / 3 + (2/3)\pi) - C_4 \tag{C.37}$$

$$v_3 = 2\sqrt{E}\cos(\vartheta / 3 + (4/3)\pi) - C_4 \tag{C.38}$$

DIMENSIONLESS FORM AND ALTERNATIVE SOLUTION OF THE PRSV EQUATION

Although the above analytic solution is direct and yields all the roots, whether real or imaginary numbers, it is not always the best approach in terms of precision and speed, mostly if the equation is to be solved repeatedly, as in applications in process simulation [34,35]. Several authors [35–39] have proposed iterative algorithms as an alternative to the analytic solution method presented above. There is also the matter of units, for example, volumes in m^3/kmol are 10^3 times smaller than in cm^3/mol, and each term in the cubic polynomial is accordingly smaller by a factor of 10^9. The precision of small, liquid-like volume roots may in particular be affected [35], and it has been advised that the cubic equation be written and solved in terms of molar densities instead of volumes [35,39]. We present in what follows an alternative solution especially applicable to the dimensionless form of the cubic EOS written in terms of the compressibility factor $z = Pv/RT$. These terms also appear in the calculation of fugacity coefficients and other dimensionless fluid properties. A thorough analysis of the numerical problems involved and a more advanced algorithm for solving cubic equations are given in Ref. [39].

A simple alternative scheme to the analytical solution is to find a first root z_I by the Newton–Raphson search:

$$z^{(n+1)} = \left[z - \frac{z^3 + C_1 z^2 + C_2 z + C_3}{3z^2 + 2C_1 z + C_2} \right]^{(n)} \tag{C.39}$$

The initial ideal-gas approximation $z^{(0)} = 1$ usually works well, but more sophisticated estimates can be generated [39]. Once converged, the value of z_I is used to "deflate" the cubic polynomial into a quadratic equation that can be solved directly for the remaining two roots. Define

$$D_1 = C_1 + z_I \tag{C.40}$$

$$D_0 = \frac{C_3}{z_I} \tag{C.41}$$

$$\Delta = \frac{D_1^2}{4} - D_0 \tag{C.42}$$

The other two roots are then:

$$z_{II,III} = -\frac{D_1}{2} \pm \sqrt{\Delta} \tag{C.43}$$

If $\Delta < 0$, there is only one real root, z_I, and two complex conjugate roots. If $\Delta = 0$, say $|\Delta| < 10^{-3}$, there are three real roots, at least two of them are equal. If $\Delta > 0$, there are three distinct real roots. Volumes can then be obtained in any desired set of units as $v = zRT/P$.

HOW TO CALCULATE THE SATURATION PRESSURE OF A PURE COMPOUND WITH PRSV EOS

This is the first question that a beginner asks when dealing with a cubic EOS and, although the answer is simple, the algebra is not so much. At saturation, the pressure of the liquid is equal to the pressure of the vapor phase and also the fugacity coefficient of the pure compound in the vapor phase (*sv*) is equal to its fugacity coefficient in the liquid phase (*sl*):

$$P^{sv}\left(\mathrm{v}^{sv},T\right)=P^{sl}\left(\mathrm{v}^{sl},T\right) \tag{C.44}$$

$$\phi^{sv}\left(\mathrm{v}^{sv},T\right)=\phi^{sl}\left(\mathrm{v}^{sl},T\right) \tag{C.45}$$

with $P(T,\mathrm{v})$ given by Equation C.1, and $\phi(T,\mathrm{v})$ obtained from Equation 13.14 (see also Problem 13.3) as:

$$\ln\phi = z-1-\ln\left[z\left(1-\frac{b}{\mathrm{v}}\right)\right]-\frac{a_c\alpha(T_r)}{2bRT\sqrt{2}}\ln\left[\frac{\mathrm{v}+b\left(1+\sqrt{2}\right)}{\mathrm{v}+b\left(1-\sqrt{2}\right)}\right] \tag{C.46}$$

Parameters a_c and b and the constants required to calculate the cohesion parameter are obtained from Table C.1 for PRSV and for PRSV2 with constants from [1–3].

Although it is not an easy task, the set of Equations C.44 and C.45 can be solved by iterating simultaneously on the two saturated phase volumes. However, it is more usual to turn the problem into a single-variable search, that is, to iterate on values of pressure, solving Equation C.5 for the saturated volumes at each assumed pressure, and checking whether the equality of fugacities is satisfied. The first step is to estimate an initial value of the saturation pressure with an equation that does not require knowing the saturation volumes. After the first estimate, a typical trial and error procedure is followed until the equality of fugacity is satisfied. A good initial estimate is important to ensure that the cubic polynomial does have three real and physically meaningful ($\mathrm{v} > b$) roots. For this first estimation of the saturation pressure, we follow the path proposed by Kolasinska [14] and use the Antoine equation written as:

$$\log_{10} P^s = A_A - \frac{B_A}{C_A + T_r} \tag{C.47}$$

Applying this equation to the critical point, the normal boiling point and the reduced temperature 0.7, where the saturation pressure is directly obtained from Equation 5.38, with the value of the acentric factor ω calculated from C.9, Kolasinska [14] obtained:

$$A_A = \frac{B_A}{C_A + \theta} \tag{C.48}$$

$$B_A = \frac{(C_A+\theta)(C_A+1)}{1-\theta}\log_{10} P_c \tag{C.49}$$

$$C_A = \frac{D_A - 0.7}{1 - D_A} \tag{C.50}$$

with

$$D_A = \frac{(0.7-\theta)\log_{10} P_c}{(1-\theta)\left(\log_{10} P_c - \omega - 1\right)} \tag{C.51}$$

and

$$\theta = \frac{T_b}{T_c} \tag{C.52}$$

where T_b is the absolute temperature of the normal boiling point, θ is dimensionless, and pressures P^s and P_c are in atmospheres, and ω is the acentric factor, which can be calculated from Equation C.9 if not otherwise available.

As in the previous section, the calculation of saturation pressures can also be performed in dimensionless terms. Given the reduced temperature T_r and an initial estimate of the corresponding reduced saturation pressure P_r^s, parameters A_P and B_P are computed from Equations C.6–C.12 as applicable. Solution of Equation C.14 then yields the compressibility factors of the saturated liquid and vapor, which are tested to check whether they satisfy the equality of fugacity coefficients, given now by the dimensionless equivalent of Equation C.17:

$$\ln\phi = z - 1 - \ln(z - B_P) - \frac{A_P}{2B_P\sqrt{2}}\ln\left[\frac{z + B_P(1+\sqrt{2})}{z + B_P(1-\sqrt{2})}\right] \tag{C.53}$$

Iteration proceeds, for example, by the Newton–Raphson search, with Equation C.16b rewritten as:

$$F(P_r^s) = \ln\frac{\phi^{sv}}{\phi^{sl}}, \quad F'(P_r^s) = \frac{z^{sv} - z^{sl}}{P_r^s}$$

where we have used a reduced version of Equation 13.7a for the differential change of the fugacity coefficients with pressure at constant temperature. Here, too, it is important to start from a good initial estimate. Equation C.18 can be recast in reduced form as:

$$\log_{10} P_r^s = E_A - \frac{B_A}{C_A + T_r} \tag{C.54}$$

Applying this equation at $T_r = 1,\ P_r^s = 1,\ \log_{10} P_r^s = 0$, and at $T_r = 0.7,\ \log_{10} P_{r,T_r=0.7}^s = -(1+\omega)$, we get:

$$B_A = E_A(C_A + 1) \tag{C.55}$$

$$E_A = -\frac{10}{3}(1+\omega)(C_A + 0.7) \tag{C.56}$$

We note that setting $C_A = 0$ would yield:

$$\log_{10} P_r^s = -\frac{7}{3}(1+\omega)\left(\frac{1}{T_r} - 1\right)$$

which is known as the Wilson or "shortcut" vapor pressure equation [40–42]. Following Kolasinska [14], however, we incorporate normal boiling point data, $T_{r,b} = T_b/T_c = \theta,\ P_{r,b}^s = 1/P_c(\text{atm})$ to get the third constant,

$$C_A = \frac{F_A\theta - 0.7}{1 - F_A} \tag{C.57}$$

with

$$F_A = -\frac{3}{10} \times \frac{\log_{10} P_{r,b}^s}{(1+\omega)(1-\theta)} \tag{C.58}$$

It should be emphasized that Equations C.47 and C.54 are wholly equivalent, the only difference being that $E_A = A_A - \log_{10} P_c$.

An alternative estimation method for P_r^s [15] is based on the fact that, as shown in Equations C.10 and C.53, compressibility factors and fugacity coefficients depend only on the dimensionless parameters A_P and B_P. Hence, specification of α / T_r uniquely determines P_r^s / T_r, regardless of the particular form of the $\alpha(T_r)$ function or the number of fluid-specific parameters that it may contain. Values of P_r^s / T_r computed a priori over a wide range of values of α / T_r, sufficient for all practical purposes, have been correlated by the empirical expression [15]:

$$\ln\left(P_r / T_r\right) = \sum_{k=1}^{4} c_k X^{(k+1)/2} + \sum_{k=5}^{10} c_k X^{k-2} \; ; \qquad X = \frac{\alpha}{T_r} - 1$$

This formula is highly accurate and supplies pressure estimates that lead to convergence in a few steps, sometimes even just one iteration. The coefficients are given in the original reference [15].

Example C.1 Use the PRSV EOS to calculate the vapor pressure of n-hexane from the triple to the critical point. Data on this compound are [12]: critical temperature $T_c = 507.43$ K; critical pressure $P_c = 30.338$ bar=29.941 atm; normal boiling temperature $T_b = 342.09$ K; acentric factor $\omega = 0.3077$; and triple point temperature $T_{tp} = 177.84$ K. The PRSV parameters for n-hexane are $\kappa_0 = 0.79879$, $\kappa_1 = 0.12382$.

Solution: From Equations C.48–C.52 and the above data we get $\theta = 0.67416$, $A_A = 4.4485$, $B_A = 2.9183$, $C_A = -0.018145$, $D_A = 0.69446$. Using Equation C.47, pressure estimates are generated at reduced temperatures from $T_r = 1$ down to $T_{tp}/T_c = 0.3505$. Equation C.15 is solved for the compressibility factors of the saturated phases, and the fugacity coefficients follow from Equation C.53. The pressure estimates are corrected using Newton–Raphson iteration until convergence is achieved to better than 10^{-6} relative error. The results are collected in Table EXC.1 below and plotted as $\log_{10} P^s$ against $1000/T$ for ease of visualization in Figure EXC.1, where the initial estimates and PRSV pressures are compared with reference values from [12]. Equation C.18 is seen to give excellent estimates from the critical to the normal boiling temperature, with increasing deviation on extrapolation to the lowest temperature, requiring more iterations for convergence of the pressures. The PRSV EOS, on the other hand, yields excellent predictions down to the triple point temperature.

TABLE EXC.1
Estimated and PRSV-Calculated Vapor Pressures of *n*-Hexane

		Equation C.47	PRSV			
T/K	T_r	P^s/bar	$v^{sl}/(m^3/kmol)$	$v^{sv}/(m^3/kmol)$	P^s/bar	Iterations
507.43[a]	1.00	30.34	0.4275	0.4275	30.34	1
456.69	0.90	13.96	0.1956	1.835	14.28	3
405.94	0.80	5.268	0.1598	5.213	5.502	3
355.20[b]	0.70	1.494	0.1418	17.94	1.551	3
342.09[c]	0.6742	1.013	0.1385	26.18	1.040	3
304.46	0.60	0.2746	0.1309	95.41	0.2615	3
253.72	0.50	0.02499	0.1235	1167	0.01804	5
202.97	0.40	6.483E−04	0.1183	71212	2.370E−04	8
177.84[d]	0.3505	4.708E−05	0.1163	1580938	9.353E−06	14

[a] Critical point.
[b] Temperature for acentric factor.
[c] Normal boiling point.
[d] Triple point.

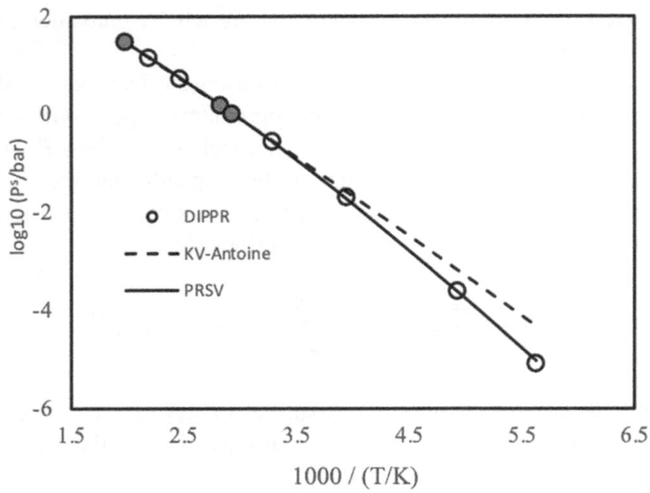

FIGURE EXC.1 Vapor pressure of n-hexane. Dashed line: Antoine Equation C.47 with Kolasinska–Vera parameterization [14]. Continuous line: PRSV EOS with parameters from Table C.3. Circles: Reference values from vapor pressure equation in [12]; the three points used for fitting Equation C.18 are shown as filled circles.

HOW TO CALCULATE SATURATED LIQUID MOLAR VOLUMES WITH THE PRSV EOS

It has long been known that the cubic equations of state do not give reliable values of the molar volumes or enthalpies of pure compounds. In fact, the molar volumes obtained above do not correspond to the real values of the saturated molar volumes at the temperature of the calculation. They are just values that help to evaluate the fugacity coefficient of the pure compound and give a very accurate value of the vapor pressure. In order to obtain more realistic saturated liquid volumes, Proust et al. [16] proposed to use:

$$\mathrm{v}^{sl} = \left(v_{\mathrm{PRSV}}^{sl} + D_{\mathrm{v}}\right)\left[1 - \exp\left(A_{\mathrm{v}} - \frac{B_{\mathrm{v}}}{T_r^3}\right)\right] \quad \text{(C.59)}$$

where D_{v} is a displacement parameter given in Table C.1 for some compounds of industrial interest. The temperature correction factor in brackets extends the use of the correlation to temperatures close to the critical point. The parameters A_{v} and B_{v} are given in Table C.2 for families of typical compounds. For compounds not listed in Table C.1, it suffices to know one single value of liquid density to calculate the value of the displacement D_{v}. As the temperature correction factor is given for classes of similar compounds, for estimates of liquid densities at other temperatures it is only necessary to identify to what family the compound belongs.

HOW TO CALCULATE ENTHALPIES AND HEATS OF VAPORIZATION WITH THE PRSV EOS

From exact thermodynamics, the difference between the enthalpy of a pure compound fluid h and the enthalpy of the ideal gas h^* at the same temperature, that is, its *residual enthalpy*, is given by:

$$h - h^* = P\mathrm{v} - RT + \int_{\mathrm{v}}^{\infty}\left[P - T\left(\frac{\partial P}{\partial T}\right)_{\mathrm{v}}\right]d\mathrm{v} \quad \text{(C.60)}$$

TABLE C.1
PRSV D_v-Parameters for Pure Compounds

Compound	$D_v \times 10^3$ (m³/kmol)	Compound	$D_v \times 10^3$ (m³/kmol)	Compound	$D_v \times 10^3$ (m³/kmol)
Methane	4.050	Benzene	2.951	Ethanol	−3.957
Ethane	4.450	Toluene	0.764	1-Propanol	−1.319
Propane	4.708	o-Xylene	−0.261	2-Propanol	−3.196
n-Butane	4.974	Ethylbenzene	−0.664	Isobutanol	−0.877
n-Pentane	3.551	Methyl chloride	0.305	1-Butanol	−0.571
n-Hexane	0.823	Chloroform	5.889	1-Pentanol	1.110
n-Heptane	−0.745	1,2-Dichloroethane	1.435	1-Hexanol	5.416
n-Octane	−4.488	Nitrogen	4.056	Acetone	−9.661
n-Decane	−11.749	Oxygen	3.402	Butanone	−6.176
n-Dodecane	−20.204	Sulfur dioxide	0.526	Diethyl ether	1.253
n-Tetradecane	−55.068	Carbon monoxide	4.107	Ethyl acetate	−5.200
Acetylene	1.557	Carbon dioxide[a]	1.661	Propionic acid	−5.305
Ethylene	3.897	Chlorine	2.505	Ammonia	−3.164
Propylene	5.208	Hydrogen sulfide	2.977	Water	−3.566
Styrene	−0.750	Hydrogen chloride	−0.384	Acetic acid	−18.031
Cyclohexane	5.979	Methanol	−7.341	Acetonitrile	−33.772

[a] At the triple point temperature.

Reproduced from [16], with permission. Copyright John Wiley & Sons, Inc., 2015.

TABLE C.2
Parameters for the Volume Correction

Class	Typical Compounds	A_v	B_v
1	Nonpolar, slightly polar nonassociated (hydrocarbons, halogenated compounds, amines, N_2, O_2, SO_2, CO_2, Cl_2, H_2S, HCl)	0.808	2.575
2	Polar and associated compounds (alcohols, ketones, ethers, esters, organic acids other than acetic acid, ammonia)	0.983	2.392
3	Three special compounds: water, acetic acid, and acetonitrile	0.685	1.894

Reproduced from Ref. [16], with permission. Copyright John Wiley & Sons, Inc., 2015.

For the PRSV EOS, then

$$h - h^* = P\text{v} - RT - \left(1 - \sqrt{\frac{T_r}{\alpha}} F(T_r)\right) \frac{a_c \alpha}{2b\sqrt{2}} \ln \frac{\text{v} + \left(1+\sqrt{2}\right)b}{\text{v} + \left(1-\sqrt{2}\right)b} \tag{C.61}$$

with

$$F(T_r) = \kappa_1 (0.7 - T_r)\left(1 - \sqrt{T_r}\right) - 2\kappa_1 \sqrt{T_r}\,(1 - T_r) - \kappa \tag{C.62}$$

The dimensionless version of Equation C.61 is:

$$\frac{h-h^*}{RT}=z-1-\left(1-\sqrt{\frac{T_r}{\alpha}}F(T_r)\right)\frac{A_P}{2B_P\sqrt{2}}\ln\frac{z+\left(1+\sqrt{2}\right)B_P}{z+\left(1-\sqrt{2}\right)B_P} \tag{C.63}$$

The heat of vaporization is directly obtained as:

$$\Delta h^s=\left(h-h^*\right)^{sv}-\left(h-h^*\right)^{sl} \tag{C.64}$$

The PRSV EOS gives accurate values of the vapor pressure of pure compounds as a function of temperature. As the heat of vaporization is directly related by the Clausius–Clapeyron equation to the slope of the saturation pressure with respect to temperature, Equation C.64 gives reliable values of heats of vaporization [16] and Equations C.61–C.63 can also be used to calculate enthalpy changes.

HOW TO CALCULATE FUGACITY COEFFICIENTS IN MIXTURES WITH PRSV

To calculate the fugacity coefficient of a compound in a mixture it is necessary to introduce mixing rules. For gas phase mixtures, commonly used mixing rules for cubic EOS have the form:

$$b=\sum_k y_k b_k \tag{C.65}$$

and

$$a=\sum_i\sum_j y_i y_j a_{ij} \tag{C.66}$$

with

$$a_{ii}=a_{c,i}\alpha_i\left(T_{ri}\right) \tag{C.67}$$

and

$$a_{ij}=\left(1-\delta_{ij}\right)\sqrt{a_{ii}a_{jj}} \tag{C.68}$$

where, in one way or another, the binary parameter δ_{ij} is adjusted to fit known experimental data. In many cases the value of δ_{ij} is close to zero so the value of $1-\delta_{ij}$ is close to unity. For mixtures of compounds belonging to the same homologous series, or of compounds of similar structure, the value of the adjustable parameter is closely independent of composition. For gas mixtures of very dissimilar compounds or for liquid mixtures, experience has shown that the binary parameter is a function of the composition of the mixture. For a component i in a multicomponent mixture, the fugacity coefficient expression obtained with the above mixing rules, where δ_{ij} is independent of composition, has the form:

$$\ln\phi_i=\frac{b_i}{b}(z-1)-\ln\left[z\left(1-\frac{b}{v}\right)\right]-\frac{a}{2b\sqrt{2}}\left[\frac{2}{a}\sum_k y_k a_{ik}-\frac{b_i}{b}\right]\ln\left[\frac{v+b\left(1+\sqrt{2}\right)}{v+b\left(1-\sqrt{2}\right)}\right] \tag{C.69a}$$

or, in dimensionless terms:

$$\ln\phi_i=\frac{B_{P,i}}{B_P}(z-1)-\ln\left(z-B_P\right)-\frac{A_P}{2B_P\sqrt{2}}\left[\frac{2}{A_P}\sum_k y_k A_{P,ik}-\frac{B_{P,i}}{B_P}\right]\ln\left[\frac{z+B_P\left(1+\sqrt{2}\right)}{z+B_P\left(1-\sqrt{2}\right)}\right] \tag{C.69b}$$

In this expression, z, A_P, and B_P are given by Equations 5.16, C19, and C.20, respectively and the cross parameters $A_{P,ik}$ are defined by the same combining rule of Equation C.66.

Very many mixing rules have been proposed. Here we only comment in detail on the two-parameter extension of the mixing rule presented above and also on what we think is a consistent method to combine the PRSV EOS with expressions for the excess Gibbs energy of liquid mixtures. In principle, the PRSV EOS can be used in at least three ways to calculate phase equilibria. First, it can be used to calculate fugacities in the vapor phase only, using an independent expression for the excess Gibbs energy to calculate activities in the liquid phase. This is the well-known $\gamma - \phi$ approach and the simple mixing rule presented above suffices for the gas phase in most cases. An alternative to this, not used in practice, would be to use the PRSV EOS with different mixing rules (or different binary parameters) for the vapor and the liquid phase. Normally, PRSV is used with the same binary parameters and the same mixing rules for both the liquid and the vapor phase, the $\phi - \phi$ approach. For this third option, a two-binary-parameter mixing rule is necessary.

TWO-BINARY-PARAMETER MIXING RULE AND ITS EXTENSION TO MULTICOMPONENT MIXTURES

For strongly nonideal mixtures presenting an asymmetric pattern of the excess Gibbs energy, among other mixing rules, Stryjek [2] proposed to use Equation C.39 with a composition-dependent binary parameter of the form:

$$\delta_{ij} = x_i k_{ij} + x_j k_{ji} \tag{C.70}$$

with

$$k_{ij} \neq k_{ji}$$

Almost simultaneously, Adachi and Sugie [17] and also Panagiotopoulos and Reid [18] presented slightly more complex, but perfectly equivalent, forms differing only in the details. In a study predicting ternary vapor–liquid equilibrium using binary parameters only, Sandoval et al. [19] found that both the Adachi-Sugie and the Panagiotopoulos-Reid mixing rules gave better results than the direct extension of the Stryjek form, which presented the Michelsen-Kistenmacher syndrome [20] discussed further by Zabaloy and Vera [21]. Thus, Sandoval *et al.* [19] proposed the following form to extend Stryjek's two binary parameters to multicomponent mixtures:

$$\delta_{ij} = \left(1 - x_i - x_j\right)\bar{k}_{ij} + x_i k_{ij} + x_j k_{ji} \tag{C.71}$$

with

$$\bar{k}_{ij} = \frac{1}{2}\left(k_{ij} + k_{ji}\right) \tag{C.72}$$

For a binary mixture, Equation C.71 reduces to Equation C.70.

A completely general form of Equations 69a,b valid for all mixing rules is:

$$\ln \phi_i = \frac{\bar{b}_i}{b}(z-1) - \ln(z - B_P) - \frac{A_P}{2B_P\sqrt{2}}\left[1 + \frac{\bar{a}_i}{a} - \frac{\bar{b}_i}{b}\right]\ln\left[\frac{z + B_P\left(1+\sqrt{2}\right)}{z + B_P\left(1-\sqrt{2}\right)}\right] \tag{C.73}$$

where $\bar{a}_i$ and $\bar{b}_i$ are analogous to partial molar coefficients,

$$\bar{a}_i = \left(\frac{\partial(na)}{\partial n_i}\right)_{T,P,n_{j\neq i}}, \quad \bar{b}_i = \left(\frac{\partial(nb)}{\partial n_i}\right)_{T,P,n_{j\neq i}}$$

The fugacity coefficient of compound i in a binary mixture described by the mixing rules of Equations C.65, C.66, and C.70 or C.71 is given by Equation C.73 with [2]

$$\bar{a}_i = 2\left(x_i a_{ii} + x_j a_{ij}\right) - a + 2x_i x_j^2 \left(a_{ii} a_{jj}\right)^{0.5}\left(k_{ji} - k_{ij}\right), \quad \bar{b}_i = b_i \tag{C.74}$$

and the fugacity coefficient of compound i in a multicomponent mixture described by the mixing rules of Equations C.65, C.66, and C.71 is given also by Equation C.73, but with [19]

$$\bar{a}_i = 2\sum_j x_j a_{ij} - a + 2\sum_{j\neq i} x_i x_j \left(a_{ii} a_{jj}\right)^{0.5}\left[\left(x_i - x_j - 1\right)\Delta k_{ij}\right] + 2\sum_{j\neq i}\sum_{m>j} x_j x_m \left(a_{jj} a_{mm}\right)^{0.5}\left[\left(x_j - x_m\right)\Delta k_{jm}\right] \tag{C.75}$$

where

$$\Delta k_{ij} = k_{ij} - \bar{k}_{ij} \tag{C.76}$$

For a binary system the third term of the right-hand side of Equation C.75 cancels out, and it reduces to Equation C.74.

CONSISTENT METHOD TO COMBINE THE PRSV EOS WITH EXCESS GIBBS ENERGY MODELS

Considering that parameters for group methods like ASOG or UNIFAC were available to calculate activities in the liquid phase, it was desirable to have a way to combine these group methods with the PRSV EOS. As early examples, Orbey et al. [22] used UNIFAC and Tochigi [23] used ASOG.

The activity coefficient of a component of a liquid mixture was defined in terms of volumetric properties by Equation 15.1:

$$\ln \gamma_{i,0} \equiv \frac{1}{RT}\int_0^P \left[\bar{v}_i - v_i^0\right] dP \tag{15.1}$$

This definition, combined with Equation 13.9, gave:

$$\gamma_{i,0}(T,P,x) = \frac{\phi_i(T,P,x)}{\phi_i^0(T,P)} \tag{15.2c}$$

This is a result of exact thermodynamics. No assumptions or models are involved. Thus, the use of Equation (15.2c) was proposed [24] as a consistent method to combine the PRSV EOS, or any other EOS, with excess Gibbs energy functions. Notably, this simple formulation gained acceptance and made it to the front cover of the *AIChE J.* "Founders Tribute" dedicated to John M. Prausnitz [25].

Many other mixing rules are available in the literature. See, for example, the work of Zabaloy et al. for vapor–liquid equilibrium [26,27]. For the use of the PRSV EOS for liquid mixtures, Ohta et al. have presented studies of liquid–liquid equilibrium [28], of ternary vapor–liquid equilibrium [29], and even of excess enthalpies [30]. Just to name a few examples, Orbey et al. combined the PRSV EOS with UNIFAC [22] for vapor–liquid and liquid–liquid calculations using a mixing rule previously proposed by Wong and Sandler [31]. Tochigi used the PRSV EOS combined with ASOG to predict high-pressure vapor–liquid equilibrium for highly nonideal ternary mixtures [23].

TADIP TABLE OF PRSV PARAMETERS

As mentioned previously in this appendix, joint solution of Equations C.15 and C.53 for a pure fluid in vapor–liquid equilibrium leads to a mutual dependence between P_r^s / T_r and α / T_r. In Example C.1, this connection was used to obtain the vapor pressures predicted by the PRSV EOS given the substance-specific parameters κ_0 and κ_1 of Equation C.5. But the reverse calculation is also possible: given a known value of P_r^S at T_r and an initial estimate of α, A_p and B_P are calculated, Equation C.15 is solved for the compressibility factors z^{sl} and z^{sv} of the saturated phases, and the fugacity coefficients are obtained from Equation C.35 and checked for equality. Iteration on α can proceed simply by successive substitution:

$$\frac{\alpha}{T_r}=\frac{2B_c\sqrt{2}\left[z^{sv}-z^{sl}-\ln\frac{z^{sv}-B_P}{z^{sl}-B_P}\right]}{A_c\ln\frac{\left[z^{sv}+\left(1+\sqrt{2}\right)B_P\right]\left[z^{sl}+\left(1-\sqrt{2}\right)B_P\right]}{\left[z^{sv}+\left(1-\sqrt{2}\right)B_P\right]\left[z^{sl}+\left(1+\sqrt{2}\right)B_P\right]}}$$

As in the computation of saturation pressures, a good initial estimate is desirable, and this is supplied by a highly accurate approximation also developed by Soave [15]:

$$\frac{\alpha}{T_r}=1+\sum_{k=1}^{2}d_kY^{(k+1)/2}+\sum_{k=3}^{10}d_kY^{k-1};\qquad Y=-\frac{\ln(P_r/T_r)}{10}$$

Figueira et al. [11] used pseudo-experimental vapor pressure data generated from equations available in the print version of the DIPPR® database [12] to obtain $\alpha(T_r)$ for 846 pure fluids over their entire range of vapor–liquid coexistence. The values thus obtained were fitted to the PRSV Equations C.4 and C.5 to get optimal values of κ_0 and κ_1 by minimizing the sum of squared relative deviations of the predicted pressures:

$$\min_{\kappa_0,\kappa_1}\sum_k\left[\frac{P_r^s(\kappa_0,\kappa_1)-P_{r,\exp}^s}{P_{r,\exp}^s}\right]_k^2\cong\min_{\kappa_0,\kappa_1}\sum_k\left[W\frac{\alpha(\kappa_0,\kappa_1)-\alpha_{\exp}}{\alpha_{\exp}}\right]_k^2 \tag{C.77}$$

where the weighting factors W account for the statistical propagation of errors from α to P_r^s and are given from exact thermodynamics by [43]:

$$W=\frac{d\ln P_r^s}{d\ln\alpha}=1-\frac{1}{z^{sv}-z^{sl}}\ln\frac{z^{sv}-B_P}{z^{sl}-B_P} \tag{C.78}$$

The optimized parameters κ_0 and κ_1 are listed in Table C.3 to eight significant figures. They reproduce the vapor pressures of all fluids in the database with an average 2.38% and a maximum 5.87% absolute deviation [11]. It may be noted that since the entire procedure employed only dimensionless variables, it was not necessary to compute the fluid-specific constants a_c and b given by Equations C.19 and C.20. These constants, however, are required for use of the EOS at true (not reduced) temperatures and pressures; therefore, they have been included in Table C.3.

Furthermore, in the unlikely event that the PRSV parameters are needed for a compound not in Table C.3 (other than CO_2, see below), they can be estimated from an extended corresponding states correlation [11]:

$$\kappa_0=0.379368+1.459994\ \omega-0.125563\ \omega^2$$

$$\kappa_1=0.599529-1.952083\ z_c+0.080764\ \omega-0.209272\ \omega^2$$

where ω is the acentric factor and z_c is the critical compressibility factor of the compound in question. This correlation for κ_0 differs from Equation C.6 because the parameters in Table C.3 were not constrained to reproduce the acentric factor of the fluids, as this would have reduced the degrees of freedom of the statistical procedure. The difference between κ_0 predicted by the above equation and by Equation C.6 is about 1.1% on average but can be as high as 17% for some compounds.

Example C.2: Calculation of PRSV parameters for CO_2 by fitting vapor pressure data

In the course of preparing this new edition, it was noticed that CO_2 was missing from the original version of the table of PRSV parameters (Table C.4 of the first edition). Although parameters for this compound had been obtained [44], the omission happened for reasons explained below. Given the considerable importance of CO_2 in process engineering, we are aware that its parameters should be reported, and we take the opportunity to illustrate the method by which these, and all other parameters in Table C.3, are obtained from vapor pressure data of each fluid.

We now take the critical temperature and vapor pressure equation of CO_2 from the online DIPPR® 801 Sample Database [45], and rewrite that equation in reduced form as:

$$\ln\left(P_r^s\right) = -9.33237\left(\frac{1}{T_r}-1\right) - 3.86388\ln T_r + 0.222805\left(T_r^E - 1\right)$$

valid from the triple $\left(T_{r,tp} = 0.71194\right)$ to the critical point. Since $T_{r,tp} > 0.70$, we note that, strictly speaking, carbon dioxide does not have an experimental acentric factor because vapor–liquid coexistence is not possible at the "omega temperature" $T_r = 0.70$. This is why CO_2 was excluded from the development of the generalized correlation of PRSV parameters and ended up being inadvertently absent from the original version of Table C.3. Values of ω_{CO2} reported in the literature, obtained no doubt by extrapolation, differ from one source to another. The DIPPR® 801 Sample Database [45] quotes the value $\omega = 0.223621$, not entirely consistent with the above vapor pressure equation that, upon slight extrapolation beyond the triple point temperature, yields $\omega = 0.223855$. The latter value will be used for full consistency with the vapor pressures.

Table EXC.2a shows selected values obtained for CO_2 using the reduced vapor pressure data derived from the DIPPR® 801 Sample Database.

Next, the complete $\{T_r, \alpha\}$ set is correlated by the PRSV cohesion function. The original optimization procedure [11] minimized the weighted deviation function of Equation C.77. Here we use two more traditional statistics, the average absolute relative deviation (AARD), and the root mean square deviation (RMSD):

$$\text{AARD} = \frac{1}{N}\sum_{\text{all } T_r}\left|W\frac{\alpha_{\text{PRSV}} - \alpha_{\text{exp}}}{\alpha_{\text{exp}}}\right|$$

$$\text{RMSD} = \sqrt{\frac{1}{N}\sum_{\text{all } T_r}\left(W\frac{\alpha_{\text{PRSV}} - \alpha_{\text{exp}}}{\alpha_{\text{exp}}}\right)^2}$$

where N is the number of data points. In either case, given the structure of PRSV, there are two possible strategies:

- One-parameter optimization: Obtain $\kappa_0 = 0.70400845$ directly from $\omega = 0.223855$ using Equation C.6, and search only for κ_1 that minimizes the objective function.
- Two-parameter optimization: Do not use Equation C.6. Search for both κ_0 and κ_1 that minimize the objective function. This will not honor the experimental ω, but will improve the overall fit of the data.

Table EXC.2b gives the results for these 2×2 combinations. As could be expected, better statistics are obtained when both parameters are optimized. On the other hand, there is little difference between the goodness of fit with each objective function, and our very slight preference would be for the parameters obtained by minimizing the RMSD, $\kappa_0 = 0.70132798$, $\kappa_1 = 0.02149019$. In practical applications, these values could and would of course be rounded off to fewer decimal places, say $\kappa_0 = 0.7013$, $\kappa_1 = 0.0215$. They are now included in group 17 (inorganic gases) of Table C.3, with more significant figures than usual as all the other values therein, just to emphasize the precision of the computational method employed to obtain them.

TABLE EXC.2A
Reduced Vapor Pressures, Estimated [15] and Calculated α Factors of CO_2

T_r	P_r^s	α [15]	$\alpha(T_r)$	T_r	P_r^s	α [15]	$\alpha(T_r)$
0.70	0.0597	1.243167	1.243169	0.86	0.3433	1.103471	1.103467
0.72	0.0779	1.223785	1.223787	0.88	0.4074	1.088097	1.088093
0.74	0.1001	1.204991	1.204994	0.90	0.4799	1.073032	1.073030
0.76	0.1266	1.186775	1.186777	0.92	0.5616	1.058214	1.058215
0.78	0.1581	1.169118	1.169118	0.94	0.6533	1.043575	1.043579
0.80	0.1949	1.151993	1.151993	0.96	0.7561	1.029042	1.029050
0.82	0.2377	1.135371	1.135369	0.98	0.8712	1.014540	1.014551
0.84	0.2869	1.119212	1.119208	1.00	1.0000	1	1

TABLE EXC.2B
Regression Statistics for CO_2 Vapor Pressures Fitted by the PRSV Cohesion Function

	min AARD		min RMSD	
	2-par	1-par	2-par	1-par
κ_0	0.70086573	0.70400845	0.70132798	0.70400845
κ_1	0.02270210	0.03784388	0.02149019	0.03428902
AARD	1.98E−03	2.42E−03	2.00E−03	2.42E−03
RMSD	2.34E−03	2.78E−03	2.29E−03	2.74E−03

TABLE C.3
TADiP Pure Compound Parameters for the PRSV EOS

TADiP is the acronym for the Research Group on Applied Thermodynamics for Process Design (in Spanish, Termodinámica Aplicada al Diseño de Procesos) at Simón Bolívar University in Caracas, Venezuela.

Index of Compounds

Compounds in Table C.3 are classified by functional groups according to the following index. Within each group, they are listed in alphabetical order ("n" counts as initial letter). Compound names starting with numbers precede those that start with letters.

1. Alcohols
   1.1 Aromatic alcohols
   1.2 n-Alcohols
   1.3 Other aliphatic alcohols
   1.4 Polyols
2. Aldehydes
3. Alkanes
   3.1 Cycloalkanes
   3.2 Methylalkanes
   3.3 n-Alkanes
   3.4 Other alkanes
4. Alkenes
   4.1 1-alkenes
   4.2 Dialkenes
   4.3 Other alkenes

(Continued)

TABLE C.3 (*Continued*)
TADiP Pure Compound Parameters for the PRSV EOS

Index of Compounds

5. Alkylbenzenes
   - 5.1 n-Alkylbenzenes
   - 5.2 Other alkylbenzenes
   - 5.3 Other hydrocarbon rings
   - 5.4 Other monoaromatics
   - 5.5 Polyaromatics
6. Alkynes
7. Anhydrides
8. Amines
   - 8.1 Aliphatic amines
   - 8.2 Aromatic amines
   - 8.3 Other amines, imines
9. Chlorides
   - 9.1 Aliphatic chlorides
   - 9.2 Aromatic chlorides
10. C, H, halogen compounds
    - 10.1 C, H, Br compounds
    - 10.2 C, H, F compounds
    - 10.3 C, H, I compounds
    - 10.4 C, H, multihalogen compounds
11. C, H, N, O, S compounds
    - 11.1 C, H, S compounds
    - 11.2 C, H, NO_2 compounds
    - 11.3 Other monofunctional C, H, O, N
    - 11.4 Polyfunctional C, H, N, (O), halide
    - 11.5 Polyfunctional C, H, O, halides
    - 11.6 Polyfunctional C, H, O, S
    - 11.7 Polyfunctional C, H, O
    - 11.8 Polyfunctional C, H, O, N
12. Elements
13. Epoxides and Peroxides
14. Esters
    - 14.1 Aromatic esters
    - 14.2 Other saturated aliphatic esters
15. Ethers
16. Formates and acetates
17. Inorganic gases
18. Ketones
19. n-Aliphatic acids
    - 19.1 Aromatic carboxilic acids
    - 19.2 Inorganic acids
    - 19.3 Inorganic bases
    - 19.4 Inorganic halides
    - 19.5 n-Aliphatic acids
    - 19.6 Other aliphatic acids
    - 19.7 Other inorganics
    - 19.8 Other salts
20. Nitriles
21. Organic/inorganic compounds

Compound	a_c [kPa (m³/kmol)²]	b (m³/kmol)	κ_0	κ_1
		1. Alcohols		
		1.1. Aromatic alcohols		
2-Phenylethanol	3777.4531	0.11301238	1.3976121	−0.35496631
2,3-Xylenol	3328.7344	0.094222346	1.1063922	−0.064106776
2,4-Xylenol	3637.8970	0.10519980	1.0964629	0.024473524
2,5-Xylenol	3229.1262	0.093458312	1.1678710	−0.12997002
2,6-di-tert-Butyl-p-cresol	7789.5233	0.22139174	1.3258175	0.047625599
2,6-Xylenol	3564.9169	0.10405991	1.0305397	−0.051116932
3,4-Xylenol	3325.9848	0.093241699	1.1890974	−0.044438156
3,5-Xylenol	4465.6561	0.12769316	1.0759485	0.19707395
Benzyl alcohol	3184.5472	0.096259158	1.3246100	−0.040642426
Dinonylphenol	20038.462	0.46282168	1.8762113	−0.097076076
m-Cresol	3455.6144	0.10018343	1.0027092	0.24875357
Nonylphenol	8751.2751	0.23656936	1.5955112	−0.039827477
o-Cresol	3040.5558	0.089199142	0.99643372	0.084189762
p-Cresol	3113.4831	0.090418249	1.0825546	0.10514202
p-Cumylphenol	8234.6529	0.20205160	1.2940233	0.004532896
p-Ethylphenol	3867.2954	0.11045986	1.1028754	0.053675336
p-tert-Amylphenol	5966.4208	0.16257620	1.1799403	0.016727111
p-tert-Butylphenol	5105.1254	0.14232900	1.1040679	0.10244591
p-tert-Octylphenol	8124.3042	0.21732404	1.2552972	0.028012406
Phenol	2514.6609	0.074121906	1.0089691	0.034254695
		1.2. n-Alcohols		
1-Decanol	6147.8137	0.18232853	1.2376226	0.45744412
1-Dodecanol	8502.9057	0.24133212	1.2551663	0.53159853
1-Eicosanol	16030.680	0.41419973	1.6357748	0.57218529
1-Heptadecanol	13055.424	0.34696311	1.4722042	0.54683794
1-Heptanol	3967.7674	0.12849342	1.1984070	0.47200174
1-Hexadecanol	11245.301	0.30239141	1.4273216	0.50592828
1-Hexanol	3474.6722	0.11630730	1.1743997	0.27955353
1-Nonanol	5322.8952	0.16185119	1.2054542	0.51880190
1-Octadecanol	14150.758	0.37268483	1.5507037	0.45378492
1-Octanol	4716.4786	0.14791778	1.2055906	0.43568739
1-Pentanol	2809.2821	0.098077551	1.1867307	0.34811989
1-Tetradecanol	10020.020	0.27671545	1.3109096	0.57998135
1-Tridecanol	9344.6842	0.26159551	1.2379595	0.57803445
1-Undecanol	7511.9818	0.21835589	1.1795127	0.61542808
Ethanol	1377.1474	0.054588783	1.2557386	−0.11187817
Methanol	1023.2767	0.040852108	1.1675457	−0.15297483
n-Butanol	2283.3190	0.083003275	1.2257630	0.14564592
n-Propanol	1817.8969	0.069312667	1.275062853	−0.016346386
		1.3. Other aliphatic alcohols		
1-Methylcyclohexanol	3007.7492	0.10207217	1.3331638	−0.015831479
2-Ethyl-1-butanol	3116.5374	0.10995814	1.3471562	0.26628077
2-Ethyl-1-hexanol	4749.4845	0.15180285	1.0469280	0.90530461
2-Heptanol	3674.2639	0.12787218	1.4091552	−0.13025589
2-Methyl-1-butanol	2604.0437	0.094315452	1.2559180	0.61186349

(*Continued*)

(*Continued*)

Compound	a_c [kPa (m³/kmol)²]	b (m³/kmol)	κ_0	κ_1
2-Methyl-1-pentanol	3151.5778	0.11081233	1.3868641	−0.15291931
2-Methyl-2-butanol	2418.9336	0.090801068	1.0664842	0.47681481
2-Octanol	4725.0957	0.15175813	1.0843306	0.56382178
2-Pentanol	2503.6324	0.092814220	1.3130143	0.20321816
2,2-Dimethyl-1-propanol	2441.5646	0.090842393	1.2379500	0.097659048
2,6-Dimethyl-4-heptanol	4520.1911	0.15339900	1.4739630	0.0073888279
3-Methyl-1-butanol	2727.9942	0.096340859	1.1936161	0.29342910
3-Methyl-2-butanol	2630.4450	0.093777871	0.86255118	0.85029331
3-Pentanol	2521.9929	0.094349492	1.3029060	0.052476805
4-Methyl-2-pentanol	3001.3776	0.10692746	1.1821588	0.068709508
Allyl alcohol	1670.1746	0.062705924	1.2142942	−0.15098591
cis-2-Methylcyclohexanol	3270.9450	0.10901541	1.2977436	−0.28406500
cis-3-Methylcyclohexanol	3242.9516	0.10738287	1.3414774	0.0041776831
cis-4-Methylcyclohexanol	3179.2901	0.10459786	1.3514890	0.067291833
Cyclohexanol	3289.5366	0.10767962	1.1012428	0.38722080
Isobutanol	2213.8892	0.082712738	1.1601736	0.56226259
Isodecanol	5745.6229	0.18257205	1.6106318	−0.045315963
Isopropanol	1713.1452	0.068968148	1.3130071	0.0096807165
Propargyl alcohol	1628.2522	0.057448242	1.1508768	−0.24272745
sec-Butanol	2108.4546	0.080496018	1.2081003	0.24966940
tert-Butanol	2048.3856	0.082808058	1.2339148	0.40386756
trans-2-Methylcyclohexanol	3184.1044	0.10577661	1.3177754	−0.26253362
trans-3-Methylcyclohexanol	3220.4036	0.10680908	1.3339598	0.011161879
trans-4-Methylcyclohexanol	3209.2084	0.10558217	1.3468914	0.033061990
	1.4. Polyols			
1,2-Benzenediol	2462.7220	0.065963709	1.3543023	−0.56334402
1,2-Propylene glycol	2050.3545	0.067025129	1.7689859	−0.39321587
1,2,3-Benzenetriol	2471.7912	0.060942002	1.6441405	−0.64317938
1,3-Benzenediol	2741.0914	0.069250281	1.3254256	−0.013309769
1,3-Butanediol	2603.1194	0.082844972	1.826367	−0.23792323
1,3-Propylene glycol	2313.6377	0.071953595	1.8895127	−0.86473221
1,4-Butanediol	2880.2033	0.088365013	1.9389936	−0.70771683
1,5-Pentanediol	3255.7391	0.098995983	1.9973319	−0.66978444
1,6-Hexanediol	3915.4995	0.11959015	2.0301390	−0.40687037
2-Butyne-1,4-diol	2606.0034	0.076731417	1.8791104	−0.11439182
Bisphenol A	7527.4830	0.18143670	1.6647339	0.76571926
cis-2-Butene-1,4-diol	2780.3385	0.083932062	1.9054736	−0.64317668
Diethylene glycol	3857.7509	0.10602156	1.2381365	0.12212874
Dipropylene glycol	3769.3000	0.11794132	1.9383295	−0.98208594
Ethylene glycol	1758.6628	0.055796364	1.8500276	−0.75042577
Glycerol	4347.8057	0.12305942	2.0566400	−0.67389547
Hexylene glycol	3035.6154	0.10003182	1.9469537	−0.10562306
Neopentyl glycol	3080.6504	0.098042523	1.8917425	−0.17180118
p-Hydroquinone	2857.3397	0.071133324	1.3227115	−0.085643991
p-tert-Butylcatechol	5069.3021	0.13368094	1.3954447	−0.0077521208
Tetraethylene glycol	6369.0900	0.18051913	2.3543546	−0.20694433
trans-2-Butene-1,4-diol	2864.8833	0.086088047	1.9094683	−0.70850205
Triethylene glycol	4720.7113	0.13800424	2.1584085	−1.1394347
Trimethylolpropane	4063.2812	0.11727722	2.3485345	−1.0668890

(*Continued*)

(*Continued*)

Compound	a_c [kPa (m³/kmol)²]	b (m³/kmol)	κ_0	κ_1
		2. Aldehydes		
1-Decanal	6349.7386	0.19777596	1.2539582	0.063879879
1-Dodecanal	7982.0787	0.23845612	1.4121641	−0.12156634
1-Heptanal	4101.8743	0.13920283	1.0503397	0.037703657
1-Hexanal	3409.4101	0.12049908	1.0015466	−0.027325714
1-Nonanal	5561.0697	0.17781214	1.1906436	0.022206238
1-Octanal	4775.9380	0.15738021	1.1328561	0.0096494503
1-Tridecanal	8891.4654	0.25993116	1.4311685	0.070801835
1-Undecanal	7151.1119	0.21776463	1.3385100	−0.10707945
2-Ethylhexanal	4503.9524	0.15184068	1.1152635	0.018839258
Acetaldehyde	1206.4464	0.053553781	0.7771264	0.12790222
Acrolein	1612.1025	0.065196635	0.82463387	−0.084975116
Benzaldehyde	3302.8971	0.097250822	0.82047774	0.081628669
Formaldehyde	798.01999	0.040025489	0.78503569	−0.092696316
Glyoxal	1317.0873	0.054449312	0.97019251	0.030978116
Isobutyraldehyde	1980.9833	0.079956890	0.89674647	−0.010838000
Methacrolein	2090.5272	0.080716623	0.73957949	0.039560879
n-Butyraldehyde	2185.8588	0.085201226	0.84392390	0.093135004
n-Propionaldehyde	1672.6626	0.069009631	0.81263770	−0.021573833
p-Tolualdehyde	4197.3902	0.12305718	1.0092089	0.036824060
trans-Crotonaldehyde	2440.6427	0.087468390	0.86815816	0.057018288
Valeraldehyde	2772.5712	0.10241321	0.91702877	0.066827024
		3. Alkanes		
		3.1. Cycloalkanes		
1,1-Dimethylcyclohexane	3758.8535	0.13011901	0.71596732	0.11576958
1,1-Dimethylcyclopentane	2735.1622	0.10232430	0.76291624	0.049335320
cis-1,2-Dimethylcyclohexane	3951.0907	0.13338897	0.71651742	0.14438800
cis-1,2-Dimethylcyclopentane	2929.4382	0.10607270	0.75962522	0.059067834
cis-1,3-Dimethylcyclohexane	3758.2827	0.13009925	0.72032252	0.13678178
Cis-1,3-dimethylcyclopentane	2781.1802	0.10329055	0.76969344	0.018263420
cis-1,4-Dimethylcyclohexane	3846.3868	0.13159091	0.71385102	0.15099432
cis-Decahydronaphthalene	4807.1897	0.14008207	0.78079605	0.13939813
Cyclobutane	1341.1568	0.059672036	0.64955475	−0.0064370216
Cycloheptane	3020.8312	0.10229559	0.72686916	0.067069769
Cyclohexane	2374.3440	0.087776381	0.68946108	0.038600844
Cyclopentane	1839.1438	0.073541484	0.65814630	0.065107850
Cyclopropane	903.29378	0.046454437	0.56538105	0.055790235
Ethylcyclohexane	3856.8583	0.12956642	0.72967925	0.15077781
Ethylcyclopentane	3018.0785	0.10844377	0.73965203	0.14489438
Methylcyclohexane	2971.3645	0.10626707	0.68774525	0.17819218
Methylcyclopentane	2373.2388	0.091152462	0.69114508	0.12456638
n-Butylcyclohexane	5462.2289	0.16758190	0.78520629	0.20665874
n-Propylcyclohexane	4596.8702	0.14717786	0.75597038	0.18037089
n-Propylcyclopentane	3830.2481	0.12998482	0.76633852	0.15479468
trans-1,2-Dimethylcyclohexane	3823.8978	0.13126041	0.72012993	0.12067994
trans-1,2-Dimethylcyclopentane	2806.3693	0.10382094	0.74573598	0.10601635
trans-1,3-Dimethylcyclohexane	3846.0061	0.13161089	0.71611508	0.15786375
Trans-1,3-dimethylcyclopentane	2809.9434	0.10398136	0.72436527	0.16154295

(*Continued*)

(*Continued*)

Compound	a_c [kPa (m^3/kmol)^2]	b (m^3/kmol)	κ_0	κ_1
trans-1,4-Dimethylcyclohexane	3745.8127	0.12988730	0.72248944	0.11191860
trans-Decahydronaphthalene	5124.7819	0.15264061	0.73805425	0.20351006
3.2. Methylalkanes				
2-Methylheptane	3993.2937	0.14601771	0.90596104	0.081592372
2-Methylhexane	3267.7930	0.12608361	0.82469338	0.13834835
2-Methyloctane	4754.2372	0.16581002	0.98124570	0.060507045
2-Methylpentane	2598.4021	0.10687997	0.76894140	0.073970634
3-Methylheptane	3992.4459	0.14494297	0.88013717	0.12761574
3-Methylhexane	3241.1098	0.12391392	0.82589318	0.094135029
3-Methyloctane	4700.7146	0.16299884	0.96395480	0.065004550
3-Methylpentane	2570.4589	0.10427803	0.70955648	0.23608391
4-Methylheptane	3942.3667	0.14361662	0.88747576	0.12022021
4-Methyloctane	4660.1233	0.16227877	0.96582751	0.050735709
Isobutane	1454.5432	0.072929041	0.61593026	0.13945189
Isopentane	1987.8537	0.088349448	0.69168484	0.083572804
3.3. n-Alkanes				
Ethane	605.16028	0.040546743	0.51946845	0.035633202
Methane	250.02652	0.026846754	0.39401837	−0.0034966480
n-Butane	1513.5590	0.072846645	0.66059669	0.055288924
n-Decane	5676.8304	0.18783839	1.0628541	0.073735336
n-Dodecane	7567.8366	0.23528644	1.1767082	0.077090006
n-Eicosane	17792.910	0.47469211	1.5734281	0.069258112
n-Heptadecane	12611.236	0.35189860	1.4166634	0.25194847
n-Heptane	3350.7582	0.12691803	0.86400112	0.13097652
n-Hexadecane	11642.561	0.33062623	1.3909216	0.073958472
n-Hexane	2682.7522	0.10819008	0.79879078	0.12382030
n-Nonadecane	16209.634	0.43880813	1.4873084	0.14057068
n-Nonane	4908.6517	0.16863748	1.0027846	0.015317067
n-Octadecane	14430.876	0.39624876	1.4670828	0.10038420
n-Octane	4124.5157	0.14837936	0.93566027	0.082977153
n-Pentadecane	10341.188	0.29940355	1.3381724	0.14329424
n-Pentane	2084.7613	0.09083747	0.72422164	0.12048929
n-Tetradecane	9345.3612	0.27619898	1.2831461	0.12978703
n-Tridecane	8373.1522	0.25354429	1.2364216	−0.0075047849
n-Undecane	6562.6013	0.21024290	1.1194446	0.10958350
Propane	1023.4796	0.056633293	0.58804342	0.069355822
3.4. Other alkanes				
2-Methyl-3-ethylpentane	3768.3284	0.13600307	0.83318233	0.11220805
2,2-Dimethylhexane	3786.4318	0.14093162	0.84343781	0.12468075
2,2-Dimethylpentane	3081.8792	0.12116520	0.79512405	0.046834454
2,2,3-Trimethylbutane	3031.1128	0.11677546	0.74027884	0.040322650
2,2,3-Trimethylpentane	3685.5185	0.13384055	0.79054432	0.10596082
2,2,3,3-Tetramethylpentane	4312.4064	0.14446683	0.78290552	0.086672350
2,2,3,4-Tetramethylpentane	4321.4836	0.14934276	0.79392605	0.15851460
2,2,4-Trimethylpentane	3672.6066	0.13816259	0.79364545	0.11486691
2,2,4,4-Tetramethylpentane	4369.0332	0.15648263	0.83557265	0.047220444
2,2,5-Trimethylhexane	4375.1128	0.15761071	0.89085356	0.069686789
2,3-Dimethylhexane	3827.9266	0.13903681	0.87503682	0.039989423

(*Continued*)

(*Continued*)

Compound	a_c [kPa (m^3/kmol)^2]	b (m^3/kmol)	κ_0	κ_1
2,3-Dimethylpentane	3200.9110	0.12189879	0.78459267	0.086377149
2,3,3-Trimethylpentane	3697.5636	0.13193659	0.78576993	0.073843009
2,3,4-Trimethylpentane	3732.9837	0.13489398	0.82035644	0.074093599
2,4-Dimethylhexane	3804.1613	0.14064501	0.86885320	0.049421012
2,4-Dimethylpentane	3131.3865	0.12327976	0.79859061	0.10682246
2,5-Dimethylhexane	3864.9588	0.14380209	0.87698249	0.096299502
3-Ethylhexane	3894.4512	0.14095273	0.89569496	0.041821312
3-Ethylpentane	3183.2889	0.12048998	0.83313770	0.021415977
3,3-Diethylpentane	4398.6136	0.14754803	0.86611704	−0.0030473720
3,3-Dimethylhexane	3767.0190	0.13716539	0.81634050	0.12131257
3,3-Dimethylpentane	3108.4773	0.11858833	0.75180510	0.070230532
3,4-Dimethylhexane	3803.7548	0.13684721	0.86479082	0.037678248
Neopentane	1865.0510	0.087984107	0.66433810	0.017404661
		4. Alkenes		
		4.1. 1-Alkenes		
1-Butene	1419.2404	0.069217177	0.62766637	0.14979282
1-Decene	5544.5711	0.18387837	1.0360903	0.072006985
1-Dodecene	7191.9381	0.22400804	1.1665569	0.078771076
1-Eicosene	15373.637	0.40804238	1.5448676	0.18291081
1-Heptene	3232.8706	0.12312964	0.83526192	0.10907262
1-Hexadecene	11241.446	0.31861633	1.3850916	−0.048310908
1-Hexene	2557.6128	0.10383923	0.76404784	0.11398095
1-Nonene	4772.9252	0.16463794	0.95693462	0.12457744
1-Octadecene	12941.573	0.35405318	1.4329027	0.32868103
1-Octene	3980.9281	0.14377746	0.89544624	0.12206380
1-Pentene	1920.3225	0.084549252	0.66972773	0.19475786
1-Tetradecene	9140.7908	0.27030913	1.2574763	0.14094369
1-Tridecene	8174.5266	0.24782315	1.2338412	−0.031450505
1-Undecene	6324.8555	0.20286773	1.0888575	0.14740966
Ethylene	501.04665	0.036312657	0.50127555	0.05105929
Propylene	916.47576	0.051415825	0.57073782	0.078028419
		4.2. Dialkenes		
1,2-Butadiene	1385.0623	0.063836546	0.74551122	−0.080954304
1,2-Pentadiene	2078.3847	0.085062663	0.61503312	0.19863267
1,3-Butadiene	1320.6149	0.063531978	0.65732315	0.048193173
1,4-Pentadiene	1938.8670	0.082831503	0.52074824	0.24829997
cis-1,3-Pentadiene	2103.8342	0.086276795	0.60369696	0.20841027
Cyclopentadiene	1578.0881	0.063695144	0.68445465	−0.14529401
Isoprene	1922.8411	0.081298225	0.61172142	0.15576166
Propadiene	898.76649	0.046781229	0.57504748	0.14011472
trans-1,3-Pentadiene	2111.9592	0.086436776	0.55621700	0.24724884
		4.3. Other alkenes		
2-Ethyl-1-butene	2620.8389	0.10474985	0.72099161	0.11524390
2-Ethyl-1-hexene	3392.6959	0.12095284	0.91329420	−0.040058389
2-Methyl-1-butene	1967.6008	0.086589868	0.71730110	0.091569556
2-Methyl-1-pentene	2571.8671	0.10380627	0.71862436	0.15315716
2-Methyl-2-butene	2054.8433	0.089277259	0.76355882	0.055974887
2-Methyl-2-pentene	2642.7483	0.10521453	0.73494641	0.13690857

(*Continued*)

(*Continued*)

Compound	a_c [kPa (m³/kmol)²]	b (m³/kmol)	κ_0	κ_1
2,3-Dimethyl-1-butene	2453.5958	0.10041904	0.69794888	0.13233294
2,3-Dimethyl-2-butene	2736.9231	0.10688440	0.72206709	0.15933617
2,4,4-Trimethyl-1-pentene	3676.8938	0.13606267	0.76874699	0.10744525
2,4,4-Trimethyl-2-pentene	3741.4945	0.13721258	0.76380443	0.18411163
3-Methyl-1-butene	1823.3116	0.082846562	0.68236740	0.098472275
4-Methyl-1-pentene	2410.5859	0.099454392	0.69621073	0.15331867
4-Methyl-cis-2-pentene	2442.8035	0.10017769	0.71881315	0.14001697
4-Methyl-trans-2-pentene	2464.9012	0.10068037	0.73379090	0.13221479
cis-2-Butene	1414.8288	0.066468979	0.66131489	0.098690607
cis-2-Heptene	3356.5799	0.12511451	0.81159071	0.032706082
cis-2-Hexene	2633.0507	0.10503279	0.78032989	0.034776042
cis-2-Pentene	1959.1054	0.084236002	0.70447356	0.13948331
cis-3-Heptene	3307.2070	0.12417893	0.77284990	0.18146432
Cyclohexene	2286.7465	0.083503164	0.66434675	0.14941548
Cyclopentene	1697.0139	0.068495253	0.64620803	0.090678880
Isobutene	1388.3781	0.067985834	0.64429294	0.10073019
trans-2-Butene	1422.9686	0.067935343	0.69042305	0.010787968
trans-2-Hexene	2631.4364	0.10496839	0.74775644	0.13817411
trans-2-Octene	4078.9903	0.14466381	0.87026053	0.024263103
trans-2-Pentene	1954.3480	0.084130441	0.71289567	0.093177115
trans-3-Octene	4038.0738	0.14396118	0.87534380	0.034028625
trans-4-Octene	4020.5183	0.14358546	0.86496050	0.074392244
	5. Alkylbenzenes			
	5.1. n-Alkylbenzenes			
Benzene	2046.1605	0.074483967	0.68506759	0.062034848
Ethylbenzene	3345.5798	0.1109302	0.79976174	0.094566787
n-Butylbenzene	4741.5392	0.1468915	0.92591435	0.10507815
n-Decylbenzene	10090.526	0.2742217	1.3153224	0.0097323093
n-Dodecylbenzene	11975.879	0.3165218	1.4057441	0.23257127
n-Hexylbenzene	6468.9478	0.1896537	1.0528018	0.033166983
n-Nonylbenzene	9176.1936	0.2534121	1.2583675	−0.0035988938
n-Pentylbenzene	5590.1309	0.1682519	1.0125630	−0.073349094
n-Propylbenzene	4039.2871	0.1294817	0.85165106	0.10889984
n-Tridecylbenzene	12906.623	0.3373136	1.4775618	0.23303326
n-Undecylbenzene	11058.974	0.2962133	1.3509162	0.15980916
Toluene	2698.6040	0.093315668	0.74845373	0.084609266
	5.2. Other alkylbenzenes			
1,2,3-Trimethylbenzene	4033.0074	0.12419310	0.91004185	−0.044546018
1,2,4-Trimethylbenzene	4111.1344	0.12960240	0.92766148	−0.053917890
1,2,4,5-Tetramethylbenzene	4902.3289	0.14858850	1.0027764	−0.047857355
3-Ethyl-o-xylene	5074.8679	0.15272104	0.89692161	0.11511670
Cumene	3923.7554	0.12721912	0.83005984	0.15741982
Isobutylbenzene	4393.7736	0.13829521	0.91934303	0.0074921716
m-Cymene	4657.2224	0.14505899	0.87029905	0.068777169
m-Diethylbenzene	4820.0907	0.14877320	0.87838812	0.13520073
m-Diisopropylbenzene	6039.7827	0.18069588	0.90739454	0.13445387
m-Ethyltoluene	4519.6880	0.14516095	0.84640454	0.11931100
m-Xylene	3410.7318	0.11311241	0.84226904	0.034174997
Mesitylene	4083.4797	0.13110783	0.94425092	0.011589387

(*Continued*)

(*Continued*)

Compound	a_c [kPa (m³/kmol)²]	b (m³/kmol)	κ_0	κ_1
o-Cymene	4729.3530	0.14619307	0.86500523	0.077574280
o-Diethylbenzene	4894.5542	0.14994076	0.87136793	0.11682717
o-Ethyltoluene	4411.2965	0.13863352	0.81792280	0.10163870
o-Xylene	3373.5706	0.10951593	0.82388628	0.044600673
p-Cymene	4756.6355	0.14902872	0.89337505	0.12991662
p-Diethylbenzene	4882.3264	0.15184845	0.95436285	0.026106151
p-Diisopropylbenzene	6128.1437	0.18200895	0.94206643	0.16088596
p-Ethyltoluene	4409.7974	0.14096780	0.84064119	0.12542512
p-Xylene	3517.1433	0.11679093	0.83040471	0.019601599
sec-Butylbenzene	4715.4738	0.14520690	0.75429246	0.28027395
tert-Butylbenzene	4624.3251	0.14337963	0.75753761	0.23843242
5.3. Other hydrocarbon rings				
1-Methylindene	4516.3235	0.13146579	0.86365775	0.050880053
1,2,3-Trimethylindene	6068.5583	0.17105349	0.99466506	−0.036145798
1,2,3,4-Tetrahydronaphthalene	4516.6342	0.12834383	0.84086957	0.098896237
2-Methylindene	4272.7396	0.12783017	0.85281631	0.048306179
alpha-Pinene	4579.7780	0.14828949	0.81486128	0.024979991
beta-Pinene	4707.5270	0.14981831	0.82732938	0.13013014
Bicyclohexyl	6566.3249	0.18482937	0.97552062	−0.019335902
Indene	3884.9773	0.11572169	0.84523814	0.069539096
Vinylcyclohexene	3249.4977	0.11101264	0.79131312	0.14320684
5.4. Other monoaromatics				
alpha-Methylstyrene	4016.7100	0.12568278	0.85453375	0.026262519
Cyclohexylbenzene	6032.2753	0.16591714	0.92222112	0.22702774
m-Divinylbenzene	4852.0610	0.14348391	0.91453593	0.0027090368
p-Methylstyrene	4150.7783	0.12772942	0.83154458	0.10873647
Styrene	3337.4410	0.10539539	0.71026689	0.18538206
5.5. Polyaromatics				
1-Methylnaphthalene	5149.2253	0.13648506	0.88379322	0.031750442
1-n-Decylnaphthalene	14756.973	0.35155005	1.2737470	0.0054251448
1-n-Nonylnaphthalene	13563.1276	0.32691526	1.2417437	0.0091505710
1-Phenylnaphthalene	8651.0669	0.20851870	1.1289246	0.025658341
1,1-Diphenylethane	7088.5718	0.18717165	1.0310671	0.028585251
1,2-Diphenylethane	7261.7737	0.19051586	1.0646812	0.012439095
2-Methylnaphthalene	5643.8539	0.15176588	0.86601589	0.11942678
Acenaphthene	6565.9059	0.16729425	0.92321721	0.070964965
Anthracene	8307.9555	0.19474349	1.0831790	−0.29527534
Biphenyl	5159.2396	0.13376689	0.89622487	0.10941505
Chrysene	12684.376	0.26513647	1.2234185	0.013330917
Diphenylacetylene	7547.1535	0.18562775	0.92995161	0.040867808
Diphenylmethane	6342.9651	0.16901067	1.0286317	0.046811926
Fluoranthene	9972.4084	0.22549375	1.1900777	−0.31611267
Fluorene	5121.6237	0.12046793	0.87460108	0.031142057
m-Terphenyl	7659.6805	0.16948154	1.1650858	−0.027198465
Naphthalene	4451.5411	0.12172749	0.80320350	0.050534105
o-Terphenyl	6457.3422	0.14831446	1.0375328	0.028976008

(*Continued*)

(*Continued*)

Compound	a_c [kPa (m³/kmol)²]	b (m³/kmol)	κ_0	κ_1
p-Terphenyl	8152.9095	0.18018064	1.1240736	0.028212152
Phenanthrene	8245.7569	0.19411937	1.0724456	–0.12758150
Pyrene	10625.2949	0.23229945	1.0928389	–0.093348371
Tetraphenylethylene	18303.1639	0.37605370	1.3845765	–0.0082589990
trans-Stilbene	7843.4168	0.19573770	1.0623701	0.11942697
Triphenylethylene	12409.963	0.27968403	1.2187175	0.015086834
6. Alkynes				
Acetylene	488.84754	0.03244552	0.65862824	–0.021336446
Dimethylacetylene	1480.9272	0.062081680	0.56837264	0.25148826
Ethylacetylene	1253.7020	0.057886543	0.72759068	0.076269391
Methylacetylene	910.58847	0.046308211	0.69012441	0.072257551
7. Anhydrides				
Butyric anhydride	4847.7861	0.15524786	1.2202788	0.16604294
Maleic anhydride	2251.4721	0.063901985	1.1503399	–0.39468163
Phthalic anhydride	4187.2755	0.10832734	1.3603023	–0.84423779
Propionic anhydride	3911.8685	0.12953252	1.1961093	–0.11777679
Succinic anhydride	3111.4660	0.078510409	1.1244250	–0.27794003
Trimellitic anhydride	6136.2866	0.14109079	1.7796532	–0.094248163
8. Amines				
8.1. Aliphatic amines				
1-Aminoheptane	4086.9654	0.13778290	1.0925863	–0.021862401
di-n-Butylamine	3755.2410	0.12649535	1.0801879	0.18779859
Diethylamine	2117.7915	0.087268904	0.81297377	0.044729915
Diisopropylamine	2703.9594	0.10577876	0.79668306	0.66463218
Dimethylamine	1162.7048	0.054365773	0.79039080	0.16622220
Dodecylamine	8154.5462	0.23975827	1.4342054	–0.20599954
Ethylamine	1175.7853	0.052747681	0.78714949	0.11133117
Isobutylamine	1980.7633	0.078900671	0.89823606	0.017552039
Isopropylamine	1553.0821	0.067355568	0.77401877	0.18826950
Methylamine	788.49766	0.037520144	0.78220827	0.11418555
n-Butylamine	2130.7845	0.081977104	0.85323906	0.072252275
n-Decylamine	6371.1980	0.19664849	1.3073976	–0.15681712
n-Hexylamine	3368.1207	0.11822304	1.0387718	–0.060299274
n-Nonylamine	5601.9043	0.17690647	1.2304668	–0.099007710
n-Octylamine	4817.79434	0.15724026	1.1763967	–0.13125845
n-Pentylamine	2719.5150	0.10027242	0.96191467	–0.012840982
n-Propylamine	1681.5519	0.069243756	0.79661792	0.060363680
Piperazine	2324.8174	0.074567778	0.97110119	0.027596767
sec-Butylamine	2092.5394	0.083260719	0.79257734	0.035670577
tert-Butylamine	1927.6165	0.081516974	0.77732667	0.049637857
Tetradecylamine	9905.6831	0.28064018	1.5472622	–0.27075069
Tri-n-butylamine	7279.8222	0.23132253	1.3417221	–0.0045711459
Triethylamine	2980.3809	0.11396704	0.84242656	0.0032910651
Trimethylamine	1452.6990	0.068615164	0.66743979	0.099917910
Tripropylamine	4787.7964	0.16965502	1.3384916	0.006675618

(*Continued*)

(*Continued*)

Compound	a_c [kPa (m³/kmol)²]	b (m³/kmol)	κ_0	κ_1
	8.2. Aromatic amines			
2-Methylpyridine	2783.7436	0.091731957	0.77739915	0.11066797
2,4,6-Trimethylpyridine	4055.3624	0.12708653	0.91638720	−0.0035983491
2,6-Diethylaniline	4641.6276	0.14009532	1.6634638	−0.054729063
2,6-Dimethylpyridine	3248.4175	0.10657230	0.87984174	0.087207730
3-Methylpyridine	3000.7246	0.095202744	0.76937642	0.11131135
4-Methylpyridine	2832.1886	0.089695750	0.81738286	0.010460029
Aniline	2896.9123	0.084808855	0.92371528	0.13135722
Benzylamine	3417.6743	0.10232345	0.96682747	0.031307198
Dibenzopyrrole	7828.1107	0.17818876	1.0882548	−0.14985598
Diphenylamine	6703.1904	0.16789684	1.1233050	−0.078441183
Hydrazobenzene	6414.7172	0.16574307	1.3209133	0.0089063040
Isoquinoline	4053.2850	0.10327460	0.79865696	0.057421375
m-Phenylenediamine	4141.4986	0.10285214	1.1444133	−0.19763062
m-Toluidine	3821.9070	0.11028718	0.97517602	−0.012796565
n-Methylaniline	2995.4824	0.087375803	1.0415383	−0.12713380
n,n-Diethylaniline	5465.4426	0.15932035	0.99539034	0.11854472
n,n-Dimethylaniline	4115.8063	0.12257063	0.95746593	−0.058398731
n,n-diphenyl-p-phenylenediamine	11193.281	0.25282048	1.5695984	−0.040663287
o-Ethylaniline	4190.5540	0.12180969	1.0402785	0.025401387
o-Phenylenediamine	3719.0010	0.097444716	1.0783265	0.034740170
o-Toluidine	4047.2323	0.11931302	0.97708590	0.16955824
p-Aminodiphenyl	6424.1184	0.16090684	1.1461666	0.025291840
p-Aminodiphenylamine	7428.3351	0.17532967	1.3410122	0.00062096110
p-Phenylenediamine	3867.5254	0.099426726	1.1382081	0.026671923
p-Toluidine	3830.8213	0.11309612	1.0480644	−0.081468336
Phenylhydrazine	3727.2141	0.10022654	1.1055395	−0.088014016
Pyridine	2155.6861	0.071156060	0.72291745	0.091423562
Quinaldine	6383.1191	0.16898046	0.78000766	0.48201131
Quinoline	4067.4508	0.10641805	0.83193722	0.15010735
Toluenediamine	4768.3690	0.12136595	1.1698590	0.055713020
	8.3. Other amines, imines			
1,3-Diphenyltriazene	7989.2224	0.19347767	1.2474316	0.014595097
Allylamine	1559.6614	0.063200714	0.85271825	0.040151102
Dicyclohexylamine	6786.7662	0.18844232	1.0876654	−0.055947529
Diethylene triamine	3428.9762	0.10380083	1.3317590	−0.39373502
Ethylenediamine	1765.3009	0.060918204	1.0586569	−0.092123129
Ethyleneimine	1330.5738	0.050704644	0.67843194	0.24470382
Hexamethylenediamine	4229.5496	0.13054602	1.2822895	0.010338717
Indole	4611.1375	0.11944391	0.89444768	0.27388875
n-Methylcyclohexylamine	3491.0215	0.11485375	0.93464115	0.040002553
n-Methylpyrrolidine	2277.3503	0.084732533	0.71522169	0.035795195
p-Aminoazobenzene	8386.9226	0.19569786	1.2626582	0.012590822
Piperidine	2399.6420	0.082662111	0.73246427	0.080821396
Propyleneimine	1632.8648	0.063165158	0.75326956	0.049017132
Pyrrole	2088.9409	0.066818833	0.78213460	0.16728841
Pyrrolidine	1821.0981	0.065546231	0.76308365	0.16497406
Tetraethylenepentamine	7486.0807	0.19792314	1.9899364	−0.11046364

(*Continued*)

(*Continued*)

Compound	a_c [kPa (m³/kmol)²]	b (m³/kmol)	κ_0	κ_1
		9. Chlorides		
		9.1. Aliphatic chlorides		
1-Chloropentane	3048.7582	0.10983928	0.84325952	0.0044550753
1,1-Dichloroethane	1703.3703	0.066648507	0.72398484	0.049619531
1,1-Dichloroethylene	1415.2297	0.060084568	0.74314992	0.034343292
1,1,1-Trichloroethane	2184.9493	0.082040423	0.69333475	0.070119383
1,1,1,2-Tetrachloroethane	3059.3126	0.10032803	0.72642782	0.13947838
1,1,2-Trichloroethane	2613.4536	0.088838523	0.74454571	0.076753698
1,1,2,2-Tetrachloroethane	3227.0775	0.10238415	0.73254081	0.31241555
1,2-Dichloroethane	1901.6704	0.069367400	0.77586388	0.0033469306
1,2-Dichloropropane	2418.9024	0.086537700	0.75370110	0.017773642
1,2,3-Trichloropropane	3472.2944	0.10898130	0.82387663	0.045753824
1,3-Dichloro-trans-2-butene	3185.0408	0.10546529	0.73225598	0.19282551
1,4-Dichloro-trans-2-butene	3482.6480	0.11032149	0.86196443	0.044292512
1,4-Dichlorobutane	3598.3685	0.11487635	0.84673984	0.040988880
1,5-Dichloropentane	4356.4042	0.13446142	0.93647090	0.027886580
2-Chloropropene	1533.7794	0.065662594	0.61202990	0.020379697
3-Chloropropene	1774.5865	0.070630188	0.61804941	0.0078959732
3,4-Dichloro-1-butene	2848.0346	0.098949364	0.81683421	0.039612571
Carbon tetrachloride	2153.2881	0.079202171	0.65810098	0.059990409
Chloroform	1637.4026	0.062466869	0.69937426	0.074815004
Chloroprene	2045.8775	0.079744980	0.67150756	0.021585083
cis-1,2-Dichloroethylene	1699.4314	0.065989685	0.74397947	0.026520219
Dichloromethane	1349.4050	0.054144603	0.65173483	0.16168954
Ethyl chloride	1227.3592	0.054559021	0.66728552	0.040568965
Hexachloro-1,3-butadiene	6111.0501	0.16876434	0.61886762	0.45300524
Hexachlorocyclopentadiene	5850.4407	0.16048439	0.90794151	0.15486681
Isopropyl chloride	1660.2822	0.069479404	0.60777142	0.39415110
Methyl chloride	818.5718	0.040242556	0.60388742	0.024419777
n-Butyl chloride	2385.2798	0.090896691	0.75552756	0.11589664
n-Propyl chloride	1746.1399	0.071017372	0.69688330	0.095832467
Pentachloroethane	3773.1485	0.11610884	0.71977811	0.16114569
Propargyl chloride	1745.0352	0.066006997	0.60484107	0.038957074
sec-Butyl chloride	2196.0424	0.086321622	0.78811355	−0.031077508
tert-Butyl chloride	2086.0404	0.084197228	0.66418016	0.051294027
Tetrachloroethylene	2705.3306	0.089291822	0.69294266	0.10126193
trans-1,2-Dichloroethylene	1572.2691	0.063335356	0.75886448	−0.061517268
Trichloroethylene	2100.1708	0.075266470	0.70405877	0.014657104
		9.2. Aromatic chlorides		
1-Chloronaphthalene	5727.4865	0.14930607	0.88733613	0.16922787
1,2,4-Trichlorobenzene	4465.9027	0.12605332	0.89680159	−0.087923471
2,4-Dichlorotoluene	4451.3377	0.12920653	0.88509609	0.057206258
Benzotrichloride	5139.4284	0.14270210	0.76017815	0.041111857
Benzyl chloride	3869.8444	0.11543897	0.83201729	0.042321707
Benzyl dichloride	4630.4400	0.12962475	0.85518909	−0.11341992
Hexachlorobenzene	7482.7006	0.18560405	1.13768038	0.12606337
m-Dichlorobenzene	3633.0933	0.10870142	0.78808071	0.018297238
Monochlorobenzene	2790.5940	0.090307158	0.74654906	0.028144324

(*Continued*)

(*Continued*)

Compound	a_c [kPa (m³/kmol)²]	b (m³/kmol)	κ_0	κ_1
o-Chlorotoluene	3527.5152	0.11003937	0.81337350	0.053919954
o-Dichlorobenzene	3855.2138	0.11190317	0.69697201	0.18406882
p-Chlorotoluene	3540.8710	0.10978657	0.82261160	0.083760975
p-Dichlorobenzene	3641.3974	0.10882260	0.79071942	0.065305905
10. C, H, Halogen Compounds				
10.1. C, H, Br compounds				
1-Bromobutane	2266.7368	0.080391164	0.77779123	0.11453937
1-Bromonaphthalene	5813.2500	0.14436929	0.91174173	0.041450827
1-Bromopropane	1735.0591	0.065267714	0.76734549	−0.034738973
1,1,2,2-Tetrabromoethane	4607.1128	0.11441545	0.59312880	0.52708317
1,2-Dibromoethane	2485.8993	0.078244353	0.66919076	0.054167724
2-Bromobutane	2198.5000	0.079346257	0.77614127	0.029406445
2-Bromopropane	1623.7905	0.062459917	0.72084723	−0.035210241
Bromobenzene	3140.8808	0.095909686	0.74150429	0.058264429
Bromoethane	1289.8074	0.052390160	0.71397619	−0.0091853261
Dibromomethane	1645.7476	0.055119472	0.67540318	0.064931635
Methyl bromide	861.99431	0.037772050	0.65299472	−0.070380509
p-Bromotoluene	3531.9732	0.10340065	0.83851674	−0.038020147
Tribromomethane	2515.5783	0.073962509	0.61055425	0.12964783
Vinyl bromide	984.83892	0.042607603	0.77462430	−0.32610635
10.2. C, H, F compounds				
1,1-Difluoroethane	1037.9149	0.054939276	0.76293871	−0.012124777
1,1-Difluoroethylene	650.1436	0.043937613	0.58597066	0.041912703
1,1,1-Trifluoroethane	1007.8093	0.059562322	0.74580479	0.030043818
Benzotrifluoride	2974.1371	0.10771981	0.78656636	0.088027038
Carbon tetrafluoride	437.53220	0.039356064	0.64999212	0.0018623429
Decafluorobutane	2026.6844	0.10734659	0.91558442	0.028158147
Difluoromethane	670.54306	0.039026611	0.77116880	0.029719974
Ethyl fluoride	885.65756	0.048290170	0.70791907	−0.062754594
Fluorobenzene	2182.2961	0.079733146	0.73969740	0.032694424
Hexafluorobenzene	2574.4857	0.10195531	0.94728273	0.0063861614
Hexafluoroethane	910.00178	0.063599574	0.73040640	0.10359936
Hexafluoropropylene	1505.7675	0.083732320	0.67738538	0.068237049
Methyl fluoride	527.48889	0.033976539	0.68716232	−0.057331558
Octafluoro-2-butene	2113.4831	0.11033054	0.79185326	0.047841882
Octafluorocyclobutane	1715.3618	0.090384319	0.89047275	0.039238509
Octafluoropropane	1415.5632	0.083951848	0.85368655	−0.11351326
Tetrafluoroethylene	752.42763	0.050244456	0.70873689	0.068877119
Trifluoromethane	584.66451	0.040029336	0.76000922	0.039509414
Vinyl fluoride	646.70703	0.040372130	0.64320722	−0.16843792
10.3. C, H, I, compounds				
Diiodomethane	3236.6864	0.088667214	0.57387768	0.22755254
Ethyl iodide	1663.0406	0.060662882	0.67812989	−0.0051828372
Iodobenzene	3636.5087	0.10319106	0.73214161	0.082031988
Isopropyl iodide	2062.1064	0.073007295	0.70481823	−0.010371280
Methyl iodide	1195.7425	0.046343282	0.66115212	−0.086030288
n-Propyl iodide	2211.2593	0.076307638	0.75238670	−0.089021888

(*Continued*)

(*Continued*)

Compound	a_c [kPa (m³/kmol)²]	b (m³/kmol)	κ_0	κ_1
	10.4. C, H, multihalogen compounds			
1-Chloro-1,1-difluoroethane	1288.9019	0.064299448	0.73259104	–0.16845853
1,1,2-Trichlorotrifluoroethane	2166.8393	0.091003456	0.75843566	0.052313342
1,1,2,2-Tetrachlorodifluoroethane	2794.2343	0.10377536	0.80235301	0.055508491
1,2-Dibromotetrafluoroethane	2232.3268	0.093648110	0.74610909	0.018972871
1,2-Dichlorotetrafluoroethane	1702.9055	0.083198417	0.73976650	0.10709837
2-Chloro-1,1-difluoroethylene	1137.1747	0.058096971	0.70377218	0.037080295
Bromochlorodifluoromethane	1347.5338	0.064708333	0.63855297	0.072571636
Bromochloromethane	1437.6351	0.052817338	0.70489493	0.042253562
Bromotrichloromethane	2339.6394	0.079005866	0.66119156	0.025938629
Bromotrifluoroethylene	1317.3808	0.062403739	0.63846315	0.043065145
Bromotrifluoromethane	922.0936	0.055473757	0.62535640	0.059071939
Chlorodifluoromethane	849.57495	0.047076637	0.68175459	0.15350398
Chloropentafluoroethane	1253.4148	0.072630435	0.74167008	0.035455245
Chlorotrifluoroethylene	1116.3040	0.060249640	0.69118084	0.21475426
Chlorotrifluoromethane	743.00038	0.050352687	0.64206246	–0.0090716394
Dibromodifluoromethane	1354.7936	0.058000036	0.67319593	–0.061534056
Dichlorodifluoromethane	1139.7861	0.060590156	0.62456046	0.099439002
Dichlorofluoromethane	1242.2222	0.056292124	0.69134332	0.0016072868
Halothane	2187.1338	0.085905438	0.51854669	0.30441548
p-Chlorobenzotrifluoride	3790.2998	0.12905717	0.91587269	0.034490433
Trichlorofluoromethane	1586.8071	0.068913114	0.65043218	0.061939584
	11. C, H, N, O, S compounds			
	11.1. C, H, S compounds			
Benzothiophene	4323.9582	0.11735272	0.81046971	0.043204381
Diethyl disulfide	3364.9001	0.10725565	0.86052535	0.030003758
Diethyl sulfide	2475.9482	0.090939479	0.80098249	0.024135906
Dimethyl disulfide	2166.3786	0.073155127	0.74361724	0.044663647
Dimethyl sulfide	1478.4696	0.060144084	0.63724449	0.072243952
Ethyl mercaptan	1430.1143	0.058630381	0.64148602	0.082089195
Methyl mercaptan	964.65227	0.042005115	0.58478758	0.18776185
n-Butyl mercaptan	2577.9072	0.092712447	0.76062339	0.11275636
n-Dodecyl mercaptan	8989.7828	0.25409357	1.2916485	–0.095469034
n-Hexylmercaptan	3963.5830	0.13019159	0.90553495	0.056847731
n-Pentyl mercaptan	3257.1938	0.11146165	0.83956866	0.054683578
n-Propylmercaptan	1961.4340	0.074884501	0.70202346	0.11149194
Phenyl mercaptan	3173.6298	0.094258404	0.75942512	0.20497960
tert-Butyl mercaptan	2186.2935	0.084414223	0.65982446	0.094358058
tert-Octyl mercaptan	4806.3005	0.15686513	0.82755909	0.034696588
Tetrahydrothiophene	2443.7560	0.079133095	0.65957564	0.12411374
Thiophene	1856.5827	0.065577721	0.65994924	0.10371980
	11.2. C, H, NO_2 compounds			
1-Nitropropane	2659.2957	0.089948576	0.95573244	–0.076449231
2-Nitropropane	2508.5548	0.086421180	0.91676358	–0.096056203
2,4-Dinitrotoluene	6160.8164	0.15488056	1.3672416	–0.71090600
2,6-Dinitrotoluene	5493.2526	0.14598959	1.3969156	–0.0083118060
3,4-Dinitrotoluene	6589.9091	0.16015864	1.3941784	–0.0073235598

(*Continued*)

(*Continued*)

Compound	a_c [kPa (m³/kmol)²]	b (m³/kmol)	κ_0	κ_1
m-Dinitrobenzene	5333.9956	0.13559379	1.3225690	0.0090607399
m-Nitrotoluene	4482.5195	0.12497098	1.0758846	−0.33371368
Nitrobenzene	3712.8684	0.10567292	1.0053108	−0.10450033
Nitroethane	2155.1689	0.074372034	0.90647300	−0.07151607
o-Dinitrobenzene	5695.8517	0.14026221	1.3286568	0.010298334
o-Nitrotoluene	4329.9744	0.12306537	1.0604948	−0.14173868
p-Dinitrobenzene	5278.8638	0.13452653	1.3317130	−0.0071327734
p-Nitrotoluene	4504.0825	0.12523092	1.1364876	−0.40697642
11.3. Other monofunctional C, H, O, N				
Cyclohexyl isocyanate	3686.2999	0.11917084	1.1162716	0.033052091
Methyl isocyanate	1552.3296	0.062903613	0.63739245	0.045790383
n-Butyl isocyanate	2966.5446	0.10687733	0.97414355	0.031335588
Phenyl isocyanate	3268.7897	0.10322741	1.0008073	−0.17814072
Toluene diisocyanate	5548.9214	0.15407214	1.1624697	0.063558379
11.4. Polyfunctional C, H, N, (O), halide				
1-Chloro-2,4-dinitrobenzene	6000.2031	0.15088543	1.3874004	−0.0044978098
1,2-Dichloro-4-nitrobenzene	5046.2891	0.13623414	1.1375561	0.026672155
3-Nitrobenzotrifluoride	4993.5664	0.15320327	1.1235395	0.021771208
3,4-Dichloroaniline	4909.3024	0.12557779	1.0452990	0.034619238
4-Chloro-3-nitrobenzotrifluoride	5449.6309	0.16256461	1.2090581	−0.035625130
m-Chloroaniline	3884.7390	0.10585343	0.96125778	0.090019597
m-Chloronitrobenzene	4341.6222	0.11973777	1.0762473	−0.19902248
o-Chloroaniline	3588.4696	0.10170800	0.97552703	−0.041666560
o-Chloronitrobenzene	4521.3205	0.12222286	1.0675917	−0.20666206
p-Chloronitrobenzene	4483.4251	0.12216675	1.0696122	0.047834585
11.5. Polyfunctional C, H, O, halides				
2-Chloroethanol	1827.4156	0.063924082	1.3018544	−0.41065883
Acetyl chloride	1426.4704	0.057462178	0.79945754	0.081824941
alpha-Epichlorohydrin	2398.7112	0.080469476	0.76209761	0.15019934
Benzoyl chloride	3783.6635	0.11108688	0.98127798	−0.13268202
bis(Chloromethyl)ether	2314.2756	0.081793643	0.84787400	0.047500026
Carbonyl fluoride	483.93642	0.033343785	0.80198736	0.1268412
Chloroacetaldehyde	1812.0647	0.066813427	0.85666285	0.047273446
Chloroacetic acid	2605.7168	0.077729545	1.1429189	0.048235544
Chloroacetyl chloride	2084.3284	0.073413014	0.89027363	0.078541361
Dichloroacetaldehyde	1953.6478	0.072033798	0.87869362	0.043361502
Dichloroacetyl chloride	2300.4937	0.081306548	0.91023863	0.028648327
Ethyl chloroformate	1812.0480	0.072972766	1.5166836	−0.026740606
Hexafluoroacetone	1417.7884	0.081237389	0.90401330	0.11574155
Isophthaloyl chloride	5656.0813	0.15070840	1.2674364	0.017693856
m-Chlorobenzoyl chloride	4538.5227	0.12828001	1.0212183	−0.056623118
m-Chlorophenol	3190.2349	0.089552593	1.0582744	−0.28753934
Methyl chloroacetate	2530.3028	0.086298704	0.99718789	0.055190425
Methyl chloroformate	1626.4885	0.063397879	0.94298599	0.036694094
o-Chlorobenzoic acid	4918.5154	0.12708430	1.3013590	0.0039672154
o-Chlorophenol	2884.9190	0.087460687	0.99792426	−0.36328295
p-Chlorophenol	3281.4821	0.090990639	1.0573930	−0.23697911
Trichloroacetaldehyde	2289.4309	0.082920541	0.85985452	0.044738838

(*Continued*)

(Continued)

Compound	a_c [kPa (m³/kmol)²]	b (m³/kmol)	κ_0	κ_1
Trichloroacetic acid	3128.1841	0.093043686	1.1525970	0.055019479
Trichloroacetyl chloride	2686.1534	0.093166941	0.87910466	0.025427066
Trifluoroacetic acid	2345.5204	0.097705651	1.1147483	–0.078213188
11.6. Polyfunctional C, H, O, S				
3-Methyl sulfolane	4980.4996	0.12474809	0.98037871	0.031024710
Carbonyl sulfide	707.33894	0.038212076	0.52624200	0.064791637
di-n-Butyl sulfone	7317.0418	0.19521951	1.3316847	0.0021959431
di-n-Propyl sulfone	5944.9999	0.15944480	1.1883972	0.032165327
Sulfolane	3197.9073	0.077079913	0.94237060	0.024432281
11.7. Polyfunctional C, H, O				
2-(2-Ethoxyethoxy)ethanol	4019.3490	0.13014326	1.5588855	–0.22407809
2-(2-Methoxyethoxy)ethanol	3550.1123	0.11531467	1.5116417	–0.15029013
2-Butoxyethanol	3508.3923	0.11965750	1.4530104	–0.20645545
2-Ethoxyethanol	2414.3905	0.086831697	1.3431685	–0.10232741
2-Methoxyethanol	1999.6757	0.072554358	1.3482114	–0.21410771
Citric acid	5640.8378	0.14042836	2.6824919	–0.39615614
Diethylene glycol ethyl ether acetate	5667.1527	0.17571305	1.3613476	–0.32027769
Diethylene glycol monobutyl ether	5284.4653	0.16535081	1.5145987	0.23841343
Diglycolic acid	4804.1953	0.11989190	1.8172800	–0.11044920
Ethyl acetoacetate	4027.9094	0.12818929	1.1700672	–0.16139209
Ethyl lactate	2832.5652	0.098579276	1.4581123	–0.77861293
Furfural	2451.8155	0.076366951	1.0120984	–0.12197605
Furfuryl alcohol	2358.4243	0.076363865	1.3921231	–0.26299568
Glycolic acid	1649.6256	0.054800902	1.7884171	–0.11073781
Hydroxycaproic acid	4973.9872	0.13428221	1.9126828	–0.11592734
Lactic acid	2027.0679	0.067339613	1.7567190	–0.085424482
Methoxyacetic acid	3023.2941	0.089533463	1.2393917	0.11848404
Methyl acetoacetate	3514.2462	0.11201603	1.1079688	0.017428755
Methyl salicylate	3785.7656	0.11051437	1.1360756	0.34215209
p-Methoxyphenol	3650.5252	0.098552849	1.1408396	0.024608856
Salicylaldehyde	2925.4984	0.088038775	1.2433335	–0.48527491
Tartaric acid	4177.0801	0.10323465	2.8481211	–0.53600569
Tetrahydrofurfuryl alcohol	2770.5168	0.088724377	1.3231888	–0.32111670
Vanillin	4730.4017	0.12458336	1.4236148	–0.026639145
11.8. Polyfunctional C, H, O, N				
1-Amino-2-propanol	2112.0365	0.070390829	1.4504471	–0.35288772
2-Pyrrolidone	3212.4634	0.083003430	0.99878459	0.033642366
3-Amino-1-propanol	2413.2822	0.076093305	1.5088299	–0.026854360
3-Methoxypropionitrile	3519.1522	0.11287569	1.0446425	–0.10345078
8-Hydroxyquinoline	4503.5855	0.11695404	1.1158712	0.031010468
Acetanilide	5743.6886	0.14246887	1.1737524	0.0025324270
Acetone cyanohydrin	3112.5055	0.098443916	1.3773837	–0.73392663
bis(Cyanoethyl)ether	7079.6320	0.18502566	1.4204766	0.019745540
Dehydroabietylamine	13791.814	0.32703455	1.4021430	–0.0091061545
Epsilon-caprolactam	4307.0243	0.10935164	1.0557071	0.020892266
Formanilide	4767.7933	0.12397260	1.1457701	0.025676928
l-Glutamic acid	6032.8110	0.13933783	1.9516167	–0.16858266
Lactonitrile	2588.7466	0.082387553	1.4697344	–0.021051493
Lysine	6041.5089	0.15058624	1.7349738	–0.10914624

(Continued)

(*Continued*)

Compound	a_c [kPa (m³/kmol)²]	b (m³/kmol)	κ_0	κ_1
Methyl cyanoacetate	3916.3440	0.11665601	1.1521583	0.022162572
Methyl diethanolamine	3740.7893	0.11290589	2.0669461	−0.13483684
Monoethanolamine	1928.6426	0.061860598	1.4477034	−0.21966904
Morpholine	2261.1517	0.074872833	0.87874008	0.034459321
n-Aminoethyl ethanolamine	3442.4519	0.10092424	1.7747197	−0.075539317
n-Methyl-2-pyrrolidone	3431.5481	0.096991696	0.89861870	0.035036314
n-Methylformamide	2924.9995	0.083018248	0.96680736	0.036516683
n,n-Dimethylacetamide	3395.4820	0.10559870	0.90333759	0.039040866
n,n-Dimethylformamide	2998.7957	0.094847444	0.85487690	0.13014625
Nitroglycerine	5086.4225	0.15306876	1.9149758	−0.095169695
o-Nitroanisole	5144.8392	0.13463194	1.1683604	0.019277578
Oxazole	1536.0541	0.056738752	0.72283726	0.037000604
p-Dimethylaminobenzaldehyde	7100.7215	0.17464743	1.1214359	−0.27101896
p-Phenetidine	5034.9430	0.13664893	1.1579228	0.020735895
Pentaerythritol tetranitrate	6416.5541	0.19423979	2.2471341	−0.30655400
Triethanolamine	8123.9451	0.21123957	1.7944963	−0.13684108
12. Elements				
Argon	146.99772	0.019939742	0.37414086	0.0029062082
Bromine	1049.6354	0.036770298	0.56728738	−0.0011322542
Chlorine	705.80912	0.034624065	0.48735538	0.075796202
Deuterium	27.958434	0.014918684	0.15918813	−0.059285425
Fluorine	125.89156	0.017851849	0.46282373	0.0015521138
Helium-3	2.9981803	0.018535862	−0.36700701	−0.15714048
Helium-4	3.7413227	0.014723290	−0.22409080	−0.13537160
Hydrogen	26.492357	0.016339067	0.051526807	−0.11738005
Mercury	595.90532	0.007028470	0.11805156	−0.022005780
Neon	23.487290	0.010825126	0.31024345	0.024655119
Nitrogen	148.62678	0.024119318	0.43743964	0.010707724
Oxygen	150.43964	0.019915536	0.40832681	0.020801738
13. Epoxides and peroxides				
1,2-Epoxybutane	1995.8558	0.077647318	0.73762517	0.014905147
1,2-Propylene oxide	1491.5234	0.063290839	0.75270077	0.068254857
1,3-Propylene oxide	1488.8405	0.058590590	0.67249097	0.038812753
1,4-Dioxane	2122.1611	0.073981527	0.77243968	0.041781418
2,5-Dihydrofuran	1687.2894	0.063704974	0.71522067	0.049168186
Ethylene oxide	967.1134	0.042184093	0.68750738	0.012995286
Furan	1378.1845	0.057538878	0.67315670	0.091589917
Tetrahydrofuran	1772.6274	0.067156201	0.71024113	0.041312022
Trioxane	1989.2831	0.067397330	0.85927304	0.047841607
14. Esters				
14.1. Aromatic esters				
Benzyl acetate	4830.4317	0.14141380	1.0303396	0.056424738
Benzyl benzoate	8145.2037	0.20326899	1.2504283	0.014857767
Dibutyl phthalate	11139.9201	0.29188655	1.5499898	0.21565362
Diethyl phthalate	7789.2252	0.21056269	1.4005204	0.052544349
Dimethyl phthalate	6672.4211	0.17825336	1.2860778	−0.12449116
Dimethyl terephthalate	6781.3149	0.17975445	1.2611592	−0.33446437
Dioctyl phthalate	17395.991	0.44166926	1.9019842	0.025851568
Ethyl benzoate	4784.1144	0.14025849	1.0637752	−0.14755105

(*Continued*)

(Continued)

Compound	a_c [kPa (m³/kmol)²]	b (m³/kmol)	κ_0	κ_1
Methyl benzoate	4228.9156	0.12487596	0.97468747	0.031451080
14.2. Other saturated aliphatic esters				
Dibutyl sebacate	14358.1528	0.38257835	1.8655572	−0.20101677
Diethyl malonate	4780.3554	0.14980629	1.2247330	−0.049599112
Diethyl oxalate	4270.5601	0.13528055	1.1778395	0.018126655
Diethyl succinate	5419.0899	0.16802173	1.3935650	0.027684410
Dihexyl adipate	14112.247	0.37651636	1.8213368	−0.033515161
Ethyl isovalerate	3847.5576	0.13391455	0.98024756	−0.021859326
Ethyl propionate	2824.2551	0.10585084	0.92997999	0.11176239
Ethylene glycol diacetate	4364.5717	0.13677650	1.0881057	0.81965576
Gamma-butyrolactone	2904.8818	0.080439074	0.92601492	−0.16443755
Glyceryl triacetate	6797.8774	0.19759853	1.5202189	−0.024120604
Isobutyl isobutyrate	4408.2310	0.14984798	0.95510575	0.12383530
Isopentyl isovalerate	5823.8983	0.18709293	1.1973192	−0.36308778
Methyl dodecanoate	9198.3741	0.26437118	1.3234423	0.11429634
Methyl n-butyrate	2816.0113	0.10392400	0.90576228	0.12581828
Methyl propionate	2247.5025	0.086679415	0.87566154	0.078826526
n-Butyl n-butyrate	4722.0861	0.15686867	1.0717582	0.018934418
n-Butyl propionate	4111.2973	0.14163659	1.0394833	0.045970120
n-Butyl stearate	16578.8345	0.44406206	1.7596807	−0.045961461
n-Propyl n-butyrate	4003.9954	0.13793998	1.01913712	−0.037815918
n-Propyl propionate	3391.5485	0.12007517	0.92670194	0.050682799
Vinyl propionate	2561.6704	0.096009376	0.86025894	0.0016081910
15. Ethers				
1,2-Dimethoxyethane	2352.5045	0.089789838	0.87437233	−0.019487712
Acetal	3105.5388	0.11746885	1.0093134	0.078745965
Anethole	5691.2641	0.16108440	1.0669792	0.029967840
Anisole	3116.5919	0.099395053	0.90724769	0.0067410215
Benzyl ethyl ether	4449.6633	0.13754734	0.99593703	−0.031577730
Butyl vinyl ether	2900.9670	0.11075441	0.92694131	0.044499615
di-n-Butyl ether	4372.4217	0.15400292	1.0431002	0.040259132
di-n-Hexyl ether	7532.9898	0.23427423	1.3624551	−0.0028267190
di-n-Propyl ether	2941.7473	0.11345435	0.87539258	0.11721086
Dibenzyl ether	7456.5144	0.19638028	1.2452426	0.080565570
Diethyl ether	1909.4888	0.083726389	0.76760851	0.14440991
Diisopropyl ether	2755.0548	0.11274567	0.87264874	−0.064554698
Dimethyl ether	965.17439	0.049365129	0.66710318	0.0098302824
Diphenyl ether	5876.8170	0.15761613	1.0457127	−0.16939040
Divinyl ether	1594.8502	0.070489095	0.80020021	0.036192868
Ethyl propyl ether	2345.0083	0.095930716	0.80765157	0.24986080
Ethyl vinyl ether	1753.2416	0.075508194	0.76946362	0.040335995
Methyl ethyl ether	1356.6407	0.063412100	0.67488632	0.68013086
Methyl tert-butyl ether	2281.8834	0.093936143	0.77022242	0.047964831
Methyl vinyl ether	1317.0667	0.061675031	0.72605707	0.060751760
Methylal	1848.3387	0.078701104	0.80297845	0.038067161
n-Butyl ethyl ether	2983.6923	0.11498536	0.94121183	0.031523589
Phenetole	3866.2089	0.12225409	0.97494659	0.031529239
Tetraethylene glycol dimethyl ether	7945.1758	0.23062025	1.6931668	−0.42429078

(Continued)

(*Continued*)

Compound	a_c [kPa (m^3/kmol)^2]	b (m^3/kmol)	κ_0	κ_1
	16. Formates and acetates			
2-Ethylhexyl acetate	5952.5789	0.19062828	1.2614381	0.00039260867
Allyl acetate	2651.8717	0.097078651	0.93693961	−0.12445373
Ethyl acetate	2248.1324	0.087913218	0.90207766	0.055381792
Ethyl formate	1735.3270	0.069848779	0.78589032	0.055433003
Ethylidene diacetate	3921.2828	0.12636810	1.0587419	0.29894956
Isobutyl formate	2493.9604	0.092564659	0.95118738	−0.12899941
Isopentyl acetate	4010.2986	0.13700390	0.97422068	0.036069623
Isopropyl acetate	2555.3570	0.097196883	0.89521687	−0.0035489757
Methyl acetate	1745.5688	0.070482839	0.83818916	0.061918731
Methyl formate	1284.9215	0.053969993	0.73105842	0.097081783
n-Butyl acetate	3409.4015	0.12046756	0.96676661	0.051625813
n-Butyl formate	2811.2928	0.10291468	0.94609348	−0.067112430
n-Pentyl acetate	4039.8853	0.13824546	1.0748246	0.024651643
n-Propyl acetate	2854.8407	0.10633501	0.92558516	0.12423781
n-Propyl formate	2292.6621	0.087204885	0.82953845	0.055333415
tert-Butyl acetate	2961.7165	0.11120646	0.87398446	0.048305412
Vinyl acetate	2050.8044	0.080089576	0.87571466	−0.030224144
Vinyl formate	1592.6367	0.065444087	0.78366914	0.15321973
	17. Inorganic gases			
Arsine	671.01844	0.036813631	0.39095266	0.042992568
Carbon dioxide	396.18119	0.026650380	0.70132798	0.021490195
Carbon disulfide	1202.9106	0.044594089	0.54625774	−0.019869276
Carbon monoxide	159.83481	0.024607306	0.45098475	0.039868702
Diborane	702.39723	0.049598375	0.52781942	0.095435474
Disilane	1148.7332	0.054414981	0.52717691	−0.035910443
Hydrazine	915.40685	0.028680338	0.83283219	0.0051948669
Hydrogen sulfide	490.08867	0.026849253	0.52175647	0.045558612
Nitric oxide	157.44303	0.017884311	1.1966190	0.058053276
Nitrogen dioxide	582.04962	0.027625787	1.5312032	−0.79316762
Nitrogen trifluoride	381.23074	0.033360579	0.58053421	−0.0091212901
Nitrosyl chloride	672.55400	0.031233195	0.82236572	−0.22636125
Nitrous oxide	416.20319	0.027512469	0.59455708	0.15322754
Ozone	386.09791	0.030271920	0.66687090	0.079545135
Perchloryl fluoride	793.48025	0.044075732	0.62541153	0.064745462
Phosgene	1145.4199	0.051515315	0.65746421	0.090821885
Phosphine	509.85398	0.032127704	0.43265394	0.0035315482
Silane	475.11290	0.036049472	0.48420872	0.016652831
Sulfur dioxide	746.20839	0.035450134	0.73328102	0.093364814
Sulfur trioxide	929.65995	0.038757733	1.0007079	0.67478535
Tetrafluorohydrazine	815.90611	0.053972571	0.70497870	−0.30616308
	18. Ketones			
2-Heptanone	4048.6982	0.13547728	0.96698716	0.088062742
2-Hexanone	3289.0558	0.11465135	0.94780030	0.040867249
2-Pentanone	2693.2460	0.098227767	0.88849954	−0.089041441
5-Methyl-2-hexanone	3838.2389	0.13068946	0.97898941	0.089556900
Acetone	1733.5551	0.069804917	0.82114231	−0.017586622
Acetophenone	4040.9197	0.11796285	0.98587807	−0.065552953

(*Continued*)

(Continued)

Compound	a_c [kPa (m³/kmol)²]	b (m³/kmol)	κ_0	κ_1
Acetylacetone	2893.4151	0.098355192	1.0812389	0.032638569
Benzophenone	6994.0164	0.17539593	1.1501039	−0.17301392
Camphor	5299.9172	0.15296986	0.84536409	−0.010016609
Cyclohexanone	3147.5914	0.10237813	1.0193956	−0.20223933
Cyclopentanone	2128.7215	0.069586909	0.90264751	−0.11635843
Diethyl ketone	2655.8330	0.09688569	0.88104494	−0.0072863217
Isophorone	4851.6616	0.13885691	0.94921118	−0.16908669
Ketene	751.30436	0.041552506	0.57363314	0.069469146
Mesityl oxide	3338.5125	0.11386357	0.84877108	0.10548778
Methyl ethyl ketone	2199.9697	0.084069848	0.83933590	0.0084544073
Methyl isobutyl ketone	3149.0830	0.11277865	0.88985029	0.18470690
Methyl isopropyl ketone	2516.4044	0.093119007	0.78436109	0.36916243
19. n-Aliphatic Acids				
19.1. Aromatic carboxilic acids				
Isophthalic acid	8138.41965	0.16538406	1.794398509	−0.134193535
o-Toluic acid	4618.38633	0.12584424	1.292968353	0.010838362
p-Toluic acid	4892.58155	0.12952143	1.297555235	−0.008631397
Phthalic acid	5104.23613	0.13056411	1.792846681	−0.12699126
Pyromellitic acid	8015.89277	0.18368914	2.656669216	−0.505780886
Terephthalic acid	9931.31966	0.18259750	1.790947037	−0.139377479
19.2. Inorganic acids				
Hydrogen bromide	488.69005	0.027537879	0.48121536	0.13349968
Hydrogen chloride	398.67837	0.025129873	0.57793434	−0.046943151
Hydrogen fluoride	1036.1812	0.045980802	0.92137589	−0.50671496
Hydrogen iodide	688.62073	0.033246883	0.42983469	0.082471874
Perchloric acid	3263.4499	0.10583532	0.45721839	0.12540479
19.3. Inorganic bases				
Ammonia	460.26755	0.023218914	0.74545487	−0.012061178
19.4. Inorganic halides				
Antimony trichloride	4134.4449	0.10655645	0.62641676	0.11139889
Boron trichloride	1650.6602	0.074739523	0.60002024	0.17733657
Boron trifluoride	428.43586	0.033604294	1.0108459	−0.044808936
Dichlorosilane	1437.7143	0.065525341	0.62969367	0.082773087
Silicon tetrachloride	2236.8459	0.090284071	0.72261540	0.057156476
Sulfuryl chloride	2037.6484	0.076509574	0.64107352	0.18165662
Titanium tetrachloride	2745.5820	0.088063669	0.77923727	−0.10523728
Trichlorosilane	1742.2479	0.074431620	0.68817984	−0.012170739
19.5. n-Aliphatic acids				
Acetic acid	1955.2632	0.067506565	1.0216356	−0.17966872
Formic acid	1439.2961	0.050781460	1.0492666	−0.52864419
n-Butyric acid	2847.2519	0.092778914	1.2180017	−0.014442586
n-Decanoic acid	7095.2627	0.20363948	1.5495994	0.69737178
n-Dodecanoic acid	8752.2381	0.24400914	1.6823928	0.22405382
n-Heptanoic acid	4883.9928	0.14697692	1.3705469	0.27719333
n-Hexadecanoic acid	12623.382	0.33288717	1.7637494	0.76607256
n-Nonanoic acid	6364.8572	0.18527481	1.4958342	0.22841814
n-Octanoic acid	5660.3961	0.16738779	1.4345819	0.25973482

(Continued)

(*Continued*)

Compound	a_c [kPa (m³/kmol)²]	b (m³/kmol)	κ_0	κ_1
n-Tetradecanoic acid	10684.839	0.28922060	1.7474202	0.065771850
Pentadecanoic acid	11668.979	0.31173613	1.7288166	0.68778397
Propionic acid	2549.0796	0.086363354	1.1344665	−0.034737088
Stearic acid	14948.838	0.382863260	1.8133525	−0.27017975
Valeric acid	3518.8555	0.11061231	1.2443862	0.075962748
19.6. Other aliphatic acids				
2-Methylbutyric acid	3362.6133	0.10701607	1.2081557	0.0059066559
Acrylic acid	2112.0131	0.070275594	1.1356440	−0.14560751
Azelaic acid	8079.4619	0.20386591	1.9242514	−0.13829119
cis-Crotonic acid	2815.9062	0.089062923	1.1818168	0.022801532
Glutaric acid	5128.7720	0.13005375	1.6598487	0.051204460
Isobutyric acid	2893.7510	0.097212014	1.2465735	−0.13187224
Isovaleric acid	3267.2164	0.10545609	1.2813614	−0.0035128126
Linoleic acid	13431.545	0.35465598	1.9232537	−0.098878848
Maleic acid	3779.9353	0.10006631	1.7182121	−0.086617539
Methacrylic acid	2792.4506	0.088870488	1.0472514	0.045461466
Sebacic acid	8946.8357	0.22464404	1.9634932	−0.25428501
Succinic acid	4334.6773	0.11005373	1.7134879	−0.10089981
trans-Crotonic acid	2957.0770	0.090859729	1.1895622	0.10397067
19.7. Other inorganics				
Deuterium oxide	604.05272	0.019197567	0.90091414	−0.047223017
Hydrogen peroxide	777.20957	0.021782568	0.89040053	−0.063312378
Water	603.34119	0.019078950	0.86975396	−0.062787920
19.8. Other salts				
Potassium chloride	21144.339	0.12469461	0.17378161	0.31508072
20. Nitriles				
1.4-Dicyano-2-butene	6109.2715	0.16558672	1.3044724	0.0063192095
Acetonitrile	1938.6867	0.07272704	0.86591088	−0.17479070
Acrylonitrile	2019.4879	0.07724502	0.86564622	−0.13055942
Adiponitrile	6815.2753	0.17857284	1.3121771	0.0024393034
Benzonitrile	3674.3322	0.10751446	0.88943360	−0.0054126211
cis-Dicyano-1-butene	5120.6849	0.15164673	1.3112027	0.0029824494
Cyanogen	851.50508	0.04354592	0.77998536	0.0064955920
Cyanogen chloride	1074.6603	0.04897878	0.82991437	0.0085544278
Glutaronitrile	6154.1275	0.16104334	1.1923210	−0.22240893
Hexanenitrile	4192.8430	0.13793242	1.0677150	−0.23000489
Hydrogen cyanide	1231.4180	0.05518297	0.96227219	−0.34796116
Isobutyronitrile	2684.8287	0.09724139	0.86939331	0.037120775
Methacrylonitrile	2498.1932	0.09227824	0.81216017	−0.14037982
Methylglutaronitrile	6047.5416	0.16678539	1.2697683	0.0026382615
n-Butyronitrile	2829.6282	0.09944957	0.90127759	−0.011749574
Propionitrile	2402.7240	0.08711640	0.83358725	−0.017768606
Succinonitrile	5293.1032	0.14067038	1.1640566	0.025905582
trans-Crotonitrile	2782.5987	0.09717085	0.94462581	−0.11770785
trans-Dicyano-1-butene	5088.1113	0.15111947	1.3016477	0.0050616206
Valeronitrile	3521.5237	0.11950782	0.94047244	0.03220135
Vinylacetonitrile	2779.6098	0.097398892	0.92320370	−0.073845497

(*Continued*)

(*Continued*)

Compound	a_c [kPa (m³/kmol)²]	b (m³/kmol)	κ_0	κ_1
	21. Organic/inorganic compounds			
Diethyl carbonate	3103.4112	0.11025540	1.0413448	0.052649200
Dimethyl sulfate	3519.6866	0.095020613	0.51660430	0.40377576
Ethylene carbonate	2914.0387	0.075483365	0.97328705	0.13253627
Tri-n-butyl borate	8781.0766	0.24179890	0.66491664	0.021304369

REFERENCES

1. Stryjek, R., and Vera, J. H. 1986. An improved cubic equation of state. In *Equations of State: Theories and Applications,* ACS Symposium Series 300. eds. K. C. Chao, and R. L. Robinson. Washington, DC: American Chemical Society, pp. 560–570.
2. Stryjek, R., and Vera, J. H. 1986. PRSV: An improved Peng-Robinson equation of state for pure compounds and mixtures. *Can. J. Chem. Eng.* 64(2): 323–333; PRSV: An improved Peng-Robinson equation of state with new mixing rules for strongly nonideal mixtures. *Can. J. Chem. Eng.* 64(2): 334–340. ERRATA: *Can. J. Chem. Eng.* 67: 523. 1989.
3. Stryjek, R., and Vera, J. H. 1986. PRSV2: A cubic equation of state for accurate vapor-liquid equilibria calculations. *Can. J. Chem. Eng. 64*: 820–826; ERRATA: *Can. J. Chem. Eng. 67*: 523. 1989.
4. Peng, D.Y., and Robinson, D.B. 1976. A new two-constant equation of state. *Ind. Eng. Chem. Fundam.* 15: 50–64.
5. Soave, G. 1972. Equilibrium constants from a modified Redlich-Kwong equation of state. *Chem. Eng. Sci.* 27: 1197–1203.
6. Redlich, O., and Kwong, J. N. S. 1949. On the thermodynamics of solutions. V. An equation of state. Fugacities of gaseous solutions. *Chem. Rev.* 44(1): 233–244.
7. Zabaloy, M. S., and Vera, J. H. 1998. The Peng-Robinson sequel - An analysis of the particulars of the second and third generations. *Ind. Eng. Chem. Res.* 37(5): 1591–1597.
8. Mathias, P. M. 1983. A versatile phase equilibrium equation of state. *Ind. Eng. Chem. Process Des. Dev.* 22: 385–391.
9. Proust, P., and Vera, J. H. 1989. PRSV The Stryjek-Vera modification of the Peng-Robinson equation of state. Parameters for other compounds of industrial interest. *Can. J. Chem. Eng.* 67: 170–173.
10. Olivera-Fuentes, C. 2017. Appendix C: Working with the PRSV EOS in Vera, J. H., and Wilczek-Vera, G. *Classical Thermodynamics of Fluid Systems. Principles and Applications.* Boca Raton: CRC Press.
11. Figueira, F. L., Lugo L., and Olivera-Fuentes, C. 2007. Generalized parameters of the Stryjek-Vera and Gibbons-Laughton cohesion functions for use with cubic EOS of the van der Waals type. *Fluid Phase Equilib.* 259: 105–119.
12. Daubert, T. F., and Danner, R.P. 1988. *Physical and Thermodynamic Properties of Pure Chemicals-Data Compilation.* Washington: Hemisphere.
13. Wilczek-Vera, G., and Vera, J. H. 2015. Understanding cubic equations of state: A search for the hidden clues of their success. *AIChE J.* 61(9): 2824–2831.
14. Kolasinska, G., and Vera, J. H. 1986. On the prediction of heats of vaporization of pure compounds. *Chem. Eng. Commun.* 43: 185–194.
15. Soave, G. 1986. Direct calculation of pure-compound vapour pressures through cubic equations of state. *Chem. Eng. Sci.* 31: 203–207.
16. Proust, P., Meyer, E., and Vera, J. H. 1993. Calculation of pure compound saturated enthalpies and saturated volumes with the PRSV equation of state. Revised k1 parameters for alkanes. *Can. J. Chem. Eng.* 71: 292–298.
17. Adachi, Y., and Sugie, H. 1986. A new mixing rule-modified conventional mixing rule *Fluid Phase Equilib,* 28: 103–118.
18. Panagiotopoulos, A. Z., and Reid, R. C. 1986. A new mixing rule for cubic equations of state for highly polar, asymmetric systems. *ACS Symposium Series.* 300: 571–582.
19. Sandoval, R., Wilczek-Vera, G., and Vera, J. H. 1989. Prediction of ternary vapor-liquid equilibria with the PRSV equation of state. *Fluid Phase Equilib.* 52: 119–126.

20. Michelsen, M. L., and Kistenmacher, H. 1990. On composition-dependent interaction coefficients. *Fluid Phase Equilib.* 58: 229–230.
21. Zabaloy, M. S., and Vera, J. H. 1996. Identification of variant and invariant properties in the thermodynamics of mixtures: tests for models and computer codes. *Fluid Phase Equilib.* 119: 27–49.
22. Orbey, H., Sandler, S. I., and Wong, D. S. H. 1993. Accurate equation of state predictions at high temperatures and pressures using the existing UNIFAC model. *Fluid Phase Equilib.*, 85: 41–54.
23. Tochigi, K. 1995. Prediction of high-pressure vapor-liquid equilibria using ASOG. *Fluid Phase Equilib.* 104: 253–260.
24. Wilczek-Vera, G., and Vera, J. H. 1989. A consistent method to combine the PRSV EOS with excess Gibbs energy models. *Fluid Phase Equilib.* 51: 197–208.
25. "Founders Tribute dedicated to John M. Prausnitz". 2015. *AIChE J.* 61(9): 2673–3144.
26. Zabaloy, M. S., Brignole, E. A., and Vera, J. H. 1999. A conceptually new mixing rule for cubic and non-cubic equations of state. *Fluid Phase Equilib. 158–160*: 245–257.
27. Zabaloy, M. S., Brignole, E. A., and Vera, J. H. 2002. A flexible mixing rule satisfying the ideal-solution limit for equations of state. *Ind. Eng. Chem. Res.* 41(5): 922–930.
28. Ohta, T., Todoriki, H., and Yamada, T. 2004. Representation of liquid-liquid equilibria at low and high pressures using the EOS-G E mixing rules. *Fluid Phase Equilib.* 225: 23–27.
29. Ohta, T. 1991. Prediction of multicomponent vapor-liquid equilibria using the PRSV and PRSV2 equations of state with the Huron-Vidal mixing rules. *Thermochimica Acta.* 185: 283–293.
30. Ohta, T. 1992. Prediction of ternary excess enthalpies using the PRSV and PRSV2 equations of state. *Thermochimica Acta*, 202: 51–63.
31. Wong, D. S. H., and Sandler, S. I. 1992. A theoretically correct mixing rule for cubic equations of state. *AIChE J.* 38: 671–680.
32. Cardano, G. 1545, Artis Magnae Sive de Regulis Algebraicis. English translation by T. Richard Witmer published as Cardano, G. 1993. *Ars Magna or the Rules of Algebra.* New York: Dover.
33. Finlayson, B. A., and Biegler, L. T. 2019. Mathematics. In *Perry's Chemical Engineers' Handbook,* 9th ed., eds. D. W. Green and M. Z. Southard. New York: McGraw-Hill Education.
34. Zhi, Y., and Lee, H. 2002. Fallibility of analytic roots of cubic equations of state in low temperature-region. *Fluid Phase Equilib.* 201: 287–294.
35. Deiters, U. K. 2002. Calculation of densities from cubic equations of state. *AIChE J.* 48 (4): 882–886.
36. Edmister, W. C., and Lee, B. I., *Applied Hydrocarbon Thermodynamics*, 2nd ed., Vol. I. Houston, TX: Gulf Publishing Co.
37. Gosset, R., Heyen, G., and Kalitventzeff, B. 1986. An efficient algorithm to solve cubic equations of state. *Fluid Phase Equilib.* 25: 51–64.
38. Olivera-Fuentes, C. 2003. The optimal solution of cubic equations of state. *Lat. Am. Appl. Res.* 23: 243–256.
39. Deiters, U. K., and Macías-Salinas, R. 2014. Calculation of densities from cubic equations of state: Revisited. *Ind. Eng. Chem. Res.* 53: 2529–2536.
40. Wilson, G. M. 1968. A modified Redlich-Kwong EOS. Application to general physical data calculations. Paper No. 15C presented at the American Institute of Chemical Engineers 65th National Meeting: Cleveland, OH.
41. de Hemptinne, J.-C., Ledanois, J.-M., Nougin, P., and Barreau, A. 2012. *Select Thermodynamic Models for Process Simulation*, Paris: Technip.
42. Elliott, J. R., and Lira, C. T. 2012. *Introductory Chemical Engineering Thermodynamic*s, 2nd ed., Upper Saddle River, NJ: Prentice-Hall
43. Figueira, F., Rodríguez, A. J., Stamateris, B. C., and Olivera-Fuentes, C. G. 2006. An energy-based cohesion function for cubic equations of state. *Polish J. Chem.* 80: 81–97.
44. Figueira, F. L. 2005. Development of equations of state of van der Waals type for polar and nonpolar pure fluids (in Spanish). MSc Thesis, Simón Bolívar University, Caracas.
45. *DIPPR$^{(r)}$ 801 Sample Database.* https://dippr.aiche.org/SampleDb (last accessed February 15, 2023).

Appendix D
Working with ASOG-KT

ABSTRACT

This appendix gives the essential tools for estimating the activity coefficients of compounds in organic and also in nonelectrolyte aqueous mixtures. The logarithm of the activity coefficient is given by a structural contribution, which depends only on the number and size of the groups forming each molecule in the mixture, and a contribution arising from the interactions between the groups of the different molecules. An extensive list of group interaction parameters prepared by Gmehling and Tochigi is reproduced in a rearranged format.

As discussed in Chapter 28, the Analytical Solution of Groups, ASOG, is based on studies carried out in the early sixties at Shell Research Co., Emeryville, California, by G. M. Wilson, C.H. Deal, and E. L. Derr, among others. It basically recognizes the importance of the behavior of mixtures of functional groups characteristic of the homologous series. Although the basic elements of the method were clearly set forward by Wilson and Deal in 1962 [1], its use only became possible with the publication of parameters for the group interactions by K. Kojima and K. Tochigi [2]. For this reason, the ASOG method used with the Kojima–Tochigi parameters is known as ASOG-KT. Variations of ASOG such as PRASOG [3] are not included here.

The use of ASOG-KT is simple, but a word of warning is required. It is necessary to remember that not all interactions between groups are physical in nature. The most common cases of *non-physical interaction* are those of the hydroxyl (–OH) group that reacts with the carboxyl group (–COOH) and the amino (–NH_2) group that reacts with the ketone (–CO–) group.

BASIC EQUATIONS

According to ASOG, the activity coefficient of component *i* of a liquid phase mixture is given by the sum of a combinatorial term and a term representing the group interactions:

$$\ln\gamma_i = (\ln\gamma_i)_{combinatorial} + (\ln\gamma_i)_{interactions} \tag{D.1}$$

For the combinatorial term ASOG uses

$$(\ln\gamma_i)_{combinatorial} = (1 - \frac{r_i}{\sum_k x_k r_k} + \ln\frac{r_i}{\sum_k x_k r_k}) \tag{D.2}$$

Where x_k is the mole fraction of compound *k* in the liquid mixture and r_i is the number of atoms different from hydrogen atoms in molecule *i*. The contribution of the interactions between groups is given by:

$$(\ln\gamma_i)_{interactions} = \sum_k \nu_{ki}[\ln\Gamma_k - \ln\Gamma_k^{(i)}] \tag{D.3}$$

where ν_{ki} is the total number of atoms different from hydrogen in all the groups of type *k* of molecule *i*. Only for three groups, this counting is done differently:

DOI: 10.1201/9781003431985-43

a. the value of ν_{ki} is considered to be 1.6 for water
b. the group saturated methine (CH) is considered to be 0.8 of methylene (CH_2) groups
c. the group quaternary carbon (C) is considered to be 0.5 methylene (CH_2) groups.

The methyl (CH_3) and the methylene (CH_2) groups are considered to be the same group.

The group activity coefficients Γ_k are given by the Wilson Equation, namely,

$$\ln\Gamma_k = 1 - \ln\sum_l X_l A_{kl} - \sum_l \frac{X_l A_{lk}}{\sum_m X_m A_{lm}} \tag{D.4}$$

and

$$\ln\Gamma_k^{(i)} = 1 - \ln\sum_l X_l^{(i)} A_{kl} - \sum_l \frac{X_l^{(i)} A_{lk}}{\sum_m X_m^{(i)} A_{lm}} \tag{D.5}$$

In the liquid mixture the (mole) fraction of group k is given by:

$$X_k = \frac{\sum_j x_j \nu_{kj}}{\sum_l \sum_m x_m \nu_{lm}} \tag{D.6}$$

and in pure compound i,

$$X_k^{(i)} = \frac{\nu_{ki}}{\sum_l \nu_{li}} \tag{D.7}$$

In both cases,

$$A_{kl} = \exp(m_{kl} + \frac{n_{kl}}{T}) \tag{D.8}$$

TABLES OF PARAMETERS

Parameters m_{kl} and n_{kl} [K] for 341 group pairs formed by 43 groups were reported by K. Tochigi *et al.* [4] and in an extension and revision was presented by Tochigi and Gmehling [5]. These values are reproduced by permission here in re-arranged tables. We have kept the group number assigned to each of the groups in the original publication of ASOG-KT [4], which included the first 31 groups, and also in the extension that followed [5]. With the purpose of avoiding an excessive number of blank spaces, we have reorganized the table of parameters and presented it in separate sections. Interactions for groups whose number does not appear in Table D.1 should be found in Tables D.2–D.6. The nomenclature for the groups presented in Tables D.1–D.3 is quite simple, as it follows the one used for the homologous series in organic chemistry: CH_2 includes the groups CH_3, CH_2, CH, and C (with the different way of counting the number of the carbon atoms indicated above); CO, COO, CHO, COOH, CN, and CON stand for the groups ketone, ester, aldehyde, carboxylic acid, nitrile, and amide, respectively. Formic acid is a group by itself: HCOOH. The group OH in an alcohol is differentiated from the group GOH in a polyalcohol. The groups: NH_2, NH, and N denote primary, secondary, and tertiary amine groups while NO_2 is the group in a nitro compound. The prefix *Ar* indicates a group that is part of an aromatic ring like ArCH, or that is linked to an aromatic ring like in ArCl, ArF, $ArNH_2$, $ArNO_2$, or ArOH for aromatic chloride, fluoride, amine, nitro compound or a phenol, respectively. The prefix Cy indicates a group in a cyclo compound; while CCl_4 and CS_2, carbon tetrachloride and carbon disulfide are groups by themselves. The groups CCl_2 and CCl_3 are carbon dichloride and carbon trichloride. For Tables D.4–D.6, we have kept the designation of groups used in the original tables and indicated their meaning in Table D.7.

HOW TO READ PARAMETERS IN TABLE D.1

Due to the format used in the original tables of parameters, also used here in Table D.1, it is necessary to pay special attention to the order of reading the values of the parameters for groups k and l. For the m_{kl} and n_{kl} parameters of the *leading* group k is found down in the column of the left-hand side of the table and of the *secondary* group l is found in the heading (upper) line of the table. For interaction of the groups OH and Cl, for example, the parameters are:

$m_{OH/Cl} = -1.7798$ and $n_{OH/Cl} = 0.1$; $m_{Cl/OH} = -0.6453$ and $n_{Cl/OH} = 0.2$

In Tables D.2–D.6 we have indicated clearly what the groups k and l are, and listed separately the values of m_{kl} and n_{kl} from the values of m_{lk} and n_{lk}.The nomenclature used for the new groups included in the extension of the method is reproduced (with permission) in Table D.7.

System Hexane-Methanol

As our mind works better with numbers than with symbols, the easiest way of using ASOG is to assign numbers to the groups, write all equations, and only in the final step go back to replace the group numbers by their symbols.

Let us consider the prediction of VLE at 60°C for the system hexane (1)–methanol (2). To do this, we need only two groups (CH_2) and (OH). Hexane has 6 (CH_2) groups as the methyl (CH_3) and the methylene (CH_2) groups are not distinguished by the ASOG method. Methanol has one (CH_2) and one (OH) group. Let us denote (CH_2) as the group #1 and (OH) as #2. Table D.1 shows in detail how to use Equations (D.1)–(D.8) for this system.

Example D.1: Step-by-Step Calculation for the Hexane(1)–Methanol(2) system

Part I. Calculations for compounds in the mixture.

Hexane (i=1) at mole fraction x_1	**Methanol (i=2) at mole fraction x_2**
$r_1 = 6, k = 1(CH_2), \nu_{11} = 6, \nu_{21} = 0$	$r_2 = 2, k = 1(CH_2), \nu_{12} = 1, k = 2(OH), \nu_{22} = 1$
$\sum_k x_k r_k = 6x_1 + 2x_2$, then using eqn. (D.2)	
$(\ln\gamma_1)_{combinatorial} = \left(1 - \frac{6}{6x_1 + 2x_2} + \ln\frac{6}{6x_1 + 2x_2}\right)$	$(\ln\gamma_2)_{combinatorial} = \left(1 - \frac{2}{6x_1 + 2x_2} + \ln\frac{2}{6x_1 + 2x_2}\right)$
Using Equation D.3	
$(\ln\gamma_1)_{interactions} = 6(\ln\Gamma_1 - \ln\Gamma_1^{(1)})$	$(\ln\gamma_2)_{interactions} = (\ln\Gamma_1 - \ln\Gamma_1^{(2)}) + (\ln\Gamma_2 - \ln\Gamma_2^{(2)})$
Where $\Gamma_1^{(1)}$ is Γ_1 calculated for x_1=1	Where $\Gamma_k^{(2)}$ is Γ_k calculated for x_2=1

Part II. Calculations of X_k and A_{kl} for the groups constituting the system, Equations D.4 and D.6.

(CH$_2$) group, $k = 1$		(OH) group, $k = 2$	
$X_1 = \dfrac{6x_1 + x_2}{6x_1 + 2x_2}$		$X_2 = \dfrac{x_2}{6x_1 + 2x_2}$	
$\ln\Gamma_1 = 1 - \ln(X_2A_{12} + X_1) - \left(\dfrac{X_1}{X_1 + X_2A_{12}} + \dfrac{X_2A_{21}}{X_1A_{21} + X_2}\right)$		$\ln\Gamma_2 = 1 - \ln(X_1A_{21} + X_2) - \left(\dfrac{X_1A_{12}}{X_1 + X_2A_{12}} + \dfrac{X_2}{X_1A_{21} + X_2}\right)$	
$A_{kl} = \exp\left(m_{kl} + \dfrac{n_{kl}}{T}\right)$	(D.8)		
A_{12}		A_{21}	
m_{12}	n_{12} [K]	m_{21}	n_{21} [K]
−41.2503	7686.4	4.7125	−3,060

Note: Values of parameters are taken from Table D.2.

The results of these calculations are presented in Figure D.1.

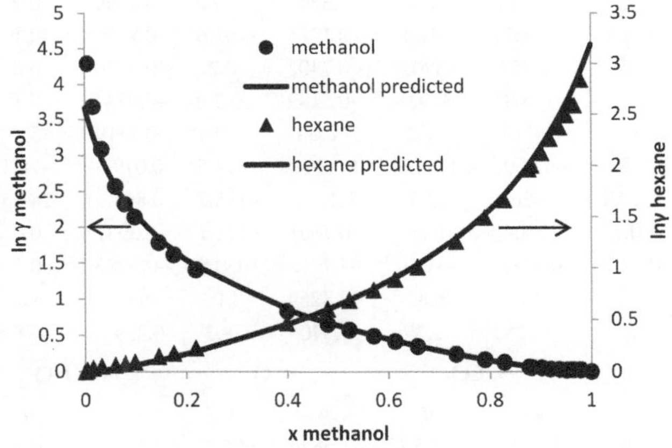

FIGURE D.1 Comparison of values of the activity coefficients predicted by the ASOG method with the experimental data [6] for the system hexane (1)—methanol (2) system at 60°C.

TABLES OF PARAMETERS

Tables D.1–D.7 contain the ASOG parameters for the temperature range 298.15–423.15 K. Explanation of symbols: NMP, C_5H_9NO; furfural, $C_5H_4O_2$; DMSO, $(CH_3)_2SO$; ACRY, $CH_2{=}CH_2$ group; GOH, OH group in glycols.

TABLE D.1
ASOG Group Interaction Parameters (298.15–423.15 K)[a]

l	1. CH_2		2. C=C		3. ArCH		4. CyCH		5. H_2O	
k	*m*	*n*	*m*	*n*	*m*	*n*	*m*	*n*	*m*	*n*
1. CH_2	0.0	0.0	0.7767	−94.4	−0.7457	146.0	0.1530	2.1	−0.2727	277.3
2. C=C	−0.4816	−58.9	0.0	0.0	−0.0622	−140.0	−1.0732	263.4	0.8390	−331.0
3. ArCH	0.7297	−176.8	−0.0744	88.8	0.0	0.0	−0.3288	156.3	n.a	n.a
4. CyCH	−0.1842	0.3	1.2487	−347.0	0.5301	−251.0	0.0	0.0	n.a.	n.a.
5. H_2O	0.5045	−2382.3	−9.5958	498.6	n.a	n.a	n.a.	n.a.	0.0	0.0
6. OH	4.7125	−3,060	−0.4867	−751.8	−0.5859	−939.1	0.56308	−3221.4.	−5.8341	1582.5
9. CO	−1.7588	169.6	2.8184	−1212.8	−0.4021	−216.8	0.0319	−350.3	0.3198	−91.2
10. O	0.7666	−444.0	−0.8602	476.9	−2.4476	562.6	0.1546	0.2	−3.2419	1037.9
11. CHO	−1.1266	0.2	−4.9564	1355.3	−0.5546	−0.1	−1.2628	0.1	−5.0228	1562.0
12. COO	−0.3699	162.6	−0.1323	114.2	−0.1541	97.5	−0.0991	2.4	−2.5548	659.9
16.CN	1.2569	−990.6	−1.4116	−48.1	−0.1163	−379.9	−4.5090	478.7	−0.2016	−85.6
19.N	0.1993	2.8	0.1398	72.4	2.324	−733.0	−0.4799	0.2	1.9691	−304.6
21.NO_2	−1.4089	228.5	−1.0057	141.4	−0.1225	−161.6	−0.8600	0.1	0.7062	−341.5
23.Cl	−1.2497	0.3	2.2753	−690.8	−0.7402	−0.2	−0.7970	0.0	n.a.	n.a.
24 CCl_2	0.2849	−151.1	−1.0081	379.0	−0.5189	0.0	−0.0744	0.1	0.0430	−335.3
25 CCl_3	−0.1134	41.1	0.4263	−1.2	0.2511	1.0	−0.2819	3.9	−3.2238	670..1
26 CCl_4	0.6926	−358.5	−3.5607	1161.7	0.8304	−374.5	0.0193	−78.1	3.7046	−1414.1
27 ArCl	3.1729	−520.4	−7.268	53.7	2.4031	−195.3	0.8643	243.5	n.a.	n.a.
28 ArF	−0.5051	−0.1	n.a.	n.a.	−0.7661	−111.1	−0.3937	0.0	n.a.	n.a.
29 Br	−7.0135	1842.7	0.5317	−419.0	−3.7012	1050.8	−0.6985	0.2	n.a.	n.a.
30 I	−2.0131	2.8	n.a.	n.a.	−1.8259	0.1	n.a.	n.a.	n.a.	n.a.
31.CS_2	−0.0033	6.0	−0.1252	2.3	−1.3705	279.3	0.3599	−178.5	−2.2182	15.4

l	6. OH		9. CO		10. O		11. CHO		12. COO	
k	*m*	*n*	*m*	*n*	*m*	*n*	*m*	*n*	*m*	*n*
1. CH_2	−41.2503	7686.4	2.6172	−865.1	−1.3836	606.4	0.1147	0.1	−15.2623	515
2. C=C	−4.1886	566.7	−1.093	367.8	−0.0489	−407.5	3.358	−1057.2	−2.4963	−31.6
3. ArCH	2.2682	−1111.5	0.9273	−185.8	−0.4061	370.9	0.1448	0.1	−0.5812	−249.3
4. CyCH	−11.9939	−2231.6	0.8476	−281	−0.4055	0.1	0.1354	0.1	−2.0465	10.5
5. H_2O	1.4318	−280.2	0.0585	−278.8	−0.3108	369.2	7.98	−2720.9	−2.4686	565.7
6. OH	0.0	0.0	−0.7262	2.9	0.4251	−474.9	−1.7642	0.1	0.0583	−455.3
9. CO	0.3283	1.3	0.0	0.0	−0.266	−0.1	−1.4943	176.3	−2.5152	489.5

(Continued)

TABLE D.1 (*Continued*)
ASOG Group Interaction Parameters (298.15–423.15 K)[a]

l	1. CH_2		2. C=C		3. ArCH		4. CyCH		5. H_2O	
k	*m*	*n*	*m*	*n*	*m*	*n*	*m*	*n*	*m*	*n*
10. O	−1.2619	380.7	0.265	0.2	0.0	0.0	−0.4918	0.0	−7.8816	−0.3
11. CHO	0.9824	−0.1	0.3871	72.9	0.4562	0.1	0.0	0.0.	−1.1887	0.0
12. COO	−0.0296	2.6	−0.1212	180	1.005	0.7	0.5393	0.1	0.0	0.0
16. CN	0.6616	−230.8	−0.2673	0.	n.a.	n.a.	n.a.	n.a.	−0.5146	0.0
19. N	2.6807	−115.6	0.8531	0.6	n.a.	n.a.	n.a.	n.a.	n.a.	n.a.
21. NO_2	2.8755	−916.7	0.0617	1.9	0.5306	0.1	n.a.	n.a.	0.0238	0.1
23. Cl	−0.6453	0.2	−0.7939	0.0	0.5451	0.3	−0.2703	1.1	−2.3239	0.1
24. CCl_2	−0.6986	0.2	0.1845	0.1	0.2922	−0.1	n.a.	n.a.	−0.9944	0.2
25. CCl_3	−2.2978	605.7	0.3823	3.1	0.7693	−0.1	n.a.	n.a.	−16.8658	−3637.1
26. CCl_4	−9.7985	2539	0.8583	−252.5	0.4234	−9.0	n.a.	n.a.	−2.8851	−27.1
27. ArCl	1.4497	0.1	1.5244	0.0	−11.339	137.0	n.a.	n.a.	1.9759	2.1
28. ArF	−0.9964	0.5	n.a.	n.a.	−4.3765	−0.2	n.a.	n.a.	n.a.	n.a.
29. Br	0.7523	−589.5	0.1539	−280.2	n.a.	n.a.	n.a.	n.a.	0.6142	−280.1
30. I	−0.8586	−421.2	−0.8678	−0.1	0.2383	0.0	n.a.	n.a.	−1.6441	0.0
31. CS_2	0.9985	−4056.7	−0.1141	−88.2	0.0461	0.1	n.a.	n.a.	−1.5267	0.1

l	16. CN		19. N		21. NO_2		23. Cl		24. CCl_2		25. CCl_3	
k	*m*	*n*	*m*	*n*	*m*	*n*	*m*	*n*	*m*	*n*	*m*	*n*
1. CH_2	−0.0786	10.4	−14.6389	5.9	−0.3228	−70.8	0.2909	0.3	0.2805	−96.9	0.2352	−119.8
2. C=C	0.7720	−186.5	−4.9244	−14.9	0.7995	−355.5	−1.7799	218.2	2.2117	−769.4	−0.5531	8.2
3. ArCH	−0.2142	176.4	−3.2954	728.4	−0.407	131.4	0.391	0.2	0.439	0.0	−0.2599	1.9
4. CyCH	1.4239	−278.4	−1.3776	0.1	−0.4029	0.2	0.0409	0.1	−0.0065	−0.1	0.1390	1.9
5. H_2O	−0.3984	−107.2	−3.8611	1497.3	−2.1995	162.5	n.a.	n.a.	−4.0265	577.5	−3.4302	−811.2
6. OH	−1.9168	166.7	−2.1231	231.6	0.4399	−821.2	−1.7798	0.1	−2.4298	0.2	2.8040	−1898.7
9. CO	0.3168	0.0	−3.4855	−0.6	0.0734	0.8	0.5934	−0.1	−0.185	−0.1	−1.118	−101.3
10. O	n.a.	n.a.	n.a.	n.a.	−0.3103	0.0	−0.9684	0.3	0.688	0.1	0.2244	0.1
11. CHO	n.a.	n.a.	n.a.	n.a.	n.a.	n.a.	0.1527	36.2	n.a.	n.a.	n.a.	n.a.
12. COO	0.0960	0.0	n.a.	n.a.	−0.1814	0.0	0.8513	0.0	0.3375	0.1	−0.4238	380.9
16. CN	0.0	0.0	−4.3706	1400	−0.7931	−0.6	0.7529	0.0	0.4827	−1.3	−0.4294	0.0
19. N	−0.3426	−301.5	0.0	0.0	n.a.	n.a.	−2.357	368.5	1.2659	−0.4	I 6569	−0.2
21. NO_2	0.5731	0.2	n.a.	n.a.	0.0	0.0	2.3999	−659.6	n.a.	n.a.	n.a.	n.a.
23. Cl	−0.7719	0.1	1.4946	−471.8	−0.9281	111.5	0.0	0.0	−0.7878	−0.3	−0.7114	0.1
24 CCl_2	−1.6043	0.4	−1.1286	0.8	n.a.	n.a.	0.5065	0.2	0.0	0.0	0.1852	57.6
25. CCl_3	0.0081	0.1	0.0056	−0.2	n.a.	n.a.	0.4207	0.0	−0.535	24.2	0.0	0.0
26. CCl_4	n.a.	n.a.	−3.0713	503.1	−1.0435	229.9	0.0817	0.1	0.073	2.1	0.0195	24.1
27. ArCl	1.0743	0.1	1.6708	−0.2	2.5829	−236.2	1.9183	0.2	3.8955	−854.7	n.a.	n.a.
28. ArF	n.a.	n.a.	−0.5503	11.6	n.a.	n.a.	n.a.	n.a.	n.a.	n.a.	n.a.	n.a.
29. Br	n.a.	n.a.	n.a.	n.a.	0.1008	2.3	0.7114	0.0	n.a.	n.a.	−0.9142	10.0
30. I	n.a.	n.a.	n.a.	n.a.	−1.4827	−0.1	n.a.	n.a.	−0.2628	0.0	−2.7033	694.7
31. CS_2	−0.5953	0.2	n.a.	n.a.	−2.0453	344.4	0.3352	0.0	n.a.	n.a.	−1.3831	269.1

(*Continued*)

TABLE D.1 (*Continued*)

ASOG Group Interaction Parameters (298.15–423.15 K)[a]

l	26. CCl_4		27. ArCl		28. ArF		29. Br		30. I		31. CS_2	
k	*m*	*n*	*m*	*n*	*m*	*n*	*m*	*n*	*m*	*n*	*m*	*n*
1. CH_2	−0.3917	227.9	−0.0904	−414.4	0.3653	0.1	−10.7665	−1628.9	0.3598	−1.1	−0.3104	11.8
2. C=C	4.1529	−1370.5	−0.0676	−86.3	n.a.	n.a.	−5.4517	453.5	n.a.	n.a.	−0.0706	1.1
3. ArCH	−0.5769	270.4	0.1546	−469.3	0.3856	98.6	−9.6405	502.6	0.5627	0.1	−0.2253	128.7
4. CyCH	0.0103	55.07	4.1582	−1746.2	0.1225	0.1	−2.4054	0.3	n.a.	n.a.	−0.2922	114
5. H_2O	−11.816	215.5	n.a.	n.a.	n.a.	n.a.	n.a.	n.a.	n.a.	n.a.	−1.8482	12.8
6. OH	5.9993	−3241.0	−3.5499	0.1	−7.3618	0.5	1.67	−1217.7	0.5339	−836.5	5.3401	−2977.0
9. CO	0.6643	−536.3	−1.1822	0.6	n.a.	n.a.	−4.9986	1582.6	−0.0874	0.1	−0.0999	−247.9
10. O	−2.2383	631	0.7735	−121.3	0.5774	−0.6	n.a.	n.a.	−1.3801	0.0	−0.3254	0
11. CHO	n.a.	n.a.	n.a.	n.a.	n.a.	n.a.	n.a.	n.a.	n.a.	n.a.	n.a.	n.a.
12. COO	−0.5166	253.3	−4.8012	−0.4	n.a.	n.a.	−5.4226	1582.6	−0.0622	0.0	−0.2434	0.1
16. CN	n.a.	n.a.	−1.8328	0.1	n.a.	n.a.	n.a.	n.a.	n.a.	n.a.	−1.2178	0.1
19. N	−0.7196	346.8	1.3450	−0.1	−2.1126	10.8	n.a.	n.a.	n.a.	n.a.	n.a.	n.a.
21. NO_2	−0.436	−79.1	−4.7073	15.9	n.a.	n.a.	−0.792	2.1	0.3076	0.2	−1.5109	244.8
23. Cl	−0.6544	0.1	−5.5731	0.4	n.a.	n.a.	−6.0004	1.1	n.a.	n.a.	−0.9223	0.0
24. CCl_2	−0.4902	92.2	−62.946	723.4	n.a.	n.a.	n.a.	n.a.	−0.0730	0.1	n.a.	n.a.
25. CCl_3	−0.0063	−43.9	n.a.	n.a.	n.a.	n.a.	−0.2827	4.2	−0.4744	130.2	−0.2393	138.5
26. CCl_4	0.0	0.0	−1.4077	225.8	0.119	0.1	−2.5136	0.2	0.3080	0.1	0.7329	−255.8
27. ArCl	2.7659	−478.0	0.0	0.0	n.a.	n.a.	1.2242	0.8	n.a.	n.a.	n.a.	n.a.
28. ArF	−0.2123	0.0	n.a.	n.a.	0.0	0.0	n.a.	n.a.	n.a.	n.a.	n.a.	n.a.
29. Br	−0.3827	0.2	0.1938	−0.5	n.a.	n.a.	0.0	0.0	0.3107	74.6	n.a.	n.a.
30. I	−1.2747	0.3	n.a.	n.a.	n.a.	n.a.	0.9094	−441.6	0.0	0.0	n.a.	n.a.
31. CS_2	−0.7212	218.8	n.a.	n.a.	n.a.	n.a.	n.a.	n.a.	n.a.	n.a.	0.0	0.0

[a] Reproduced and modified from [4] and [5] by permission of The Society of Chemical Engineers, Japan. Copyright of the Society of Chemical Engineers, Japan, 2015.

TABLE D.2

ASOG Group Interaction Parameters for Groups 7, 8, and 13–15 (298.15–423.15 K)[a]

ArOH: OH Group in Aromatic Compounds. GOH: OH Group in Glycols

l	7 ArOH		8 GOH		13 COOH		14 HCOOH		15 CON	
k	*m*	*n*	*m*	*n*	*m*	*n*	*m*	*n*	*m*	*n*
1. CH_2 k,l	−9.5152	0.7	−7.1548	64.0	9.7236	−3797.5	0.2365	−95.3	1.0067	−378
l,k	−3.8090	0.5	−17.925	102.0	−10.9719	4022.0	−0.0721	−264.8	−1.3137	−103.2
2. C=C k,l	11.708	−4898.9	n.a.	n.a	n.a.	n.a.	n.a.	n.a.	0.9654	−323.4
l,k	−6.4189	1634.6							−1.4194	116.2
3. ArCH k,l	−5.8576	1.0	−7.6896	−7.5	1.4405	−429.9	0.1022	12.0	−1.1344	162.6
l,k	−2.6414	0.6	−0.9602	5.6	−0.2256	−213.7	−0.3000	−232.2	0.5928	−252.7
4. CyCH k,l	−14.4465	317.8	n.a.	n.a.	n.a.	n.a.	n.a.	n.a.	0.5728	−236.6
l,k	1.2180	−1955.2							−0.1198	−397.7
5. H_2O k,l	0.7186	−320.2	−7.9975	2397.4	−0.4492	7.4	−0.6400	423.3	−08550	513.7
l,k	0.8632	−357.2	2.2157	−450.2	−2.1113	797.7	1.5229	−921.5	−1.2225	159.3

(*Continued*)

TABLE D.2 (*Continued*)
ASOG Group Interaction Parameters for Groups 7, 8, and 13–15 (298.15–423.15 K)[a]
ArOH: OH Group in Aromatic Compounds. GOH: OH Group in Glycols

l	7 ArOH		8 GOH		13 COOH		14 HCOOH		15 CON	
k	*m*	*n*	*m*	*n*	*m*	*n*	*m*	*n*	*m*	*n*
6. OH k,l	−0.2115	−0.2	0.8185	−1318.4	3.8786	−1712.0	n.a.	n.a.	0.1081	−23.7
l,k	0.9971	0.1	0.8906	15.5	1.7000	−664.5			0.3059	−58.8
9. CO k,l	−12.348	363.0	n.a.	n.a.	1.0434	−626.0\	n.a.	n.a.	−0.6028	−133.4
l,k	−4.3851	2209.5			1.8864	−543.0			−0.7286	388.4
10. O k,l	n.a.	n.a.	−4.2085	2478.6	3.9356	41.4	n.a.	n.a.	n.a.	n.a.
l,k			−5.3133	1673.0	−4.8241	1433.0				
20. $ArNH_3$ k,l	0.4097	0.0	n.a.	n.a.	n.a.	n.a.	n.a.	n.a.	n.a.	n.a.
l,k	0.1214	0.0								
23. Cl k,l	−24,647	188.4	n.a.	n.a.	n.a.	n.a.	0.0373	−274.0	n.a.	n.a.
l,k	−46.055	201.3					−4.0112	915.5		
25. CCl_3 k,l	n.a	n.a	n.a	n.a	−0.3039	3.7	−0.1506	3.3	n.a	n.a
l,k					−0.6070	168.3	−0.5650	2.1		
26. CCl_4 k,l	−36.680	1301.4	n.a.	n.a.	5.2636	−2014.2	n.a.	n.a.	0.2965	4.9
l,k	−0.9416	−1006.5			−5.0606	1679.3			−1.9351	5.0

TABLE D.3
ASOG Group Interaction Parameters for Groups 17, 18, 20, and 22 (298.15–423.15 K)[a]

l	17 NH_2		18 NH		20 Ar NH_2		22 Ar NO_2	
k	*m*	*n*	*m*	*n*	*m*	*n*	*m*	*n*
1. CH_2 k,l	0.6435	−159.8	−12.3803	−785.1	−9.2417	1.2	−3.0243	854.1
l,k	1.1005	−346.7	0.2778	−274.9	−12.764	1.8	4.4726	−1571.0
2. C=C k,l	−0.0924	5.0	−1.8432	7.2	−4.3664	10.5	−0.7495	7.2
l,k	−1.3899	4.6	−0.5681	12.0	−9.8620	55.7	0.031	−4.4
3. ArCH k,l	0.9525	−268.4	−2.3433	454.7	−2.3077	−1799.9	3.7681	−985.1
l,k	−0.6231	−183.8	−0.5387	34.7	0.3858	−832.8	−4.857	1152.7
4. CyCH k,l	−0.1458	165.3	−2.4289	0.3	n.a.	n.a.	−70.4174	619.0
l,k	−2.1740	−212.2	−0.7841	0.3			−4.4362	1354.8
5. H_2O k,l	4.2174	−1082.2	−0.5958	489.1	n.a.	n.a.	n.a.	n.a.
l,k	4.4468	−1847.2	−2.6892	919.5				
6. OH k,l	0.7008	0.0	−3.5886	1484.0	−0.9734	0.1	−7.9764	0.2
l,k	0.3741	−0.1	5.9417	−1834.8	−0.2167	0.1	0.4607	0.6
9. CO k,l	n.a.	n.a.	n.a.	n.a.	−5.0439	−0.7	n.a.	n.a.
l,k					0.0578	1.6		
12. COO k,l	n.a.	n.a.	0.6579	0.0	−5.1419	0.8	n.a.	n.a.
l,k			−1.2709	0.0	−0.4415	1.2		
17. NH_2 k,l	0.0	0.0	−0.8988	0.0	n.a.	n.a.	n.a.	n.a.
l,k			0.6143	0.1				

(Continued)

TABLE D.3 (*Continued*)
ASOG Group Interaction Parameters for Groups 17, 18, 20, and 22 (298.15–423.15 K)[a]

l	17 NH_2		18 NH		20 Ar NH_2		22 Ar NO_2	
k	*m*	*n*	*m*	*n*	*m*	*n*	*m*	*n*
18. NH k,l	0.6143	0.1	0.0	0.0	n.a.	n.a.	n.a.	n.a.
l,k	−0.8988	0.0						
19. N k,l	n.a.	n.a.	−12.4922	231.6	n.a.	n.a.	n.a.	n.a.
l,k			2.6204	−769				
20. $ArNH_3$	n.a.	n.a.	n.a.	n.a.	0.0	0.0	−1.5563	2.1
k,l							0.0849	−0.4
l,k								
21. NO_2 k,l	n.a.	n.a.	n.a.	n.a.	n.a.	n.a.	0.5023	0.0
l,k							−1.3655	0.3
22. $ArNO_2$	n.a.	n.a.	n.a.	n.a.	0.0849	−0.4	0.0	0.0
k,l					−1.5563	2.1		
l,k								
23. Cl k,l	n.a.	n.a.	n.a.	n.a.	−9.3156	1.1	n.a.	n.a.
l,k					−2.6502	0.8.		
26. CCl_4 k,l	1.4075	−305.3	−6.9598	1261.6	−16.841	129.9	0.0442	5.0
l,k	−1.6955	−34.3	1.5993	−511.7	−37.843	220.3	−0.7072	1.6
27. ArCl k,l	0.0936	4.7	−34.169	13.5	n.a.	n.a.	n.a.	n.a.
l,k	−0.8748	3.2	−1.6453	548.5				

TABLE D.4
ASOG Group Interaction Parameters for Groups 32–35 (298.15–423.15 K)
Furfural: $C_5H_4O_2$; ACRY: $CH_2{=}CH_2$ Group

l	32 Pyridine		33 Furfural		34 ACRY		35 Cl(C=C)	
k	*m*	*n*	*m*	*n*	*m*	*n*	*m*	*n*
1. CH_2 k,l	0.3012	−137.6	−0.7638	114.8	0.5607	−4.6	1.1392	−171.2
l,k	0.0106	−39.0	0.6961	−241.2	−1.4541	9.1	−1.9268	227.5
2. C=C k,l	−0.0156	2.4	−0.1147	−47.1	n.a.	n.a.	0.9875	−20.9
l,k	−0.1709	2.2	−0.0993	16.7			−2.4412	31.8
3. ArCH k,l	0.2311	−5.9	−0.6636	78.0	0.0965	89.6	0.1170	182.9
l,k	−0.2436	−12.3	0.5358	−91.7	−0.5230	−41.6	−0.2341	−250.5
4. CyCH k,l	0.6196	−210.1	n.a.	n.a.	n.a.	n.a.	0.9884	−129.7
l,k	−0.4535	83.6					−1.9390	327.9
5. H_2O k,l	−0.6486	5.4	−0.6920	−262.3	−0.8319	1.8	n.a.	n.a.
l,k	−0.5628	3.0	−0.1105	−80.2	−0.8244	7.7		
6. OH k,l	−3.5249	862.8	−2.8028	241.5	−2.1018	9.8	−9.3298	56.5
l,k	1.2271	−704.9	2.5170	−817.9	−0.5398	6.8	−1.7019	35.5
9. CO k,l	−0.6859	8.7	−0.9171	144.5	n.a.	n.a.	0.7515	−140.8
l,k	0.4533	−43.4	2.6700	−794.0			−0.4612	−73.3
12. COO k,l	n.a.	n.a.	−0.4431	6.5	n.a.	n.a.	0.473	1.4
l,k			0.0487	−6.4			−2.5822	4.0
16. CN k,l	−0.5626	1.5	n.a.	n.a.	−0.2184	3.4	1.0854	−130.9
l,k	0.0608	4.2			−0.1420	1.9	−11.477	156.3
23. Cl k,l	n.a.	n.a.	n.a.	n.a.	n.a.	n.a.	−0.9831	611.2
l,k							−81.178	611.17

(*Continued*)

TABLE D.4 (*Continued*)
ASOG Group Interaction Parameters for Groups 32–35 (298.15–423.15 K) Furfural: $C_5H_4O_2$; ACRY: $CH_2{=}CH_2$ Group

l	32 Pyridine		33 Furfural		34 ACRY		35 Cl(C=C)	
k	*m*	*n*	*m*	*n*	*m*	*n*	*m*	*n*
24. CCl_2 k,l	0.0115	1.6	n.a.	n.a.	n.a.	n.a.	n.a.	n.a.
l,k	0.0393	3.2						
25. CCl_3 k,l	−0.4599	3.3	n.a.	n.a.	n.a.	n.a.	0.5687	5.1
l,k	0.5132	4.8					−1.0044	−4.8
26. CCl_4 k,l	0.2895	−0.6	0.207	4.4	0.2042	34.9	−0.0045	166.4
l,k	−0.4234	−15.4	−0.6327	0.2	−0.7635	−11.1	−0.8124	48.1
31. CS_2 k,l	n.a	n.a	n.a	n.a	n.a	n.a	0.2582	2.4
l,k							−0.5219	2.8

TABLE D.5
ASOG Group Interaction Parameters for Groups 36–39 (298.15–423.15 K)[a] DMSO: $(CH_3)_2SO$; NMP: C_5H_9NO

l	36 DMSO		37 NMP		38 C≡C		39 SH	
k	*m*	*n*	*m*	*n*	*m*	*n*	*m*	*n*
1. CH_2 k,l	1.1476	−340.7	−0.4505	62.5	−1.3791	478.8	0.6493	−90.7
l,k	−0.8839	−93.8	0.2854	−127.4	1.2210	−563.2	−1.0967	−203.0
2. C=C k,l	0.3592	−95.8	−1.5635	365.1	−0.1650	3.8	n.a	n.a
l,k	−0.4881	−41.6	2.1023	−639.4	−0.0150	1.5		
3. ArCH k,l	0.4226	−59.9	1.0714	−230.2	0.2346	−46.8	0.2005	7.4
l,k	−0.4264	−41.9	−1.5972	368,8	−0.5775	114.2	−0.9086	−2.7
4. CyCH k,l	1.1713	−253.5	0.6173	−182.3	n.a	n.a	−2.6058	316.6
l,k	−2.3215	−37.7	−0.3014	−34.0			2.2088	−821.8
5. H_2O k,l	−0.0058	−181.9	−0.4469	9.7	n.a	n.a	n.a	n.a
l,k	−0.3146	321.6	−1.4178	421.2				
6. OH k,l	−1.2893	−185.6	0.5490	−407.2	0.6644	−797.6	−2.3877	12.3
l,k	0.1689	275.2	−0.2521	56.3	1.1483	−895.2	−1.5491	11.9
7. ArOH k,l	n.a	n.a	−0.4991	0.0	n.a	n.a	n.a	n.a
l,k			1.221	0.0				
9. CO k,l	0.3498	−17.2	n.a	n.a	−0.2678	1.9	0.0602	3.6
l,k	−0.5942	31.2			0.0865	2.0	−0.5705	1.9
10. O k,l	n.a	n.a	n.a	n.a	n.a	n.a	0.0751	1.1
l,k							0.0055	1.8
11. CHO k,l	n.a	n.a	1.5366	−1094.2	n.a	n.a	n.a	n.a
l,k			−5.4464	1447.3				
12. COO k,l	0.1131	3.4	n.a	n.a	n.a	n.a	n.a	n.a
l,k	−0.6784	3.3						
16. CN k,l	n.a	n.a	n.a	n.a	−0.783	280.9	0.0864	−0.3
l,k					−0.3102	−109.4	−0.7913	3.0
17. NH_2 k,l	n.a	n.a	n.a	n.a	n.a	n.a	−0.3014	1.6
l,k							0.1122	2.6
19. N k,l	0.7100	1.6	n.a	n.a	n.a	n.a	n.a	n.a
l,k	−0.3215	−1.5						

(*Continued*)

TABLE D.5 (*Continued*)
ASOG Group Interaction Parameters for Groups 36–39 (298.15–423.15 K)[a] DMSO: $(CH_3)_2SO$; NMP: C_5H_9NO

l	36 DMSO		37 NMP		38 C≡C		39 SH	
k	*m*	*n*	*m*	*n*	*m*	*n*	*m*	*n*
21. NO_2 k,l	n.a	n.a	n.a	n.a	0.1228	0.6	n.a	n.a
l,k					−0.4273	−3.0		
24. CCl_2 k,l	0.4910	24.2	n.a	n.a	n.a	n.a	n.a	n.a
l,k	−0.5936	5.8						
25. CCl_3 k,l	0.3529	154.0	−1.0846	−182.2	n.a	n.a	n.a	n.a
l,k	−1.2394	119.5	0.5672	175.9				
26. CCl_4 k,l	0.2529	21.0	n.a	n.a	n.a	n.a	n.a	n.a
l,k	−0.8684	−16.5						
27. ArCl k,l	n.a	n.a	1.711	16.5	n.a	n.a	n.a	n.a
l,k			−0.8995	−18.9				
37. NMP k,l	n.a	n.a	0.0	0.0	0.7622	−17.1	n.a	n.a
l,k					−0.6922	−39.7		
40. DMF k,l	n.a	n.a	n.a	n.a	0.0021	3.7	2.8472	−805.0
l,k					0.1352	1.2	0.3152	−219.6

TABLE D.6
ASOG Group Interaction Parameters for Groups 40–43 (298.15–423.15 K)[a]

l	40 DMF		41 EDOH		42 DEG		43 Sulfolane	
k	*m*	*n*	*m*	*n*	*m*	*n*	*m*	*n*
1. CH_2 k,l	−0.1442	112.0	0.4982	−132.8	0.9532	−186.0	0.9906	−326.9
l,k	1.5415	−794.2	−4.3182	121.3	−7.5046	63.1	−1.4478	247.4
2. C=C k,l	−1.2730	452.7	n.a.	n.a.	−0.0696	−3.0	n.a.	n.a.
l,k	2.2580	−862.2			−0.6872	13.0		
3. ArCH k,l	1.629	−394.0	0.7659	−100.1	−0.3610	−5.7	−0.3259	−11.4
l,k	−2.8986	724.5	−10.598	208.8	0.0648	8.8	−0.4483	−30.0
4. CyCH k,l	−1.6170	511.4	n.a.	n.a.	n.a.	n.a.	n.a.	n.a.
l,k	2.8862	−1057.6						
5. H_2O k,l	0.2704	−235.2	−0.5995	141.1	n.a.	n.a.	n.a.	n.a.
l,k	−1.4479	455.7	−0.3272	51.8				
6. OH k,l	−0.9392	254.1	n.a.	n.a.	n.a.	n.a.	n.a.	n.a.
l,k	−0.6461	−251.6						
9. CO k,l	−0.2038	1.0	n.a.	n.a.	n.a.	n.a.	n.a.	n.a.
l,k	0.0236	17.4						
21. NO_2 k,l	n.a.	n.a.	0.0240	3.3	n.a.	n.a.	n.a.	n.a.
l,k			−1.0633	3.3				
23. Cl k,l								
l,k								
26. CCl_4 k,l	−0.2233	114.9	n.a.	n.a.	n.a.	n.a.	n.a.	n.a.
l,k	−1.9153	163.4						
29. Br k,l	1.7273	−332.3	n.a.	n.a.	n.a.	n.a.	n.a.	n.a.
l,k								
38. C≡C k,l	0.1352	1.2	n.a.	n.a.	n.a.	n.a.	n.a.	n.a.
l,k	0.0021	3.7						

(*Continued*)

TABLE D.6 (*Continued*)
ASOG Group Interaction Parameters for Groups 40–43 (298.15–423.15 K)[a]

l	40 DMF		41 EDOH		42 DEG		43 Sulfolane	
k	*m*	*n*	*m*	*n*	*m*	*n*	*m*	*n*
39. SH k,l	0.3152	−219.6	n.a.	n.a.	n.a.	n.a.	n.a.	n.a.
l,k	2.8472	−805.0						
40. DMF k,l	0.0	0.0	−0.0851	1.8	n.a.	n.a.	n.a.	n.a.
l,k			0.0321	2.1				

[a] Reproduced and modified from [4] and [5] by permission of The Society of Chemical Engineers, Japan. Copyright of the Society of Chemical Engineers, Japan, 2015.

TABLE D.7
Values of ν_{ki} and r_i for More Complex Groups[a]

Group	Compound: General Name	Compound: Name	Compound: Formula	ν_{ki}	r_i
32. Pyridine	Pyridines	Pyridine	C5H5N	Pyridine: 6	6
		3–Methyl pyridine	C6H7N	CH2: 1, Pyridine: 6	7
33. Furfural		Furfural	C5H4O2	Fufural: 7	7
34. ACRY		Acrylonitrile	CH2=CHCN	ACRY: 4	4
35. Cl(C = C)	Chlorides (C = C)	Trichloroethylene	Cl2C=CHCl	C=C: 2,Cl(C=C): 3	5
36. DMSO		Dimethylsulfoxide	(CH3)2SO	DMSO: 4	4
37. NMP		N–methylpyrrolidone	C5H9NO	NMP: 7	7
38. C:C	Alkynes	1–Hexyne	CH3CH2CH2CH2C≡CH	CH2: 4, C≡C: 2	6
		2–Hexyne	CH3CH2CH2C≡CCH3	CH2: 4, C≡C: 2	6
39. SH	Thiols	Methanethiol	CH3SH	CH2: 1, SH: 1	2
		Ethanethiol	CH3CH2SH	CH2: 2, SH: 1	3
40. DMF		Dimethylformamide	(CH3)2NCHO	DMF: 5	5
		Diethylformamide	(CH3CH2)2NCHO	CH2: 2, DMF: 5	7
41. EDOH		1,2–Ethanediol	(CH2OH)2	EDOH : 4	4
42. DEG		Diethyleneglycol	HOCH2CH2OCH2CH2OH	DEG: 7	7
43. Sulfolane		Sulfolane	C4H8O2S	Sulfolane: 7	7

[a] Reproduced and modified from Refs. [4] and [5] by permission of The Society of Chemical Engineers, Japan. Copyright of the Society of Chemical Engineers, Japan, 2015.

REFERENCES

1. Wilson, G.M. and Deal, C.H. 1962. Activity Coefficients and Molecular Structure. *Ind. Eng. Chem.* 1: 20–23.
2. Kojima, K. and Tochigi, K. 1979. *Prediction of Vapor-Liquid Equilibria by the ASOG Method.* Tokyo: Kodansha-Elsevier.
3. Tochigi, K. 1998. Prediction of vapor-liquid equilibria in non-polymer and polymer solutions using an ASOG-based equation of state (PRASOG). *Fluid Phase Equil.* 144: 59–68.
4. Tochigi, K., Tiegs, D., Gmehling, J. and Kojima, K. 1990. Determination of new ASOG parameters. *J. Chem. Eng. J.* 23: 453–463.
5. Tochigi, K. and Gmehling, J. 2011. Determination of ASOG parameters – Extension and Revision. *J. Chem. Eng. J.* 44: 304–306.
6. Koennecke, H. G. 1969. Ergebnisse von Dampfflüssigkeits-Phasengleichgewichtsuntersuchungen am System n-Hexan-Methylcyclohexan-Methanol bei 60°C. *J. Prakt. Chem.* 311: 974–982.

Section VI

Solutions to End-of-Chapter Problems

To learn and then have occasion to practice what you have learned—is this not satisfying?

Confucius

DOI: 10.1201/9781003431985-44

Presentation of Section VI

This section contains detailed solutions to end-of-chapter problems. It was conceived and prepared as an integral part of the second edition, but due to its length, it was decided to exclude it from the book and ebooks and make it available as a single downloadable file accessible at the book's web page: www.routledge.com/9781032557342.

DOI: 10.1201/9781003431985-45

Index